New Trends in Instrumentation for Hypersonic Research

NATO ASI Series

Advanced Science Institutes Series

A Series presenting the results of activities sponsored by the NATO Science Committee, which aims at the dissemination of advanced scientific and technological knowledge, with a view to strengthening links between scientific communities.

The Series is published by an international board of publishers in conjunction with the NATO Scientific Affairs Division

A Life Sciences **B Physics**	Plenum Publishing Corporation London and New York
C Mathematical and Physical Sciences **D Behavioural and Social Sciences** **E Applied Sciences**	Kluwer Academic Publishers Dordrecht, Boston and London
F Computer and Systems Sciences **G Ecological Sciences** **H Cell Biology** **I Global Environmental Change**	Springer-Verlag Berlin, Heidelberg, New York, London, Paris and Tokyo

NATO-PCO-DATA BASE

The electronic index to the NATO ASI Series provides full bibliographical references (with keywords and/or abstracts) to more than 30000 contributions from international scientists published in all sections of the NATO ASI Series.
Access to the NATO-PCO-DATA BASE is possible in two ways:

– via online FILE 128 (NATO-PCO-DATA BASE) hosted by ESRIN, Via Galileo Galilei, I-00044 Frascati, Italy.

– via CD-ROM "NATO-PCO-DATA BASE" with user-friendly retrieval software in English, French and German (© WTV GmbH and DATAWARE Technologies Inc. 1989).

The CD-ROM can be ordered through any member of the Board of Publishers or through NATO-PCO, Overijse, Belgium.

Series E: Applied Sciences - Vol. 224

New Trends in Instrumentation for Hypersonic Research

edited by

A. Boutier
ONERA,
Châtillon, France

Springer Science+Business Media, B.V.

Proceedings of the NATO Advanced Research Workshop on
New Trends in Instrumentation for Hypersonic Research
ONERA Le Fauga-Mauzac Center, France
27 April – 1 May 1992

Library of Congress Cataloging-in-Publication Data

New trends in instrumentation for hypersonic research / edited by A. Boutier.
p. cm. -- (NATO ASI series. Series E, Applied sciences ; vol. 224)
ISBN 978-94-010-4807-1 ISBN 978-94-011-1828-6 (eBook)
DOI 10.1007/978-94-011-1828-6
1. Aerodynamics, Hypersonic--Research--Instruments. I. Boutier, A. (Alain) II. Series: NATO ASI series. Series E, Applied sciences ; no. 224.
TL574.F5N45 1992
629.132'306'072--dc20 92-33756

ISBN 978-94-010-4807-1

Printed on acid-free paper

TABLE OF CONTENTS

PREFACE

Because of the growing interest in hypersonic flows, the AGARD Fluid Dynamics Panel initiated a sub-working group on instrumentation for hypersonics in 1989. This sub-group, chaired by J. WENDT (VKI - Belgium), was composed of: A. BOUTIER (ONERA - France), K. BÜTEFISCH (DLR - Germany), R. CATTOLICA (SANDIA Lab. - USA), V. CLINE (AEDC - USA), A. GIRARD (ONERA - France), R. McKENZIE (NASA Ames - USA), S. OCHELTREE (NASA Langley - USA) and G. SMEETS (ISL - Franco-German Inst.). As a result of several meetings, the idea came to organize this workshop, 27th April - 1st May 1992, on "New Trends in Instrumentation for Hypersonic Research", at Le Fauga-Mauzac ONERA center, in France, where the new hot-shot arc-heated facility F4, as well as a new conference building, were recently completed.

This workshop has been organized in close connection with the AGARD FDP Symposium to be held in Torino 4 - 8 May 1992 on "Theoretical and Experimental Methods in Hypersonic Flows": the main conclusions and ideas expressed by the papers and during the discussions of this workshop are reported in session 7 and have been presented in Torino in a special Instrumentation session.

As chairman of this workshop, I express many thanks to the Organizing Committee composed of Karl BÜTEFISCH, André GIRARD, Stewart OCHELTREE and John WENDT for their very constructive help, leading to a meeting that was recognized to be very fruitful for all the participants.

This workshop is a NATO Advanced Research Workshop: the NATO Science Committee has in fact decided to support this workshop, facilitating in this way the presence of many contributors; the whole Organizing Committee is deeply indebted to this very precious help and expresses many thanks to Prof. L. SERTORIO of NATO for his cooperation.

I want also to express many thanks to those who have largely contributed to the success of this NATO ARW by their financial aid: NASA, ESA (European Space Agency), DRET (Direction des Recherches, Etudes et Techniques), CNES (Centre National d'Etudes Spatiales), Région Midi-Pyrénées and DLR for encouraging a large participation of its scientists.

Finally this NATO ARW provided to all the participants a good opportunity for high quality scientific exchanges and resulted in the establishment of a set of recommendations for future research in instrumentation for hypersonics.

A. BOUTIER
Chairman

A. Boutier (ed.), New Trends in Instrumentation for Hypersonic Research, ix.

REQUIREMENTS FOR FACILITIES AND MEASUREMENT TECHNIQUES TO SUPPORT CFD DEVELOPMENT FOR HYPERSONIC AIRCRAFT

WILLIAM L. SELLERS III and DOUGLAS L. DWOYER
Experimental Methods Branch
Mail Stop 170
NASA Langley Research Center
Hampton, Virginia 23665-5225

ABSTRACT. The design of a hypersonic aircraft poses unique challenges to the engineering community. Problems with duplicating flight conditions in ground based facilities have made performance predictions risky. Computational fluid dynamics (CFD) has been proposed as an additional means of providing design data. At the present time, CFD codes are being validated based on sparse experimental data and then used to predict performance at flight conditions with generally unknown levels of uncertainty. This paper will discuss the facility and measurement techniques that are required to support CFD development for the design of hypersonic aircraft. Illustrations are given of recent success in combining experimental and direct numerical simulation in CFD model development and validation for hypersonic perfect gas flows.

1. Introduction

Over the last decade there has been a resurgence of interest in hypersonics. In the United States, much of that interest is due to the research and development of single-stage to-orbit (SSTO) vehicle concepts such as the National Aerospace Plane (Fig. 1). The NASP is required to spend a large portion of its flight time at high Mach numbers and to maneuver throughout its flight envelope. The design of such a vehicle involves extremely complicated physics and some of the critical design issues are indicated in Figure 2. The flow is three-dimensional and dominated by viscous effects. There is a strong coupling between the external flow around the vehicle and the propulsion system internal flow field. A knowledge of the boundary layer state is critical to determining the flow entering the engine inlet. Many of the flow interactions result in reduced performance or extremely high heating rates. At high altitudes, portions of the vehicle may be in transitional (low-density) flow and very little is understood about the prediction of the aerodynamics characteristics of complex configurations in this regime. Compounding the problem for the vehicle designer is the lack of an extensive experimental data base.

The general design features of a hypersonic air breathing vehicle were described in reference 1. The vehicle requires a highly integrated airframe and propulsion system to avoid excessive drag penalties. The propulsion system is based on a supersonic combustion ramjet (Scramjet) and the entire vehicle becomes part of the propulsion system. The forebody provides compression of the flow before it enters the engine inlet, which reduces the compression requirements of the inlet and the size of the engine module. The aftbody of the vehicle serves as an extension of the engine nozzle and the design of the aft body is complicated due to highly three-dimensional flow. In addition, off-design operating conditions could induce boundary-layer separation, and the inability to predict the thrust vector could cause severe trim penalties. Exit velocities in the nozzle are so high, that a knowledge of the thermodynamic state and chemical species exiting the combustor are important since the flow in the nozzle will not be in equilibrium. Therefore, the chemical reactions and their rates are vital to the nozzle design. Due to the favorable pressure gradient in the nozzle, the flow might relaminarize and that process is poorly understood.

A. Boutier (ed.), New Trends in Instrumentation for Hypersonic Research, 1–10.

Figure 1. Artist view of the NASP single-stage-to-orbit concept.

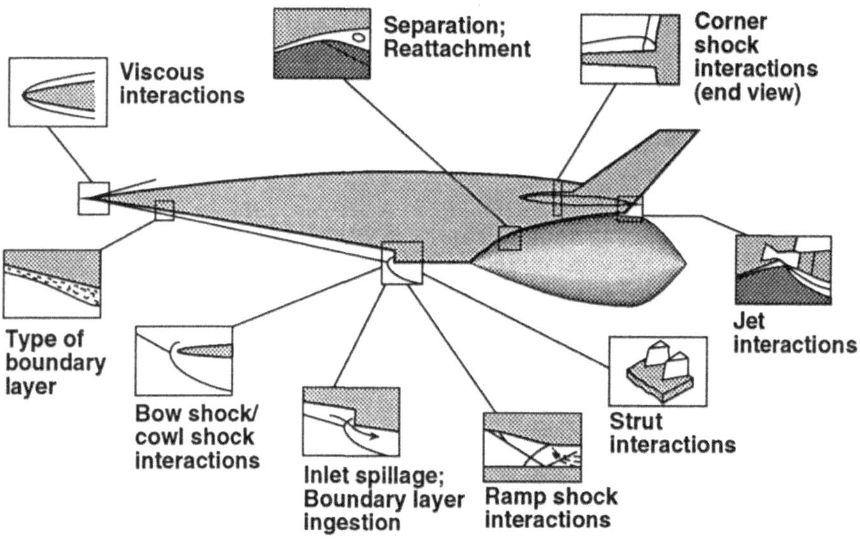

Figure 2. Hypersonic design issues

The injection of fuel (hydrogen) and its mixing with air is a critical factor for engine performance and size. The velocities in the combustor remain very high which means mixing and combustion must be rapid [2]. The residence time of the fuel in the combustor duct is on the order of microseconds. At high hypersonic speeds, injection of the fuel must be parallel to the flow since it provides a significant portion of the exit momentum of the combustor. Parallel injection of the fuel, however, results in reduced mixing due to a reduction in the convective Mach number of the fuel. Innovative concepts are required to increase mixing efficiency and reduce heat transfer.

The hypersonic flight regime can be divided into two broad categories with Mach 10 usually chosen as the dividing line. Flight below this speed involves typical high-speed fluid dynamics problems such as, pressure distributions, skin friction and heat transfer. Flight above Mach 10 brings the additional complications of molecular dissociation, ionization, rate kinetics of real gases and the problems associated with low-density flows.

2. Hypersonic Test Requirements

In the 50's and 60's hypersonic research was directed toward the development of re-entry vehicles. During this period there was a large growth in the number of hypersonic facilities. Toward the end of the Apollo program many of the facilities where mothballed or scrapped. The stringent test requirements for an air-breathing hypersonic vehicle will strain the capabilities of the few remaining facilities. There have been numerous studies that have examined the capabilities of ground-based facilities in relation to the test requirements of vehicles such as the NASP, Hermes and Sanger [3-7]. In most cases, these studies have divided the facilities into categories such as: continuous flow, intermittent or blowdown, and impulse facilities. Lukasiewicz provides an excellent comparison and historical review of the various types of hypersonic facilities [8].

To obtain reliable performance predictions from ground test facilities requires accurate simulation of flight conditions or scaling of the data based on similarity parameters. In the low-speed regime simulation of flight conditions is practical, but in the high-speed regime that is not the case. The hypersonic vehicle designer is faced with a larger number of similarity parameters because of the wide range of flight conditions that are encountered with an air breathing vehicle. In general the key test parameters are Mach, Reynolds, and Knudsen numbers and total enthalpy. Mach number is required to simulate compressibility effects, and Reynolds number is required to simulate the proper boundary layer characteristics. Knudsen number is used in flows involving low-density conditions and due to the complexity of the combustion process, the list of similarity parameters increases (e.g. Prandtl, Stanton, Eckert numbers). To satisfy the real-gas and propulsion test requirements will require full simulation of flight velocity, altitude, atmospheric environment and vehicle size.

There should be no facility generated inputs in the engineering data base. The contamination of the test gas with heater byproducts, is a concern because of their effect on transition, turbulence, and combustion. Acoustic noise and turbulence has been shown to have a major effect on transition measurements in the supersonic and low hypersonic regime and our understanding of their effects at high Mach number is limited. Facilities must also provide the proper length and time scales for hypersonic aircraft. This constitutes a major problem for ground-based testing of a air breathing hypersonic propulsion systems because the required length scales span essentially the full length of the vehicle.

It is interesting to compare the hypersonic test requirements with the existing ground- based test facilities. There are numerous ways to compare facilities based on whether the interest is to compare Reynolds number, Mach number or enthalpy. Figure 3 depicts the trajectory or flight corridor of a typical SSTO vehicle in terms of altitude and velocity. The capabilities of typical ground-based test facilities are also indicated. It is clear from the figure that we can only duplicate a small portion of the flight envelope below approximately Mach 10. Even in this range, however, the facilities have short run times that may limit their applicability. Wendt [7] further states that although many tunnels may appear adequate when comparing Mach and Reynolds capability, many important characteristics are not simulated. He states that short run-times result in much poorer measurement accuracies than in longer run-time, low-Mach number tunnels. Productivity is low and cost per data point is high in facilities of this type. Many of the facilities do not simulate the proper chemistry and may contain contaminants.

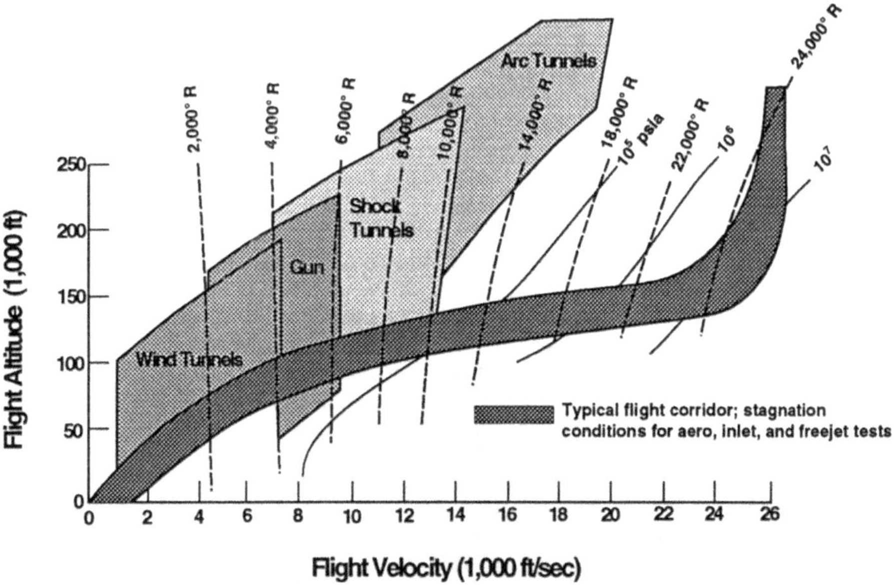

Figure 3. Comparison of ground-based test facilities

In general, aerodynamic simulation requires correct boundary layer simulation (laminar, transitional, or turbulent), wind tunnel flow quality comparable to flight, and the duplication of the correct chemistry. If we were to construct a facility for the upper portion of the flight envelope by isentropic expansion of a gas, it would require stagnation temperatures on the order of 25,000 °R and pressures on the order of 10 million psia. In fact, no ground-based facility exists, or can be expected to be built in the near future, of sufficient size to fully simulate hypersonic flight [3]. The lack of experimental facilities puts increased demands on CFD to provide the required design information and conversely CFD places demands on the experimental facilities for validation data.

There are examples, such as in the development of high-speed linear stability theory, where flight data can fill the gaps in our knowledge. There were many debates on the inconsistencies between linear stability theory and the transition data obtained in conventional wind tunnels. Not only were the magnitude of the transition data in question, but the trends were also incorrect. In the early 1980's, there was an F-15 flight experiment to investigate transition on cones. The flight data indicated much better agreement with the results from the linear stability theory. At the same time, a lengthy research effort went into the development of low-disturbance or "quiet" high-speed wind tunnels. Final closure to the problem occurred when the quiet tunnels became operational and proved without a doubt, that the discrepancy was due to facility generated tunnel noise. Interaction between all three simulation approaches (ground test, CFD and flight test) is a requirement for the high-Mach number regime.

3. CFD Development

CFD methods are progressing rapidly toward the prediction of the three-dimensional flow field about complex configurations. There are a variety of algorithms and techniques applicable to air breathing hypersonic aircraft [9-11]. Thomas [11] provides an excellent discussion of CFD requirements for hypersonic aircraft and includes a description of the physical models, algorithms, and advanced applications. The vehicle designer uses multiple codes, where each one approximates the physics of the flow in a slightly different manner. It is important to realize, however, that since the flow is viscous with three-dimensional interactions, the designer is generally forced to the Navier-Stokes equations to model the proper physics.

There are several items which are pacing CFD development. The most obvious is computer power. Even with the tremendous advances in computer speed and memory capacity, there are still limitations on the problems that can be solved without resorting to simplifying assumptions and modeling. The lack of sufficient computer power results in pacing items such as: algorithms, grid generation, flow modeling and code validation. There is a critical need for improvements in flow modeling. This is especially true in the hypersonic regime with the added complexity of chemical kinetics. Even with all the limitations, the present generation of CFD codes are able to predict flow fields where little experimental data exists, and that is making the issue of CFD code validation increasingly important.

A greater understanding of the fundamental physics of fluid dynamics is required to improve the modeling in CFD codes. Transition and turbulence modeling are key issues in CFD development. It is important to be able to predict not only the location of boundary layer transition, but also the extent of transitional flow. Inaccurate predictions can result in significant differences in vehicle performance, stability and weight. There are many critical issues regarding the behavior of turbulence at high Mach numbers. These include the reduction in turbulent mixing, the amplification of turbulence through shocks and the interaction of turbulence and chemical kinetics. Gas chemistry models need to be improved to provide better predictions of combustion, wall catalysis and combustion product expansion through the nozzle. An improved understanding of low-density flows is required for the high altitude portion of the flight envelope.

3.1 CFD CODE VALIDATION

Thomas [11] discusses the steps that are required to validate a CFD code in the hypersonic regime. The code is validated over several incremental ranges throughout the flight envelope. This is due to the fact that the physics that needs to be resolved changes in each range. The three steps that are involved in the calibration process include:

(a) Consistency and internal checks. In this step the mass, momentum and energy conservation are compared against available exact solutions.

(b) Code to code comparisons. In this step a code is compared to other established codes and algorithms.

(c) Code to experiment comparisons. This is the final step and consists of detailed comparisons of the calculated results against well-defined experiments.

There is a big gap however, between the data that is available and the full simulation that is required. Some of the data may be supplied by flight experiments but this data is limited because of the differences in flight trajectories and the available instrumentation. The majority of flight data is limited to surface measurements. A new approach to hypersonic CFD development is shown schematically in figure 4. Codes are presently being developed using experimental data in a range of flow conditions where partial simulation (Mach and Reynolds numbers) is possible.

The new approach would utilize the capabilities of Direct Numerical Simulations (DNS) to add detailed flow field data that is missing from the existing experimental data base. The DNS would also supply the data that is required to develop new models for the CFD codes in the parameter range that is not covered by ground-based facilities. The new flow models would then be calibrated/validated against flight data.

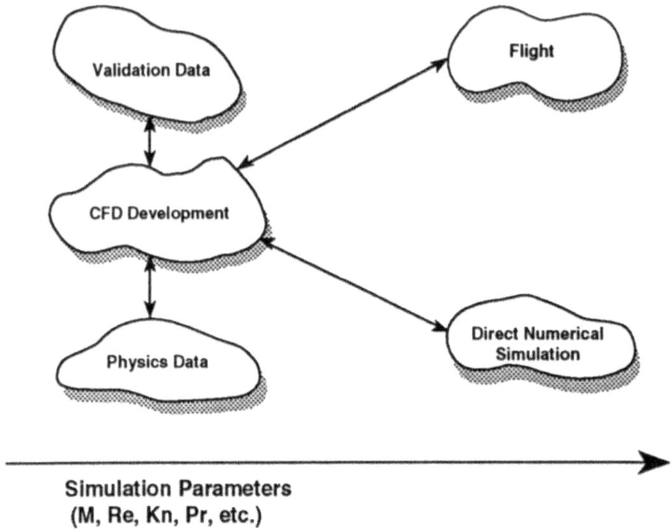

Figure 4. Future hypersonic CFD code validation process

Direct Numerical Simulations solve the Navier-Stokes equations essentially without the use of approximations or modeling. They were until recently limited to very low Reynolds and Mach numbers. They are computationally expensive and typically require 100 to 1,000 hours of time on a supercomputer. Therefore, only a few benchmark cases are typically calculated. The DNS have already proven useful in helping to analyze hypersonic flows and explain inconsistencies in experimental data. The calculations by Pruett [12] demonstrated the ability of DNS to predict the "precursor" effect in hypersonic boundary layer transition. This effect occurs when transition originates near the edge of the boundary layer and then propagates toward the body. Figure 5 shows a schlieren photograph from the wind tunnel experiment of Fischer and Weinstein [13] of a 2.87 deg half-angle cone, in helium flow at Mach=13.6. The sketch in Figure 6 depicts the details that were found in the schlieren photograph. The spreading angle of the transitional flow was approximately 0.85 deg. The cone was tested over a wide range of Mach numbers and the spreading angle was found to range from about 0.5 deg to 1.0 deg.

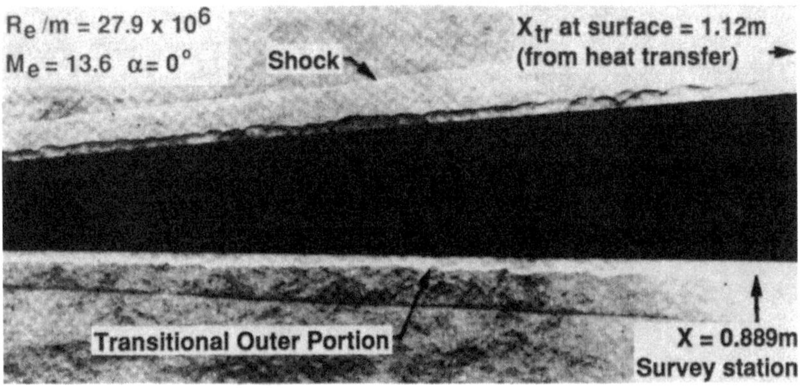

Figure 5. Spark Schlieren photograph of a 2.87° cone at Mach=13.6

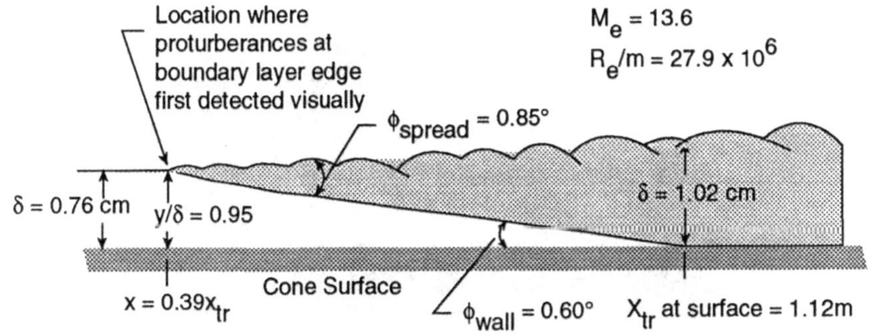

Figure 6. Growth of turbulence in a hypersonic boundary layer [13]

The results of the DNS calculations, are shown in figure 7, for the same cone geometry at Mach=6.8. The spatially reconstructed Reynolds stress ($\overline{\rho u'u'}$) clearly indicate the precursor effect. The spreading angle for this case was 0.7 deg which is in qualitative agreement with the experimental data. More importantly, the DNS calculations showed that the peaks in the Reynolds stress near the critical layer were due to secondary instabilities.

The DNS have also been able to explain the appearance of "rope-like" waves that appear at the edge of high-speed boundary layers. The structures have been reported experimentally by Potter and Whitfield [14] and Demetriades [15]. The test case was a hollow cylinder in a Mach 4.5 flow. The results from the DNS were used to generate a numerically simulated schlieren photograph (Fig 8) for comparison with the experimental data [12]. It was derived from the wall-normal density gradient from the DNS. The numerical schlieren photographs agree remarkably well with what was is seen in the experimental schlieren data. The "rope-like" waves were shown to have a scale about twice the boundary layer displacement thickness. The DNS simulations have shown that the "rope-like" appearance is really the artifact of the two-dimensional projection of a three-dimensional structure comprised of staggered lambda vortices. These lambda vortices are representative of a sub harmonic secondary instability. In the two-dimensional or integrated projection of the experimental schlieren, the rope has the apparent wavelength equal to that of a primary disturbance, which is incorrect. Therefore, the DNS was able to duplicate two previously unexplained phenomena from high-speed transition experiments. Due to the tremendous detail that is provided by the DNS, the calculations also demonstrated that neither of the two phenomena could be explained by linear stability theory alone.

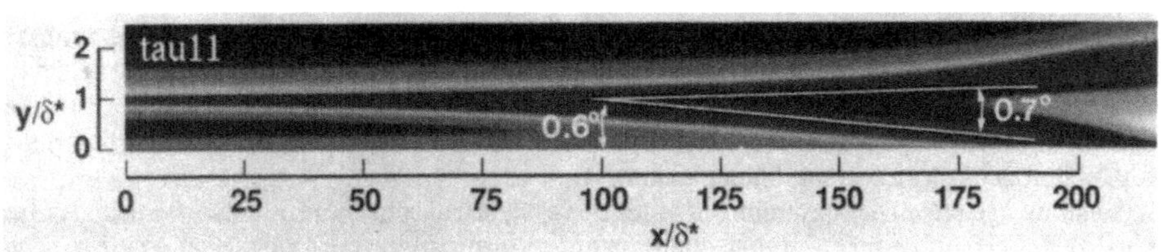

Figure 7. DNS spatially-reconstructed Reynolds stress for a cone at Mach 6.8

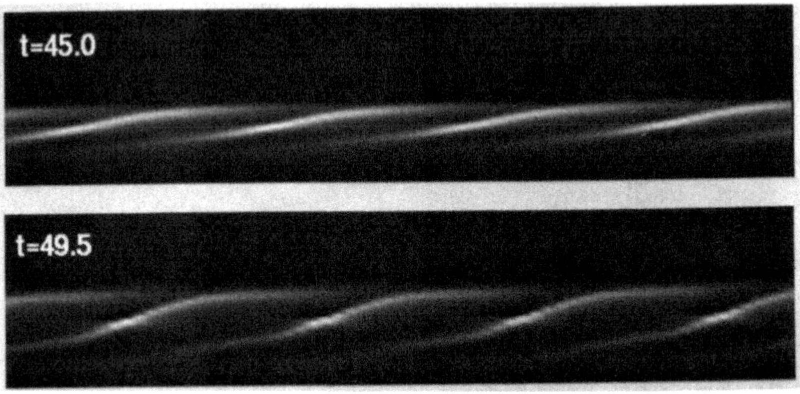

Figure 8. Numerically generated "Schlieren" Images of "rope-like" structure for a cylinder at Mach 4.5

4. Measurement Requirements

Advanced facilities and instrumentation are required to supply the highly-detailed experimental data needed for CFD code development. Particular attention must be paid to obtaining flow field surveys and experimental boundary condition measurements in order for the results to be useful for code validation [11]. The measurements required to support CFD development have been divided into two categories; data to support flow modeling; and data to support code calibration/validation.

4.1 FLOW MODELING

There is a critical need for the measurement of fluctuating flow field properties to support the development of transition, turbulence and chemistry models. To improve our ability to predict boundary layer transition, measurements of the transition location and the extent of transitional flow are important. Measurements of the initial disturbances in the flow field are important to improve the understanding of boundary layer receptivity. The initial disturbances are typically low-amplitude ($p' / p < 0.1\%$) and high frequency ($\approx$500 kHz). Not only are measurements of the amplitude and spectral characteristics of the disturbances important, but also of the wave orientation and growth rates.

Measurements of the correlation between the fluctuating flow field properties are required to support the development of turbulence and chemistry models. One area that has been neglected, is the correlation between velocities, species and temperature. These data are required for the new models for turbulent chemical kinetics [16]. These models make extensive use of probability density functions of these correlation parameters.

Accurate and reliable measurements of reaction rates, transport properties and thermodynamic properties are required for the high-temperature region of interest. At the present time these parameters are extrapolated with unknown certainty from existing data.

4.2 CODE CALIBRATION/VALIDATION

As mentioned by Thomas[11], code validation requires several steps. There is an additional interaction during the final step in which the code is compared with experimental data. The code developer compares his results globally and then in detail through different levels of experimental data. In the case of reacting flows, one of the first comparisons would be with the distribution of the major chemical species in the flow (H_2O, H_2 and O_2). Other global properties such as surface pressures, skin friction, and heating rates would then be examined. Detail comparison would then be made between the primary fluid dynamic properties in the flow field. This would require measurements of the mean and fluctuations of pressure, temperature, density, velocity and Mach number. The minor species (OH, O, H, and NO) would then be examined next. Nitric Oxide would be the least important because the majority of the codes neglect NO in their chemistry models because of the complexity involved. Comparisons with integrated parameters such as combustion and mixing efficiencies would be examined last. It is important to note that this hierarchy is from the perspective of the CFD code developer. A vehicle designer would, in all probability, have a totally different priority, because the primary interest would be vehicle performance and the parameters that influence its prediction.

There is a continual debate over the relative importance of point versus global measurements. The fact of the matter is, that both types of measurements are important. Instantaneous and time-averaged 2D planar measurements are valuable in providing considerable detail and for understanding the structures in the flow. They must be used and interpreted wisely in 3D flow fields. Planar and line type measurements also provide the ability to measure spatial correlations. Point measurements (2D and 3D) still provide accurate measurements of the statistical properties in the flow (e.g. Reynolds stress and higher correlations). At this point in time, the debate and emphasis should be on providing any quantitative measurements versus the merits of global versus point measurements.

The measurement requirements discussed above represent a significant challenge to the experimentalist. To be successful it requires a change in philosophy from the traditional way experimental research is conducted. The experiments are typically building blocks, involving different levels of physics at each step. They are **designed and conducted** in concert with the code developers. It requires a close interaction between the two individuals. The experimentalist must fully document the flow through detailed measurements of the surface, flow field and boundary condition parameters. Most importantly, the experimentalist must provide **quantitative** data with a specified measure of uncertainty

5. Concluding Remarks

The combination of test and measurement requirements far exceeds the capabilities of existing facilities. Because of this limitation, hypersonic code development requires a new approach. This approach would combine the use of flight data and direct numerical simulations to obtain the critically needed data. Their is no fundamental difference between the mechanics of CFD code validation in other speed regimes and in hypersonic flows. The process is made more difficult due to the flow quality requirements, limited optical access in hypersonic facilities, and the harsh flow environment. The hypersonic regime has the additional requirement to study the physics of chemical kinetics, low-density flows and boundary layers with extensive transition regimes. All of these factors drive the measurement technology in the hypersonic regime to advanced non-intrusive techniques and the development of high-frequency instrumentation.

6. References

1. Henry, John R. and Anderson, Griffin Y.: "Design Considerations for the Airframe-Integrated Scramjet", NASA TMX-2895, December, 1973.

2. Billig, F. S.: "Research on Supersonic Combustion", AIAA Paper 92-0001, Aerospace sciences Meeting, Reno, NV, January 1992.

3. "Review of Aeronautical Wind Tunnel Facilities", Committee on Assessment of National Aeronautical Wind Tunnel Facilities, National Research Council, National Academy Press, 1988.

4. "Hypersonic Technology for Military Application", Committee on Hypersonic Technology for Military Application, National Research Council, National Academy Press, 1989.

5. Thomas, Scott R. and Guy, R. Wayne: "Scramjet Testing From Mach 4 to 20 - Present Capability and Needs for the Nineties", AIAA 90-1388, AIAA 16th Aerodynamic Ground Testing Conference, Seattle, WA, June 1990.

6. Wittliff, C. E.: "A Survey of Existing Hypersonic Ground Test Facilities - North America", In *Aerodynamics of Hypersonic Lifting Vehicles,* AGARD CP-428, pp. 1-1, 1-8.

7. Wendt, J. F.: "European Hypersonic Wind Tunnels", In *Aerodynamics of Hypersonic Lifting Vehicles*, AGARD CP-428, pp. 2-1, 2-24.

8. Lukasiewicz, J.: "Experimental Methods of Hypersonics", Marcel Dekker, Inc., New York, 1973.

9. Dwoyer, Douglas, L. and Kumar, Ajay: "Computational Analysis of Hypersonic Airbreathing Aircraft Flow Fields", AIAA 87-0279, AIAA 25th Aerospace Sciences Meeting, Reno, NV, January, 1987.

10. McClinton, Charles, R.: "CFD Support of NASP Design", AIAA 90-3252, AIAA/AHA/ASEE Aircraft Design, Systems and Operations Conference, Dayton, OH, September, 1990.

11. Thomas, J. L.; Dwoyer, D. L. and Kumar, A.: "Computational Fluid Dynamics for Hypersonic Airplanes", Workshop on Hypersonic Flows for Reentry Problems, Antibes, France, January 22-25, 1990.

12. Pruett, C. D. and Zang, T. A.: "Direct Numerical Simulation of Laminar Breakdown in High-Speed, Axisymmetric Boundary Layers", AIAA 92-0742, 30th Aerospace Sciences Meeting, Reno, NV, January, 1992.

13. Fischer, M. C. and Weinstein, L. M.: "Cone Transitional Boundary-Layer Structure at Me=14", AIAA Journal, Vol. 10, No. 7, 1972, pp. 699-701.

14. Potter, J. L. and Whitfield, J. D.: "Boundary-Layer Transition Under Hypersonic Conditions", In *Recent Developments in Boundary Layer Research: Part III, AGARDograph 97,* May 1965, pp. 1-61.

15. Demetriades, A.: "Boundary Layer Instability Observations at Mach Number 7", J. Appl. Mech., ASME, Vol. 99, No. 1, 1977, pp. 7-10.

16. Girimaji, Sharath, S.: "A Simple Recipe for Modeling Reaction-Rates in Flows with Turbulent-Combustion", AIAA 91-1792, AIAA 22nd Fluid Dynamics, Plasma Dynamics & Lasers Conference, Honolulu, Hawaii, June, 1991.

HERMES MEASUREMENT NEEDS IN HOT FACILITIES

Marc RAPUC
DASSAULT AVIATION
78, quai Marcel Dassault
92210 SAINT-CLOUD
FRANCE

ABSTRACT. At hypersonic speed, real gas phenomena take place in the shock layer. The previous space shuttle flight experiments have shown important aerothermodynamic characteristics deviations induced by these effects. Two new facilities, F4 and HEG, have been built in the frame work of the HERMES program in order to get experimental data in real gas conditions. Such tests require an important effort in terms of measurements. One needs to know, the incoming flow quality and conditions but also its behaviour at the model wall. These measurements have to take into account the severe conditions of the shots (very short time duration, high level of temperature, wide range in level of pressure...). They are required for post processing data analysis and validation of numerical codes. The paper describes the different parameters to be measured.

1 - Flight conditions

At hypersonic speed, the flow surrounding the vehicle can't be considered as a perfect gas one. The increment in temperature that happens when crossing the bow shock induces different real gas phenomena such as :

- excitation of vibrational degree of freedom
- dissociation of molecules
- ionization
- ...

Then, different chemical processes take place in the shock layer according to velocity gradient and chemical reaction rates.

Furthermore, the wall properties (catalycity, emissivity,...) influence the chemical process in the boundary layer and the wall temperature.

A. Boutier (ed.), *New Trends in Instrumentation for Hypersonic Research*, 11–23.

Fig. 1 shows that such a complex flow occur during the main part of the reentry trajectory, when the speed exceeds 3 km/s. At the highest speed both nitrogen and oxygen dissociations take place while the vehicle flies the viscous interaction regime, characterised by low Reynolds number and strong shock/boundary layer interaction, and enters the hottest part of the trajectory.

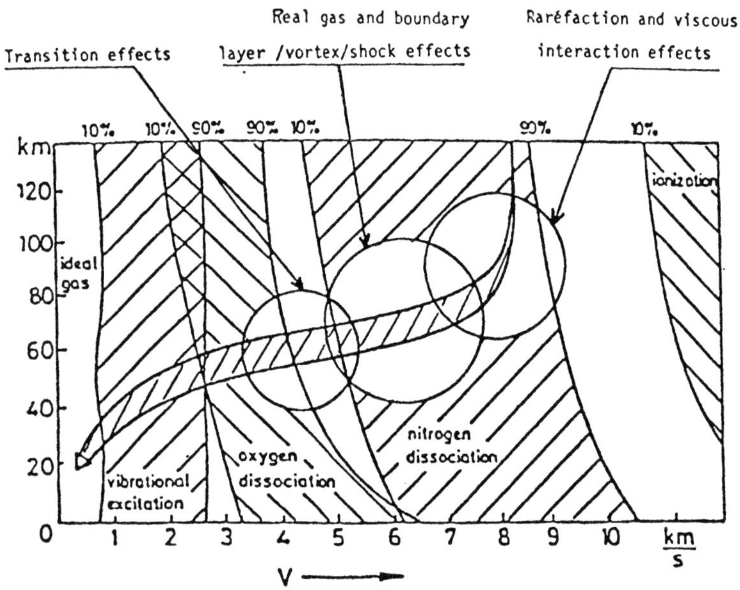

Fig. 1 - Typical flow conditions along reentry

The previous space shuttle flight experiments, US Orbiters and Buran, have shown that in comparison with characteristics obtained à M = 8 in perfect gas conditions the real gas effects influence the aerothermodynamics.

An example is given in fig. 2, with the trimmed body flap deflection angle of Columbia's first flight. Above M = 10, an important unpredicted pitch-up moment required an extra 8° deflection angle of the body flap in order to trim the vehicle. Thus the thermal limitation was reached on the body flap.

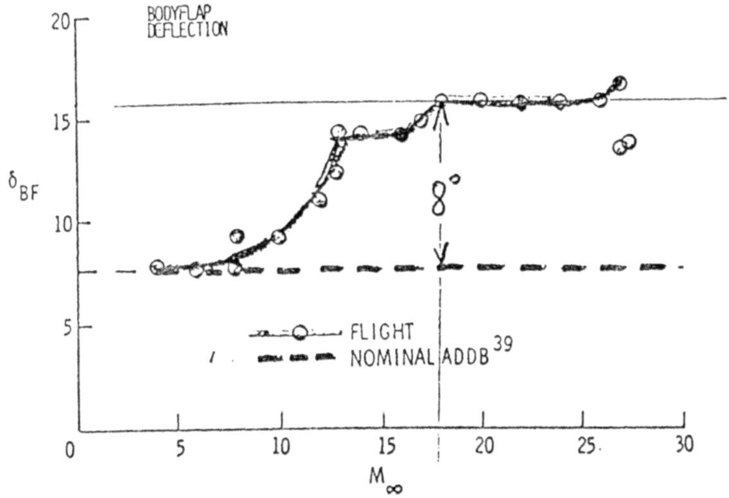

Fig. 2 - STS.1 body flap deflection

Hermes design is evaluated through a set of data including nominal characteristics and uncertainties. These uncertainties, have to be carefully studied and can't remain too large otherwise there won't be any feasible vehicle.

Therefore, the proposed HERMES methodology takes into account the analysis of the above mentioned real gas effects.

2 - Ground simulation

Looking at the two ways available to predict flight conditions, wind tunnel experiments or computations, it appears that both have to be used and cross-checked in order to overcome their different shortcomings.

It is impossible to duplicate in wind tunnel all the similarity parameters at the same time (Mach number, Reynolds number, stagnation enthalpy, wall temperature, chemical kinetic process...). In addition, after the expansion in the nozzle the high enthalpy flows are in non-equilibrium conditions.

The representativity of the tests remains therefore partial.

The approach by computation gives more hopes. But particular attention has to be paid to mesh influence and chemical kinetic model influence.

So there is a need for code validation.

Within the HERMES program, two new high enthalpy facilities have been developed in order to get experimental data at high velocity conditions and validate Euler and Navier-Stokes codes in non-equilibrium chemical condition.

These two facilities are :

- an arc heated type, F4 at ONERA-FAUGA, FRANCE
- a free piston driven shock tunnel, HEG at DLR - GOTTINGEN, GERMANY

Fig. 3 shows the facility domains in the (ρ . L, V) diagram. These similarity parameters are generally used for real gas effect analysis since the velocity is closely linked with the total enthalpy and the density times the model length correlate the binary reactions.

It comes then that these two wind-tunnels are complementary :

- F4 covers the range from 3 to 5,5 km/s that corresponds roughly to oxygen dissociation domain. The test durations are from 20 to 100 ms.

- HEG covers the highest speed from 5 to 8 km/s that corresponds to nitrogen dissociation in addition to oxygen dissociation. The test duration is about 1 ms.

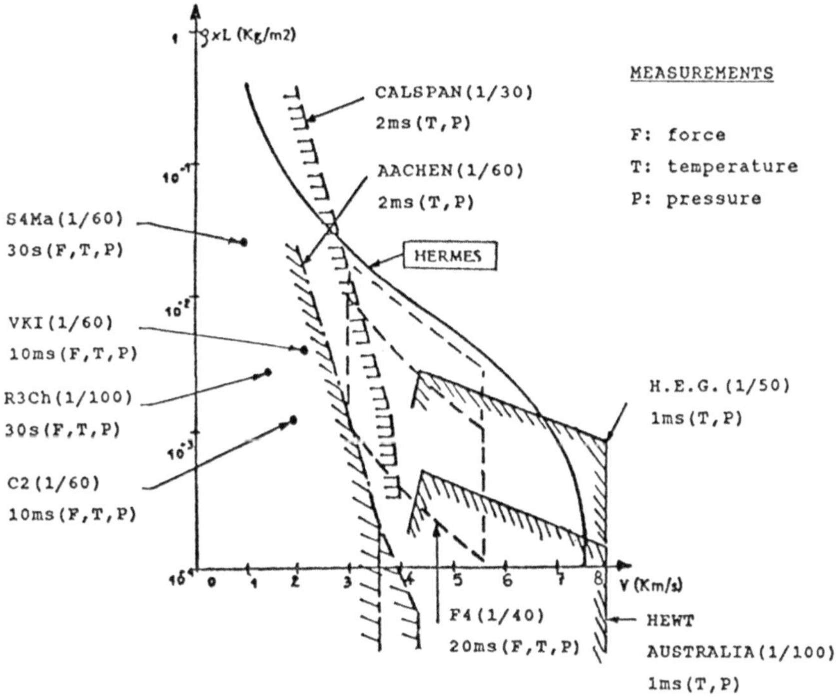

Fig. 3 - Wind-tunnel domains

These facilities will be used during the definition and qualification phases of HERMES design in three different ways.

- data base extension to real gas conditions
- sensibility studies of aerodynamic parameter variations in hot conditions.
- validation of numerical codes.

But the success of these challenging objectives relies on the availability of different measurement techniques that are mandatory to identify the flow and understand the aerodynamic and aerothermodynamic phenomena.

3 - Measurements

The measurements deal with :

- the flow conditions
- the aerodynamic and thermal loads on the model.

The flow conditions have, of course, to be known in order to check wether the test provides the desired aerodynamic parameters. But they are also needed to define the different coefficients (Stanton, pressure coefficient, force coefficient) that are more useful for the aerodynamicist than the measured data (heat flux (watt/m^2), pressure (Pa), force (N)).

In the case of high enthalpy facilities, since the flow is in non-equilibrium at the nozzle exit some specific characterisation of its thermodynamic and chemical state are also required.

The different parameters to be evaluated are now discussed according to the measurement locations in the tunnel, and the facility types (arc heated, shock tunnel).

Four main locations can be defined :

- in the reservoir
- in the test section
- in the shock layer
- at the model wall

3.1 - Measurement in the reservoir

The requirements here deal with the knowledge of the stagnation conditions in terms of :

- Hi stagnation enthalpy
- Pi stagnation pressure

Due to the high enough level of pressure in the reservoir, thermodynamic equilibrium assumption is valid.

The difficulties are due to the very short time duration (some milliseconds) of the shots. If, it is possible to measure the pressure level, it remains impossible to make a direct measurement of temperature. Different approaches have then to be considered according to the tunnel types.

In the hot shot type (F4), pressure measurement and equilibrium assumption are sufficient since the heating process is performed at constant and known density. The density value gives the second parameter that define the thermodynamic state of the gas.

In the shock tunnel type (HEG), the shock speed in the tube allows to get the enthalpy level specially when the tube is operated at Taylored condition (without multiple shock reflections or expansion in the test gas).

Of course, crosscheck of these standard measures or computed parameters would be appreciated.

The range of variation of these parameters is presented in fig. 4.

3.2 - Measurement in the test section

One specific item concerns the quality of the flow in the test section after the very fast expansion in the nozzle.

The nozzles are countoured in order to minimise the flow angularity variation and the longitudinal gradient that comes with conical ones, but have to be calibrated.

The homogeneity must be checked in the volume where the model is going to be placed in terms of pitot pressure and enthalpy. This can be achieved using specific rakes equiped with a sufficient number of probes (pressure probes and hemispheres with thermocouples for instance - see after). These rakes must be shifted in the longitudinal direction.

These measurements give also an important information about the available core at different flow conditions and help to define the model size.

Let us come now to the flow condition measurements. The parameters that have to be evaluated in order to characterise the flow can be put into two categories according to :

- first priority : ρ density, V velocity, Hi total enthalpy,Hf frozen enthalpy.
- second priority : T temperature, P pressure, M Mach number, Re Reynolds number

This classification comes from the following analysis.

The first priority parameters are those required for the different data reduction, they need the highest accuracy.

As a matter of fact, we have :

- dynamic pressure = $1/2 \cdot \rho \cdot V^2$
- stanton coefficient = $Q/\rho . V ((Hi - Hf) - Hw)$
- with Q = heat flux
  Hw = enthalpy at the wall

A first evaluation of these parameters can be obtained through pitot pressure and heat flux on sphere measurements since :

- pitot pressure is nearly proportional to dynamic pressure
- heat flux on sphere is linked to total enthalpy through Fay-Riddell formula with different assumption on equilibrium or frozen state.

Some other measurements like :

- mass flow probe ($\rho . V$)
- optical measurement (ρ , V)

can also be added, when it is possible, in order to increase the confidence of the parameter evaluations.

As a general rule it can be said that when it is possible it is worthwhile to get the same information from different ways in order to cross-check the results. For instance, when total enthalpy can be evaluated both from reservoir and test section measurements, it is the best way to operate.

Some computations of the expansion in the nozzle, in non-equilibrium conditions, even in one-dimension, may also be considered as another evaluation of the test section parameters.

The range of variation of these parameters is presented in Fig. 4.

The second priority parameters are more difficult to measure since they are related to static temperature that is more sensitive to flow non-equilibrium conditions. They less influence the data reduction process but are useful as inputs for code computations and data correlation.

When they can't be measured, one can take computed values from nozzle expansion as explained above. However, some difficulties may come from non-equilibrium processes, like vibrational relaxations, that influence significantly the temperature and Mach number values.

The previous discussed parameters are mainly needed for comparison of different experimental results and correlation studies. But when one wants to compare the test outputs with computation other information appear mandatory.

First of all, since the flow is non-equilibrium in the nozzle an appropriate flow field characterization in terms of species concentration, molecular vibrational temperature and translational temperature is needed.

Probably, inputs for the vehicle computation will come from outputs of axisymetric nozzle computation including a non equilibrium chemical model.

This nozzle modelisation has therefore to be validated by different specific measurements in the test section.

These measurements are required to assess the degree of non-equilibrium of the flow.

They can be measurements of :

- species concentrations O_2, NO, N , O, N_2
- vibrational and rotational temperature of N_2, O_2
- electronic density

Different spectroscopic techniques have to be tested in actual conditions.

Here, the measured values can be integrated along the optical beam since the distribution should be nearly uniform.

Some information concerning the pollution by different particles can also be obtained with these spectroscopic techniques and taken into account in the analysis process.

These measurements have to be performed for the different nozzles and stagnation pressures. Thus the range of variation of density is very large (see fig. 4) and different techniques would probably have to be proposed to cover all the cases.

F4	HEG
Stagnation conditions	
$200 < Pi\ (b) < 2000$	$500 < Pi\ (b) < 2000$
$20 < {}^{Hi}/_{RT} < 180$	$180 < {}^{Hi}/_{RT} < 350$
Test section conditions	
$2500 < V(m/s) < 5300$	$5300 < V\ (m/s) < 7400$
$10^{-4} < \rho\,(kg/m^3) < 2\,10^{-1}$	$10^{-3} < \rho\ (kg/m^3) < 10^{-2}$

Fig. 4 - Range of variation of different parameters

3.3 - Measurements in the shock-layer

Once the computation on the nozzle is checked, one must compare the flow field distribution around the model itself.

The first information to compare can be the bow shock position from the model. Such a measurement can be achieved by strioscopy or luminous photographs.

In addition some measurements behind the bow shock would be interesting.

These measurements can be :

- density gradient by holographic interferometry
- NO concentration
- electronic density

These measurements have to be local ones, since the flow is highly non uniform behind the bow shock.

3.4 - Measurements on the model

Once the incoming flow is well characterized, let us think about the measurements to be performed to identify the behaviour of such a flow at the wall.

The first point can be to check that the pollution level is low enough to avoid severe damages to the model or its heat flux equipment. Some preliminary tests in actual conditions will answer this questions.

Then three types of measurements have to be considered:

- force measurements
- pressure measurements
- heat flux measurements

It has also to be pointed out that model material has a great influence on the flow through the boundary layer behaviour and therefore on the measured values.

Different models have to be considered :

- metallic models that are the current technology derived from cold wind-tunnel.
- insulating models in order to increase the wall temperature and get a more appropriate wall temperature to recovery temperature ratio.
- catalytic or non catalytic models (or coatings) in order to see the influence of chemical process on measured heat fluxes.

3.4.1 - Force measurements

Force measurements are very useful to identify the real gas effects on the vehicle general balance (see fig. 2). But the very short time duration of the shots make this task rather difficult.

The mean aerodynamic values can't be directly extracted from the balance dynamic response due to not enough oscillation periods and important inertia forces. But, after inertia forces compensation taking into account accelerometer measurements, six components aerodynamic forces can be obtained.

With a sufficiently high system stiffness, 10 ms shots should be enough to perform force measurements. It means that if F4 is well suited for this technique, its feasibility hasn't been yet demonstrated in HEG.

The other difficulty remains in the necessity to have a set of balances well adapted to the different dynamic pressure conditions of the tunnel.

3.4.2 - Pressure measurements

This type of measurement is rather usual and only few comments have to be made.

These measurements will be used to :

- analyse some other measurements like heat flux measurements
- compare wind-tunnel results with computed ones.

The main difficulties concern the range of sensibility that must be covered. The pressure level will change with stagnation pressure and nozzle expansion ratio but also with model angle of attack and probe location on the model.

3.4.3 - Heat flux measurements

As in any wind-tunnel, one needs two types of heat flux measurements :

- thermal mapping to get a total heat flux cartography
- discrete measurements (thermocouple, thin films...) for more accurate measurements or different locations on the model.

If the short time duration of high enthalpy runs doesn't lead to specific problems when using thermocouple techniques for instance since their response time is very short, thermal mapping are more difficult to obtain.

The best way seems to perform infra-red measurements but the need for a sufficient number of images in order to compute heat fluxes from the temperature history requires a specific high speed camera. This activity is on the way at FAUGA.F4 facility with a 400 images/s camera.

The different techniques used have to be complementary in order to get a complete heat flux distribution of the model with a sufficient accuracy in the hottest points such as nose, leading edge, canopy, deflected control surfaces.

4 - CONCLUSION

High enthalpy tests are very specific and require the developpment of special techniques for their different operations (model manufacturing, measurement, post-processing data analysis...).

They remain with flight experiments the only way to validate the different new codes that are or will be available soon.

An important effort in measurement techniques is therefore needed in order to get a sufficient understanding of the tests.

They represent an important challenge for HERMES and other future project designs.

5 - References :

- G. de RICHEMONT : "Existing and new hypersonic facilities required for HERMES spacecraft testing". ICAS 90 Sweden.

- E. HIRSCHEL, A. KOC, S. RIEDELBAUCH : "Hypersonic flow past radiation-cooled surfaces". AIAA - 91 - 5031.

- G. KOPPENWALLNER, Ph. SAGNIER, G. EITELBERG, K. HANNEMANN : "Philosophy of high enthalpy wind-tunnel utilization". HTG Report 4-91 or CNES TA Report H-NT-0-2025 DLR.

- PAPIRNYK, SAGNIER, LOURME, BUTEFISH : "Measurement techniques currently in use or in development within the HERMES program : Synthesis of the presentations made at the DLR/Cologne meeting on June 25-27, 1990. ONERA H-NT-0-2032-ONER.

Hot Experimental Technique: a New Requirement of Aerothermodynamics

E.H. Hirschel
Messerschmitt-Bölkow-Blohm GmbH
Postfach 801160
8000 München 80
Germany

KEYWORDS/ABSTRACT. Hypersonic flows/hot vehicle surfaces/experimental simulation/hot experimental facilities/hot model surfaces/hot measurement techniques.

The flow past hypersonic vehicles causes large heat loads on the vehicle surfaces. It is shown that the heat loads and the flow properties are strongly coupled, if radiation cooling is employed. Similar phenomena occur in interacting flows, and in non-equilibrium viscous flows. The large demands on ground-facility simulation of hypersonic flows make new ways in experimental work necessary. A Hot Experimental Technique, each for three different typical measurement problems, is proposed. Details of the technique, regarding the tunnel type, model and measurement techniques are discussed. A stepwise approach to the involved problems is suggested.

1. INTRODUCTION

Worldwide several hypersonic vehicle projects and technology programmes are pursued. The vehicles encompass pure reentry vehicles like HERMES, and single- or two-stage-to-orbit-vehicles like X30, SÄNGER etc.. Common to all, although with different consequences, are aerothermodynamics with more or less strong viscous effects, real-gas effects, and coupled to them large heat loads. These effects influence performance, flight quality, controllability, propulsion efficiency (thrust minus drag), materials and structure concepts (cold primary structure with TPS, hot primary structure, depending on Mach-number range etc.).

Heat loads (temperatures, heat fluxes) often are not so much a concern of the aerodynamicist, who makes flux measurements at cold-surface models in the wind tunnel. They concern more the materials and structure man. In hypersonic research experiments usually are made with laminar flow past cold-surface models. Theoretical work mostly assumes adiabatic walls.

A. Boutier (ed.), New Trends in Instrumentation for Hypersonic Research, 25–39.

The very large design problems especially of hypersonic vehicles with airbreathing propulsion have led to a new assessment of the phenomena of aerothermodynamics: in many cases a strong coupling exists between the thickness of the viscous layer, the flow state (laminar or turbulent), interaction effects (shock/vortex/viscous layer), catalytic effects, emissivity (radiation cooling), the surface temperature and the heat flux in the surface. This holds for both external (fuselage, wing, ...) and internal flows (inlet, nozzle, ...) [1].

Depending on the vehicle trajectory the coupling may exist between different groups of these effects. During reentry the largest heat loads usually occur at about 70 km altitude, where the flow is still laminar. Airbreathing propulsion will be employed at altitudes where the flow is turbulent on large parts of the vehicle surface, with all the consequences also for the propulsion system (RAM and/or SCRAM). The aerothermodynamic designer is confronted with three basic problems: 1.) an insufficient understanding of the effects, partly regardless whether they are coupled or not, 2.) the limitations of ground-simulation facilities, 3.) the limitations of computation methods, especially that of their flow-physical and thermodynamic models.

In this paper it is shown, that severe shortcomings in ground-facility simulation exist, because some aerothermodynamic properties of hypersonic vehicles are coupled to the thermal state of the vehicle surface. The paper addresses with the Hot Experimental Technique (HET) a possible remedy. First, in the following Chapter 2, the present aerothermodynamic design philosophy is sketched with regard to the three basic problems mentioned above and the HET solution perspective. Chapter 3 illustrates with results of some numerical simulations where the simulation problems lie, and shows the range of coupled effects to be covered. In Chapter 4 finally the demands on a future HET in its different forms are discussed in some detail.

2. THE PROBLEM AND THE PERSPECTIVE OF ITS SOLUTION

Hypersonic aerothermodynamics, much more than classical aerodynamics, suffers from the restriction of ground-simulation facilities to fulfil at the same time the most important similarity parameters Mach number, Reynolds number, total enthalpy, Damköhler numbers, etc.. The current approach to overcome this problem is, somewhat simplified, to use "transfer models" to extrapolate from small-scale wind-tunnel simulations with limited similarity to full-size free-flight conditions. Transfer models are scaling laws and/or the methods of computational aerothermodynamics, which came to maturity, as it seems, during the last years.

This sounds good, and the success of reentry aerothermodynamics, although with some restrictions [2], seems to support this approach. Apart from the fact that computational aerothermodynamics was not employed extensively during the design of the U.S. Space Shuttle, transfer models in their present state are not sufficiently developed to fulfil the needs of future hypersonic vehicle design. This statement bases on three facts:

1. The next generation of hypersonic vehicles will be airbreathing systems for orbital launch and other, perhaps military purposes. These vehicles are from the view of aerothermodynamics on major parts of their trajectories viscous-effects dominated. This is obvious when considering the lengths of these vehicles, which are of the order of one-hundred meters. The flow path from tip to tail, which leads through the propulsion system, sees thick turbulent boundary layers with a weak forcing of transition laminar-turbulent on the forebody. The main problem is propulsion integration (forebody, boundary-layer diverter (with regard to inlet stability?), inlet, engine, nozzle, expansion ramp) with its demand of sufficiently large thrust minus drag.

2. Transfer models need on the one hand a representative data base (presently basically from the wind tunnel), and on the other hand validation. If the ground-simulation facility is not sufficient for the aerothermodynamic design, it will also not be sufficient to validate transfer models to the needed degree. The way out seem to be down-scaled hypersonic experimental vehicles. Such vehicles, however, first of all, must fly themselves, and in addition, they cannot be used as "flying wind tunnels", in order to make large numbers of detailed measurements. They can "only" be the final - and decisive - tool in a large technology effort [3].

3. To the similarity parameters mentioned above some wall-temperature relation must be added. The community of aerothermodynamicists seems only lately to become aware of the consequences of the fact, that hypersonic vehicles predominantly have radiation-cooled surfaces [4]. In the frame of this paper it can only be pointed to the fact that the materials and structure concept of an airframe initially is based on the radiation-adiabatic temperature as a conservatively estimated materials constraint. Where radiation cooling is not possible, for instance in the inlet, etc., the recovery temperature is the design parameter, which leads then to cooling requirements and so on.

These three facts, together with the partly strong coupling of aerothermodynamics effects mentioned in the introduction (for illustration see next chapter) make necessary a technology development approach in aerothermodynamics where the Hot Experimental Technique is a key element in order

- to improve the understanding of aerothermodynamic phenomena (attached and separated flow, interaction phenomena, heat loads etc. at hot surfaces),

- to obtain a more representative experimental design data base,

- to improve the flow-physical (transition laminar-turbulent, turbulence) and thermodynamic (finite-rate effects, transport properties, catalytic effects) modelling of transfer models (especially of numerical methods).

Of course, finally experimental vehicles of some kind are necessary to validate this technology approach, because also the Hot Experimental Technique is a ground-simulation approach, which is not fully representative for an advanced flying hypersonic system with its tremendous design risks.

3. ILLUSTRATION OF WALL-TEMPERATURE EFFECTS BY MEANS OF NUMERICAL STUDIES

In the following the coupling of aerothermodynamic effects with thermal surface states (hot-surface aerothermodynamics) is illustrated with results of numerical studies. The discussion is restricted to the major points, the reader is referred to the references for details.

Fig. 1 shows the result of a Navier-Stokes study of a forebody boundary layer [5]. Considered are laminar and turbulent flow (transition assumed at $x/L = 0.1$), perfect gas and equilibrium real gas, and the effect of radiation cooling. The recovery temperature is nearly constant in x/L, the influence of the boundary-layer state (laminar or turbulent) is weak. Real-gas effects are not strong, but appreciable. The impact of radiation cooling is very large. Very surprising is the strong influence of the boundary-layer state on the radiation-adiabatic temperature, which is approximately ten times larger than for the recovery temperature. This is a numerical result. The real-flight transition location cannot be predicted today to a sufficient degree and accuracy. The accuracy of any turbulence model in presence of especially the large wall-normal temperature gradients, and hence inverse density gradients in the radiation-cooling case cannot be assessed. Wind-tunnel data representative for this case are not known. They would require radiation-adiabatic equilibrium of the model surface, representative Mach number and Reynolds number (no boundary-layer tripping), a proper flow quality (quiet tunnel), and a representative total enthalpy.

At the OMS-pod of the U.S. Space Shuttle larger heat loads than predicted were found during the first flights [2]. They were attributed to "vortex scrubbing". Fig. 2 shows the radiation-adiabatic temperature (upper curve) found with a Navier-Stokes solution on a delta wing with laminar flow [6]. The distribution at the cross-section $x/L = 0.99$ exhibits peaks (A_1, A_2, A_3) and valleys (S_1, S_2). The peaks lie at attachment lines (A_1: primary attachment line at the windward side) and the valleys at separation lines. "Vortex scrubbing" essentially is an attachment-line heating effect in presence of radiation cooling, and also

a)

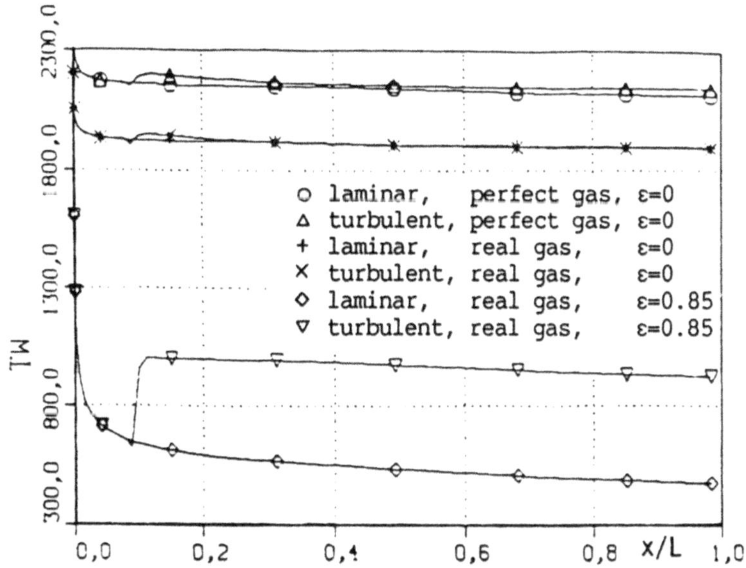

b)

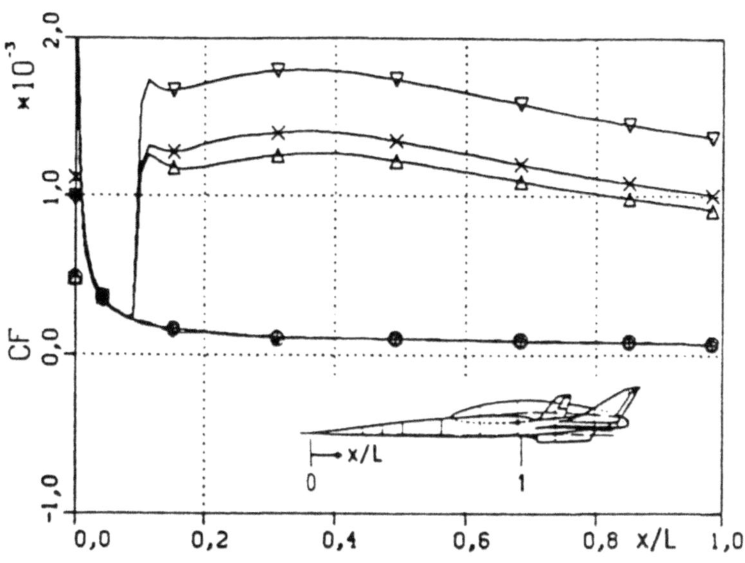

Fig. 1 Influence of the state of the forebody boundary layer (laminar/turbulent), real-gas effects and surface radiation on a) wall temperature, b) skin friction, [5] (lower symmetry line of SÄNGER forebody, $M_\infty = 6.8$, $Re_L = 1.22 \cdot 10^8$, $L = 55$ m, $\alpha = 6°$)

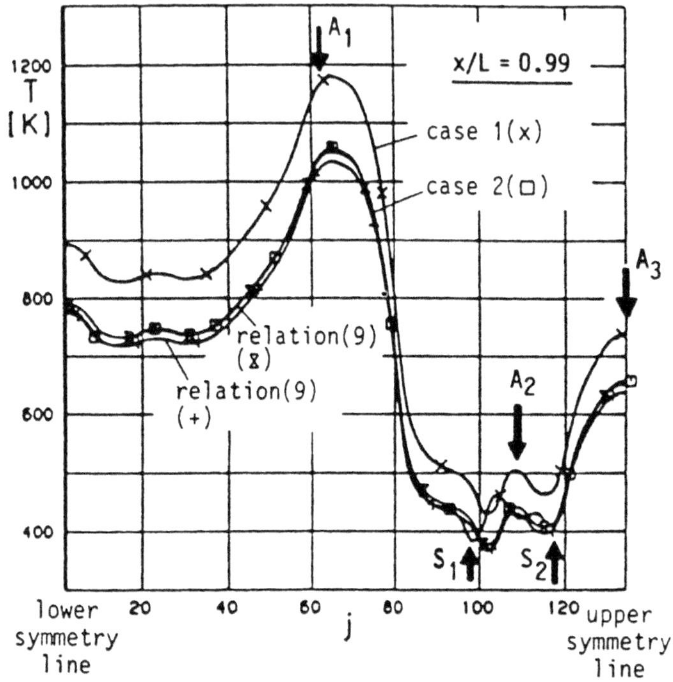

Fig. 2 Radiation-adiabatic temperature at the surface of a delta wing with attachment-line heating, [6] (cross section x/L = 0.99, $M_\infty = 7.15$, $Re_L = 5 \cdot 10^5$, L = 13 m, $\alpha = 15°$, laminar flow, $\varepsilon = 0.85$)

of heat flux into the wall. The radiation-adiabatic temperature is scale and Reynolds-number dependent, in contrast to the recovery temperature. The effect in Fig. 2, but also the decrease of the radiation-adiabatic temperature with increasing x/L in Fig. 1 can be explained in detail today [7], but an experimental verification is not possible with the available wind-tunnel techniques.

A Navier-Stokes study of the shock/boundary-layer interaction in an hypersonic inlet revealed a strong influence of the wall temperature on the strength of the interaction [8], Fig. 3. This can be explained in the following way: at the cold wall the flow density is larger than at the hot wall. Therefore the momentum flux is larger at the cold wall and hence the interaction weaker, if similar pressure fields are assumed [1]. Of course, also here the quality of the turbulence model cannot be assessed. In any case the inlet in reality has hot walls according to the materials and structure concept, probably with a certain degree of active cooling. In view of the high demands on inlet efficiency, a cold-wall simulation in a ground facility will not give the right answers. Even the size of the simulation error can only be estimated by numerical studies with questionable turbulence models.

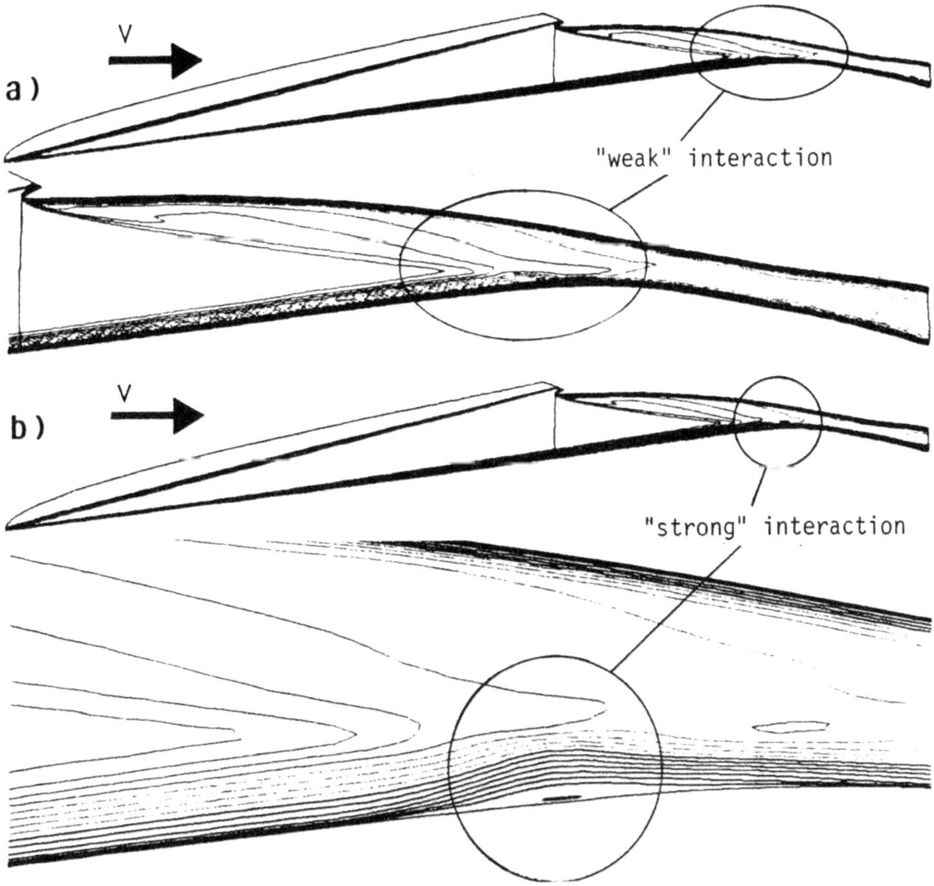

Fig. 3 Hypersonic inlet flow with a) cold wall: T_w = 303 K, b) hot wall: $T_w = T_{adiabatic} \approx$ 600 K [8], (M_∞ = 7.4, Re/L = $8.86 \cdot 10^6 m^{-1}$, turbulent flow), © AIAA – used with permission

Fig. 4 finally shows results of a numerical study of the viscous flow in chemical non-equilibrium along the lower symmetry line of the HERMES configuration [9]. Fig. 4.a gives the radiation-adiabatic wall temperature and the mass fraction of atomic Nitrogen at the wall. A non-catalytic wall was assumed. The results for a typical wind-tunnel situation are shown in Fig. 4.b. The model size is 1:10, the free-stream is in a weakly dissociated frozen state, the wall is cold (500 K have been chosen in order to achieve a more stable solution). Obviously, the mass fraction of atomic Nitrogen behaves completely different in the two cases. This holds also for the other air species and flow properties. Regarding the modelling of such flows, including the transport properties of multicomponent gases, the validation of transfer models from wind-tunnel data is at least very problematic. A hot model wall, if possible, in radiation-adiabatic equilibrium, would enhance very much the value of experimental data.

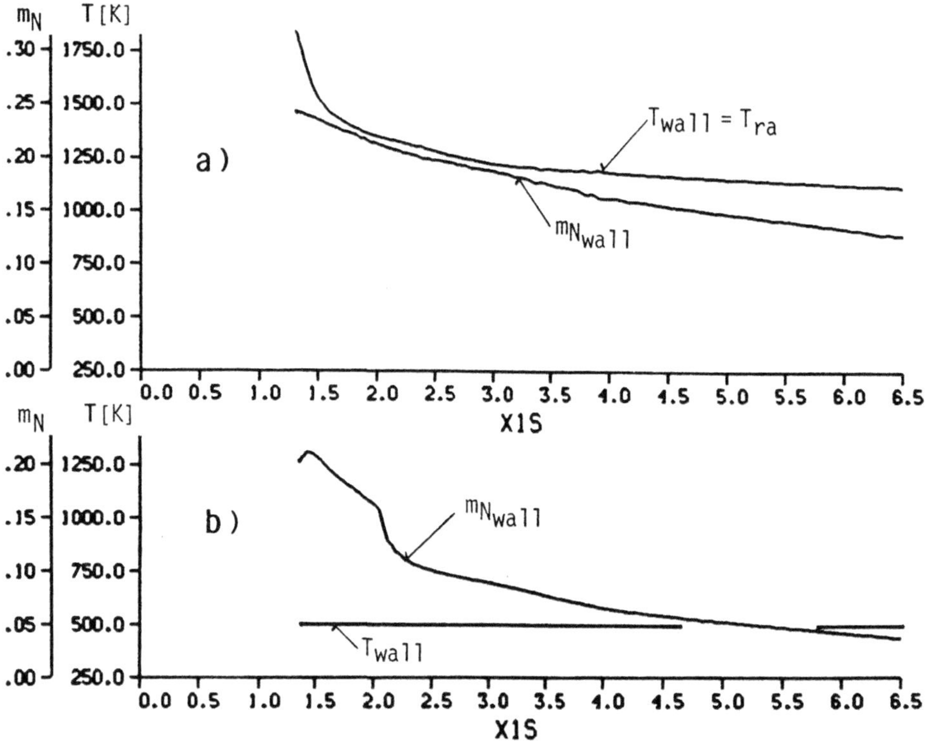

Fig. 4 Comparison of wall atomic-mass fraction at a) free flight with radiation adiabatic wall, b) wind-tunnel situation, 1:10 model, weakly dissociated frozen freestream, cold wall, [9] (lower symmetry line of HERMES, $\alpha = 30°$, $M_\infty = 20$, $Re/L = 1.21 \cdot 10^5 m^{-1}$, air, non-catalytic wall, laminar flow)

In addition to the effects discussed above, hypersonic vehicles at low speed after a high-speed heat-up (predominantly reentry vehicles) consistently seem to have a lift to drag ratio higher than predicted. First estimates show that the hot wall in free-flight has a considerably lower viscous drag than in the typical cold-wall wind-tunnel situation. Here possibly lies another need for a low-speed hot-wall HET.

4. HOT EXPERIMENTAL TECHNIQUE AS A NEW REQUIREMENT

In view of the design problems and risks of hypersonic vehicles much better ground simulation capabilites are needed than available today. This holds already for the development of experimental vehicles as forerunners of operational vehicles. The few illustrating examples of the preceding chapter show that an exploratory approach to the problem of hot-surface aerothermodynamic simulations is overdue. The proposed Hot Experimental Technique (HET) addresses the needs as they are perceived today. These may get partly redefined, when first actual results are available. These are soon needed, in order to start larger efforts in building an operational HET for both research and project work.

The name "Hot Experimental Technique" stands for a ground simulation technique with different constituent parts, namely

- the simulation facility (wind tunnel, etc.),
- the hot-model technique,
- the hot-instrumentation technique,
- transfer models (scaling methods, numerical methods).

In view of the examples given in the preceding chapter, three different forms of the HET are proposed:

1. External-Flow HET,
2. Internal-Flow HET,
3. High-Enthalpy External Flow HET.

In the following the features of these three forms are discussed, which of course, have hot-surfaces in common, and also some of the constituent parts.

1. External-Flow HET

   - Simulation objectives: The External-Flow HET applies to flow past airbreathing hypersonic vehicles in the RAM/SCRAM propulsion regime. The typically long and slender configurations are radiation cooled, and may partly be actively cooled at higher speeds. Attached, and interacting/separating flows are to be simulated with model surfaces in radiation-adiabatic equilibrium or near-equilibrium. Of particular interest is the flow at the lower side of the forebody of these vehicles (at rather low angles of attack $\alpha = 6°$ to $8°$), which is the onset flow of the inlet area (propulsion integration, boundary-layer diverter, inlet performance, etc.).

   - Research use: Basic research on transition laminar-turbulent, turbulence, interacting flows, separating flows.

   - Industrial use: Measurement of heat loads, drag, inlet onset flow, control surface efficiency, general aerodynamic coefficients.

   - Flow parameters: A duplication or near-duplication of the free-flight Mach number, Reynolds number and the total enthalpy appears to be possible for vehicles with lengths up to 50 to 80 m at Mach numbers $M_\infty \lesssim 5$ to 8 (2.5 km/s), altitudes up to 35 km.

- Facility type: The above flow parameters can be realized in a Ludwieg tunnel, with measurement times of 0.1 to 0.3 s. These times can be much longer, if no heating of the tunnel wall and a re-radiation of heat occurs, which would affect the radiation-adiabatic equilibrium of the model surface (large plenum chamber and/or LN2?). Such a tunnel could be employed also for the Internal-Flow HET (see below). The Ludwieg tunnel in principle can be considered as a quiet tunnel [10], however, for transition studies further improvements over the present state appear to be necessary. A near-duplication of the Reynolds number is necessary especially for industrial measurements. For the high boundary-layer edge Mach numbers present at slender configurations at low angle of attack no reliable tripping technique is known. The diameter of the test section should be 1 to 2 m for industrial applications, for exploratory work present tunnels are large enough.

- Model technique: The model surface must come into radiation-adiabatic equilibrium in the running times of 0.1 to 0.3 s defined above. If the tunnel type allows it, much longer running times are available (see above: Facility type). This appears possible with pre-test heat conditioning by either external heating or surface-sheet heating. Of course, the model surface must have the specified emissivity coefficient. A problem with regard to the necessary equilibrium state can be the large tangential temperature changes at the location of laminar-turbulent transition (Fig. 1), which might warrant longer running times (see above). Although in reality the transition zone will be smeared out, large tangential gradients are expected, which are also of much concern for structure and materials considerations.

- Measurement technique: The hot model surface, and also the associated internal model heating, point to large heat loads for measurement gauges, balances etc.. Surface mounted gauges should attain the unfalsified radiation-adiabatic temperature of the surrounding model surface. In any case the time history must be measured. This seems to be unrealistic with the present techniques and their development potential. Therefore for flowfield measurements non-intrusive opto-electronic techniques (IR, LDA, LIF, etc.) seem to be the candidate techniques. Force measurements etc. need appropriate balance techniques.

- Transfer models: The radiation-adiabatic surface temperature for instance, is scale dependent [7]. In any case a scaling of these data to free-flight conditions is necessary. Transfer models in form of scaling laws or the numerical methods of aerothermodynamics need to be employed also for other data. The HET will enhance via basic research measurements the flow-physical modelling of the numerical methods (transition criteria, turbu lence models, interaction phenomena, etc.), and hence appropriate transfer models for industrial use can be designed.

2. Internal-Flow HET

- Simulation objectives: Internal flows stands here basically for inlet flows of airbreathing hypersonic vehicles, but also for flows in inlet ducts, nozzles, etc.. The internal part of the inlet etc. cannot be radiation cooled. Depending on the flight-speed range either uncooled structures (T_r< 1500 K?), or actively cooled structures will be employed.The simulation problem is that of viscous flow with strong shock and embedded vortex interactions [1]. Special problems are the onset flow with a thick turbulent boundary layer stemming from the forebody, and bleed needs to alleviate interaction phenomena, and the prevention of inlet unstart.

- Research use: Basic research for inlet-flow transfer-model development including bleed flow (large transfer problem!) probably can be made to a certain extend with the External-Flow HET, which allows a better access to the flow. However, for special viscous interaction problems, and for performance studies (unstart, buzz, etc.) the Internal-Flow HET is necessary.

- Industrial use: Performance measurements, bleed studies, studies of off-design behaviour, distortion, unstart, measurement of aerodynamic loads and heat loads, all at real inlet configurations.

- Flow parameters: A duplication or near-duplication of the onset flow (M, Re, H_0) appears to be possible for flight speeds in the RAM and in the SCRAM range up to 3 to 4 km/s.

- Facility type: Because the inlet flow needs to be started short-time facilities like for the External-Flow HET cannot be employed. Running times of the facility of about 2 to 3 s appear to be the minimum, however, for certain inlet tests longer times up to 60 s might be necessary. Because no radiation cooling is employed, re-radiation of walls is no issue. However, internal hot-spot situations might be influenced by internal radiation, if high-emissivity internal surface materials are used. A problem, however, is the flow past the external ramp(s), because these are at least radiation-cooled in reality. If the loss of total enthalpy in the wall-near flow at the ramps due to either cooling means is of concern for the internal flow behaviour, a proper simulation is necessary (see the remarks on the Facility type of the External-Flow HET).

- Model technique: In order to reach the representative steady thermal state of the external and internal inlet surfaces, either a temperature pre-conditioning or longer test times than 2 to 3 s are necessary. Because of the high total enthalpy of realistic inlet flows very large heat loads result for the inlet-model, which in the extreme must be cooled like the final flight model. Whether such extreme conditions are necessary cannot be guessed at this time.

- Measurement technique: Depending on the speed regime the heat loads on instruments are much higher than for the External-Flow HET, because the internal surfaces will be much hotter than external surfaces. Another problem is the limited access for opto-electronic measurement techniques because of the highly loaded model. On the other hand in the internal part of the inlet possible hot gauges need not to have in addition surface emissivity properties like for external flow.

- Transfer models: In view of the extreme heat-load problems of the Internal-Flow HET the numerical methods of aerothermodynamics as the primary transfer model play an all-important role in hypersonic inlet design. The necessary flow-physical models can be predominantly derived from the work on the External-Flow HET. The Internal-Flow HET must then be used to validate the transfer model(s) with global performance data.

3. High-Enthalpy External-Flow HET

- Simulation objectives: High-enthalpy aerothermodynamics typically present during reentry flight pose simulation problems, which cannot be tackled with the ordinary External-Flow HET. The problems are mainly connected to the simulation facilities presently used or developed for such flows, the piston driven shock tubes (Stalker tubes) [11]. The measurement times in these facilities are extremely short O (100μs) down to O (10μs) for very high speeds. Although the surface temperatures of reentry vehicles are only of O (2000K) (conservative estimation) due to radiation cooling [7], their proper simulation is necessary for the assessment of thermodynamic and of possible catalytic-surface recombination effects (heat loads) and of the efficiency of control surfaces (interacting and separating flows, hot spots).

- Research use: Basic research in non-equilibrium aerothermodynamics, transport properties, catalytic-surface recombination in the presence of hot (not necessarily) radiation-adiabatic equilibrium surfaces for transfer-model developments.

- Industrial use: Configurational aerothermodynamics (limited because of frozen non-equilibrium test-section flow), heat loads, control-surface efficiency.

- Flow parameters: Total enthalpy equivalent to speeds up to · km/s (measurement times!), large total pressure in order to reduce frozen non-equilibrium test-section flow effects.

- Tunnel type: Free-piston shock tunnel, test-section diameter O (1 m).

- Model technique: Pre-test heat conditioning, but not necessarily in order to get radiation-adiabatic equilibrium or near-equilibrium.

- Instrumentation technique: In principle with the same demands and constraints as for the External-Flow HET, but at higher temperature levels of O (2000 K). Important are spectroscopic techniques, as they are in use or developed today.

- Transfer models: As above, but with modelling needs in finite-rate effects, transport properties, and surface catalycity effects.

5. CONCLUSIONS AND A LOOK AT THE NEXT STEPS

The development especially of next generation airbreathing winged space-transportation systems faces enormous aerothermodynamic challenges. Major problems are posed by the integration of the propulsion system and the airframe, the prediction of heat loads and aerodynamic properties. As was shown, the major problems lie in viscous effects which are strongly coupled to the thermal surface state.

While in flight external and internal vehicle surfaces attain high temperatures (conservative estimations: radiation adiabatic temperature, and recovery temperature, respectively), the present ground-simulation techniques essentially work with cold-model surfaces.

The ground-simulation capabilities must be improved in order to get more realistic design data, and in order to improve and validate transfer models (scaling laws, numerical methods of aerothermodynamics). A way to improve ground simulation is the Hot Experimental Technique. It was shown how the demands are on this technique in three major simulation classes.

The discussion is based on a still rather small number of numerical studies. Definite directions of work cannot be defined yet. Elements of the Hot Experimental Technique are already there. Hot hypersonic tunnels, even high enthalpy tunnels exist. "Hot" measurement techniques have been employed during the first U.S. Space-Shuttle flights in order to measure wall pressures, temperatures and heat fluxes. They are also employed in turbo-engine design.

In general, however, the old hypersonic thinking in terms of thin laminar viscous layers at cold surfaces which may or may not interact with shocks prevails in research, and partly also in industry. This must give way to a facing of the real problem of thick, laminar, transitional and turbulent viscous layers at hot surfaces. These surfaces may for design purpose considered to be for free flight in radiation-adiabatic equilibrium (external flows), or at recovery-temperature level (internal flows), with active cooling according to flight speed. Interaction effects of importance are those of shocks and vortices with these thick viscous layers.

The approach to the Hot Experimental Technique with the goal of an employment in the design of future hypersonic transportation systems should consist of the following major steps:

1. Studies of simulation needs with numerical methods by both research and industry:

   Investigation of reference configurations
   - at free-flight,
   - in cold-model tunnel situation,
   - in Hot-Experimental Technique situation.

   Even with the presently available numerical methods with insufficient flow-physical models major effects can be studied. Parametrical studies can give qualitative and quantitative answers to what to expect in terms of a possible HET. At the same time, however, basic pilot experiments should be conducted, also in order to give insights into the effects.

2. Pilot work on hot-model and hot-measurement techniques in available hot hypersonic tunnels.

3. Development of pilot HET facilities for research work (small tunnels), and for limited configurational work on hypersonic experimental vehicles (medium size tunnels).

4. After establishment of the value and the potential of the HET, latest after the first flights of experimental vehicles, development of a large HET facility for the development of operational hypersonic transportation systems.

This certainly is an ambitious proposal, which needs a large combined effort of the aerothermodynamic community, certainly on an international level. Step 1 should need two years until sufficiently deep insights are generated. Step 2 should be made simultaneously. Ideally work on a pilot HET facility (step 3) should start now, too, but latest after the two years of steps 1 and 2. A large HET facility should then be available in eight to ten years.

Acknowledgement: The author wishes to thank his colleagues at Dassault St.Cloud, Deutsche Airbus Bremen, DLR Göttingen and Köln, Dornier Luftfahrt Friedrichshafen, MBB München, MTU München and ZARM Bremen for many fruitful discussions, ideas and material, which went into this paper.

6. REFERENCES

[1] E.H.Hirschel: "Viscous Effects". Space Course Aachen 1991, Paper No.12, 1991. Also MBB/FE202/S/PUB/441, 1991.

[2] J.P.Arrington, J.J.Jones (eds.): "Shuttle Performance: Lessons Learned". NASA CP 2283, 1983.

[3] E.H.Hirschel, H.Grallert, J.Lafon, M.Rapuc: "Acquisition of an Aerothermodynamic Data Base by Means of a Winged Experimental Reentry Vehicle". ZFW, Vol.16, No.1, 1992, pp. 15 - 27. Also MBB/FE202/S/PUB/0461/A, 1991.

[4] E.H.Hirschel: "Heat Loads as Key Problem of Hypersonic Flight". To appear in ZFW, 1992.

[5] M.A.Schmatz, R.K.Höld, F.Monnoyer, Ch.Mundt, H.Rieger, K.M.Wanie: "Numerical Methods for Aerodynamic Design II". Space Course Aachen 1991, Paper No.62, 1992. Also MBB/FE211/S/PUB/442, 1991.

[6] S.Riedelbauch: "Aerothermodynamische Eigenschaften von Hyperschallströmungen über strahlungsadiabate Oberflächen". Doctoral Thesis, Technical University München, 1991.

[7] E.H.Hirschel, A.Koç, S.Riedelbauch: "Hypersonic Flow Past Radiation-Cooled Surfaces". AIAA-Paper 91-5031, 1991.

[8] A.Brenneis, K.M.Wanie: "Navier-Stokes Results for Hypersonic Inlet Flows". AIAA-Paper 91-2472, 1991.

[9] Ch.Mundt: "Calculation of Hypersonic, Viscous Non-Equilibrium Flows Around Reentry Bodies Using a Coupled Boundary-Layer/Euler Method". AIAA-Paper 92-2856, 1992.

[10] P.Krogmann: "An Experimental Study of Boundary-Layer Transition on a Slender Cone at Mach 5". AGARD-CP-224, 1977, pp. 26-1 to 26-12.

[11] W.H.Beck, G.Eitelberg, T.J.McIntyre: "The High Enthalpy Shock Tunnel in Göttingen (HEG)". Proc. 3rd Aerospace Symposium Braunschweig, 26 - 28 August, 1991, Springer Verlag, Heidelberg, 1992.

EXPERIMENTAL VALIDATION OF HYPERSONIC VISCOUS FLOW MODELS

B. AUPOIX
CERT/ONERA
2, Avenue E. Belin
B. P. 4025
31055 Toulouse Cedex FRANCE

ABSTRACT. As new hypersonic vehicles design relies upon CFD, both the numerics and the physical models require a fine validation to optimize the vehicle. The emphasis will be put here on viscous flows, mainly boundary layer flows. Two key problems are addressed, namely laminar high-enthalpy re-entry flows and hypersonic turbulent boundary layers. A brief overview of the state-of-the-art in both high-enthalpy flows and turbulence modelling is first given to identify needs and relevant information which could be extracted from experiments.

For re-entry flows, maximum heat loads occur in a non-equilibrium, laminar regime. Boundary layer flow predictions are ruled by the wall catalytic efficiency. An effort has been done to correctly model the chemistry, at least for temperatures relevant to boundary-layer flows. Unknowns still exist for high-temperature chemistry, as well as for thermal relaxation and for the transport properties. Thermal relaxation effects are larger in high-enthalpy wind-tunnel flows than in re-entry flows. It is however shown that wind-tunnel experiments can supply interesting data for model validation.

For the end of the hypersonic re-entry regime or for future hypersonic air-breathing airplanes, the flow is turbulent. Very little is known up to now about hypersonic turbulence and extensions of incompressible models as well as some specific hypersonic models behave poorly. There is a dire need for new, well documented, hypersonic turbulent flow experimental data.

1. Foreword

The design of a space shuttle or of an air-breathing hypersonic airplane nowadays relies upon CFD. Attention is paid as well to the wall pressure distribution to predict the stability and manoeuver ability as to the viscous effects which determine the wall heat fluxes and the viscous drag. Models are required, e.g. to represent the high-enthalpy effects or the turbulence effects as well as to predict the onset of the laminar/turbulent transition. This paper will focus on the validation of the models used in viscous flows computations, mainly boundary layer flow computations. This study will be restricted to air flow modelling and will not deal with combustion problems. Moreover, the laminar/transition problem will not be discussed and attention will be devoted to high-enthalpy flows in the laminar regime in section 2 and to turbulence modelling in section 3. In both

A. Boutier (ed.), New Trends in Instrumentation for Hypersonic Research, 41–50.

parts, a brief overview of the state-of-the-art will first be presented and validation needs and possibilities will be discussed.

2. High-enthalpy flow model validation

2.1. High-enthalpy flow modelling problems

During a major part of the re-entry, the flow is in thermal and chemical non-equilibrium. In order to describe such a flow, models are required for the thermodynamic properties of the species and of the mixture, for the transport properties of the species and of the mixture, for the chemical and thermal relaxation processes. At last, boundary conditions have to be prescribed, which may require models (e.g. for the wall catalytic efficiency) and/or hypothesis (e.g. for the wall temperature).

When boundary layer flows are considered, the key parameter is the wall catalytic efficiency. This is shown on figure 1 where the wall heat flux along the space shuttle windward centerline is given for an altitude of 71.29 km and an upstream velocity of 6730 m s^{-1} for both a catalytic and a non-catalytic wall. Flight measurements of the wall heat fluxes under these conditions during flight STS–2 are given as a reference but should not be compared as entropy layer swallowing effects are not accounted for in the computations which are moreover performed over an axisymmetric analog to the space shuttle windward centerline. The heat release on a catalytic wall can lead to an increase in the wall heat flux up to a factor of three when compared to a non-catalytic wall (see e.g. [2] and references herein).

The wall temperature is known to affect the wall heat flux, mainly on low catalytic efficiency walls, but its influence is moderate [2].

A few years ago, the chemical model was considered as the second key parameter. Computations of boundary layer flows exhibited a significant scatter according to the reaction rate data set used, despite a good consensus on the chemical reactions which are likely to occur in dissociated air as well as on the relative importance of these various reactions [2].

It must be reminded that reaction rate data are always represented using Arrhenius (or non-Arrhenius) forms. These forms have only limited theoretical grounds : the ratio of the forward and backward reaction rates is the equilibrium constant which can be derived from thermodynamics and will not exactly fit Arrhenius laws ratio. Moreover, even if there is presently a sort of "convergence" of the experimental reaction rate data, these data are available only over a restricted temperature range, up to about 4000 K. In re-entry flows, temperatures of about 20000 K can be encountered downstream of strong shock waves so that chemical reaction rates have to be extrapolated up to these temperatures with the help of Arrhenius fits !

Recent results of boundary layer computations (A. Daiss, unpublished) show very small discrepancy using different (recent) reaction rate data sets. Larger scatter could be

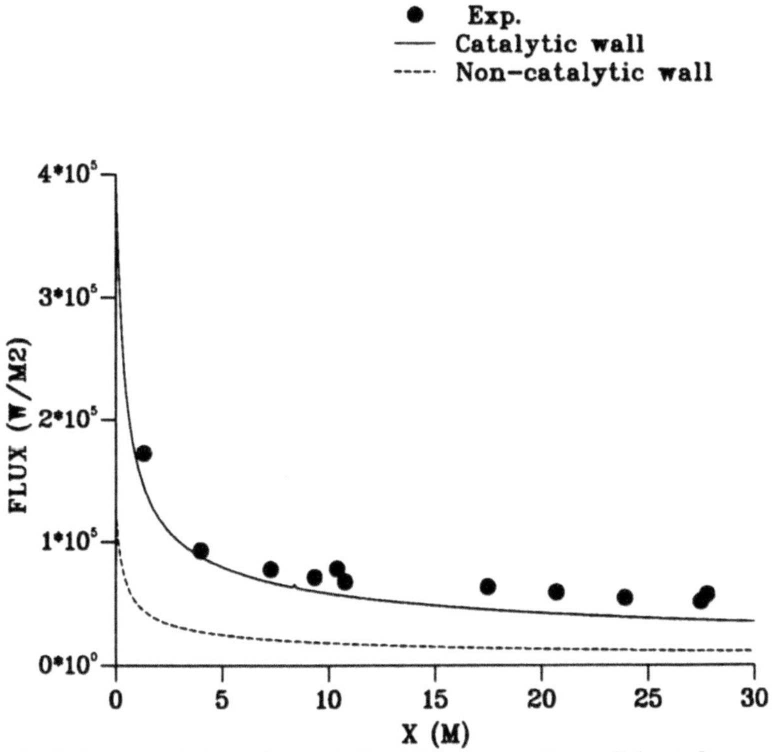

Figure 1: Influence of the wall catalytic efficiency on the wall heat flux prediction

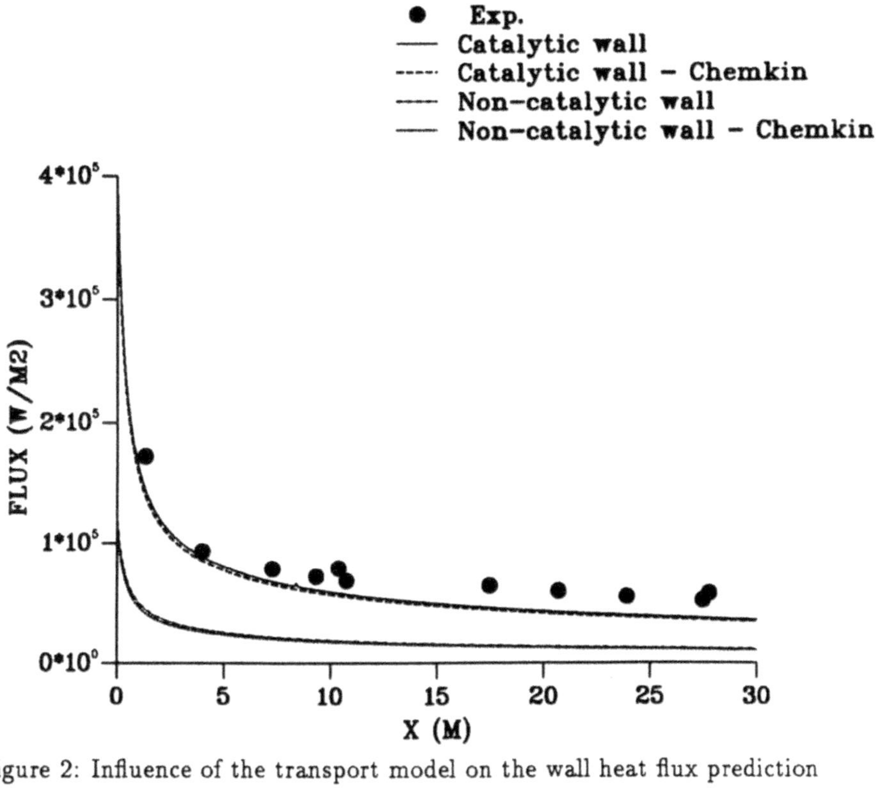

Figure 2: Influence of the transport model on the wall heat flux prediction

observed if higher temperatures were achieved.

The situation is not so good when thermal non-equilibrium is considered. There is presently a large effort being done in this field, and consequently a variety of models yielding somewhat different results, as well concerning the thermal relaxation process itself as the thermal non-equilibrium/chemistry coupling.

Concerning mixture transport properties, gas kinetic theory [12] yields models for transport properties of individual species together with mixing rules for equilibrium flows. However, the interaction potentials between species or the collision integrals have to be prescribed. There is a large scatter in these data which can have some importance, as shown on figure 2 where two data sets for the dissociated air species interaction potentials yield a 8 % difference in the wall heat flux on a non-catalytic wall for the same 71 km altitude test case and up to 20 % at higher altitudes. Moreover, mixture transport properties are complex to compute so that there is a great effort to derive simpler mixing rules which can also introduce some errors. At last, the influence of non-equilibrium on transport properties is generally not accounted for.

Finally, the thermodynamic properties of species are known with good accuracy and approximate expressions are available with enough accuracy for CFD approaches. Similarly, the mixing rules to compute the thermodynamic properties of a mixture from the thermodynamic properties of each species are quite simple and can be found in handbooks of thermodynamics or hypersonics flows [16].

2.2. Validation with respect to experimental data

Two main sources of experimental data can be used to validate models for high-enthalpy flows, namely flight data and wind-tunnel data.

Due to the large expansions required in hypersonic flows, high-enthalpy nozzle flows are frozen from both a thermal and a chemical point of view. Therefore, thermal non-equilibrium effects are larger in experiments than in re-entry flows and may play an important rôle when comparing CFD and wind-tunnel results. Hence, flight conditions cannot be duplicated in wind tunnels. Moreover, due to the reduced size of models, if upstream flow characteristics are roughly reproduced, the relative importance of advection and relaxation time scales, i.e. a Damköhler number, is not conserved. This can be partly circumvented by duplicating the density times length product. It ensures the above Damköhler number conservation, but only for binary reactions (dissociation and shuffle reactions), not for the other processes. At last, many high-enthalpy wind-tunnel flows are polluted and the incidence of this pollution is unknown.

The key problem with flight data is an instrumentation one. What measuring device can fly ? Up to now, mainly wall pressures and wall heat fluxes have been obtained. Very few in-field data are available, such as a global measure of electron density through radio-attenuation measurements. Electron-beam techniques could be used to measure density profiles.

When wall values are considered, wall pressure is quite insensitive to the chemistry and of course to the transport. On the other hand, wall heat fluxes are first ruled by the wall catalytic efficiency. As an illustration, the wall heat fluxes measured on a metallic flat plate in a dissociated air flow [7] were accurately computed by Eldem and myself (unpublished) using a perfect gas assumption as the wall is catalytic so that all the dissociation energy is released at the wall. On a metallic (highly catalytic) surface, the wall heat flux is quite independent of the chemical model while it is weakly affected by the transport. This is a low sensitivity test for the transport model. If one could be able to accurately estimate the wall catalytic efficiency, the lower it is, the more the wall heat flux is sensitive to the chemical and transport models.

Therefore, in-flow measurements are desirable to validate high-enthalpy flow models. Simple optical controls can give first information. The shock stand-off distance is a good candidate to check chemical and thermal relaxation models. A complementary information about iso-density contours can be deduced from Schlieren pictures [13] and can also be used to validate chemical and thermal relaxation models.

Finner measurements, such as flow temperature, mass fractions or light emission measurements are difficult to perform in free-flight but easier (quite still difficult) in wind-tunnels. In the inviscid part of the flow, e.g. behind a shock-wave, they can provide very useful information to validate chemical and thermal relaxation models.

It seems more difficult to extract relevant information to validate the transport model. In wind-tunnels, boundary layers over models are thin and cannot be easily probed while very few in-flow measurement techniques can fly. An alternative could be to probe nozzle wall boundary layers. Moreover, the transport models are quite complex as they rely upon collision integral data and only a global check of the viscosity, thermal conductivity or diffusion predictions could be achieved.

3. Turbulence model validation

3.1. Present status of hypersonic turbulence knowledge

For air-breathing hypersonic airplanes, turbulence modelling plays a major rôle to characterize the boundary layer at the engine inlet, to optimise the combustion in the scramjet or to study the turbulent flow in the nozzle.

It is well known that, as the turbulent Mach number increases, the turbulent motion becomes compressible. A striking example of the importance of compressible turbulence is the modification of the mixing layer spreading rate versus the convective Mach number as shown on figure 3. No steady flow computation, with an incompressible turbulence model is able to predict such an evolution.

Another well-known example of the importance of experimental data and of the present turbulence modelling status is the prediction of the skin-friction coefficient evolution on an adiabatic flat plate versus Mach number. Before experimental data were available, various theories yield a large variety of predictions (figure 4). Once experimental data have been obtained, theories tried to reproduce them (figure 5). It must however be pointed out that all present turbulence models are still far from being able to reproduce the observed

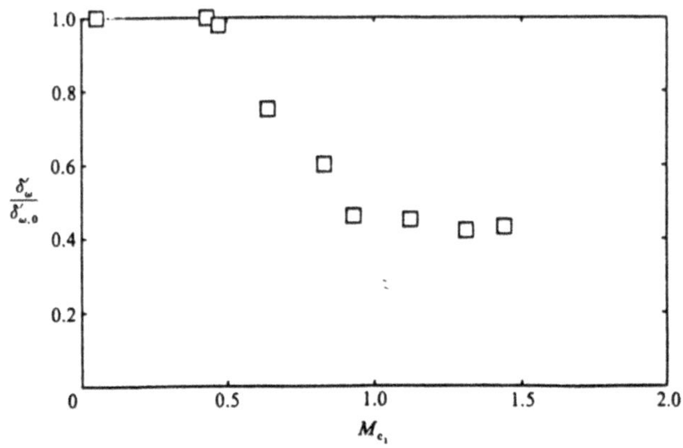

Figure 3: Evolution of the mixing layer spreading rate versus convective Mach number from [14]

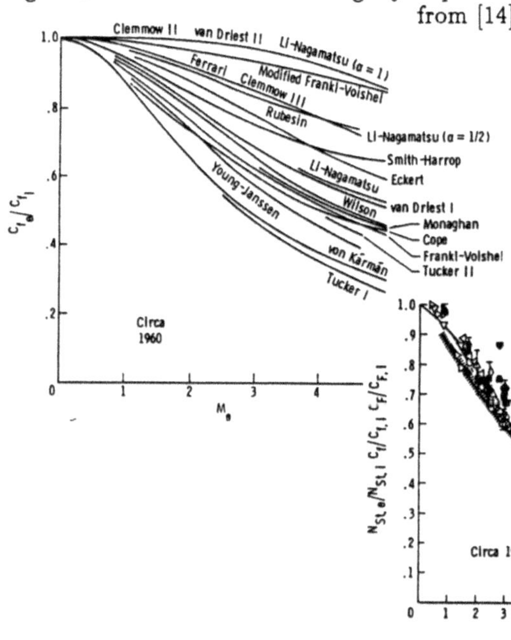

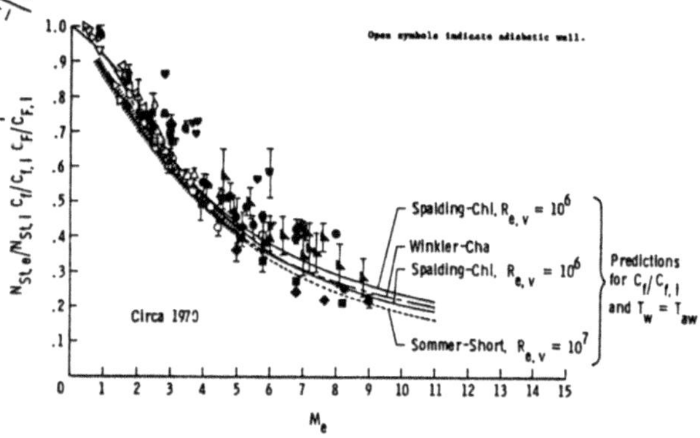

Figure 4: Skin-friction coefficient predictions on an adiabatic flat plate versus Mach number (60's status)

Figure 5: Skin-friction coefficient predictions on an adiabatic flat plate versus Mach number (70's status)

from [5]

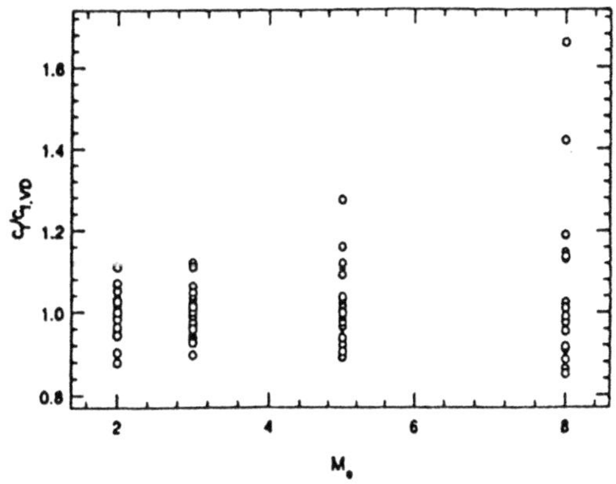

Figure 6: Skin-friction coefficient on an adiabatic flat plate versus Mach number, normalised by "van Driest II" formula values
from [4]

experimental behaviour, as shown on figure 6 which gives the scatter of turbulence model results obtained during the last Stanford turbulence workshop.

Our recent hypersonic boundary layer studies [1, 3] support Rubesin's conclusions [15] : in hypersonic boundary layer flows, the near wall region turbulence is in equilibrium with the mean motion and can be described with simple, even incompressible, turbulence models as pointed out earlier by Bushnell [6] while compressibility and strong non-equilibrium effects can occur in the outer region so that more complex models are required there.

Two main problems have to be solved to improve turbulence modelling. The first one is the near wall modelling as our knowledge of the turbulence structure in the wall region is poor, even in incompressible flows. Some characteristic behaviours are identified, on the one hand, very close to the wall and, on the other hand, in the logarithmic region. They are used to tune models but quite no information is presently available in the buffer layer. The use of ad-hoc wall damping functions is a proof of the lack of knowledge in the wall region. The second problem is the extension of incompressible turbulence models to compressible turbulence. The problem was tackled mainly by Rubesin's group at NASA Ames in the late seventies and the eighties with emphasis on shock/boundary layer interaction problems. Only recently, a major interest in compressible turbulence has occurred, as well for astrophysics as for the mixing layer and other hypersonic flow problems. Various models are presently emerging but require validation. As an illustration, several models, using very different fundamental arguments, are able to reproduce the variation of the mixing layer spreading rate shown on figure 3. The use of very different arguments shows that efforts are surely still required to understand the real physics.

Turbulence models development and validation rely upon analysis of and comparisons with basic experimental data. We only consider boundary layer flows as we think it is presently the best way to test turbulence models in an economical way without facing numerical problems. Only wind-tunnel data will be considered here, as experiments are easier, but not so easy, to perform at "moderate" Mach numbers, i.e. up to Mach numbers about 10. Let us however mention that flight experiments on experimental vehicles could also be achieved, providing both wall and in-flow measurements, as pressure and temperature surveys of the boundary layer are possible in this Mach number range. Such experiments could give access to higher Reynolds numbers than wind-tunnel experiments.

An important survey effort of available compressible turbulent boundary layer experimental data has been performed by AGARD [8, 9, 11, 10]. Analysis of the available data shows that there is a real lack of "good" data for hypersonic flows.

Our definition of "good" data is close to the one proposed in [9, page 2]. It means that the flow is well enough documented so that it can be computed without any "ad-hoc" hypothesis. This leads to the following requirements :

- Boundary layer hypothesis are fulfilled, mainly there is no significant normal pressure gradient or this pressure gradient is documented.
- Boundary conditions, namely wall pressure and temperature are measured at close intervals. Also, departures from two-dimensionality, in a two-dimensional flow must be quantified as well as free-stream conditions such as the turbulence level. At last, the end of the transition regime has to be estimated.
- Wall data, i.e. wall skin friction and wall heat fluxes are measured at various stations as they are the values which are used for vehicle design so that the real goal is to reproduce them with models.
- Boundary layer profiles are measured at various stations, if possible at the same station as the wall values, in order to provide both initial profiles to start computations with and stations to compare predictions with experiments after a significant boundary layer development. Independent measurements of mean flow properties are required to avoid the use of an hypothetical velocity/temperature relation. Independent measurements of pressure, temperature and density can give an error estimate on these quantities.
- Turbulence measurements, such as the Reynolds stresses, the turbulent heat flux or the density variance are also expected.
- At last, measurements error estimates will be appreciated.

Direct numerical simulations of compressible turbulence are still limited to low Reynolds number and presently provide information on simple flows such as homogeneous turbulence or mixing layers, thanks mainly to CTR and ICASE efforts. They can presently yield some guidelines for the modelling effort but not supply wind-tunnel data which are still the only way to investigate more complex flows and/or higher Reynolds numbers. However, it must be pointed out that these numerical simulations give access to quantities that play an important rôle in turbulence modelling such as pressure/velocity correlations which are very difficult to measure. Moreover, they are presently the only way to have accurate turbulence data very close to the wall.

4. Conclusion

After a period of great belief in "numerical" wind-tunnel, the present hypersonic modelling status tends to show that there is still information to extract from "physical" wind-tunnels to improve our knowledge and modelling of the high-enthalpy flows and turbulence effects. This is however far from an easy task as the important quantities are difficult to measure and less information can be extracted from the easily available quantities.

It is the author's belief that simple flow experiments in high-enthalpy wind tunnel could give important contribution to the improvement of high-enthalpy flow models as in-flow measurements can be performed to validate chemical and thermal relaxation models. Wall heat flux data may be more a way to estimate material catalytic efficiency than to improve models. Nozzle wall boundary layer surveys could also be a way to check transport model and improve boundary layer modelling.

Concerning turbulence modelling, there is a dire need of good experiments to validate models. New, well documented, turbulent flow (boundary layer but also mixing layer, wakes ...) experiments are required to improve turbulence models. Direct numerical simulations of compressible, high Reynolds number, turbulent boundary layer flows seem presently out of reach but could provide interesting data in the near future.

The author wishes to acknowledge Dr C. Eldem, F. Leclère and A. Daiss for their contributions in reacting boundary layer computations, O. Rouzaud, M. Marty and E. Desmet for turbulence models studies.

References

[1] D. Arnal and B. Aupoix. Hypersonic boundary layers : Transition and turbulence effects. In *First European Symposium on Aerothermodynamics for Space Shuttle*. ESTEC – Noordwijk, May 28–30 1991.

[2] B. Aupoix. An introduction to real gas effects. In *Special Course on Aerothermodynamics of Hypersonic Vehicles*. AGARD–FDP VKI Lecture Series – AGARD Report 761, 10 May – 3 June 1988.

[3] B. Aupoix and J. Cousteix. Analysis of turbulence models for hypersonic boundary layers. In *Eighth Symposium on Turbulent Shear Flows - Munich*, pages III–2, September 9–11 1991.

[4] P. Bradshaw, B. E. Launder, and J. L. Lumley. Collaborative testing of turbulence models. AIAA Paper 91–0215 29th Aerospace Science Meeting Reno, Nevada, January 7–10 1991.

[5] D. M. Bushnell. Simple compressible strains. In S. J. Kline, B. J. Cantwell, and G. M. Lilley, editors, *Volume II - Taxonomies, Reporter's Summaries, Evaluation, and Conclusions*, pages 812–836. 1980–81 AFOSR–HTTM–Stanford Conference on Complex Turbulent Flows, 1981.

[6] D. M. Bushnell and D. J. Morris. Shear–stress, eddy–viscosity, and mixing-length distributions in hypersonic boundary layers. Technical Memorandum TM X–2310, NASA, August 1971.

[7] R. A. East, R. J. Stalker, and J. P. Baird. Measurements of heat transfer to a flat plate in a dissociated high-enthalpy laminar air flow. *Journal of Fluid Mechanics*, 97(4):673–699, 1980.

[8] H. H. Fernholz and P. J. Finley. A critical compilation of compressible turbulent boundary layer data. AGARDograph AG–223, AGARD, June 1977.

[9] H. H. Fernholz and P. J. Finley. A critical commentary on mean flow data for two–dimensional compressible turbulent boundary layers. AGARDograph AG–253, AGARD, May 1980.

[10] H. H. Fernholz, P. J. Finley, J. P. Dussauge, and A. J. Smits. A survey of measurements and measuring techniques in rapidly distorted compressible turbulent boundary layers. AGARDograph AG–315, AGARD, May 1989.

[11] H. H. Fernholz, P. J. Finley, and V. Mikulla. A further compilation of compressible turbulent boundary layer data with a survey of turbulence data. AGARDograph AG–263, AGARD, November 1981.

[12] J. O. Hirschfelder, C. F. Curtiss, and R. B. Bird. *Molecular Theory of Gases and Liquids*. John Wiley and Sons, 1954.

[13] H. G. Hornung. 28th Lanchester memorial lecture – Experimental real-gas hypersonics. *L'Aéronautique et l'Astronautique*, 133(6):11–21, 1988.

[14] D. Papamoschou and A. Roshko. The compressible turbulent shear layer : An experimental study. *Journal of Fluid Mechanics*, 197:453–477, 1988.

[15] M. W. Rubesin. Turbulence modeling for aerodynamic flows. AIAA Paper 89–0606 27th Aerospace Science Meeting Reno, Nevada, January 9–12 1989.

[16] W. G. Vincenti and C. H. Kruger, Jr. *Introduction to Physical Gas Dynamics*. Robert E. Krieger Publishing Company, Malabar, Florida, 1965.

COMPUTATION AND EXPERIMENTATION OF FREE PISTON SHOCK TUNNELS - EXAMPLE OF THE MARSEILLE FACILITY

R. Brun, Y. Burtschell, L. Dumitrescu, M. Dumitrescu, L. Houas,
L. Labracherie and D. Zeitoun
Université de Provence
Centre de Saint -Jérôme,
Lab. SETT, Dept MHEQ, Case 321
13397 Marseille, France

ABSTRACT. A numerical analysis of the hypersonic free piston shock tunnel in construction in Marseille is presented. Since the few wind tunnels of this type actually operational (Australia) suffer of performance losses when compared to expected data given by the analytical theory (Stalker, 1967), an axisymmetric numerical simulation is performed by a finite volume method which allows to emphasize the unsteady phenomena which restrict the expected performances. Taking into account these effects, it is possible to obtain simulations of the complete wind tunnel with initial conditions allowing to realize high enthalpy flows characteristic of the reentry phase of the Hermes european shuttle. Preliminary results of pressure measurements in the compression chamber are also presented and compared with computational data.

1. Introduction

Actually, the development of free piston shock tunnels is observed due to the theoretical possibility of obtaining higher performances than those obtained in other types of facilities for which combustion or heated light gases (Lukaziewicz, 1973), are used in the driver section. Thus, a new high enthalpy hypersonic free piston shock tunnel is in construction in Marseille (Université de Provence) which must be able to produce high stagnation enthalpy flows. The existing facilities have partially confirmed these predictions but some deficiencies have appeared concerning for example the values of pressure behind the reflected shock wave which are far from attaining the calculated values (Stalker and

A. Boutier (ed.), New Trends in Instrumentation for Hypersonic Research, 51–64.

Hornung, 1970). An analysis of the wave phenomena in the whole facility may explain these deficiencies and may constitute a basis for an optimization of this type of facility.

First experimental results of pressure measurements at the end of the compression chamber of the facility are also presented: They confirm the comptutational analysis exposed here after for the so-called Phase I (diaphgram 2 closed, Fig.2), showing the pressure increase, the influence of the wave system and also strong pressure oscillations due to resonant effects typical of the geometry of the connection between compression and shock tube chambers.

2. Computational analysis

2.1 PRINCIPLE AND GEOMETRY OF THE FACILITY

In order to illustrate the following description, a theoretical (x,t) diagram and a scheme of the apparatus are presented in Fig.1.

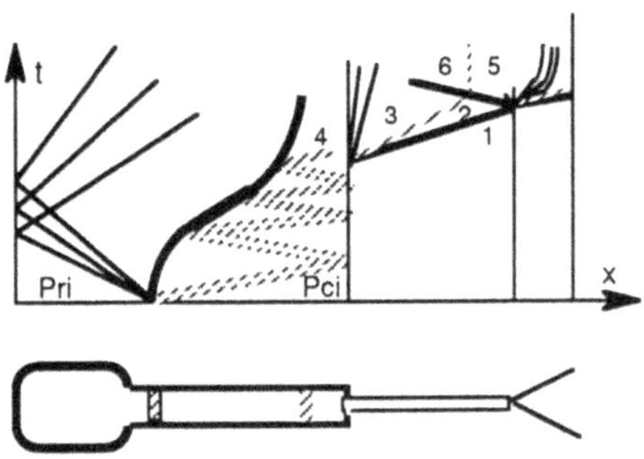

Figure 1 : Theoretical (x,t) diagram and scheme of the apparatus

A compression chamber of 7 m long and 20 cm diameter contains helium compressed by a 60 Kg piston moved by high pressure air (maximum 350 bar) coming from an air reservoir of 3 m long and 34 cm diameter and separated from the compression chamber by a double diaphragm system. When the pressure in front of the piston becomes high enough, a downstream diaphragm is broken, and a shock wave propagates in an air filled shock tube of 6 m long and 7 cm diameter. This tube is followed by a converging diverging axisymmetric nozzle of about 1,13 m long (6 mm throat and 400 mm exit diameter) in which air, under high pressure and high temperature conditions, is expanded so that a Mach number about 10 is obtained in the test section of the exit nozzle.

In a first step, a steady computation of different quantities inside the shock tube has been performed with several assumptions, in particular equilibrium for the chemical state of the air inside the shock tube, and tailored conditions for the contact surface. Afterwards, an analytical analysis (Stalker, 1967) allows to obtain relations between the rupture parameters and the initial conditions of the compression system : In this analysis it is assumed that pressure increase inside the compression tube and air expansion in the reservoir are isentropic and uniform all along these chambers ("ideal theory"). However, these hypotheses are not realistic for high velocities of the piston, and an unsteady quasi one-dimensional numerical simulation of the whole compression system has been made by Burtschell et al. (1989) : One conclusion of this analysis is that the initial pressure of the tank air obtained analytically is not sufficient to reach the necessary rupture conditions. However this kind of simulation does not allow to obtain easily the values of the stagnation conditions as a function of rupture conditions and the cross section changes are not correctly taken into account. Thus it has been necessary to realize a two dimensional simulation of the whole free piston shock tunnel in order to foresee correctly its performances.

2.2 NUMERICAL METHOD AND MESH

The finite volume numerical method (Colas, 1989), is applied with a decentred scheme allowing the computation of strong discontinuities. An explicit unsteady method (Desideri and Dervieux, 1988), with the conservative law form of the Euler equations in the spatially two-dimensional axisymmetric case has been used with a Van-Leer flux decomposition. In Fig.2, the mesh is presented with the different parts constituting the facility as well as the piston and the positions of the 3 diaphragms. The dimensions are those of the Marseille facility. The numerical simulation has been operated in three different stages corresponding respectively to the compression phase (diaphragm 2 closed), the shock wave propagation (diaphragm 3 closed) and the expansion in the nozzle.

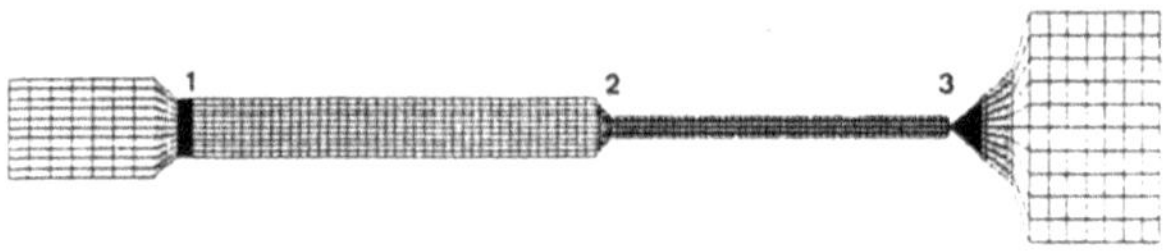

Figure 2 : Used axisymetric mesh of the whole free piston shock tunnel

2.2.1 *Phase I : Phenomenological Analysis and results*. The piston displacement inside the compression tube is itself divided in two phases. The first one corresponds to the acceleration of the piston (initially at rest), under the influence of the pressure of the tank air. During this phase the piston gains its kinetic energy. The second phase, during which the piston continues to compress helium, corresponds to the restitution of this kinetic energy. It is during this last phase that the rupture pressure of diaphragm 2 (P4), which is a function of the chosen incident Mach number, is reached with a theoretical piston velocity (Up) at least equal to U4, steady flow value given by the ideal theory. If the diaphragm rupture between the compression tube and the shock tube is not allowed, the pressure inside the compression chamber becomes maximum when the piston velocity is equal to zero. At this instant, the piston bounces towards its initial position.
The simulations presented here correspond to future experimental cases : In these cases, the initial air tank pressure is maximum, which gives the highest possible incident Mach number for a given initial air shock tube pressure, and a comparison is made, versus Mach number, between ideal and simulated values of the rupture piston velocity. Thus, in Fig.4, the optimum point of good running is obtained with an initial air pressure of 1.5 bar for an incident Mach number about 8.

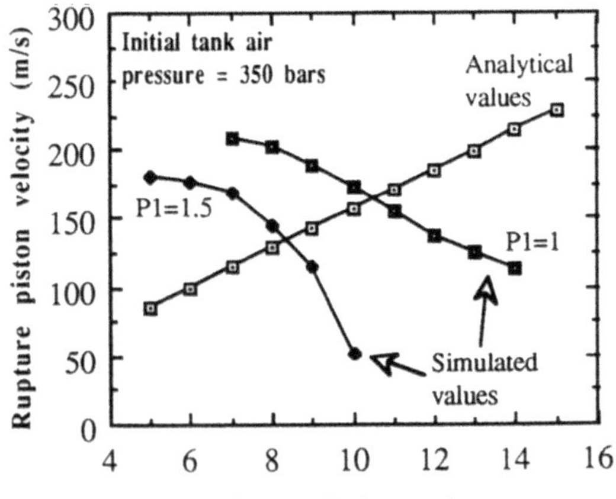

Figure 3 : Ideal and simulated values of the rupture piston velocity versus Mach number

This result also deduced from the method of the characteristics (Burtschell, 1990) validates both numerical methods, each one with respect to the other one. Moreover, for a lower air pressure (1 bar), a value of 10 for the incident Mach number is reached, which is the value corresponding to stagnation conditions (6500 K and 1200 bar) allowing to simulate in the test section flight conditions of the reentry phase of Hermes.

Thus, the initial conditions are chosen with P1=1bar (Fig.1) for the following two-dimensional numerical simulation, whereas the rupture conditions are respectively P4=688bars and Up=158 m/s (minimum rupture piston velocity).

During its displacement, the moving piston generates waves which appear in Fig.4 representing an (x,t) velocity diagram along the axis of the closed compression system.

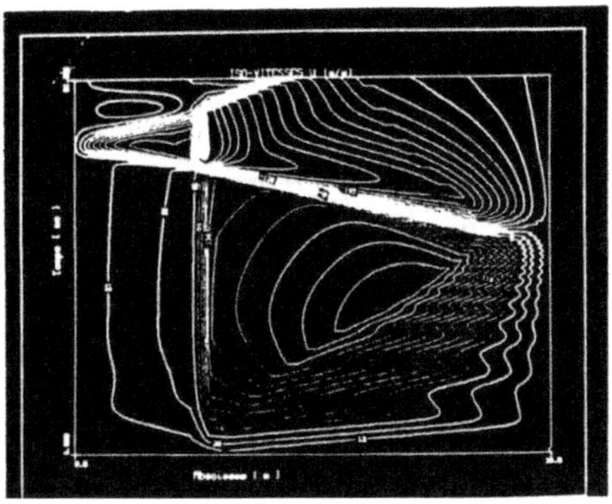

Figure 4 : (x,t) Velocity diagram along axis of the closed compression system

The piston displacement versus time may be noted as well as the compression waves travelling from the piston to the end of the compression system and vice versa. The velocity of these waves is faster when the helium temperature increases. On the contrary, on the rear face of the piston, the expansion waves are slower and slower because of temperature decrease of the tank air. The same, in Fig.5 which represents an (x,t) axis pressure diagram, the oscillatory moving piston is visible with an upstream shock wave and reflection at the section variation between the reservoir and the compression tube.

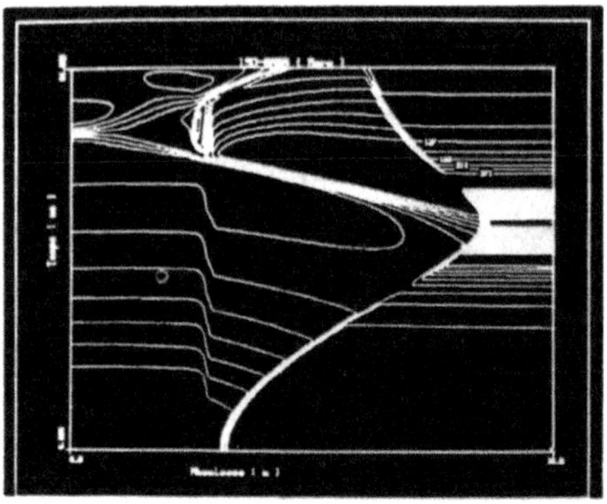

Figure 5 : (x,t) Axial pressure diagram along the closed compression system.

The aim of the next discussion is the comparison of the performances of the compression system obtained during this numerical simulation with the ideal values. If we refer to the analytical ideal computations (Burtschell, 1990), the maximum pressure numerically obtained, with an initial theoretical tank air pressure requiring an overestimation, is 1300 bars. However, the maximum speed of the piston is 238 m/s, that is 7% above the ideal value, as it is seen in Fig.6 which represents versus time the axis velocity for 10 points of the mesh equally divided between the front of the piston and the second diaphragm. Moreover, some oscillations are observed which represent in fact the compression waves. These oscillations disappear when the piston bounces and generates expansion waves.

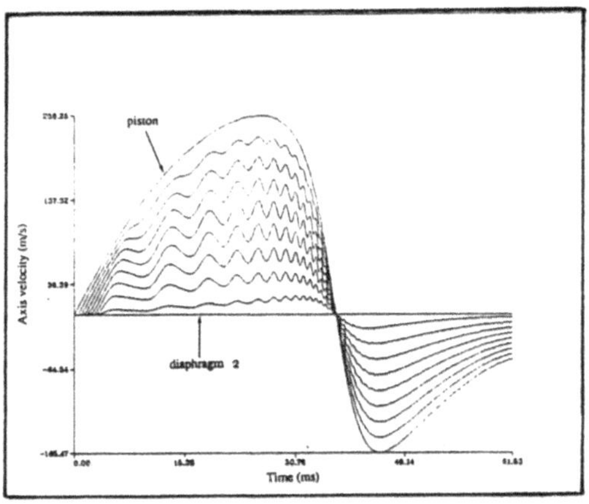

Figure 6 : Axial velocity, versus time, for 10 points of mesh equally divided between the front of the piston and the second diaphragm

Thus, the rupture values are reached only with an initial ideal tank air pressure greater than the ideal value. Then it appears that the wave systems observed in Fig.2 strongly modify the performances of the facility. This is easily explained by the fact that, when the piston moves, the pressure on its rear face falls quickly, and of course is less than assumed ideally. The opposite process is created on the upstream piston face : Thus, the accelerating phase is less important than expected. However, the differences are more important for the high Mach numbers : This comes from the fact that the first expansion waves in the tank gas are independent of this Mach number. The result is an increase of the depression, at the rear face of the piston with the Mach number and consequently with the piston velocity. Then it is better, as far as possible, to use relatively weak volumetric compression ratios which gives results close to isentropic processes.

The instant of the rupture occurs 32.8 ms after the starting of the piston. The velocity is then 171m/s, 7% over theoretical value whereas the rupture pressure is 688 bars which is in agreement with the ideal computations.

2.3.2 *Phase II - Phenomenological Analysis and results*. The following results concern the second phase corresponding to the instantaneous opening of the diaphragm 2 and to the shock wave propagation along the shock tube up to the diaphragm 3 which is assumed to remain closed. In Fig.7, an (x,t) axis velocity diagram after diaphragm 2 rupture is presented : The incident and reflected shock waves as well as the piston displacement and the pressure gradient at the section variation are cleary visible. The small number of isovalues inside the air of the tank and behind the reflected shock wave proves that the pressure is quasi constant in these regions. After the opening of the diaphragm 2, the piston compression continues because the expansion waves coming upstream are relatively weak, which is in agreement whith the ideal computation.

This result is confirmed (Fig.8) by the evolution of the axis pressure versus time, for 10 points of mesh equally divided between the front of the piston and the beginning of the section variation between the compression tube and the shock tube. Independently of the oscillations which are an effect of wave reflections, it is to be noted that the mean pressure inside the compression tube is approximately constant until the arrival of the reflected shock wave which occurs 3.2 ms after the rupture diaphragm. Until this instant, the total pressure variation has been about 11% around the rupture value (7% above and 4% below).

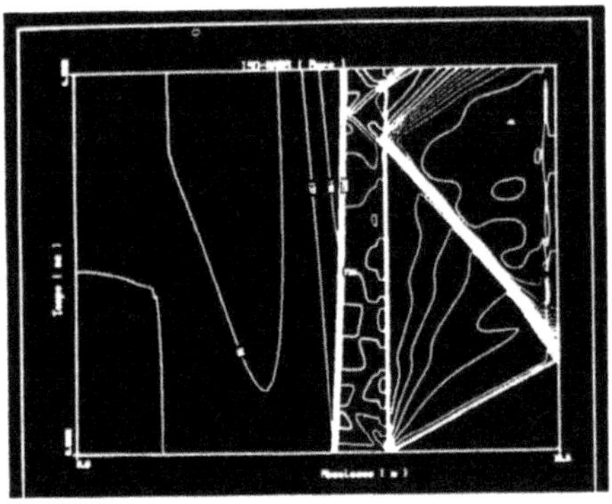

Figure 7 : (x,t) Axial velocity diagram after diaphragm 2 rupture

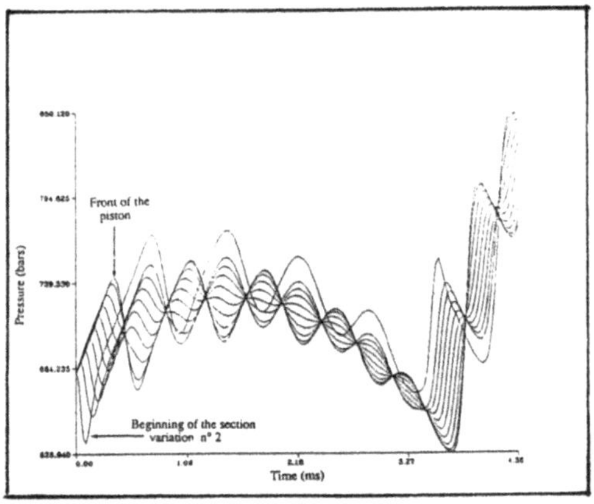

Figure 8 : Evolution of the axis pressure versus time, for 10 points of mesh equally divided between the front of the piston and the beginning of the section variation 2

With the same conditions as above, the evolution of the axis velocity is given in Fig.9. It is interesting to note that the piston speed is not sensitive to these oscillations and decreases monotically. In Fig.10 the temporal pressure evolution is presented at four different abscissa along the shock tube, at the beginning of the section variation, at the diaphragm position, at 36 cm after it and at the end of the shock tube : At all stations, the pressure is constant and especially the stagnation pressure at the end of the shock tube, where it is 7% below the theoretical value. Thus, it seems that the ideal value of the piston velocity (at least equal to U4) is necessary to obtain correct flow conditions. In order to confirm this fact, a computation has been performed with a rupture piston velocity which is 1/3 of the ideal value : In this case the piston displacement is no more able to compensate the pressure loss of the driver gas and a quick pressure decrease reaching 50% of the rupture pressure is observed only 3 ms after the diaphragm opening.

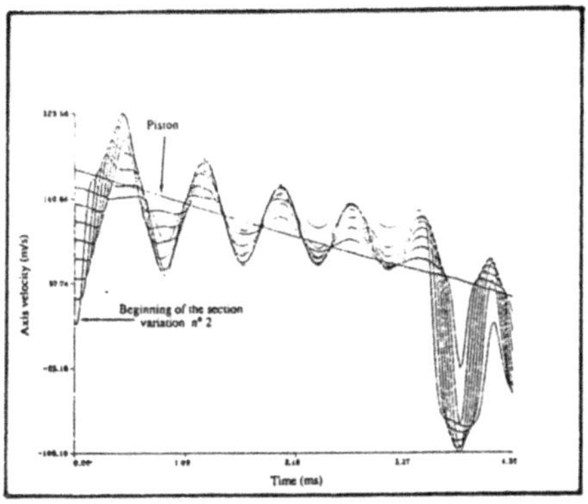

Figure 9 : Evolution of the axis velocity versus time, for 10 points of mesh equally divided between the front of the piston and the beginning of the section variation 2

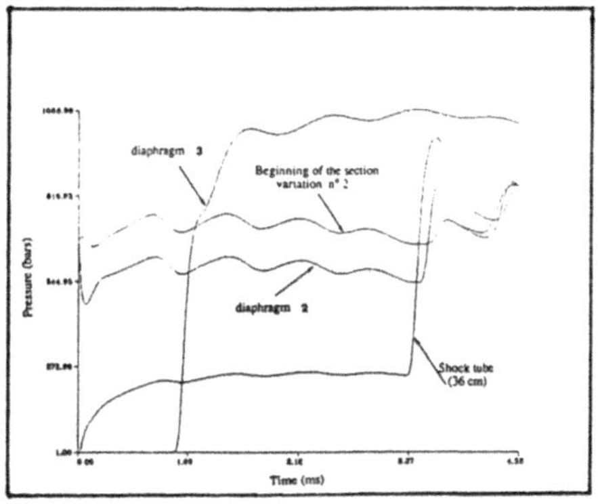

Figure 10 : Temporal pressure evolution at four different abscissa along the shock tube

In order to summarize results concerning this second phase, Table 1 supplies the relative discrepancies, about hot flow velocity and stagnation pressure, between numerical and ideal values. For all the simulations the rupture pressure is equal to the ideal value : The estimation of the accuracy of the method is within 5%, value found for a simple shock tube. Moreover, the disastrous effect of a discontinuous section between a long high pressure chamber and shock tube may be noted. At the diaphragm opening instant and for the same length of the HP chamber, comparisons are made about the interest to have a piston velocity at least equal to the ideal value (last two schemes of Table 1).

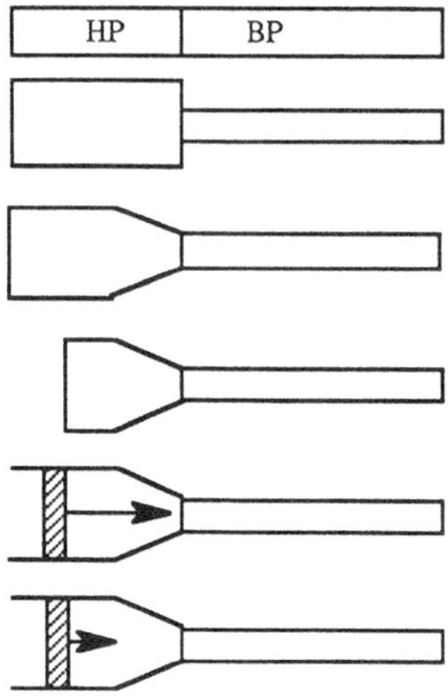

$\frac{\Delta V_2}{V_2}$	$\frac{\Delta P_5}{P_5}$
-4%	-5%
-21%	-55%
-6%	-8%
-5.5%	-23%
-4.5%	-7%
-6.5%	-27%

Table 1 : Relative discrepancies, about hot flow velocity and stagnation pressure, between numerical and ideal values

2.3.3 *Phase III - Phenomenological analysis and results.* The third phase includes the arrival of the incident shock wave on diaphragm 3 inducing its rupture, and consequently generating the flow inside the nozzle. This flow consists of a starting phase and a steady phase. The first one is visible in Fig.11 which gives the evolution of the axis velocity along the free piston shock tunnel, the flow tending to steady conditions. The first phase lasts 0.7 ms after the instantaneous diaphragm opening which occurs 33.7 ms after piston departure. The steady phase continues during at least 2 ms, time at which the simulation has been stopped, as seen in Fig.12 which presents the temporal evolution of the axis velocity at the nozzle test section. This last period is, in fact, composed of two phases which correspond first to the hot air flow and then to the helium contaminated flow. However, problems due to numerical diffusion on species concentrations have not allowed to specify the duration of each flow phase. While the theoretical profile of helium concentration at the nozzle exit must give a step from 0 to 1, we have here a smooth profile of the helium arrival and at the present time, it is not possible to evaluate the duration of the hot flow.

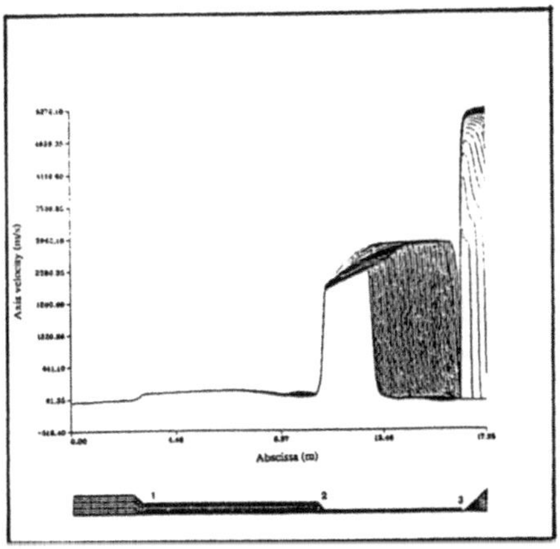

Figure 11 : Evolution of the axis velocity along the free piston shock tunnel

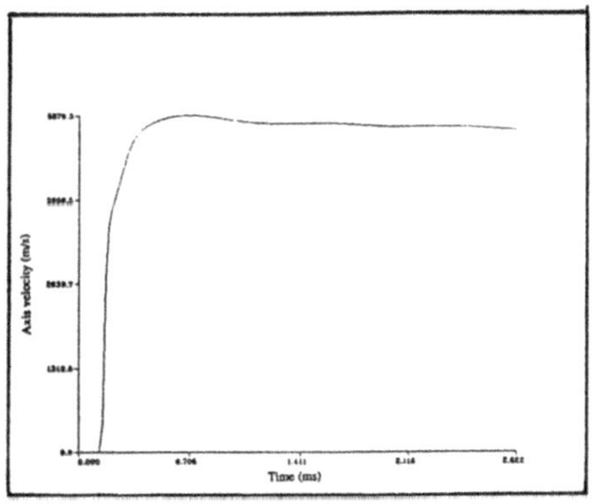

Figure 12 : Temporal evolution of the axis velocity at the nozzle test section

3. First experimental comparisons

Some cases corresponding only to the phase I have been examined experimentally by measuring the pressure close to the end of the compression tube. Experimental conditions correspond to a tank pressure of 105 bar (N2 and then He). Thus a comparison is presented on Fig. 13 and 14 between experimental and computational pressure (N2 case): A qualitative agreement may be observed for the wave system and for the values of the obtained pressures. The agreement is much better when helium is used as the test gaz and Fig.15 and 16 represent the experimental and computational pressure evolution in this case, including the rupture of the diaphgram 2 but without shock tube flow: The drop in pressure corresponds to this rupture, and superimposed oscillations are characteristic of the resonance effects due to the particular geometry of the connection between compression and shock tube chambers: they represent strong pressure variations which may disturb the rupture process.

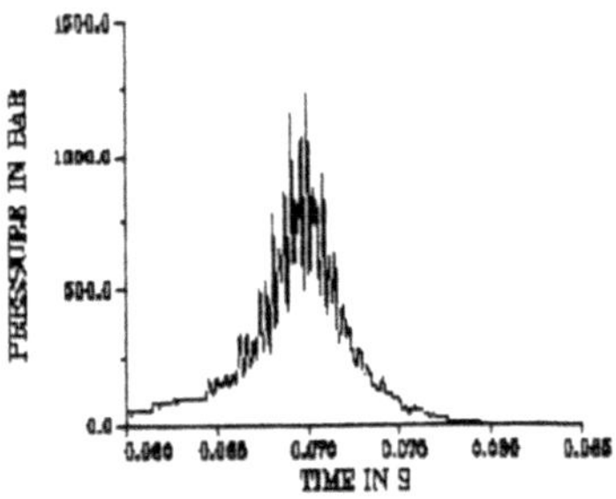

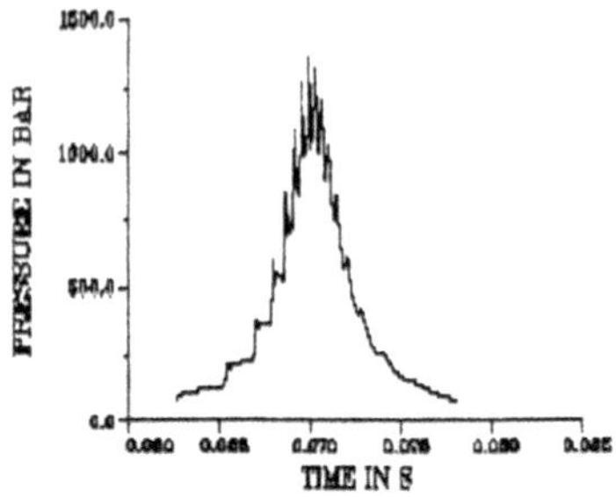

Figure 13 : Experimental pressure evolution at the closed end of the compression chamber

Figure 14 : Numerical simulation of the pressure evolution

(Case N2)

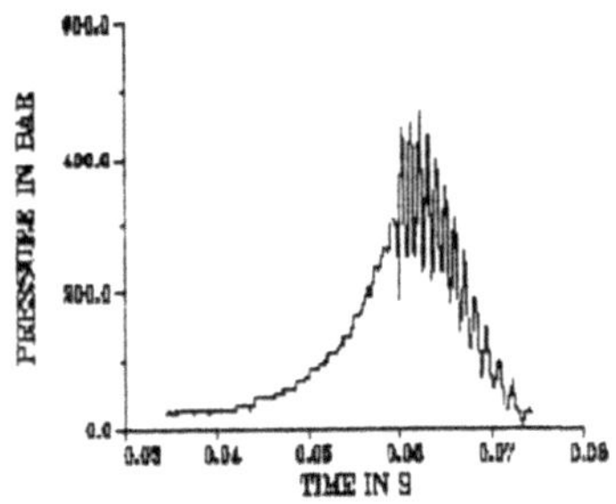

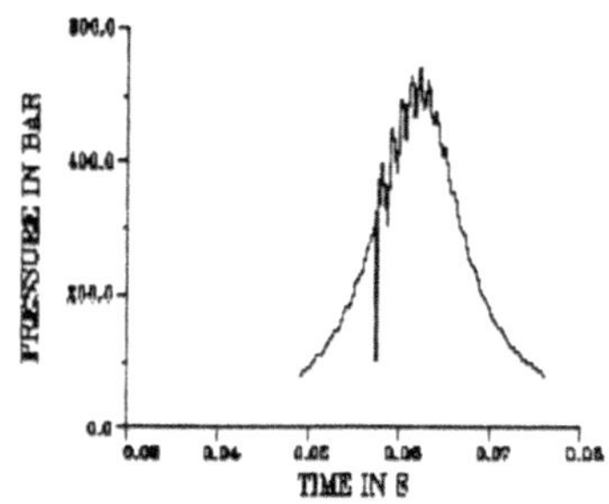

Figure 15 : Experimental pressure evolution at the end of the compression chamber with rupture of the diaphragm

Figure 16 : Numerical simulation of the pressure evolution

(Case He)

4. Concluding remarks

In order to optimize the use of a free piston shock tunnel, quasi one-dimensional and axisymmetric numerical simulations have been performed. As a first conclusion, the strongly unsteady character of the compression phase has been cleary observed. Therefore in order to obtain ideal rupture conditions, the air tank pressure needs to be greater than ideal values. Then, in order to obtain a compression as uniform as possible, it is necessary to use some reasonable compression ratio (about 10). Moreover, at the diaphragm opening instant, it has been found that the rupture piston velocity must be at least equal to its ideal value in order to prevent a too important decrease of the driver gas pressure and then to obtain the expected stagnation conditions behind the reflected shock wave. About the expansion in the nozzle, the starting and steady phases are described. However, problems due to numerical diffusion have not allowed to specify the duration of the useful flow.

The improvement of the results requires a finest modelization of the numerical flux in order to restrict the numerical diffusion and it would be also necessary to take into account the finite opening time of the diaphragms and the boundary layer effects.

From the experimentalpoint of view, pressure measurements along the shock tube, nozzle and test section will be made in the next future before beginning local concentration and temperature measurements by Raman diffusion methods.

5. References

Burtschell Y.(1990) 'Performances, dimensionnement et simulation numérique d'une soufflerie hypersonique à choc réfléchi à piston libre', Thèse de doctorat, Université d'Aix-Marseille I

Burtschell Y., Colas P., Gubernatis P., Zeitoun D., Imbert M., Houas L. and Brun R. (1989) 'Numerical simulation of a free piston shock tunnel', 17th I.S.S.W., Bethlehem, U.S.A.

Colas P. (1989) 'Ecoulements hypersoniques eulériens : modèles physiques et numériques, application aux ondes de choc et aux tuyères bidimensionnelles', Thèse de doctorat, Université d'Aix-Marseille I

Desideri J.A. and Dervieux A. (1988) 'Compressible flow using unstructured grids ,. Von Karman Institute Lecture Series 1988-05

Lukaziewicz J. (1973) 'Experimental methods of hypersonic', Gasdynamics series vol.3, Dekker

Stalker R.J. (1967) 'A study of a free piston shock tunnel', AIAA Journal, Vol.5, No.12

Stalker R.J. and Hornung H.G. (1970) 'Two developments with free piston drivers', 7th I.S.S.W.

GROUND TEST FACILITIES AND INSTRUMENTATION FOR AEROTHERMAL AND AERO-OPTICAL STUDIES OF HYPERSONIC INTERCEPTORS

M.S. HOLDEN
Calspan-UB Research Center
P.O. Box 400
Buffalo, New York 14225

ABSTRACT. In this paper we describe the hypervelocity facilities and instrumentation for the aerothermal and aero-optical studies of hypersonic interceptors. We first review the major fluid dynamic phenomena that control the flowfield environment around an interceptor. The key role of duplicating flowfield velocity for the simulation of both fluid dynamic and real-gas effects is discussed. The design and operating characteristics of the Large Energy National Shock tunnel (LENS) facility is then described, and the test conditions that will be obtained for endo-interceptor studies are discussed. The surface and flowfield instrumentation that will be used to determine the aerothermal characteristics of the flowfield over the hypersonic vehicle is described. Finally we discuss the aero-optic instrumentation which is used to evaluate the optical degradation through the shock layer surrounding the vehicle.

1. Ground Test Simulation of Flows Around Hypersonic Interceptors

The key phenomena and major parameters that must be simulated over an interceptor seeker head are illustrated in Figure 1. On the relatively small seeker-head configurations, a key parameter that must be defined and controlled is the position of boundary layer transition. Not only is the heat transfer distribution strongly influenced by boundary layer transition, but so is the mean flowfield structure. Also, the distributions of mean and fluctuating densities across the boundary layer are strongly dependent on the position of transition. Transition itself is dependent on the freestream Mach number and Reynolds number, on the seeker-head configuration, and on the coolant blowing rate.

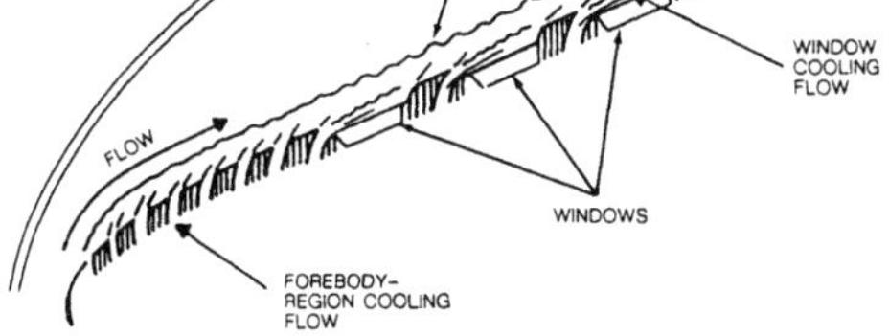

Figure 1 MAJOR PARAMETERS CONTROLLING APERTURE PERFORMANCE ASSOCIATED WITH FLOWFIELD PROPERTIES AND WINDOW CONFIGURATION OF A SEEKER HEAD

The flowfield structure over the forebody and, subsequently, over the windows is controlled principally by the velocity of the external flow and the coolant injection parameters for the forebody and windows. To simulate correctly coolant/flowfield mixing from the fluid dynamic viewpoint, it is important to duplicate the local convective Mach number of the injectant and the local Reynolds number of the flow.

A. Boutier (ed.), New Trends in Instrumentation for Hypersonic Research, 65–74.

In simulating the major aero-optic characteristics of the flowfield, it is important that the density and the temperature characteristics of the flowfield be replicated. To achieve this, it is necessary that both the Reynolds number and the total enthalpy of the freestream be duplicated, together with the detailed injection characteristics of the forebody and window cooling. It is of key importance to duplicate the flowfield velocity so as to obtain the correct density profiles through the shock layer—and, hence, the correct aero-optic distortion. This is illustrated in Figure 2, where are shown typical density and velocity distributions through the shock layer generated in flight, in high-enthalpy shock tunnels, and in state-of-the-art high Mach number blowdown facilities. Basically, the magnitude of the density difference across the viscous shock layer that controls the aero-optic distortion is directly related to the velocity. The convective Mach number across the shear layer, which is a key parameter in controlling the rate of mixing between the shock layer and the coolant flow, is also strongly influenced by velocity. Conventional low-velocity hypersonic facilities are capable of obtaining only 1/10 of the velocity and density ratios expected in flight.

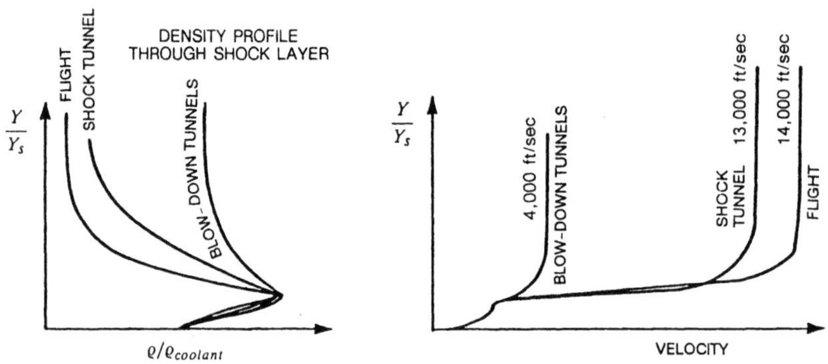

Figure 2 IMPORTANCE OF VELOCITY SIMULATION IN GROUND TEST FACILITIES TO ESTABLISH THE CORRECT SHOCK-LAYER DENSITY PROFILES OVER AN INTERCEPTOR

2. Characteristics of the Large Energy National Shock (LENS) Tunnel

2.1. DESIGN FEATURES OF LENS

LENS is a large, high-pressure but conventional, shock tunnel that is based on an upgrade of the existing 48" shock tunnel. The LENS has been constructed specifically for the purpose of generating hypervelocity flows to simulate the flow over small endo-interceptors traveling at speeds up to 18,000 ft/sec. It incorporates a number of features that have been evolved over the past 25 years of shock tunnel development and testing at Calspan. In the design of the facility, particular care was taken in developing a contamination-free hypersonic flow and in isolating the loads associated with operation of the tunnel from the model support system and the aero-optic instrumentation.

The LENS design is based on an 8-inch-diameter shock tube, driven by a 12-inch-diameter driver. The tunnel layout configuration is shown in Figure 3. The driver section of the tunnel is heated to 750^{0}F , and, because it is ordinarily operated with hydrogen, it has a stainless steel liner to prevent hydrogen embrittlement. The driver and heater section of the tunnel are mounted on a carriage, which recoils with the tunnel as it is fired. The firing of the tunnel is controlled by a double-diaphragm rig via which the pressure in the intermediate chamber is vented to fire the tunnel.

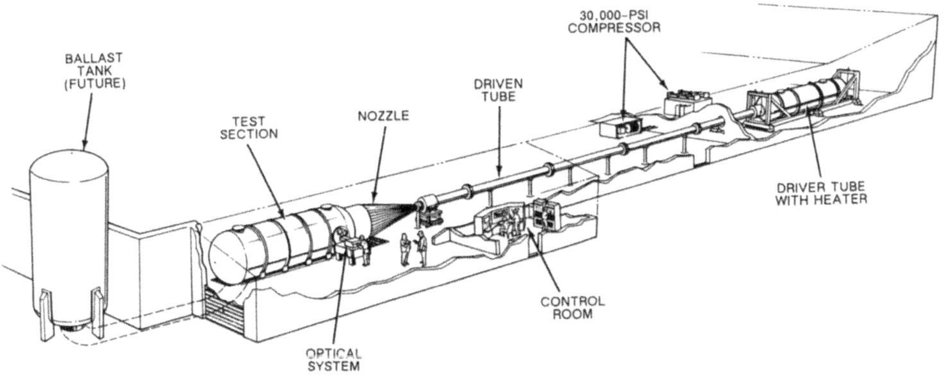

Figure 3 PHYSICAL LAYOUT OF LARGE ENERGY NATIONAL SHOCK (LENS) TUNNEL

Because flowfield chemistry and cleanliness are of prime importance in this facility, all components that come in contact with the test gas are constructed from stainless steel. A centerbody apparatus, which employs a fast-acting plug valve to close the nozzle throat, is used to terminate the flow once the useful slug of air has been exhausted from the end of the shock tube. A Mach 11 to 18 contoured fiberglass nozzle is used in the tunnel. However, to handle the large loads generated by the recoil of the driver, a metal corset, which surrounds the fiberglass nozzle, is used to prevent overloading the fiberglass. Because of the large volume of gas contained in the high-pressure driver, a large test section volume is required to prevent the pressure level in the test section from exceeding that necessary to fracture the schlieren windows. If the centerbody valve fails to operate, then additional volume, provided by a ballast tank which is coupled to the test section through a large-volume pipe isolated through a diaphragm, is used to prevent overpressurization of the test section.

2.2. SIMULATION CAPABILITIES OF LENS

The large dimensions of the LENS driver and driven tubes allow test times of up to 18 milliseconds for operation at low incident Mach numbers, with contamination-free test times of the order of 3 to 4 milliseconds at 14,000 ft/sec. The current operating range with the existing nozzle is Mach 11 to 18, for stagnation temperatures up to 14,000°R in a reflected shock mode, and up to 18,000°R for non-reflected mode.

The velocity/altitude performance of the LENS facility is shown in Figure 4, which compares the envelopes for LENS and for the 48-inch shock tunnel within which they are capable of simultaneously duplicating velocity (i.e., enthalpy), Mach number, and density as obtained in hypersonic flight. Also shown is the LENS envelope for test conditions in which the important parameters to be duplicated are density-altitude and flight velocity (which two parameters then ensure duplication of freestream momentum and energy). The operation, performance, and instrumentation of the LENS facility are described in greater detail in Reference 1.

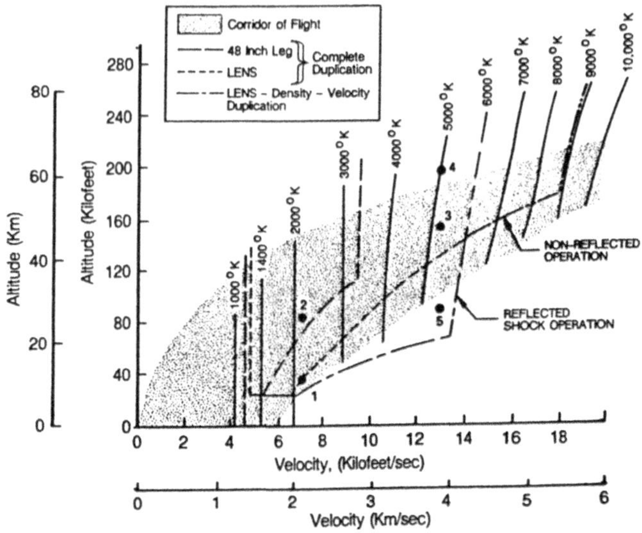

Figure 4 VELOCITY/ALTITUDE PERFORMANCE OF LENS FACILITY FOR GROUND TEST DUPLICATION

3. Instrumentation for Aerothermal and Aero-Optical Studies

3.1. SURFACE AND FLOWFIELD INSTRUMENTATION

3.1.1. *Review of Measurement Techniques.* In selecting the surface and flowfield instrumentation, particular emphasis was placed on measurements of the fluctuating density and temperature fields, as these properties directly control the aero-optic characteristics of the flowfield. In terms of wall quantities, we have in hand excellent means of measuring surface pressure, heating rate, and skin friction. Heating-rate measurements using "thin films" (thin-film resistance thermometers) also provide key information on the position of transition. Mean flow measurements under low-enthalpy conditions can be made with pitot pressure and total temperature intrusive probes. It is also possible to obtain direct measurements of concentration under these conditions. Holographic interferometry and velocimetry are possible to obtain measurements of density or velocity. On the 2-D models, holographic interferometry can be used to obtain flowfield density information. At this point, however, it is questionable whether laser doppler velocimetry (LDV) techniques are capable of measuring at the high velocities we anticipate in our experimental studies; and laser-induced fluorescence (LIF), while an extremely important potential technique, is yet to be used successfully in large-scale facilities.

For flowfield fluctuation measurements, very few definitive techniques are available. High-frequency pitot pressure measurements have been made and will be useful in these experiments, to determine principally turbulent scale size. Density measurements can be made with electron beams, and this technique would be employed to determine density fluctuations across the shock layer.

3.1.2. *Surface Instrumentation.* Thin-film heat transfer instrumentation has the capabilities of measuring both mean and fluctuating heat transfer rates up to frequencies of 1 MHz. High-frequency Kulite pressure gages are capable of accurately providing the mean pressure and the fluctuating level up to typically 300 kHz. Skin friction instrumentation using floating-element balances is capable of measuring up to 5 kHz in frequency response, which basically must be regarded as a mean measurement. Thin-film and calorimeter heat transfer elements are illustrated in Figures 5 and 6, and a description of these techniques can be found in Reference 1. The skin friction gage which is used in earlier programs in the shock tunnel is shown in Figure 7. The gage is based on a single-component force balance and contains extensive accelerometer compensation.

ure 5
IN-FILM HEAT TRANSFER
STRUMENTATION

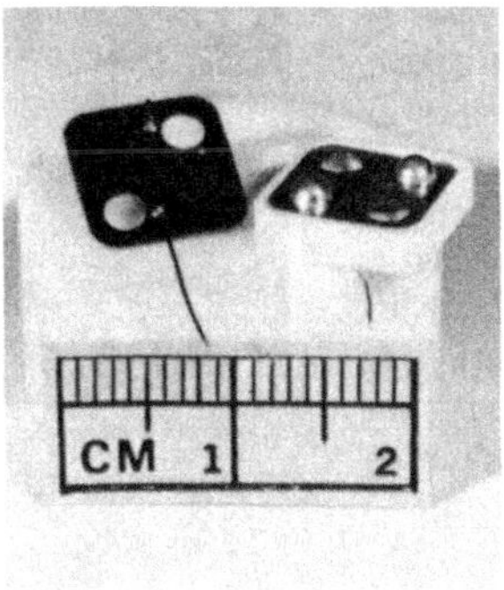

Figure 6
TYPICIAL CALORIMETER INSTRUMENTATION

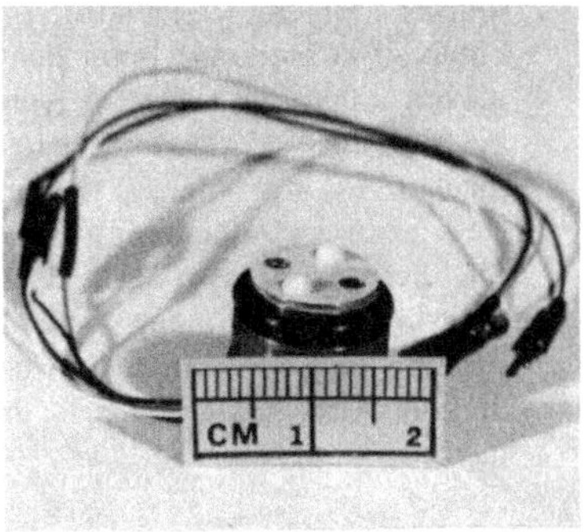

Figure 7
TYPICAL SKIN FRICTION GAGE INSTRUMENTATION

3.1.3. *Flowfield Instrumentation.*

Intrusive Instrumentation—Three sets of instrumentation are used to probe the flowfield under low-enthalpy conditions. These are mean-frequency pitot probes, total temperature probes, and thin-film probes. A typical probe assembly is shown in Figure 8.

The Kulite-based high-frequency pitot probes have a frequency response of 300 kHz. The total temperature gages and the thin-film gages will give us the ability of measuring fluctuations of up to 1 kHz and 500 kHz, respectively. Both the pitot pressure and thin-film instruments can be used to obtain turbulent scale size in the mixing region of the flowfield.

Figure 8 PITOT PRESSURE AND TOTAL TEMPERATURE SURVEY RAKE

Non-intrusive Instrumentation—We will be using a number of non-intrusive flowfield measurement techniques during the experimental studies. Later, we hope to incorporate more techniques, such as LIF, as these techniques mature. As mentioned above, we will use schlieren photography and holographic interferometry to determine the distribution of the density in the flowfield over the models. This will enable us to determine the boresight shift for direct comparisons with the boresight-shift data obtained from line-of-sight measurement through the window.

We have used the electron-beam technique in earlier shock tunnel studies (Ref. 2) to obtain mean and fluctuating density measurements across turbulent boundary layers in hypersonic flow. Static temperature measurements can also be obtained at the same point that the density measurements are made. A typical electron-beam setup, in which measurements were made across a 6° cone, is shown in Figure 9. The mean and fluctuating densities across the boundary layer are determined by measuring the mean and fluctuating intensities of emitted light at the measurement stations across the boundary layer.

Figure 9 ELECTRON BEAM FOR BOUNDARY LAYER DENSITY MEASUREMENT

3.2. AERO-OPTIC INSTRUMENTATION

3.2.1. *Introduction.* To evaluate an optical system associated with an interceptor, we must determine the refractive and radiative environment generated by the hypersonic flow over the window. Non-uniform and turbulent flows over the window induce refractive effects that degrade the optical performance of an infrared sensor looking through the flow. The optical wavefront distortion blurs, shifts, and jitters the target image on the focal plane of the sensor, thus degrading line-of-sight data sent to the guidance and control system. Also, radiative emissions from high-temperature air and coolants reach the sensor, raising background levels and degrading performance. The optical evaluation of the sensor is based on direct measurements of refractive and radiative levels produced in the flow. The radiative environment achieved in the high-temperature flow over the seeker head produces an increased background level on the focal plane and absorbs incident target radiation. These radiation levels are expected to depend on incident gas velocity and density, coolant gas, and seeker head design. Design considerations for these systems are discussed in Ref. 3.

3.2.2. *Model Support System.* Design of the support system for the aero-optical studies is one of the most important aspects of the construction of the aero-optic evaluation facility. The requirement that neither the model nor the optical bench move more than 10 microradians during the run time places a stringent demand on this design. The basic sting system is not attached to the test section of the tunnel and, thus, is not subjected to the large acceleration loads associated with the tunnel recoil. The lower half of the support system is mounted on two massive rods that pass through the test section via four sets of rolling seals and are attached to the concrete floor beneath the test section. The massive (40,000-lb) upper half of the support system is suspended on four rubber blocks such that this system has a natural frequency of under 20 hertz and, thus, will totally isolate the model attached to it from recoil loads for the 4- to 18-millisecond run time of the tunnel. Fairings, which are attached to the test section of the tunnel, are employed to prevent the model support structure from being exposed to the airloads, leaving only the model and the source plate exposed in the airflow. The optical bench system, which contains the optical components used for holographic and infrared evaluations of seeker-head performance, is mounted inside the model. The entire optical bench system is at atmospheric pressure and temperature conditions. Figures 10 through 13 show the major components of the model support and optical bench system, and the entire system mounted in the shock tunnel.

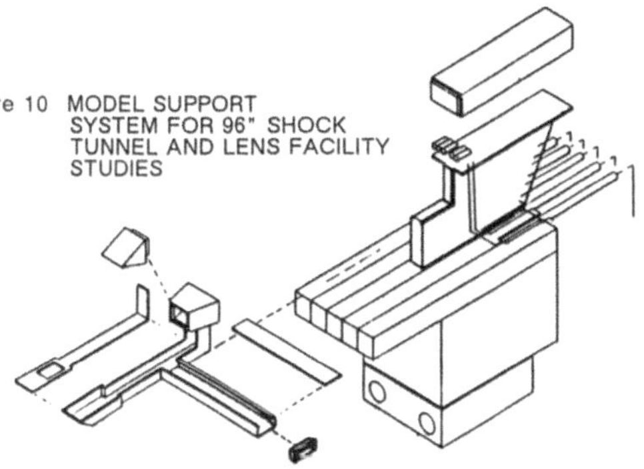

gure 10 MODEL SUPPORT SYSTEM FOR 96" SHOCK TUNNEL AND LENS FACILITY STUDIES

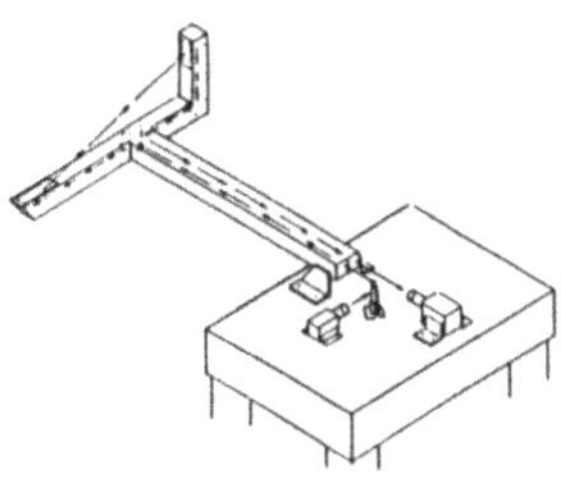

Figure 11 SCHEMATIC DIAGRAM OF OPTICAL BENCH

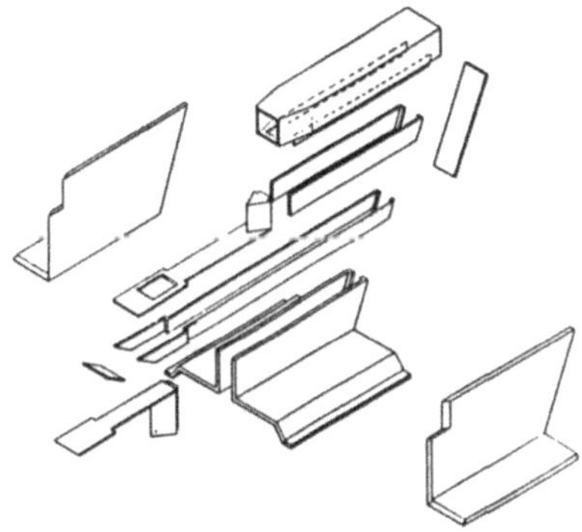

Figure 12 MODEL AND SUPPORT FAIRINGS

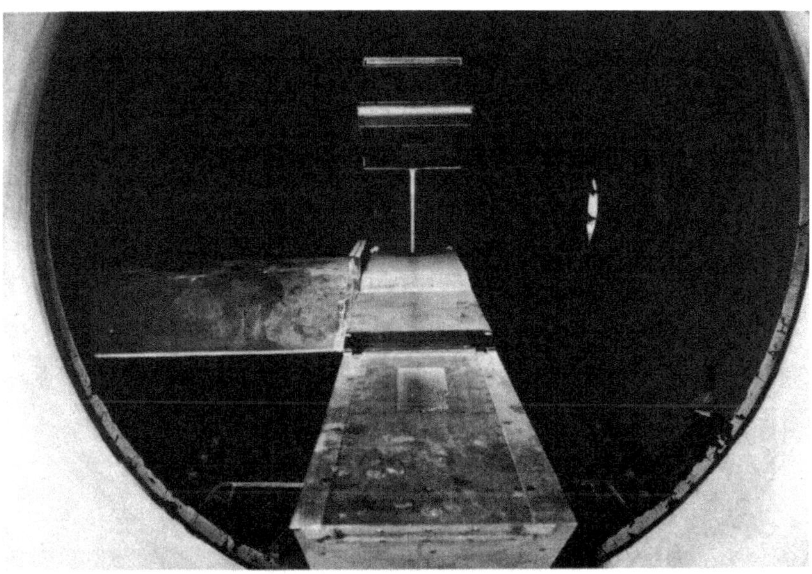

Figure 13 MODEL SUPPORT/OPTICAL BENCH SYSTEM INSTALLED IN SHOCK TUNNEL

3.2.3. *Refractive Instruments.* The refractive degradations are measured with a combination of wavefront-distortion and image-formation diagnostics. The wavefront distortion produced by the flow over the seeker head can be measured directly with holographic interferometry. Pulsed laser holography is a practical and accurate means for interferometric measurements of wavefront distortion. With attendant specification of the seeker aperture and wavelength, the focal-plane intensity distribution can be obtained through Fourier analysis. The alternative to wavefront phase measurement in the input aperture is to measure intensity in the focal plane. This is accomplished by focusing the beam onto a camera. This direct image-formation approach has the advantage of producing a real image (i.e., with actual lens elements) and recording real intensity distributions. This bypasses the analysis intensive step used in holographic interferometry of high-resolution phase measurements and two-dimensional Fourier transformation. The distinct disadvantage of the direct image formation approach is that the optical degradation is wavelength and aperture dependent; therefore, this evaluation is less comprehensive than the holographic interferometry diagnostic. The holographic interferometry diagnostic provides direct wavefront distortion measurements and, through analysis, provides image blur, modular transfer function, and point spread function data for the wavelength and aperture size of interest that is in the UV, visible, or infrared. The imagery diagnostic provides direct measurement of boresight error (BSE), jitter, and image blur, in the visible and IR bands.

Aero-optic holographic interferometry provides visual and quantitative measurements of the optical effects of coolant turbulence. Direct measurement of optical wavefront distortion is achieved through interference with a plane optical wave. In its most simple form, holographic interferometry is accomplished by recording both the plane (flow-off) and the distorted (flow-on) wavefronts on the same holographic plate. When reconstructed, such holograms reproduce both wavefronts, and the resulting interference depicts the flow-on wavefront distortion as shown in Figure 12.

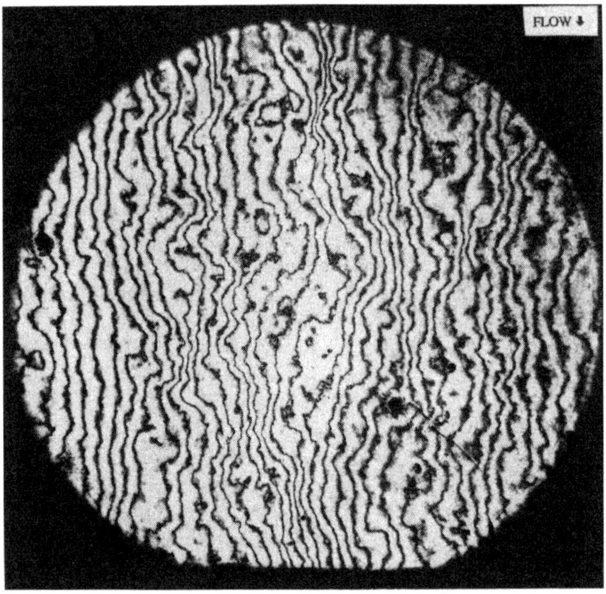

Figure 14 HOLOGRAPHIC INTERFEROGRAM

Vibration is tolerated between these steps by recording the plane and distorted waves on separate holograms. The two waves are then superimposed with the incorporation of a double-plate hologram reconstruction device. In reconstruction, the relative hologram positions are adjusted to achieve either the infinite or the finite fringe configuration. The interference fringes depict the changes produced by the flow. This approach is known as single-pulse holographic interferometry, since only one flow-on exposure is used.

In double-exposure or double-pulse holographic interferometry, two holograms are superimposed in rapid succession in the recording step on a single plate or film. Since both holograms are recorded in the presence of flow, the interferogram fringes depict the change in the turbulence-induced distortion between the pulses. This complicates interpretation of the interferogram. However, the fact that each pulse traverses a turbulent volume possessing statistics that are stationary in time is the basis for the quantitative interferometric analysis. Since the mean flow in which the turbulence is embedded is steady in time, the turbulence statistics such as velocity and density correlations are time invariant. Hence, for optical waves that traverse such media, the spatial statistics of the distortion are also time invariant. The RMS distortion of the difference wavefront is obtained by interferometric analysis.

3.2.4. *Imagery.* The invisible and infrared wavelength imagery system is an integral component of the aero-optic instrumentation. The imager field-of-view is coaxial with the holography system and operates simultaneously. For the visible systems, a pulsed light source and a telescope are used to form the source image on the CCD detector array. Various forms of 1- and 2-dimensional scanning are implemented to expose a large number of images during the short-duration flow. The focal plane is used to obtain the required intensity distribution accuracy and resolution. Further, we will utilize a wideband dynamic range video digitizer to record the CCD exposures. The visible imager also implements a line-of-sight detector for BSE/Jitter measurement. For the IR imager, a black-body source is directed through the flow, and an image is formed on an indium antimonide (InSb) focal-plane array (FPA). Figure 13 shows imagery data taken from shock tunnel studies.

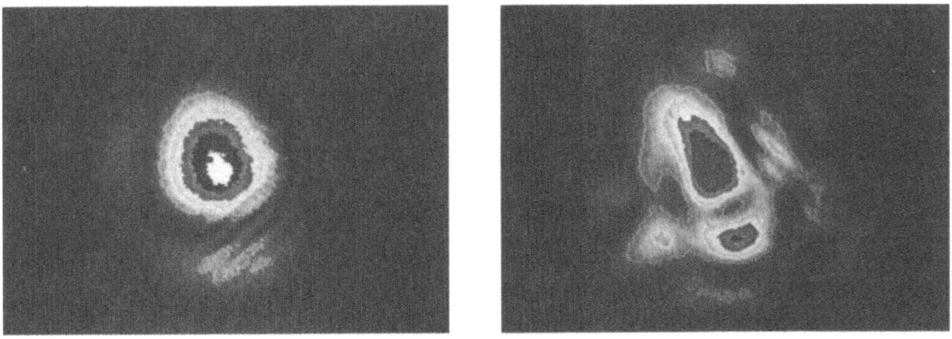

A B

Figure 15 IMAGERY DATA: (A) BEFORE FLOW IMAGE ATTENUATION TO COMPENSATE FOR ENERGY LOSS DURING FLOW, (B) IMAGE DURING FLOW

3.2.5. *Radiative Instruments.* The seeker-head sensor must look through the hot, turbulent flow over the seeker-head window. Because it is hot, and because it contains infrared-active chemical species, the flow will emit infrared radiation that can interfere with the operation of the sensor. In addition, the simple presence of these chemical species means that infrared radiation from a target will be absorbed by the flow at certain wavelengths. To predict and optimize the performance of a seeker-head sensor, the designers will need to know what wavelength band(s) to use to minimize these problems, what the levels of the emission and absorption are over these bands, and how these effects vary with seeker-head and window cooling design. Available mathematical models of these effects are subject to large uncertainties.

The Spectroscopy and Radiometry system makes measurements to address these issues. The radiation emitted by the flow is directly measured by two instruments. The relatively broad spectral bandwidth radiometer measures the emission over an operating band selected by the individual seeker-head contractor. This measurement is made with high temporal resolution, thus producing data not only on the quantitative level of the emission, but also its time history and fluctuations. The spectrometer measures the emission over a number of relatively narrow spectral bands simultaneously, thus producing an emission spectrum of the flow. This measurement is made at a relatively low temporal resolution, i.e., it is a time-averaged spectrum. Both instruments view the flow against a background emitter that is alternately switched between two temperatures. This modulates the infrared background radiation, which provides a signal from which one can deduce changes in transmission through the flow.

4. Conclusions

Hypervelocity facilities and surface and flowfield instrumentation have been developed to perform aerothermal and aero-optical studies of hypersonic interceptors. The major aerothermal phenomena, which must be duplicated to obtain a true ground test simulation of aero-optical phenomena, are discussed. The design and operating characteristics of the Large Energy National Shock (LENS) tunnel facility has been described. Surface and flowfield instrumentation which are being used to evaluate the aerothermal and aero-optical characteristics of hypersonic interceptors are reviewed.

5. References

1. Holden, M.S. (1991) 'Large Energy National Shock Tunnel (LENS) Description and Capabilities.'

2. Harvey, J., Bergman, R.C., and Holden, M.S. (1989) 'An Experimental Study of Hypersonic Turbulence on a Sharp Cone,' AIAA-89-1866, AIAA 20th Fluid Dynamics, Plasma Dynamics and Lasers Conference.

3. Karaguleff, C., Craig, J., Parker, R., and Navaz, H.K. (1992) 'Design Considerations for an Infrared Camera System for Aero-Optical Window Evaluation,' AIAA-92-2821, AIAA SDIO Interceptor Technology Conference.

INTERFEROMETRY IN RAREFIED GAS FLOWS

G.SMEETS
French-German Research Institute Saint-Louis
Gasdynamics Department
Boîte postale 34
F 68301 SAINT LOUIS CEDEX, France

ABSTRACT. A survey is given on optical interferometry using a Wollaston prism differential interferometer. With a number of selected examples, its potential, especially in the field of low density flow diagnostics, is demonstrated. In the first part, differential interferometry with spark light sources is treated and is shown to be a very effective means for high sensitive flow visualization. Finally, the laser differential interferometer is described originating from a combination of the differential interferometer with collimated light beams from a CW laser. Of special interest for applications to rarefied flows is its capability of resolving optical paths changes down to 10^{-5} of the light wavelength. This is three orders of magnitude beyond the resolution limit of classical interferometry based on fringe shift analysis.

1. INTRODUCTION

Optical interferometry has been widely used in aerodynamic and gasdynamic research for more than 5 decades now. Its interest results from the easy access to gas density ρ by a rather simple set of relations connecting ρ to the optical path variation of light passing through the gas, δl, detectable by a change of the interference pattern.

The optical path of a light ray from its source to a screen or detector surface is given by the integral:

$$l = \int n \cdot ds \qquad (1)$$

l is affected by refractive index changes δn along the light path resulting from changes of the gas density ρ according to the linear relationship between δn and $\cdot \delta\rho$:

$$\delta n = K \cdot \delta\rho \qquad (2)$$

where K depends on the gas and also slightly on the light wavelength λ. For air, this so- called "Gladstone-Dale constant" has the value: $K = 0.2274\ cm^3/g$ (at $\lambda = 0.5097\ \mu m$). Gladstone-Dale constants at different wavelengths and for other gases are given in (Merzkirch 1987). For high temperature gases, values of gas refractivities can be found in (Alpher 1958, 1959 and Wettlaufer 1972).

A. Boutier (ed.), New Trends in Instrumentation for Hypersonic Research, 75–85.

For interferometric measurements in high enthalpy tunnel flows it is a very favourable occurrence that for dissociating air, K is approximately constant up to temperatures of about 5000 Kelvin, when nitrogen markedly dissociates. This is because the specific refractivities for atomic and molecular oxygen are very close to another and the percentage of CO molecules is comparatively low in equilibrium air.

In a two-beam interferometer, a light bundle is divided into two coherent bundles by a beam divider. After their recombination at the exit of the interferometer, the bundles interfere. The irradiance H at each point of a screen or photographic film depends on the optical path difference Δl of the two light rays from the light source to this point and is given by:

$$H = H_0 \cos^2 [\pi \cdot \Delta l / \lambda] \qquad (3)$$

Equations (1) to (3) connect the detectable quantity H to flow density ρ. In classical interferometry, data reduction is done by quantifying the shifting of fringes on interferograms. This procedure when applied together with high speed photography overcomes the difficulties from the great vibrational sensitivity inherent to all absolute interferometers as for example the Mach Zehnder. As the fringe shifting can - according to equation (3) - only be resolved to about 1/25 of the fringe spacing, the limit for resolving or detecting optical path changes $\delta\Delta l$ from fringe shifts is 1/25 λ.

A very powerful alternative to "absolute" interferometers are "differential" interferometers, in which the two interfering light bundles both pass through the phase object having only a small displacement (shear) to each other. In the following, differential interferometers based on Wollaston prisms as the beam splitting elements will be treated. These interferometers have a number of important advantages:

- they are relatively unsensitive to vibrations and to imperfections of the optical components used,
- are very easy to handle,
- because of the low sensitivity to vibrations and optical imperfections, adjustment to an interference fringe slope at infinite fringe spacing becomes feasable enabling much higher resolving capability to small optical path changes.

When a collimated beam of continuous laser light (as for example from a He-Ne laser) is sent through a two beam interferometer, the irradiance modulation by interference, H in equation (3) can be directly monitored with a photodiode placed at the corresponding point of the image plane. This type of laser interferometers enable rather sensitive recordings of the time history of the optical path difference.

When this technique is used with differential interferometers, a powerful tool is created capable of quantitative measurements in rarefied gas flows as for example in high enthalpy tunnels.

DIFFERENTIAL INTERFEROMETER

In a differential interferometer, the two interfering light bundles, instead of being totally separated, both pass through the phase object with only a small lateral shift relative to each other. Among a variety of possible optical components that may be used for beam separation, the Wollaston prism has become the favourite.

The Wollaston prism interferometer was originally developed for the use in interference microscopes (Françon 1952, Nomarski 1954, 1956). Since 1954 its application was extended to flow diagnostics, first at ONERA (Renet 1954, Chevalerias 1957, Philbert 1958), a little later at ISL/Saint-Louis (Oertel 1960, Oertel 1961, Smeets 1964) and at the Ernst-Mach-Institut/Freiburg (Merzkirch 1965).

A Wollaston prism is a double prism made of a birefringent crystal like quartz or calcite. The optical axes of the two prism parts have orientations perpendicular to each other. A light beam passing the Wollaston prism is split into two partial beams. The splitting angle ε depends on prism angle α and the difference between the refractive indices n_{ex} and n_{or}:

$$\varepsilon = 2 \cdot \alpha \cdot |n_{ex} - n_{or}| \qquad (4)$$

At $\lambda = 0.589\ \mu m$, the values of $|n_{ex} - n_{or}|$ for Quartz and Calcite are 0.0111 and 0.1720 respectively.

The principle arrangement of a differential interferometer based on Wollaston prisms is explained in Fig. 1. Every individual light ray passing the first Wollaston prism is split by an angle ε into two rays of perpendicular linear polarization. If the original ray was of circular polarization or polarized linearly under 45° with respect to the optical axes of the prism, the partial beams are coherent and have equal intensities. By a subsequent objective lense or spherical mirror of focal length f, arranged exactly at a distance f behind the Wollaston prism, the rays become parallel with a separation distance (shift or shear of the differential interferometer):

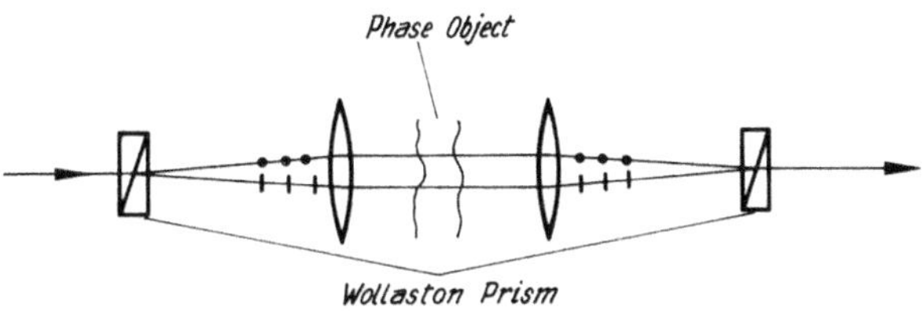

Fig. 1. Differential interferometer based on Wollaston prisms.

$$e = f \cdot \varepsilon = 2 \cdot f \cdot \alpha \cdot |n_{ex} - n_{or}| \qquad (5)$$

After having passed through the phase object, the rays are recombined in the second part of the interferometer which is exactly symmetrical to its first part. The partial rays are joined in the second Wollaston prism and then have a common path to the screen. Because of their perpendicular linear polarization, they can only interfere after having been polarized by a polarization filter in a common direction - preferentially 45°. If the Wollaston prisms are arranged exactly at the focal points of the objective lenses, the interferograms (without a phase object) show no fringes. When the prisms are shifted along the optical axis, parallel and equidistant fringes in a direction normal to the beam separation appear. The interferogram in Fig. 2 was taken in the rarefied flow of a shock tunnel around a blunted cone with a 1.2 cm shear in the vertical direction. Due to the low density of the flow, the interference fringes are only shifted by a fraction of an interfringe spacing.

In rarefied flows, applicability of interferometry normally ends

Fig. 2. Interferogram of a rarefied flow in a shock tunnel around a blunted cone at 1.2 cm shear.

when the fringe shifts become too small for a quantitative evaluation or even for its visibility. However, by adjusting the interferometer to infinite fringe spacing and to the center part of a slope between a bright and a dark interference fringe, this limit may be extended by at least one order of magnitude.

It follows from equation (3) that when adjusting the interferometer to one of the optical path differences:

$$\Delta l = \lambda/4 \pm n\cdot\lambda/2, \quad (6)$$

small changes of the optical path difference by a weak phase object, δl, lead to changes of the irradiance at the screen given by:

$$H/H_0 = \pm\, \pi\cdot\delta l/\lambda \quad (7)$$

This linear relationship between δl and dH is, of course, only valid within the linear part of the interference fringe slope, which limits the optical path changes to $|\delta l| \leq 0.1\cdot\lambda$.

With the increased sensitivity, however, the slightest (unavoidable) imperfections of the optics - also representing a weak phase object and introducing optical path changes of the same order - become important and lead to erroneous shading of the background. A Wollaston prism interferometer offers a simple possibility of clearing the spotty background by means of complementary double exposure. As is explained in Fig. 3, an additional Wollaston prism is arranged in front of the differential interferometer and the double image of two separate spark light sources is formed in the plane of the first Wollaston prism of the interferometer. By adapting the lateral distance of the light sources it can be obtained that the two inner light bundles exactly merge and pass through a diaphragm. They are perpendicularly polarized and lead to complementary interference on the screen.

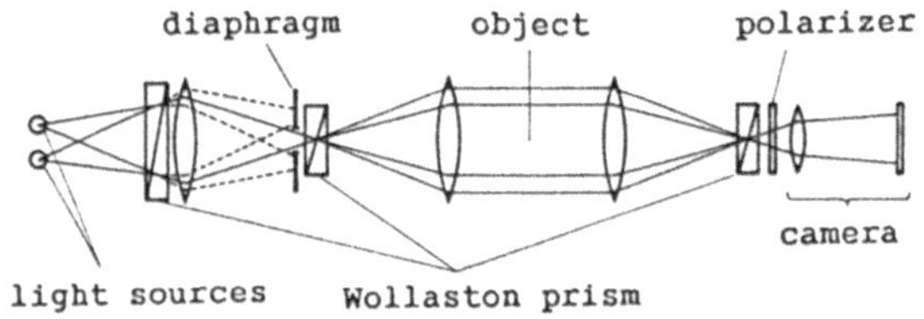

Fig. 3. Differential interferometer for complementary double exposure.

The light bundle having a polarization direction perpendicular to that of the analysing polarization filter in front of the screen leads to the complementary irradiance:

$$H = H_0 \sin^2 [\pi\cdot\Delta l/\lambda] \quad (8)$$

With the adjustment of the interferometer to an optical path difference (6), a change of the irradiance complementary to that of equation (7) results (see Fig. 4):

$$dH/H_0 = \mp \pi \cdot \delta l/\lambda \qquad (9)$$

As an example of a very weak phase object, Fig. 5 shows a series of spherical pressure waves generated by an encapsuled spark gap. The resultant fringe shift (upper picture) is not greater than 1/20 of the

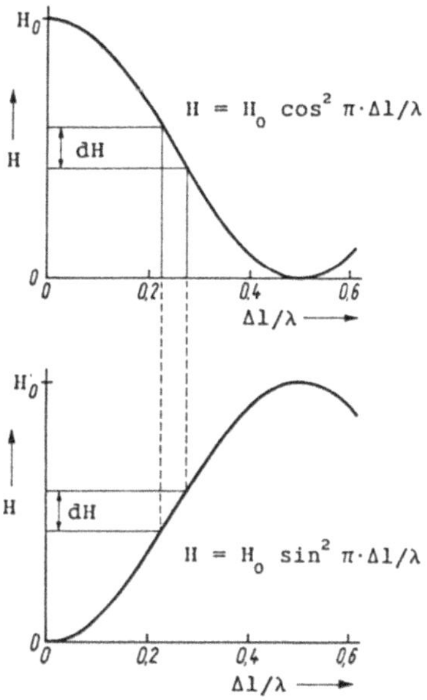

Fig. 4. Complementary change of the irradiance used with double exposure interferometry.

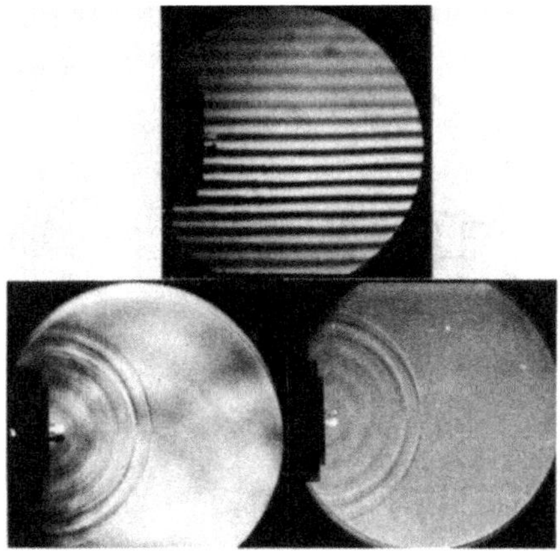

Fig. 5. Spherical pressure waves generated by an encapsuled spark gap.

fringe spacing. By adjusting the interferometer to infinite fringe spacing and to the center part of a slope between a bright and a dark interference fringe, the pressure waves become visible. By complementary double exposure - one of the exposures at a time before or after the event - the object appears on a uniform background. On high contrast photographic film, a visibility limit for weak phase objects as low as $\lambda/500$ can be achieved (see Smeets 1972). Fig. 6 shows a double exposure interferogram of the bow shock wave at a blunted cone in a low density ($\rho = 1.5\cdot 10^{-6}$ g/cm^3) hypersonic (M = 9) flow in a shock tunnel.

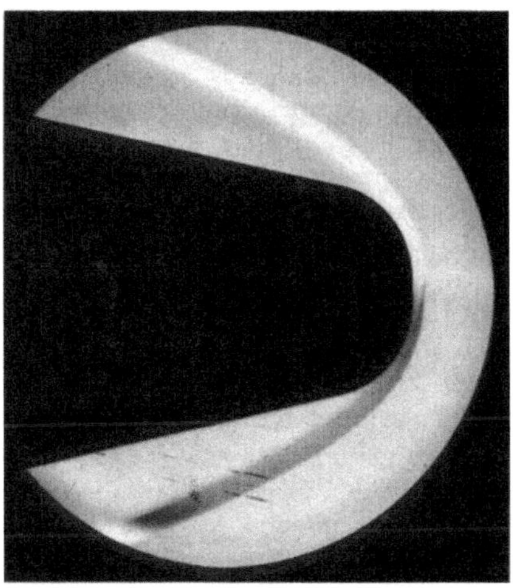

Fig. 6. Low density ($\rho = 1.5\cdot 10^{6}$ g/cm^3) hypersonic (M = 9) flow around a cone.

LASER DIFFERENTIAL INTERFEROMETER

Using collimated beams from CW lasers (as for example from a He-Ne laser) in combination with Wollaston prism differential interferometers, leads to a class of laser interferometers being very effective tools for hypersonic flow diagnostics (see Smeets 1972 and Smeets 1977).

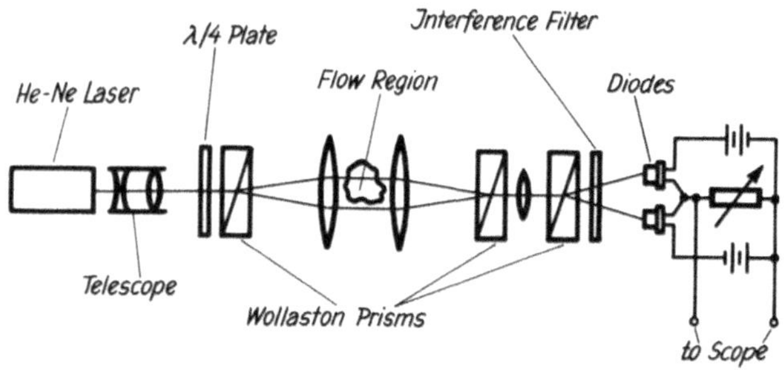

Fig. 7. LDI using two separate diodes for complementary interference detection.

If the polarizing filter behind the interferometer is replaced by an additional Wollaston prism, two beams of complementary interference are created which are sent to separate photodiodes. By means of the simple circuit shown in Fig. 7, a difference current is formed. It follows from equations (7) and (9), that the increment of the current, di, is linearly related to an optical path change, δl, by:

$$di/i_0 = \pm\, 2\cdot\pi\cdot\delta l/\lambda \qquad (10)$$

so that the signal is amplified by a factor 2 and its sensitivity to laser power fluctuations is much reduced.

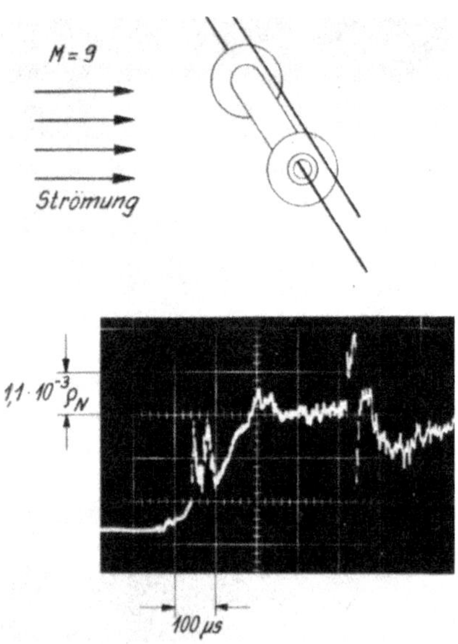

Fig. 8. Free stream density measurements in a hypersonic shock tunnel.

In cases when vibration sensitivity becomes a problem as in extremely rough and noisy invironments, autostabilisation can be attained by different means, for example by arranging a Pockels cell in series with the interferometer (see Smeets 1975).

For free streem density measurements in a hypersonic shock tunnel, a device illustrated in Fig. 8 has been used. A small tube, 10 mm in diameter and 10 cm in length, closed at both ends by glass windows, was positioned perpendicular to the flow. In order to prevent the gas of the stagnation region in front of the cylinder from passing over the windows, knife edged rings were fixed to both ends of the tube. One of the laser beams was sent through the tube and thus shielded from the flow. The recorded signal in Fig. 8 gives information on time and duration of the short stationary flow period of a nonreflected shock tunnel and also on the absolute density within this period.

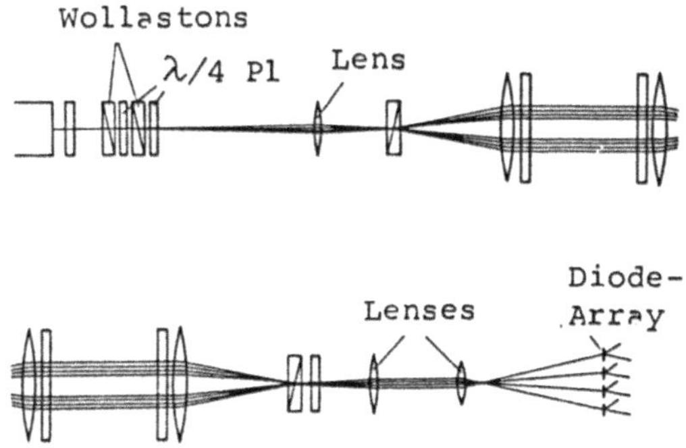

Fig. 9. Multiple beam LDI.

Using a set of additional Wollaston prisms having suitably graded prism angles, a multiple beam system (see Fig. 9) can be realized. The original laser beam can be multiplied in such a way that a series of parallel and equidistant beam pairs is produced in the phase object. Simultaneous recording of the signals can easily be performed by a linear diode array. In this way, up to 8 beampairs have been realized and were found to be practical.

This technique was first applied to the investigation of a boundary layer behind a shock wave passing over a cylinder in a shock tube (the geometry is explained in Fig. 10). Fig. 11 shows a series of 8 signals from 8 adjacent laser beams. The corresponding reference beams pass outside the boundary layer and are not shown in the figure. At the low Reynolds number, the boundary layer remains laminar. For the stationary part of the laminar boundary layer, the optical path data - using suitable nondimensional coordinates - could be made to coincide into one experimental optical path profile (see Fig. 12). From this, the nondimensional flat plate density profile could be derived by means of the Abel inversion (see Smeets 1973).

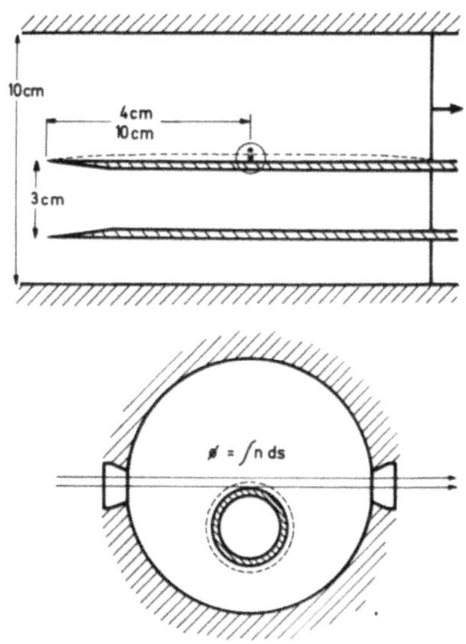

Fig. 10. Model arrangement inside the shock tube.

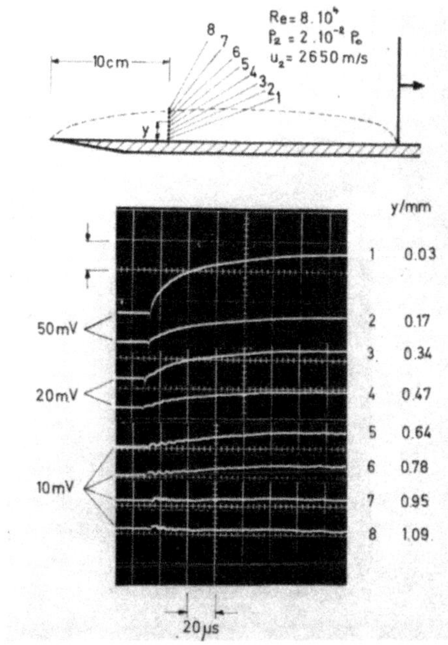

Fig. 11. 8 simultaneous signals from a shock boundary layer.

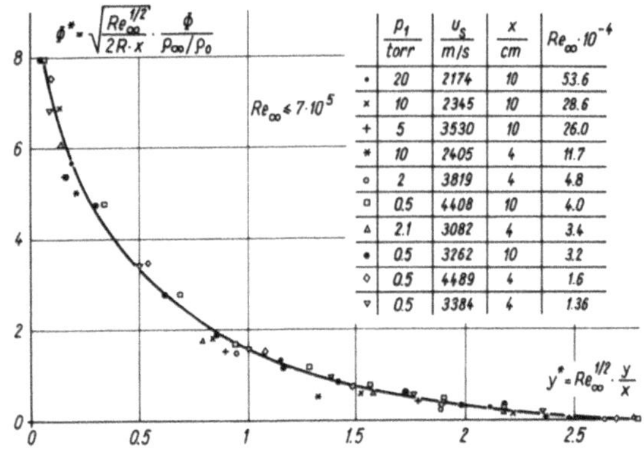

	p_1 / torr	u_s / m/s	x / cm	$Re_\infty \cdot 10^{-4}$
•	20	2174	10	53.6
×	10	2345	10	28.6
+	5	3530	10	26.0
*	10	2405	4	11.7
○	2	3819	4	4.8
□	0.5	4408	10	4.0
△	2.1	3082	4	3.4
●	0.5	3262	10	3.2
◇	0.5	4489	4	1.6
▽	0.5	3384	4	1.36

Fig. 12. Optical path profile of the stationary boundary layer.

Applying the multiple beam arrangement for shock tunnel experiments, the density distribution within the hypersonic flow field around a blunted cone could be determined. 8 laser beam pairs were arranged in a plane perpendicular to the cone axis. They passed close to the cone through its shock layer and were focussed to about 0.05 mm within the flow field. The 8 reference beams pass outside the shock layer. A set of simultaneous signals is given in Fig. 13. They show the time dependence of the hypersonic flow generated in the shock tunnel. From each signal only the constant value over the time period of stationary flow condition is taken for further calculation. The density profiles during the stationary period of the tunnel were calculated by the Abel inversion. At the small (10°) angle of attack, the limited sector of the flow at the windward side of the cone that is traversed by the laser beams was regarded as approximately axisymmetric. A number of profiles determined at different positions x/R is given in Fig. 14. They

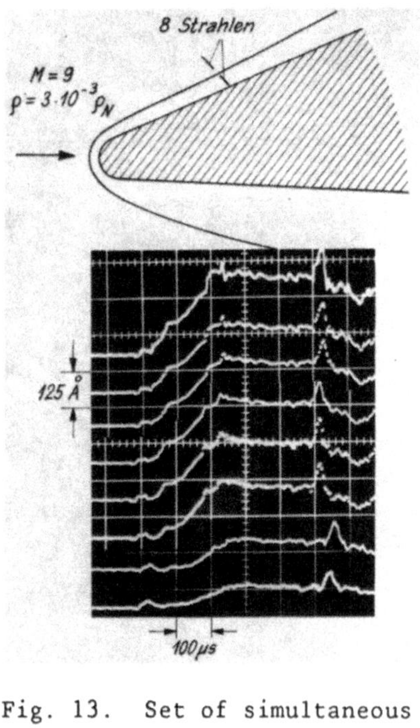

Fig. 13. Set of simultaneous signals from the shock layer.

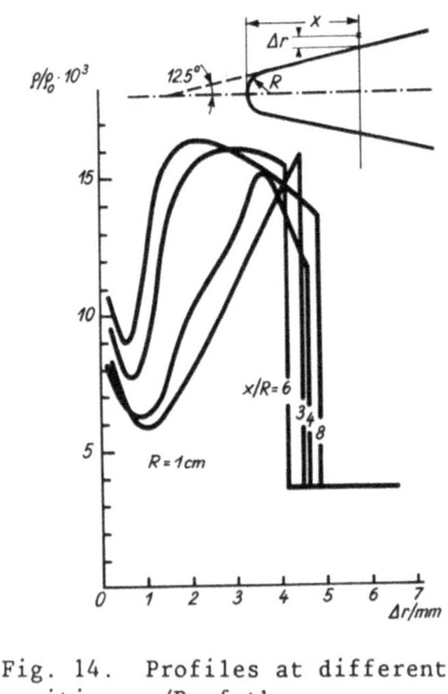

Fig. 14. Profiles at different positions x/R of the cone.

show how the high entropy gas from the stagnation region - represented by the density minimum close to the wall - is cooled (presumably by heat conduction through the boundary layer).

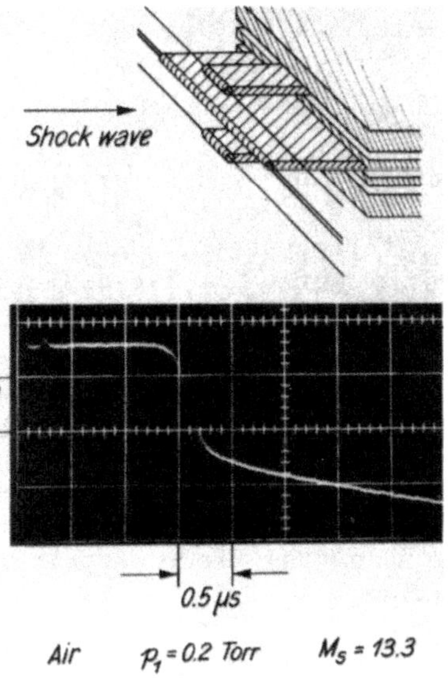

Fig. 15. Density profile of a shock wave.

An experiment for recording the density profile of a shock wave of Mach number 13 running into air of initial pressure of 0.2 Torr is explained in Fig. 15. The four laser beams from 2 beam pairs are focussed to about 0.1 mm in the center of a 10 cm in diameter shock tube and are adjusted exactly perpendicular to the tube axis. The beams are protected by a system of capillary tubes having an inner diamater of 1 mm and a length of 10 mm and 48 mm, respectively. The difference of the signals from the two beam pairs is recorded. By this, the disturbances equally occurring at any of the capillary tube ends, do not affect the difference signal. The experiment demonstrates the advantage of laser differential interferometers as to cancel the effects of perturbations including the unknown contributions from the boundary layers at the test chamber windows. It is also an example of an experiment requiring a rather high sensitivity together with a good spatial and temporal resolution. Resolving the translational and rotational structures within shock fronts has formerly been only possible by the electron beam absorption technique.

An optical system as outlined in Fig. 16 composed of two laser differential interferometer for λ = 633 nm and λ = 1152 nm fused into

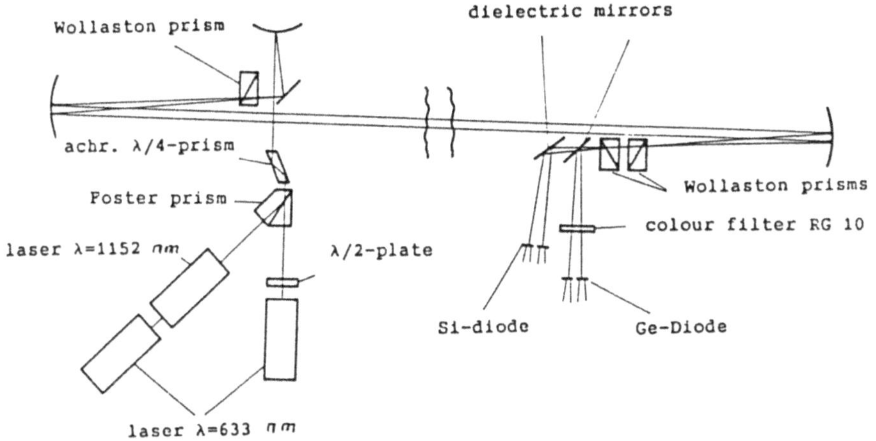

Fig. 16. Optical system for simultaneous recording of refractive index variations at two wavelengths.

one was used. Because of the strong wavelength dependence of free electrons (dispersion), from the two simultaneous signals, both, Argon gas density and electron number density can be determined.

For further applications of the laser differential interferometer to high speed flows see Smeets 1977 and 1982 and Oertel and Oertel 1989.

CONCLUDING REMARKS

By different examples, the possibilities of optical interferometric techniques for low density flow diagnostics were demonstrated.

The Wollaston prism differential interferometer was shown to have the capability of high quality visualization of flow fields even at lower densities. On double exposure interferograms, the bow shock at a blunted cone in a hypersonic (M = 9) flow having a density as low as $\rho = 1.5 \cdot 10^{-6}$ g/cm^3 could be visualized.

Laser interferometry has pushed the limitations of classical interferometers much further. Optical path changes as small as $10^{-5} \cdot \lambda$ can be detected in this way. For highest spatial resolution, the laser beams can be focussed to a few hundreds of a millimeter and the time resolution is only limited by the time constant RC of the diode circuit which may be only a small fraction of a microsecond.

REFERENCES

Alpher, R. A., White, D. R. (1958) 'Interferometric measurement of electron concentration in plasmas', Phys. Fluids 1,5, pp. 452-3.

Alpher, R. A., White, D. R. (1959) 'Optical refractivity of high temperature gases: I. Effects resulting from dissociation of diatomic gases. II. Effects resulting from ionisation of monatomic gases', Phys. Fluids 2,2, pp. 153-69.

Chevalerias, R., Latron, Y., Veret, C. (1957) 'Methods of interferometry applied to the visualisation of flows in wind tunnels', J. Opt. Soc Ann. 47,8, pp. 703-6.

Françon, M. (1952) 'Interférence par double réfraction en lumière blanche', Revue d' Optique, 31, 2.

Merzkirch, W. (1965) 'A simple schlieren interferometer system', AIAA J. 3, pp. 1974-76.

Merzkirch, W. (1974) 'Flow Visualization', Academic Press, New York and London, 2nd edition, Academic Press, Orlando (1987).

Merzkirch, W. (1987) 'Techniques of flow visualization', AGARD-AG 302.

Nomarski, G. (1955) 'Microinterféromètre différentiel à ondes polarisées', J. de Phys. 16, pp. 95-135.

Nomarski, G. (1956) 'Remarques sur le fonctionnement des dispositifs interférentielles à polarisation', J. de Phys. 17, pp 15-35.

Oertel, H. (1960) 'High speed photography of hypersonic phenomena by a schlieren interferometric method', Proc. 5th Int. Congr. on High Speed Phot., Wash. DC.

Oertel, H. (1961) 'Ein Differentialinterferometer für Messungen im Hyperschallstoßrohr', ISL T 17/61.

Oertel, H. sen, Oertel, H. jun. (1989) 'Optische Strömungsmeßtechnik', Braun Verlag, Karlsruhe.

Philbert, M. (1958) 'Emploi de la strioscopie interférentielle en aérodynamique', Rech. Aéron. 65, pp. 19-27.

Renet, C. (1954) 'Strioscopie quantitative en soufflerie', ONERA N.T. 23.

Smeets, G. (1964) 'Differentialinterferometer zur Beobachtung und Ausmessung von Grenzschichten', ISL T 25/64.

Smeets, G. (1972) 'Laser interferometer for high sensitivity measurements on transient phase objects', IEEE-Transact.-AES 8.2, pp. 186-190.

Smeets, G., George, A. (1972) 'Interferograms of high optical quality by double exposure', Proc. 10th Int. Congr. on High Speed Phot., Nice, pp. 244-47.

Smeets, G., George, A. (1973) 'Investigation of shock boundary layers with a laser interferometer', Proc. 9th Int. Shock Tube Symp., Stanford.

Smeets, G. (1975) 'Laserinterferometer mit Phasennachführung', ISL R 135/75.

Smeets, G. (1977) 'Flow diagnostics by laser interferometry', IEEE Transact.-AES, 13.2, pp. 82-90.

Wettlaufer, D., A. (1972) 'An interferometric determination of the specific refractivities of the nitrogen and oxygen atoms', UTIAS Techn. Note No.175 AFOSR-TR-72-0411.

HOLOGRAPHIC INTERFEROMETRY ON THE HIGH ENTHALPY SHOCK TUNNEL IN GÖTTINGEN (HEG)

G. EITELBERG, B. FLECK, T. J. McINTYRE
Institute for Experimental Fluid Mechanics
German Aerospace Research Establishment (DLR)
Bunsenstraße 10, 3400 Göttingen, Germany.

ABSTRACT. A holographic interferometer for the high enthalpy shock tunnel, HEG, is described. The system relies on recording a flow and a no-flow hologram on a thermoplastic film allowing reconstruction of each beam and thus the interferogram after an HEG run. From this interferogram it is possible to extract the shock shape around an aerodynamic model mounted in the test section and also the density distribution between the shock and the model. Results of initial tests on HEG are presented.

1. Introduction

The free - piston driven shock tunnel, HEG, has been built to investigate high enthalpy hypersonic gas flows. Of particular interest is the characterisation of chemically reacting flows at atmospheric reentry conditions [1]. Such an aerodynamic testing facility requires dedicated instrumentation for the utilization of its full potential. This instrumentation includes surface pressure and heat transfer sensors, laser induced fluorescence (LIF) [2] and interferometry for flow visualisation and for quantitative measurement of density variations [3].

Figure 1 shows the setup of the shock tunnel. It consists of an air supply, the compression tube within which a heavy piston is free to move, the shock tube initially containing the test gas, the nozzle and the test section. The facility run starts with the acceleration of the piston (of mass up to 730 kg), when air from the air reservoir is released to enter the space behind the piston. As the piston moves, it compresses the helium ahead of it, until the first diaphragm separating the driver gas from the driven gas bursts at a given design pressure. The design pressures range up to 2000 bar, current operation has been up to 500 bar with nitrogen as test gas.

The bursting of the primary diaphragm causes a shock to propagate through the driven gas. The tailored shock speed is about 4 - 5 km/s so that it reaches the end of the driven tube in approximately 3 - 4 ms. The shock is reflected at the end of the tube and heats and compresses the driven gas up to

Table 1. Design conditions for the operation of HEG

Parameter	Free Stream	Frozen	Equilibrium
Density (x 10^{-3} kg/m^3)	1 - 10	6 - 60	10 - 1000
Temperature (K)	800 - 2600	10000 - 20000	5000 - 10000
Pressure (mbar)	5 - 50	500 - 3500	600 - 3600
Velocity (km/s)	5 - 8	0.5 - 1.0	0 - 0.8

A. Boutier (ed.), New Trends in Instrumentation for Hypersonic Research, 87–96.

12000 K. There a second diaphragm ruptures immediately and the heated test gas expands through the nozzle reaching a velocity of between 5 and 8 km/s. The design range of operating conditions for HEG is shown in Table 1 with values given for conditions in the free stream flow, for directly behind the bow shock formed ahead of a blunt model ('frozen') and for equilibrium conditions which may be obtained further back in the flow. The total test time is of the order of a millisecond.

2. Optical Arrangement

Figure 2 shows the optical arrangement for holographic interferometry on the tunnel. The light source is a Q-switched ruby laser with an output energy of 50 mJ. The pulse duration is about 30 ns and the coherence length is about 1 m. The linearly polarized laser output is divided into two collimated beams, one of which is expanded to a maximal diameter of 300 mm and passed through the test section of the HEG. The other collimated beam is diverted around the test section and used as the reference beam. The two beams are passed to the receiving unit consisting of a two-reference beam module and a holographic camera which uses a thermoplastic film [4]. The receiving table is situated some 5 m away from the transmitting unit as necessitated by the dimensions of the facility and the alternative use of another measuring technique, laser-induced fluorescence. The two-reference beam module is used to record two successive holograms with separate reference beams. This allows later reconstruction of the two beams forming an interferogram which can then be processed using the phase shifting technique as described by Dändliker and Thalmann [5].

For recording during an HEG test, a no flow hologram is recorded immediately before the run. During the run, the flow hologram is recorded by triggering the ruby laser from a pressure sensor located upstream from the test section. The time between the two recordings must be kept as short as possible as the diffraction efficiency of the first hologram is observed to drop rapidly over a period of several minutes. One possible explanation of this phenomenon could be the limited lifetime of the "latent picture" due to diffusion processes.

Afterwards, both beams are simultaneously reconstructed using a cw He-Ne laser, the light of which is divided into two beams and passed through the holographic film at angles similar to the two recording beams. Because of the presence of two holograms on the film, four separate beams are reconstructed, two of which are allowed to interfere. As no scattering elements such as a speckle plate are used in the ray path, the angular separation of the other two beams must be large enough so that they do not overlap with the interferogram. This leads to an angle between the two recording reference beams of approximately 4°. Also, the angular dependence of the diffraction efficiency of the thermoplastic material requires an angle between the two reference beams and the object beam of approximately 34°. Hence the angle of the first reference beam to the test beam was chosen as 32° and the angle of the second reference beam to the test beam as 36°. For reconstruction these angles are adjusted to account for the difference between the wavelength of light from the Ruby and He-Ne lasers.

The interferogram is recorded on a CCD camera connected to a computer. A mirror mounted on a piezo-electric crystal is used to adjust the phase of one reference beam with respect to the other allowing recording of three phase shifted interferograms at 0°, 120° and 240°. A software package [6] is then used to evaluate these interferograms yielding the wanted phase distribution. The determination of the density variation can then be performed dependent of the geometry and constituents of the flow.

3. Estimation of the resolving power of the interferometer

Investigations have shown that the resolution of the interferometer is limited essentially by the resolving power of the CCD camera. The camera used has a CCD matrix with the dimensions of 8.8 x

6.6 mm^2 and pixel size of 11 x 13 μm^2. That means that the test section with a diameter of 300 mm has to be imaged on the CCD matrix with a diameter of about 6 mm. Therefore the scale of reduction is 50. Under the assumption that two pixels per structure are needed, the maximal resolved imaged structure size is about 30 µm. This corresponds to a possible resolution in the test section of about 1.5 mm. Experimental investigations confirm this value. The resolving power of this optical arrangement can be increased by decreasing the object area imaged onto the CCD.

4. Results

Experiments have been completed in HEG for diaphragm burst pressures of 100 and 500 bar with test gas nitrogen. This represents only one quarter of the maximal operating conditions but already provides sufficient density variations for measurements with holographic interferometry. As an aerodynamic model, a 90 mm diameter cylinder with length 280 mm has been mounted in the test section, as shown in figure 2, providing a suitable two-dimensional object for both examining the flow obtained in the test section as well as testing the holographic interferometry arrangement. The two - dimensionality can be seen in the flow luminosity picture, taken from above the cylinder, shown in figure 3.

An infinite fringe interferogram for one such HEG run is shown in figure 4. Evident ahead of the cylinder is a bow shock across which the density jumps. Although the quality of the interferogram is not good, an adjustment of the alignment of the reconstruction beams provides a finite fringe interferogram which can than be processed by the software package OPTOCAT to give the phase distribution and then filtered to enhance the contrast as shown in figure 5. The influence of the finite fringes can be removed to provide a phase distribution between shock front and model for the region behind the shock above the plane of symmetry, as shown in figure 6. The improved quality of this phase distribution image over that seen in the original interferogram shows the value of the phase shifting technique as applied in holographic interferometry.

Having obtained the phase distribution behind the shock, it is then necessary to determine the fringe shift across the shock in order to obtain absolute density changes. The shock is seen as a discontinuity in the interferogram and it is not always possible to determine from the interferogram alone, which fringes match across the shock. However, both the density before the shock and the density change across the shock are known with sufficient accuracy from numerical modelling of the flow to uniquely assign a fringe order to each fringe behind the shock. Since this involves assigning integer numbers to fringes no uncertainties are introduced by the process. The density distribution, ρ, can then be calculated from the phase by using the Gladstone - Dale relationship

$$n - 1 = K\rho \qquad (1)$$

where n is the refractive index and K is the Gladstone-Dale coefficient for the gas. For a density change, Δρ, this reduces to

$$\Delta n = K\,\Delta\rho \qquad (2)$$

or when more than one gas species is present

$$\Delta n = \Sigma_i K_i \Delta\rho_i \qquad (3)$$

The Gladstone-Dale coefficients for N and N_2 are given by Alpher and White [7] as

$$K_N = 3.05 \times 10^{-4}\ m^3/kg,\quad K_{N2} = 2.38 \times 10^{-4}\ m^3/kg.$$

The relationship between the fringe shift ΔN and the refractive index change due to a phase object is given by

$$\Delta N = \frac{1}{\lambda} \int [n(x,y,z) - n_0]\, dz \qquad (4)$$

which for the case of a two-dimensional object reduces to

$$\Delta N = \frac{L}{\lambda} \Delta n \qquad (5)$$

Combining equations (2) and (5) produces the final result

$$\Delta \rho = \frac{\lambda}{L} \frac{\Delta N}{K} \qquad (6)$$

where L in this case is the length of the cylinder.

Table 2. Calculated conditions for the HEG test section for the interferogram shown in figure 4.

Region	Mass fraction N (mol/g)	Mass fraction N_2(mol/g)	Density ρ (kg/m^3)
Nozzle exit	3.35×10^{-3}	$3,4 \times 10^{-2}$	$2,3 \times 10^{-3}$
Frozen bow shock	3.35×10^{-3}	$3,4 \times 10^{-2}$	$1,7 \times 10^{-2}$
Equil. bow shock	2.07×10^{-2}	2.53×10^{-2}	$2,8 \times 10^{-2}$

Table 2 summarizes the calculated conditions for the 500 bar HEG shot during which the interferogram in figure 4 was recorded. From these values, one can calculate the approximate fringe jump across the shock to be $\Delta N \simeq 1,5$ allowing the matching of the fringes. The density variation over the interferogram can than be calculated using equation (6) as given in figure 7. It shows, as expected, a large density increase across the shock and along the stagnation streamline. The density increase is smaller as one looks futher away from the horizontal plane of symmetry of the cylinder. Furthermore, assuming the free-stream density as given in table 2, an absolute density for every point in the flow field can be determined. For the flow along the stagnation streamline immediately behind the shock and on the cylinder surface the following densities have been determined

$$\rho_{shock} = 1.43 \times 10^{-2} kg/m^3$$
$$\rho_{cylinder} = 2.63 \times 10^{-2} kg/m^3.$$

These values agree reasonably well with the result of the numerical flow calculation for the frozen bow shock and the equilibrium bow shock as presented in Table 2.

The complete distribution along the stagnation streamline is shown in figure 8. The initial jump corresponds to the shock followed by a region where the velocity decreases and the density increases. A small error has been introduced into the calculation from the Gladstone - Dale coefficient. For all points behind the shock, the equilibrium value has been used whereas in reality there will be a continuous change from the frozen to the equilibrium case as the gas relaxes. This error is always less than 10 % but could perhaps be reduced by introducing a simple model relating the N and N_2 concentrations to the total density throughout the relaxation zone.

As well as providing a quantitative measure of the density variations in the flow, the interferogram provides a recording from which the shock shape can be measured. This shock standoff is shown in figure 11 for two shots in HEG compared with a curve calculated as if the chemistry of the gas was always frozen and a curve for the case of always in equilibrium. As can be seen, the standoff is inside that for a frozen gas indicating that nonequilibrium effects in the form of nitrogen dissociation are being observed.

5. Conclusions

A holographic interferometry system has been constructed and initial tests completed on HEG. The results have allowed the determination of the density distribution between the shock front and the model as well as the shape of the shock around the model. The use of a thermoplastic film provides the advantage that the film can be immediately developed without being removed from its position and reconstruction can be performed directly after an HEG shot. A disadvantage is that the quality of the interferogram is perhaps not as good as that provided by Mach-Zehnder interferometry for example, but the ability to utilize the phase shifting technique alleviates this problem.

Further work involves improving the quality of the interferogram together with incorporation of prior knowledge about the flow into the data reduction.

Acknowledgements

The authors acknowledge the assistance of Mr. S. Kortz for providing the calculations on the shock standoff distance.

The financial contribution of the ESA/CNES toward the HEG and its instrumentation is gratefully acknowledged.

References

[1] W. H. Beck, G. Eitelberg, T. J. McIntyre, J. P. Baird, J. Lacey & H. Simon, "The High Enthalpy Shock Tunnel in Göttingen (HEG)", 18th Int. Symposium on Shock Waves, Sendai, Japan, 1991.

[2] W. H. Beck, DLR - Nachrichten 63 Mai 1991, p. 24 - 28.

[3] C. M. Vest, "Holographic Interferometry"; John Wiley&Sons 1979.

[4] Betriebsanleitung Prozeß - Kamera TPC 200 Sales by: Steinbichler Optotechnik GmbH.

[5] Dändliker & Thalmann, Opt. Eng. 24 (1985) 824 - 31.

[6] B. Breuckmann, Technisches Messen - tm 57 (1990) 389 - 94.

[7] R. A. Alpher & R. R. White, Physics of Fluids 2l(1959) 153.

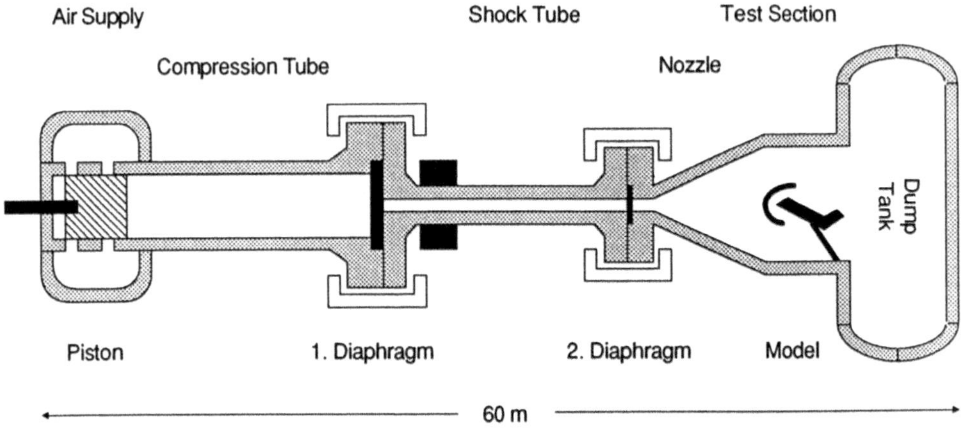

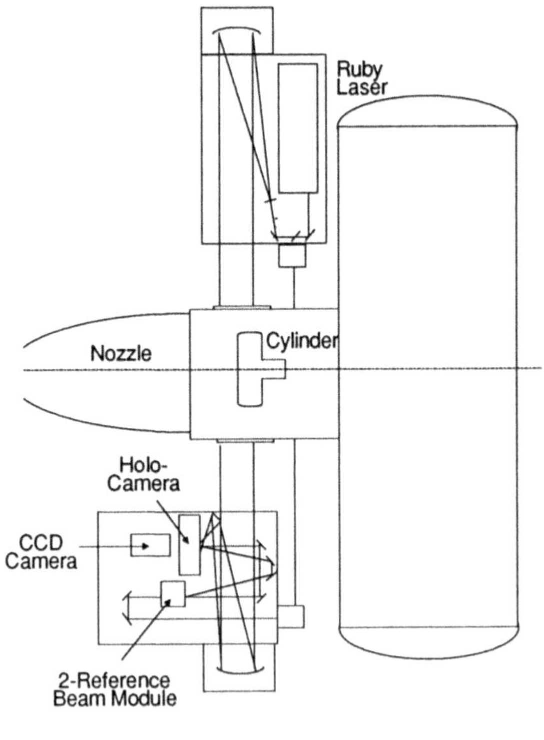

Fig. 2. Holographic interferometry layout around HEG.

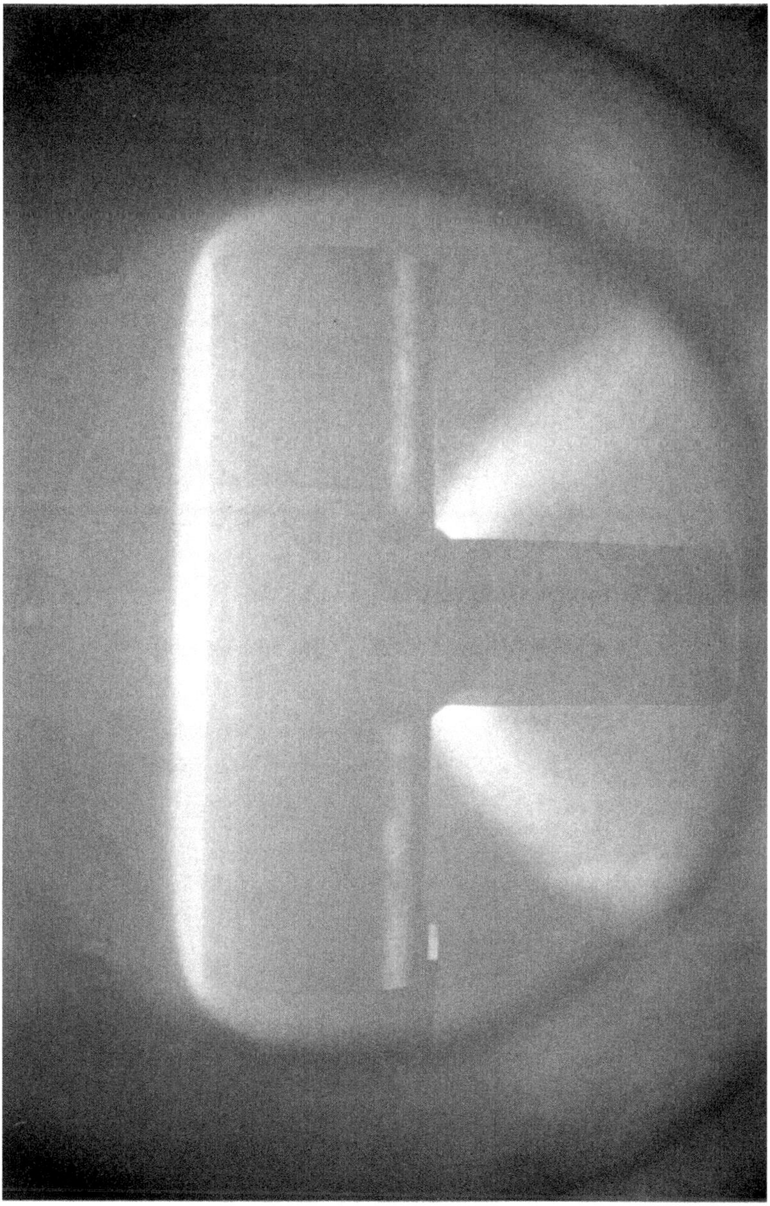

Figure 3. Luminosity photograph of flow over a cylinder

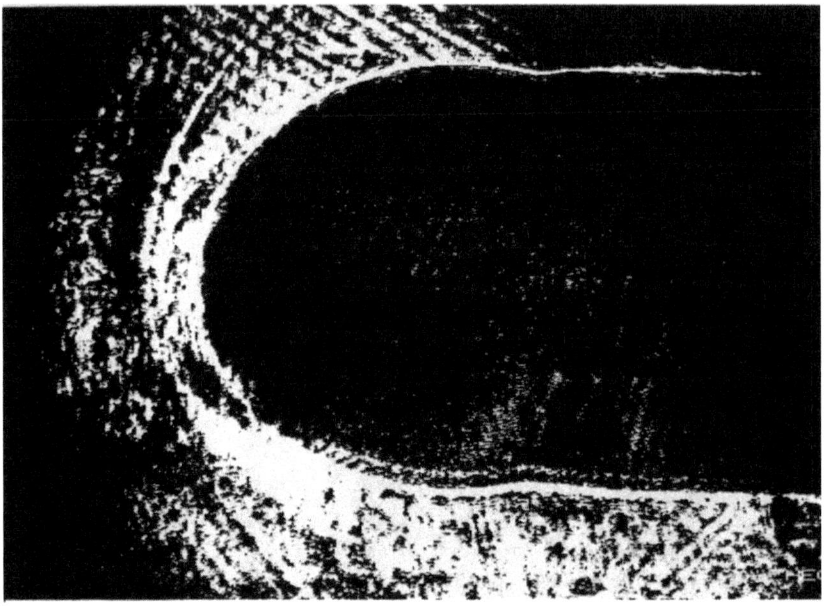

Figure 4. Infinite fringe interferogram of flow over a cylinder

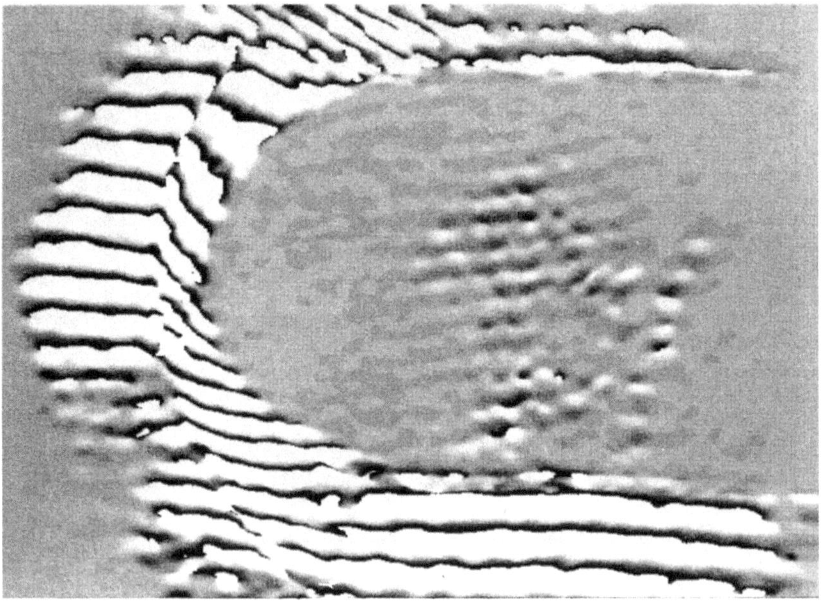

Figure 5. Phase extracted from the interferogram

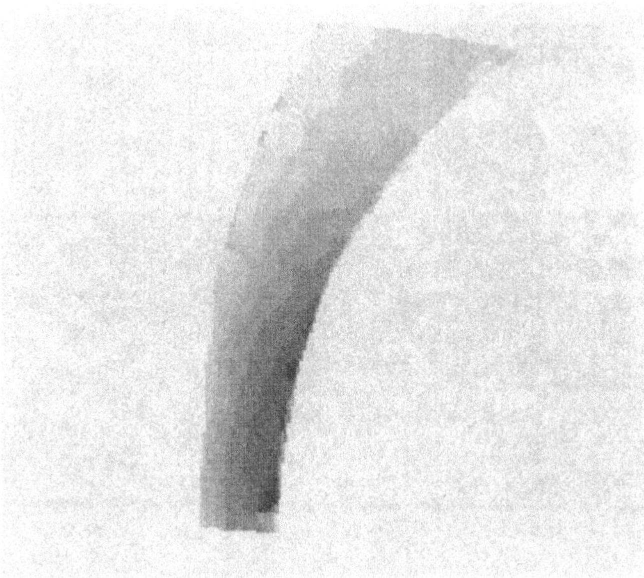

Figure 6. Phase distribution between the shock front and the model.

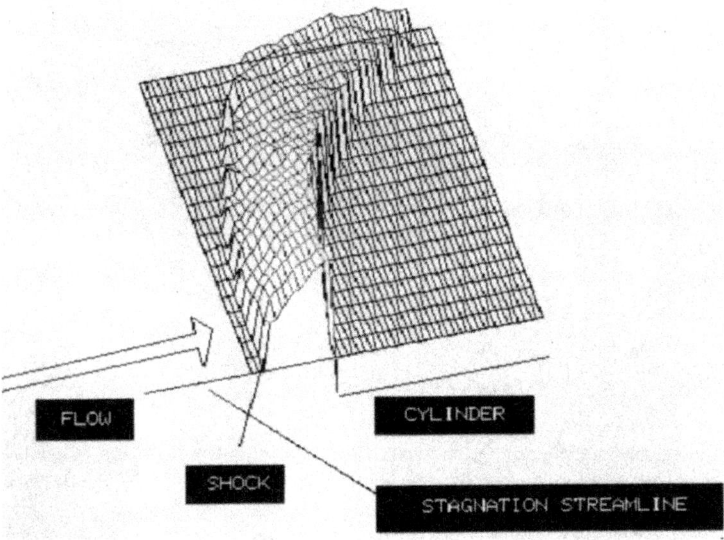

Figure 7. Three-dimensional representation of the phase distribution

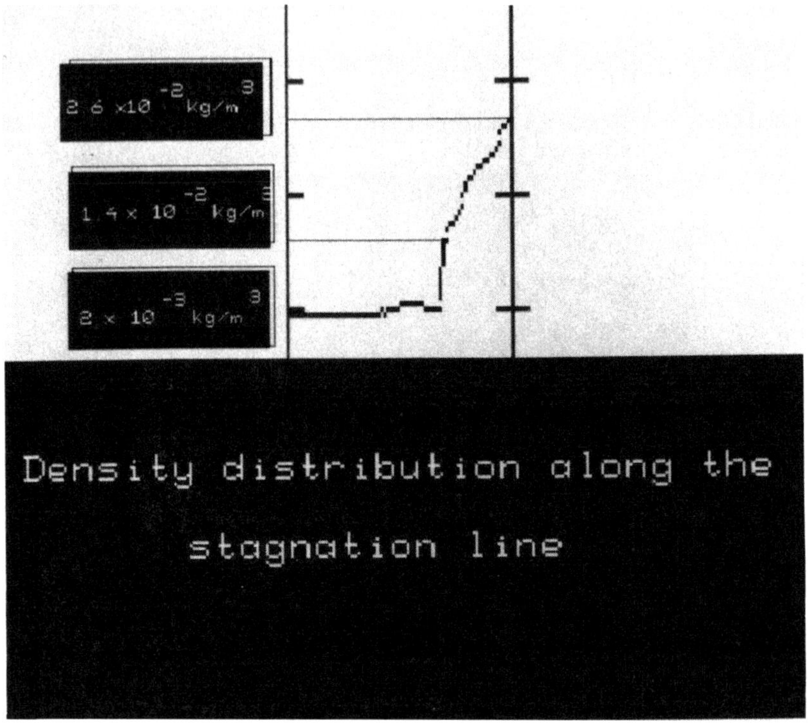

Figure 8. Density distribution along the stagnation streamline.

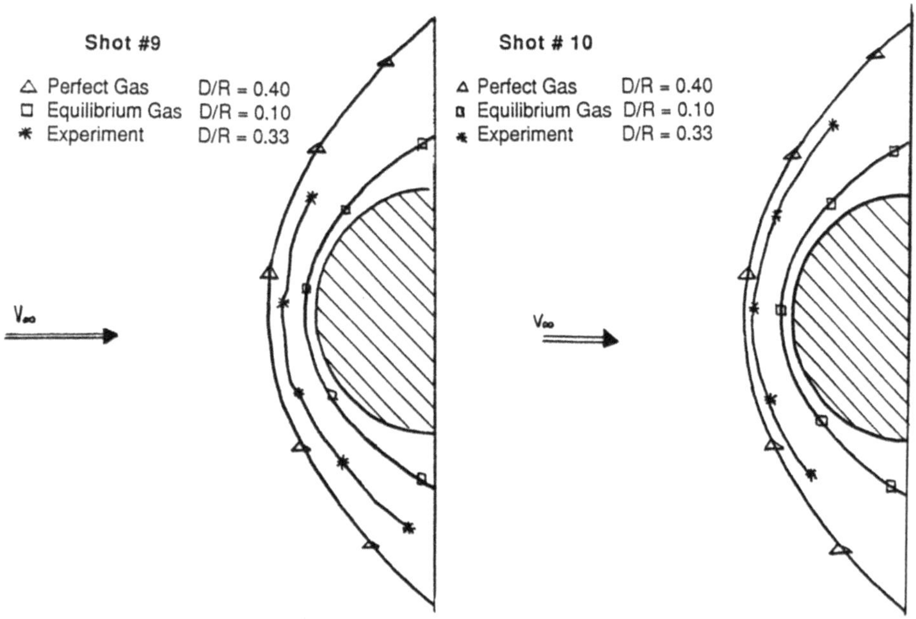

Figure 9. Measured and calculated shock shapes on a cylinder

THE USE OF HOLOGRAPHIC INTERFEROMETRY FOR FLOW FIELD DIAGNOSTICS

W. CHARLES SPRING, III
WILLIAM J. YANTA
KIMBERLY GROSS
CARLOS A. LOPEZ
NAVAL SURFACE WARFARE CENTER, DAHLGREN DIVISION,
WHITE OAK DETACHMENT, SILVER SPRING, MARYLAND, 20903-5000

ABSTRACT. The use of interferometry as a viable tool for aerodynamics research has been through several cycles of popularity in the last five decades. It has always held great potential for providing an enormous amount of information about a flow field, but the greatest difficulty has been the retrieval of data from the interferograms. With the ready availability of high-speed desk-top computers with large memory capacity, and high-resolution digital video image processing and storage, the technology now exists to do the data reduction from interferograms effectively and efficiently. This paper will describe the recent applications of holographic interferometry at the U. S. Naval Surface Warfare Center, Dahlgren Division (NSWCDD), and will present details of the development of phase-shifting analysis to make flow field calculations from these interferograms.

1. Holographic Interferometry

Holography itself is the recording of a unique interference pattern on a photo-sensitive plate such that this pattern can be reconstructed to regenerate the original optical wave. (See Vest, Reference 1, for background on holography.) In much the same way as with the interferometer, the output from a laser light source is split into two beams. The first, the object beam, is reflected off or transmitted through the object; the second, the reference beam, is directed around the object. Both beams are then recombined in the plane of a photo-recording medium (usually a film plate). Because of the nature of the interference pattern, both the amplitude and the phase of the light waves are recorded. When the hologram is re-illuminated with a coherent light (or white light in some unique cases) that is a duplicate of the reference beam, the recorded interference pattern will diffract some energy into a beam that is a replica of the object beam. This reconstructed object beam contains all of the amplitude and phase information of the original object beam, and hence looks and acts like the original beam. This is why holography has been called "lensless photography", and is why it can produce visually dramatic three-dimensional images.

Since the reconstructed wave contains phase information, as well as amplitude, images simultaneously reconstructed from two separate holograms can be made to interfere with each other, in much the same way as the two beams of an interferometer. This is the basis of the holographic interferometry used at NSWCDD to determine flow field densities during wind tunnel tests. One hologram is taken through the tunnel test cell before the flow is started, and a second one after the supersonic flow is established. The two holograms are reconstructed together and the optical path length differences between them, caused by the density changes in the air flow, are seen as interference patterns.

A. Boutier (ed.), New Trends in Instrumentation for Hypersonic Research, 97–112.

To construct holograms of wind tunnel flow fields at NSWCDD, a Z-type schlieren system (Fig. 1) is used. This optical system is the standard technique for routine flow visualization. The normal light source is replaced with pulsed ruby laser, which provides a high-intensity, short duration, collimated, coherent beam of light. A partial beamsplitter at the output of the laser splits off a small fraction (typically about 10%) of the energy, which is used to create the hologram reference beam. This beam is directed over the wind tunnel toward the film plane. The remaining laser energy is imaged into the schlieren system, which produces a large diameter collimated beam passing through the tunnel test cell and then is focused near the film plane. In the film plane both beams are recombined to form the hologram.

The special holographic quality ruby laser (Korad) used as the light source is Q-switched to provide very short duration exposures, on the order of 20 nanoseconds (2×10^{-8} seconds). These short exposure times freeze the flow field density patterns, and also freeze the minute interference patterns that form the holograms. The laser also has mode selectors that increase both the temporal and spatial coherence, so that high-quality holograms can be made. The temporal coherence is improved by using a tuned etalon in the laser cavity. Spatial coherence is increased by inserting an aperture that reduces the beam diameter, and hence the dispersion angle, of the output beam. These modifications reduce the output energy of the laser by about two orders of magnitude. However, there is still enough light for well-exposed holograms, and the improved coherence is critical.

The ruby laser is constructed such that a small helium-neon laser can be aimed through it for aligning the laser system with the schlieren optics. Provisions are also made for monitoring the peak power and total energy of the Q-switched pulses with oscilloscopes. This is important for assuring proper operation of the laser.

As mentioned earlier, two separate holograms are constructed for each test - the reference hologram, taken with no flow in the tunnel, and the object (or scene) hologram, taken with the tunnel at the flow conditions to be measured. In an optical sense, these two holograms are the two separate legs of a classical interferometer. After the plates are processed they are reconstructed with a helium-neon laser (for video digitizing) or an argon laser (for making large negatives). When these plates are reconstructed simultaneously, the optical path length differences between the two are visible as interference fringes (Fig. 2). The interference pattern can be changed by moving one of the holograms with respect to the other, as shown in Fig. 3.

Also, the scene hologram can be reconstructed by itself to form either a shadowgraph or a schlieren image. Shadowgraph and schlieren are much more common methods of flow visualization, but are less quantitative than interferometry. Thus, holography offers the advantage that all three of the classical flow visualization techniques can be employed on the same test.

Another advantage of holographic interferometry is that the interference patterns can be generated in the controlled environment of the optics lab at a convenient time after the test, instead of in the relatively hostile, real-time environment of modern hypervelocity wind tunnels. This places fewer demands on the optical set-up and data reduction techniques.

2. Data Reduction

The basic information that needs to be extracted from the interferograms is the optical path length differences between the object and reference images. The fringe shifts across the field are a direct measure of these path length differences. The classical way to measure this is to count, or track, fringes. The straight-line fringes in the free-stream region outside the shock wave, where the gas density is known, are used as reference fringes. The distance the fringes are shifted within the flow field is directly related to the unknown gas density in these areas. In a complex flow field, however, the task of determining these fringe shifts for every point in the field is tedious and time consuming if manually counting fringes is the only method of extracting the data. For years it was this limitation that prevented interferometry from becoming a practical measurement technique.

The key to the data reduction technique that has been developed at NSWCDD for analyzing flow-field interferograms is the fact that the two holograms are recorded on separate plates. The dual plate method, developed by Havener and Radley[2], was applied to the NSWCDD tunnels originally by Hannah and Havener[3]. All work described in this paper uses the dual plate technique. This method allows the interference pattern to be continuously varied during the reconstruction. (The double exposure technique, which is the other common method of holographic interferometry, fixes the interference pattern in place at the time of the exposures.) The dual plate method leads directly to the scheme called phase-shifting interferometry. Although not the first to use phase-shifting interferometry, NSWCDD has developed the technique for analysis of flow field interferograms where sharp density gradients are encountered, such as at supersonic shocks.

The principal of phase-shifting interferometry is that a map of the relative phase of the interference pattern can be calculated from three or more intensity maps of the field. The condition is that the background fringes of the intensity maps must have a constant, known phase shift between them. This is why the ability to shift the fringe pattern during the reconstruction of a pair of holograms of the same test conditions is important. Also required for this technique are a computer-based high-resolution video digitizing system with the memory and software capability to do image processing and large array calculations.

The current data reduction capabilities at NSWCDD include a dual plate holder, developed by Havener and Radley[2], used during reconstruction of the holograms. The apparatus provides multiple degrees of freedom to move one plate with respect to the other. The plates are initially aligned to reconstruct an infinite fringe interferogram (see Fig. 2), then the object plate is translated horizontally until the desired background fringe spacing is obtained.

The reconstructed interferograms are imaged into a high-resolution video camera, which has gain and contrast adjustments to control the level and range of the output signal. Though it will be shown later that the phase calculations are relatively insensitive to local intensity and contrast variations, this ability to adjust the video signal is important to optimize the images obtained from a variety of holograms. The ideal signal utilizes the full 8-bit intensity range of the digitizer without clipping any values from either the high or low ends of the scale.

The video output is then fed into a PC-based image digitizing system (FrameGrabber from Imaging Technology Corp.). The result is a 512x512 pixel array of digitized intensity values representing the interference pattern of the flow field.

To capture the images required for the phase shifting methods, the object plate is driven laterally small fractions of an inch by an accurately controlled piezoelectric micrometer that has been calibrated to produce $\pi/2$ phase shifts in the interference pattern. Typically, five images are digitized for each pair of holograms analyzed. They represent phase shifts of 0, $\pi/2$, π, $3\pi/2$, and 2π radians (the 0 and the 2π images will be identical). Once these images have been digitized, they can be digitally filtered or processed in a number of ways, depending on the quality of the input images and the desired output product. Also, the data reduction routines, developed at NSWCDD, can be run which will calculate phase distributions and density values of the flow field.

3. Data Reduction Theory

As shown by Creath[4], phase shifting interferometry utilizes three or four (depending on the reduction algorithm selected) digitized images of the intensity fields from the reconstructed interferograms, with $\pi/2$ phase changes between each of the images. The intensity of each of the points (pixels) in the field is:

$$I = I_0 [1 + \gamma \cos(\phi)] \qquad (1)$$

where:

I = intensity of any given point within the field
I_0 = the DC, or background, intensity
γ = modulation, or contrast, of the fringes, and
ϕ = the phase of the wavefront

The intensity, I, of each point in the field is detected and digitized by the video camera and PC frame grabber. The other three values, I_0, γ, and ϕ, are unknown, hence the need for at least three digitized images. When the three (or four) intensity fields are recorded with a phase shift of $\pi/2$ between them, three (or four) simultaneous equations for the intensity of any given point in the field (from Eq. 1) are as follows:

$$I_1(x,y) = I_0(x,y)\{1+\gamma\cos[\phi(x,y)+\pi/4]\} \qquad (2)$$
$$I_2(x,y) = I_0(x,y)\{1+\gamma\cos[\phi(x,y)+3\pi/4]\} \qquad (3)$$
$$I_3(x,y) = I_0(x,y)\{1+\gamma\cos[\phi(x,y)+5\pi/4]\} \qquad (4)$$
$$I_4(x,y) = I_0(x,y)\{1+\gamma\cos[\phi(x,y)+7\pi/4]\} \qquad (5)$$

These can be solved for the phase ϕ at that point as follows, where the phase is in radians:

$$\phi(x,y) = \tan^{-1}\left[\frac{I_3(x,y) - I_2(x,y)}{I_1(x,y) - I_2(x,y)}\right] \qquad (6)$$

The general form of Eqs. 2-5 can be solved, but if the phase shifts between the separate intensity measurements are 90° ($\pi/2$) this simplified form results. Also, to solve the general form, the phase shifts between each intensity map must still be known, so it is more expedient to pre-select the equally spaced shifts and use the simpler form.

Equation 6 describes a particular algorithm, termed the "three-bucket" method, since three intensity maps are required to calculate the phase map of the field. Several other algorithms have also been examined, namely the "four-bucket", which uses four intensity maps with phase shifts of $\pi/2$, the "three-by-three", which averages two three-buckets, and the Carré method, which uses four intensity maps with unknown (but assumed to be linear) phase shifts.

For the four-bucket method, which uses four of the intensity maps, the equation for ϕ becomes:

$$\phi(x,y) = \tan^{-1}\left[\frac{I_4(x,y) - I_2(x,y)}{I_1(x,y) - I_3(x,y)}\right] \qquad (7)$$

The three-by-three method averages two three bucket phase calculations, the first using I_1, I_2, and I_3, and the second using I_2, I_3, and I_4 .

These three methods, by virtue of the addition of arbitrary phase shifts introduced into the mathematics by the intensity map phase shifts, will produce different calculated phase values from the same set of intensity maps. Since the desired input for the density calculations is phase shift relative to the free stream (or other reference), all calculated phase values are normalized to zero in the free stream, or to some other reference value.

The Carré method uses four intensity maps having a phase shift that is not known, but is assumed to be linear. The actual phase shift is a fourth unknown, but can be calculated from the four intensity equations. This method has not yet been integrated into the NSWCDD data reduction programs.

All of these methods produce phase values that are modulo 2π, with a range from $-\pi$ to $+\pi$, which means that a plotted curve of phase values vs. distance (radius) will be discontinuous. To produce a continuous phase distribution curve, the calculated phase values are stacked on each other at the discontinuities. This is done with software, by comparing the phase values calculated for adjacent pixels. If the phase difference is greater than π, then 2π is added or subtracted to make the difference less than π. This technique produces smooth phase distribution curves. Typical results for the initially calculated phase and the final phase distributions are shown in Figures 4 and 5. These results are for the flow around a blunt cone which will be described in more detail later in the paper.

4. Calculation of Density Distribution

Once the phase has been determined, the density distributions may then be calculated. The complexity of the calculations will depend upon the geometry of the phase object (flow field). For a simple 2-D channel-type flow it can be assumed that the density is constant along the light path through the field of view. The refractive index n and density ρ, at a place where the fringe shift, or phase, is N(y) is given by:

$$N(y)\lambda = (n-n_0)L = K(\rho-\rho_0)L \qquad (8)$$

where:

$n-1 = K\rho$
n_0 = refractive index of known gas
ρ_0 = density of known gas
λ = wavelength of constructing laser
K = the Gladstone-Dale constant
L = pathlength through flow field

Solving for the density ρ, then:

$$\rho = \rho_0 + N(y)\lambda/KL \qquad (9)$$

or normalizing to the known density:

$$\frac{\rho}{\rho_0} = 1 + \frac{N(y)\lambda}{\rho_0 KL} \qquad (10)$$

For axisymmetric flows, the problem is more complex. The phase distribution can be related to changes in the optical path lengths and density for a refractionless, axisymmetric flow field by:

$$N(y)\lambda = 2\int_y^R \frac{r\,f(r)}{\sqrt{(r^2-y^2)}}\,dr \qquad (11)$$

where:

N(y) = fringe order number
R = boundary of flow field
r and y = independent geometric variables
f(r) = radial index of refraction

The fringe order number (phase shift non-dimensionalized by 2π), N(y), is the functional description of the changes in the optical path length through the flow field. The function f(r) is the radial distribution of the index of refraction in the phase volume. It is referenced to some known condition n_0 (usually the free stream index of refraction) by the equation:

$$f(r) = n(r) - n_0 \qquad (12)$$

Since the index function f(r) in Equation 11 is the desired quantity, the integral must be inverted resulting in:

$$f(r) = \frac{-\lambda}{\pi}\int_r^R \frac{dN/dy}{\sqrt{(y^2-r^2)}}\,dy \qquad (13)$$

The density ρ_r can then be calculated from this index function, using the Dale-Gladstone relationship $(n=1+K\rho)$ as follows:

$$\rho(r) = \frac{1}{K}[f(r) + (n_0-1)] \qquad (14)$$

A variety of methods are used to solve the above equations for the density distribution. In the analysis of interferometric data, the fringe order number N(y) is known only at a finite number of discrete locations and must be inverted numerically. In view of the fact that a derivative must be computed for the experimental data, large errors may develop in the resulting density values. One approach is to use an analytical representation of N(y) by curve fitting N(y) with a polynomial either locally or over the entire region. Many investigators have used this technique successfully (see for example Ref. 5).

A second method has proven to be more practical at NSWCDD. This method divides the flow region into a series of discrete annular elements (rings) of constant width Δr as shown in Figure 6 (also see Ref. 1). Since the output from the digitization results in a large number of discrete, equally spaced values of the phase N(y), this method lends itself particularly well to the present set-up. This is because the phase value determined at each pixel can represent the value for each ring.

Different assumptions can also be made about the density in each of the annular elements, including linear or higher order variation. The simplest is to assume that the density is constant, hence a uniform refractive index exists in each element. This results in a system of equations given by:

$$\sum_{k=i}^{I-1} A_{k,i} f_k = \left(\frac{\lambda}{2\Delta r}\right) N_i \qquad (15)$$

where the coefficients $A_{k,i}$ are given by:

$$A_{k,i} = \sqrt{(k+1)^2 - i^2} - \sqrt{(k^2 - i^2)} \quad \text{and } f_k = n(r_k) - n_0.$$

Equation 15 represents a set of simultaneous, linear, algabraic equations which must be solved for the unknown values of f_k, the change in index of refraction. The density is then calculated by:

$$\rho_k = \frac{1}{K}[f_k + (n_0 - 1)] \qquad (16)$$

Instead of assuming the index of refraction is constant in each zone, one can approximate it by a linear function in r, which results in the following equations:

$$\lambda N_1 = \Delta r f_1 A_{i,i} + \Delta r \sum_{k=i+1}^{I-1} f_k (A_{k,i} - A_{k-1,i}) \qquad (17)$$

where $f_k = n(r_k) - n_0$ as before and $A_{k,i}$ is now defined by:

$$A_{k,i} = (k+1)\sqrt{(k+1)^2-i^2} - k\sqrt{(k^2-i^2)} - i^2 \log\left\{\frac{k+1+\sqrt{(k+1)^2-i^2}}{k+\sqrt{(k^2-i^2)}}\right\}$$

A third approach assumes the fringe shift is a linear function of y^2, which results in a technique for directly computing f(r) without solving a system of equations. This results in

$$f_i = \frac{2\lambda}{\pi \Delta r} \sum_{k=i}^{I-1} (N_k - N_{k+1}) A_{k,i} \qquad (18)$$

In this case, $A_{k,i}$ is defined by:

$$A_{k,i} = \frac{\sqrt{(k+1)^2-i^2} - \sqrt{(k^2-i^2)}}{2k+1}$$

One of the difficulties with the annular ring method is that the method breaks down when step discontinuities in density occur. Such a case would be the region near a shock wave, and especially near a strong bow shock. A technique was proposed by Weyl[6] and revised by Bennett, et.al.[7], to account for this density jump.

This method results in the following equation for the density:

$$\rho = \rho_s + \frac{2\lambda}{\pi k} \sum_{k=i}^{I-1} \left\{\frac{(N_k - N_{k-1}) + \Delta r \rho'_2 B_k}{A_{k,i}}\right\} \qquad (19)$$

where $A_{k,i} = \sqrt{(k+1)^2-i^2} + \sqrt{(k^2-i^2)}$ and $B_k = \sqrt{I^2-(k+1)^2} - \sqrt{(I^2-k^2)}$

Also $\rho'_2 = 2\rho_2 K/\lambda$, with ρ_s being the density behind the shock, and ρ_2 the step discontinuity at the shock wave. More details on these methods can be found in Reference 5. Results from all of these methods will be presented later.

5. Applications of Holographic Interferometry

The first real test of the capabilities of the phase shifting technique was for a two dimensional flow field in a scramjet inlet investigation in a Mach 4 flow (Fig. 7). This test provided a good evaluation of the technique because several other flow diagnostics methods were also used, including Laser Doppler Velocimetry (LDV) and pressure probes. This allowed the results from the various methods to be compared. This work has been presented by McArthur, et. al.[8] Basically, the dual-plate holograms were reconstructed to form the real-time interferograms and a series of phase-shifted images was digitized. These images were then converted to phase maps, from which densities were calculated. The calculated densities were then compared to densities that had been measured in the inlet with Pitot probes. These results are shown in Figure 8, and indicate a very good agreement between the physical and optical measurements.

The next challenge for the phase shifting technique was the axisymmetric flow field around a blunt cone (see Fig. 3). This problem was selected because it was a relatively straight-forward case and both experimental and computational data (including data from hand-plotted interferograms) was available to compare the phase shifted calculations. The Mach number was 2.5 for these tests, and, again, several other techniques, including LDV and Pitot probes were used for comparisons. This program provided the opportunity to compare several different methods of calculating the phase distributions from the intensities (namely the three-bucket, the four-bucket, and the three-by-three methods).

Also, density values were calculated from the phase distributions using several different assumptions for the nature of the density in each annular element. The simplest assumption used was that the density is a constant, which produces density values through the flow field using Equation 16. Densities were also calculated from the results of Equation 17 (where the density is assumed to be linear with radius, r) and from Equation 18 (assuming the relationship is as r^2). A fourth set of density values was calculated using Equation 19, which accounts for the step discontinuity in density at the shock wave.

The results, presented in detail by Yanta, et. al.[9], show that the phase shifting technique does indeed calculate flow field densities that compared very closely to densities measured experimentally by other methods and to densities predicted by computational fluid dynamics (CFD) codes. Figure 9, for example, shows the comparison between the phase shifted densities and densities calculated using a three-dimensional Euler solver by Hsieh and Priolo[10]. These measurements were taken in an axial plane three inches behind the nose tip. This graph (Fig. 9) is typical of those taken at several axial locations, and all show good agreement except in the region of the shock, a problem that will be addressed later.

Figure 10 shows virtually no difference in the phase distributions calculated using the three-bucket, the four-bucket, and the three-by-three methods. Also, Figure 11 shows little difference in the density calculations using the four assumptions for the nature of the density in each of the annular rings. It should be noted that the general case might not show such good agreement for these methods, and future work will not ignore the possibility that one method may become preferred.

The next steps involved trying the phase-shifting methods that worked well for the blunt cone on cases that presented more difficult circumstances - namely where either much smaller or much larger density gradients are to be measured. To investigate these cases, the flow field around a sharp cone at low Mach number and an indented nose shape were examined.

The sharp cone in low Mach number (3.0) flow presents a complication because the density changes across the shock wave are quite small, which means that the fringe shifts will also be small (see Fig. 12). Preliminary evidence indicates that the methods developed so far will need to be refined for these cases. Possibly a scheme that averages several phase maps will provide a greater signal-to-noise discrimination that will increase the confidence in the phase distributions determined under these conditions. Similar situations could occur in high Mach number, low Reynolds number flows. Measurement of small density gradients will be necessary for practical analysis of simulated re-entry conditions.

The indented nose shape (Fig. 13) also presents problems because of the opposite situation - that is the density changes are so great that the interference pattern contains a very large number of finely spaced fringes, requiring a high-resolution system for digitizing. Figure 14 shows the results of the densities calculated for the indented cone flow field using the phase shifting technique, again compared to CFD predictions as calculated by Hsieh (Ref. 11). These results show that all the fringes must be resolved in the digitized intensity images in order for the phase distributions to be calculated correctly. If all the fringes are not resolved, the phase distributions are reduced by 2π for each lost fringe.

A higher resolution digitizing system is one solution to this problem. Another solution is to digitize several enlarged sections of the interferogram so that the fine detail can be resolved, then let the computer put the sections back together to determine the total phase distribution. New equipment will soon be obtained by NSWCDD that will make available a high-resolution camera and digitizing capability. This can be used in conjunction with the computer to digitize areas of the image in high resolution that can then be processed and recombined at lower resolution to provide phase data for the entire image. This should be helpful in overcoming some of the problems in areas where the present system has limitations, such as in the area of a strong shock wave or the recirculation regions of the indented cone.

Another serious problem in using interferometry to calculate flow densities is the underlying assumption that the gas is refractionless. Since the calculations of the gas density involve determining the index of refraction of the gas, it is known that changes in this index are going to cause refraction of the light passing through the gas. This is the principle on which the schlieren and shadowgraph techniques are based.

As is shown by the data in Figure 9, there is a discrepancy between the phase values calculated from the interferograms and those values predicted by CFD in the region near the shock. It has been shown by Lopez in Ref. 12 that diffraction/refraction are the main contributors to this error. Ref. 12 modelled the optical systems and the wind tunnel using the theories of wave optics. The idea is best explained by assuming the wind tunnel behaves like a "lens". Fresnel or Fraunhofer diffraction integrals were applied, depending upon whether the wind tunnel density field behaved as a long-focal-length or short-focal-length lens. The inviscid radial density distribution was calculated for the flow over the blunt cone and the theoretical phase distribution was then determined. The results from these calculations are shown in Fig. 15. Shown in the figure are the predicted effects of diffraction as light propagates from the phase object to the observation plane. The curve labeled "00" is the phase distribution calculated from the CFD densities for a plane at the edge of the phase object. The curve labeled "15" is the predicted phase distribution 15 feet away from the phase object, including the diffraction effects. This curve is compared to the experimental phase values measured by Yanta, et. al, in Ref. 9. In these experiments the film plane was located approximately 15 feet from the model plane. It is seen that the results which include diffraction agree very well with the experimental values. There is still some small discrepancy near the shock wave where density gradients are most severe. This discrepancy may then be due to refraction effects. Work currently underway is investigating methods for correcting the phase data to include these diffraction and refraction effects.

6. Conclusions

This paper has presented some of the recent theoretical and experimental work at NSWCDD that have provided major improvements in the use of holographic interferometry for wind tunnel flow field analysis. The application of phase shifting techniques, combined with high-resolution video digitizing and powerful desk-top computers, has overcome one of the severe limitations in the use of interferometry as a practical flow diagnostic tool. These techniques have greatly increased the speed, density (number of data points), and accuracy of the flow density calculations made from the interferograms. Experimental data presented here show good correlation between the results of the phase shifting interferometric techniques and other methods of measuring the densities. The theoretical investigation indicates a reasonable approach for improving the density calculations in areas of the flow field where discrepancies occur.

7. Acknowledgements

The authors would like to acknowledge the many fruitful discussions with Dr. Andrew A. Wardlaw concerning the data reduction processes involved in this work. We also are grateful to the NSWCDD Independent Research Program, which has provided most of the funding for the recent work implementing the phase shifting data reduction technique and for the theoretical analysis of the diffraction corrections. Earlier work on the dual plate development had been sponsored by Naval Air Systems Command.

BIBLIOGRAPHY

1. Vest, C. M., Holographic Interferometry, Wiley & Sons, Inc., New York, N. Y., 1979.

2. Radley, R. J., Jr. and Havener, A. G., "The Application of Dual Hologram Interferometry to Wind Tunnel Testing", AIAA Journal, Vol. 11, Sept. 1973, p. 1332.

3. Hannah, B. W. and Havener, A. G., "Applications of Automatic Holographic Interferometry," Internation Congress on Instrumentation in Aerospace Simulation Facilities proceedings, Ottawa, Canada, Sept. 1975.

4. Creath, K., "Phase-Measurements Interferometry Techniques, "Progress in Optics XXVI, Edited by E. Wolf, Elsevier Science Publishers B. V., 1988, p. 350-393.

5. Havener, A. G., "Holographic Measurements of Transition and Turbulent Bursting in Supersonic Axisymmetric Boundary Layers," AIAA Paper 83-1724, presented at AIAA 16th Fluid and Plasma Dynamics Conference, July 12-14, 1983, Danvers, Mass.

6. Weyl, F. J., "Analytical Methods in Optical Examination of Supersonic Flow, NAVORD Report 211-45, Bureau of Ordnance, Navy Dept., December 11, 1945.

7. Bennett, F. D., Carter, W. C., and Bergdolt, V. E., "Interferometric Analysis of Airflow about Projectiles in Free Flight", Journal of Applies Physics, Vol. 23, Number 4, April 1952, pp 453-469.

8. McArthur, J. C., Yanta, W. J., Spring, W. C., III, and Gross, K. U., "Laser Holographic Interferometric Measurements of the Flow in a Scramjet Inlet at Mach 4", AIAA Paper #89-0043, presented at AIAA 27th Aerospace Sciences Meeting, January 1989, Reno, Nev.

9. Yanta, W. J., Spring, W. C., III, Gross, K. U., and McArthur, J. C., "Phase-Measuring Laser Holographic Interferometry For Use in High Speed Flows", presented at the 13th International Congress on Instrumentation in Aerospace Simulation Facilities, Gottingen, West Germany, 18 September 1989.

10. Hsieh, T. and Priolo, F. J., "Generation of the Starting Plane Flowfield Over a Spherically Capped Body", NAVSWC TR 84-484, May 1985, NAVSWC, Silver Spring, Md.

11. Hsieh, T., "Calculations of Flowfield About Indented Nosetips", NSWC TR-286, 23 August 1982, NAVSWC, Silver Spring, Md.

12. Lopez, Carlos A., "Diffraction Corrections of Holographic Interferometric Images", AIAA Paper 91-0564, Presented at the 29th Aerospace Sciences Meeting, Reno,Nev., Jan. 1991.

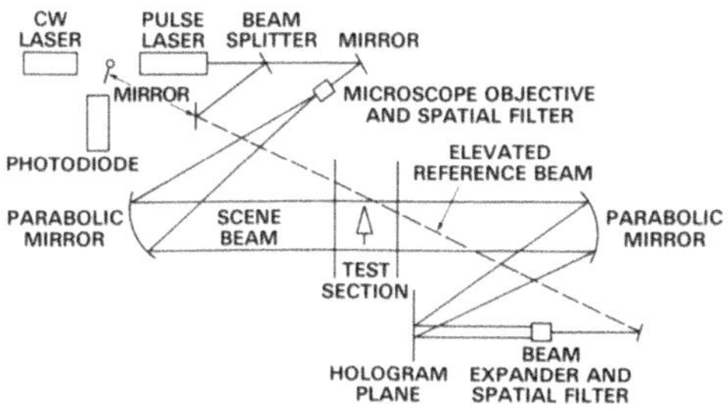

FIG. 1: SCHEMATIC OF Z-TYPE SCHLIEREN SYSTEM FOR HOLOGRAM CONSTRUCTION

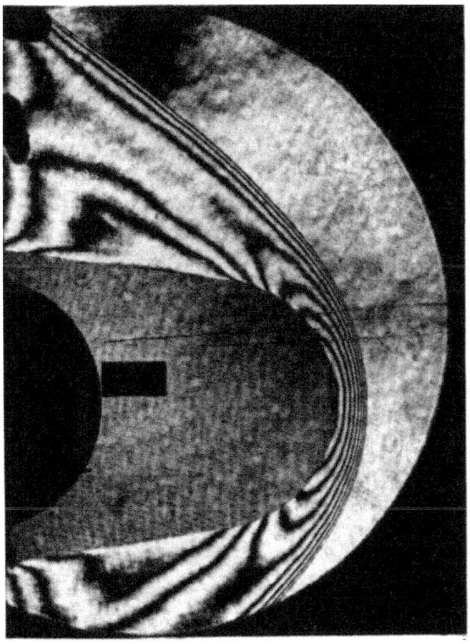

FIG. 2: INFINITE FRINGE INTERFEROGRAM OF BLUNT CONE AT MACH 2.5

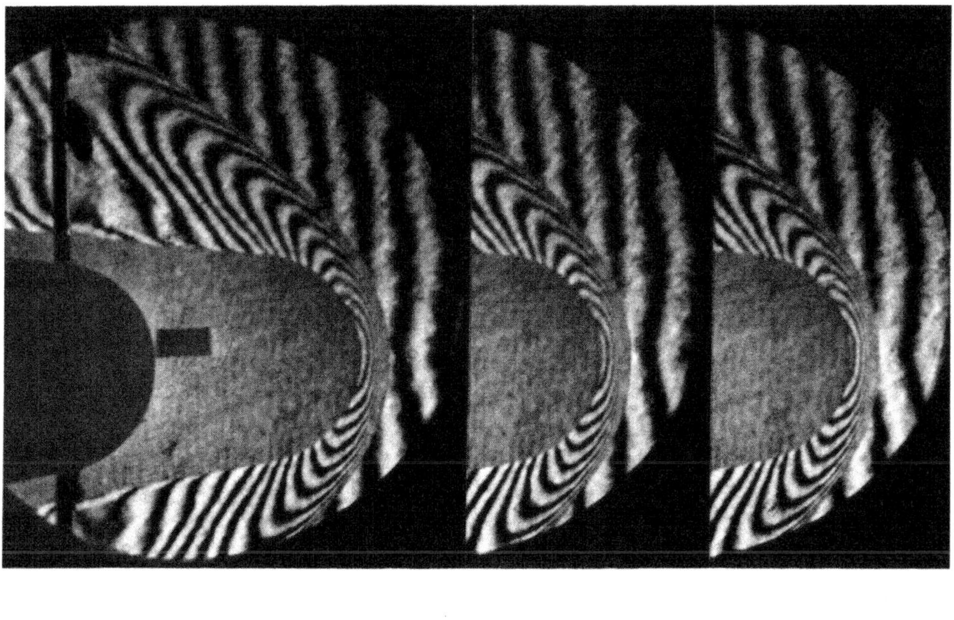

FIG. 3: PHASE-SHIFTED DUAL PLATE INTERFEROGRAMS

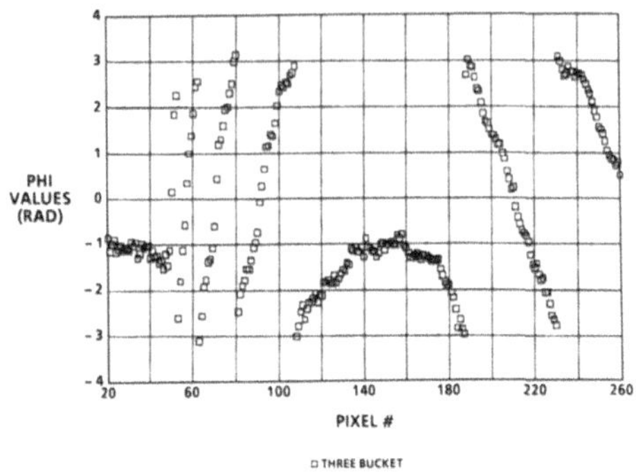

FIG. 4: PHASE JUMP TECHNIQUE

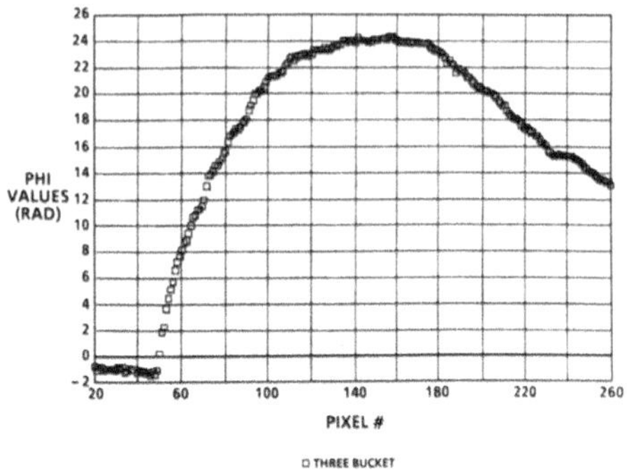

FIG. 5: CONTINUOUS PHASE DISTRIBUTION

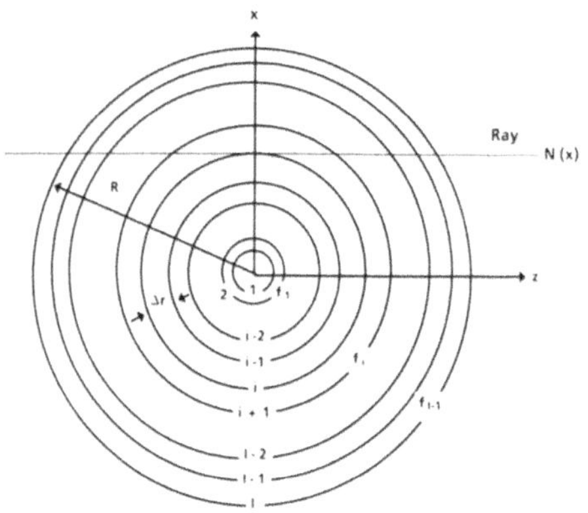

FIG. 6: PHASE OBJECT DIVIDED INTO DISCRETE ANNULAR RINGS

FIG. 7: FINITE FRINGE INTERFEROGRAM OF SCRAMJET INLET

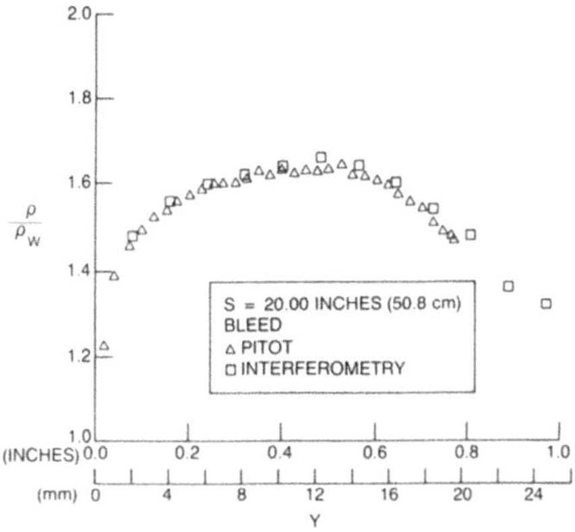

FIG. 8: DENSITY DISTRIBUTION IN SCRAMJET INLET

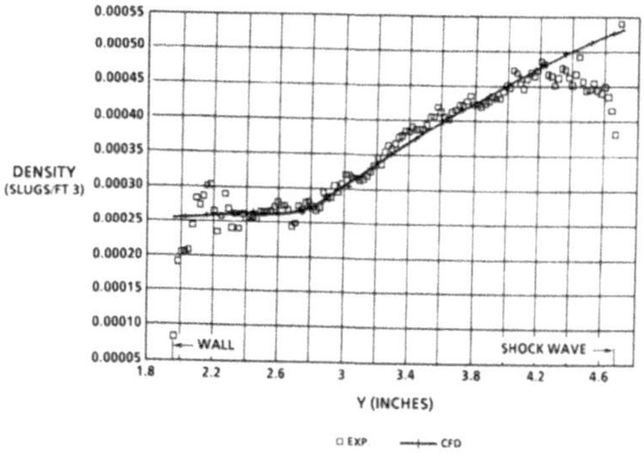

FIG. 9: MEASURED DENSITY VS.CFD SOLUTION FOR BLUNT CONE

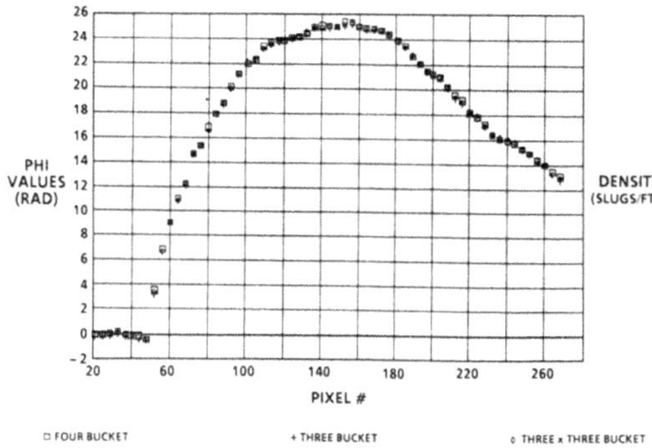

FIG. 10: PHASE DISTRIBUTION COMPARISONS

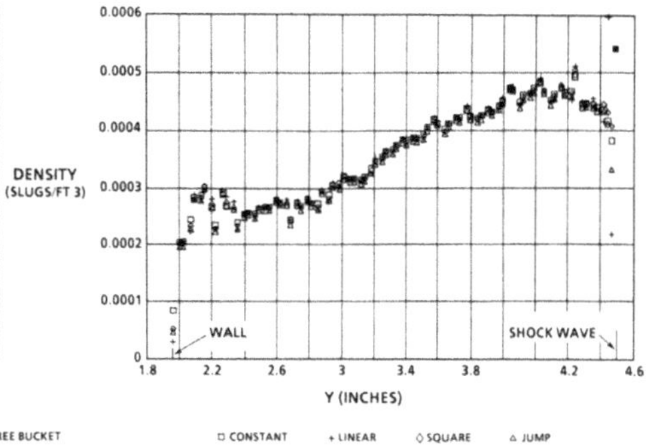

FIG. 11: DENSITY DISTRIBUTION COMPARISONS

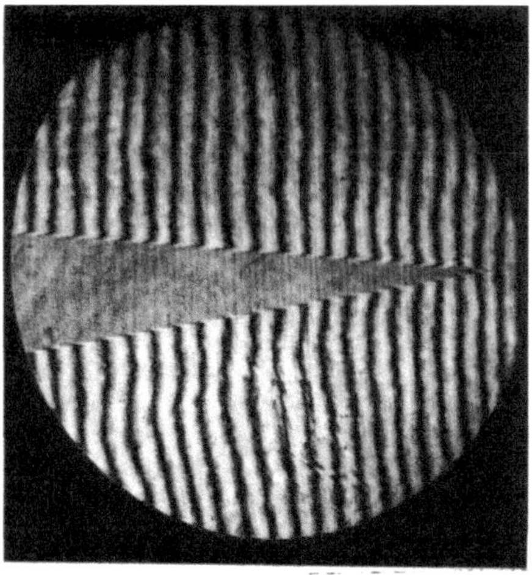

FIG. 12: FINITE FRINGE INTERFEROGRAM OF SHARP CONE AT MACH 3

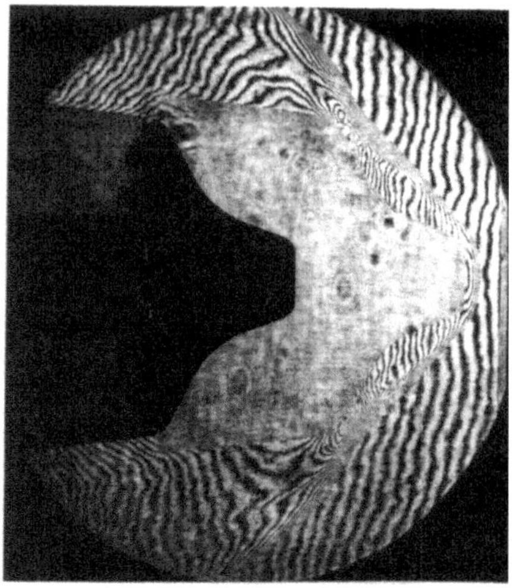

FIG. 13: FINITE FRINGE INTERFEROGRAM OF INDENTED NOSE SHAPE AT MACH 5

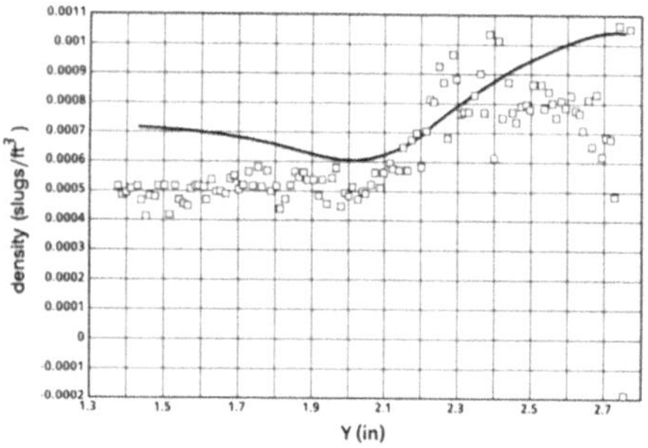

FIG. 14: MEASURED DENSITY VS. CFD FOR INDENTED NOSE SHAPE

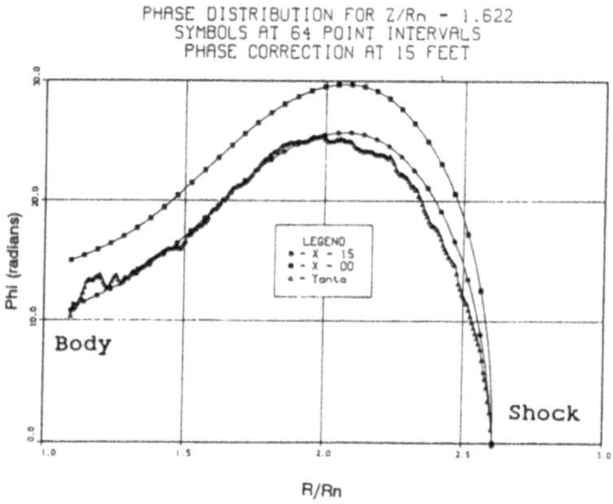

FIG. 15: THEORETICAL PHASE DISTRIBUTION FOR CORRECTED AND UNCORRECTED DIFFRACTION EFFECTS

MULTIPASS HOLOGRAPHIC INTERFEROMETER FOR THE HIGH ENTHALPY HYPERSONIC WIND TUNNEL F4

J. Surget and G. Dunet
Office National d'Etudes et de Recherches Aérospatiales
BP 72, 92322 Châtillon Cedex, France

SUMMARY. This paper describes a multipassage principle to increase the interferometric sensitivity. It presents the holographic set-up built for testing the method at the R5 low density hypersonic wind tunnel of ONERA Chalais-Meudon center. Interferograms of air flows ($M = 10$ and density $3.9\ 10^{-4}$kg m^{-3})obtained for 2, 4 and 6 passages are shown. Finally we describe the multipass holographic interferometer designed for the high enthalpy arc heated hypersonic wind tunnel F4 of Le Fauga-Mauzac ONERA center.

1. Introduction

The high enthalpy arc heated hypersonic wind tunnel F4 is equipped with the schlieren apparatus described in the paper 2F of this workshop. This large optical field device (fifty centimeters in diameter) provides schlieren or phase contrast pictures of aerodynamic phenomena in the totality of the flow surrounding the model under study. But a correct interpretation of the schlieren figures requires the knowledge of the generated gas flow density. This information can be obtained from schlieren visualization of the flow in the vicinity of two-dimensional or axisymetric models. However, density evaluation is more easily and consequently more precisely achievable using an interferometric technique. For this reason, it is planned to build a holographic interferometer to complete the F4 visualization equipment. A fifteen centimeter optical field diameter is sufficient for this apparatus which will be applied to small cylindrical or spherical models.

2. Principle of the holographic multipass interferometer

Single pass holographic interferometry, currently used for supersonic gas flow studies, can be also applied to visualize hypersonic flows generated in shock tunnels [1]. However it is not enough sensitive to enable accurate measurement of the very low densities existing in the flows generated by the F4 nozzle, ranging from 10^{-3}down to 6.10^{-5}kg.m^{-3}.Consequently the apparatus is designed to achieve improved sensitivity thanks to multipassages of the test beam through the aerodynamic phenomenon: the distortion of the test light wavefront, due to the gas index gradients, increases each time the beam crosses the flow [2,3].

A. Boutier (ed.), New Trends in Instrumentation for Hypersonic Research, 113–122.

3. First multipass holographic interferometer

A first holographic set-up has been built in order to test the practical possibilities of application of the multipass process [4]. Various experiments have been conducted successively at the optical laboratory and at the R5Ch low density hypersonic wind tunnel of ONERA Chalais-Meudon center [5].

3.1. OPTICAL DIAGRAM OF THE SET-UP

As indicated in figure 1, the laser beam is separated by the optical splitter S1: the two light flux transmitted and reflected by S1 propagate respectively along the test and the reference paths of the holographic device.

3.1.1. *Test path.* The beam transmitted by S is expanded by the lens L1, partially reflected by the semireflecting plane mirror S2, and then collimated by the lens O. The parallel beam emerging from O successively passes through the semireflecting plane mirror S3, the test chamber TC (closed by windows W) and is then reflected back by the plane mirror M1 (S3 and M1 are both adjusted at normal incidence). The reflected beam crosses again TC and reaches once more S3.

If the light crossing S3 is used for hologram recording by the photoplate H, the sensitivity is only doubled compared to that corresponding to a single pass optical system. But the use of light several times reflected by S3 and M1, and consequently having crossed 4, 6, 8 ...times TC, makes it possible to multiply the sensitivity by the same factor.

However H receives actually at once all the multireflected beams due to the action of the optical cavity S3–M1. Therefore, it is necessary to select only one of them. This selection is achieved by adjustment of the length of the reference path of the device. Indeed, hologram recording of a given test path is only possible using a reference path of equal length. In fact, the length adjustment tolerance is equal to about half the coherence length of the laser (a few centimeters for argon ion laser).

3.1.2. *Reference path.* The part of the laser light reflected by S1 is successively reflected by two rows of rectangular prisms down to the plane mirrors M2 and M3. The beam is then enlarged by the L2–L3 lens collimator, and ultimately impinges on the photoplate H. The distance between the left and right rows of prisms is equal to the distance between S3 and M1. So, the displacement of the mirror M2 from point P1 to points P2 or P3 makes it easy to equalize the reference pathlength to the testpath length corresponding respectively to one, two or three double passages through TC.

3.1.3. *Passage number selection.* The method for path number selection only founded on the limited coherence length of the laser light presents a drawback: the test light flux corresponding to the selected number N of double passages (useful flux ϕ_u) is in any case inferior to the useless test light flux ϕ_p due to the unselected beams. The unwanted part of the test beam light lowers the microfringe contrast of the hologram, and consequently reduces its image reconstruction efficiency. This problem is illustrated in figure 2, which shows S3, M1 and only one incident light ray, for a clearer statement. Although normally parallel to M1, S3 is traced slightly tilted for making visible the successive reflected rays.

For three double passages, the useful flux is less than 10% of the total flux, as indicated in the table associated to the figure. In spite of this unfavorable ratio, interferograms can be obtained for N = 3, which corresponds to an increase of sensitivity by a factor six (see figures 7 and 8).

To multiply the sensitivity by 8, 9, 10 etc... using passage numbers greater than 3, the unuseful test light flux must be eliminated, or at least deeply reduced. For this aim, two techniques can be used. The first one is a spatial filtering. In this case, S3 is slightly tilted (as represented in figure 2) producing an angular difference between test beams when they are successively reflected. Thus other beams than the selected one are eliminated, owing to a spatial filter set in the focal plane of the collimating lens O (figure 1). With this method, the large stray light flux ϕ_p is entirely suppressed, but the light rays successively crossing the test section are no longer parallel each other, which induces a slight displacement of object edges in the reconstructed holographic image.

The second technique consists in a time filtering. Whereas both continuous wave and pulsed lasers can be used in the two previous techniques, in this second technique we take advantage of the short duration of the light pulse delivered by a pulsed ruby laser. Indeed this method is based on the value of transit time T of the light into the optical cavity of the device (path S3-M1-S3). Using a pulse duration t smaller than T, the different beams corresponding to the various light path numbers, successively fall upon the photoplate H at time intervals T-t. Thus a fast light shutter (as a Pockels cell for instance) selects the test beam corresponding to the chosen number of test chamber crossings. On the reference beam, no shutter is required; the optical length has just to be adjusted to the test path length.

This selection principle is illustrated by figure 3 for a 30 ns laser pulse duration and a gap of 5 m between S3 and M1, similar to the time filtering configuration of the holographic device planned for the wind tunnel F4 (alinea 4). The shutter must be operated during the 3.3 ns separating two successive H illuminations.

3.2. EXPERIMENTAL SET-UP AND RESULTS.

The holographic set-up built for multipass interferometry testing is presented on figure 4. Its optical diagram corresponds to figure 1. The test chamber of the R5Ch wind tunnel of ONERA Chalais-Meudon center appears on the right side of the picture. It hides the plane mirror M1 of the set-up which is located behind. Figure 5 shows a close-up view of the L1-L2 lens collimator area. Figure 6 presents the left prism row P_L. The main characteristics of the set-up are as follows:

* light source: continuous wave 2 watts argon laser,
* optical field: 145 mm in diameter,
* optical cavity (interval S3-M1): about 2.1 m due to the space required by wind tunnel test chamber T_C,
* distance between the left and right prism rows P_L and P_R: 2.1 m (like S3-M1 interval as indicated in 3.1.2),
* optical path length of both test and reference paths:
  8.5 m for only one double passage through T_C,
  12.7 m for two double crossings,
  16.9 m for three.

First tests of the multipass technique have been carried out in laboratory, using an optical glass plate with a surface defect for air gradient effect simulation. Only the first and second number pass selection methods (paragraph 3.1.3) have been tested, considering the continuous wave laser used.

Due to the poor resolution provided by the second selection method, the best interferograms were obtained using the first method. Figure 7 presents a set of interferograms corresponding respectively to one, two and three double crossings of the test beam through the glass plate. The sensitivity, indicated by the number of fringes, clearly increases proportionately with the number of crossings.

Three double passages are practically the upper limit: according to figure 2 indications, the stray light already very important (93.7 %) becomes too high for four double passages (96.9 %). Thereby contrast and luminosity of the reconstructed interferograms decrease too much.

The interest of the multipass technique being demonstrated, the device was then installed at the R5 wind tunnel of ONERA Chalais-Meudon center [4] to visualize a low density gas flow (density about 4.10^{-4} $kg.m^{-3}$ and Mach number 10). Various models such as a cylinder and a sphere were tested. But the most significant results were obtained using a wedge (15° angle and 150 mm spread). Figure 8 shows three interferograms recorded respectively for one, two and three double crossings like interferograms of figure 7. However, the infinite fringe interferogram adjustment, used for the glass plate examination, is replaced here by a finite fringe adjustment, well adapted to visualize very small modifications of the optical path due to very low density gradients of the medium: the test light wavefront distortion, due to the aerodynamic phenomenon, is about twenty times inferior to the one created by the surface defect of the glass plate.

On figure 8, the distortion of the original fringe pattern, due to the shock wave, increases with the number of light passages through the wind tunnel test chamber: just visible on 8a, it clearly appears on 8c and becomes easily measurable.

4. F4 holographic multipass interferometer

For the F4 wind tunnel, in spite of an optical principle similar to the apparatus used at R5 wind tunnel, the holographic set-up has been designed on complementarity with the schlieren device. Only a few optical and mechanical components of the schlieren device must be disassembled for allowing the interferometric installation; therefore only four optical elements are located between the two spherical schlieren mirrors. In the same way, the schlieren window W is also used to introduce laser light into the test chamber, and to return it towards the photoplate for hologram recording.

Such an arrangement of course prevents to use simultaneously the schlieren apparatus and the interferometer. However, visualization method exchange requires only one or two hours.

Figure 9 shows the optical diagram of the F4 interferometer. As previously indicated, only the mirror M4, the objective lens O, the semireflecting mirror L0 and the mirror M5 are introduced into the test chamber and installed on the optical axis of the schlieren device. The optical field diameter is 145 mm in diameter. A continuous wave argon laser (2 watts) or a pulsed ruby laser (1 joule) can be used for hologram recording. The optical configuration represented on figure 9 is suitable for the number pass selection method only founded on the limited coherence length of the laser light. The interval L0-M5 (optical cavity) is 1.8 m long, and the total optical length for both test and reference arms is 11.2 m long for one double crossing, 14.8 m long for two double crossings, and 18.4 m long for three double crossings.

For a larger number of crossings, the time filtering method must be applied using a pulsed ruby laser. In this case, the optical cavity LO-M5 must be enlarged to 5 m (by

M5 displacement), corresponding to the laser pulse duration (30 ns). This time selection was previously described in paragraph 3.1.3, and illustrated by figure 3.

The optical pathlengths corresponding respectively to the two F4 optical configurations and to the R5Ch configuration (for comparison) are indicated in the table of figure 10.

5. Conclusion

The multipass holographic interferometric technique has been successfully tested at the hypersonic wind tunnel R5Ch of Chalais-Meudon ONERA center; a similar apparatus adapted to the high enthalpy hypersonic facility F4 of Le Fauga-Mauzac ONERA center has been designed and will be built to equip the facility by the end of 1992.

References

[1] Fleck B., Mc Intyre T.J., Eitelberg G. (1991) "Holographic interferometry on the new High Enthalpy Shock Tunnel in Gottingen (HEG)", 3rd French-German congress on applications of holography, Saint Louis (F).

[2] Ostrosky Yu I., Bustov M.M., Ostrovskaya G.V. (1980) "Interferometry by holography", Spring-Verlag, Berlin, Heidelberg, New-York, 178.

[3] Trolinger J.D. (1988), AGARDograph 296, Laser applications in flow diagnostics, 93.

[4] Surget J., Dunet G. (1990) "Multipass holographic interferometry for low density gas flow analysis", 19th International congress on High Speed Photography and Photonics", Cambridge (G.B).

[5] Chanetz B., Délery J., Nicout D., Pot T., Quelin C. (1990) "La soufflerie hypersonique à faible nombre de Reynolds R5Ch", RT ONERA n° 11/4362 AN.

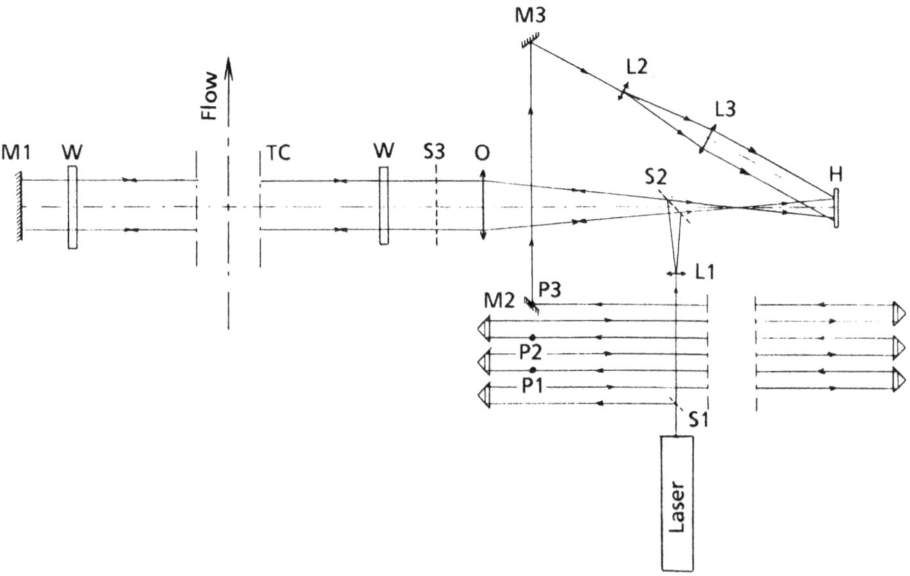

Fig. 1. Optical diagram of the multipass holographic set-up.

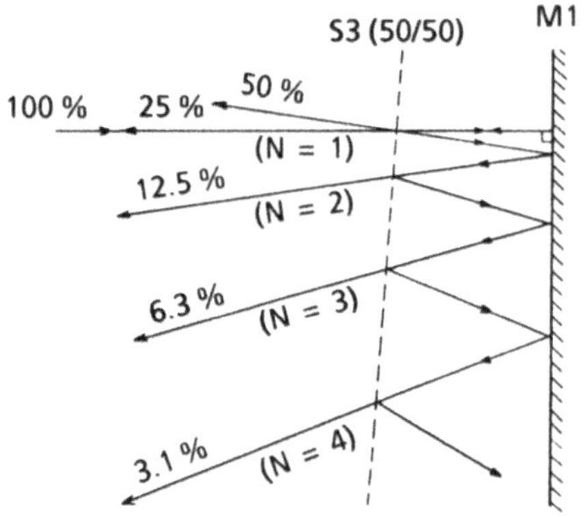

N	ΦU	ΦP	ΦP/ΦU
1	25 %	75 %	3
2	12.5 %	87.5 %	7
3	6.3 %	93.7 %	15
4	3.1 %	96.9 %	31
5	1.6 %	98.4 %	63

Fig. 2. Multipass system.

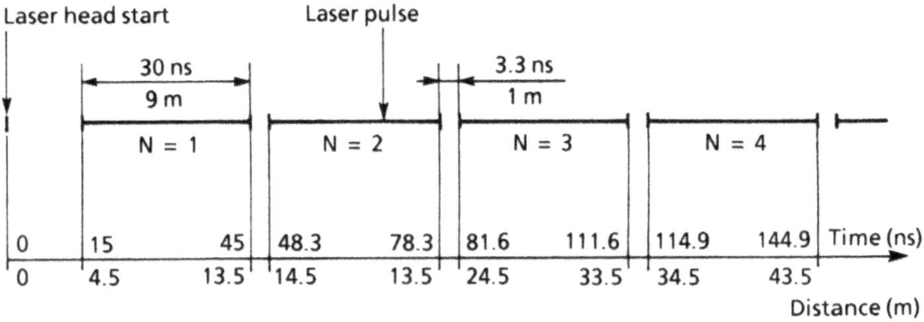

Fig. 3. Temporal selection.

Fig. 4. General view of the holographic set-up installed in front of the R5Ch wind tunnel.

Fig. 5. The L2 L3 lens collimator area.

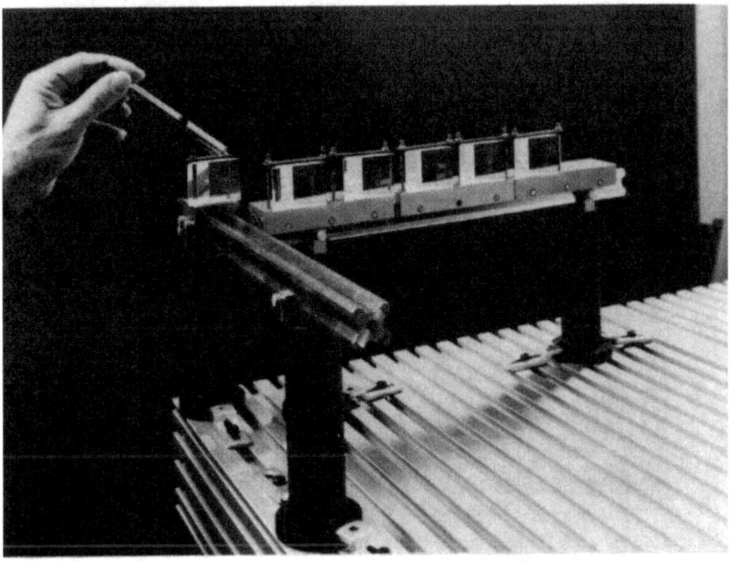

Fig. 6. The left side prism row.

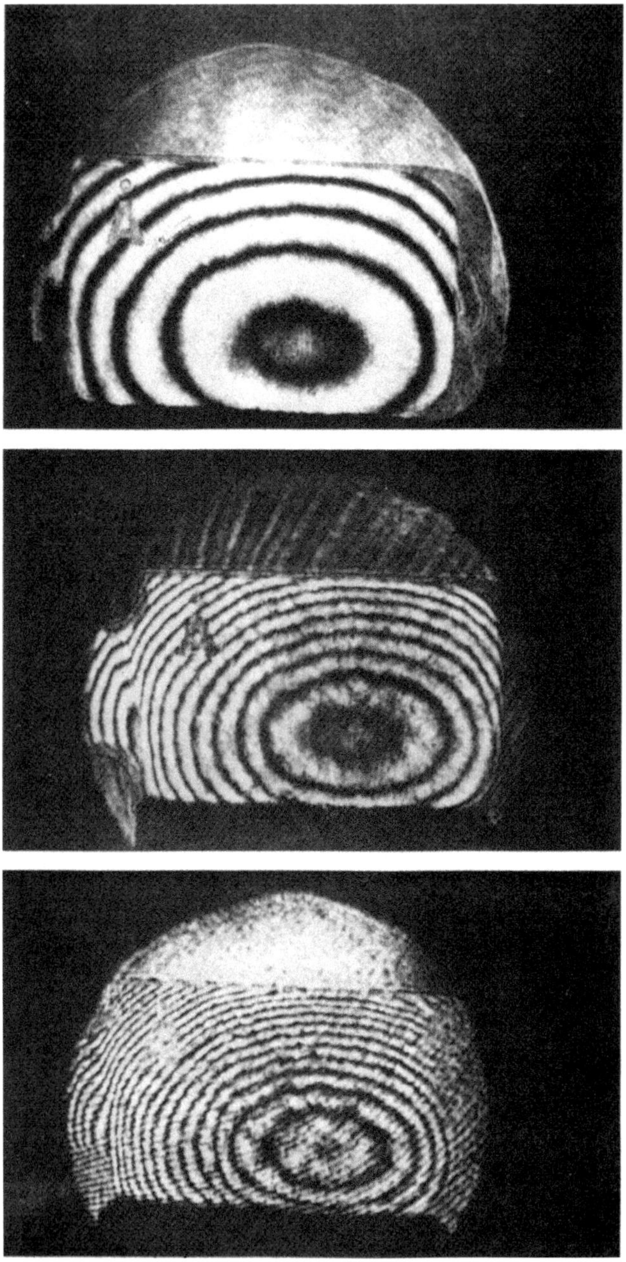

Fig. 7. Infinite fringe interferograms of a glass plate obtained by means of the first multipass technique:
a) N = 1
b) N = 2
c) N = 3.

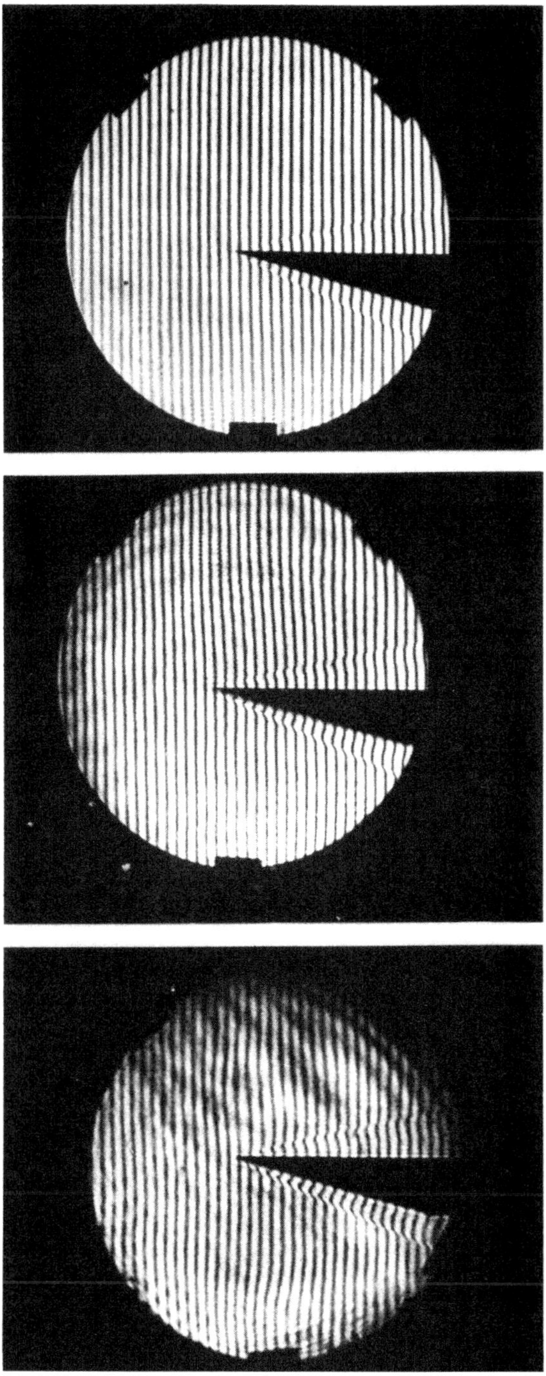

a) N = 1

b) N = 2

c) N = 3

Fig.8. Finite fringe interferograms recorded in the R5Ch wind tunnel.

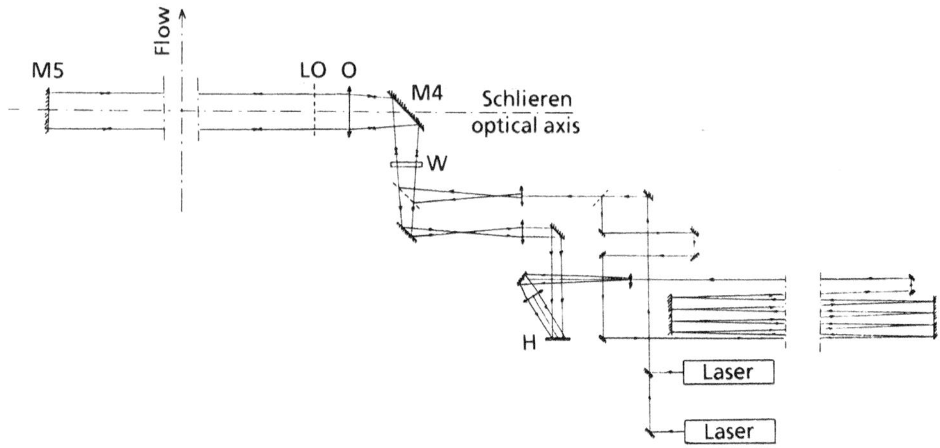

Fig.9. Optical diagram of the multipass holographic interferometer designed for the F4 wind tunnel.

N	R5 config.		F4 1st config		F4 2nd config.	
	LC	LT	LC	LT	LC	LT
1	4.2	8.5	3.6	11.2	10	17.5
2	8.4	12.7	7.2	14.8	20	27.6
3	12.6	16.9	10.8	18.4	30	37.6
4					40	47.6
5					50	57.6
7					60	67.6
8					70	77.6

Fig.10. Optical pathlengths for R5 and F4 configurations
N: numbers of double crossings,
LC: pathlength into the optical cavity (m),
LT: total optical pathlength (m).

RESONANT HOLOGRAPHIC INTERFEROMETRY - A MULTIPOINT, MULTIPARAMETER DIAGNOSTICS TOOL FOR HYPERSONIC FLOW

NATO Advanced Research Workshop
New Trends in Instrumentation
for Hypersonic Research
April 27-May 1, 1992

JAMES D. TROLINGER
MetroLaser, 18006 Skypark Circle 108
Irvine, CA 92714
(714) 553-0688

and

RONALD K. HANSON, BRANDON YIP, AND BRETT BATTLES
Stanford University
M. E. Department
Stanford, CA 94305
(415) 723-1745

ABSTRACT. This paper describes experimentation with a new type of flow diagnostics referred to as Resonant Holographic Interferometry Spectroscopy (RHIS). This technique combines the power of holography with the species selectivity of spectroscopy to provide three dimensional images of the density of selected species in complex flows. The technique is particularly suitable to study high speed mixing processes, to measure minor species in the presence of major species, and to observe major species in a heterogeneous low pressure environment. Both experiments and modeling are being conducted to establish the feasibility of RHIS in measuring the concentration of water vapor, oxygen, and OH in high speed propulsion applications. It is expected that in addition to the species concentration, the resonant holographic technique has the potential of providing temperature, pressure, and flow velocity.

1. Introduction

The increasing use of short duration hypersonic testing has generated a considerable effort in the search for effective multipoint, multiparameter diagnostics methods. Planar laser induced fluorescence (PLIF), flow tagging, and RELIEF methods have received a considerable amount of attention to satisfy these requirements in recent years, but considerable difficulty still exists in accurately quantifying the results. Another method referred to here as resonant holographic interferometry spectroscopy (RHIS) is now raising considerable interest because of its potential for instantaneous three dimensional diagnostics. RHIS combines the power of holography with the species selectivity of spectroscopy resulting in a potential to provide three dimensional images of the concentration of selected species in complex flows.

To exploit the full diagnostics potential of holography in a general flow, developments are required in four different technologies. **Figure 1** shows the relationship of each of the technologies the overall diagnostics hierarchy.

A. Boutier (ed.), New Trends in Instrumentation for Hypersonic Research, 123–134.

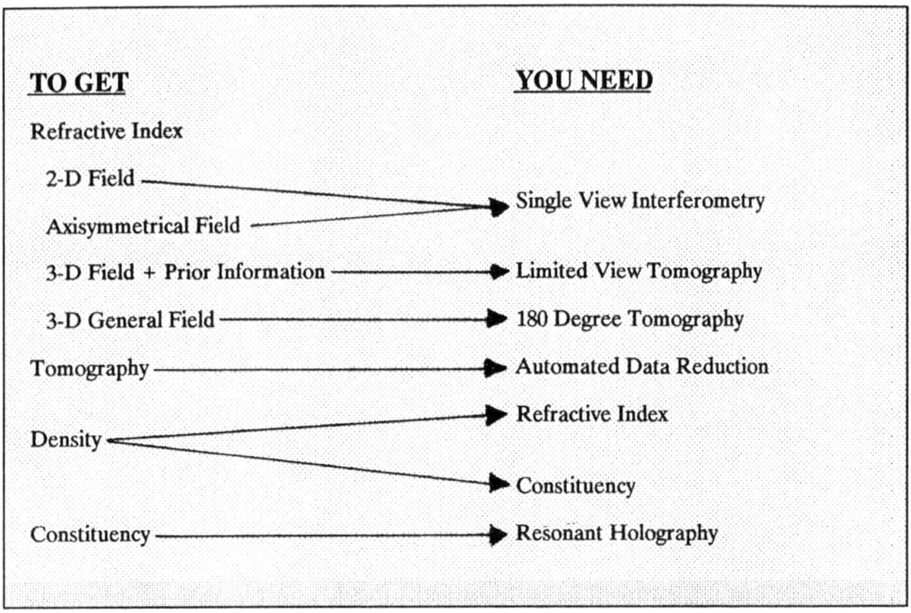

Figure 1. The Development Hierarchy for Holographic Flow Diagnostics

Holographic interferometry provides a measure of the integrated refractive index (optical pathlength) through the field of interest. Then **tomography** converts the integrated data provided by holographic interferometry to point data. **RHIS** then converts refractive index data to species density data. RHIS also greatly enhances sensitivity and accuracy. Finally, **automated interferogram data reduction** makes the handling of the large amounts of data required to perform the first three tasks possible.

Supplementing the basic framework illustrated in **Figure** 1 are other advanced techniques which further enhance the information gathering power of holography. The features of these techniques are summarized in **Table 1**. Although some of these have been known for many years, recent refinements and applications have made them more useful and important. Using a single recording plate to superimpose multiple holograms which are separable later (multiplexing), and which contain different types of data, can greatly enhance the information-potential and measuring capability of an instrument[1]. Multiple pulses of different pulse separation, multiple wavelengths, and multiple angles of view stored with different reference waves are a few examples of such information[2-4]. Tomography has been used to extract three dimensional data from complex subsonic flows[5]. Phase shift interferometry has been demonstrated to improve sensitivity of holographic interferometry by over an order of magnitude[6-8]. The automated holographic data reduction techniques which are required here have been demonstrated by a number of investigators, for example references [9-13].

TABLE 1

Advanced Holographic Techniques Supporting Flow Diagnostics

TECHNIQUE	FEATURES
1. **Phase shift and heterodyne interferometry**	Improves sensitivity of interferometry and simplifies data reduction.
2. **Multiplex holography**:	Allows simultaneously storage and independent reconstruction of different types of wavefront information by using different reference waves. Provides a method to implement phase shift and heterodyne interferometry, optical subtraction and correlation of two image conditions.
3. **Sandwich holography**:	Allows different wavefronts to be stored and reconstructed separately and directly compared optically during reconstruction. Further simplifies phase shift and heterodyne interferometry, optical subtraction and correlation of two image conditions.
4. **Multiple wavelength recording**:	Allows additional information about the subject to be stored (i.e. different variables that affect the refractive index).
5. **Particle image displacement velocimetry**:	Provides three dimensional velocity fields and enhances flow diagnostics.
6. **On-line systems-Thermoplastic and TV holography**:	Provides quicker access to data.
7. **Fiber optic holography**:	Improves optical access, simplifies path matching, and ruggedizes the optical system.
8. **Holographic optical elements**:	Reduces weight and volume and simplifies optics.

Even though progress has been made each of these areas, the practical integration of more than one or two of them into a single instrument has not been achieved. Such integration and optimization is the overall purpose of the present effort. A system design which has incorporated much of the power of holographic interferometry is described in the following paragraphs.

2. A Flow Diagnostics System Design

A single pass, diffuse lighted system has been designed and is shown in **Figure 2**. The object beam passes through a frosted plano-convex condensing lens which diffuses and scatters it over the field.

The purpose of the condenser is to improve efficiency and brightness, especially at the edge of the diffuser. The receiver lens is comprised of two matched condenser lenses. These lenses serve to increase the field of view which can be captured by the 4x5 inch glass plate. The hologram should be at the focal point of the lens pair for Fourier transform holography.

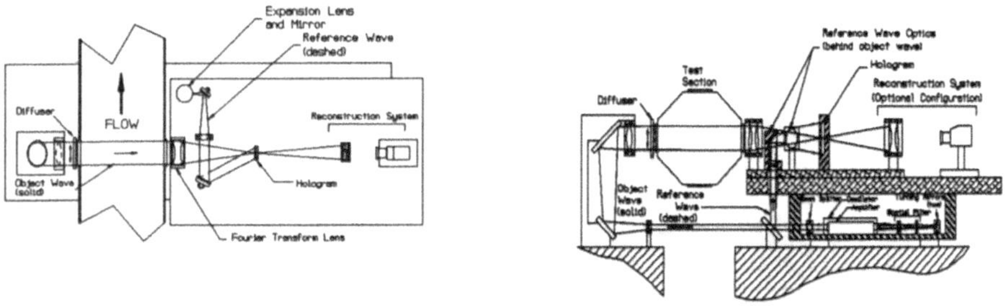

Figure 2. Conceptual Holographic System Assembly

The ultimate use of the diagnostics is to understand and quantify the spatial and temporal distributions of the species concentration, and other thermophysical parameters such as temperature, pressure, and velocity. Holography provides the optimum way to break up the flow field into its more basic components to help simplify its interpretation. This can be done by producing the following series of holograms for comparison:

1. **Reference holograms:** Holograms without flow to record the effects of all of the test facility optics. Holography provides a simple method to subtract background effects from the measurement. Reference holograms can be recorded on the same plate that is to be used later for recording data, but with a different reference wave. However, the most practical way to store reference data is to record it in a separate plate and later use sandwich holography to make the comparison.
2. **Data holograms:** Holograms that record wavefronts for comparison with reference waves or for comparison with other data waves. The different waves may be stored in the same plate with different reference waves, or in different plates for sandwich holography.
3. **Double pulsed holograms**: Holograms that record two different time separated waves and allow time differential interferometry. Recording the two wavefronts with different reference waves makes phase shift or heterodyne interferometry possible also. The greatest power in this method lies in its ability to display only the changes in the field, subtracting out all other steady effects.

Pulsed facilities are characterized by long, expensive setup times and short run times. Under these circumstances a significant savings can be realized by maximizing the amount of holographic data obtained in each run. Recording two or more types of holograms simultaneously in a single test is often of great value and actually reduces the number of tests required. If, for example, both the time independent and temporal changes in the field are of interest, the number of tests required is reduced by simultaneously recording both double pulsed and double plate holograms. First, a reference plate is recorded. Then the two pulses are stored with two different reference waves in a second plate. The second plate provides double pulsed holography when used alone, and when sandwiched with the reference plate, it provides two absolute interferograms of the flow field.

Except for RHIS, each of the above methods has seen a relatively wide application and has previously been described. To provide a preliminary demonstration of RHIS, OH density was measured in this study using a tunable UV laser. The measurements were conducted with a ring dye laser operating near 308 nm on laminar flames produced by propane and acetylene burners. In the present work the two holograms were obtained sequentially, the first at resonance and the second off resonance. The disadvantage of this scheme is that to maintain sensitivity to the resonant species, the flame must remain steady for the two exposures. Ultimately RHIS must simultaneously capture two holograms using two different wavelengths, either from two lasers or from a two-wavelength laser. This allows RHIS to measure turbulent flames and other transient phenomena.

3. Resonant Holographic Interferometry Spectroscopy

The potential benefits of performing interferometry and holographic interferometry near resonance have been recognized for many years[14-16] but suitable lasers and test gases have not been available except for limited applications in combustion and plasmas. Craig[17] appears to have been first to recognize that *"The obvious conclusion here is that resonant and non resonant holograms must be recorded simultaneously. In this case, the contributions from nonresonant species to the resonant and nonresonant frequency phase shifts exactly cancel, and the resulting differential phase shift is directly proportional to the resonant specie, OH; and accurate subtraction can be accomplished with holographic interferometry."* This study did not, however, demonstrate this fact experimentally.

Sirota, et al[18] compared RHIS with other spectroscopic measurements and concluded that the method could be extremely competitive. Subsequently, experimental demonstrations have been completed[19,20].

Laser spectroscopic methods for measuring temperature, pressure, velocity, and species concentration are directly applicable to RHIS[21-28].

Innovative schemes have been proposed and a few relatively simple experimental demonstrations have led to measurement of minor species in the presence of major species. In the vicinity of an absorption line both the real and imaginary parts of the index of refraction can increase by orders of magnitude (**Figure 3**) resulting in a proportional increase of the optical path length. Therefore, when interferometry is done with a laser tuned to near resonance, enhancement of the sensitivity of the interferometry to the resonant species will be achieved. Of even greater significance, is that the effects of all non resonant species can be removed entirely if the reference wave for the interferometer is produced by passing a second wave, tuned slightly off resonance, through the same flow field simultaneously with the resonant wave and then storing both waves holographically. When the two waves are reconstructed, their phase difference (and therefore the resulting interferogram) is due to the resonant species only, since the refractive index of the non-resonant species is essentially the same for both waves.

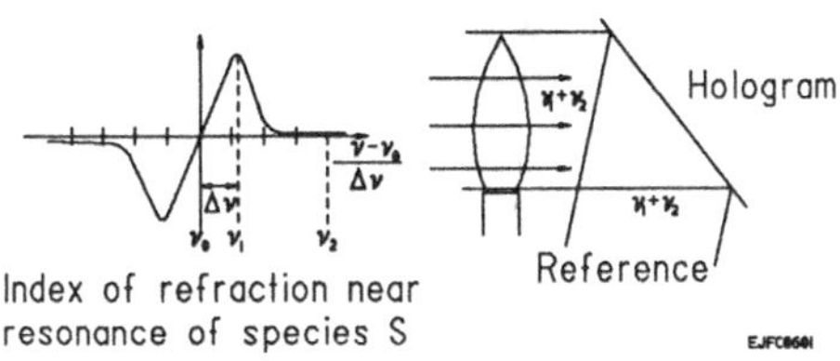

Figure 3. Holographic recording at frequencies ν_1 (near resonance) and ν_2 (off resonance)

Figure 4 provides a comparison of the anticipated results of RHIS (4a) and ordinary interferometry (4b) of a flame. We assume that the concentration of OH lies in an optically thin shroud near the edge of the flame. For finite fringe interferometry the fringes will remain straight except near the edge of the flame where the optical pathlength through the shroud is significantly larger. The anticipated fringe shifts computed later in this paper are less than one fringe.

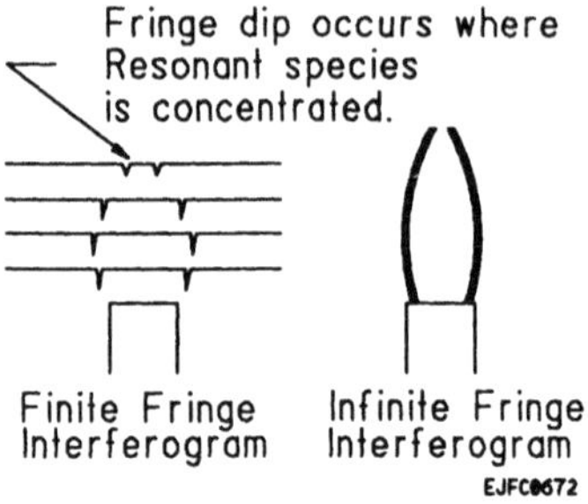

Fig.4a) Resonant holographic interferogram of species S

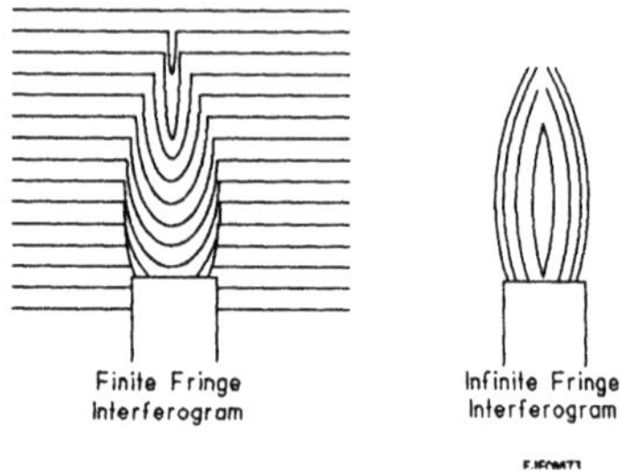

Fig. 4b) Ordinary holographic interferogram

The expected infinite fringe interferogram is also included in the figure, showing a darker region near the edge of the flame where OH is concentrated. For ordinary interferometry, **Figure 4b** shows a significantly larger fringe shift for finite fringe interferometry throughout the flame (or more fringes for infinite fringe interferometry). This is a consequence of the overall change in density in the flame which is measured by ordinary interferometry.

Since the resonance linewidth and position also depend on temperature, pressure and flow velocity, RHIS also has the potential of providing measurements of these parameters, in a way similar to that used in the other spectroscopic techniques.

Unlike laser induced fluorescence, RHIS is not subject to quenching problems since the species remain in the ground level state. Furthermore, RHIS may ultimately be implemented with low power, inexpensive lasers, thus opening the road for an inexpensive yet very sophisticated spectroscopic system. RHIS also offers some significant advantages over absorption spectroscopy, both in sensitivity and in data- handling capability. RHIS is a holographic interferometric technique which may be combined with tomography to provide a three-dimensional characterization of the index of refraction which can be converted to density.

Many processes of interest are characterized by the simultaneous presence of multiple species with various concentrations. In combustion processes involving air, the partial pressure of nitrogen can be many orders of magnitude larger than the partial pressure of minor species. Therefore, the phase change experienced by the object beam will be primarily due to nitrogen, even if the laser is tuned near resonance of the minor species. So the ability to optically subtract the effects of a major species is extremely important for many applications.

In describing the principles of RHIS, we adopt the nomenclature of Sirota and Christiansen[5].

Elementary dispersion theory indicates that both the real and imaginary components of the index of refraction experience dramatic changes in the vicinity of an absorption line. The real part (n) induces phase changes to a wave front, while the imaginary part (a) causes beam attenuation. The latter is used in laser absorption spectroscopy. Holography makes use primarily of the real component which near an absorption line is given by:

$$n-1 = K f N_i \, (\nu-\nu_0)/[\nu_0 \, (\nu-\nu_0)^2+\Delta\nu^2],$$

where ν_0 is the central frequency, $\Delta\nu$ is the line width of the probed species, N_i is the number density of species "S", f the oscillator strength of the particular absorption line, and $K = 2.24 \times 10^{-14}$ cm when all the frequencies are given in cm^{-1}. Equation (1) contains variables which yield thermophysical parameters. N_i provides direct information on the plasma density of selected species, ν_0 can be Doppler shifted by the flow velocity, and $\Delta\nu$ depends on temperature and pressure.

The holographic interferometer records an integrated phase change which is manifested by a series of fringes which are separated by half the wavelength of light. The actual phase shift is:

$$\Delta\phi = 2\pi/\lambda \int_c (n-1)\, dl,$$

where l is the pathlength traveled by the object beam.

Modern interferometric data interpretation techniques allow a resolution of somewhere between 0.01 and 0.001 fringes depending on the noise level of the hologram. This would allow the measurement of about 10 to 100 ppm of hydroxyl in a 10 cm span.

The complex portion of the index of refraction contributes to attenuate the laser beam. This attenuation is given by:

$$\alpha(\nu, l) = \int \alpha(\nu)\, dl,$$

where

$$\alpha(\nu) = 4\pi K f N_i \Delta\nu / [(\nu - \nu_0)^2 + \Delta\nu^2]$$

This attenuation may offer additional information about the parameters of interest. The attenuation of the beam must be balanced with the phase shift to optimize the holographic recording. Sufficient light must emerge from the test section to record a hologram with adequate signal/noise.

The above equations are for a laser with infinitesimal linewidth. Practical lasers have a linewidth, which is convoluted over the absorption linewidth. Under this scenario, the fringe shift must be integrated over the laser linewidth to obtain the total shift.

4. Preliminary Experiments

Each of the concepts described above in the holocamera design, except for RHIS, were tested in MetroLaser's laboratory and were applied in pulsed hypersonic facilities during the past two years. A simple experiment series was designed by the authors to test RHIS with existing hardware in the High Temperature Gas Dynamics Laboratory at Stanford University and to provide preliminary design information for implementation of RHIS in the system. Since the desired two wavelength, pulsed laser was not available, a method for applying a tunable, CW laser to achieve similar results was developed. The high coherence of the selected laser provided a capability to perform precision measurements that may not be practical with commercially available pulsed lasers and provided many of the answers required for the final system design.

The laser was tuned to a known wavelength at or near a predetermined resonance. A hologram was then produced with the burners operating. The laser was then tuned to a different wavelength and a second hologram was superimposed upon the first. The resulting holographic interferogram shows the phase variations caused by the difference in refractive index at the two wavelengths and also shows phase variations caused by changes in pathlength resulting from motion or changes in the flow field with time (such as turbulence and instabilities). To evaluate the stability of the flame with time, multiple exposure holograms were recorded at the same wavelengths. All of the holograms produced in this experiment were of the double exposure type.

These experiments were designed to probe for OH in propane-air and oxyacetylene flames. The specific rovibronic transition probed is the $R_1(5)$ feature in the A <- X (0,0) system centered at 32625.58 cm^{-1}. This system was chosen for its well characterized nature and the fact that it is well isolated from adjacent transitions.
The laser used in these preliminary experiments is a commercially available ring dye laser (Spectra Physics Model 808 with Rhodamine 590 dye) tuned to operate at approximately 610 nm. An intracavity temperature-tuned, doubling crystal (deuterated ADA) generates a 4 - 8 mW UV beam at 306 nm. Various intracavity devices (etalons and birefringent filters) are used to generate an extremely narrow linewidth for the output beam. The use of such a narrow laser line source insures that there is no overlapping interferences from adjacent spectral features within the bandwidth of the laser. Typical FWHM linewidths for the UV beam are on the order of 0.0005 cm^{-1}. In comparison, OH linewidths at atmospheric pressure and room temperature are typically 0.3 cm^{-1}, or 10 GHz. The narrow linewidth (long coherence length) facilitates matching of the object and reference beam pathlengths, and also permits a maximal resonant phase shift from OH when the wavelength is optimally tuned. At the higher temperatures in flames, an increase in the linewidths of these features would be expected.

Within the laser cavity are a pair of galvanometer-driven Brewster plates, whose function is to change the effective cavity length by small amounts while keeping the beam alignment constant. These plates are rapidly scanned by control electronics (Spectra Physics Model 481B) which yields a single-mode linear modulation of the output wavelength. The frequency scans used in this work were close to 3 cm^{-1} (60 Ghz) in length. The staring frequency was set so that the laser scanned across the A <- X (0,0) $R_1(5)$ rovibronic transition of OH was centered at 32625.58 cm^{-1}. This state was chosen for its well characterized nature and the fact that it is well isolated from any adjacent transitions.

A reference flame was employed, using an in-line beam splitter, to monitor the actual output frequency of the laser. Since the entire OH molecular lineshape could be resolved, the laser could be set up to operate in a fixed frequency, non-scanning mode at the line-center resonance, the half power points from line center, or completely off resonance. At the half power points from the absorption maximum, the resonant perturbation of the refractive index is a maximum, thus the largest phase change in the incident wavefront would be realized.

Tuning the laser from on to off-resonance was accomplished manually and typically required 4 - 6 seconds. A manual shutter was used to block the beam path between on and off-resonance hologram exposures. However, the Brewster plates could be scanned at rates up to 3 kHz, so that the technique could be applied to high speed or some turbulent flows by use of a shutter synchronized with the scanning rate of the laser.

Figure 5 shows the holocamera recording system layout. The beam emerging from the laser is split into object and reference beams which then enter identical optical systems that expand, spatial filter, and collimate the beams at approximately two inches in diameter. Quartz was used for most of the optical components to accommodate the UV wavelengths. In the object beam leg, the two burners were inserted one above the other to provide simultaneous viewing of the two types of flames. An imaging lens was used to image the burners at the film plane (image plane holography) so that white light reconstruction could be used.

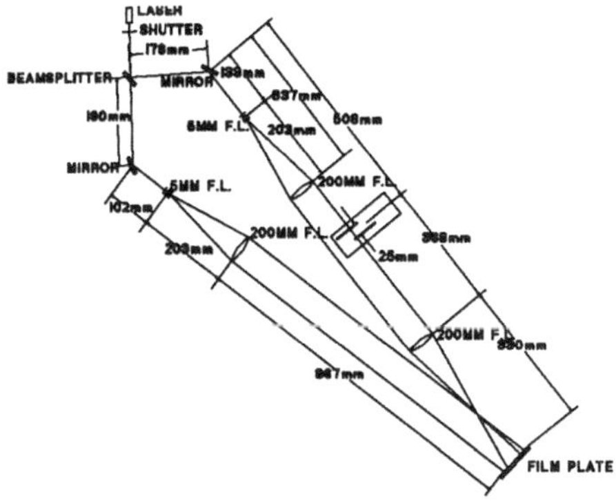

Figure 5. Experimental RHIS System.

The recording system was located on a vibration isolated table to prevent motion from occurring between exposures.

One of the purposes of the test was to evaluate different types of recording materials since little data exists on the use of holography recording materials in the UV. Three materials which were selected were Kodak 125, Kodak 120, and Agfa Gavaert 8E56. These materials gave similar results and were considered satisfactory. Film speeds, though not measured precisely, appeared to be similar in the UV to that published in the visible range.

Many variations of the double exposure holograms were recorded including (1) on and off resonance, (2) both off resonance, (3) both on resonance, and (4) symmetrically placed at maximum and minimum values of refractivity where the greatest sensitivity can be expected. Some of the multiple exposures were made with flame on and flame off for comparison of the flow field with no flow field.

5. Estimated Optical Pathlength Increase

The two combustion sources used in this work were axisymmetric jet flames. One was a propane-air torch, which is the larger torch in the figures. The other was an oxyacetylene torch. The flames were adjusted to near stoichiometric operating conditions. Typical flame conditions in either torch were a temperature of 2000 K and OH concentrations of between 750 and 1000 ppm.

These types of burners are well suited for these experiments since they are of manageable size, exhibit good spatial and temporal stability, and possess both a core region of unburned gases where no OH is present, and a well-marked flame zone where peak OH levels occur. Little flame fluctuation was observed in the 4-6 seconds between exposures.

The expected number of fringes in the holographic interferogram due to the resonant enhancement of the refractive index by OH can be estimated by applying equations 1 and 2. When two waves of wavelength k_1 and k_2 are holographically recorded and then superimposed during reconstruction, the phase difference between the two waves will be

$$\Delta\phi = [2\pi l/\lambda_0][n(\lambda_2)-(n(\lambda_1)]$$

$$= [Ne^2 l f\lambda_0{}^2/8\pi\epsilon_0 mc^2][1/(\lambda_2-\lambda_1)-1/(\lambda_1-\lambda_0)]$$

Putting in l = 5mm, k_0 = 306 nm, $k_1 << k_0$, $k_2 = k_0 + 0.0015$ nm, N = 2.5 x 10^{15} cm^{-3} for 1000 ppm of OH at 2000K and f ~ 4 x 10^{-4} we obtain $\Delta\phi$ ~ 0.4 radians, which corresponds to 6% of a fringe due to OH effects in the holographic interferogram.

6. Experimental Results

Figure 6 shows the holographic interferogram of the flame and no flame, showing a flow field strength of about ten fringes. The presence of turbulence and minor species is buried in the large number of fringes.

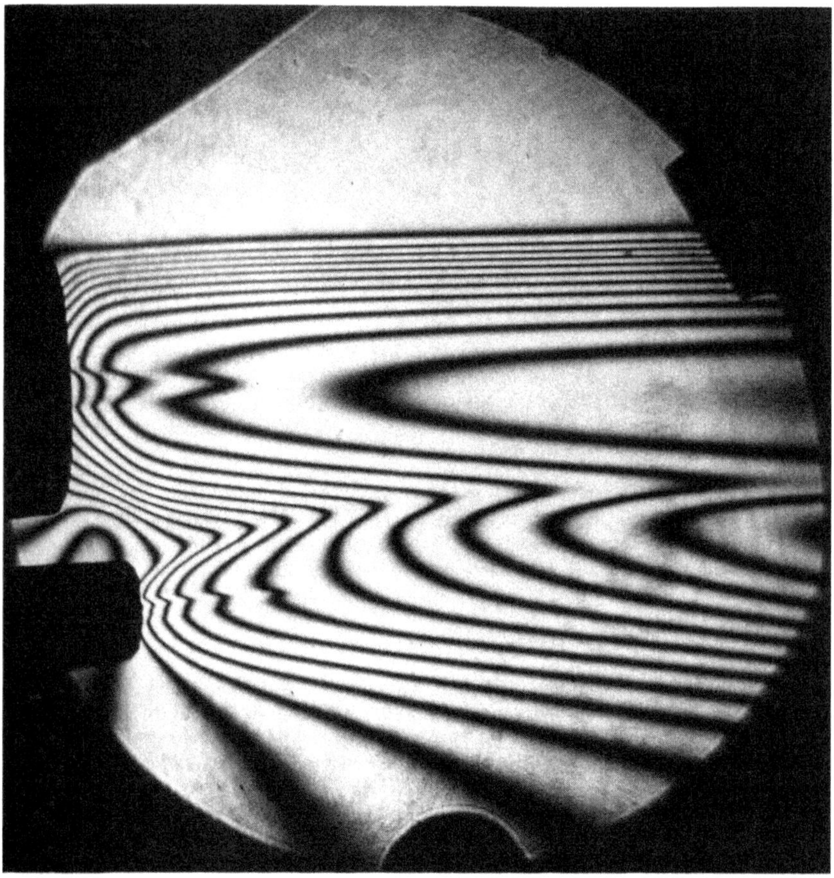

Figure 6. Holographic Interferogram Comparing a Flame with the Background

Figure 7 shows a double exposure holographic interferogram of the flames produced with the laser tuned off-resonance, made to evaluate the transient effects. If the flame was perfectly stable and laminar, the interferometric data should show a uniform blank field since the two recorded wavefronts would be identical. The turbulent region in the flame center is clearly visible as expected in the propane burner (the larger of the two). No flow is visible outside this region until a distance of about two nozzle diameters when the evolving and spreading fringe suggests less the presence of turbulence. To see the detailed structure of the turbulence requires that the recordings be short compared to the time constants in the flow. Since these recordings were relatively long, the turbulence effects are averaged. No flow is visible in the smaller acetylene flame.

Figure 8 shows the resonant holographic interferogram. In this case the two holograms were obtained at two different frequencies (resonant and off-resonant) and the fringes here correspond to OH as well as the transient effects already demonstrated on **Figure** 7. The presence of OH is suggested by the resonant interferogram by the fringes near the edge of the flame. This can be seen near the edge of the flames in both burners. The optical pathlength has increased by a small but observable amount (less than one wavelength). A small amount of convective flow is observable near the burner. This is caused by convection over the hot burner itself. Convection of this type required that the experiments be done with the burners facing horizontally so that the effects were dominant only near the burner. While the results are qualitative and preliminary, they are very encouraging.

7. Future Plans

Future plans include modeling RHIS for water vapor and oxygen and to experimentally verify its performance in laboratory experiments. Some of the questions being addressed are:

1. What is the sensitivity increase in the absorption band of the interrogated species?
2. What is the minimum detectable density per unit pathlength near resonance and off resonance?
3. How can RHIS be combined with tomographic techniques to yield quantitative 3-D maps of the density profile?
4. What schemes should be considered to measure temperature?
5. What schemes should be considered to measure pressure?
6. What schemes should be considered to measure velocity?

Many of the above questions will be answered by modeling the spectroscopic holography and by performing selected experiments to verify primarily the ability to measure density. It is expected that pulsed and/or cw IR diode lasers can ultimately act as an illumination source.

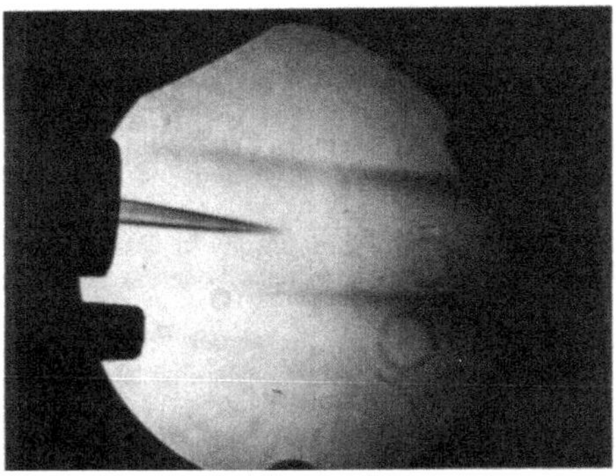

Fig 7. Holographic Interferogram flames with themselves at a later time with two non resonant wavelengths.

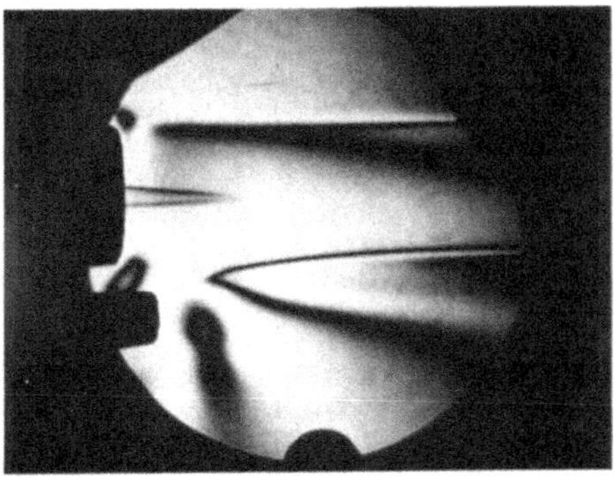

Fig. 8. Holographic Interferogram comparing the (RHIS) comparing the flames with themselves, using resonant plus non resonant wavelengths.

8. References

1. Trolinger, J.D., and Havener, G.A., "A Renaissance in Holographic Flow Diagnostics, AIAA Paper no. AIAA 91-0566, 29th Aerospace Sciences Meeting, Reno, NV, 7-10 January 1991.
2. Ballard, B., "Double Exposure Holographic Interferometry With Separate Reference Beams", Applied Physics, Vol. 39, p. 4846, 1986.
3. Surget, J., "Two Reference Beam Holographic Interferometry For Aerodynamic Flow Studies," Communication presented in Conference Internationale sur les Applications de l'Holographic et le Traitement Optique des Donnes, Jerusalem, Israel Aug 23-26, 1976.
4. Trolinger, J.D., "Applications of Generalized Phase Control During Reconstruction to Flow Visualization Holography," Applied Optics, Vol. 18, No. 6, March 15, 1979.
5. Modarress, D., Tan, H., and Trolinger, J.D., "Tomographic Reconstruction of Three-Dimensional Flow Fields Over Air Foils", AIAA Paper 85-0479, 23rd Aerospace Science Meeting, Reno, NV, 1985.

6. Danliker, R., Ineichen, and Mottler, F., "High Resolution Hologram Interferometry by Electronic Phase Measurement," Opt. Commun. Vol. 9, p. 412, 1973.
7. Trolinger, J.D., Craig, J., and Dean, P., "Advanced Holographic Diagnostic Method For 3D Hypersonic Flow Fields," AIAA Paper 88-4653-CP, AIAA/NASA/AFWAL Sensors & Measurement Techniques For Aeronautical Applications Conference, Atlanta, GA., September, 1988.
8. Yanta, W.J., Spring, W.C., Gross, D.U., and McArthur, J.C., "Phase Measuring Laser Holographic Interferometer For Use In High Speed Flows," ICIASF 13TH Congress on Instrumentation In Aerospace Simulation Facilities, September, 1989.
9. Takeda, M., Hideki, I., and Kobaysahi, S., "Fourier-Transform Method of Fringe Analysis For Computer-Based Topography and Interferometry," J. of The Optical Society of America, Vol. 72, No. 1, pp. 156-160, January 1982.
10. Havener, G., Kirby, D., "Axisymmetric Density Measurements Reduced From Holographic Interferograms of M-6 Flow: Application of 1D FFT Interferometry," AIAA Paper in preparation for presentation at the AIAA 30TH Aerospace Science Meeting, Reno, NV, 1992.
11. Becker, F. and Yu, Y., "Digital Fringe Reduction Techniques Applied To The Measurement Of 3D Transonic Flow Fields," Optical Engineering, Vol. 24, No. 3, pp 429, 1985.
12. Kim, R.C. and Rubin, L., "Diffuse Holographic Interferometry System For Rocketdyne Hypersonic Shock Tunnel," In "Laser Interferometry: Quantitative Analysis of Interferograms," SPIE Proceedings, edited by R.J. Pryputniewicz, p. 1162, 1989.
13. Trolinger, J.D., Hess, C.F., Craig, J.E., and Azzazy, M., "Non-intrusive Diagnostics for National Aerospace Plane Testing," Paper No. AIAA-88-2051 of the AIAA 15th Aerodynamic Testing Conference, San Diego, California, 18-20 May 1988.
14. Bershader, D., Prakash, S., and Huhn, H., "Improved Flow Visualization By Use of Resonant Refractivity," Experimental Diagnostics in Gas Phase Combustion Systems, AIAA Progress in Aeronautics and Astronautics, 53, edited by B.T. Zinn, 1977.
15. Dreiden, G.V., et al., "Plasma Diagnostics By Resonant Interferometer And Holography," Sov. J. of Plasma Physics, Vol. 1, pp 256, 1975.
16. Mayinger, F., and Panknio, W., "Holographic Two Wavelength Interferometry For Measurement of Combined Heat And Mass Transfer," Combustion Measurements, R. Goulard Ed., Academic Press, New York, pp 270, 1976.
17. Craig, J.E., Azzazy, M., and Poon, C.C., "Resonant Holographic Tomography for Detection of Hydroxyl Radicals in Reacting Flows" AIAA-84-0202, AIAA 22nd Aerospace Sciences Meeting, Reno, NV, January, 1984.
18. Sirota, J.M., and Christiansen, W.H., "Flow Diagnostics By Resonant Holographic Interferometry," AIAA Paper 90-1550, AIAA 21st Fluid Dynamics, Plasma Dynamics and Lasers Conference, Seattle, WA, June, 1990.
19. Rubin, L., and Swain, D., "Near Resonant Holography" Optics Letters, 7, p. 526, 1991.
20. Trolinger, J.D., Yip, B., Battles, B., and Hanson, R.K. "Hydroxyl Density Measurements in a Flame with Resonant Holographic Interferometry", AIAA Paper no. AIAA 92-0582, 30th Aerospace Sciences Meeting & Exhibit, Reno, Nevada, Jan. 6-9, 1992.
21. Philippe, L.C. and Hanson, R.K., "Tunable Diode Laser Absorption Sensor for Temperature and Velocity Measurements of O_2 in Air Flows", AIAA-91-0360, 29th Aerospace Sciences Meeting, January 7-10, 1991.
22. Hanson, R.K., "Combustion Diagnostics: Planar Imaging Techniques", presented at 21st Symposium (International) on Combustion, Munich, Germany, August, 1986.
23. Seitzman, J.M., Kychakoff, G., and Hanson, R.K., "Instantaneous Temperature Field Measurements Using Planar Laser-Induced Fluorescence," Opt. Lett. 10, No. 9, 439, 1985.
24. Chan, C., and Daily, J.W., "Measurement of Temperature in Flames Using Laser-Induced Fluorescence Spectroscopy of OH," Applied Optics 19, No. 12, 1963 (1980).
25. McDaniel, J.C., Hiller, B., and Hanson, R.K., "Simultaneous Multiple-Point Velocity Measurement Using Laser-Induced Iodine Fluorescence," Optics Letters 8, No. 1, 1983.
26. Trolinger, J.D., "Laser Instrumentation for Flow Field Diagnostics," AGARDograph No. 186 (1974).
27. Cattolica, R.J., "OH Rotational Temperature From Two-Line Laser Excited Fluorescence," Appl. Opt. 20, 1156, 1981.
28. Zizak, G., Omenetto, N., and Winefordner, J.D., "Laser Excited Atomic Fluorescence Techniques for Temperature Measurements in Flames: A Summary," Opt. Eng. 23, No. 6, p. 749, 1984.

SCHLIEREN DEVICE and SPECTROSCOPIC MEASUREMENTS in F4

D. Bize*, G. Dunet, M. Philbert, A. Roblin, J. Surget
* ONERA Centre du Fauga - Mauzac 31410 NOE France
ONERA Optics depart. BP 72 92322 CHATILLON France.

In F4, a new high enthalpy arc heated hypersonic facility of the Fauga-Mauzac ONERA Center, the two initial optical instruments installed to characterize the flow are:

- a schlieren device, to visualize the whole aerodynamic phenomena surrounding any model set in the test section.
- a spectroscopic apparatus to analyze the light emitted from the arc chamber through the nozzle or the light around the model, just down-stream the shock-wave.

1 Schlieren device

As most of the ONERA wind tunnel facilities, F4, which is the new high enthalpy hypersonic facility built at CFM - ONERA center, disposes of a schlieren device to visualize the aerodynamic phenomena around the model, set in the flow. With a modification of the knives we can transform the device in phase contrast detection and increase the sensitivity.

The principles of this technique are well known [1]. We just describe the possibilities of the density measurements in F4 experimental conditions.

1.1 Experimental device

We chose for F4 a Z schlieren device. The configuration of this device is shown on the figure 1.

The flow field is 500 mm, so the apertures of the optics have to be set at the same dimensions and to limit the wave front distorsion by the aberrations, the focal length is 5 metres.The mirrors are polished with 0.1" arc of precision. With this optical configuration, the total optical length is 22.5 m. To avoid perturbations by air convection, the most of the optical length is set in the vacuum chamber during the run.

The emission and reception set up allow several possibilities.

- Emission set up (fig.2)

There are 3 commutable sources:

-a continuous Hg vapor source (HBO, 200 W)
-a snapshot source (FX 12)
-a continuous halogen source.

- Reception set up (fig.3)

Several options are offered

-a high speed video camera up to 10.000 images/s (NAC E10)
-a CCD video camera
-a HASSELBLAD camera (2000 F CW, 6x6, 500 mm)
-a NIKON camera (F3, 24x36, 85 mm)

The mirrors and the schlieren knives are monitored from the controle room.

A. Boutier (ed.), New Trends in Instrumentation for Hypersonic Research, 135–149.

1. 2 Threshold of the detection

For phase objects with high gradient like shock waves, the detection threshold of the schlieren technique is achieved for a difference path of the optical length near $\lambda/50$ (λ wavelength of the source).
The minimum optical different path can be written

$$\Delta_{min} = K.L.\Delta\rho_{min} = \frac{\lambda}{50}$$

with : K the Gladstone - Dale constant (Air : $K = 2.27\ 10^{-4}kg.\ m^{-3}$)
L length of the optical path in the gas flow
$\Delta\rho$ density minimum detectable, supposed constant along L
Here L = 300 mm and λ = 0.5 mm

$$\Delta\rho = 1.5.10^{-4}kg.m^{-3}$$

The figure 4 locates the threshold in the diagram ρV for F4 with Pi = 500 bar. The set up with the phase contrast (ref.1) increases the sensitivity by a factor 10.

1.3 Experimental results

1.3.1 *Measurements at R5 CH*

To verify the sensitivity of this method, tests have been made in the R5 Ch facility of ONERA at Chalais - Meudon [2], with a Z schlieren device of 200 mm of field of view. The R5 Ch characteristics are:

- nozzle : Mach 10
- Ti = 1100 K
- Pi = 2.5 bar
- $\rho i = 0.8\ kg.m^{-3}$

In the flow:

- Tv = 52 K
- Pv = 0.06 mbar
- $\rho v = 4.10^{-4}\ kg.m^{-3}$

Two models have been tested:

- a cylinder (ϕ = 30 mm, length = 150 mm)
- a sphere (ϕ = 60 mm)

The figure 5 shows a schlieren picture with the knives parallel to the gas flow. For this run the density in the flow was ;

$$\rho_v = 7.5.10^{-4}kg.m^{-3}$$

Assuming a value of 6 for the ratio of the densities on both sides of the shock wave, the variation of the density which occurs with the aerodynamic phenomena is equal to 5ρ, so:

$$\Delta\rho = 3.75.10^{-3}kg.m^{-3}$$

For a crossed length L = 150 mm, this $\Delta\rho$ gives a variation of the optical length :

$$\delta = K.L\ .\Delta\rho\ = 0.13\ \mu m$$

Or a phase shift of $\lambda/4$. This value is about 10 times higher than the visibility threshold ($\lambda/50$).
The figure 6 shows a picture of the gas flow around the sphere of 60 mm obtained with the phase contrast set up.
The density in the flow was :

$$\rho_v = 5.10^{-4} kg.m^{-3}$$

and

$$\Delta\rho\ = 2.5.10^{-3} kg.m^{-3}$$

With the axisymetric configuration, the optical length in front of the model is difficult to technique. If we take 1 cm for the shock layer crossed by the beam, we obtain :

$$\delta = K.L\ .\Delta\rho\ = 6.10^{-3}\ \mu m$$

or a phase shift of $\lambda/100$
This value is above the threshold for the schlieren device but it is still higher than the threshold for the phase contrast.
This results confirm the possibilities of the F4 visualisation techniques.

Schlieren : $\lambda/50 = 0.01\ \mu m$

Phase contrast : $\lambda/500 = 0.001\ \mu m$

or

Schlieren : $L_{cm}.\Delta\rho_{kg.m^{-3}} = 4.5\ 10^{-3}$

Phase contrast : $L_{cm}.\Delta\rho_{kg.m^{-3}} = 4.5\ 10^{-4}$

1.3.2 *Measurements at F4*

The schlieren technique has been the first optical device installed at F4. The measurement visualized the gas flow and the shock waves around the probes for the aerodynamic calibration. Only the schlieren configuration has been used. The pictures have been taken with the 6x6 HASSELBLAD and the snapshot source FX12.
The pictures 7 and 8 show the shock waves around the probes. The conditions of the run are :

- gas : nitrogen
- Hi/RT = 37
- Pi = 520 bar
- $\rho = 1.3\ 10^{-3}\ kg.m^{-3}$

The pictures 9 and 10 show the optical conditions with higher enthalpy with respectively nitrogen and air at low density.

nitrogen
- Hi/RT = 95
- Pi = 270 bar
- $\rho = 4.10^{-4}\ kg.m^{-3}$

air
- Hi/RT = 98
- Pi = 286 bar
- $\rho = 5.10^{-4}$ kg.m^{-3}

We can see that with high enthalpies, the light coming from the hot gas is important in the layer of the shock wave. This light gives a saturation of the pictures around the probes and hides the shock wave and the schlieren effect. This light can be very important and can disturb the record, even with the high speed video camera (5000 images/s) and a short exposure time. An improvement, which is studying now, is to use like source a monochromatic source (HeNe laser (632.8 nm), 15 mW) and in front of the detector a filter at the same wavelength to cut all the stray light coming from the shock layer.

2 Spectrometric measurements

2.1 Experimental device

An U.V - visible emission spectroscopic instrument has been set up in F4. It is composed of Czerny - Turner monochromator (focal length = 275 mm, aperture f / 3.8) which is coupled with a linear photodiodes array detector piloted by an optical multichannel analyser. The detector is made up of 1000 single elements (height = 2.5 mm, width = 2.5 µm) and of 700 of them are intensified.Three gratings can be used, leading to the following characteristics:

grating number	grooves / mm	spectral domain (nm)	dispersion (nm / element)
1	150	400	0.60
2	600	100	0.15
3	1200	50	0.075

The optical signal is sent to the monochromator entrance slit through a bundle of optical fibers.Two geometric configurations have been used. In the first one (axial measurement), a small probe set up in the flow, is used to collect the light emitted by the arc chamber through the nozzle without any optical conjugaison. In the second case (transverse measurement), an U.V-visible objective (focal length = 105 mm, aperture f / 4.5) is used to image a small part of the flow on the optical fibers entrance. The probed volume is either located in the free flow or behind the shock layer created by a circular impactor (ϕ = 3 cm). For each run, the wavelength calibration is performed with a low pressure Hg discharge lamp. On the contrary, the spectral response correction has been carried out only for some runs, using a tungsten ribbon lamp. For each run, the acquisition is triggered few milliseconds before the beginning of the arc. Morever, spectra are usually acquired every 11.6 ms and the time exposition is choosen between 0.1 and 10 ms depending on the arc chamber.

2.2 Preliminary results

2.2.1 *Axial measrements*

This configuration has been tested to verify if the spectral emission from the arc chamber could be compared with a black body curve. The figure 11 shows a corrected spectrum recorded during the run 131 (Nitrogen, reduced enthalpy = 40, arc chamber pressure = 200 bar). Due to the presence of various emission and absorption lines, the spectral signature cannot be easily assimilated to a black body curve. Morever, the measurements are perturbated by the gas in the free flow and behind the shock layer produced by the probe itself. Impureties such as sodium and potassium can be seen in absorption around 590 and 770 nm. In the following, the axial configuration has been used without spectral correction in order to probe the entire flow and to detect various emitting or absorbing species. The figue 12 shows an example of spectrum recorded during the run 198 (Nitrogen, reduced enthalpy = 75, initial arc chamber = 750 bar). Copper arising from the electrodes erosion during the electric arc gives rise to UV (324.7, 327.4 nm) and visible (510.5, 515.3, 521.8 nm) emission lines. Three bands around 358, 386 and 417 nm have been assigned to Δv = -1,0 and +1 sequences of CN violet system (359.0 - 388.3 - 421.6nm). The presence of this species in the flow has been confirmed by using the grating n° 3 around the $\Delta v = 0$ sequence of the violet system. CN comes mainly from impureties containing carbon and from the coating (silastene) used as electrodes isolation in the arc chamber.It has already been detected in the seventies, in ARC 2 facility at ONERA [3].

2.2.2 *Transverse mearurements*

Transverse measurements have been performed looking at behind the shock layer produced by a circular impactor. The figure 13 shows one of the spectra recorded during the run 236 (Nitrogen, reduced enthalpy = 101, arc chamber pressure = 300 bar), using the grating n° 3 around $\Delta v = 0$ sequence of CN violet system. The CN spectrum may be perturbated by Fe lines which are clearly seen around 373 nm. The next spectrum has been recorded in the same spectral region during the run 234. This run has the same initial conditions except that nitrogen is replaced by air. Emission from CN is no longer obvious but atomic lines (Fe,Mn,Ni) are still present.

Spectra recorded during air runs using the grating n° 1 have show the presence of excited OH giving rise to the well know A - X system. An example of a such emission, given by the figure 15, was recorded during the run 268 (Air, reduced enthalpy = 55, arc chamber pressure = 222 bar). OH A - X system and Cu doublet are clearly seen and no another emitting species are detected. Figure 16 shows another spectrum taken during the run 265 (Air, reduced enthalpy = 54, arc chamber pressure = 754 bar). OH emission can be guessed but is greatly perturbated by surrounding atomic emissions from impureties such as Ni, Al and Fe. These last two runs give rise to different spectra, however it must be mentionned that in contrary of the run 268, the electric arc was still on when the run 265 began. This fact can explain the presence of metallic pollutions such as Fe, Ni, and Al detected in one case and not in the other one.

Finally, spectra have been recorded in the same domain with nitrogen bursts. The presence of exited OH has not yet been clearly detected in the cases. On the contrary, main features about atomic emission are the same as in the runs with air.

3. Conclusion

The two first optical devices set up in F4 let us to record the optical aerodynamic phenomena of the first runs. The sensitivity of the schlieren device is good for densities more than $10^{-4}kg.m^{-3}$. However for high enthalpies, the stray light hide the schlieren effect. A monochromatic apparatus will let us to record with more sensitivity.
The spectroscopic device gives a lot of information, but the analyse is not easy. The axial test is severe, because the flow is emitting during all his trajectory from the nozzle to the probe. The transversal device gives interesting spectra in the shock layer. In the future when the pollution in the flow will be lower, we will be able to record the gas concentration and analyse his temperature.

4. References

- 1. Boutier.A, Philbert.M, Surget.J, Veret.C (1978) "Procédés optiques de mesures en aérodynamique" R 2160 Technique de l'ingénieur - 4.
- 2. Chanetz.B, Delery.J, Nicout.D, Pot.T, Quelin.C (1990) "Soufflerie hypersonique à faible nombre de Reynolds : R5Ch" R.T. ONERA 11/4362
- 3. Chevalier.J.P, Carega.F (1970) "Silastene coating for pollution reduction in a hot shot wind tunnel inspection by spectroscopy" Recherche Aerospatiale 1970/1

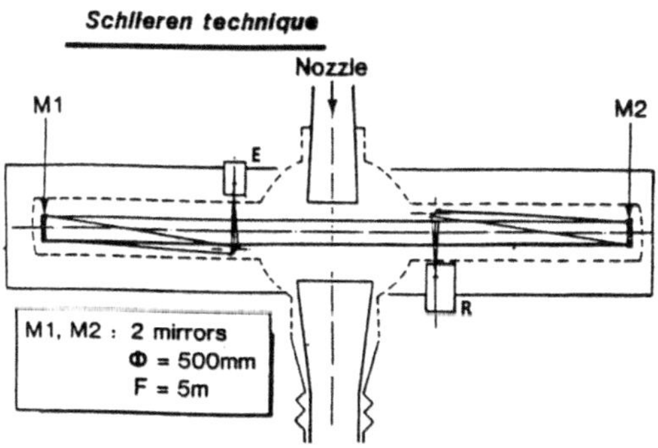

Fig. 1
Z Schlieren configuration

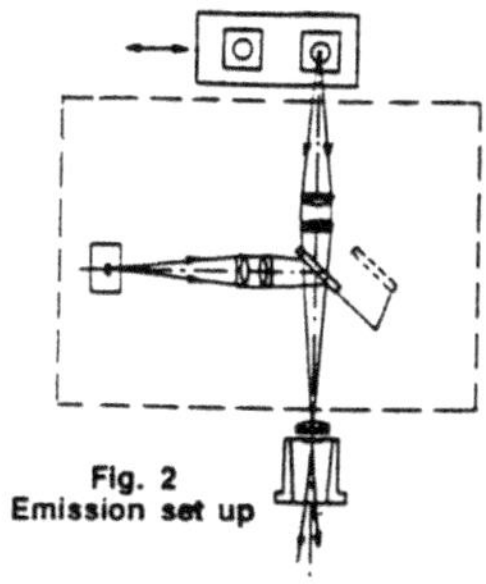

Fig. 2
Emission set up

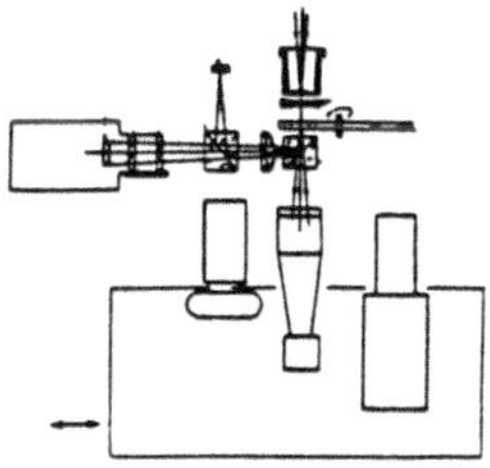

Fig. 3
Reception set up

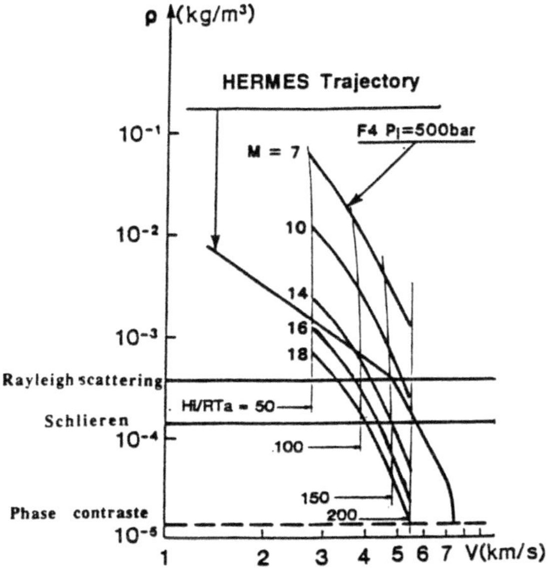

Fig. 4
Sensitivity domains versus the density

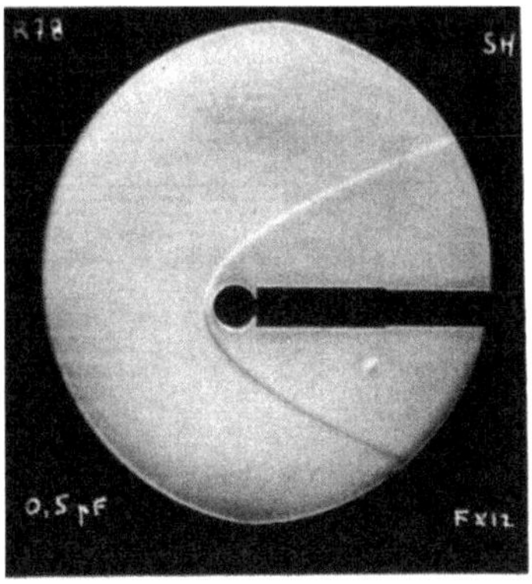

Fig. 5
Shock wave around a cylinder

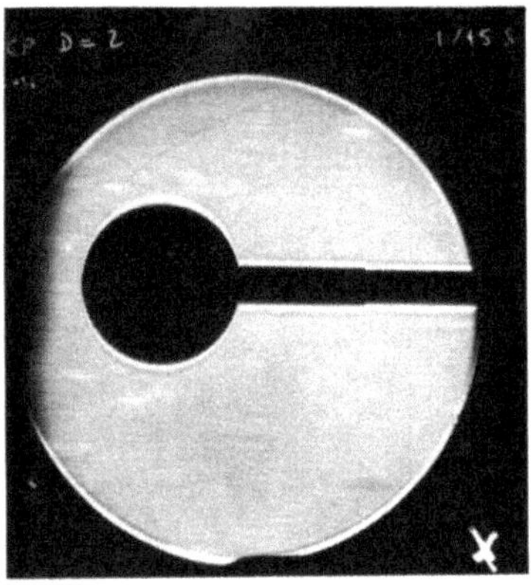

Fig. 6
Shock wave around a sphere

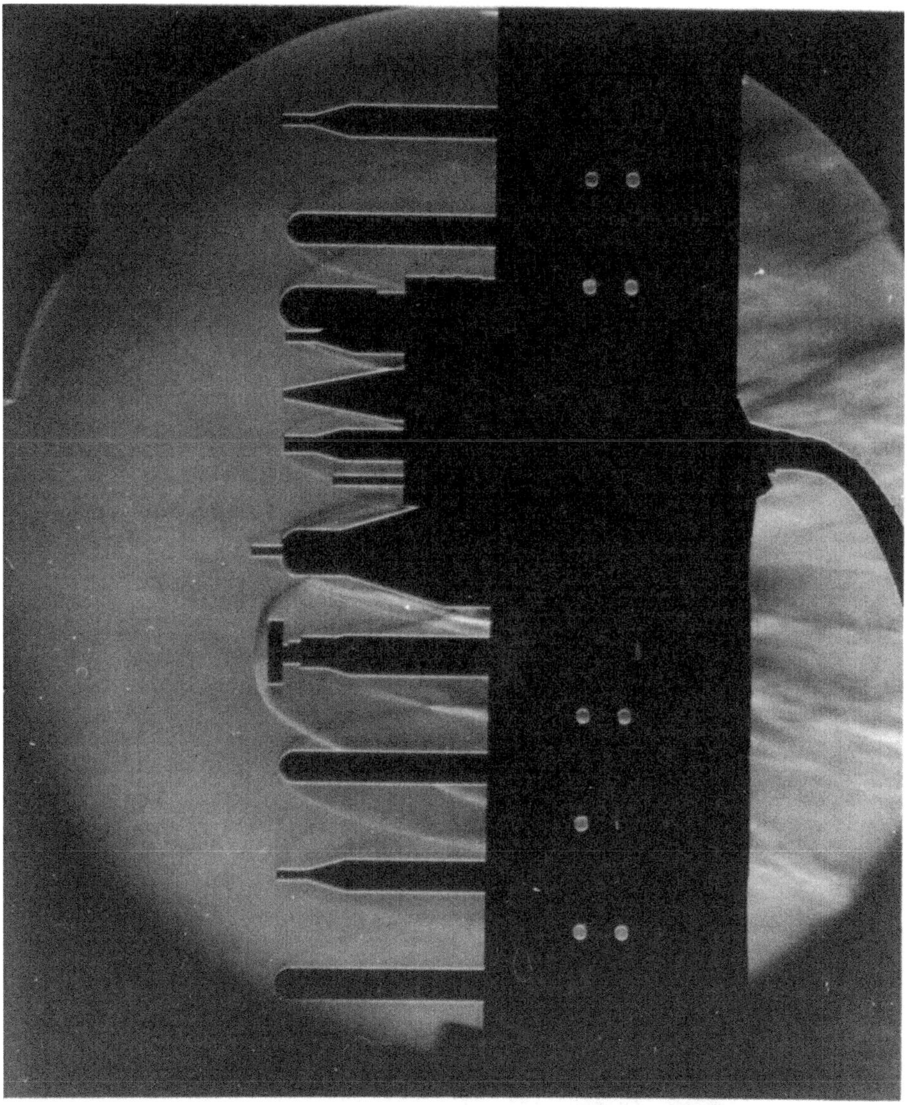

RAFALE 160

Fig. 7
Shock waves around probes in F4

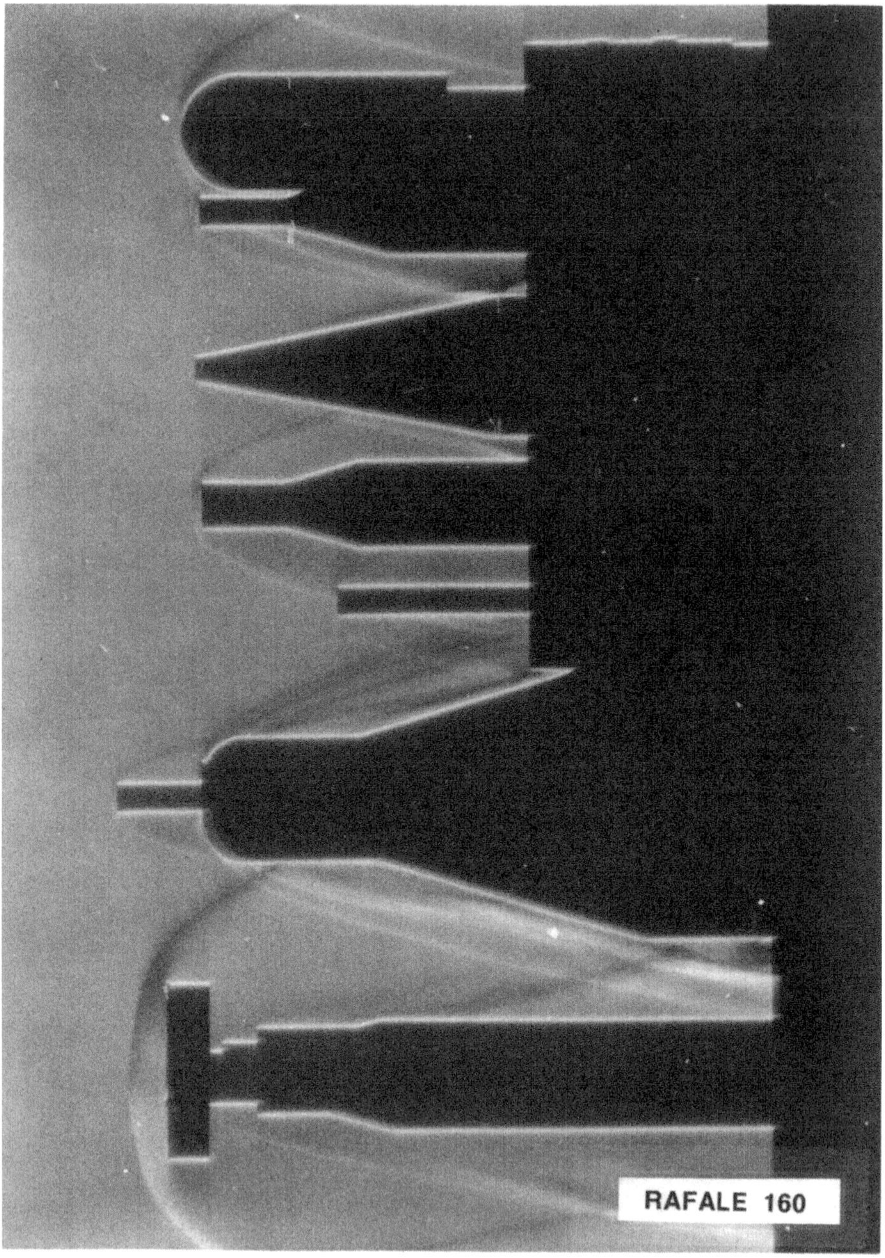

Fig. 8
Shock wave around impactor in F4

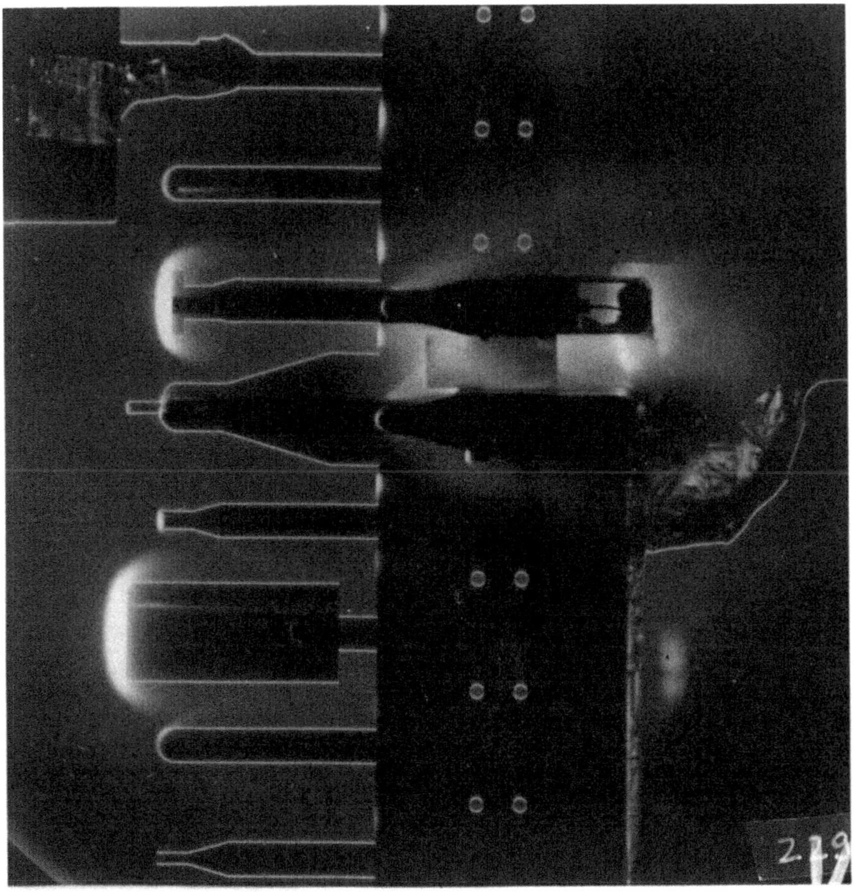

Fig. 9
Light around probes in high enthalpy run with nitrogen

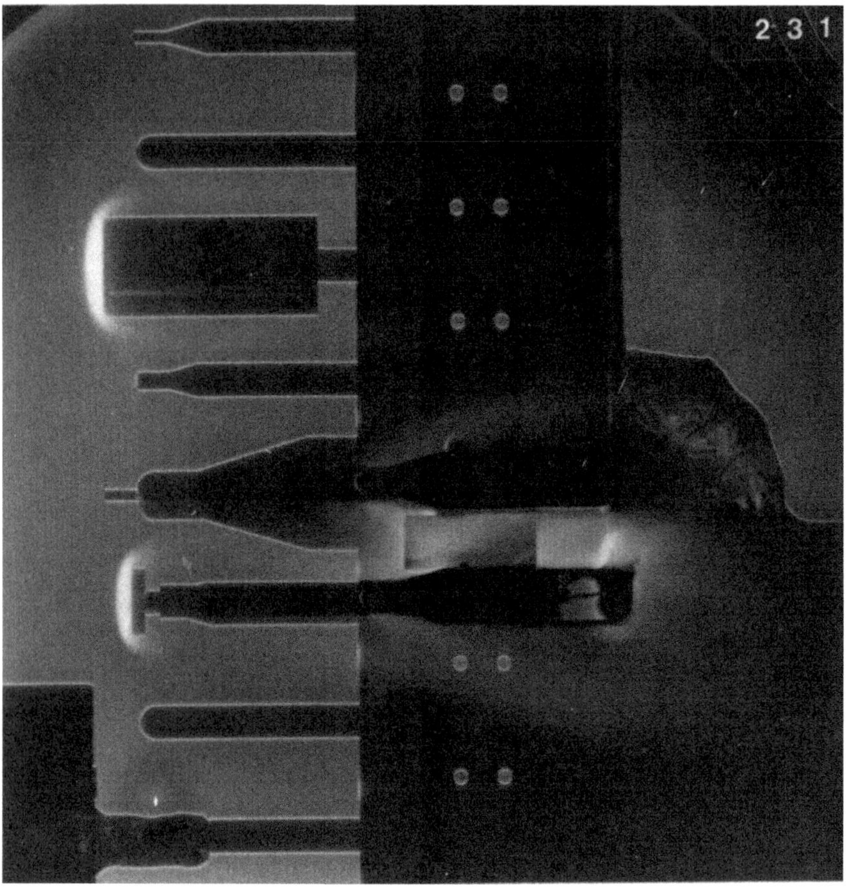

Fig. 10
Light around probes in high enthalpy run with air

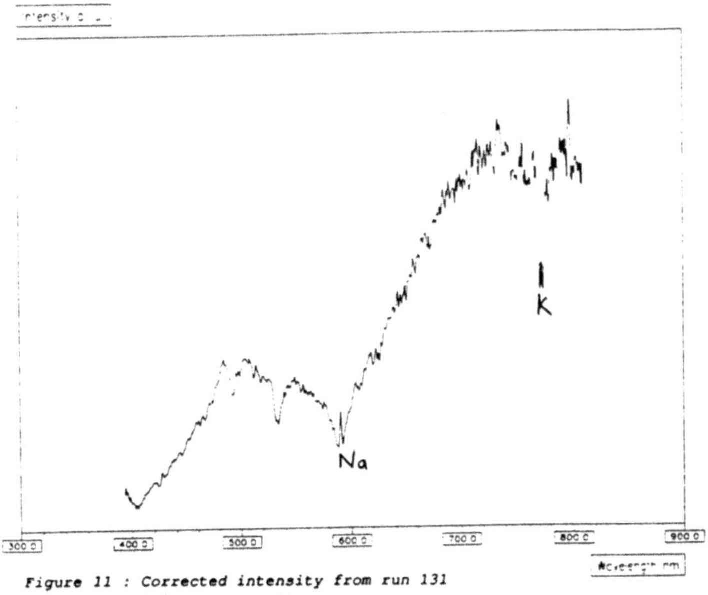

Figure 11 : Corrected intensity from run 131
(Nitrogen-axial measurement)

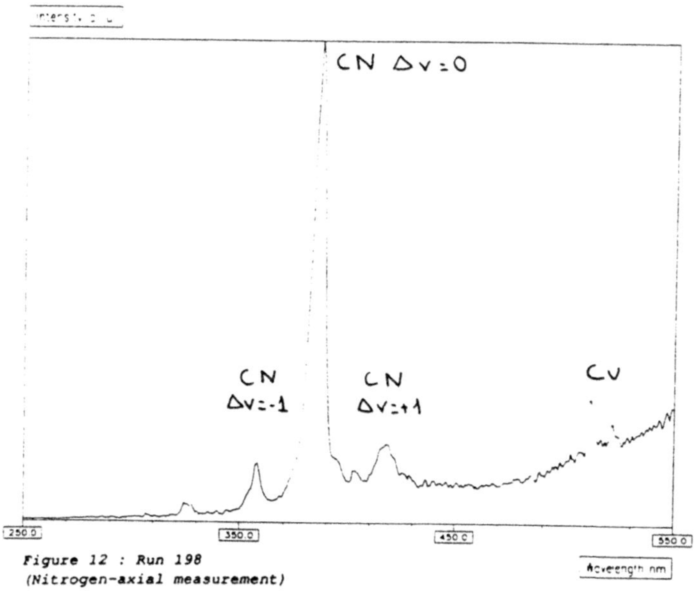

Figure 12 : Run 198
(Nitrogen-axial measurement)

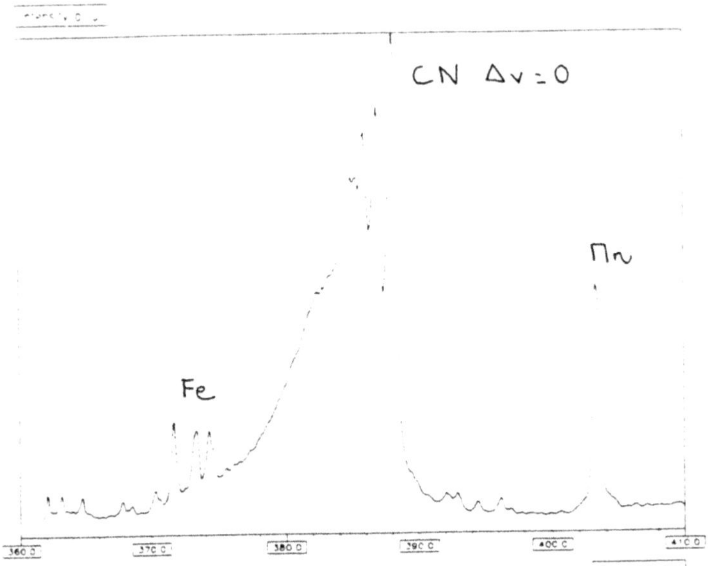

Figure 13 : Run 236
(Nitrogen-axial measurement)

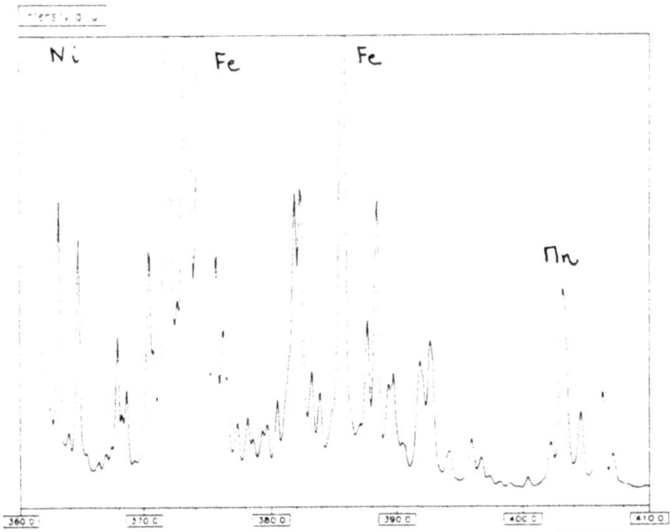

Figure 14 : Run 234
(Air-transverse measurement)

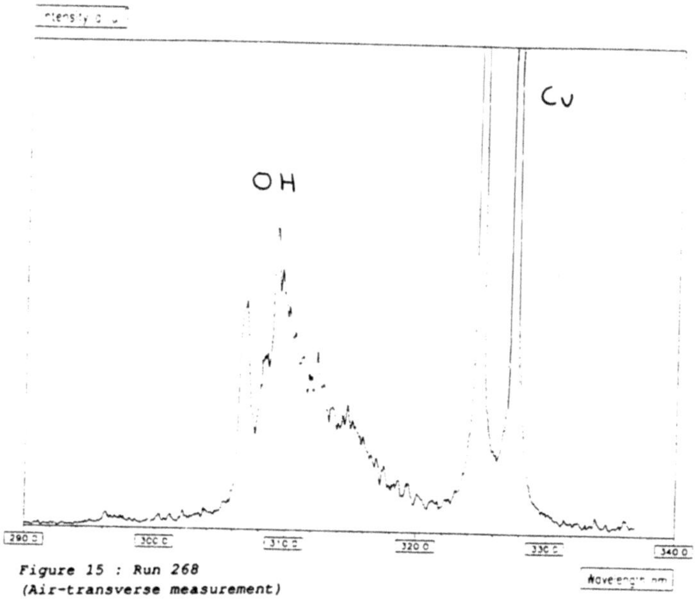

Figure 15 : Run 268
(Air-transverse measurement)

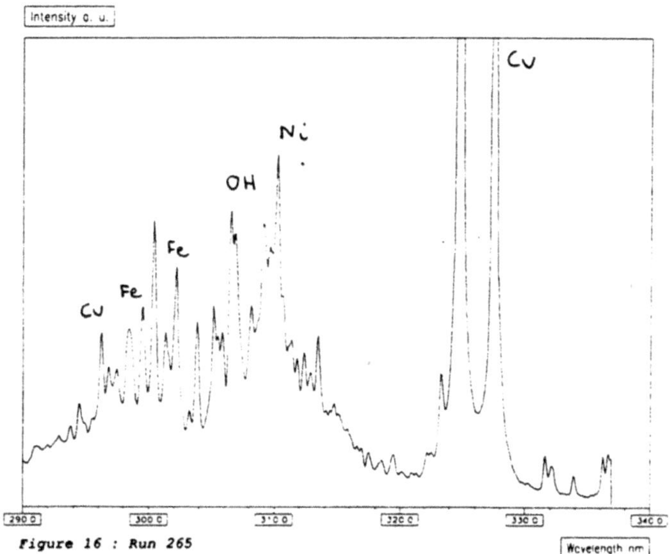

Figure 16 : Run 265
(Air-transverse measurement)

CONSIDERATIONS FOR THE APPLICATION OF THE INFRARED EMISSION ABSORPTION TECHNIQUE TO HYPERSONIC PROPULSION DIAGNOSTICS*

C. C. Limbaugh and D. G. Brown
Sverdrup Technology, Inc./AEDC Group
Arnold Engineering Development Center
Arnold Air Force Base, TN 37389 (USA)

Abstract. The Infrared Emission/Absorption Technique (E/A) is a line of sight (LOS) diagnostic technique for determining the steady state static temperature and species partial pressure of high speed hot combustion gases. The technique is based on coupling the appropriate radiative transfer model (either line-by-line or broadband) with a geometry model for the unknown region and performing a conventional onion-peel type inversion. Results of a recent application are described. Issues for applying the technique to the environment typical of hypersonic propulsion devices are briefly examined and the current state of development as applied to the E/A technique is described. It is recommended that application of the E/A be extended to the short wavelength IR H_2O spectrum for robust application to hypersonics diagnostics. This will require laboratory determination of the locations and line strengths of the hot H_2O spectral lines in the victinity of 1.0-μm.

1. Introduction

Determination of the properties of the working gases in propulsion systems being developed for hypersonic flight presents a challenging environment to the experimenter. The unusually harsh environment characteristic of the propulsion system gases leads one to consider nonintrusive instrumentation for diagnostic measurements. Nonintrusive techniques that have been used successfully in hostile environments similar to those in hypersonic propulsion systems are so-called line of sight (LOS) techniques. These nonintrusive techniques typically measure either the emitted radiation, absorbed radiation, or both, within a defined spectral bandpass. The bandpass may be in any spectral region and may be either very narrow, encompassing a single linewidth, or very broad, encompassing many linewidths and spectral lines. One such technique that has been used successfully in the infrared spectral regions is the Infrared Emission/Absorption Technique (E/A),[1-3] developed at the Arnold Engineering Development Center (AEDC). The E/A technique uses a wide bandpass encompassing a large number of spectral lines with a band model formulation for the radiative transfer model and uses a radial inversion to determine plume properties.

The E/A technique has been applied successfully in a variety of environments to determine the properties of gases typical of the effluent of turbine engines and liquid-propellant rockets. The instrumentation used is robust, and, in one application, data have been successfully acquired from the Space Shuttle Main Engine exhaust with the detector and source placed within approximately 20 m of the engine. Current applications of the technique have been limited to the H_2O and CO_2 molecules in relatively high-speed flow (up to Mach 6), with static pressures between 0.02 atm and 0.5 atm and static temperatures below 2,000 K.

*The research reported herein was performed by the Arnold Engineering Development Center (AEDC), Air Force Systems Command. Work and analysis for this research were done by personnel of Sverdrup Technology, Inc./AEDC Group, operating contractor of the AEDC propulsion test facilities. Further reproduction is authorized to satisfy needs of the U. S. Government.

A. Boutier (ed.), New Trends in Instrumentation for Hypersonic Research, 151–160.

Geometries have been cylindrical. A model for simple rectangular geometries has been developed but there has been no requirement for application.

The following discussion outlines very briefly the elements of the technique. A recent example of an inversion is given, and current concerns and work being accomplished to define those physical phenomena still required for other applications are described. Based on the presently understood attribute of hypersonic propulsion systems, work that will enhance the application of the E/A to their diagnostics is identified.

2. Elements of Technique

The essential elements of the measurement and analysis procedure for implementing an emission-absorption (E/A) line of sight (LOS) technique are illustrated in Fig. 1. The measurements of the emitted and absorbed radiation from the unknown gas path are acquired along several LOS across the extent of the unknown region. The physical properties (temperature and partial pressure of the IR active species) are assumed to be in isoproperty contours so that the entire region can be represented by a series of similarly shaped homogeneous regions. A radiative transfer model which includes a functional description of the physical properties along the LOS is used to calculate a predicted emission and absorption along an LOS. The calculations are compared to the measured values, and adjustments to the properties along the LOS are made until the calculated and measured values agree within an acceptable level. The process proceeds in conventional radial inversion or onion-peel fashion at the outer edge with only one homogeneous zone in the LOS. With these outer zone properties determined, the properties in the outer isoproperty zone are defined for all other lines of sight. The next inward LOS only has the properties at the interior zone unknown and they can be determined iteratively in the same fashion. The process proceeds inward to the centerline until the properties within all isoproperty contours have been determined.

The radiative transfer models used for LOS techniques depend upon the characteristics of the gas path and instrumentation being used. In the most fundamental formulation, and one which is the starting point for single spectral line techniques, the emission and transmission is modeled by the familiar solution to the Einstein formulation.As the instrument bandpass

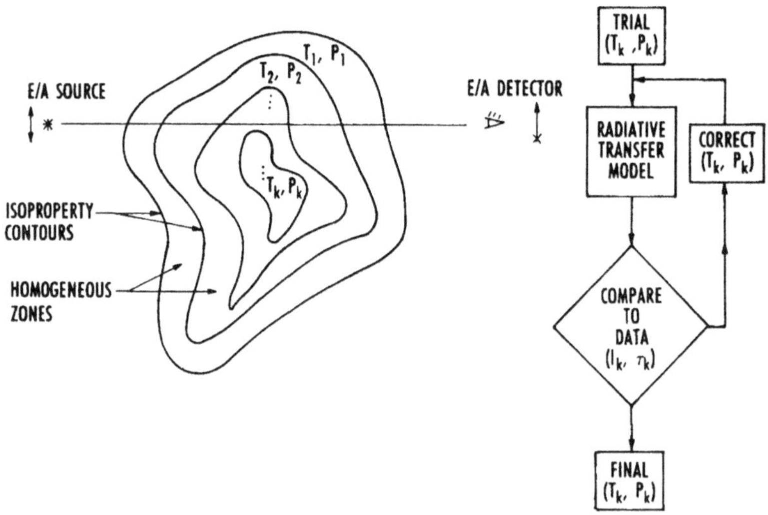

Fig. 1. E/A LOS inversion technique schematic.

increases, the contribution from each additional line is simply added to the solution. This situation is typical of single isolated atomic lines or diatomic molecules. Robust instrumentation with broad spectral response is typical for LOS nonintrusive diagnostics in hostile environments. When the bandpass is large enough to encompass many spectral lines, the computational load requirement to integrate over all contributing spectral lines becomes overwhelming, and simpler models are used. This is more likely to be typical of diagnostics of the hypersonics environment.

When the gas is a thermal equilibrium, the solution to the radiative transfer problem can be expressed

$$I_{\Delta v} = \int_{\Delta v}\int_{x} \varepsilon(v,\xi)\exp^{-\int_{\xi} k(v,\eta)\,d\eta}\, d\xi dv \qquad \ln\tau_{\Delta v} = \int_{\Delta v}\int_{x} k(v,\xi)\,d\xi\,dv \tag{1}$$

where the ε is the familiar emission coefficient

$$\varepsilon(v,x) = \sum_i \frac{h\,v_i}{4\pi} N_i^{(u)}(x)\,A_i^{(u\ell)}\,L(x, v - v_i) \tag{2}$$

and k is the absorption coefficient

$$k(v,x) = \sum_i \left[B_i^{(\ell u)} N_i^{(\ell)}(k) - B_i^{(u\ell)} N_i^{(u)}(x)\right] I_{\Delta v}\, L(x, v - v_l) \tag{3}$$

The summations in the ε and k equations are over all lines that, with central frequency v_i, have a contribution at v given by the line shape function $L(v - v_i)$. The N's, A's, and B's in the ε and k equations are the usual number densities and transition probabilities. With a large number of lines in the instrument bandpass, the use of Eqs. (1)-(3) becomes impractical because of the summations and evaluations for all the contributing lines. This is exacerbated when the model is used in a radial inversion, since this iterative technique will require a very large number of calculations at each of the trial solutions on the way to the final solution choice. In this situation, a considerable improvement in the work required to evaluate the radiative transfer is realized with band model theory.

For the band model theory approach, the formal solution to the radiative transfer equation [Eqs. (1)-(3)] is expressed in equivalent form (assuming local thermal equilibrium):

$$I_{\Delta v} = \int_{\Delta v}\int_{x} N_b[v, T(\xi)]\,\frac{\partial\tau}{\partial\xi}\, d\xi dv \qquad \ln\tau_{\Delta v} = \int_{\Delta v}\int_{x} k(v,\xi)\,d\xi dv \tag{4}$$

where N_b is Planck's black body function. The continuous variation of the gas physical properties along the line of sight is replaced by numerical integration based on approximating the continuously inhomogeneous path by a series of smaller homogeneous zones:

$$I_{\Delta v} = \sum_i N_b(v, T_i)(\tau_{i-1} - \tau_i) \qquad \ln\tau_{\Delta v} = \sum_i F(K_i, \gamma/d_i) \tag{5}$$

The functional form of F is dependent upon certain assumptions and observations about the probability distribution of the band line strengths, line spacing, and width of the spectral bandpass, and is molecule specific. In the formulation γ is the usual line width parameter that enters the line shape function in Eqs. (1) - (3), and band model parameters K and 1/d are the mean absorption coefficient and average reciprocal line spacing for the band pass. The parameters for inhomogeneous path are accomplished by establishing path-averaged values through, for example, the Curtis Godsen equation. The elements of the band model formulation for molecular radiative transfer are described in a variety of sources (e. g., Refs. 4 and 5).

The band model parameters K and 1/d are required for this approach and can be derived from the fundamental line data required for the line-by-line calculations , Eqs. (1) - (3), or can be determined with correctly designed experiments and instrumentation of wide bandpass, e.g., Reference 6.

3. Example

To illustrate the technique using the band model approach, Fig. 2 shows radiance and transmittances in the 2.7-µm H_2O band recently acquired from a medium-sized (thrust class reater than 5,000 lbf) liquid-propellant rocket engine. Instrumentation for acquiring the data included a bandpass filtered spatially scanning radiometer which acquired data at 25 scans/sec and digitized them to approximately 75 positions/scan. The same instrument is capable of up to 1,600 scans/sec with an attendant improvement in the average signal to noise. A heated plate provided the continuum source for the absorption measurement on the opposite side of the plume. Approximately 400 scans each of emission and absorption were acquired for the data. These data were used with a cylindrical geometry model consistent with the rocket nozzle and codes that were being used to analyze the nozzle flow. The results of the inversion are shown in Fig. 3. Figure 3a is the static temperature and Fig. 3b is the H_2O partial pressure. Included on the plots are predictions of the temperatures and partial pressures by a codes suite that includes a model of the O/F variations across the combustion chamber. The E/A analysis that was conducted here was based on the assumption of local thermal equilibrium. The E/A technique does give relatively good agreement with the codes predictions,

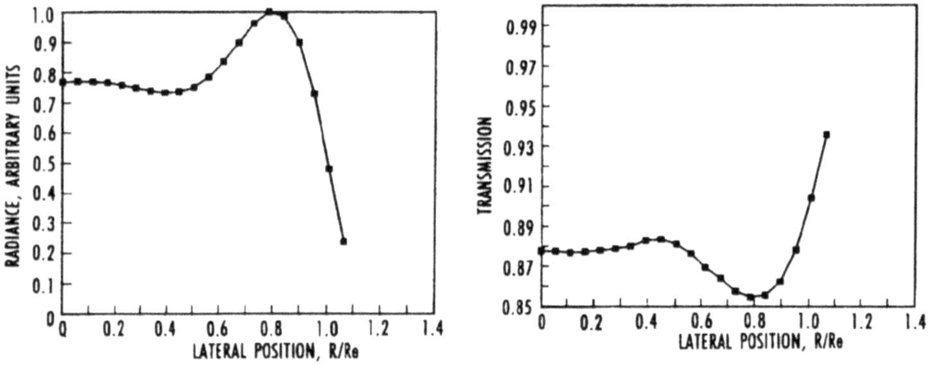

Fig. 2. Example of radiance and tranmission profiles.

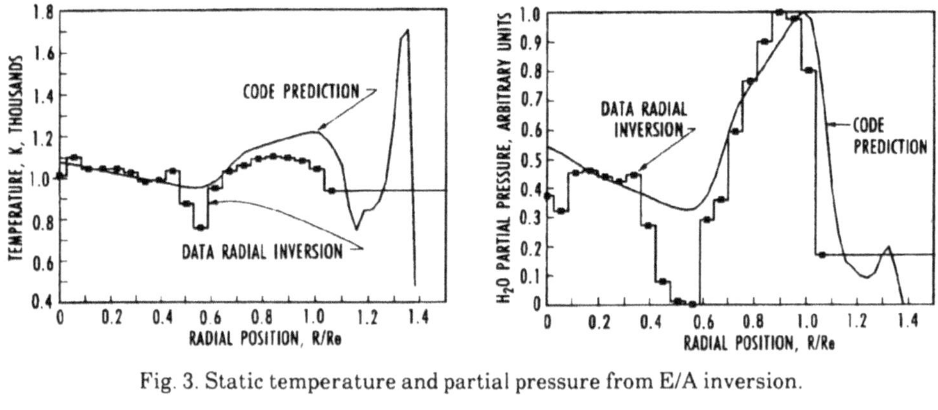

Fig. 3. Static temperature and partial pressure from E/A inversion.

although there is some disagreement in the detailed structure of the distribution between the data and the codes predictions. In this case, additional work towards modeling the O/F distribution in the combustion chamber is required for the codes to model the data accurately.

4. Line-By-Line Considerations

Implicit in the application of the E/A technique in the above example are several underlying assumptions within the radiative transfer model that have a direct bearing on the efficacy of the approach with potential implications for application to hypersonics diagnostics. Specifically, the assumption is that the gas exhibits at least local thermodynamic equilibrium (LTE) and that the spectral location of the emitted/absorbed line radiation from different regions of the gas is at the same central frequency throughout the gas. Since the band model theory is based on local thermodynamic equilibrium, results must be interpreted in this light.

In the example above, H_2O was the IR active molecule and, parenthetically, is the most likely molecule for IR E/A diagnostics in the hypersonics developments. The collision number for thermal equilibration of hydrogenic molecules (OH and H_2O) is of the same order as the collision number for thermal equilibration of the rotational distribution of molecules, so H_2O is generally considered to be in thermal equilibrium at the conditions of that data. It is, however, possible over the hypersonic flight path to generate conditions in some portion of the gas path that relaxation effects of these molecules cannot be discarded out of hand. In such applications, temperatures derived from the E/A analysis are indicative of the vibrational temperature, and additional considerations are required. Vibrational relaxation effects are especially important if the IR active molecule is carbon- or nitrogen-bearing (NO, CO, CO_2). For these molecules, computational techniques are available for determing the effect of the relaxation.[7]

A second complication to be considered in the hypersonic environment is the effect of the frequency shift of the individual spectral lines in the infrared band resulting from the Doppler shift of the gas molecules. The phenomenological representation of the effect of the Doppler shift on the radiation of a single specific spectral line is shown in Fig. 4. For example, Fig. 4 shows the relative line positions for a simple model of two zones, one zone moving toward the observer and one zone moving away from the observer, the velocity component along the LOS equal to that of a 15 deg expansion at Mach = 3. The lines are significantly displaced from each other, and the composite line shape will be appreciably affected. This effect can be significant on the radiative transfer but it is not presently included in conventional band model theory analysis, although work is progressing on this enhancement.[9,10]

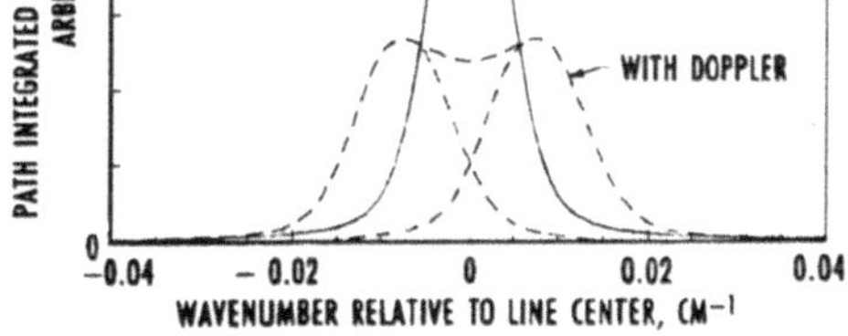

Fig. 4. Effects of line-of-sight Doppler shifts on absorption lines.

A quantitative illustration of the detailed effect of the Doppler shift on the line radiation is shown in Fig. 5, taken from Ref. 8, for CO_2 spectral lines from 2,324 cm^{-1} to 2,326 cm^{-1}. This portion of the spectrum was chosen because of the high density of the lines, it includes both optically thick and optically thin lines. Although the specific example here is for CO_2, the physical phenomena are important for all species in hypersonic combustor exhaust. The calculations are accomplished for a simple conical expansion to an area ratio of 50, illustrated schematically on the figure. Figure 5a. shows the result for the LTE distribution with no velocity shift as a baseline for comparison. Figure 5b shows the result for the LTE distribution with the Doppler shift. The LTE calculations (Fig. 5a) are for a single homogeneous zone across the exit plane of the nozzle. The Doppler shift case, Fig. 5b, is modeled by three zones, each of 50-cm length across the exit plane of the nozzle and all having the same gas dynamic properties and vibrational state distributions as the single LTE zone. The outer zones have a radial component of velocity corresponding to the 15-deg radial component towards and away from the observer, as indicated on the figure.

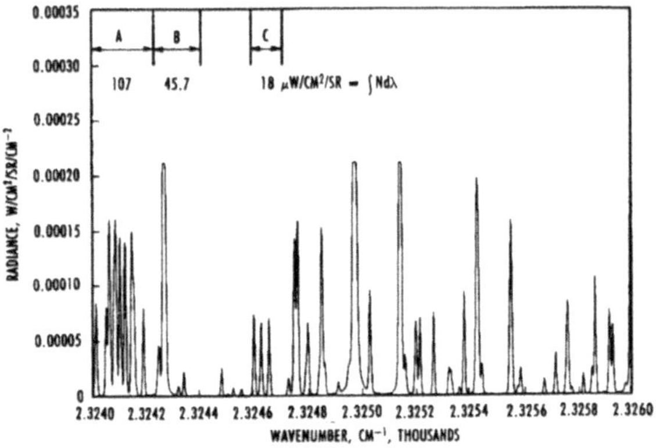

a. LTE

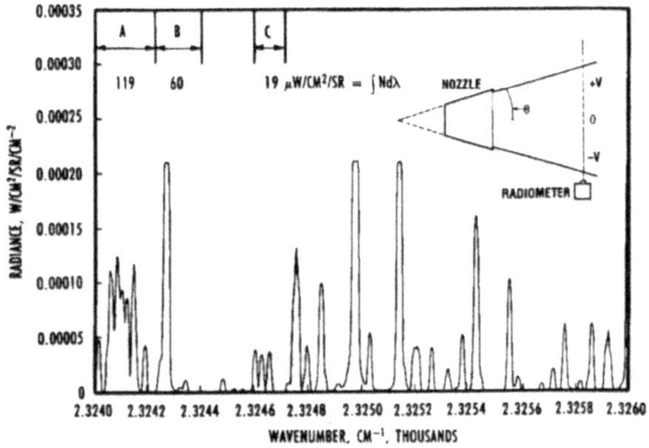

b. LTE and Doppler

Fig. 5. Calculated high resolution spectra with LTE and Doppler shift effects.

Examining and comparing details of the spectra of Figs. 5a and 5b illustrates the effects of the Doppler shift and allows one to infer some of the consequences for the broadband radiative transfer. One sees immediately that the spectral details are broadened between Figs. 6a and 6b. This is a natural result, since one effect of the Doppler shift is simply to broaden the spatially integrated spectral lines. This results in the loss of spectral detail, as well as a change in the radiance levels of individual lines and integrated bands. Spectral lines that are well separated and optically thin will show a small increase in the radiance, whereas optically thick or significantly overlapping lines will show greater effect. This is shown on the figures by noting the integrated radiance across the A, B, and C bandpasses. The first region includes several closely spaced and overlapping spectral lines and the effect of the Doppler shift is to virtually merge the lines together. The second region encompasses only an optically thick line, and the third region includes three optically thin lines that remain separated with the Doppler shift included. The inclusion of the Doppler shift causes an increase of about 10 percent for the first region, about 30 percent for the second optically thick region, and about 6 percent for the third near optically thin region.

5. Broadband Calculations

The effects of the Doppler shift phenomena described above can be modeled on a first principles basis, as illustrated, since the fundamental descriptions of the spectral line and emission and absorption are well known. A complete accounting of the spectral positions and the strengths of each individual line in the instrument bandpass is required and then calculation is amenable to application of Eq. 1. However, incorporating such an *ab initio* approach to the broadband E/A inversion problem will require considerable computational power and significant calculation times and can quickly become impractical. Including the Doppler shift effects into the band model theoretic approach is required for practical applications.

Efforts are underway to modify the conventional band model theory to include velocity effects.[9,10] The effect of the Doppler shift on the line strength of a single spectral line is extrapolated to the broadband absorption coefficient. The effect of the Doppler shift on the line width is accounted for through the average line spacing in the band model formulation. Results are illustrated in Fig. 6. The curve marked LTE is a conventional broadband LTE calculation. The solid curve is a measured spectrum, and the dashed curve is a line-by-line prediction with Doppler shift and vibrational relaxation. The dot-dash curve is a band model calculation-based LTE calculation adjusted NLTE conditions. The broadband correction obviously does a better job of representing the data than if the effects are ignored. The work is in its preliminary stages, but it appears to have promise for providing the band model corrections needed. Correct accounting of the combined effects of the Doppler shift on the width of the integrated line as well as the line spacing, is expected to improve these comparisons.

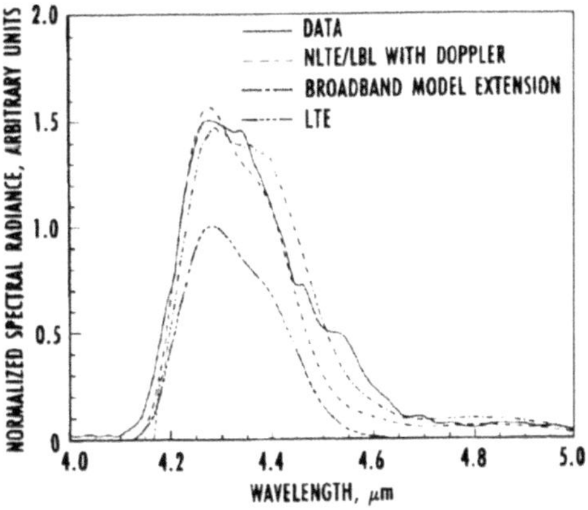

Fig. 6. CO_2 4.3-μm band spectral radiance, with band model extension.

6. Extensions/Applications to Hypersonics

To use an LOS technique to infer physical properties along the line of sight, it is necessary to describe in some fashion how the properties in one region relate to the properties in another region. Although previous applications of E/A have been for combustion gases with circular geometry, unreported work has developed a rectangular geometry model, consistent with proposed combustor geometries. In principle, any environment which has isoproperty contours is a candidate for E/A analysis. For those portions of the hardware which cannot be viewed to have axisymmetric isoproperty contours, these contours can be developed from CFD codes predictions. Multiple valued unknown properties along the LOS are allowed; all that is required are additional measurements at different wavelengths or bandpasses. However, in practice, this has not been successfully accomplished. Extensions to other geometries must be validated.

Hypersonic flight development presents a range of conditions in static temperatures from below 300 K to above 3,000 K and static pressures from fractions of an atmosphere to several hundred atmospheres, depending upon the position on the vehicle and the flight and altitude regime. Temperature in the combustor may range above 3,000K with static pressures a few tenths of an atmosphere. The ground test facilities will be highly energetic, and inlet gas flows are typically the effluent of a highly energetic gas generator or arc-heated air. The combustor for the propulsion system will probably be burning hydrogen as a fuel and using the surrounding atmosphere as the source of oxygen. Thus, the typical gaseous environments will be made up, in no particular order, of N_2, H_2O, CO_2, CO, NO_x, NO, OH, and perhaps, H, H_2, O, and O_2. Of these, only the H_2O, CO_2, CO, NO_x, NO and OH are infrared active, and thus are candidate molecules for IR E/A work. Even though some of these species may be present only in small concentrations they are candidate species for a correctly designed E/A experiment, and by virtue of their spatial distribution, can provide information about the combustion process. The H_2O molecule will always be a major consituent and is the most likely molecule for E/A diagnostics.

At the temperatures and pressures to be expected in the combustor, the very high lying transitions from H_2O are likely to be populated to significant levels and provide radiation into the short-wavelength IR. In addition to the usual radiation at 2.7 μm, the band normally previously for past applications of IR E/A, there are other bands at shorter-wavelength IR that should be accessible in the extreme conditions of the combustor. Of special interest in this context are those bands near 1 μm. Although these short wavelength transitions are generally not observed in expanded plumes, there can be measurable levels at higher temperatures and pressures.

To illustrate that these transitions can be significant, Fig. 7 is an example of a near IR spectrum acquired from the SSME plume during ground testing. The SSME is fueled by liquid oxygen and hydrogen. The spectra were obtained from the Mach disc approximately 3/4 diameter downstream of the nozzle exit. The static pressure in the disc is about 2 atm, temperature is about 3,000 K and the size of the disc is about 2/3 m in diameter. The short wavelength portion of the spectrum was obtained with an unintensified silicon detector array. The H_2O bands commencing at about 0.75 μm are weak on this scale, which is set to show the stronger bands near 0.92 - 0.98 μm. The conclusion to be drawn from the spectrum is that there should be significant radiation in the very short IR wavelengths for measurement with uncooled silicon or germanium detector arrays and thus is a wavelength region that can lend itself to successful IR E/A determination of gas path properties.

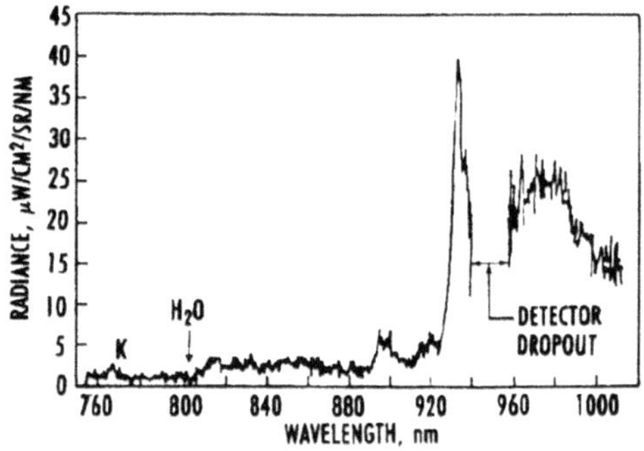

Fig. 7. Near IR spectrum from SSME exhaust.

In addition to the use of economical and rugged detectors, short-wavelength IR measurements will allow the use of quartz fibers for optical path routing. These detector arrays do not have to be cryogenically cooled and, with quartz fibers to direct the optical path, a rugged installation can be designed. A conceptual schematic which might be appropriate for installation in a combustor that might be used for both ground test and flight is shown in Fig. 8. In this concept, the detector and broadband source are installed at a convenient place in the installation, and fiber-optic cables are used to direct the field of view to the appropriate location on the combustor walls. The source shutter and detector electronics are coupled so that the detector can alternately acquire emission and absorption data.

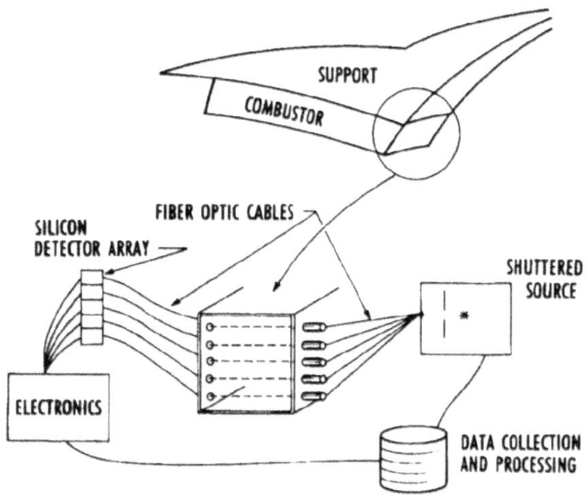

Fig. 8. E/A concept for combustor diagnostics.

The accessibility of these wavelength regions with modern instrumentation is illustrated in Fig. 9 (R. L. Moyers, unpublished data, Feb. 1992). The figure shows an H_2O absorption spectrum between 0.89-µm and 1.0-µm obtained with a simple silicon detector array mounted on a grating spectrometer. The gas path is a simple cell about one meter long in a furnace. The ability of the simple instrumentation to resolve the spectral details is apparent, as is the change in the spectra with the temperature. It is obvious that this spectral region is rich in spectral detail, and the accessibility to the silicon detector technology presents an opportunity to design diagnostic and performance evaluation instrumentation.

7. Conclusions

The IR E/A technique will have its greatest utility in diagnostics of hypersonics systems when it is used to develop steady-state averaged profiles in gas paths that can be represented by a set of isoproperty contours. Considerable detail in the temperature and specie partial pressure profiles can be obtained when the technique is appropriate and may be useful for steady-state diagnostics.

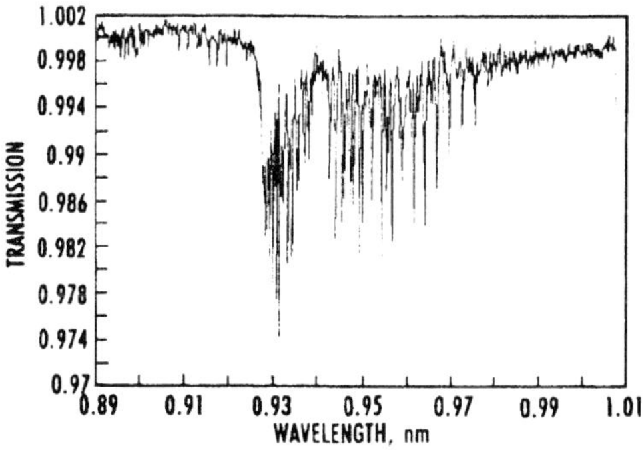

Fig. 9. H_2O transmission spectrum in 0.9 - 1.0 µm spectral range.

The geometries used with the IR E/A to this date have been circular. In principle, there is no difficulty in extending the technique to other geometries; indeed, geometry packages for rectangular systems have been developed but not validated. The extension to o ther geometries is viewed generally as being based upon codes predictions of the shapes of these expected steady-state contours. Subsequent inversions are consistent with these predictions.

Current work in the E/A technique includes considerations for vibrational relaxation and the effect of Doppler shift for the 4.3-µm band of CO_2 and the 2.7-µm band of H_2O. This preliminary work shows that the physical effect of the Doppler shift can be significant in the E/A work, and effective broadband calculations for including the Doppler shift effects must be completed for applications to the hypersonic environment. The importance of the vibrational relaxation of CO_2 is well known, and the tools to include it are available. Vibrational relaxation is probably not an issue for H_2O except at very low pressures.

The high combustor temperatures and the high concentrations of H_2O make the H_2O molecule the molecule of choice for most perceived applications of the IR E/A. This molecule similarly lends itself to band model approach to calculate the radiative transfer. Band model parameters exist for the 2.7-µm band of H_2O, but additional spectral line information is needed at the elevated temperatures to validate the broadband approach for the Doppler shift. Additionally, the H_2O and combustor conditions make the shorter wavelength IR extremely attractive from an instrumentation point of view because the spectral region near 1 µm allows the use of uncooled silicon or germanium detector arrays. The band strengths of the H_2O are needed for these short bands, however, to construct the appropriate efficient band models.

Extending the current E/A technology into the short-wavelength IR range of H_2Owill present some challenges for the development of the radiative transfer model. Since the expectations for the temperatures in the combustor are likely 3,000 K or more, line strengths or band model parameters, and broadening data will be required for the radiative transfer model. A recent survey of the literature (A. A. Mason, private communication, March 1992) shows that the database of fundamental band parameters at these short wavelengths is limited, and measurements at elevated temperatures have only recently been made.[11] Exploitation of these short wavelengths will require enhancement of these data. For broadband radiative transfer, band model parameters can be developed without going to the spectral details,[6] but the development of the band model parameters from the spectral line details is far more satisfying and, ultimately, more productive and accurate. Having the individual spectral band strengths has the additional advantage that, when appropriate, the NLTE and Doppler shift considerations will require the detailed state data to validate broadband results.

The use of the short wavelength H_2O bands opens opportunities for monitoring of the combustor both in the flight environment and the ground test environment. This is possible because the near-IR makes possible the use of fiber optics for optical path guides and economical, uncooled silicon or germanium detector arrays. This type of hardware lends itself to robust designs that can withstand extremely hostile environments. However, this approach has not been accomplished previously and the concept needs development and validation.

8. References

1. L. E. Brewer and C. C. Limbaugh, "Infrared Band Model Technique for Combustion Diagnostics". *Applied Optics*, Vol. 11, May 1972, p. 1200.

2. Limbaugh, C. C. "The Infrared Emission Absorption Method for Temperature and Species Partial Pressure Determination in Flames and Plumes." Chapter 5, *Infrared Methods for Gaseous Measurements. Theory and Practice.* Optical Engineering Series, Vol. 7, Edited by Joda Wormhoudt. Marcel Dekker, Inc., New York. 1985.

3. Limbaugh, C. C. Brown, D. G. and Zaccardi. V. A. "Experimental Determination of Plume Properties in Full Scale Hydrogen-Oxygen Rockets." AIAA-89-5240, AIAA 25th Joint Propulsion Conference, Monterey, CA, July 10-14, 1989.

4. Ludwig, C. B., Malkmus, W., Reardon, J. E., and Thomson, J. A. L. *Handbook of Infrared Radiation from Combustion Gases.* (Ed. by R. Goulard and J. A. L. Thomson) NASA SP-3080, National Aeronautics and Space Administration, Scientific and Technical Information Office, Washington, DC, 1973.

5. Ludwig, C. B., Malkmus, W., Walker, J., Slack, M., and Reed, R. A. "Standard Infrared Radiation Model." AIAA-81-1051, AIAA 16th Thermophysics Conference, Palo Alto, CA, June 23-25, 1981.

6. Phillips, W. J. "Band Model Parameters for the 2.7-μm Band of H_2O". *J. Quant. Spect. and Rad. Transfer*, Vol. 43, No. 1, Jan. 1990, pp. 13-31.

7. Limbaugh, C. C. "Gas Diagnostics for High-Area-Ratio Rockets". AIAA-85-1082, AIAA 20th Thermophysics Conference, Williamsburg, VA, June 19-21,1985.

8. Limbaugh, C. C., Hiers III, R. S., and Phillips, W. J. "Spectral Radiative Transfer for the 4.0- to 5.0-μm Bands of CO_2 With Mild Vibrational Relaxation and Doppler Shift." AIAA-90-1782, AIAA/ASME Joint Thermophysics and Heat Transfer Conference, Seattle, WA. June 18 - 20, 1990.

9. Drakes, J. A., Hiers III, R. S., and Reed, R. A. "Doppler Shift Effects on Infrared Band Models." *Journal of Thermophysics and Heat Transfer*, Vol 6, No.1, Jan.-Mar., 1992, pp. 44-47.

10. Drakes, J. A. and Limbaugh, C. C., "Detailed Examination and Extensions to Infrared Band model Calculations Including Doppler Shift Effects." AIAA-91-1430, AIAA 26th Thermophysics Conference, Honolulu, HI, June 24-26, 1991.

11. Chevillard, J. P., et al. "$H_2{}^{16}O$ Line Positions and Intensities Between 9,500 cm^{-1} and 11,500 cm^{-1}". *Canadian Journal of Physics, 16*, November, 1989, pp. 1065-1084.

MOLECULAR ELECTRONIC RESONANCE STATE TECHNIQUES*

W. H. McGregor and R. P. Howard
Sverdrup Technology, Inc./AEDC Group
Arnold Engineering Development Center
Arnold Air Force Base, TN 37389 (USA)

Abstract. Resonant electronic (A←X) absorption transitions in high-enthalpy gas flows carry with them information on density and thermodynamic state. By spectral examination of the absorption (or re-emission as fluorescence) of a suitable source the information can be extracted. In this paper the utility of the approach for diagnostics of hypersonic flow properties is examined. Some important aspects of the method requiring special attention are discussed, including high-resolution characterization of the resonant lamp source, treatment of Doppler shifts, collisional broadening, nonequilibrium vibrational state population, and modeling of the spectrometer response function. Finally, some approaches for application to hypersonic propulsion are discussed.

1. Introduction

This paper is directed toward the determination of properties of a gas flow that might either be a test medium or an engine exhaust for a hypersonic vehicle. Maximum temperatures for the test gas thus correspond to the reservoir temperature of the heater used to produce a hypersonic flow, and the chemical state of the gas is controlled by the method of heat addition. If the test facility uses an arc heater, the gas, originally ambient air, is composed of a nonequilibrium mixture of O_2, N_2, O, N, NO, Ar, CO_2, CO, CN, ions, electrons and impurities such as Cu, Ni, and oxides. Moreover, in the expansion from reservoir to high-altitude, high-speed conditions, certain of the species relax slowly so that vibrational populations are different than for equilibrium. The gas flow which must be diagnosed thus stresses the tools available. If the high-temperature gas is produced by combustion of some of the oxygen with a fuel, followed by make-up oxygen - so-called vitiated air - then the products produced by the 15,000° or 20,000° electrons in the arc may no longer be present, but the added free radicals of combustion (OH, NH, CH, CN, ---) as well as the water vapor (H_2O) and carbon oxides now must be dealt with.

A few of the species present in such flows lend themselves to resonance absorption diagnostics. In this technique a resonance (A→X) emission is produced by a source, is absorbed (X→A) by the test gas, and is detected by some receiver, Fig. 1. The amount of absorption at any wavelength is directly related to the density of the species along the path and its temperature. Re-emission of the photons absorbed constitutes fluorescence. Finding the absorption relationship constitutes the major physics problem for this technique.

*The research reported herein was performed by the Arnold Engineering Development Center (AEDC), Air Force Systems Command. Work and analysis for this research were done by personnel of Sverdrup Technology, Inc./AEDC Group, operating contractor of the AEDC propulsion test facilities Further reproduction is authorized to satisfy needs of the U. S. Government.

A. Boutier (ed.), *New Trends in Instrumentation for Hypersonic Research*, 161–170.

The source traditionally used for the resonance absorption (RA) is a gas discharge tube containing the specie of interest. Capillary DC discharges through which a mixture of NO with argon flows is typical as a source of NO gamma band radiation. The question "why not a continuous source such as a deuterium lamp?" is answered simply by the increased sensitivity due to correlation of the source and absorber spectral lines. For larger absorber densities, the continuum may play a role. A second question "Why not a laser tuned to a particular A→X line?" is a better one and will be answered only by experience. The tunable laser approach deals with one line at a time; the RA approach with many lines. The complexity of the full-up RA method is thus much greater than the tunable laser approach.

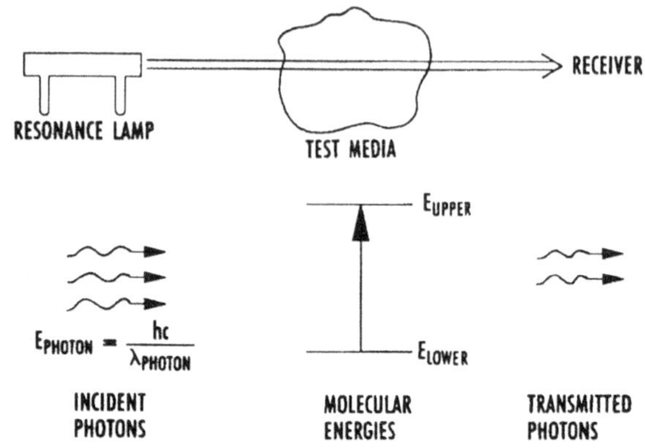

Fig. 1. Illustration of resonance absorption diagnostic technique.

It is this complexity of the modeling required to correctly execute the RA approach that this paper will address. Much attention must be paid to detail. The rewards for a correct application of the method are considerable, as witnessed by application to radiative mechanisms research[1], emissions from supersonic aircraft engines,[2,3] arc-heated test facilities,[4] and unpublished measurements on liquid-propellent rocket exhausts. Future applications to the testing of hypersonic vehicles either in ground test facilities or in flying test beds require further scrutiny.

The measured quantity for a gas flow is, of course, the line-of-sight transmission

$$\tau_v = \frac{I_v}{I_v^0} = e^{-\int_0^L k_v dx}$$

where the subscript v means that the equation is valid only at the frequency v.

2. The Physics

The intrinsic absorption of a single line (upper state n, lower state m) at a point within a media is expressed by the absorption coefficient, k, given by[5]

$$k_{nm} = \frac{c^2}{8\pi v_0^2} A_{nm} N_m \frac{g_n}{g_m}\left(1 - \frac{g_m}{g_m}\frac{N_n}{N_m}\right) \quad (1)$$

where:

$$k_{nm} = \int_{\Delta v} k_{v(nm)} dv$$

h, c = Plank's constant, velocity of light

ν_o = line center frequency, sec^{-1}

A_{nm} = Einstein transition probability, sec^{-1}

$N_n N_m$ = population density of states n, m

$g_n g_m$ = statistical weights of states n, m

The above expression holds generally for single lines, but care and conciseness are required to express the coefficients for a molecule. The transition probability now becomes

$$A_{nm} = A_o q_{v'v''} S_{j'j''}(Br) \tag{2}$$

where A_0 is the overall transition probability (or lifetime, $A_0 = 1/\tau_0$), $q_{v'v''}$ is the normalized overlap integral between the upper v′ and lower v″ state , or Frank-Condon factor, and $S_{j'j''}$ is the normalized relative rotational line transition probability, or Hönl-London factor for the branch (Br) of concern. The other factor that bears consideration in expression is the sum of states, or partition function, arising from the Boltzmann expression,

where now:

$$N_n = N_o \frac{g_n e^{-E_n/kT}}{\sum_{\text{all } i} g_i e^{-E_i/kT}} \tag{3}$$

$$E_i = E_u + E_v + E_r$$

$$g_i = (g_u)(g_r)$$

For the molecules of interest in hypersonic flow which have a resonance transition, the electronic states are generally messy. For the molecules of principal concern, NO, OH, and O_2, the following state designations hold

	upper state	lower state	λoo (nm)
NO	$A^2\Sigma$	$X^2\Pi$	227
OH	$A^2\Sigma$	$X^2\Pi$	306
O_2	$B^3\Sigma_u$	$X^3\Sigma_g$	204

In all three cases many transitions are allowed, and finding a single, isolated line for use in laser methods is difficult. Moreover, in practical RA applications the spectral resolution may be such that many lines are included within the resolution element.

Another principal factor of consideration for the RA method must be the spectral line profile, k_v. In general, the combined Doppler and Collision broadening must be considered and, for classical Lorentz collision broadening, a Voigt profile is used.

$$k_v = k_o \frac{a}{\pi} \int_{-\infty}^{+\infty} \frac{e^{-y^2} dy}{a^2 + (\omega - y)^2} \tag{4}$$

where:

k_o = the absorption coefficient at line center

a = the ratio of the collisional to Doppler half-widths ,

$= \sqrt{\ln 2} \, \frac{\Delta_c v}{\Delta_D v}$

$\Delta_c v = \frac{Z_c}{\pi}$ or. $\frac{\sigma_c \overline{V} n}{\pi}$

Z_c = effective collision frequency for broadening

σ_c = cross section for broadening

$\overline{V}$ = mean velocity

n = density of broadening molecules

$\Delta_D v = 2\left(\frac{2kT\ln 2}{Mc^2}\right)^{\frac{1}{2}} v_o$

y = variable of integration

$\omega = 2\sqrt{\ln 2} \, \frac{v - v_o}{\Delta_D v}$

Of course, the source profile as well as the absorber profile, must be taken into account in the radiative transfer model.

The equation of transfer through a homogeneous path of molecular radiation within some Δv frequency of interest can be written as follows:

$$\tau_{\Delta v} = \frac{\sum_j \int_{\Delta v} g(v) I_j^0 (v, T_s) \left[\exp\{- \sum_i k_i(v,n,T,p)L\} \right] dv}{\sum_j \int_{\Delta v} g(v) I_j (v, T_s) \, dv} \tag{5}$$

where

$\tau_{\Delta v}$ = transmission within the frequency interval Δv

$g(v)$ = instrument response function

$I_j^0 (v, T_s)$ = source intensity for each line j within Δv with Doppler broadening at temperature T_s

$k_i(v,n,T,p)$ = absorption coefficient of each line i contributing to absorption of line j at a frequency v as a function of the density, pressure, temperature, and velocity of the gas

L = path length

In most cases the path will not be homogeneous, so that the density, pressure, and temperature will vary along the path x. Moreover, the velocity of the gas along the line of sight will not be zero, but may be quite high, leading to Doppler shift of the line. This velocity effect can be quite important since the absorption and source lines are now displaced and a given single line may not even be absorbed. See Fig. 2 for an illustration of this point.

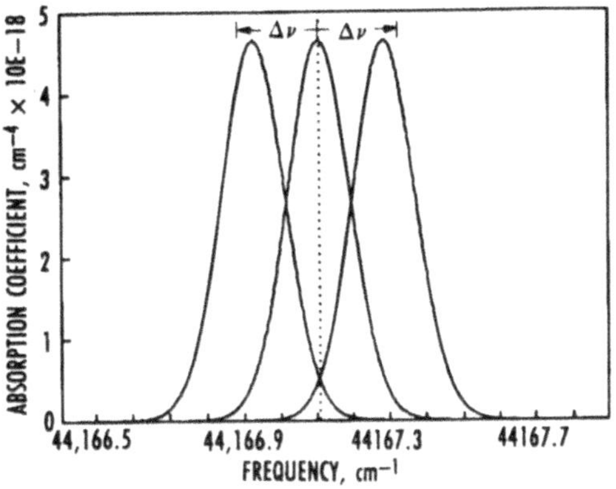

Fig. 2. Illustration of the effect of doppler shifts on the P_2 (21.5) absorption line.

The absorption now must be written as an integral along the path,

$$\exp\left[-\int k_i(v', n, \rho, T)_x \, dx\right]$$

where $v' = v\,(1 - \frac{V}{c})^{-1}$

and the velocity v can be either away from, or toward the observer (+ or -) and is also path dependent.

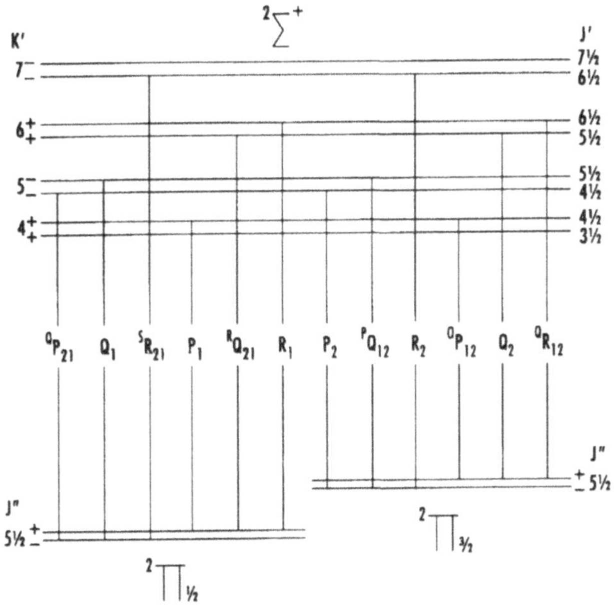

Fig. 3. Transition diagram for $A^2\Sigma^+ - X^2\Pi$ transitions.

We have introduced quite a complex model and have yet to discuss the spectra of the molecule of interest. For NO, a $A^2\Sigma \leftrightarrow X^2\Pi$ transition, all 12 allowable rotational transitions must be considered as shown in Fig. 3. The Π state transitions from Hunds' case a to case b as the rotational quantum number increases so that the Hönl-London factors for the transitional case must be included. The theoretical line positions, although predicted within a fraction of a wave number by the Hill and Van Fleck formula,[6] are still not predicted well enough for the actual case, especially when Doppler shifts must be considered. To build a precise numerical model tabulated, measured line positions of Deesi[7] must be used. The inversion of the integrated measurements to local properties will await the discussion of the apparatus since it, too, must be a part of the modeling.

3. Apparatus

An arrangement that has produced nice results for NO in axisymmetric flows is shown in Fig. 4. The source lamp and spectrometer are remotely located and the light is transmitted through optical fibers to opposing translators. Figure 5 illustrates the spectral features of the measured transmission. A profile is produced (Fig. 6) which may be inverted via an onion peeling procedure (Fig. 7) to a radial profile of molecular density and temperature (Fig. 8).

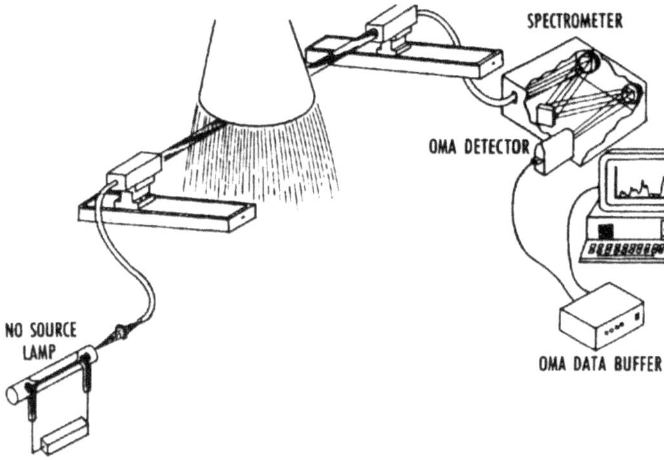

Fig. 4. Experimental arrangement for resorance absorption application to field measurments.

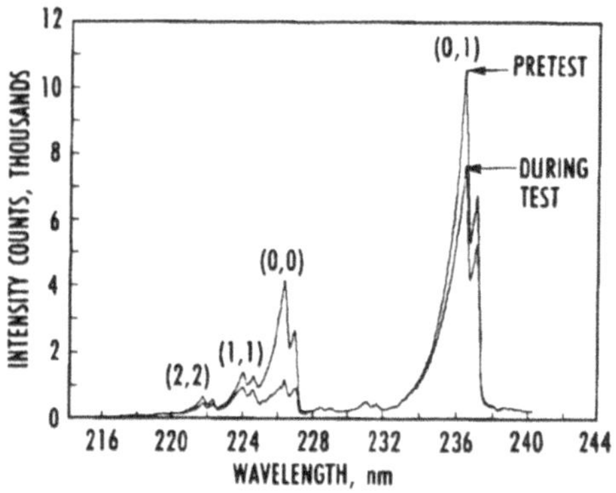

a. Spectral transmission

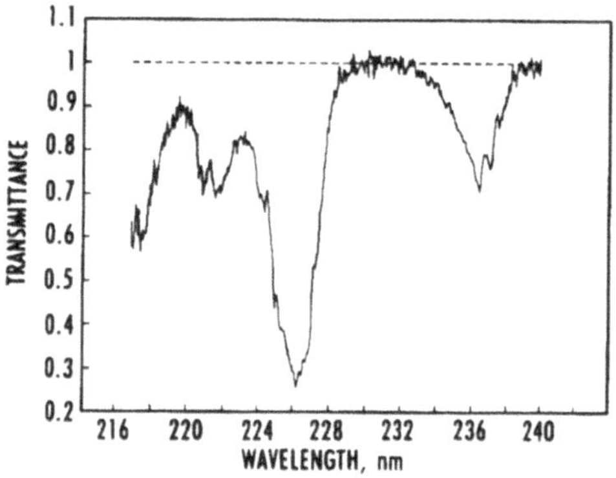

b. Spectral transmittance (ratio of curves in a)

Fig. 5. Spectrum measurements of transmission of resonance lamp radiation of NO through gas generator flow.

To achieve the results cited above a great deal of laboratory work had to be accomplished. First, the source had to be examined using a high-resolution (0.004 nm, in this case) spectrometer, resulting in an intensity distribution of the lines in both the (0,0) and (0,1) bands (Fig. 9). Then the entire instrument function g(v) had to be established. The spectrometer in this case utilized a linear intensified, silicon detector with a multi-channel analyzer electronics. The spectrometer slit width was chosen to correspond to a single pixel, but as can be seen in Fig. 10, leakage to several adjacent pixels occurs so that this distribution function must be included in the model. Finally, a calibration is performed in which the transmission through a cell of known concentration and controlled temperature is measured. The single greatest unknown in the physics model is the collisional broadening parameter (α). Although some attempts have been made to measure this parameter, we blatantly use it as an adjustment parameter to force model and laboratory results to agree. Perhaps surprisingly, a single value of α, adjusted for temperature and pressure, gives consistent results over a wide range of densities, pressures, and temperatures.

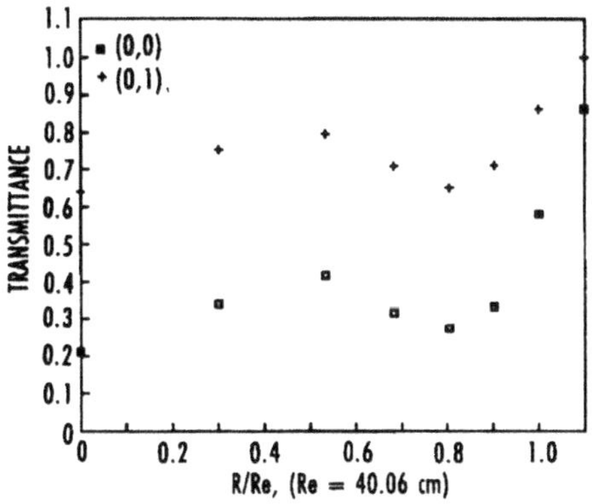

Fig. 6. Transverse profile of transmittance of the (0,0) and (0,1) bands of the gamma system of NO through the gas generator flow.

4. Inversion

The procedure for inverting the integrated line-of-sight measured transmission to local properties is the most controversial element of this procedure. In the axisymmetric case, Fig. 7, the ever-present question "What if the flow isn't axisymmetric?" or "What if there are in homogeneities?" is answerable only by a statement of uncertainty peculiar to each application.

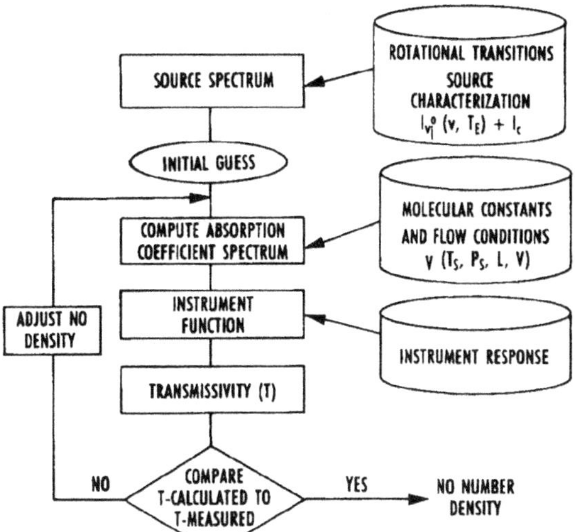

$$T = \frac{\int_{\Delta\nu} g(\nu)\,[I^{o}_{\nu j}(\nu, T_E) + I_c]\, e^{-\sum_j k^{o}_{\nu j}(T_s, P_s, \nu)\, N_{NO} L}\, d\nu}{\int_{\Delta\nu} g(\nu)\,[I^{o}_{\nu j}(\nu, T_E) + I_c]\, d\nu}$$

Fig. 7. Illustration of procedure to invert transverse transmittance profiles to radial profiles of space density.

The axisymmetric case has been the one most studied because of applications to jet engine and rocket exhausts and to arc-jet flows. However, the peeling approach to inversion can be applied to any case for which constant property zones can be specified: rectangles, ellipses, etc. Current designs of scramjet engines tend to be rectangular with fuel injections along one, or opposing

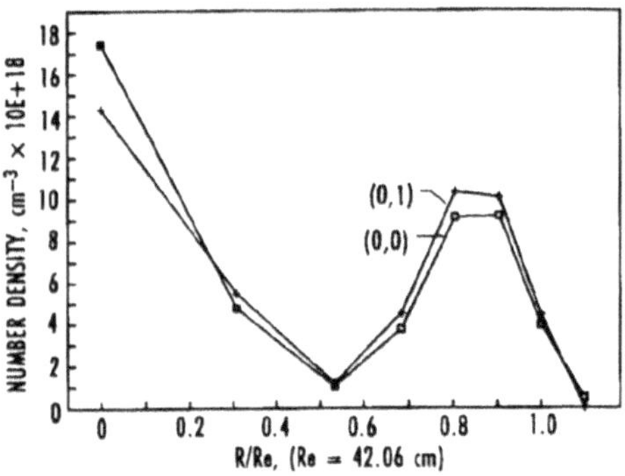

Fig. 8. Radial profile NO density from the (0,0) and (0,1) bandhead transmittances.

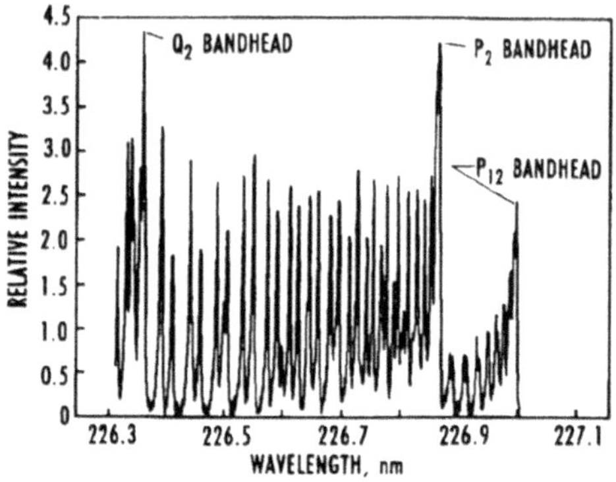

Fig. 9. High resolution (0.0045 nm) spectrum of the NO gamma (0,0) band system produced by a capillary discharge lamp.

walls. Integrated path methods can be applied in such cases. One proposed configuration would use a diagnostic channel (Fig. 11) to accomplish the measurement, again utilizing fiber optics to place the source and receiver in a quiescent environment. The success of such a measurement depends upon defining the cells of near-constant properties.

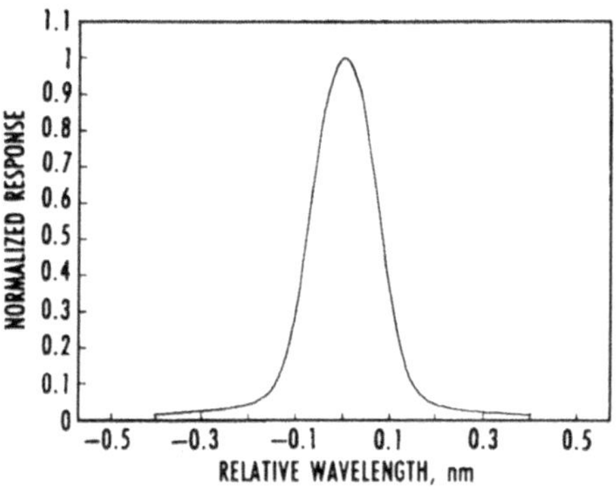

Fig. 10. Spectral response of intensified linear array detector used in spectrometer.

5. Applications

One application will be discussed briefly to illustrate the current status of the technique and to show the effect of ignoring some of the important physics. The data and subsequent inversion shown in Figs. 5, 6, and 8 will illustrate several points. The application was made to a flow produced by the expansion of gas generated by storable propellent combustion through an axisymmetric nozzle. The temperature profile used in the inversion was calculated. The nozzle had a rapidly expanding contour so that the radial velocity at the measurement point reached about one km/sec. Shown in Fig. 12 is the result of ignoring the Doppler F

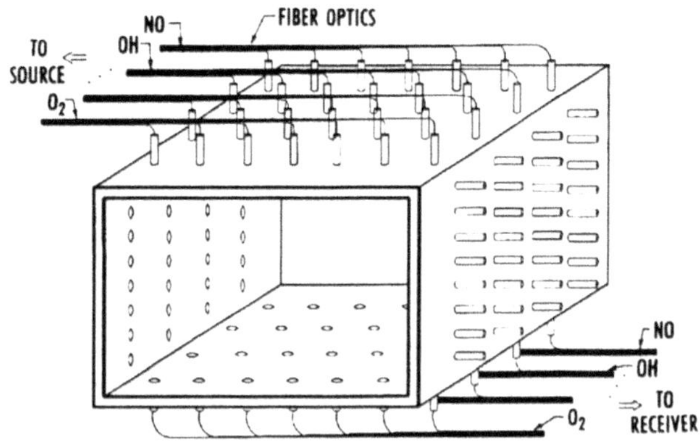

Fig. 11. Illustration of diagnostic channel for rectangular duct flows.

shift resulting from this velocity. Also shown is the density obtained from the 0,0 band measurements, assuming an equilibrium vibrational distribution. Still another effect is shown by ignoring the effect of temperature on the broadening coefficient. Quite clearly, attention must be paid to such detail if the results are to be consistent.

The method was also applied to determine the OH density profile of the same flow. Vitiated air heaters produce unknown amounts of NO and OH which must be taken into account when scramjet propulsion devices are evaluated. Similiar comments can be made for OH as for the NO application – attention to detail is necessary. Also, laboratory calibration of the OH technique relies on being able to specify the OH concentration in a flat-flame or diffusion burner, usually burning hydrogen and air. The equilibrium calculation of both temperature and OH concentration is the foundation of the calibration, and a certain amount of faith is required.

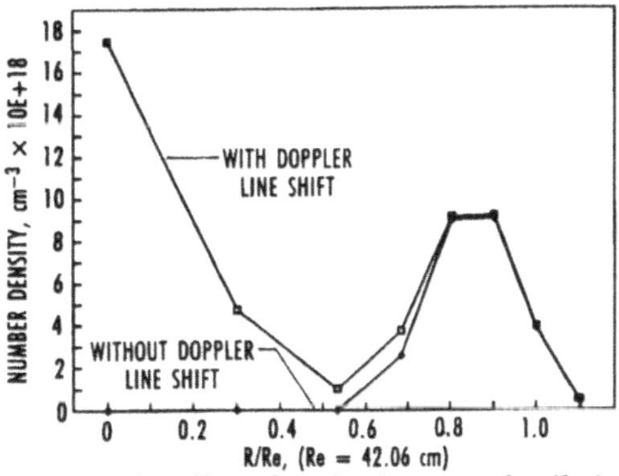

Fig. 12. The effect of neglecting some details in the radiative transfer model.

The application of the NO measurement in an arc-jet heated flow [4] produced a nice result – the vibrational population of the v = 1 state. In this application the flow was expanded from a reservoir to Mach 6. The ratio of measured population of the v = 1 state to the equilibrium value was about 1,000, an unexpectedly high value. If only the v = 0 state had been considered in the diagnostics, an error in number density of some 10 percent would have been introduced.

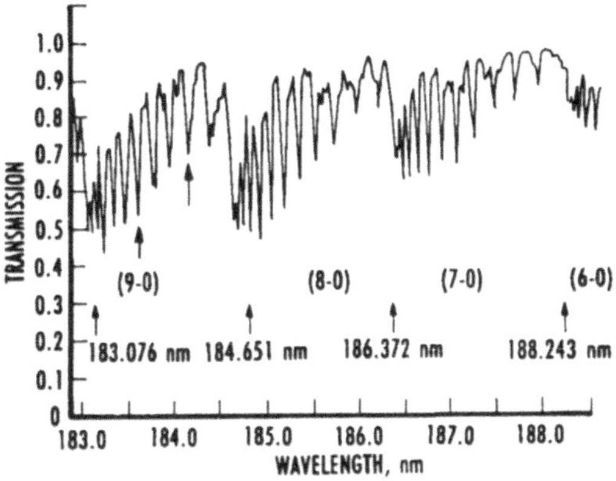

Fig. 13. Example absorption spectra of the Schuman-Runge lands from a continuoum source.

An application currently being pursued is the "calibration" of a vitiated air facility to be used in prototype scramjet development. NO and OH concentrations will be measured by the RA method and O_2 concentration by a kindred method, using a continuum source. The objective of the O_2 measurement is to determine whether the make-up percentage of O_2 is correct for simulation of air. The absorption of the (9,0) band of the Schumann-Runge system is being used. An example absorption spectra from the laboratory is shown in Fig. 13. The major anticipated problem here is interference from products of combustion such as CO, NO unburned fuel components.

Applications to scramjet testing have been proposed in which "diagnostic channels" like that of Fig. 11 are made a part of the test engine. The opposed fiber optic ports can then be used for resonance absorption using different sources (resonance lamps, lasers, continuum lamps, etc.) or emission spectroscopy. The arrangement shown in Fig. 11 would be especially useful in determining the extent of mixing and combustion by measuring the OH emission and/or absorption. This rugged approach to diagnostics is useful in ground testing, but can also be used for flight test beds or free-flight tests.

6. Future Efforts

The robust methods discussed herein do not produce point measurements within flows. Inversion methods used to infer point values of properties are subject to errors when insufficient information is available to specify zones of homogeneity. Turbulence effects also become difficult (if not impossible) to detect using line-of-sight measurements. Yet, these methods probably offer more reliable information for experiment evaluation than beam scattering methods. For this reason, further development in a few areas is believed warranted.

First, the engineering of diagnostic channels is prerequisite to deployment. The materials to be used must be studied and the fiber-optic installation and protective design must be determined. This development will be time consuming and require testing and validation before an engine designer will feel safe in incorporating it into a test article or prototype.

Secondly, and again an engineering problem, source and receiver components must be made more rugged and be miniaturized to be generally useful. This is especially true if these techniques are to be made flightworthy. The relay racks and optical benches currently used just will not do.

Finally, some physics always needs improvement. The collisional broadening in molecular electronic transition has never been pursued intensively. Yet, for the resonance techniques to be successful, research in this area is required, especially on such molecules such as NO, OH, and O_2. Extension to molecules such as NH may also lead to a good diagnostic since the Q-branch of the A-X transition at about 336 nm offers a dense line spectra that can be modeled almost as a continuum.

7. References

1. Davis, M. G., McGregor, W. K., Jr., and Mason, A. A. "Determination of the Excitation Reaction of the OH Radical in H_2-O_2 Combustion." AEDC-TR-69-95 (AD-695471), October 1969.

2. Few, J. D., McGregor, W. K., and Glassman, H. N. "Resonance Absorption Measurements of NO Concentration in Combustor Exhausts". AIAA Progress in Astronautics and Aeronautics-Experimental Diagnostics in Gas Phase Combustion Systems, B. T. Zinn, ed., AIDD Press, Princeton, NJ, 1977.

3. Dodge, L. G., Colket, M. B., III, Zabielski, M. F., Dusek, Jr., and Seery, D. J. "Nitric Oxide Measurement Study: Optical Calibrations". United Technologies Research Center, December 1979.

4. Howard, R. P., Dietz, K. L., McGregor, W. K., and Limbaugh, C. C. "Nonintrusive Nitric Oxide Density Measurements in the Effluent of Core-Heated Airstreams". AIAA 21st Fluid Dynamics, Plasma Dynamics and Lasers Conference, AIAA-90-1478, June 18-20, 1990, Seattle, WA.

5. Mitchell, A. C. G. and Zemansky, M. W., "Resonance Radiation and Excited Atoms." Cambridge University Press, 1971.

6. Hill, E. L., and Van Fleck, J. H. Phys. Rev. 32, 250, 1923.

7. Deezsi, I. "A Recent Rotational Analysis of the τ-Bands of the NO Molecule". Acta Physica, Vol. 9, 1957, p. 125.

LOCAL NON-INTRUSIVE MEASUREMENTS OF THE INLET AND EXHAUST FLOWS IN GROUND TESTING A HYPERSONIC ENGINE*

RONALD H. KOHL
Sverdrup Technology, Inc. / AEDC Group
Mail Stop 900
Arnold AFB, TN 37389
United States of America

ABSTRACT. The freejet ground testing of full-sized hypersonic, technology-demonstration engines presents unprecedented challenges even at relatively low equivalent flight Mach numbers (≤ 4). The required quantitative knowledge of the inlet and exhaust gas flows cannot be achieved with techniques used in the past, and local non-intrusive measurement techniques must be developed and used. A generic hydrogen-air engine is used as the focus in presenting test-result requirements, the implications of these requirements on the development of candidate measurement techniques, the development path that these techniques must follow and why, and the techniques whose development has been begun by AEDC and the status of these. Some comments are added regarding the additional development problems the testing of hypersonic hydrocarbon-air engines will bring.

1. Introduction

The freejet ground testing of complete, full-sized hypersonic, technology-demonstration engines presents unprecedented challenges even at relatively low equivalent flight Mach numbers (≤ 4) with a relatively long time spent at each test condition (30 sec).

Using the techniques of the past, the required quantitative measurements of the exhaust gas flow cannot be accomplished, and those of the inlet can only be achieved at great cost and risk. At the exhaust this is due primarily to intrusive sampling effects on both instantaneous and time-averaged chemical composition, leading to an inability to obtain combustion efficiency at the exhaust plane and any but the time-averaged local fuel-to-air ratio. At the inlet this is primarily due to the preclusion of the use of in-and-out probe rake(s) at the start of the test condition due to the expected close coupling of the facility nozzle and the (only slightly smaller) engine inlet. This close coupling at the inlet is required to put all of the inlet in the test rhombus of the facility nozzle.

---

* The research reported herein was performed by the Arnold Engineering Development Center (AEDC), Air Force Systems Command. Work and analysis for this research were done by personnel of Sverdrup Technology, Inc., AEDC Group, operating contractor for the AEDC propulsion test facilities. Further reproduction is authorized to satisfy the needs of the U. S. Government. Cleared for public release, AEDC/PA 92-139.

A. Boutier (ed.), New Trends in Instrumentation for Hypersonic Research, 171–184.

1.1 NON-INTRUSIVE MEASUREMENTS

The problems cited above imply that non-intrusive measurement techniques must be developed and used. A few key introductory remarks concerning these are given here.

1.1.1. *Set of Measured Parameters.* Non-intrusive techniques interact with the molecules of the gas flow, usually the molecules of a particular constituent. Those techniques suitable for measuring density therefore measure the number density, n_i, of one or more constituents, i, the number density being stated in number/cm^3 or Amagats. (One Amagat is the number density relative to that of a gas at 300 K and 1 atmosphere pressure.) All viable techniques for the accurate measurement of the number density in a single volume element moving with the gas flow must, to date, be calibrated in place, a mixed blessing.

Non-intrusive techniques usually interact with those molecules in a certain internal energy level. The fraction of those molecules present which are in that energy level is a function of the temperature, T. Such a temperature measured by a non-intrusive technique is the temperature of the particular internal-energy distribution, which is equal to the static temperature of the gas in a number of situations. (Accurate measurement of the temperature of a single volume element moving with the gas flow precludes, to date, the use of the Doppler width corrected for pressure broadening.)

The form of the perfect gas law used is $P = nkT$, where n is the total number density, k is Boltzmann's constant and P is the static pressure. The link to stagnation quantities (P_0 and T_0) is the local value of the non-intrusively measured velocity, $\underline{v}$. As shall be seen the vector velocity, $\underline{v}$, is required.

The essentially complete set of non-intrusively measured parameters is n_i, T and $\underline{v}$.

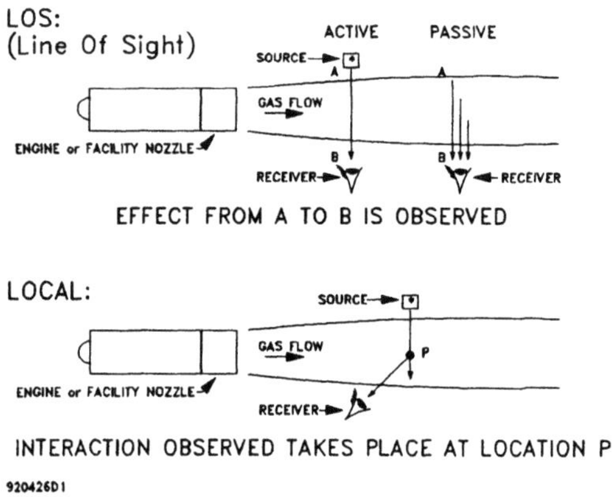

Figure 1. The definitions of line-of-sight and local non-intrusive techniques.

1.1.2. *Local Non-Intrusive Techniques.* There are two types of non-intrusive techniques: Line-of-Sight and Local, as shown in Figure 1. Line-of-Sight techniques, where the receiver receives radiation influenced by the gas-flow properties all along a line through the flow, are the simplest to implement, and there are many such techniques in use and/or development at AEDC. However, a number of reasons, including the need for profiles in what can be expected to be locally complex and/or rapidly varying flows, especially in the exhaust, and the lack of a spectrally accessible, linearly absorbing transition in two key molecules, N_2 and H_2, direct one to seek local techniques to develop.

Because they are local, local techniques usually have certain characteristics. Consider the lower portion of Figure 1. The interaction at point P, which causes radiation to be observed at the receiver, is an interaction that exists all along the radiation beam from the source as it crosses the flow. If the radiation observed by the receiver is to be characteristic of the point P only, the interaction in question must be weak, so that it does not remove energy of any significance from the source beam as it traverses the flow. But for measurements of some precision, the number of photons received at the receiver from the point P must be large or photon statistics will ruin the precision of the measurement. To obtain large photon numbers at the receiver due to a weak interaction at the point P requires the use of a very powerful source; for this reason pulsed, high-power lasers will be typically used. A receiver array, not just a single receiver as shown in Figure 1, will be used so that at each pulse of the laser beam, the interaction at many points all along the beam, not just at P, will be observed. Finally we note that a particular radiative-molecular interaction used successfully in a laboratory with a flow 1 cm wide, and which, for example, allows the source beam to exit the flow with 95% of its original energy intact, will be completely unacceptable for use as a local technique in a flow 1.5 m in width at the same flow conditions, where not even 0.1% of the original source-beam energy will exit the flow.

1.2. TEST CONDITIONS

In this paper a generic hydrogen-air engine is used as the focus. This full-scale engine is assumed to have inlet and exhaust flows of from 1 to 1.5 m in width and in height, over which measurements are to be made within the 30-sec time interval assumed spent on any one test condition. The flow presented to the inlet is dry, clean air (not vitiated) consistent with the flight Mach number (≤ 4).

The flow conditions in the inlet and exhaust are shown in Table I. Note that there will be no requirement to measure the number density of OH or NO.

Table I. FLOW CONDITIONS

	INLET	EXHAUST	
		Combusted	Uncombusted
STATIC TEMP.	217 to 300 K	500 to 1900 K	200 to 800 K
NUMBER DENSITY (Static Press.)	0.05 to 1.1 Amagats (0.05 to 1.2 Atm)	0.03 to 0.1 Amagats (0.1 to 0.6 Atm)	0.1 to 0.3 Amagats (0.1 to 1 Atm)
VELOCITY (Mach Number)	170 to 1080 m/s (0.5 to 4)	1420 to 2220 m/s (1 to 3)	280 to 1700 m/s (1 to 3)
CONSTITUENTS (Numbers are mole fractions)	Dry Air	H_2O: 0.1 to 0.35 H_2: 10^{-5} to 0.26 O_2: 10^{-7} to 0.1 N_2: 0.4 to 0.7 OH: 10^{-5} to 10^{-2} NO: 10^{-5} to 5×10^{-3}	All Mixtures: From H_2 Only to Air Only

The total flow span, including both the flow of interest (at the center) and the surrounding flow is 1.5 to 3.0 m. (At the inlet the surrounding flow is air at the same conditions. At the exhaust the surrounding flow is air at conditions between those at the inlet and those in an uncombusted exhaust airflow.) Therefore, the distance of the furthest point of interest from the closest optics position is 1.25 to 2.25 m.

2. Test-Result Requirements

Throughout this paper, each requirement is stated in terms of two approximate limits: the first, specifically called the "requirement," is the condition which, if not met, calls into question the value of making the measurement. The second, called the "target," has been set such that going beyond the target will give very little additional value in the use of the measurement.

2.1. TEST PARAMETERS REQUIRED

Flow-related parameters are required to be determined on two types of imaginary surfaces in the flow.

2.1.1. *On Reference Planes.* The first type of imaginary surface consists of cross-flow planes called reference planes. The flow conditions on a reference plane are the upstream conditions of the flow used in ground tests of a downstream section of the airframe and/or propulsion system. The test parameters required to be known as local parameters on reference planes in a freejet hydrogen-air engine test are:

- Vector Velocity
- Density or Static Pressure
- Static Temperature (inlet especially)
- Gas Composition (exhaust* especially)
- Core-Flow Turbulence (exhaust* especially).

(In implementation, measurement of the core-flow turbulence will be limited to the probability distribution function of the vector velocity.)

2.1.2. *On Control-Volume Surfaces.* The second type of imaginary surface on which flow parameters are required consists of the surface elements which are part of the closed surface surrounding the engine control volume and which are in the engine gas flows, the inlet and exhaust flows especially. (The control volume encloses the engine and is used to describe the engine test results, particularly the force accounting.) The test parameters required to be known on the control-volume surfaces in the inlet and exhaust flows are listed in the lower part of Table II in priority order. Above them are listed the parameters to be measured, which can be seen to be complete by comparison. (In Table II, $\hat{n}$ is the outward-pointing unit normal at any point on the control-volume surface, m_i is the appropriate molecular weight of flow-constituent i, and A indicates any area of the control-volume surface through which flow is passing.

---

* Also at the inlet in vitiated-air facilities.

The square brackets, [ ], surrounding the measured parameters n and T in the upper part of Table II indicate that in order to measure n in the exhaust and, in some if not all the measurement techniques useable in the inlet, in order to measure n in the inlet, a temperature T must also be measured.)

Table II. CONTROL-VOLUME FLOW PARAMETERS

	INLET	EXHAUST
MEASURED	*Local*: $\left[\begin{array}{c} n_j, \text{ for a major constituent } j \\ T \end{array}\right]$ $\underline{v}$	*Local*: $\left[\begin{array}{c} n_i, \text{ for all major constituents, } i \\ T \end{array}\right]$ $\underline{v}$
REQUIRED	*Local Mass Flow* $= \rho\, \underline{v}\cdot\hat{n}$ $\rho = n_j\, \Sigma_i\, m_i \left(\frac{n_i}{n_j}\right)_{air}$ *Total Mass Flow* $= \int_A \rho\, \underline{v}\cdot\hat{n}\, dA$ *Force Contributions* $\left\{ \begin{array}{l} \int_A \rho\, \underline{v}\, \underline{v}\cdot\hat{n}\, dA \\ \int_A P\, \hat{n}\, dA, \quad P = nkT \end{array} \right.$ *Moment Contributions* { *Substitute* $\int_A \underline{r}\times$ *for* $\int_A$ *in the two Force Contribution expressions.*	*Local Fuel/Air Ratio or Equivalence Ratio*, ϕ *Local Combustion Efficiency*, η_c *Force Contributions,* *Moment Contributions* *and* *Total Mass Flow* (*all as at left but with* $\rho = \Sigma_i\, m_i\, n_i$)

To the inlet test-parameter list in Table II should be added the local velocity direction (for inlet function) and the local static temperature (for ignition effects in operability testing).

2.2. TIME AND SPACE REQUIREMENTS

Flow cross-section coverage-time and spatial-resolution requirements for both the inlet and exhaust are shown in Figure 2.

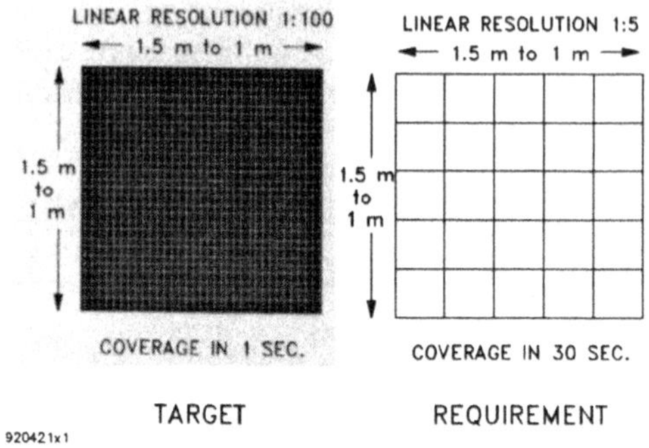

Figure 2. Flow cross-section interrogation requirements.

2.3. TEST-PARAMETER UNCERTAINTIES REQUIRED

Approximate required test-parameter uncertainties were only obtained, and that with some difficulty for the control-volume point of view. These on the inlet end are identical to the inlet-reference-plane requirements. Exhaust-reference-plane requirements have yet to be championed.

The required and target test-parameter uncertainties are given in Table III and Figures 3 and 4. Here the "uncertainty" is the total uncertainty, the square root of the sum of the approximate precision and bias, where the precision uncertainty is approximately 1 standard deviation in a random distribution of values obtained for a result and the estimated bias is appropriately sized.

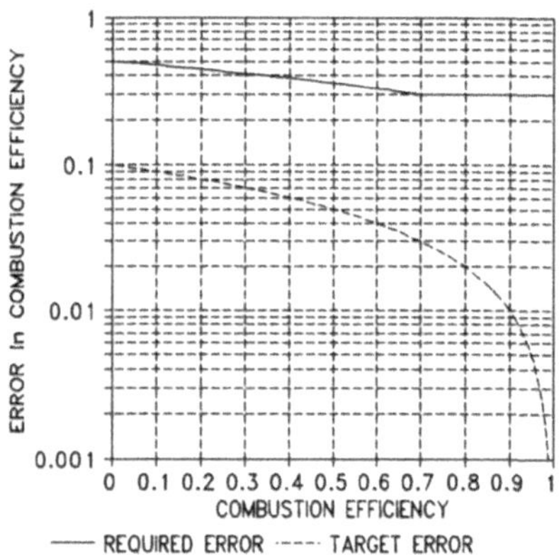

Figure 3. Required and target errors in the combustion efficiency.

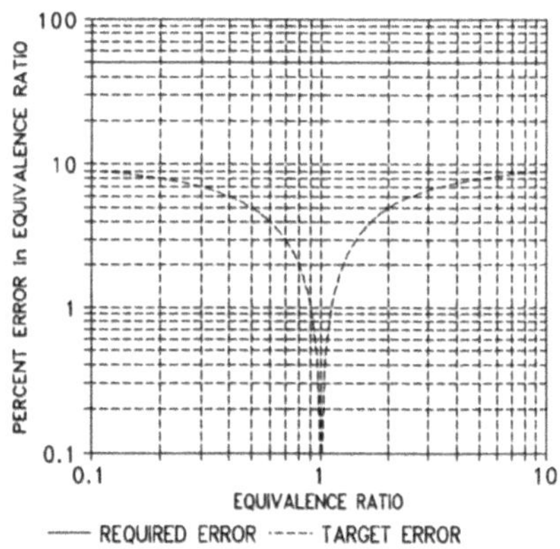

Figure 4. Required and target relative errors in the equivalence ratio. (Required relative error is 50%.)

Table III. TEST-PARAMETER UNCERTAINTIES
T = **Target** **R** = **Requirement**

TEST PARAMETER	**T**	**R**	**Remarks:**
INLET:			
Local Mass Flow	1%	8%	
Total Mass Flow	1%	8%	
Force Contributions:			
Momentum Transport	1%	11+%	Direction: **T**=5°, **R**=28°
Pressure	*	**	* 1%, M<1; M^2%, M>1. ** Better than estimates from wall pressures. Direction as above.
Moment Contributions (Total)	1%	11+%	
Local Velocity Direction for Inlet Function	1/2°	5°	
Local Static Temperature for Ignition Effects	1%	10%	
EXHAUST:			
Local Equivalence Ratio, ϕ	*	*	* See Figure 4.
Local Combustion Efficiency, η_c	*	*	* See Figure 3.
Force Contributions:			
Momentum Transport	1%	11+%	Direction: **T**=5°, **R**=28°
Pressure	*	**	* 6% at all Mach numbers (M). ** Better than estimates from wall pressures. Direction as above.
Moment Contributions (Total)	1%	11+%	
Total Mass Flow	1%	8%	Only for closure check of other measurements.

3. Implications

3.1. PRIORITY OF DEVELOPMENT OF PARAMETER MEASUREMENT

Of those test-result parameters listed in Table II, the local, exhaust-plane equivalence ratio and combustion efficiency are most important in the testing of hydrogen engines. These are obtained essentially from the measured, major-constituent (H_2O, H_2, O_2 and N_2) exhaust number densities, n_i, which also require the measurement of the relevant temperature(s).

Without prior thought, one would assign the force and moment contributions from the exhaust plane as the next priority. However, the development state and the current outlook for development of the measurement of vector velocity in the exhaust appears to be much poorer than that of the density and vector velocity in the inlet under these test conditions. In addition, the net contribution to force and moment from both inlet and exhaust planes can be measured in the test-cell with sufficient accuracy using the current state of the art that indirect determination of the exhaust plane contributions, via the vector relationship *exhaust contribution = net contribution + inlet contribution*

should be much more accurate and sure than direct determination. For these reasons the force and moment contributions of the inlet -- and the measured inlet number density and vector velocity -- are assigned second priority. The *direct* determination of exhaust-plane force and moment contributions -- and the measurement of the local vector velocity in the exhaust (also required for the exhaust reference plane) -- are then assigned third priority.

In summary, the list of parameters to be measured and the list of parameters for which local, non-intrusive measurement techniques need to be developed for freejet testing of hydrogen-air engines are, in priority order:

(1) In the Exhaust:
Local n_i (for all major constituents i) and T ,

(2) In the Inlet:
Local $\underline{v}$, n_j (for one major constituent j) and T ,

(3) In the Exhaust:
Local $\underline{v}$.

3.2. MEASUREMENT-PARAMETER UNCERTAINTY REQUIREMENTS

Given in Table IV are the uncertainty requirements for the measured parameters which arise from the test-parameter uncertainty requirements of Table III as influenced, in some indicated cases, by foreknowledge of a likely measurement technique.

3.3. DEVELOPMENT DIRECTIONS

3.3.1. *Measurement of All Parameters: n_i, T, $\underline{v}$.* The requirements for accurate measurements, rapid coverage of the flow cross section (< 30 sec), moderate spatial resolution and no low f-number optics imply the use of a pulsed laser beam directed across the flow, with simultaneous point measurements made all along the beam, across the entire flow, at each laser pulse, with the beam position traversed sideways, across the flow at right angles to itself, between pulses.

3.3.2. *Measurement of Velocity, $\underline{v}$.* The requirements for high accuracy and measurement of the full three-dimensional velocity vector over the free surfaces of the control volume (over which the velocity vector must be integrable) imply a marker or flow-tagging technique. The requirement for the three-dimensional velocity vector implies the marks must be points; the points being where two marking laser beams cross. The paragraph above implies that, as an initial development direction, many points should be marked simultaneously along a laser beam.

As the components of a marker-technique measurement are time and displacement, if the specific technique allows the mark (excitation) to last long enough, uncertainties of 1% to 1/4% may be common.

3.3.3. *Measurement of Number Density, n_i.* If using laser-induced fluorescence or spontaneous Raman scattering to measure a number density n_i, the relation between the signal and the number density is of the form

$$\text{Signal}_i = C\, f(T)\, n_i \quad . \tag{1}$$

Table IV. MEASUREMENT-PARAMETER UNCERTAINTIES
T = Target R = Requirement

MEASUREMENT PARAMETER	T	R	Remarks:
INLET:			
$\|\underline{v}\|$	0.25%	1%	Given, from expected performance of marker technique (RELIEF)
Direction of $\underline{v}$	0.5°*	10°*	From inlet function
[* Note, already	0.1°	0.4°	at **T** and **R** $\|\underline{v}\|$ uncertainties with RELIEF.]
n_j	0.7%	**	From pressure force for $M \leq 1$ (** not done)
"	1%	8%	From mass flow (more restrictive than momentum transport, given $\underline{v}$ uncertainties)
T			Aside from any requirement for n_j determination:
"	0.7%	**	From pressure force for $M \leq 1$ (** not done)
"	1%	10%	From ignition effects
EXHAUST:			
$\|\underline{v}\|$	0.6%	6.5%	From momentum transport, assuming equal relative uncertainties in ρ and $\|\underline{v}\|$
Direction of $\underline{v}$	5°	28°	From momentum transport
n_i	$(f_i\sqrt{N})^{-1}$ ×0.6%	$(f_i\sqrt{N})^{-1}$ ×6.5%	From ρ, from momentum transport as with $\|\underline{v}\|$. N = number of contributing constituents, f_i = <u>mass</u> fraction in constituent i. Note this is an arbitrary assignment of uncertainty to the n_i which happens to work.
"	?	?	From ϕ and η_c, assignment of error is just too arbitrary to be useful. For ϕ, η_c and ρ, apply the uncertainty requirements to the uncertainties arising from the candidate techniques to put requirements directly on the techniques.
T	-	-	No requirement apart from n_i determination. (Internal energy determination in 2nd generation.)

The constant C includes the efficiency of all optical elements and requires calibration in place in the test cell, automatically, before and after each run. The constant C is obtained from

$$(\text{Signal}_i)_{cal} = C\, f(T_{cal})\, (n_i)_{cal} \tag{2}$$

with the subscript "cal" indicating the calibration and T_{cal} and $(n_i)_{cal}$ known. (The calibration procedure must insert no additional optical elements or change existing ones, so the calibration gas mixture, at whatever calibration temperature is used, will be at atmospheric pressure.)

Equations (1) and (2) imply uncertainty in the initial test-cell determinations of n_i of about 5%, with hard work and experience perhaps bringing it down to the vicinity of 2.5% eventually.

3.3.4. *Measurement of Temperature, T.* Where temperature is required for its own sake, it is the static temperature, the molecular translational temperature, that is required. The use of non-intrusive interactions imply the measurement of molecular rotational or vibrational temperatures. Under the conditions of interest the static (translational) and rotational temperatures are expected to be equal.

Given equilibrium with the static temperature, there will be less error in the determination of static temperature if the vibrational temperature is measured rather than the rotational temperature. The vibrational temperature is expected to be in equilibrium with the static temperature in the exhaust in constituents whose vibrational energy is thermal (as opposed to chemical). At the lowest Mach number values in the inlet the vibrational temperature approaches equilibrium, but unlike a vitiated air-flow facility, the dry air of the test conditions here ($M \leq 4$) mean that the vibrational temperature will be significantly higher than the static temperature at the higher Mach numbers. Rotational temperature measurement will have to be used where the vibrational temperature is not in equilibrium and to confirm that the vibrational temperature is in equilibrium where that is expected to be the case.

3.3.5. *Validation of Measurements.* Aside from velocity obtained by marker technique, the validation of the use of a measurement technique at a given flow condition in a given geometry will be by blind comparison with the results of simultaneous measurement of the same parameter by an entirely different technique.

4. General Development Path

Because of the differences, especially in length scales, between the application in a large ground-test facility and the typical, initial laboratory demonstration of a local, non-intrusive technique, another laboratory demonstration will usually have to be done with application to a large ground-test facility in mind as the first step in a chain leading to more or less routine use in such a facility, as indicated in the lower part of Figure 5.

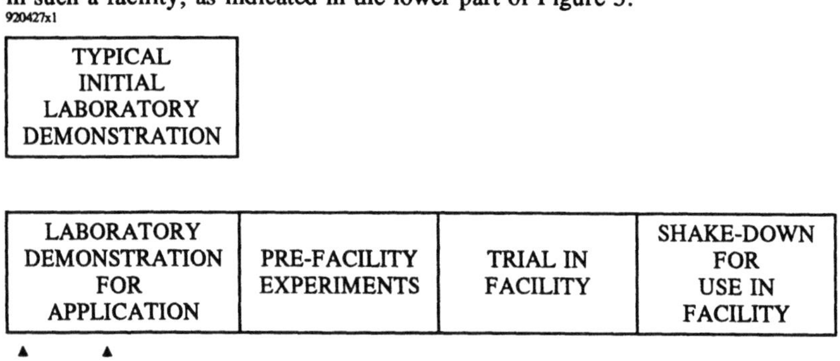

Figure 5. Development path for application in large ground-test facilities.

4.1. LABORATORY DEMONSTRATION FOR APPLICATION

Much can be borrowed from the typical, initial laboratory demonstration, but even where a fair amount of ultimate capability in the test cell has to be sacrificed in order to bring initial costs down to where a laboratory development/demonstration can be pursued, the laboratory demonstration for application must focus on the ultimate use of the technique in a test cell. At the conclusion of such a laboratory demonstration for application,

- a feasibility calculation for the facility use will have been done and found to be favorable;
- the molecular transitions and resulting wavelengths will be known to be appropriate for the facility geometry and conditions;
- any unknown parameters will have been measured (may require measurements in the facility or at other locations, e.g., to obtain laser-beam/particle interaction);
- the apparatus used may have been different from that expected to be used in the facility, but the performance in the facility can be calculated or scaled to a high degree of confidence based on the measurements made in the laboratory demonstration;
- the apparatus envisioned for facility use is, or can, by reasonable conception, be expected to be made appropriate for facility use.

4.2. PRE-FACILITY EXPERIMENTS

In the pre-facility experiments,

- equipment and software expected to be used in the facility are obtained and/or fabricated,
- these are checked out for facility use (in the laboratory, research cell, or piggy-backed onto other scheduled use of test cells) under all facility conditions (all aspects may not be checked simultaneously, but care must be taken that the end result is the same),
- and any modifications found to be necessary or prudent are made and the equipment and software are checked out again.

The two remaining steps of Figure 5 are straight forward.

The development projects which have been begun by AEDC are, for the most part, at the stage of development indicated by the first arrowhead at the bottom of Figure 5. Of the two projects at the second arrowhead, laser-induced fluorescence of oxygen for inlet density and Raman scattering at 248 nm for exhaust density, only the latter still appears potentially useful in large-scale groundtest facilities.

5. Technique Development

5.1. TECHNIQUES

The local, non-intrusive techniques, the development of which have been begun by AEDC for the test cell use cited here, are given in Table V below. Immediately after the technique the application considered here is given as well as the parameters to be obtained. This is followed by the organization which is doing or did the development work to date (with a point of contact in parentheses) and the sponsor (again with point of contact in parentheses). Finally there are remarks: general remarks, a comment on the stage of development, and a comment on current or planned future development activity.

Table V. TECHNIQUES FOR WHICH DEVELOPMENT HAS BEEN BEGUN BY AEDC

MEASUREMENT TECHNIQUE	**APPLICATION: PARAMETERS**	**Developing Org. (P.O.C.)** **Sponsoring Org. (P.O.C.)**
REMARKS	**STAGE OF DEVELOPMENT**	**ACTIVITY**
RAMAN SCATTERING AT 532 nm	Exhaust: Equivalence Ratio and Combustion Efficiency (and Density and potential for Static Temperature) and Inlet: Density	AEDC-Sverdrup (Kohl) AEDC (Hart)
Most important for advanced-propulsion engines. 532 nm for less fluorescence; visible laser beam in facility environment. Potential for 3,000 to 7,000 measurements in 30 sec over 1.5 m by 1.5 m flow.	Calculated photon-statistical noise and scanning times and implications. Initial receiver concept. A few receiver components obtained for testing.	Planned resumption, 1 Oct 96.
RAMAN SCATTERING AT 248 nm	Exhaust: Equivalence Ratio and Combustion Efficiency (and Density and potential for Static Temperature) and Inlet: Density	MetroLaser/Vanderbilt U. (Hess/Pitz) AEDC SBIR (Mitchell/Kohl)
Began as two-photon H_2O laser-induced fluorescence, but temperature sensitivity coupled with impossibility of sufficiently accurately determining temperature of H_2O, even using Raman cells, led to necessary (and welcomed) redirection. Actual contract is still for concentration of H_2O. Each receiver is a spectrometer.	About to do blind comparison in equilibrium flames and well-characterized mixture with known H_2O content. Upon successful completion will demonstrate in test-cell-sized setup without scanning.	Ongoing, completion of phase II approx. Jan 93.
RELIEF (Oxygen Marking and Interrogation)	Inlet: Vector Velocity	Princeton U. (Miles) AEDC-Sverdrup (Kohl)
From 10,000 measurements in 30 sec over a 1.5 m by 1.5 m flow with 1/4% to 1% accuracy using marking laser pulses of 100-psec pulsewidth to 300-3600 measurements at the same specifications depending on the degree of modification of a Raman cell and using marking laser pulses of 10-nsec pulsewidth.	For 100-psec marking laser pulse, feasibility calculations are completed. For 10-nsec marking laser pulse with no modifications to Raman cell, much of non-scanning laboratory demonstration completed. [Study completed yielding parameters required to determine feasibility in other applications including flight -- sponsored by AEDC and NASA Langley.]	Planned resumption, 1 Oct 93,* preceded by equipment purchase. —— * Other uses increase priority above Raman.

Continued...

Table V. TECHNIQUES (Continued)

LASER-INDUCED FLUORESCENCE (LIF) OF OXYGEN AT 193 nm	Inlet: Density	AEDC-Calspan (Williams) assisted by AEDC-Sverdrup (Kohl) AEDC (Hart) [and NASA Langley in the past]
Temperature sensitivity of transitions requires temperature compensation. When laser oscillator-amplifier was operated narrow-band without broad, low pedestal of frequencies also output, output energy was too low for non-signal receiver noise at these flow conditions. With broad frequency pedestal, laser frequency distribution varies too much, shot-to-shot, and no signals were sensitive enough to temperature compensate.	Blind trials done which project failure for this use at these test conditions. (Development continued by AEDC-Calspan for small-facility and single-point, time-averaged facility-flow calibrations.)	None planned for engine-test use except to monitor current activity for significant signal-to-noise increase.
LASER-INDUCED FLUORESCENCE (LIF) OF OXYGEN AT 203 nm	Inlet: Density	Deacon Research (O'Keefe) AEDC SBIR (Hart/Kohl)
Investigated for use in higher densities with greater temperature sensitivity. (Development continued by AEDC SBIR phase II for small-facility and single-point, time-averaged facility-flow calibrations.)	Laboratory experiments done that could be scaled to test-cell conditions. Too many complex laser assemblies would be required to get even borderline performance at the very best of flow conditions.	None planned for engine-test use.
USE OF NITROGEN: MULTI-PHOTON APPROACHES	Exhaust: Velocity and/or Temperature	U. of Virginia (Laufer) AEDC-Sverdrup (Kohl)
Only limited possibilities, but no other way foreseen in exhaust velocity. Fair possibilities in combusted exhaust. Initial approach (spectrally inflexible N_2^+ creation and interrogation) definitely not applicable in uncombusted (air) exhaust or in inlet.	Very early in development. More basic experiments required. Knowledge of multi-photon resonant(?) spectral excitation and subsequent behavior required.	Awaiting availability of funding for continuance.
ROTATIONAL TEMPERATURE OF OXYGEN BY EXCIMER-LASER LIF	Inlet: Temperature (and possibly, as a check, Exhaust: Temperature, where O_2 is present)	U. of Virginia (Laufer) AEDC-Sverdrup (Kohl)
Results from U. of Virginia are encouraging.	Needs to be evaluated by AEDC for potential test-cell use.	Awaiting resumption of engine-test development activity by AEDC, 93?-96?

5.2. COMMENTS

5.2.1. *Exhausts, Hydrogen-Air and Hydrocarbon-Air.* The exhausts have additional problems not associated with inlet flows. These problems will not be well characterized, so the measurement techniques developed will have to be conservative in the handling of these problems. Hydrogen-air exhaust-measurement techniques must deal with exhaust-gas luminosity and exhaust flow unsteadiness. The latter implies that a single measurement at each area element in the flow cross section may not be sufficient, and that multiple measurements at each area element may have to be made in a part of the cross section or over the entire cross section to obtain a valid mean measurement. Hydrocarbon-air exhaust-measurement techniques must deal with the above problems -- with the luminosity problem probably worse than in hydrogen-air exhausts -- and, in addition, with laser-induced particle incandescence, laser-induced particle fluorescence, much worse potential for laser-induced gas fluorescence, and potentially low gas transmittance if ultraviolet wavelengths are used. If SiO_2 particles may be present at times in the hydrogen-air exhaust, then any resulting laser-induced particle incandescence and fluorescence must be confirmed to be no problem with the measurement techniques to be used in the hydrogen-air exhaust as well.

5.2.2. *Technical Aspects.* There are two parameters which fall below the others in terms of the author's confidence in developing sufficient local non-intrusive measurement techniques in the ground testing of hydrogen-air engines. These are the vector velocity in the exhaust and the static temperature at the lower temperatures of the inlet. (The confidence in the latter may be increased with more investigation.) Fortunately, these are lower priority parameters as explained above, but the author would appreciate potential candidate methods being called to his attention. The author expects that local non-intrusive methods will first be used in combination with selected current state-of-the-art approaches (perhaps static pressures in the inlet and water-cooled Mach-number probes in the exhaust, if it is steady enough) despite the short-comings of the latter. Such a combination would be used to cover the incompleteness of what will be the initia suite of local non-intrusive measurement techniques.

5.2.3. *Programmatic Aspects.* There is a dearth of support (especially relative to the cost) fo even the initial, application-oriented, laboratory-demonstration stage in the development of loca non-intrusive measurement techniques for engine testing, particularly as opposed to support fo development of measurement techniques for combustion research and development. This mean that considerable cutting will have to be done on the ultimate capability of systems to b developed in order to minimize the initial cost of the laboratory demonstration stage and so t allow development to begin in the current financial climate. This reduced capability may eve have to exist as far along the development path as the trial in the facility (See Section 4 above) It will exist until those who can fund the required research and development see the obviou benefits of increased capability in these techniques and additional funds are therefore mad available to develop it.

LASER-BASED DIAGNOSTICS FOR HYPERSONIC FLOWS

R. K. Hanson
High Temperature Gasdynamics Laboratory
Stanford University
Stanford, CA 94305

ABSTRACT. New laser-based diagnostic techniques, developed primarily within the combustion community, offer considerable promise for measurements in hypersonic flows. In this paper we overview three diagnostic methods under development at Stanford University: spectrally resolved line-of-sight absorption (LOSA), conducted with a wavelength-modulated semiconductor diode laser and applied to O_2 in a shock tube flow; spectrally resolved single-point laser-induced fluorescence (SP LIF), conducted with a rapid-tuning ring dye laser and applied to NO in a supersonic underexpanded jet; and planar laser-induced fluorescence (PLIF), conducted with a tunable pulsed laser source and intensified CCD array camera, and applied to single-shot measurements of NO in a shock tunnel. These methods have unique capabilities for nonintrusive measurements of flowfield properties such as temperature, species concentration, velocity, density and pressure, as well as quantities derived from these properties such as mass flux (product of velocity and mass density).

1. Background

Over the past decade or so, several promising laser-based diagnostics have been proposed for nonintrusive measurements in gaseous flows. Interest in combustion, in particular, has driven the development of these advanced measurement techniques. Linear methods, such as absorption and laser-induced fluorescence, are especially attractive owing to their signal strength, species specificity, and relative simplicity of equipment and data interpretation. Equally important is the fact that variations of absorption and fluorescence have been identified for sensing several different flowfield parameters of interest including species concentrations, temperature, velocity, pressure and density.

Laser-induced fluorescence, originally developed as a single-point diagnostic but extended in 1982 to simultaneous multipoint measurements in a plane (i.e., PLIF), offers the important attributes of spatial and temporal resolution, assuming use of a pulsed laser source. The PLIF method has been rapidly accepted by the combustion community owing to its value for visualizing complex reacting flowfields. For example, PLIF images of OH produced in a diffusion flame serve to locate the instantaneous position of flame zones; and PLIF images of a tracer compound can be used to quantify the extent of mixing in a turbulent jet. During the past few years, as activity in hypersonic flow has been rekindled, there has been considerable interest in

A. Boutier (ed.), New Trends in Instrumentation for Hypersonic Research, 185–194.

developing PLIF as a diagnostic for aerodynamics and propulsion applications. The ability of the PLIF method to capture data for many flowfield points simultaneously is particularly relevant to applications in pulsed and short-duration flow facilities. It should be noted, however, that LIF and PLIF suffer one particular disadvantage, namely that fluorescence data are difficult to convert to absolute species concentrations owing to various uncertainties, especially in the collisional quenching rate. Much of the current research on PLIF is aimed at developing measurement strategies which avoid these difficulties. Fortunately, good progress has been made for some variables, for example to measure temperature through the ratio of two PLIF images, and to measure velocity through PLIF monitoring of Doppler-shifted absorption. By using signal differences and ratios, many of the problems of unknown quench rates are eliminated. Another limitation of LIF and PLIF, when executed in the usual manner with spectrally broad pulsed lasers, is that an individual measurement typically yields, at best, only one flowfield property. In most cases the fluorescence signal is a function of temperature, density and mixture composition, and possibly velocity. Signal interpretation can thus be complex, and may require a combination of measurements.

Recently, spectrally resolved methods, in both line-of-sight absorption and single-point LIF, have emerged as strategies which avoid some of the problems of PLIF and also enable simultaneous determination of multiple flowfield properties. Spectrally-resolved absorption methods using cw laser sources complement pulsed laser techniques in two important ways: (1) cw absorption can be highly quantitative, easily interpreted, and free of calibration uncertainties; and (2) measurements can be made at high repetition rates or even continuously. On the other hand, absorption is a line-of-sight method and thus provides poor spatial resolution, although extension of absorption to multiple paths, together with tomography, can relieve this latter deficiency somewhat. Another potential advantage of cw laser absorption worth noting is that the low power levels available are compatible with the use of optical fibers, and hence remote location of laser sources becomes feasible. Also, in the case of semiconductor diode lasers sources, laser absorption offers high promise for flight applications as well as ground testing, since these sources are compact, rugged, and economical, and they have low power requirements. We anticipate that packaged, stand-alone instruments based on these lasers and spectrally resolved absorption spectroscopy are likely to have substantial impact in basic and applied studies of hypersonic flows in the future.

Single-point fluorescence detection of spectrally resolved absorption offers a means of providing spatial resolution while retaining most of the advantages of the absorption method. The key limitation to this strategy, of course, is that the signal levels may be weak. The concept is therefore limited to cases with sufficiently high absorption and fluorescence yields.

In the following, we provide a brief overview of three current research efforts which represent the ideas discussed above. More detailed descriptions may be found in the recent publications cited in the Reference section of this paper.

2. Tunable Diode Laser Absorption Measurements of O_2

Tunable semiconductor diode lasers offer high potential for diagnostic applications owing particularly to their compact size and low cost, and to their capability for rapid wavelength modulation. In the past two years we have explored their use for LOS absorption measurements of O_2 and H_2O in a variety of low and high temperature environments including static cells, shock tubes and laboratory flames. Here we summarize our work with O_2 aimed at providing an air mass flux sensor for use in directed supersonic air flows, as found for example in supersonic inlets. More detailed descriptions of this work may be found in Refs 1-3.

The spectral region of interest is the A-band of O_2 near 760 nm in the near-IR. A calculated absorption spectrum of the R-branch of this band at room temperature is shown in Fig. 1; the conditions correspond to a path of 5.08 m of air at 1 atm. Transitions receiving particular attention in our work are shown in a boxed region centered at 760.1 nm, although in fact we've made measurements involving about half of the lines shown. Note that absorption exceeds 1%/meter for many of the lines at these conditions, even though this is a spin-forbidden system. Such absorption levels are well within the capabilities of absorption detection schemes based on various forms of wavelength or frequency modulation.

Our approach has been based on repetitively scanning the laser in wavelength over a spectral region of about 1 cm^{-1}, at a typical rate of 10 kHz, with a superposed modulation of wavelength at a higher frequency, f, of 10 MHz. Detection then is conducted at the second harmonic of f, i.e. 20 MHz, with a low-pass electronic filter set at 500 kHz; the effective temporal resolution of the 2f signal is thus less than 0.5 microseconds, but the time interval for a complete 2f spectral scan is 100 microseconds. A schematic of the current apparatus arrangement, used in shock tube experiments, is shown in Fig. 2. Note that the beam is split twice to enable simultaneous recording of three signals: wavelength change (on the etalon channel), and transmitted intensity for two orientations through the test section of the flow under study. A more detailed view of the optical arrangement is shown in Fig. 3.

The lasers are commercial devices manufactured by Mitsubishi (cost, approximately $30), housed in a diode laser mount (ILX Lightwave) which provides thermoelectric cooling. The temperature controller and current source are also commercial products of ILX. The power level of the laser is about 3 mW, its linewidth about 0.003 cm^{-1}, and its useful operating range is 760-762 nm. The laser is single mode above a threshold current of about 50 mA; the tuning rate is about 10 GHz/mA. The detectors are simple, fast-response photodiodes. The electronics are home-built, with economical commercial components, and based on homodyne detection ideas.

A typical data set obtained in the region behind an incident shock wave (with T = 941 K, P = 0.5 atm, v = 600 m/s) is shown in Fig. 4. The spectral region selected in this case includes a group of 5 transitions. Note the shift in the spectrum for the two observation angles (90 and 60 degrees to the shock tube axis). Absorption levels ranged from 0.2% to 0.9% in the tests conducted. This 2f spectrum is best-fit to infer temperature (mostly from the ratio of peak heights), pressure (from linewidth) and velocity (from the Doppler shift). Use of two angles through the same flow eliminate potential problems with impact shifts. From pressure and temperature, density may be computed; and from density and velocity the mass flux, which was especially sought, is calculated. A sample of measured and theoretical results for mass flux, over a range of shock conditions, is shown in Fig. 5. The measurement error is typically less than 10%.

3. Tunable Ring Dye Laser Fluorescence Measurements of NO

During the past two years we have sought to extend our spectrally resolved absorption measurements of NO to single-point fluorescence. The laser is a commercial ring dye laser, pumped by an argon ion laser, which has been modified in two critical ways: (1) to incorporate intracavity frequency doubling with a BBO crystal, thereby allowing access to the 215-230 nm range; and (2) to allow rapid repetitive tuning (4 kHz) over 1 cm^{-1} intervals using a dual-rhomb tuner developed in our laboratory. The experimental arrangement now in use is shown in Fig. 6. As indicated, the focussed laser beam (0.2 mm dia, about 1 mW of power) crosses the flow at an angle, which can be varied, allowing sensitivity of the spectral scan to Doppler shifts. The LIF signal is collected with a low f-number lens at right angles to the flow. A representative single-sweep data trace is given in Fig. 7, shown here in frequency units, for observation of the $Q_2 + R_{12}(8)$, $R_2(4)$ line pair in the (0,0) band of the A←X system of NO at 226.69 nm. The data were obtained in a free jet (operating with a 5% NO in N_2 mixture) at an axial location x/D = 0.75 (the nozzle exit diam = 3.1 mm) where the computed flow properties (see Fig. 7) are about 160 K, 0.123 atm, V = 521 m/s. The data are analyzed by finding the best-fitting Voigt profiles to each line, and then extracting the temperature from the peak ratio of the two lines, followed by inferring the pressure from the Voigt a parameter of each line. The velocity is of course found directly from the Doppler shift of the line pair. Provision is made for the impact shift using known (measured in our laboratory) impact parameters. Example single-sweep LIF results for temperature along the jet axis are shown in Fig. 8. The agreement with the calculated temperature distribution is quite good except at x/D greater than 2 where rotational freezing appears to set in. Further details of this work may be found in the recent papers by DiRosa and Chang, et al., of our laboratory (e.g., see Refs. 4,5).

It is interesting that such quantitative results can be obtained with only 1 mW of laser power and that the measurements can be executed at a repetition rate of 4 kHz. Although successful scaling of this approach to a larger scale facility would require increased power levels, we believe these results confirm the potential of spectrally-based LIF measurements.

4. Planar Laser-Induced Fluorescence Imaging of NO in a Shock Tunnel

We turn now to our third diagnostics strategy, planar laser-induced fluorescence, known as PLIF, which offers the ability to simultaneously monitor the flowfield at a large number of measurement locations. The history of PLIF is well known (see, for example, Refs. 6,7) and need not be repeated here. Instead we will focus on one of our most recent activities which is aimed at monitoring velocity and temperature in a high-enthalpy, supersonic free jet produced in a shock tunnel.

The experimental facility with its associated optical and electronic components is shown in Fig. 9. For NO imaging we employ an excimer-pumped dye laser (Coumarin 450 dye), frequency-doubled to provide about 1 mJ of energy at the wavelengths of transitions in the (0,0) band of the A←X system of NO. The camera employed in these measurements is a home-built, intensified (single-plate MCP) CCD (Amperex, 512x240 pixels) with variable gatewidth typically set at 1 microsecond. A Cassegrainian reflecting lens (f/1.2, 95 mm focal length) was used to image the fluorescence from a typical flowfield region of 36.0 x 26.5 mm located on the jet centerline. The shock tube itself is 6 m in length, with a square interior cross section 7.6 cm on a side, and is pressure (helium) driven. A smooth, converging axisymmetric nozzle (located in the center of the shock tube end wall) was used with a throat diameter D = 5 mm. The jet flows into a dump tank of 0.25 m^3 volume held at a controlled ambient pressure of typically 5-10 torr. A range of mixtures containing NO diluted in argon, nitrogen or oxygen was studied in order to allow study of various degrees of vibrational nonequilibrium in the jets. The stagnation temperature and pressure were also varied, up to maxima of about 4500K and 5 atm, to provide a wide range of flow conditions in the core of the jet upstream of the Mach disc which forms at large pressure ratios.

Several pairs of NO lines were investigated as candidate choice for two-line thermometry based on fluorescence intensity ratios. The optimum line pair depended on the temperature regime of greatest interest (see Ref. 8 for details). In all cases the fluorescence signal was collected broadband and normal to the laser sheet; a 2 mm thick UG-5 Schott glass filter was used to block elastically scattered light. For temperature imaging, the laser sheet was propagated radially through the flow to minimize effects of the Doppler shift, while for velocity imaging the direction of laser illumination was varied over a range of angles. For temperature, the relatively broadband laser (0.4 cm^{-1}) was centered on the absorption line center frequency, while for velocity the laser frequency was detuned to provide a controlled offset in frequency. The center frequency and spectral distribution of the laser were monitored using a 2 cm^{-1} etalon (FSR = 20) and a 512-element linear array just prior to each experiment.

Representative image data sets for temperature and radial velocity are shown in Figs. 10 and 11 for the case of 2% NO in Ar, T_{stag} = 4200 K, P_{stag} = 3.0 atm. Note that Fig. 10 includes results based on the ratio of two single laser shots, a 5-frame average of these ratios, and also a method of characteristics calculation for this flow, all as functions of r/D and x/D. The line pair selected ($Q_2(7)$ and $Q_1(5)$) provides a reasonable compromise in temperature sensitivity over the range of temperatures monitored. In Fig. 11, the radial velocity has been extracted from PLIF data using two different algorithms: a linear algorithm useful for small velocity shifts; and a more complex nonlinear algorithm which can deal with larger shifts. The agreement found, for both temperature and velocity, is quite good. Details of this work may be found in the recent AIAA paper by Palmer et al. (Ref. 8).

5. Conclusions

Each of the three laser techniques described has advantages and disadvantages. These methods should be viewed as complementary with each other and with other measurement schemes under development in various laboratories. What has become clear in the past few years, however, is that laser-based diagnostics have reached a level of maturity such that they can be employed with confidence to obtain increasingly accurate results in both applied and fundamental research programs. Furthermore, these methods can access flowfield parameters which are, in some cases, impossible to determine with the methods utilized in past decades.

6. Acknowledgements

The author gratefully acknowledges the students and staff of the High Temperature Gasdynamics Laboratory who have contributed to the successful development of these diagnostic methods, including particularly Dr. Jerry Seitzman, Dr. Louis Philippe, Mr. Mike DiRosa, Ms. Jennifer Palmer and Mr. Brian McMillin. The work has been sponsored primarily by the U.S. Air Force Office of Scientific Research and, for the diode laser activity, by NASA.

7. References

1. L. C. Philippe and R. K. Hanson, "Laser-Absorption Mass Flux Sensor for High-Speed Airflows," Optics Lett. 24, 2002-2004 (1991).

2. L. C. Philippe and R. K. Hanson, "Laser Diode Wavelength-Modulation Spectroscopy for Simultaneous Measurement of Temperature, Pressure, and Velocity in Shock-Heated Oxygen Flows," Applied Optics, submitted December, 1991.

3. M. P. Arroyo and R. K. Hanson, "Tunable Diode Laser Absorption Technique for Detection of Water Vapor in Aerodynamic Flows," AIAA 92-0510, paper presented at 30th Aerospace Sciences Meeting in Reno, Jan. 1992.

4. M. D. DiRosa, A. Y. Chang and R. K. Hanson, "CW Dye Laer Technique for Simultaneous, Spatially-Resolved Measurements of Temperature, Pressure, and Velocity of NO in an Underexpanded Free Jet," AIAA 92-0006, paper presented at 30th Aerospace Sciences Meeting, Reno, Jan. 1992; also Applied Optics, submitted March 1992.

5. A. Y. Chang, M. D. DiRosa, D. F. Davidson, and R. K. Hanson, "Rapid Tuning CW Laser Technique for Measurements of Gas Velocity, Temperature, Pressure, Density and Mass Flux using NO," Applied Optics 30, 3011-3022 (1991).

6. R. K. Hanson, "Combustion Diagnostics: Planar Flowfield Imaging," Twenty-First Symposium (International) on Combustion, The Combustion Institute, 1677-1691 (1986).

7. R. K. Hanson, J. M. Seitzman and P. H. Paul, "Planar Laser-Fluorescence Imaging of Combustion Gases," Appl. Phys. B 50, 441-454 (1990).

8. J. L. Palmer, B. K. McMillin, and R. K. Hanson, "Planar Laser-Induced Fluorescence Imaging of Velocity and Temperature in Shock Tunnel Free Jet Flow," AIAA 92-0762, presented at 30th Aerospace Sciences Meeting, Reno, Jan. 1992.

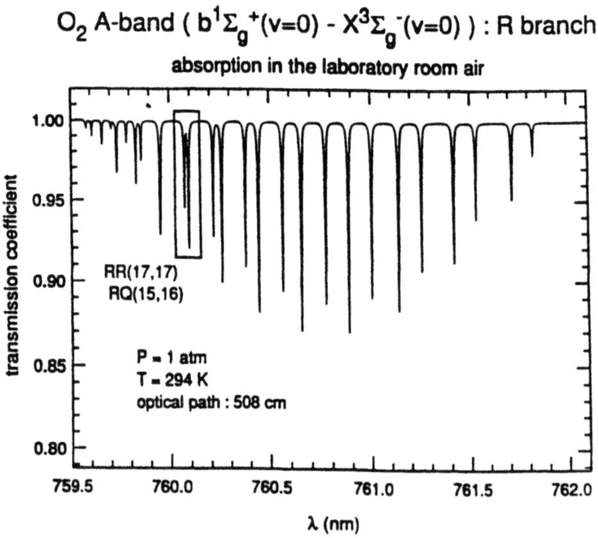

Fig. 1. Calculated absorption spectrum of O_2 A-band R-branch at 294 K.

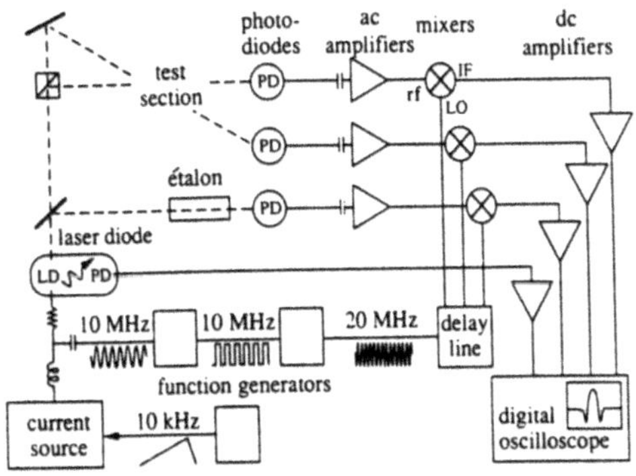

Fig. 2. Experimental apparatus setup. LD, laser diode; PD's, photodiodes.

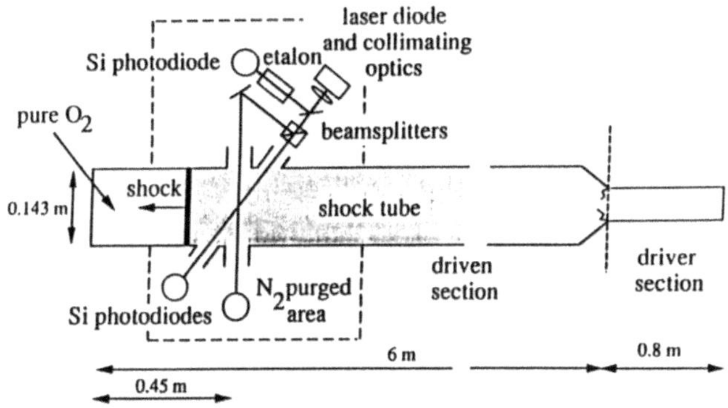

Fig. 3. Optical arrangement for mass flux measurements in the shock tube.

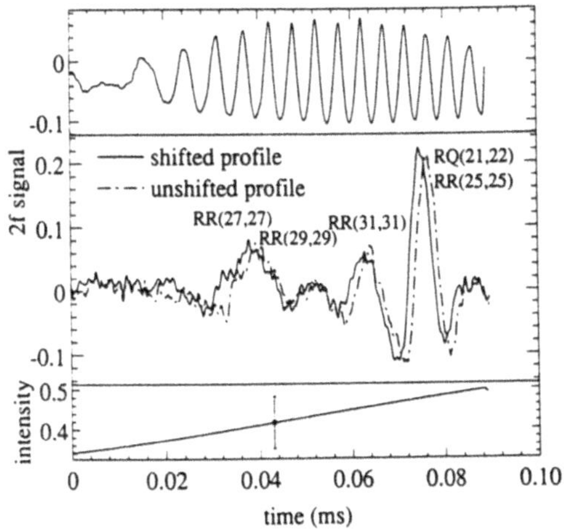

Fig. 4. 2f data for incident shock region. Top trace, étalon 2f signal; middle traces, 2f absorption profiles in the shock tube along two directions with a 30° inclined angle; bottom trace, laser emission intensity. The vertical units are arbitrary. From Ref. 1.

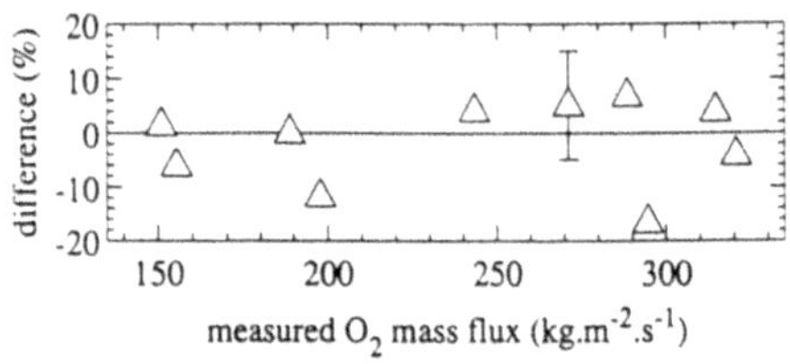

Fig. 5. Difference (%) between the measured and actual oxygen mass flux for various shock conditions. From Ref. 1.

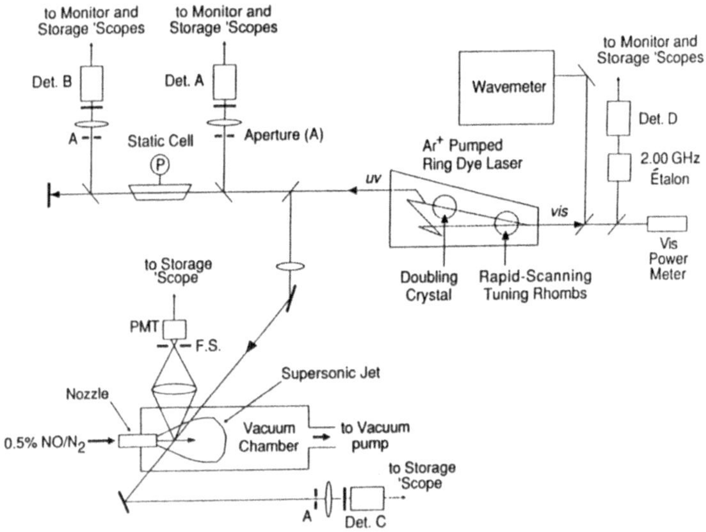

Fig. 6. Experimental arrangement for cw single-point LIF in a supersonic jet.

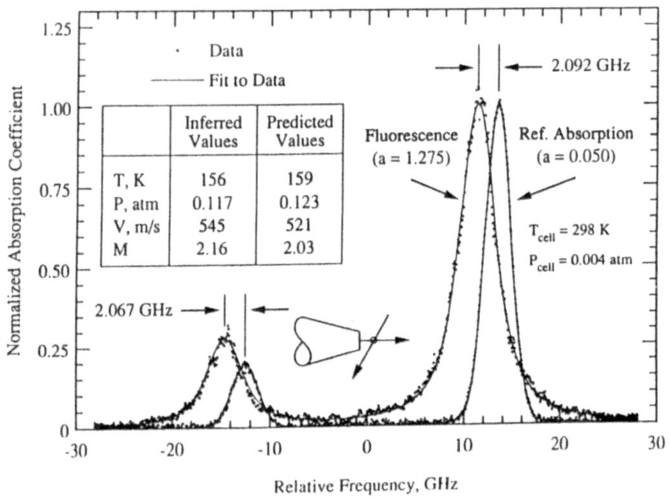

Fig. 7. Single-sweep SP LIF data at X/D = 0.75 on jet axis.

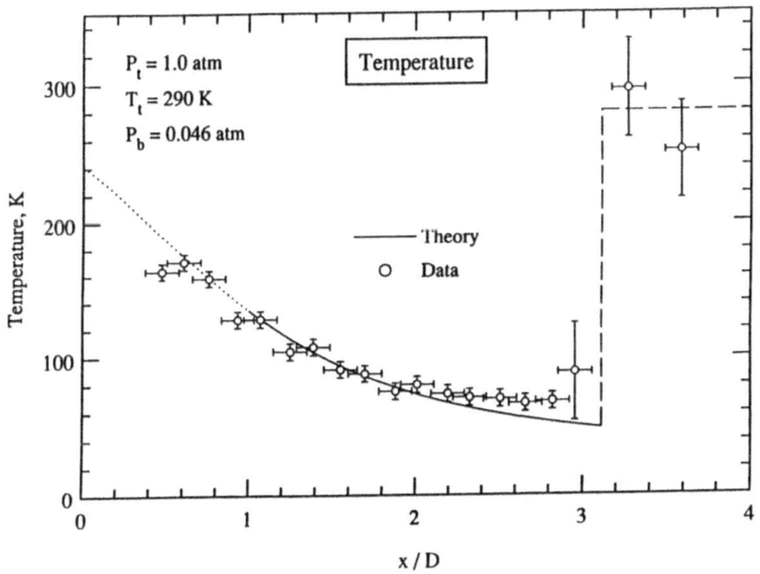

Fig. 8. Rotational temperature on jet centerline by SP LIF.

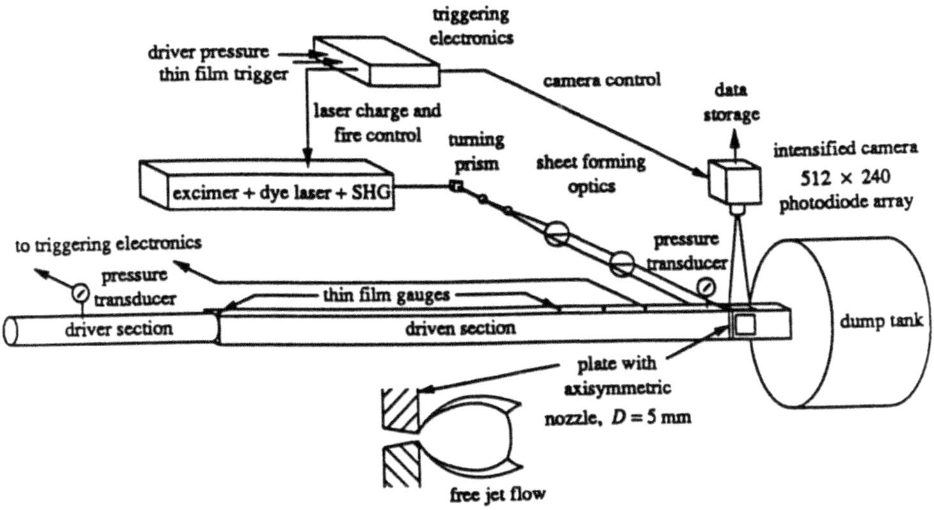

Fig. 9. Schematic of experimental facility with associated optical and electronic components for PLIF imaging.

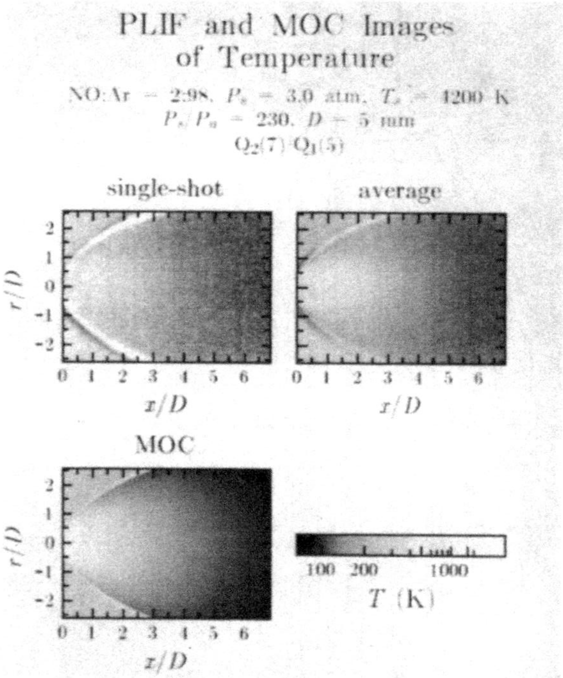

Fig. 10. PLIF and MOC results for temperature in a shock tunnel.

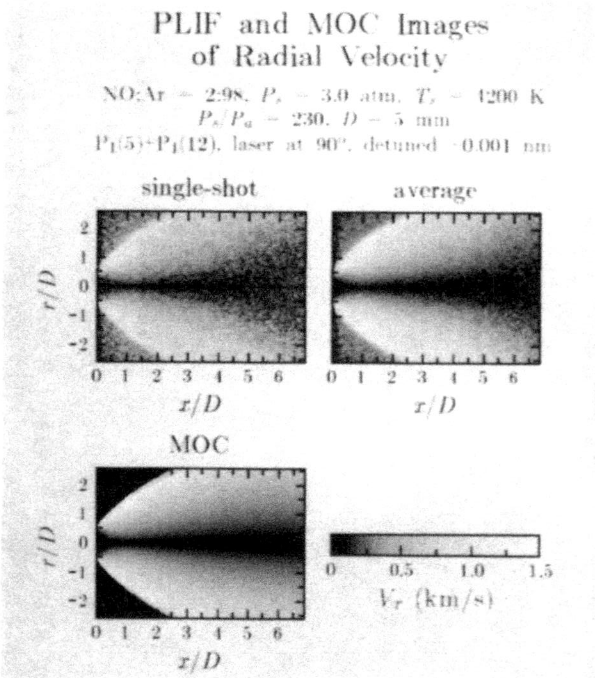

Fig. 11. PLIF and MOC results for radia velocity in a shock tunnel.

DETERMINATION OF MINORITY SPECIES CONCENTRATIONS AND TEMPERATURE IN COMBUSTION SYSTEMS BY LASER-INDUCED FLUORESCENCE

U. MEIER, I. PLATH and K. KOHSE-HÖINGHAUS
Institute of Physical Chemistry of Combustion
DLR - German Aerospace Research Establishment
Pfaffenwaldring 38-40
D-7000 Stuttgart 80
FR Germany

ABSTRACT
Laser-induced fluorescence (LIF) methods were developed for *quantitative, point-wise* measurement of minority species concentrations and temperature in laboratory-scale combustion systems. Currently, the potential of similar measurement strategies for *accurate two-dimensional, single-pulse* experiments is being investigated. In the light of some typical examples, ideas and recommendations will be outlined.

1. Introduction

A need for quantitative, correlated, time-resolved mapping of scalar and/or vector properties of - in many cases reactive - flows is common to a wide number of investigations in combustion, aerodynamics, plasma science, environmental analysis or chemical engineering. In studies of practical flames, real-time quantitative mapping of the flow field, the temperature field and the concentration distributions of some of the most important chemical constituents would be desirable. To date, a single diagnostic technique is not capable of fulfilling these requirements. LIF, however, has been demonstrated to be one of the most versatile laser-spectroscopic tools in this respect (see e.g. Refs. 1-7). Two-dimensional distributions of velocity components, pressure, temperature or concentrations of radicals as OH and CH or molecules as NO, O_2 and CO have been measured, although this could not always be done in a quantitative fashion. Some experimental approaches allow for simultaneous detection of several species, some for simultaneous measurement of different parameters; those strategies have, however, almostly exclusively been applied to point-wise measurements. In summary, part of the effort in the further development of quantitative two-dimensional LIF imaging for reactive flows should be devoted to the investigation of generally applicable strategies for accurate single-pulse concentration as well as temperature field measurements and, most importantly, to correlated multi-species or multi-parameter studies.

In this work, we outline our own approach to this task by discussing some examples for the investigation of laboratory-scale flames, directed towards both fundamental and applied issues of laser-based combustion diagnostics. In addition,

A. Boutier (ed.), New Trends in Instrumentation for Hypersonic Research, 195–204.

the potential of LIF is compared to that of other suitable techniques for some of these experiments. In the following, results of several categories of experiments shall be reported, in particular of point-wise and planar LIF applications. Furthermore, some results obtained with multi-photon techniques (multi-photon excited LIF (MPLIF), and resonance-enhanced multi-photon ionization (REMPI)) and with a coherent technique (degenerate four-wave mixing (DFWM)) shall be briefly referred to.

2. Experiment

Due to the limited space available, only a short overview over the experimental arrangements can be given, details will be reported elsewhere.

Two laser systems were available for these studies, a Nd:YAG-pumped dye laser (System 1: Spectra Physics DCR2 and PDL2 equipped with a WEX frequency doubling system and a Raman cell) and an excimer-pumped dye laser (System 2: Lambda Physik EMG150MSC and Lumonics HD500). System 1 was used for concentration measurements of a variety of radical and atomic species (OH, CH, H, O, N, CH_3) by LIF, MPLIF and REMPI, for the measurement of temperature and for the investigation of collision processes. System 2 was mainly employed in NO LIF and DFWM experiments.

For *point-wise LIF* measurements, the fluorescence was spectrally isolated using either a monochromator or interference filters. It was detected with a photomultiplier and received by either a gated integrator (Stanford Research Systems SR 250) or a transient digitizer (Tektronix 7912AD). The laser intensity was monitored by photodiodes, in some cases before and after the object under study to verify optically thin conditions. Typically, the signals from 10-100 laser pulses (and even more in the studies of collision processes) were averaged. *MPLIF* and *REMPI* measurements were performed in a quite similar fashion, the only differences being the focussing of the laser beam (typical beam waists of 0.1-0.3 mm) to ensure appropriate multi-photon excitation rates and the use of an ionization probe for the detection of charged species in the REMPI experiments.

In the *DFWM* experiments, the dye laser radiation was split into a pump and a probe beam of about the same intensity which were combined under a small angle (typically 3-5°) in the measurement volume. The pump beam was retroreflected in itself to form the third beam. The polarization of this backward pump beam was rotated by 90° with respect to the other incoming beams in order to prevent scattered light from surfaces from reaching the detector. Fluorescence was simultaneously recorded from approximately the same observation volume.

For the *two-dimensional LIF* experiments, the laser beams were shaped into sheets of appropriate dimensions (typically 0.2x20 mm) using cylindrical lenses or telescopes. The fluorescence light was detected with intensified CCD cameras (DICAM2, PCO) equipped with UV Nikkor lenses (f=105 mm, f# 4.5). Typically, gates of 100 ns duration were used. The spatial variation of the laser intensity in the sheets was monitored either independently, or - if necessary - for the same laser pulse using a third camera (Cohu 4710) which recorded the attenuated laser beam directly.

Several objects were studied depending on the purpose of the experiments. *Flat flames* of 0.01-10 bar were used in the quantitative point-wise LIF measurements of concentration and temperature. Typically, stoichiometric hydrogen and methane flames burning with oxygen or air were studied; if necessary, the equivalence ratio could be varied.

A *discharge flow reactor* served as a calibration system for the concentration measurements by multiphoton techniques. Also, it was used in the investigation of collision processes. The flow reactor was typically operated at room temperature and pressures of 1-10 mbar and provided known concentrations of radicals as OH and CH_3 or atoms as H, O and N using microwave discharges in mixtures of He or Ar with H_2 or N_2 and fast titration reactions. H and N atoms were thus produced directly,

O by reaction of N with NO, OH by reaction of H with NO_2 and CH_3 by reaction of H with CH_3I.

A *static cell* at room temperature and at pressures of 0.001-8 bar was used in the comparison of LIF and DFWM experiments with NO. It was equipped with AR-coated windows to reduce scattered light interferences in the DFWM measurements.

In the two-dimensional imaging experiments, a *turbulent hydrogen diffusion flame* was investigated for which temperature and concentration histograms of N_2, H_2, O_2 and H_2O had already been measured in single-pulse spontaneous Raman scattering experiments[8].

3. Results

3.1. MEASUREMENT OF TEMPERATURE WITH LIF

Temperature is one of the fundamental quantities for the characterization of combustion processes and of many other systems. The accurate determination of temperature and its spatial and temporal variations in a given situation is important at least in two respects: It is necessary for the comparison of experimental results with model predictions, and it is furthermore a prerequisite for quantitative concentration measurements by laser-induced fluorescence. LIF techniques are - in general - sensitive to the (temperature-dependent) population of specific quantum states and the (temperature-dependent) variations of density and chemical composition. If a temperature is not defined, as in systems with non-equilibrium rotational and vibrational quantum state distributions, knowledge of the population in specific quantum states may be required to characterize the system. The procedure for the measurement of the non-equilibrium distributions may be quite similar to that of temperature, except that information on more quantum states will be needed.

CARS is one of the most established methods for the determination of temperature in combustion systems; it is, however, in general restricted to point-wise applications. LIF is in principle capable of producing instantaneous temperature maps for frozen flow conditions. Although atomic or molecular seed materials can be added to the flow as temperature indicators, this procedure becomes questionable in high-temperature reactive flows. In combustion experiments, most attention has therefore been devoted to 'inherent' temperature indicators as OH and NO.

3.1.1. *Point-wise temperature measurements*

Using LIF in the OH (A-X) system, we have recently examined several strategies for accurate point-wise temperature measurements[9]. Linear and partially saturated excitation were investigated. A scheme with excitation of suitable isolated lines in the (1,0) band and detection of the entire (1,1 and 0,0) band fluorescence proved reliable in a wide pressure (0.001-1 bar) and temperature (300-2500 K) regime. CARS and Raman temperatures with N_2 as indicator species - measured under exactly the same conditions - yielded temperatures in excellent agreement with the LIF results. **Figure 1** shows as an example the temperature profile in a stoichiometric $H_2/O_2/N_2$ flame at 95 mbar. From **Fig.1** it is also evident that partially saturated LIF could be used further down to the burner surface at temperatures substantially lower than those measured with linear LIF. This is due to the higher detection sensitivity of saturated LIF which can exploit even the low OH concentrations formed in the initial temperature rise.

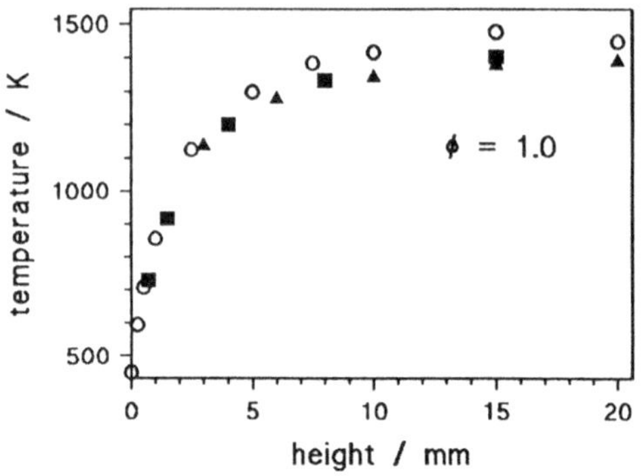

Fig.1: Temperature vs. height above the burner surface for a stoichiometric $H_2/O_2/N_2$ flame at 95 mbar.

triangles: linear excitation,
squares: non-linear excitation,
circles: CARS measurements.

Typical errors are smaller than the size of the symbols.

In an independent experiment[10], the linear LIF approach described above was extended to flames at 1-10 bar; again, CARS measurements were in good agreement with the LIF results. Simulations of the OH spectrum using measured collisional broadening coefficients showed, however, that at pressures above 10 bar, lines in the (weaker) S branch might have to be chosen because of the massive spectral overlap in other regions of the spectrum.

3.1.2. *Strategies for instantaneous two-dimensional LIF temperature measurements*

For application in combustion systems where the mole fraction of a temperature indicator is not necessarily constant, a two-line (rather than a single-line) temperature measurement seems to be the most versatile approach. Excitation is then performed out of two different quantum states of the lower electronic level, broadband fluorescence in the same wavelength region is detected (which requires a suitable temporal delay). Although not every detail of different strategies can be discussed here, several criteria for line selection for two-line temperature imaging experiments shall be given. It is obvious that isolated lines with a highly temperature-sensitive population must be chosen. To ensure good signal-to-noise ratios, the population and the transition moments should be high. Furthermore, quantum yields of the two upper levels should not be very sensitive to collisions or they should at least exhibit approximately the **same** sensitivity to collision processes. In some cases, influences of collisions on the temperature measurement can be avoided by choosing transitions to the same upper level, although this method has some limitations in terms of the available energy spacing and does not necessarily offer the highest temperature sensitivity for a given experiment.

For a typical measurement situation, as e.g. encountered in a turbulent flame, temperature and composition vary unpredictably in the observation plane. Even if this causes large spatial variations in the collisional quenching rate, a temperature measurement using two rotational transitions will not be affected as long as the collisional decay rate is not level-dependent. Consider the sensitivity of a temperature measurement to the variation in the quenching rates Q_1 and Q_2 for the levels 1 and 2:

$$dT = \frac{\partial T}{\partial Q_1} dQ_1 + \frac{\partial T}{\partial Q_2} dQ_2 \qquad [1]$$

With the ratio of the number densities in both levels $N_1/N_2 = g_1/g_2 \exp[-(E_1-E_2)/kT]$ and the fluorescence intensity for level i (i=1,2) given by $I_{flu,i} = C' N_i A_i/(A_i+Q_i)$, in the limit $Q_i \gg A_i$, the following expression for the temperature error is obtained

$$\frac{dT}{T} = \frac{kT}{E_1-E_2}\left(\frac{dQ_1}{Q_1} - \frac{dQ_2}{Q_2}\right) \qquad [2].$$

The error is thus minimized by choosing a large energy difference E_1-E_2 while keeping the difference in rotational quantum number as low as possible; the error is not sensitive to the **absolute** value of the quenching rate but to the **difference of their variations**, i.e. only to changes in the collisional environment that affect the two levels differently. Fortunately, there are some indications that dependencies of the quenching rate coefficients on rotational level in the (A) states of both OH and NO may decrease with increasing temperature, although only limited information is available[11,12]. The variation of the quenching rate with temperature and chemical composition has been modeled for the detection of OH and NO (both by their A-X transitions) in hydrogen and methane flames; details will be given in 3.2.2.

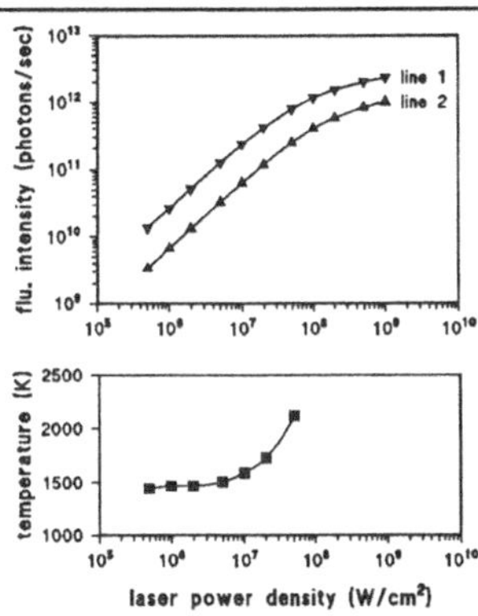

Fig.2: Calculated saturation curves for a two-line temperature measurement

top: log-log diagram of fluorescence intensity vs. laser power density for 2 transitions with a difference of a factor of 2 in Einstein coefficients; lower state populations differ also by a factor of 2

bottom: resulting temperature, applying the formalism for linear excitation

As already mentioned above, temperature measurements - at least for OH as indicator species - might profit from the enhanced signal-to-noise ratio achievable by partial saturation of the transitions. For an instantaneous temperature field measurement, this approach is, however, less suitable than for point-wise applications, as the degree of saturation might depend in a complex fashion on the local intensities and collision rates. It shall be pointed out in addition, that rather large systematic errors in the measured temperature can be caused by - erroneously - assuming an entirely linear dependence of the fluorescence intensity on laser power density. This is demonstrated in **Fig.2.** The upper panel shows in a log-log plot the calculated dependence of fluorescence intensity on laser power density for two transitions. Simulating a typical two-line temperature experiment, the populations are assumed to be different by a factor of 2, as are the Einstein coefficients. Already for very small deviations from linear excitation, the error in the temperature can be severe, as seen in the lower panel of **Fig.2.** It is therefore recommended to check very carefully for any residual saturation effects before performing the actual experiment. Of course, partially saturated measurements can be evaluated if the suitable formalism is applied (see also measurements in **Fig.1**).

Currently we are pursuing two-dimensional temperature measurements using the OH radical. Following our approach for the point-wise measurements, two transitions in the (A-X, 1,0) system are excited using both laser systems with a delay on the order of 100 ns; detection is performed broadband. For NO as indicator species, two different strategies will be applied. For high-temperature combustion systems with a substantial fraction of natural NO (~1000 ppm), the (A-X) system seems preferable, as the current data base on collision processes is much less incomplete than for the (D-X) system. The latter will be chosen in situations where the higher achievable sensitivity is necessary.

3.2. CONCENTRATION MEASUREMENTS

Similar as for the determination of temperature, several principal LIF approaches are being discussed for the measurement of absolute concentrations of important intermediates. Saturated or predissociative techniques offer elegant solutions for specific situations but are not generally applicable. The quantitative interpretation of linear LIF measurements in most combustion situations at atmospheric pressure and above is vitally dependent on the availability of information on collision rates. Also, a concentration measurement by LIF is sensitive to errors in the measurement of temperature. These two aspects shall be considered in more detail below.

3.2.1. *Point-wise concentration measurements*

3.2.1.1 OH detection with LIF

Two questions have been addressed with the quantitative detection of OH in our laboratory. Primarily premixed, flat low-pressure flames have been investigated for the examination of detailed chemical-kinetic flame models[13]. In addition, a study of flat flames at 1-10 bar has been performed to demonstrate the potential of linear LIF for quantitative measurements at high pressures in comparison with that of a non-linear technique, electronically resonant CARS[10].

3.2.1.2. NO detection with LIF and DFWM

Since DFWM as a non-linear technique might be superior to LIF in situations with high background luminosity, we have recently started a comparative study of both techniques, detecting NO in its (A-X) transition. Experiments have been performed in the pressure region of 0.001-8 bar in a cell, where different collision partners could be added, and in a welding torch producing a hot flame with a high NO mole fraction. **Figure 3** shows the (A-X,0-0) O branch for nearly unsaturated conditions in a most unfavorable situation for LIF: in a cell experiment, 200 ppm NO were detected in a CO_2 atmosphere where the NO fluorescence is rapidly quenched[12]. Whereas the DFWM signal intensity is larger than that of LIF, the detection sensitivity seems similar. In addition, the DFWM signal becomes increasingly noisy with rising pressure in spite of the increasing intensity. The potential for quantitative measurements with both techniques at high pressures is being examined.

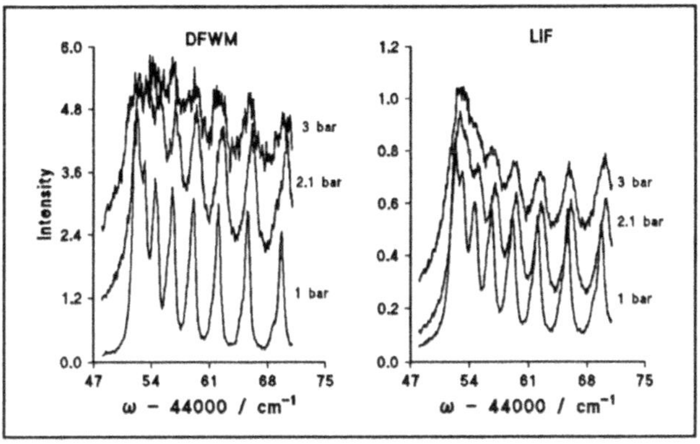

Fig.3: NO (A-X, 0-0) O branch spectra taken by DFWM (left) and LIF (right) in a mixture of 200 ppm NO in CO_2 as a function of pressure.

The intensities are arbitrarily scaled to a common maximum and offset by different amounts to make them distinguishable.

3.2.1.3. Multi-photon excitation

By multi-photon techniques, primarily some of the atomic species in flames are detectable that may also be of importance in aerodynamic sytems. We used MPLIF for the determination of quantitative concentrations of H[14], O[15] and N[16] atoms in low-pressure flames to investigate specific aspects of chemical-kinetic flame modeling. REMPI was employed for the detection of non-fluorescent species as CH_3 [17].

3.2.2. *Strategies for instantaneous two-dimensional concentration measurements*

For a typical experimental situation, as e.g. encountered in a turbulent flame, variations in temperature, density and collisional environment will be of importance for a quantitative interpretation of the fluorescence signal. Errors in the independently measured temperature field will directly influence the concentration measurement.

The sensitivity of a concentration measurement of OH or NO to errors in the temperature determination can be expressed by

$$\frac{dN_0}{N_0} = \frac{dT}{T}\left(\frac{BJ(J+1)}{kT} - 2\right) \qquad [3],$$

where B is the rotational constant and J the quantum number of total angular momentum. Assume a temperature uncertainty of 100 K for the determination of the NO concentration. Exciting J=2.5 at 2000 K leads to an error of -9.9% in the NO measurement (1000 K: -19.7%), whereas excitation of J=30.5 reduces this error to 4.4% (1000 K: -3.9%).

To get a realistic idea about the variation of the total quenching rate under typical conditions, we have modeled laminar, premixed flat H_2/air and CH_4/air flames at atmospheric pressure and at different stoichiometries. For a given temperature profile, these calculations yield the chemical composition for increasing reaction time (or height above the burner surface). Together with published temperature-dependent quenching coefficients for OH[18] and NO[12], species-specific contributions to the total quenching rate are obtained. The example in **Fig.4** which was calculated for NO (A-X) excitation in a stoichiometric H_2/air flame shows a 25% variation in the total quenching rate with temperature.

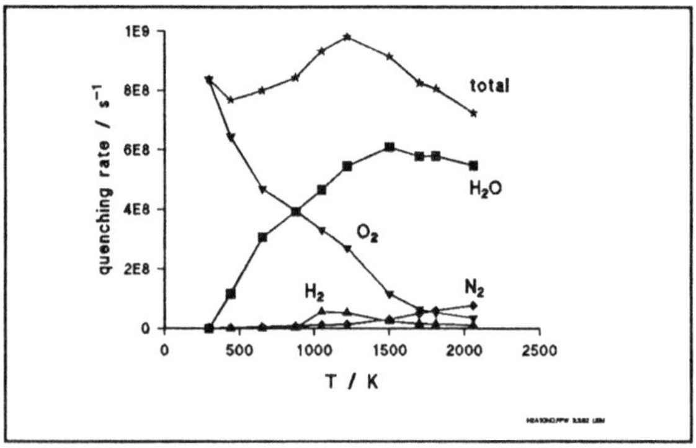

Fig.4: Total quenching rate in the reaction zone of a stoichiometric H_2/air flame as a function of temperature.

The chemical composition was calculated using the code of Warnatz[19]; the temperature-dependent, quenching coefficients were taken from Ref.12.

Compensating effects in the density, composition and temperature dependence of the individual quenching coefficients lead to this reassuring result. Changing the stoichiometry has an even smaller effect on the total quenching rate. The pattern is similar for NO detection in CH_4/air flames; it is less favorable (changes of about a factor of 2 in total quenching rate) for OH detection in both H_2/air and CH_4/air flames. This result is not too unfortunate, however, as OH - in contrast to NO - can alternatively be detected using LIPF. Again, as for the measurement of temperatures, modeling of all pertinent processes can be recommended to determine the best-suited experimental approach.

As an example from our laboratory, **Fig.5** shows the **correlated** two-dimensional fluorescence images of OH and NO in a turbulent H_2 diffusion flame where the cold H_2 flow was doped with 2800 ppm NO. The OH (1,0) $Q_2$7 line was excited; for NO, several overlapping transitions (with low J) in the (0,0) P_1 head were excited. From previous investigations with single-pulse Raman scattering[8], it seems likely that turbulent mixing of cold H_2 and hot combustion products occurs in the inner zone of the flame. Whereas from the OH measurement alone no information is obtained about the inner zone, the NO distribution is in accord with this interpretation of the Raman data.

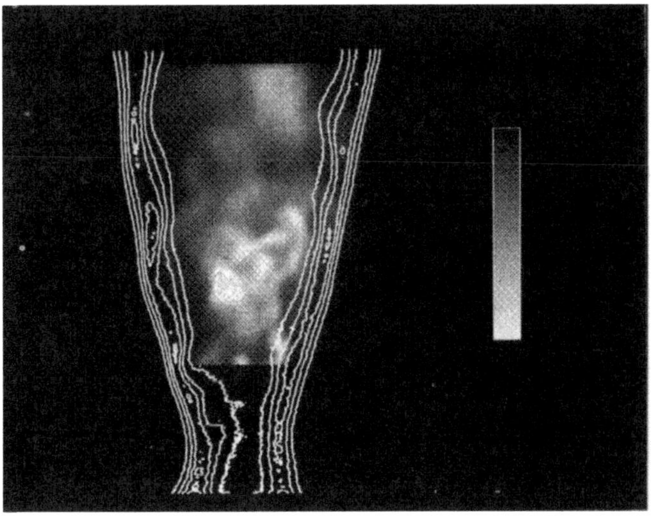

Fig.5: Simultaneously measured distributions of the fluorescence intensities of OH (contour plot) and (seeded) NO (grey scale) in a turbulent H_2/air diffusion flame.

dimension: height of OH sheet: 30 mm

contours: 6 equally spaced levels; dynamic range: approximately 50

The OH image represents the relative concentration to about 30%, taking into account the temperature sensitivity of the chosen line and the variation of the collision rate. An absolute calibration can be obtained with a flat flame of known OH (equilibrium) concentration. The NO excitation scheme is more sensitive to temperature. An instantaneous temperature field measurement, in this case obviously with NO as indicator, will be performed to address remaining questions.

4. Conclusions

Strategies for quantitative 2D LIF measurements can to a certain extent profit from ideas developed for point-wise measurements. Linear LIF as one of the most generally applicable techniques does not seem to suffer severely from variations in the collision rates even for instantaneous measurement of temperature and concentration in turbulent flames. In order to exploit the quantitative potential of two-dimensional LIF techniques for the investigation of the structure of turbulent flames, more effort should be devoted to correlated measurements of species and temperature distributions.

ACKNOWLEDGEMENTS

The authors are grateful to their colleagues in the laser diagnostics group for helpful suggestions and discussions. The cooperation of Dr.B.Attal-Trétout, ONERA, France during some of the DFWM experiments is gratefully acknowledged. We are grateful to Dr.S.Kelm for performing the flame calculations. We also wish to thank Prof.Th.Just for his continuing interest in these studies.

REFERENCES

1 Hanson, R.K. (1986) 'Combustion diagnostics: Planar imaging techniques', 21st Symposium (International) on Combustion, The Combustion Institute, Pittsburgh, USA, pp.1677-1691
2 Rotunno, A.A., Winter, M., Dobbs, G.M. and Melton, L.A. (1990) 'Direct calibration procedures for exciplex-based vapor/liquid visualization of fuel sprays', Combust. Sci. Tech. 71, 247-261
3 Andresen, P., Meijer, G., Schlüter, H., Voges, H., Koch, A., Hentschel, W., Oppermann, W. and Rothe, E. (1990) 'Fluorescence imaging inside an internal combustion engine using tunable excimer lasers', Appl. Opt. 29, 2392-2404
4 Arnold, A., Becker, H., Hemberger, H., Hentschel, W., Ketterle, W., Köllner, M., Meienburg, W., Monkhouse, P., Neckel, H., Schäfer, M., Schindler, K. P., Sick, V., Suntz, R. and Wolfrum, J. (1990), 'Laser in situ monitoring of combustion processes', Appl. Opt. 29, 4860-4872
5 Hanson, R.K., Seitzman, J.M. and Paul, P.H. (1990) 'Planar laser-fluorescence imaging of combustion gases', Appl. Phys. B50, 441-454
6 Kohse-Höinghaus, K. (1990) 'Quantitative laser-induced fluorescence: Some recent developments in combustion diagnostics', Appl. Phys. B50, 455-461
7 Chang, A.Y., Battles, B.E. and Hanson, R. K.(1990) 'Simultaneous measurements of velocity, temperature, and pressure using rapid cw wavelength-modulation laser-induced fluorescence of OH',Opt. Lett. 15, 706-708
8 Meier, W., DLR Stuttgart, private communication
9 Lawitzki, A., Plath, I., Stricker, W., Bittner, J., Meier, U. and Kohse-Höinghaus, K. (1990) 'Laser-induced fluorescence determination of flame temperatures in comparison with CARS measurements', Appl. Phys. B50, 513-518
10 Kohse-Höinghaus, K., Meier, U. and Attal-Trétout, B. (1990) 'Laser-induced fluorescence study of OH in flat flames of 1-10 bar compared with resonance CARS experiments', Appl. Opt. 29, 1560-1569
11 Jeffries, J.B., Kohse-Höinghaus, K., Smith, G.P., Copeland, R.A. and Crosley, D.R. (1988) 'Rotational-level-dependent quenching of OH ($A^2\Sigma^+$) at flame temperatures', Chem. Phys. Lett. 152, 160-166
12 Drake, M.C. and Ratcliffe, J.W. (1991) 'High temperature quenching cross sections for nitric oxide laser-induced fluorescence measurements', General Motors Research Laboratories, GMR-7426
13 Kohse-Höinghaus, K., Koczar, P. and Just, Th. (1986) 'Absolute concentration determination of OH in low-pressure hydrogen-oxygen, methane-oxygen and acetylene-oxygen flames', 21st Symposium (International) on Combustion, The Combustion Institute, Pittsburgh, USA, pp.1719-1727
14 Bittner, J., Kohse-Höinghaus, K., Meier, U., Kelm, S. and Just, Th. (1988) 'Determination of absolute H atom concentrations in low-pressure flames by two-photon laser-excited fluorescence', Combust. Flame 71, 41-50
15 Meier, U., Bittner, J., Kohse-Höinghaus, K. and Just, Th. (1988) 'Discussion of two-photon laser-excited fluorescence as a method for quantitative detection of oxygen atoms in flames, 22nd Symposium (International) on Combustion, The Combustion Institute, Pittsburgh, USA, pp.1887-1896
16 Lawitzki, A., Bittner, J. and Kohse-Höinghaus, K. (1990) 'Determination of N-atom concentrations in low-pressure premixed $H_2/O_2/N_2$ flames doped with NH_3, HCN and $(CN)_2$', Chem. Phys. Lett. 175, 429-433
17 Meier, U. and Kohse-Höinghaus, K. (1987) 'REMPI detection of CH_3 in low-pressure flames', Chem. Phys. Lett. 142, 498-502
18 Garland, N.L. and Crosley, D.R. (1986) 'On the quenching of electronically excited OH, NH and CH in flames', 21st Symposium (International) on Combustion, The Combustion Institute, Pittsburgh, USA, pp.1693-1702
19 Warnatz, J., Institut für Technische Verbrennung, Universität Stuttgart, One-dimensional laminar flame code, version of 1991

ABSORPTION, SCATTERING, AND FLUORESCENCE TECHNIQUES FOR HYPERSONIC FLOW MEASUREMENTS

R. J. Exton
Instrument Research Division
NASA Langley Research Center
Hampton, Virginia 23665, USA

ABSTRACT. This paper presents an overview of the primary nonintrusive diagnostic techniques being developed by the NASA Langley Research Center to address the validation needs of Computational Fluid Dynamic (CFD) codes. The techniques include absorption in the UV and IR, Laser Induced Fluorescence, and a number of scattering techniques including Rayleigh, spontaneous Raman, and several coherent Raman spectroscopies. Most of the techniques are highly specialized, require complex data interpretation, and can satisfy only a few of the CFD needs. For these reasons, the evolving trend in flowfield diagnostics appears to favor a mode in which the diagnostic researcher, the experimental aerodynamicist, and the CFD community jointly define experiments based on the aeronautical requirements and on available diagnostic techniques.

1. Introduction

Hypersonic Research in the United States has been brought to a focus during the past few years in response to the particular needs of the nation's National Aerospace Plane (NASP) project. Many new technologies must be developed before the first flight of the NASP X-30, the first research aircraft designed to take off from a runway and attain orbital cruise altitude and velocity using air-breathing scramjet technology. One of these technologies requires new knowledge on the fluid dynamic and thermodynamic state of the supersonic/hypersonic flowfield about the vehicle and throughout the propulsion system. The bulk of this new knowledge will be gained through ground facility testing over a wide range of Mach number and Reynolds number. The results of this testing will be used to validate CFD codes in order to develop confidence in the final aircraft design. This paper presents an overview of the primary nonintrusive diagnostic techniques being developed by the NASA Langley Research Center to address these needs.

2. Nonintrusive Diagnostic Techniques

Table 1 outlines the techniques that will be discussed. Laser spectroscopy is dominant on the list, but other non-laser techniques are also included which offer unique capabilities and possibilities for flight applications.

A. Boutier (ed.), New Trends in Instrumentation for Hypersonic Research, 205–214.

TABLE 1. Nonintrusive Diagnostic Techniques

Principle	Technique	Parameters	Configuration	Advantages/ (Constraints)
Absorption	Microwave discharge	[OH],[NO] T_r	Line of sight	Multi-channel Compact, rugged Fiber optic compatible Flight possibilities
	Tunable diode laser	$[O_2]$,$[H_2O]$ T_r,V	Line of sight	Multi-channel Compact, rugged Fiber optic compatible Flight possibilities
Scattering	Rayleigh	ρ	Line, plane	Strong signal (Clustering) (Background scattering)
	Spontaneous Raman	$[N_2]$,$[O_2]$,$[H_2O]$, $[CO_2]$,$[H_2]$,T_r,T_v	Point	Species specific (Weak signal)
	CARS	$[N_2]$,$[O_2]$, $[H_2O]$,T_r,T_v	Point	Species specific Strong signal
	RDV	P,T_t,V T_r,T_v	Point	Species specific No seeding (V) (Scanning system)
Fluorescence	LIF,PLIF, LIPF	$[O_2]$,[OH],[NO], $[H_2]$,$[H_2O]$,T_r	Line, plane	Species specific (Quenching)

T_t Translational temperature
T_v Vibrational temperature
T_r Rotational temperature
V Velocity
ρ Density
[ ] Species concentration

2.1 MICROWAVE DISCHARGE

The first two techniques are based on absorption and are line-of-sight averaged measurements over the path length, L. The first uses a microwave discharge lamp to dissociate H_2O and thus generate a low pressure electrodeless discharge containing emission from the OH radical. The emission at 310 nm is used as a background light source in order to measure the absorptivity of OH in combustors. The combustors typically operate at higher pressure (and temperature) than the discharge lamp source. The difference in line widths brought about by this difference in pressure (and temperature) ensures that the peak absorption coefficients in the combustor OH spectrum are sampled for each individual line. High resolution measurements of the absorption for each line could, in principle, be used to generate a molecular Boltzmann plot for the population distribution in the ground state and hence determine the temperature. However, in order to simplify the spectral resolution requirements, measurements are made with a low resolution spectrometer in only 8 spectral "windows" in which the absorption of a number of lines are sampled simultaneously. The system is initially characterized by measuring the spectral output of the background lamp using a high resolution monochromator and by determining the spectral shape of each "window". This information, in conjunction with a library of theoretical absorptivity values, allows one to ascertain the product [OH]·L and the temperature of the OH by "fitting" the theoretical predictions to the measurements.[1] The spectral "windows" are determined by an array of fiber optic bundles placed in the exit plane of the spectrometer. Each fiber optic output is then fed to a separate PMT. Additionally, the light directed to and from the combustor under study is carried on long fibers (10–20m), thus enabling the instrumentation to be located far from hazardous environments.

An extension of this technique is being developed to allow measurements to be made simultaneously at ten different locations. In this variant, shown in Fig. 1, the light from the background lamp is simultaneously fed to 10 separate fibers and each is chopped at a specific frequency. After each fiber optic is directed to/from the combustor, the outputs of the 10 receiver fibers are simultaneously incident on the entrance slit of the spectrometer. In this manner, the output of the spectrometer carries information from the 10 spatial positions in each of the 8 spectral "windows." The spatial information contained in the signal from each spectral "window" can be separated using FFT techniques. The technique described above can also be applied to NO in an analogous manner using the NO band systems near 227 nm.

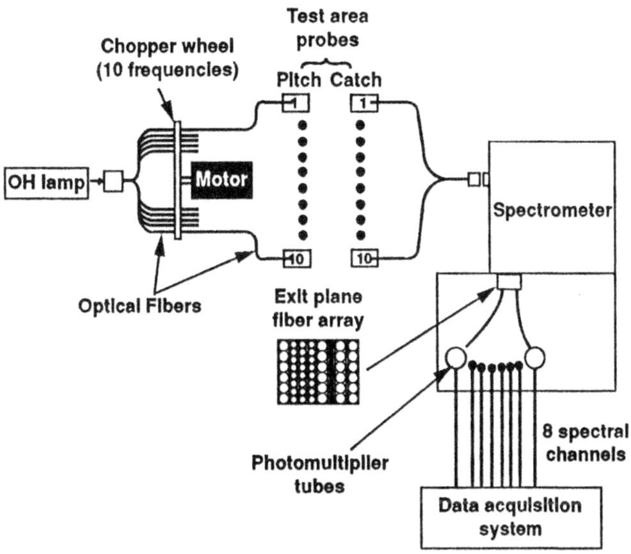

Figure 1. Schematic diagram of the ten position OH absorption system.

2.2 TUNABLE DIODE LASER

The second absorption technique operates in the near IR and employs a tunable diode laser (TDL) as a background light source [2]. This technique is based on the absorption of the O_2 atmospheric band system in the 760 to 780 nm range. Inexpensive, commercially available, GaAlAs diode lasers are employed in a frequency modulated (FM) mode to detect the weak absorption from these magnetic dipole transitions. In this particular application, the TDL is spectrally scanned over several transitions in order to measure temperature and density. The technique will also be used to measure velocity along the laser direction by measuring the Doppler shift. The instrument has been designed with a single TDL which is optically split into 8 beams. Eight fiber optic transfer systems and detectors complete the optical train in order to achieve a simultaneous multiple measurement capability. The near IR technique can also be employed to monitor other gases, such as H_2O, in an analogous manner.

2.3 RAYLEIGH SCATTERING

Several scattering techniques are being applied today that take advantage of the laser characteristics of narrow linewidth, high power and, in some cases, tunability. The first of these is Rayleigh scattering, in which the large Rayleigh cross-section generates strong signals, particularly in the UV, which can be related to total density. The strength of Rayleigh signals has allowed measurements to be made in a plane in many cases. One problem with Rayleigh scattering is the difficulty of rejecting background scattering (walls, windows, etc.) since the molecular scattering is at the same frequency as the incident light. This is particularly true in a full-size wind tunnel where it is difficult to properly control and baffle the optical train. Another difficulty with interpreting Rayleigh signals relates to the clustering of molecules brought about by the expansion (cooling) necessary to generate supersonic/hypersonic flows. Heating of the flow is routinely performed to prevent liquefaction of the air. Recent studies, however, [3] at Mach 6 show that under typical facility stagnation conditions, clustering of an unknown nature is still present at a level which completely overwhelms the molecular Rayleigh signal, making quantitative interpretation impossible. In situ, free stream measurements of the moisture content revealed a water concentration of <10 ppmv. At this level, water condensation by itself is not capable of producing the observed "Rayleigh" signals. The water and the other constituents of air, such as CO_2, may, however, provide the condensation nuclei by which sizeable clusters could be formed. This effect is still under investigation. In spite of the limitations imposed by the clustering, planar Rayleigh measurements are extremely valuable in obtaining visualization of the flow. For example, the mixing of helium into air at Mach 6 was visualized using an excimer laser sheet at 193 nm with an intensified CCD camera [4]. In this experiment, the helium (simulating H_2 fuel) was fed into a Mach 6 air flow through three "fuel" injectors, each 25 mm high by 12.5 mm wide. The laser sheet was situated at several downstream locations and viewed by a UV intensified CCD camera (Fig. 2). Visualization of the mixing was made possible by the low Rayleigh cross section for helium relative to air (lower by a factor of 80). Thus regions containing only helium appear as a dark hole relative to the surrounding air. This effect is further enhanced by the Rayleigh-sized clusters that are present in the air flow. Fig. 3 shows two consecutive (15n sec pulses at 10Hz) "snapshots" of the Rayleigh image obtained at 10 injector heights downstream from

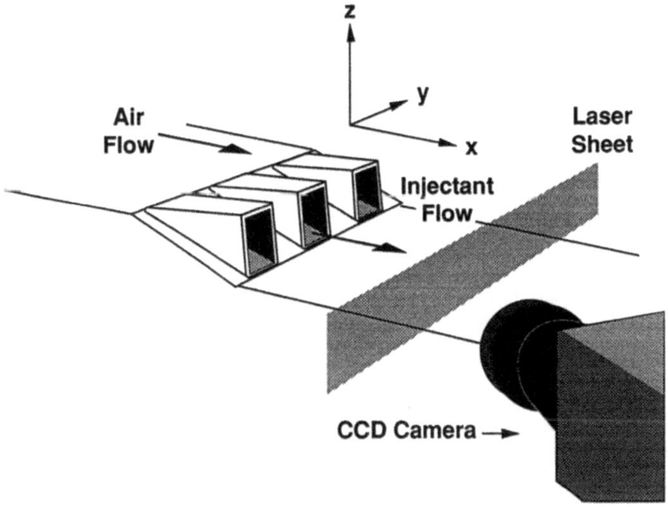

Figure 2. Experimental arrangement of the three injector model in a Mach 6 air flow. The pulsed excimer laser sheet is positioned downstream of the injectors and is viewed by a gated, intensified CCD camera.

the injectors. These data were obtained for the condition of matched helium and air static pressures. Note the different shapes for the helium "mushroom" as it lifts off the surface and mixes into the air stream. Although not evident in the individual "snapshots", videotape replays appear to show the entrainment of air into the "stem of the mushroom" brought about by vortical motion generated at the injectors. Fig. 4 shows the averaged Rayleigh flow field (22 laser shots) for the central injector and the mean density of air for the same region deduced from probe measurements. Although the clustering in the air prohibits a quantitative comparison, the qualitative agreement between the two is excellent. The individual "snapshots" also provide a qualitative assessment of the fluctuating component (unsteadiness) in the flow field, something not available from probe data. It should be noted that the temperature distribution in this flowfield (cooler on the periphery of the "mushroom") is expected to have a strong effect on the clustering and could account for some of the enhanced contrast between the helium/air interface.

2.4 SPONTANEOUS RAMAN SCATTERING

In contrast to Rayleigh scattering, Raman scattering is species specific and occurs away from the incident frequency, making it easier to reject the background. Raman cross-sections, however, are small (typically 0.1–1.0% of Rayleigh values) and the resulting signals are weak. For example, vibrational Raman measurements of N_2 were made in the low density Mach 6 free stream as a way of checking the Rayleigh clustering signal described above [3].

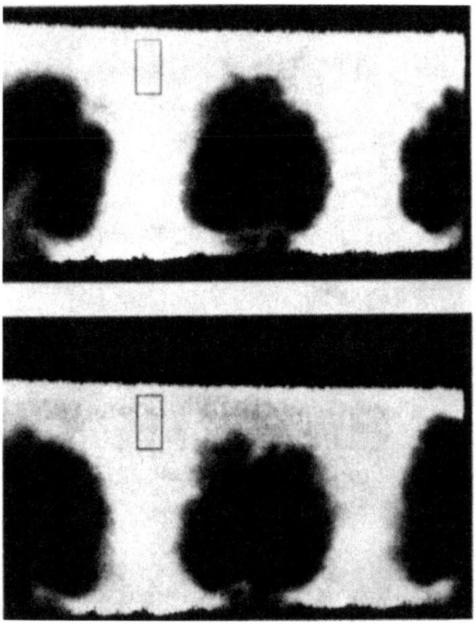

Figure 3. Two consecutive Rayleigh "snapshots" obtained at ten injector heights downstream from the injection plane.

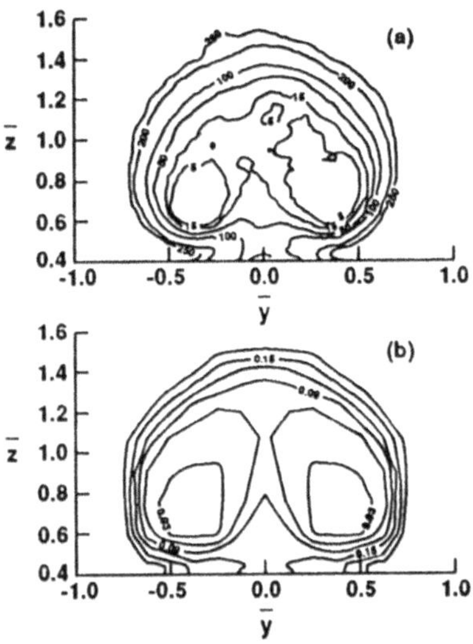

Figure 4. Qualitative comparison of the flowfield downstream of the central injector obtained from (a) averaged Rayleigh data (22 frames) converted to pressure (torr) at room temperature and (b) mean density of air (Kg/m^3) deduced from probe measurements. The coordinates are normalized to one injector height.

For these experiments, a frequency doubled (532 nm), Nd:YAG laser (60 mJ/pulse @10Hz) was employed as a light source to focus on the tunnel center line with a 40 cm focal-length lens. Practical gathering optics (f/5) focused the scattered light from a 1cm segment of the laser beam onto a double monochromator outfitted with a PMT. For this situation, signal averaging for 20 sec was required to obtain a reasonable S/N.

The use of a high power (200mJ/pulse), UV (248 nm) excimer laser to gain a ν^4 advantage in Raman cross-section has been employed to perform diagnostics on a Mach 2 supersonic combustor [5]. The atmospheric densities provided by this combustor, coupled with the opportunity to use fast gathering optics (~f/1) to focus light from a point in the combustor to the slit of a polychromator, allowed single shot measurements of the major species densities (N_2, O_2, H_2 and H_2O) to be made. By translating the combustor, simultaneous density profiles, along with local temperature measurements, provided a unique picture of the turbulent mixing and chemical reactions taking place in a supersonic reacting flow.

2.5 COHERENT ANTI-STOKES RAMAN SCATTERING (CARS)

A popular coherent Raman technique, CARS, has been developed to enhance Raman scattering through nonlinear effects. Typically, two pulsed lasers are employed such that their difference in frequency equals the Raman molecular frequency. The two laser beams are arranged in a crossed-beam (BOXCARS) mode to maximize the spatial resolution and in a spectrally "broadband" mode in order to generate temperature and density information with a single laser pulse. The CARS signal is generated at the intersection of the pump and Stokes beams and carries the diagnostic information on an efficiently controlled "laser-like" beam. The present system has been designed to simultaneously measure N_2 temperature, N_2 number density, and O_2 number density. An extension is under way to add H_2O number density as a measurable. The system has been used extensively to map the Mach 2 Hydrogen/Air combustor described earlier and the results compared favorably to computational fluid dynamics (CFD) codes. More recently, the optics have been "hardened" and the system has been demonstrated in a hostile combustion test cell environment [6].

2.6 RAMAN DOPPLER VELOCIMETER (RDV)

Another coherent Raman technique is under development that uses either Inverse Raman Spectroscopy (IRS) or Stimulated Raman Gain Spectroscopy (SRGS). In these coherent versions, two focused laser beams (one pulsed and one c.w.) interact simultaneously with the molecules in a quasi-c.w. arrangement. As in CARS, the difference in frequency of the two lasers equals the molecular Raman frequency of the molecule being probed. In this case, however, one of the lasers is scanned in frequency and the narrow laser linewidths allows a measure of the Doppler shift of the individual Raman lines. In this way, the velocity of the molecules themselves is measured (no particle seeding required), circumventing any particle lag problems and also bypassing any difficulties in seeding particles into specific regions of interest. The Doppler shift is actually obtained through the difference in frequency shift between the forward and backward scattered Raman peaks and, in addition, the two line shapes jointly determine the gas translational temperature, T, and pressure, P. In this manner, all three parameters (V,P,T) can be determined from frequency difference measurements independent of intensity or absolute frequency considerations. This technique was initially employed in a concentric IRS mode (Ar^+ probe, tunable pulsed dye amplified pump) and successfully demonstrated that the complex optical configuration would work in a major wind tunnel environment and geometry [7,8]. The system has more recently been reconfigured in the SRGS mode (tunable dye laser probe, Nd:YAG pump) in order to take advantage of the increased energy per pulse available using a narrow-line Single Axial Mode (SAM) Nd:YAG laser. The new configuration also employs crossed-beams to increase the spatial resolution.

2.7 LASER INDUCED FLUORESCENCE (LIF)

Laser Induced Fluorescence (LIF) and Planar Laser Induced Fluorescence (PLIF) techniques have gained in popularity in recent years due to the availability of tunable lasers and the opportunity they afford in increasing the detection sensitivity to specific species [9]. LIF is a two-step process in which the laser photon is first absorbed by a specific line or band, followed by fluorescence which, at low densities, is subsequently emitted with the characteristic lifetime of the upper state. At higher densities, however, collisions with other molecules occur during the lifetime, which alter the internal energy of the system and also quench the fluorescence. Quenching significantly reduces the fluorescence yield and depends collectively on the concentrations of all other species present. Since local, instantaneous information on the absolute species concentrations of all the colliding partners and the local temperature is generally not known, the interpretation of the fluorescence signals is an intractable problem. The most recent technique to circumvent the quenching problem is Laser Induced Predissociation Fluorescence (LIPF) where transitions are selected in which the upper electronic state is crossed by a repulsive state. In this situation, the upper state predissociates at a faster rate than the collision rate with the end result that collisions have little, if any, effect on the fluorescence. The negative side of LIPF is that the signal is reduced drastically since a significant proportion of the excited molecules are dissociated in the process. Nevertheless, the technique is finding widespread use and many molecular states have been identified that exhibit predissociation [10]. One of the major species of interest to aerodynamics is oxygen which is accessible through the Schumann-Runge ($B \leftarrow X$) band with an excimer laser operating at 193 nm (ArF) or 248 nm (KrF). This technique is being developed for both cold and hot flows. At the very low temperatures characteristic of hypersonic free streams, the O_2 LIF signals generated using the excimer at 193 nm are marginal due to the fact that only high rotational levels (unpopulated at low temperatures) are accessible over the limited frequency range provided by the gain profile of the excimer. For hot flows, the excitation of high rotational levels is no longer a problem, but many additional O_2 vibrational bands come into play which complicates the spectrum. Spectral interference from other molecules found in hot flows, such as NO, must also be taken into account.

The application of LIF techniques to major aerodynamic facilities today suffers from two major drawbacks: signal intensity and quantification. Many of the LIF techniques demonstrated to date use fast optics and the best lasers available but still exhibit field-of-views that are measured in millimeters or centimeters. The scaling of these techniques to meter-sized facilities severely limits their usefulness. Even where there is adequate signal, most measurements have not been quantified since factors such as the fluorescence yield are very difficult to determine [9]. Nevertheless, the species specific flow visualization attained by LIF techniques has been extremely valuable, particularly in supersonic combustion diagnostics, where the delineation of flowfield structure, reaction zones, and shock waves have helped to characterize turbulent flames. At Langley, PLIF has been used to examine OH profiles from a Mach 2 combustor. The OH "snapshots" showed flow patterns which were "frozen" by the 15n sec pulse length of an excimer laser operating at 308 nm (XeCl). The qualitative image data showed changes in the structure of the reacting jet and illustrated the localized OH concentrations occurring on a pulse-to-pulse basis.

3. Discussion

The requirements for flowfield diagnostic techniques are being driven in part by CFD solutions that are being developed to generate 3-D flowfields for complex aerodynamic shapes. This technology is maturing rapidly but still requires experimental validation. This paper has presented an overview of a number of diagnostic techniques being developed at Langley Research Center to address the validation needs of the CFD community. It would appear that no single technique can satisfy even a few of the CFD requirements, but rather each requirement must be met by a specific instrumental solution based on many factors [ex. model/test section geometry, spatial/temporal resolution, facility run time, and general flowfield characteristics (T,P,V)] . As a result of these specialized applications and the complex data interpretations required, it is questionable whether many will attain the "dedicated" instrument status that has been possible with Laser Doppler Velocimeter (LDV) systems, for example. Rather, the trend appears to favor a research mode of investigation where the diagnostic researcher works hand-in-hand with the experimental aerodynamicist and the CFD community to jointly define the experiment based on the aeronautical requirements and on available diagnostic techniques.

Finally, it should be recognized that many of the evolving diagnostic techniques are not only species specific, but, in addition, measure (and sometimes drive) the internal state distribution of the probed molecules. These measurements, at the molecular level, require a detailed knowledge of the collision physics and reaction dynamics as an integral part of their interpretation. Thus the newer diagnostics represent a revolution of sorts in that they are leading to an understanding of aerodynamic phenomena from the molecular level rather than from the traditional bulk properties approach.

4. References

1. Lempert, W.R. (1988) "Microwave Resonance Lamp Absorption Technique for Measuring Temperature and OH Number Density in Combustion Environments," Combustion and Flame 73, 89–98
2. Philippe, L.C. and Hanson, R.K. (1992) "Sensitive Diode Laser Absorption Technique For Aerodynamic Measurements", AIAA paper 92-0139 presented at Thirtieth Aerospace Sciences Meeting, Reno, NV Jan. 6–9
3. Shirinzadeh, B., Hillard, M.E., and Exton, R.J. (1991) "Condensation Effects on Rayleigh Scattering Measurements in a Supersonic Wind Tunnel", AIAA Journal 29, 242–246

4. Shirinzadeh, B., Balla, R. J., Hillard, M. E., Anders, J. B., Exton, R. J., and Waitz, I.A. (1991) "Study of Rayleigh Scattering For Visualization of Helium-Air Mixing at Mach 6", presented at The International Congress on Instrumentation in Aerospace Simulation Facilities, Rockville, MD Oct. 27–31.
5. Cheng, T. S., Wehrmeyer, J. A., Pitz, R. W., Jarrett, O., and Northam, G. B. (1991) "Finite-Rate Chemistry Effects in a Mach 2 Reacting Flow", AIAA-91-2320 presented at the Joint AIAA/ASME/SAE Propulsion Conference, Sacramento, CA. June 24–27
6. Smith, M., Antcliff, R., Cutler, A., Jarrett, O., and Northam, G. (1990) "CARS Temperature Measurements in a Hydrogen-Fueled Supersonic Combustor", AIAA-90-5260 presented to the AIAA Second International Aerospace Planes Conference, Orlando, FL. Oct. 29–31
7. Exton, R.J., and Hillard, M.E., (1986) "Raman Doppler Velocimetry: A Unified Approach for Measuring Molecular Flow Velocity, Temperature, and Pressure", Applied Optics 25, 14–21
8. Exton, R. J., Hillard, M. E., Lempert, W. R., Covell, P. F., and Miller, D. S. (1987) "Molecular Flow Velocity using Doppler Shifted Raman Spectroscopy", AIAA-87-1531 presented to the 22nd Thermophysics Conference, Honolulu, Hawaii, June 8–10
9. Hanson, R. K., Seitzman, J. M., and Paul, P. H. (1990), "Planar Laser-Fluorescence Imaging of Combustion Gases" Applied Physics B50, 441–454
10. Andresen, P., Bath, A., Groger, W., Lulf, H. W., Meijer, G., and ter Meulen, J. J. (1988) "Laser-Induced Fluorescence with Tunable Excimer Lasers as a Possible Method for Instantaneous Temperature Field Measurements at High Pressure: Checks with an Atmospheric Flame", Applied Optics 27, 365–378

SPECTROSCOPIC DIAGNOSTIC TECHNIQUES FOR THE HIGH ENTHALPY SHOCK TUNNEL IN GÖTTINGEN (HEG): PREPARATORY LIF STUDIES ON OTHER FACILITIES.

W.H.BECK[1], U.KOCH[2] AND M.MÜLLER[1]
German Aerospace Research Establishment (DLR)
Bunsenstrasse 10
3400 Göttingen
F.R. Germany

1. Institute of Experimental Fluid Mechanics
2. Central Division Windtunnels

ABSTRACT. The Laser Induced Fluorescence (LIF) technique will be used on the High Enthalpy Shock Tunnel Göttingen (HEG) to study the gas in the test section free-stream flow and between a test model and its bow shock. The LIF apparatus has been tested on several facilities in order to ascertain any likely problems which may occur in application to HEG. LIF results of the determination of the free-stream conditions (NO density and temperature) in an arc-heated wind tunnel (DLR Cologne) are presented here.

1. Introduction

The High Enthalpy Shock Tunnel Göttingen (HEG) has been built to carry out experimental studies on the influence of real gas effects on the external aerodynamics and heat loading of space vehicles (such as HERMES) reentering the earth's atmosphere. The Laser Induced Fluorescence (LIF) technique has been chosen to characterize the conditions in the test section free-stream flow and also to probe the region between a model placed in the flow and its stand-off bow shock. It is planned to measure 2d profiles of NO_{vJ} concentration (v, J = specific vibrational, rotational levels) and 1d (along a line) O-atom concentration profiles, and, if a rotational temperature is defined, to obtain 2d maps of total NO concentration and (NO) rotational temperature.

There is to this stage no published record of an application of the LIF technique to a study of the rather extreme flow conditions obtained in a high enthalpy facility. Since one can beforehand only estimate the problems likely to be encountered in such an application, and since sufficient manpower and knowhow was not available within the HEG project group to devise and construct independently an LIF apparatus, it was decided to seek outside help (LaVision 2D Meßtechnik GmbH in Göttingen) in constructing the apparatus; this apparatus was handed over to the DLR (including initial testing) in September 1991. Further, in order to evaluate the likelihood of occurence of the aforementioned problems in HEG, a detailed test program on other facilities was devised to check these problems and where relevant, to find suitable solutions.

To this end, tests were planned in:

1. A heated cell (DLR Göttingen)
2. A shock tube (University of Göttingen)
3. A vacuum wind tunnel (DLR Göttingen)
4. An arc-driven tunnel (DLR Cologne)
5. Another high enthalpy shock tunnel (RWTH Aachen).

Tests 1. - 4. are completed, evaluation is partly finished [1, 2]. Test 5. is planned for mid 1992.

LIF will be used in HEG in late 1992. Until that time (i.e. until LIF is ready), it is planned to carry out emission spectroscopy studies in HEG flows. These studies are of a qualitative nature, and serve to give an estimate of what radiating species are present between bow shock and model, without necessarily providing any information of how much is present - whether this technique can provide any quantitative data (e.g. temperatures) is not known as yet and would need to be tested. A first result in HEG has already been obtained, but it is too early yet for a definitive analysis.

A. Boutier (ed.), New Trends in Instrumentation for Hypersonic Research, 215–224.

This paper will focus on a characterisation of the free stream conditions (temperature and NO density) in the arc-driven tunnel (L2K) in Cologne.

2. The LIF Method

2.1 APPLICATION TO HEG STUDIES

To apply the LIF technique to HEG, it is important to have at least an estimate of the expected operating conditions, both in the free-stream and in front of a model. Calculations were carried out here in the Institute of Theoretical Fluid Mechanics of the DLR [3]; a summary for an enthalpy of 44.5 MJ kg^{-1} is shown in Fig. 1. Particulary noteworthy are the low concentration of NO (of order of 10^{13} cm^{-3}) and the complete dissociation of O_2 (O concentration of the order of 10^{16} cm^{-3}) in the free-stream, even though the gas temperature is only 2700 K; this exemplifies the non-equilibruim nature of the free-stream gas. Initially, efforts will be concentrated on fluorescence measurements of the species NO only.

The LIF apparatus consists of two tunable excimer lasers whose output beams are focussed to sheets and overlapped in front of a test model in HEG. The two beams are tuned to two transitions of NO. The two generated fluorescence signals are captured in 2d sequentially using two combination telescope/image intensifier/CCD camera systems. These two fluorescence signals are measured beforehand at a reference temperature, so that these four signals (2 reference, 2 experiment) can be combined, given certain conditions being met, to yield an NO rotational temperature - this is further discussed in [2], where the problem of quenching, in the form of a study being carried out with the University of Göttingen [1], is also addressed.

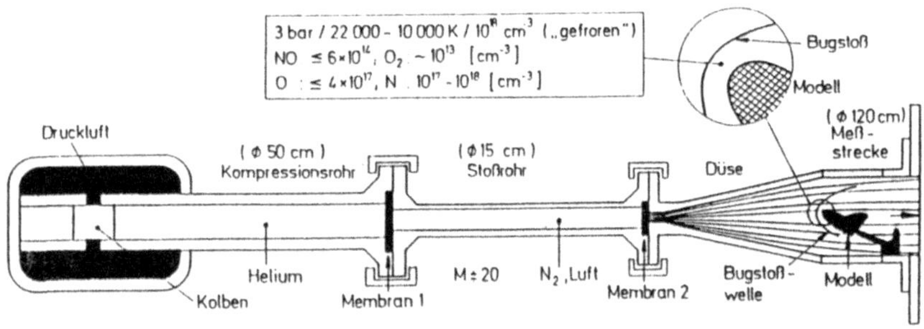

Anfang	200 bar	1-2 bar / 300 K		≤ 0.4 bar / 300 K	10^{-3} mbar
"Maximum" (später)	—	1800 bar / 4500 K		2000 bar / 13800 K / 10^{21} cm^{-3} (hinter reflektierter Stoßwelle)	40 mbar / 2700 K / 8500 ms^{-1} / 10^{17} cm^{-3} (free-stream)
			N_2	1×10^{20} cm^{-3} (12 %)	6×10^{16} cm^{-3} (56 %)
			O_2	7×10^{17} cm^{-3} ( )	1×10^{12} cm^{-3} ( )
			NO	2×10^{19} cm^{-3} (1 %)	9×10^{12} cm^{-3} ( )
			N	7×10^{20} cm^{-3} (65 %)	1×10^{16} cm^{-3} (11 %)
			O	2×10^{20} cm^{-3} (22 %)	4×10^{16} cm^{-3} (33 %)

Figure 1. Calculated conditions in HEG at an enthalpy of 44.5 MJ kg^{-1}.

2.2 THE LIF APPARATUS

A schematic of the LIF apparatus is given in Fig.2. For NO measurement (2d), the beams of the two excimer lasers are focussed to sheets and crossed in front of the model in the HEG test section. The ensuing fluorescence is captured by an image-intensified, gated camera system via a modified Newton telescope with a large light gathering power. The lasers are fired sequentially (time separation <1 μs) so that the fluorescence arising from one laser beam is captured only by its corresponding signal capturing system. Hence one obtains a "quasi-instantaneous" determination of NO density and temperature. The short flow times available in HEG (<1 ms) dictate the use of two separate lasers in this application.

Both beams also pass through a heated calibration cell with an attached monochromator and intensified camera (operating as an OMA), enabling the tuning of the laser to be checked easily. A small percentage of both beams is deflected via a beam-splitter to another camera which records both beam profiles during a run - this allows for corrections due to beam fluctuations and O_2 absorption along the lengthy (7 m) beam path to be made. This setup also incorporates an etalon, which allows the laser bandwidths to be checked before a run. Raman cells are included to widen the accessible pumping range of the ArF lasers to near 211 nm (i.e. 210-211 nm). It is further planned to modify the apparatus to use the OMA and calibration cell during the experiment to test whether the lasers have drifted off the peak absorption frequency for the desired transition.

In order to compensate, at least in part, for the complexity of the timing and control events of the laser system (gate widths and delays can be as small as 5 ns), it has been fully automated and can run remotely under computer (PC) control.

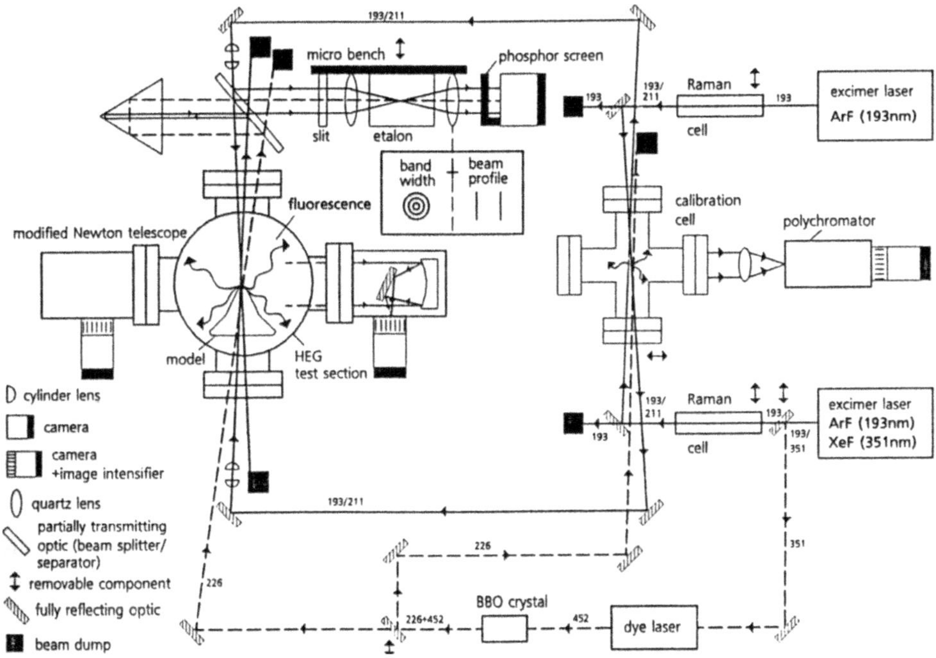

Figure 2. Block diagram of the LIF apparatus to be used on HEG.

3. First Results on Other Facilities

3.1 OVERVIEW

A brief synopsis will be given here of the results obtained in the heated cell (spectroscopy, temperature determination), the shock tube (NO quenching study) and the vacuum wind tunnel (spectra, 2d NO profiles) - the reader is referred elsewhere for further details [1, 2].

3.1.1 *Heated Cell.* Fluorescence and excitation spectra were measured using the OMA and calibration cell, and a temperature determination was carried out using a separate heated cell in place of the HEG test section shown in Fig. 2(i.e. the setup was similar to that to be used in HEG). The scatter in the data was large - this problem is being addressed anew in a study here at the DLR (but see also 3.2).

3.1.2 *Shock Tube.* In a collaboration with the University of Göttingen for studies of NO quenching, the fluorescence decay rate of laser excited NO ($R_1(26.5)$ $D \leftarrow X$) was monitored for different NO/Ar mixtures (c_{NO}) both at 300 and 1980 K behind an incident shock wave where NO densities were varied in the range 0.26 - 13.8 x 10^{-8} mol cm^{-3}. From this decay rate, a total removal rate coefficient (representing all deexcitation channels) k_{tot} was measured, and a value for k_{Ar} (quenching due to Ar) was derived. There was no great difference found between NO self-quenching and quenching by Ar - a surprising result.

3.1.3 *Vacuum Wind Tunnel.* Mixtures of NO/N_2 or NO/He (both 5%) at a pressure of 4 bar and a temperature of 1060 K were expanded through an orifice of diameter 0.7 mm into a vacuum chamber held to better than 0.03 mbar. This setup is known to lead to the following gas conditions (at a distance of 30 mm from the orifice): T_{trans} = 100 K; T_{vib} = 1000 K; T_{rot} = 100 K; $\rho = 2 \times 10^{15}$ cm^{-3}; M = 16. A model (flat plate, diameter 5 mm) was placed at a distance of 40 mm from the orifice, and 2d laser induced fluorescence images were recorded between orifice and model at several laser excitation wavelengths. Complete excitation spectra (over the tuning range of the ArF excimer laser) could also be recorded, resolved in 1d between model and orifice. These spectra showed that N_2 and He have different effects on the internal (rotational) energy distributions of NO in front of the model.

3.2 ARC-DRIVEN TUNNEL COLOGNE L2K

LIF measurements were carried out in the arc-driven tunnel L2K in Cologne. This was the first test of the LIF apparatus on a facility where the NO is generated in the normal course of operation, and has not been added to the test gas beforehand. Hence a measurement of NO fluorescence can be used to deduce a temperature and a total NO concentration, from which insight can be had into the kinetic processes occurring between model and bow shock, and those which have occurred in the nozzle leading to the free-stream conditions. The latter will be discussed here.

3.2.1 *Operating Conditions.* In L2K, a continuous dc-arc is produced in the flowing test gas (here air), which is expanded through a nozzle into an evacuated test chamber. For these LIF measurements, a model (cylinder of copper or silicon carbide with axis along the line of flow) was placed near the nozzle exit; the laser beam sheet probed the region about 30 mm upstream of the model.

The experimental setup is shown in Fig. 3. The 2 telescopes had to be installed protruding into the test section in order to obtain the optimal object-to-image separation for the telescope optics. Similar to the procedure in the vacuum wind tunnel, 2d laser induced fluorescence images upstream of the model and 1d resolved (along axis line nozzle-model) excitation spectra were recorded. Since the laser sheet had a width of ~ 30 mm, it was possible in the one image to capture fluorescence signals both in the free stream and between model bow shock.

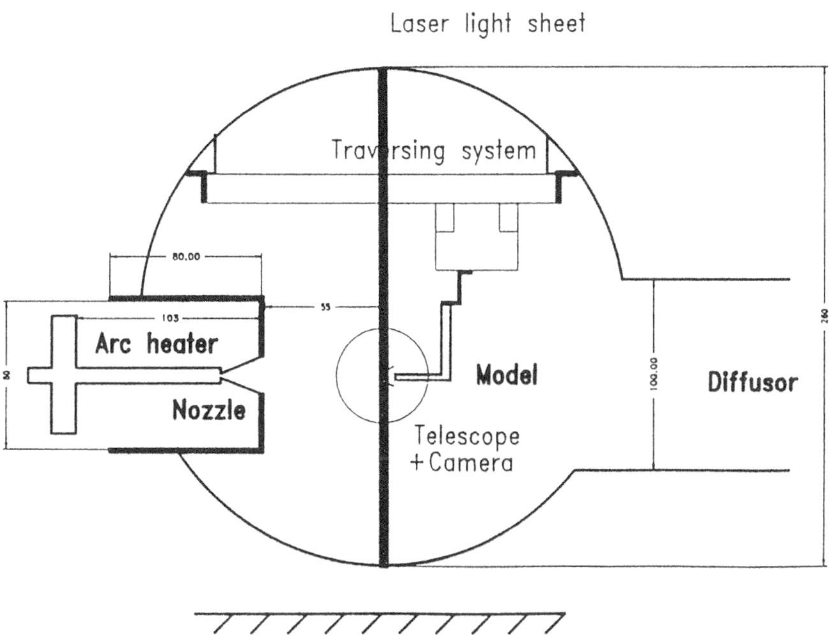

Figure 3. Block diagram of the experimental setup for LIF studies on the arc-heated tunnel L2K in Cologne.

Test conditions were:

Measured quantities:-	
- mass flow rate:	50 g s^{-1};
- stagnation pressure:	1290 mbar;
- distance between model and nozzle exit:	55 mm;
- diameter of the nozzle throat:	29 mm;
- diameter of the nozzle exit:	200 mm;
- model surface temperature:	SiC 1410 K; copper 1300 K.
Calculated quantities [see 4]:-	
- stagnation temperature:	4326 K;
- temperature in the free stream:	239 K;
- temperature behind the bow shock (frozen/equilibrium):	3122/3874 K;
- total density in the free stream:	0.1608 g m^{-3} (3.72 x 10^{15} cm^{-3});
- NO concentration in the free stream:	4.65% (1.73 x 10^{14} cm^{-3});
- pitotpressure:	11.88 mbar.

3.2.2 *Free-stream Temperature.* A typical 2d fluorescence image in front of the model with the laser at 193.04 nm (NO B $\leftarrow$ X $P_2(24.5)$ transition) is shown in Fig. 4(b). The image was taken with the laser operating in burst mode, viz. several (here 10) laser pulses were fired in rapid succession, the fluorescence signals were integrated on the CCD chip and then read off into the computer. This was necessary to be able to obtain sufficient signal in the free stream; in so doing, however, the fluorescence signals between bow shock and model were saturated (i.e. exceeded the dynamic range of the camera system) and could not be used for an analysis of this region. (A separate measurement was carried out to examine this region). Also shown in Fig. 4(a) is a 2d fluorescence image taken at the same laser wavelength as 4(b), but with static NO in the test section - this was used to determine the detection limits of the LIF system, and to calibrate the system with a known amount of NO (see 3.2.3). Further, in all 2d measurements, a laser beam profile was recorded as well.

In images similar to Fig. 4(b) at different laser wavelengths, a central narrow horizontal band was chosen, over which the signals from each pixel in a vertical column were averaged to one value, whereby the 2d image was reduced to a 1d line which was stored separately for each laser wavelength. The spatial information is preserved in a horizontal direction (along the flow), but is lost in the vertical - if the band chosen is small, the errors are negligible. The stored lines contain the spectral (excitation wavelength) and spatial (1d along flow line) information of an experiment; Fig. 5(a) shows such a spectrum of NO in the free-stream region before the bow shock in Fig. 4(b). Also shown are spectra measured in the static calibration cell (at 295 K), Fig. 5(b), and calculated at 295 K using a spectrum simulation code [5], Fig. 5(c).

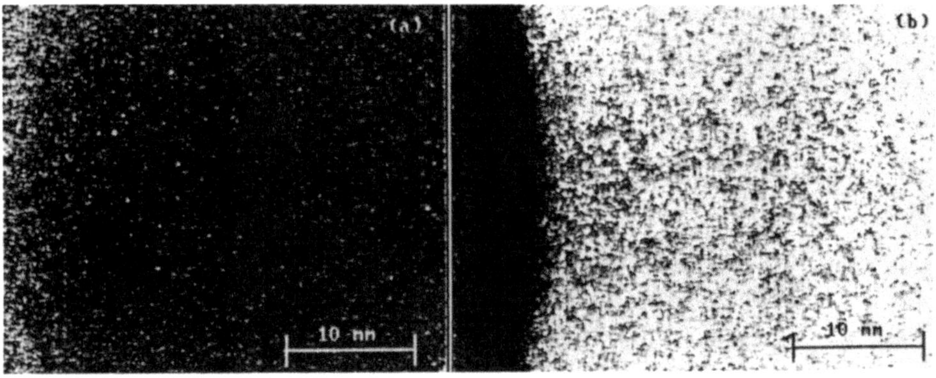

Figure 4. 2d fluorescence images in front of the model in L2K for the laser tuned to the NO B $\leftarrow$ X $P_2(24.5)$ transition (the darker the region, the higher the fluorescence signal): (a) Static NO in the test section at low pressure; (b) L2K free-stream flow (conditions as in 3.2.1).

A comparison of Fig. 5(b) and 5(c) shows that whereas a particular branch (e.g. P_2) increases with decreasing wavelength for the simulated spectrum (as expected for decreasing rotational quantum number K = 28.5 - 24.5), in the measured calibration spectrum the branch peaks near the middle. This is due to the laser locking efficiency dropping off rapidly near the edge of the tuning range, an effect which must be allowed for in a quantitative analysis of the spectra. The presence of this effect (and others) serves to emphasize the strategy which has been adopted for HEG measurements: because it is extremely difficult to account for all possible influencing factors in a quantitative measurement, a calibration is carried out and all measured spectra are analysed with reference to this calibration.

The individual line intensities S of the calibration spectrum (Fig. 5(b) were plotted logarithmically versus the energy E in the ground electronic state from which the transition arose. (For details of the method see [2]). Correction factors were applied to bring the points onto a line with slope corresponding to the calibration temperature, 295 K. A similar analysis was carried out for the free-stream spectrum (Fig. 5(a)), whereby the correction factors obtrained from the calibration were implemented - the resulting plot of ln S versus E is shown in Fig. 6. Error estimates of the points (one standard deviation) are shown. A fit to the points gives a temperature of 275 $\pm$ 20 K.

The calculated value (see 3.2.1 and [4]) is 239 K, a difference of 36 degrees. Too few experimental data are available at this stage to interpret this result. The good fit of the points to the straight line in Fig. 6 suggests that rotational equilibrium exists in NO, at least amongst the rotational levels (K = 24,5 - 31,5) which have been probed. A larger then calculated value of the temperature could be interpreted as an underestimate in the nozzle flow calculation of those processes which feed energy back into the gas, e.g. atom recombination reactions. In passing through the nozzle, the reacting gas undergoes rapid temperature and density drops; to model this correctly, kinetic rate data for the various chemical processes occurring during these changes must be known over the whole temperature range in question - this need not always be the case. One also needs to examine the aerodynamic assumptions made for the calculation - initial conditions, viscous effects, 2d effects. From the experimental point of view, it should also be stressed that improvements (refinements) are possible; the problem of ascertaining whether the laser is actually tuned to the peak of a transition can be overcome by averaging longer over those parts of the spectrum where a transition line is observed. Also the calibration procedure (determination of correction factors) can be improved.

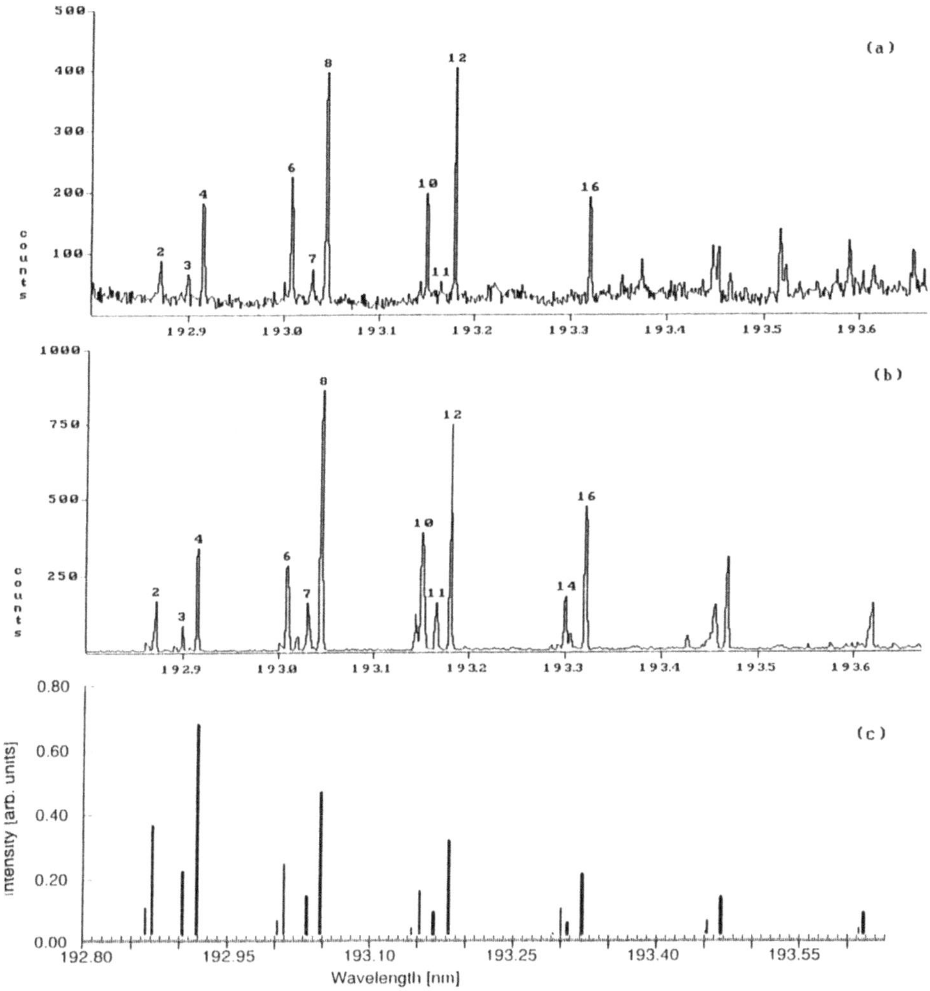

Figure 5. NO excitation spectra over the wavelength tuning range of the excimer lasers: (a) Measured in the L2K free-stream gas flow, (b) Static NO in calibration cell; (c) Simulation (calculated) at 295 K. The spectral lines are numbered and correspond to the following NO B ← X transitions:- lines 4, 8, 12, 16 - branch P_2(23.5-26.5); lines 3, 7, 11 - branch R_2(26.5-28.5); lines 2, 6, 10, 14 - branch P_1(25.5-28.5).

3.2.3 *Free-stream NO Density.* Two 2d images of the type shown in Fig. 4(b), corresponding to the 2 NO $B \leftarrow X$ transitions I ($P_2(24.5)$) and II ($P_2(26.5)$) were recorded. 2d images for both these transitions with static NO in the test section were also recorded, in every case along with a laser beam profile (using the beam profile camera - see Fig. 2) to correct for laser beam intensity fluctuations. All these images have superimposed on them the geometric shape of the laser beam intensity - for each image, the same region, lying centrally in the beam, was chosen and averaged to give a single value (with statistical error) corresponding to a mean free-stream fluorescence intensity S. This value was corrected for laser intensity fluctuation in each case using its appropriate beam profile image. The total NO density was determined using the following formula [see 2]:

$$[NO] = \frac{S}{S'} \cdot \frac{f'_i(T')}{f_i(T)} \cdot \frac{q'_f}{q_f} \cdot [NO]' \cdot a \qquad (1)$$

where S = corrected fluorescence intensity (counts);

$f_i(T)$ = Boltzmann fraction for level i with energy $E_i \equiv \frac{\exp(-E_i/kT)}{Q(T)}$ (where k = Boltzmann constant = 0.6968 $cm^{-1}K^{-1}$, Q(T) is the NO partition function at temperature T);

T = temperature (K);

q_f = quantum yield for fluorescence = $A_k/(A_k + Q)$ (where A_k = natural decay rate of excited state k, Q = quenching rate);

[NO] = NO concentration (cm^{-3});

a = c'/c (where c = detection system parameters);

and dashed (') refers to calibration (static) measurement conditions.

The static cell NO fill pressures were 1.75 Pa ([NO]' = 4.23 x 10^{14} cm^{-3}) and 4.85 Pa (1.19 x 10^{15} cm^{-3}) for the transitions I and II, respectively, T and T' were 275 (see 3.2.2) and 295 K, respectively, Q (T) and Q' (T') were 227.3 and 243.7, respectively (see [5]), E_i (I) and E_i (II) were 1189 and 1367 cm^{-1}, respectively (see [5]), and q_f and q'_f were assumed to be unity (see later). Inserting these values into equation (1) gives the following values for NO concentration:

Line I	P_2 (24.5)	[NO] = 0.9 ± 0.4 x 10^{14} cm^{-3}
Line II	P_2 (26.5)	[NO] = 1.4 ± 0.6 x 10^{14} cm^{-3}.

The calculated value for [NO] in the free-stream is 1.7 x 10^{14} cm^{-3} (see 3.2.1 and [4].)

3.2.4 *Quenching.* The values for [NO] in 3.2.3 were calculated on the basis $q_f = q'_f$ (=1). This needs to be checked for the calibration (q'_f) and free-stream (q_f) measurements.

q_f is given by: $q_f = A_k/(A_k + Q)$, where A_k = 3.16 x 10^5 s^{-1} for the NO B-state [6]. For the NO calibration conditions given beforehand, Q can be calculated from k_{NO}[NO], where k_{NO} is the B-state quenching rate constant and has a value of 1.26 x 10^{-10} cm^3 s^{-1} (see [6]). Inserting values leads to Q_I = 5.3 x 10^4 s^{-1} and Q_{II} = 1.5 x 10^5 s^{-1}, so that q'_f (I) ~ 0.9 and q'_f (II) ~ 0.7.

For the free-stream, $Q = k_{NO}[NO] + k_{N2}[N_2] + k_{O_2}[O_2] + k_O[O] = k_{NO}[NO] + k_{N_2}([N_2] + [O_2] + [O])$, where it has been assumed that quenching rates for NO B-state for N_2, O_2 and O are the same. Theoretical calculations [4] give the following values in the free-stream:

[NO] ~ 2 x 10^{14} cm^{-3}	[O] ~ 8 x 10^{14} cm^{-3}
[N_2] ~ 3 x 10^{15} cm^{-3}	[O_2] ~ 3 x 10^{14} cm^{-3}.

Taking $k_{NO} = 1.26 \times 10^{-10}$ cm^3 s^{-1} and $k_{N2} = 6 \times 10^{-13}$ cm^3 s^{-1} [6], and assuming that these values for 295 K also apply at 275 K, gives $Q = 2.8 \times 10^4$ s^{-1} and $q_f = 0.99$. This leads to:

Line I $\quad q'_f/q_f = 0.9 \quad [NO] \sim 8 \times 10^{13}$ cm^{-3}

Line II $\quad q'_f/q_f = 0.7 \quad [NO] \sim 10 \times 10^{13}$ cm^{-3}.

It can be seen that the change in [NO] when quenching is accounted for is less than the error estimates in the NO concentrations given in Section 3.2.3. The estimates show that a neglect of quenching in the free-stream leads to an error of only ~ 1%, and that for the two concentrations used for lines I and II in the calibration, the neglect of quenching introduces errors of ~10 and ~30% respectively; the value obtained from line I is therefore considered to be the more reliable.

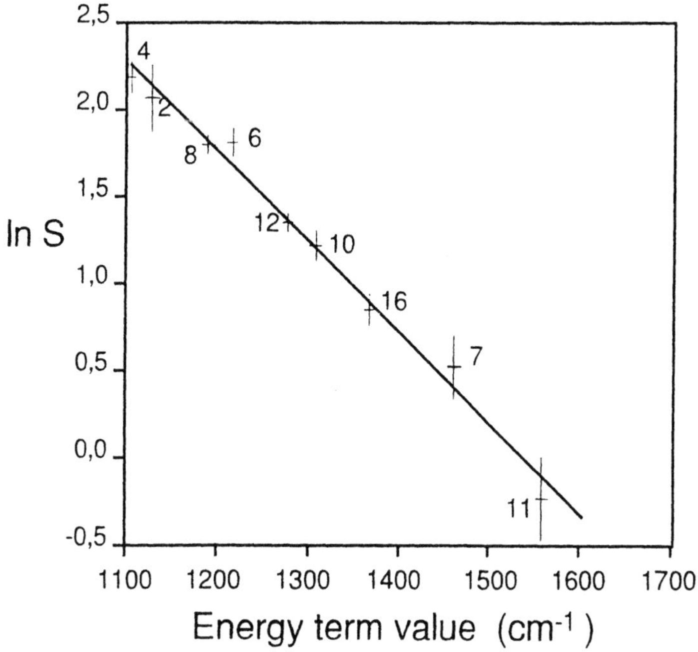

Figure 6. Plot of ln S versus E, where S = individual corrected spectral line intensities for free-stream spectrum (see Fig. 5(a)) and E = energy of vJ level in NO ground electronic state from which transition arose. Error bars correspond to one standard deviation.

5. Further Procedure

The LIF apparatus is undergoing further static tests at the DLR Göttingen; the measuring technique will be refined, the OMA will be readied for use in real-time measurement and high temperature (>500 K) spectroscopy of NO will be carried out in a static cell to prepare for implementation on the high enthalpy shock tunnel at RWTH Aachen (in June 1992) and subsequently on HEG (end of 1992). Processing of the data from L2K in the region model-bow shock will continue. Studies of NO D-state quenching in the shock tube of the University of Göttingen will be extended to study the role of O-atoms as quenching partner.

6. Acknowledgements

The authors gratefully acknowledge the assistance/collaboration of the following: G.Gundlach, W.Sader (DLR Göttingen), T.Seelemann (La Vision 2D Meßtechnik GmbH, Göttingen), P.Andresen (University of Bielefeld), A.Gülhan, D.Erdtel, J.Kalytta, P.Büscher, W.Dathe and D.Lütz (DLR Cologne). The financial support of the ESA through CNES is also gratefully acknowledged.

7. References

1. Beck, W.H., Eitelberg, G., McIntyre, T.J., Andresen, P., Seelemann, T., Hippler, H., Riedl, A. and Troe, J. (1991), "A Laser induced Fluorescence System for the High Enthalpy Shock Tunnel (HEG) in Göttingen", 18th ISSW, 21.-26. July, Sendai, Japan.
2. Beck, W.H., Dankert, C., Eitelberg, G. and Gundlach, G. (1992), "Preliminary Laser Induced Fluorescence Measurements in Several Facilities in Preparation for Application to Studies in the High Enthalpy Shock Tunnel Göttingen (HEG)", 30th ASM, 6.-9. January, Reno, Nevada.
3. Hannemann, K. (1990), "Design of an Axisymmetric, Contoured Nozzle for the HEG", DLR Research Report DLR-FB 90-04.
4. Gülhan, A. (1992), private communication of unpublished results.
5. Gundlach, G. (1992), private communication. Details of simulation program to be published as DLR Research Report (in preparation).
6. Okabe, H. (1978), "Photochemistry of Small Molecules", Wiley, New York.

ROTATIONAL TEMPERATURE AND DENSITY MEASUREMENTS IN A HYPERSONIC FLOW BY DUAL-LINE CARS

F. GRISCH, P. BOUCHARDY, M. PEALAT, B. CHANETZ, T. POT, M.C. COET
Office National d'Etudes et de Recherches Aérospatiales
BP 72, 92322 Châtillon Cedex
France

ABSTRACT. Rotational temperature and nitrogen number density are measured in a low pressure, 55 K, Mach 10 hypersonic air flow. The data are recorded using a non-linear optical technique named Dual-Line Coherent Anti-Stokes Raman Scattering (DLCARS). DLCARS allows single laser-shot spatially-resolved measurements. The flow is probed from the free stream to the vicinity of a two-dimensional model. Averaged temperature and density profiles are compared with those predicted by a Navier-Stokes solver. Good agreement is found.

1. Introduction

The flying conditions of hypersonic vehicles at altitudes between 60 and 100 km, where the pressure is very low, are characterized by large Mach numbers and low Reynolds numbers. The boundary layers are then thick and interact strongly with the inviscid flow field. The parameter used to globally qualify these effects is the viscous interaction parameter $\chi = M/\sqrt{Re_L}$. During the reentry of a 10-m-long vehicle, χ decreases from 0.1 to 0.01 while the altitude decreases from 85 to 50 kilometers. At ONERA, the blow-down wind-tunnel R5Ch was constructed to simulate this domain of values. Its flow is laminar with a unit Reynolds number of Re = 167000 m^{-1}. Equipped with a Mach 10 nozzle, the viscous interaction parameter is $\chi = 0.055$ for a characteristic length of 0.2 m. The facility is presently devoted to validating the Navier-Stokes codes developed for prediction of hypersonic laminar shock-wave/boundary layer interactions [1]. Several optical diagnostic techniques were already used to study the flow: Schlieren visualization [2], holographic interferometry [3] and, recently, electron-beam-induced fluorescence [4]. In the present paper, we report rotational temperature and N_2 number density measurements by Dual-Line Coherent Anti-Stokes Raman Scattering (DLCARS) in the free stream and in the vicinity of a two-dimensional model. The DLCARS technique has already been described [5]. It employs a single frequency pump laser and two independent narrow-band tunable dye lasers. Single shot rotational temperature is obtained by ratioing the CARS intensities of two isolated N_2 Q-lines. Once the temperature is known, the N_2 number density is deduced from one of these two intensities.

A. Boutier (ed.), New Trends in Instrumentation for Hypersonic Research, 225–234.

2. Experiment

2.1. BLOW-DOWN WIND TUNNEL

It consists of a 100 kW Joule effect heater, a contoured nozzle and a large test section followed by a diffuser and a 500 m^3 vacuum sphere. The air heater maintains a stagnation temperature of 1100 K which avoids liquefaction of the air during the expansion in the nozzle. The hypersonic contoured nozzle produces a Mach 10 flow for a stagnation pressure of 2.5 10^5 Pa. The ratio of the contour radius of curvature in the throat region to the throat radius is equal to 80, which leads to a very progressive expansion. The total nozzle length is 1.7 m, and its exit diameter is 0.355 m yielding an isentropic flow diameter of 0.2 m (i.e. discarding the boundary layer). Optical access to the hypersonic flow is allowed over a 0.470-m-long stretch between the exit of the nozzle and the diffuser. The length of the latter is equal to 7.5 m. It insures a very effective recovery of the flow, the value of the pressure at its end being almost seventy times the value at the entrance. The design of this element allows run durations comprised between 60 and 90 seconds, depending on the model obstruction.

For the above stagnation conditions, the free stream pressure, static temperature and total density are p_0= 5.9 Pa, T_0= 52.5 K and ρ_0 = 3.9 10^{-4}kg.m^{-3},respectively. Figure 1 shows the temporal evolution of the stagnation temperature, the stagnation pressure, the Mach number and p_0 during a run. The pressure and the temperature are measured by capacitive transducers and a thermocouple. Very stable conditions are obtained, allowing measurements to be averaged during the run. The Reynolds number and the Knudsen number, calculated with a characteristic length of 0.2 m are 2.9 10^4 and 2.57 10^{-3} respectively which implies that the flow is laminar and that the continuum regime is achieved.

2.2. MODEL DEFINITION

A two-dimensional aluminum model has been chosen. It is 0.15 m wide and 0.27 m long. It consists of a 0.179 m-long sharp-leading-edge flat plate, followed by a ramp tilted by 25° (Fig.2). In order to minimize side effects affecting the two-dimensional nature of the flow, the model is equipped with lateral fences covering the hinge region where a separated flow is likely to form. Effectiveness of these fences was proven by surface flow visualizations using an oil film technique.

The model can be translated while the DLCARS probe volume is fixed. The position of the latter is X and Y with the X-axis along the flow axis and the Y-axis along a direction perpendicular to the model surface (Fig.2).

2.3. DLCARS SETUP

The DLCARS technique [5] requires three pulsed lasers of frequencies ω_1, ω_2 and ω'_2. The frequency differences (ω_1-ω_2) and (ω_1-ω'_2) are tuned on resonance with the Raman frequencies ω_J and $\omega_{J'}$ of two isolated N_2 rotational lines Q(J) and Q(J') of the fundamental vibrational band where J and J' are the rotational quantum numbers of the lower states of the Raman transition.

The core of the system is an injection-seeded Q-switched Nd:YAG laser chain delivering 800 mJ of I.R. in 14 ns-long pulses at a repetition rate of 12.5 Hz. The output is frequency doubled by a KDP crystal, giving 380 mJ at 532.1 nm in 11 ns-long pulses. A second KDP crystal doubles the I.R. left from the first crystal. Some 40 mJ at 532.1 nm are produced and used to pump two narrow-band dye laser oscillators (ω_2 and ω'_2 frequencies). Part of the 380 mJ is used to pump the dye amplifiers whereas the remainder is used as the

ω_1 beams. The resonator of the ω_2 laser comprises a 2400-lines-per-millimeter holographic grating at grazing incidence, a flat rotating back mirror and an intracavity prism expander. The wavelength is selected by tilting the back mirror. The linewidth (full width at half maximum, FWHM) is 0.08 cm^{-1}. The ω'_2 resonator comprises a 2400-lines-per-millimeter grating in Littrow mount as back mirror, and two intracavity solid Fabry-Pérot etalons. They are 0.1 mm and 3 mm thick, respectively. The ω'_2 laser linewidth is 0.08 cm^{-1} (FWHM) and the frequency is tuned by tilting the grating and the etalons. The optical arrangement of the ω_2 and ω'_2 lasers is designed to ensure a good mechanical stability. The output mirror and the dye cell windows are common. After the dye amplifiers, the ω_2 and the ω'_2 laser beams pass through distinct telescopes. Then a dichroic mirror combines the ω_2 and ω'_2 beams with the ω_1 beams. At the output of the CARS bench, the energies are 1 mJ, 1 mJ, 30 mJ and 30 mJ for the ω_2 and ω'_2 beams and for the two ω_1 beams, respectively. The laser frequencies are measured with a Fizeau wavemeter (Laser Technics) and the dye lasers are tuned in resonance with the N_2 Q-branch lines.

2.4. EXPERIMENTAL CONFIGURATION

A schematic diagram of the experimental configuration is presented in Fig.3. First, the beams are focused in the testing chamber with their axis parallel to the leading edge of the model. After splitting off the anti-Stokes beams $S\omega_3$ and $S\omega'_3$ ($\omega_3 = 2\omega_1 - \omega_2$ and $\omega'_3 = 2\omega'_2 - \omega_1$) using a dichroic mirror, the laser beams are focused again in a reference cell filled with 40 Pa of air at room temperature where the reference anti-Stokes beams ($R\omega_3$ and $R\omega'_3$) are created. Figure 4 depicts the laser beam configuration incident on the focusing lens (INPUT) and the recollimating lens (OUTPUT) viewed against the direction of the beam propagation. The ω_2 and ω'_2 beams are phase matched with the two ω_1 beams in the Folded BOXCARS arrangement [6]. In this arrangement, the ω_3 and ω'_3 anti-Stokes beams are parallel and spatially separated by 10 mm. The focal lengths of the achromats must meet conflicting constraints: 1) long focal lengths are preferable in order to reduce the saturation effects which are likely to appear because powerful lasers are required to create measurable signals in low pressure media; 2) short focal lengths are preferable in order to increase the spatial resolution. The set of focal lengths 1 m and 0.75 m has been chosen for the sample and reference channels respectively as the best compromise: the power densities, maintained at acceptable values [7], are equal in the signal and reference channels. The spatial resolution at the sample probe volume is 40 mm long parallel to the leading edge and 150 μm in diameter. The sample beams are focused separately on the entrance slit of a 0.8 m focal length spectrograph used to isolate the anti-Stokes beams from the laser. The $S\omega_3$ and $S\omega'_3$ spots are separated by 10 mm. The spectrograph is equipped with a concave, 50 mm diameter, 2100 lines per millimeter holographic grating. A photodiode array (EGG1456) parallel to the slit detects $S\omega_3$ and $S\omega'_3$. The reference anti-Stokes beams ($R\omega_3$ and $R\omega'_3$) are detected in the same manner. A mini computer COMPAQ 386/33 reads the CARS data, controls the scanning procedure of the photodiode arrays and processes the data.

The software calculates the ratios Q_1, Q_2 and Q where $Q_1 = S\omega_3/R\omega_3$, $Q_2 = S\omega'_3/R\omega'_3$ and $Q = Q_1/Q_2$. First, the rotational temperature T_{rot} is inferred from the ratio Q by the following expression

$$T_{rot} = \frac{\Delta E/k}{1/2 \ln Q + \Delta E/kT_R - 1/2 \ln A} \quad (1)$$

where k is the Boltzmann constant, E(J) is the term energy of the rotational level J, ΔE is the energy difference E(J')-E(J), T_R is the reference temperature and A is a constant deduced from an experiment carried out in a gas at known temperature.

Then, once the rotational temperature is known and after calibrating the CARS signals in a cell filled with air at known pressure and temperature, the nitrogen number density ρ_{N2} is deduced from Q_1 or Q_2. The total density ρ is then deduced from ρ_{N2} knowing the N_2 mole fraction in air; no chemistry effects like O_2 dissociation are likely to appear at temperatures below 1100 K.

3. The Navier-Stokes solver

Modelling of the laminar hypersonic flow is done using a Navier-Stokes solver called HOMARD2 [7]. For a planar flow, the dimensionless time-dependent Navier-Stokes equations are written in integral form in cartesian coordinates as

$$\iint_S \partial W/\partial t \, dS + \int_{\delta S} (F_\beta - Fv_\beta) \, N_\beta \, dl = 0 \qquad (2)$$

where S is a fixed control domain, δS is the boundary of S, dS = dxdy is the surface element, N_β, (β = 1,2) is the outside normal to δS, F_β and Fv_β, (β = 1,2) are the inviscid and viscous flux terms, which are functions of (W_i), (i = 1,4) standing for ρ, ρu_1, ρu_2, ρE. ρ is the density, u_1, u_2 are the components of the velocity vector u and E is the total specific energy i.e. $E = e + u^2/2$ where e is the internal specific energy.

An implicit time-upweighted finite-volume method is used to solve the 2D time-dependent Navier-Stokes equations in the assumption of perfect gas. The inviscid fluxes are evaluated by means of a non-centered second-order total variation decreasing (TVD) scheme [8]. The viscous fluxes are taken into account by a space-centered scheme similar to the Crank-Nicolson scheme [9] where the velocity and energy space derivatives are calculated at the middle of cell sides with formulae involving six points.

The calculations are performed using a 175x50 point mesh. The points are nearly equally spaced along the X axis while the points are spaced according to a geometrical progression along the Y axis. Ten cells are required upstream of the leading edge. The boundary conditions are imposed on the upstream border Γ_1, the upper one Γ_2, the downstream one Γ_3 and on the wall Γ_4. On Γ_1 and Γ_2, a uniform flow is imposed with the conditions of the free stream. On Γ_3, the data are extrapolated from those of the cells inside the domain. On Γ_4, u = 0 and the wall temperature is 290 K. The downstream edge of the ramp is modelled with an expansion corner of deflection -20°. This ramp which does not exist on the model provides a realistic simulation of base flow effect [10].

The calculations are performed on a Cray YMP. Convergence is achieved when residues are less than 10^{-4}. Two hours of CPU calculation time are required. Figure 5 presents the calculated streamlines. It shows the front shock wave induced by the thickening of the boundary layer above the model. The recirculating bubble due to the 25° ramp appears as closed streamlines located in the hinge region. Formation of a region of separated flow is confirmed by Fig.6a which represents the calculated skin-friction coefficient versus X. Figure 6b presents a picture of the surface flow showing the skin-friction line pattern. Accumulation of oil before the ramp clearly marked the separation line at X/L=0.74 where L = 0.179 m, while the reattachment line is less visible. The separation line revealed by the visualization perfectly coincides with the prediction of the skin-friction coefficient zero-crossing. Also, the picture shows that edge effects are confined in a relatively reduced area.

4. Experimental results

4.1. ROTATIONAL TEMPERATURE AND DENSITY MEASUREMENT ACCURACIES

The N_2 Q-lines which the frequency differences (ω_1-ω_2) and (ω'_1-ω'_2) have to be tuned to, must be selected to optimize the rotational temperature and density measurement accuracies. Increasing ΔE is a priori advantageous but also deteriorates the standard deviation on Q_1 (or Q_2) which is rapidly dominated by Poisson statistics when the Q-line becomes weak. For the present study, the set Q(6), Q(8) is selected when T_{rot} is below 150 K and the set Q(6), Q(16) when T_{rot} is between 150 and 600 K. At 55 K and $\rho_0 = 3.9\ 10^{-4}$ kg.m^{-3}, i.e. the conditions of the free stream, the Q(6) and Q(8) signal strengths are 4 10^4 and 2.8 10^3 photocounts per pulse respectively. Increasing T_{rot}, the Q(6) signal decreases gently while the Q(8) signal increases rapidly. At T_{rot}= 150 K, the Q(16) signal then reaches about 4 10^2 photocounts per shot. Using these sets of Q-lines, the relative standard deviations of single shot measurements of T_{rot} and ρ are below 0.1. Averaging N laser shots gives T_{rot} and ρ with expected relative standard deviations of 0.1/ N.

4.2. RESULTS

In the free stream, 300 laser shots are averaged during a run. Averaging repetitive experiments on fourteen runs yields $\langle T_{rot} \rangle$= 54.8 K and $\langle \rho \rangle$ = 3.68 10^{-4} Kg.m^{-3}, in good agreement with the predictions, i.e. 52.5 K and 3.90 10^{-4} kg.m^{-3}, and their standard deviations are 1.2 K and 0.15 10^{-4}kg.m^{-3} respectively. The latter depend on both the accuracy of the DLCARS measurement and the reproducibility of the stagnation conditions of the runs. At the location X = 18.8 10^{-3}m (X/L = 0.1) and Y = 2.10^{-3}m, close to the leading-edge of the model, averaging over the time is forbid den because T_{rot} increases by 75 K (Fig.7) whilst the edge of the model which is more or less blunt, is heated during the run. By thermal conductivity, the first few centimeters of the model are also perturbed. At the end of the ramp (X/L = 1.2) where heat-fluxes to the wall increase, a similar phenomenon is also observed. Elsewhere, these effects do not appreciably influence the wall temperature anymore, allowing time-averaging of the measurements. Nevertheless, measurements at positions located in the vicinity of the wall are always performed at the beginning of the runs.

Rotational temperature and density profiles have been measured at five sections, located at X/L = 0.4, 0.6, 0.8, 1.0 and 1.2. The first two profiles at X/L = 0.4 and 0.6 are located upstream of the separation line (Fig. 8a, 8b).

In Fig. 8a at X/L = 0.4, the temperature and the density are those of the free stream for Y > 17.10^{-3}m. The temperature and density jumps found at height Y = 17.10^{-3}m are due to the shock created by the leading edge interaction. For Y < 10.10^{-3}m, the temperature first increases through the boundary layer due to the viscous effects which decelerate the flow and then decreases to reach the wall temperature. Meanwhile, the density decreases and then is nearly constant in the part of the boundary layer close to the wall. The maximum of the density reaches 2.5 ρ_0. At X/L = 0.6, the same behavior is observed (Fig. 8b).

Profiles located in the separated region (X/L = 0.8, 1.0 and 1.2) are shown in Figs.8c-8e. The profiles present the same features and reflect the development of the boundary layer. At X/L = 1 and 1.2, the temperature profiles remain similar although the temperature jump becomes less and less apparent. The density profiles show clearly the effects of the shock induced by the corner. More precisely, the density jump located above Y = 35.10^{-3}m is due to the leading edge shock and the density jump located at Y = 26.10^{-3}m is due to the separation shock.

Whatever the position, a good agreement is observed between experimental and theoretical temperature although T_{rot} is compared with T_{static}. Indeed, translation and rotation are expected to be in equilibrium at least for the rotational states with a rotational energy below kT_{static}. For upper rotational states, the distribution of the rotational population deviates certainly from a Boltzmann distribution as already observed [11]. Unfortunately, the sensitivity of DLCARS does not permit observation of these effects in the R5Ch flow. Concerning the density, the agreement between the experiment and the theory is also satisfactory. However, a significant discrepancy appears in Fig. 8e where the measured maximum density is 40% higher than the predicted ones. The most probable explanation is the increasing three dimensional character of the flow for positions far downstream. The curvature of the reattachment line (Fig. 6b) shows clearly these effects. Fences with better design or three dimensional modelling should improve the agreement.

5. Conclusion

DLCARS is shown to be a sensitive technique to probe molecules under the conditions encountered during the reentry flight of shuttles. For the first time, time-average temperature and density are simultaneously measured in the vicinity of a two-dimensional model and then compared to the predictions of a Navier-Stokes solver. Further works are required to improve the accuracy of single-shot measurements in order to carry out experiments in wind tunnels with short duration runs. Studies of other two-dimensional models are now in preparation. DLCARS could also provide interesting results concerning the vibrational excitation of N_2.

Acknowledgements: The authors would like to thank C. Marmignon for using the HOMARD2 code and J. Délery for helpful discussions.

References

[1] Délery J. (1988) "Shock/shock and shock wave/boundary layer interactions in hypersonic flows", AGARD-FDP-VKI special course on Aerothermodynamics of hypersonic vehicles, AR n°761.

[2] Girard A., Philbert M. et al. (1990) "Développement des moyens de mesures adaptés aux souffleries hypersoniques", report ONERA RT N° 35/3409 PN.

[3] Surget J., Dunet G. (1990) "Multipass holographic interferometry for low density gas flow analysis", 19th international congress on high speed photography and photonics, Cambridge (G.B.).

[4] Mohamed A.K. (1991) "Sondage par faisceau d'électrons des écoulements hypersoniques en rafales brèves", Thesis, university of Orsay (France).

[5] Péalat M., Lefebvre M. (1991) "Temperature measurement by single-shot Dual-Line CARS in low-pressure flows", Appl. Phys. B 53, 23-29.

[6] Prior Y. (1980) "Three dimensional phase matching in four-wave mixing", Appl. Opt. 19, 1741-1743.

[7] Hollanders H., Marmignon C. (1989) "Navier-Stokes high speed flow calculations by an implicit non-centered method", 27th AIAA Aerospace Sciences Meeting, Reno (Nevada, USA).

[8] Yee H.C. (1987) "Upwind and Symmetric Shock-Capturing Schemes" report Nasa TM 89464.

[9] Peyret R., Taylor T.D. (1982) "Computational methods for fluid flow", Springer Series in computational physics.

[10] Joulot A. (1991) "Contribution of ONERA to problem III : 2D Laminar Ramp Flows" Second Workshop on Hypersonic Flows for reentry Problems, Antibes (France).

[11] Dam N.,Liedenbaum C.,Stolte S.,Reuss J. (1987) "Rotational distributions in seeded molecular beams", Chem.Phys.Lett. 136, 73-80.

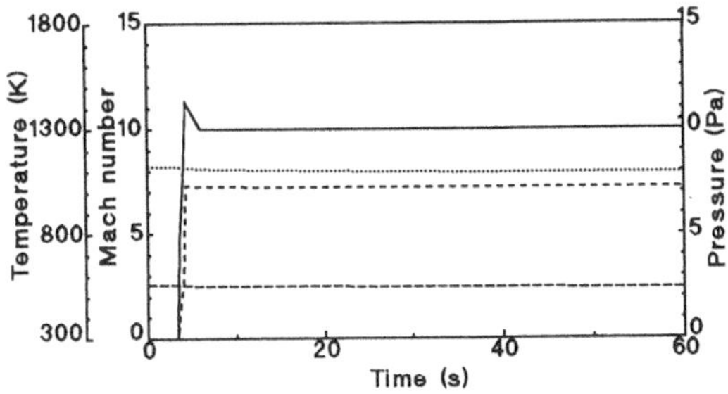

Figure 1. Temporal evolution of the stagnation pressure (— — —), stagnation temperature (•••), Mach number (—) and static pressure (- - -) measured by conventional probes. The stagnation pressure has been divided by 10^5.

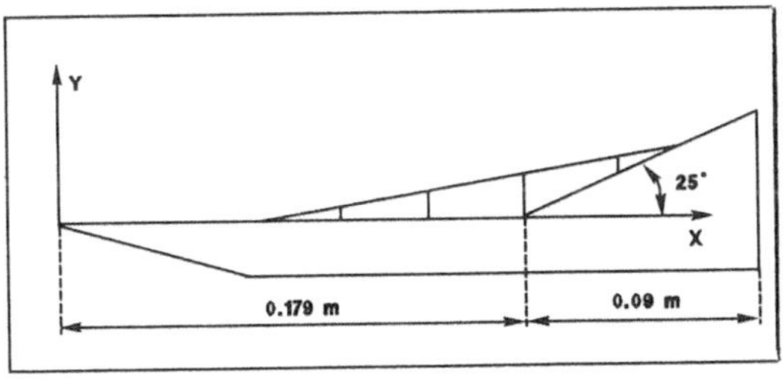

Figure 2. Sketch of the two-dimensional model with its fences and four vertical slits located at X/L = 0.6, 0.8, 1.0 and 1.2 where L = 0.179 m. The X and Y axis are also shown.

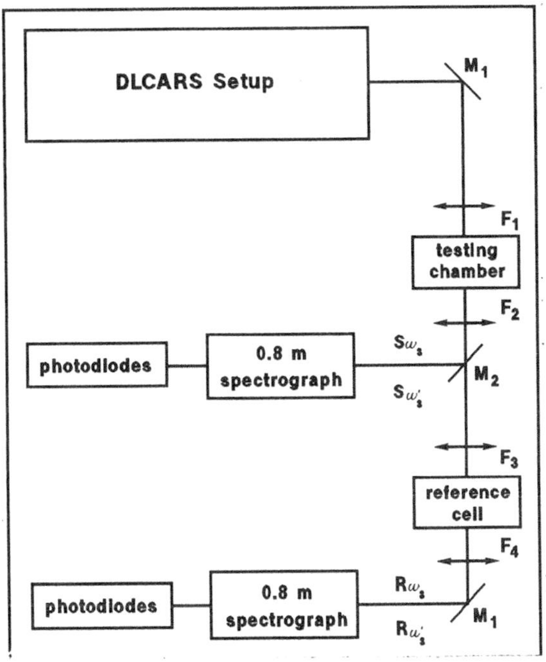

Figure 3. Schematic diagram of the DLCARS arrangement. F_1, F_2: 1 m focal length achromats; F_3, F_4: 0.75 m focal length achromats; M_1: broadband mirror, M_2: dichroic mirror.

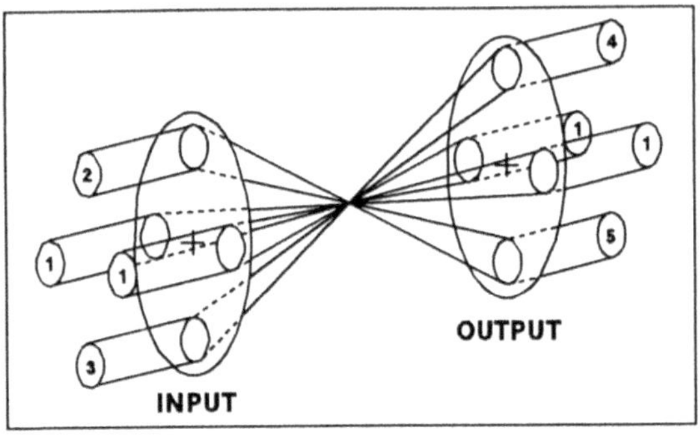

Figure 4. Laser beam arrangement for the DLCARS experiment. The beam geometry is shown in perspective on the focusing (INPUT) and recollimating (OUTPUT) achromats. 1 = ω_1 beams, 2 = ω_2 beam, 3 = ω'_2 beam, 4 = ω'_2 and ω_3 beams, 5 = ω_2 and ω'_3 beams.

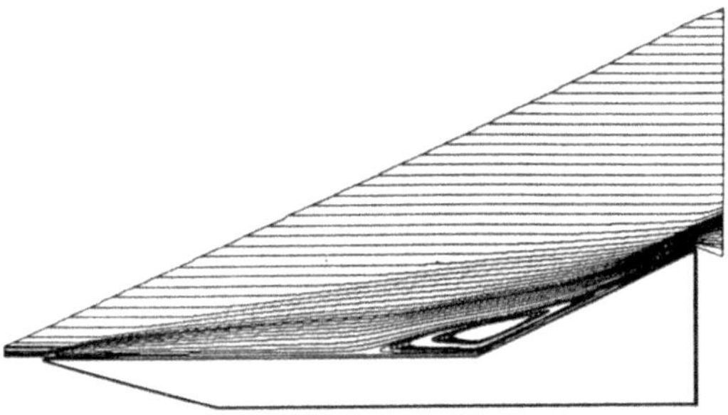

Figure 5. Streamlines obtained from the HOMARD2 code.

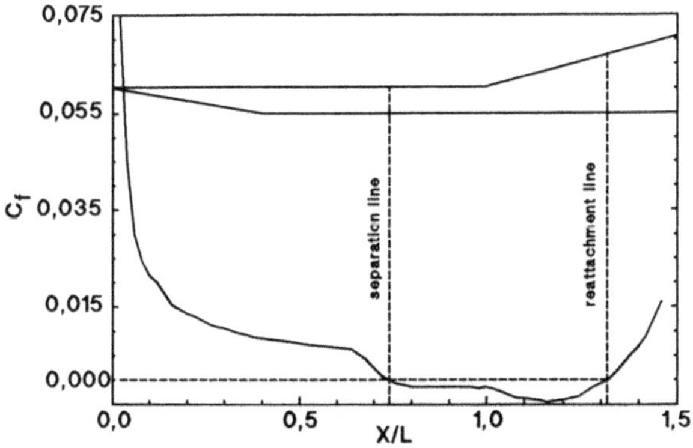

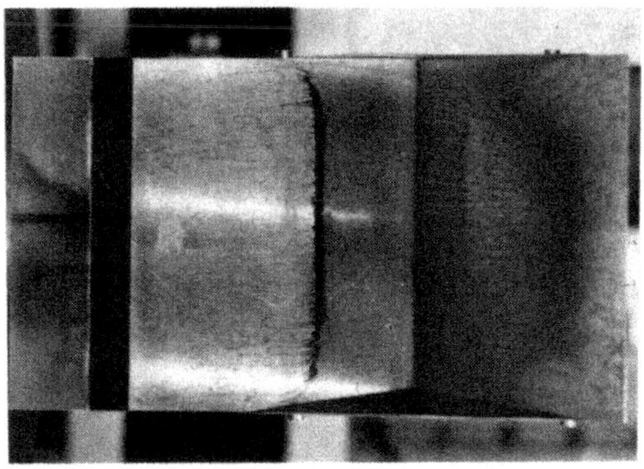

Figure 6. Skin-friction coefficient: a) calculated using the HOMARD2 code; b) view of the oil-wetted surface during a run.

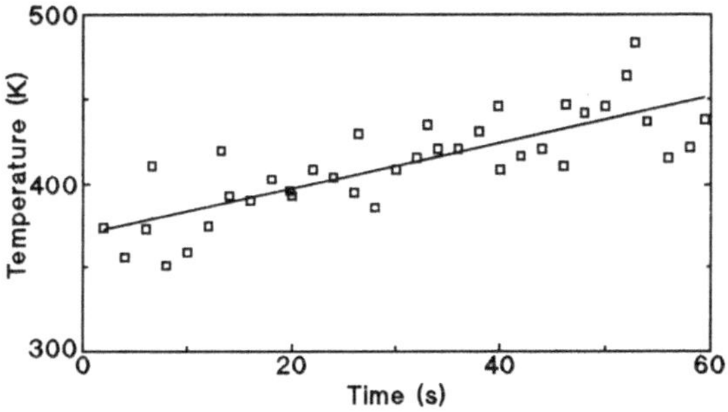

Figure 7. Temporal evolution of the rotational temperature at the position X = 18.8 10^{-3} m (X/L = 0.1), Y = 2. 10^{-3}m. Each point is the average of 20 laser shots viz a 1.6s averaging. The relative standard deviation of the temperatures is 0.04. The straight line is obtained by a least squares routine.

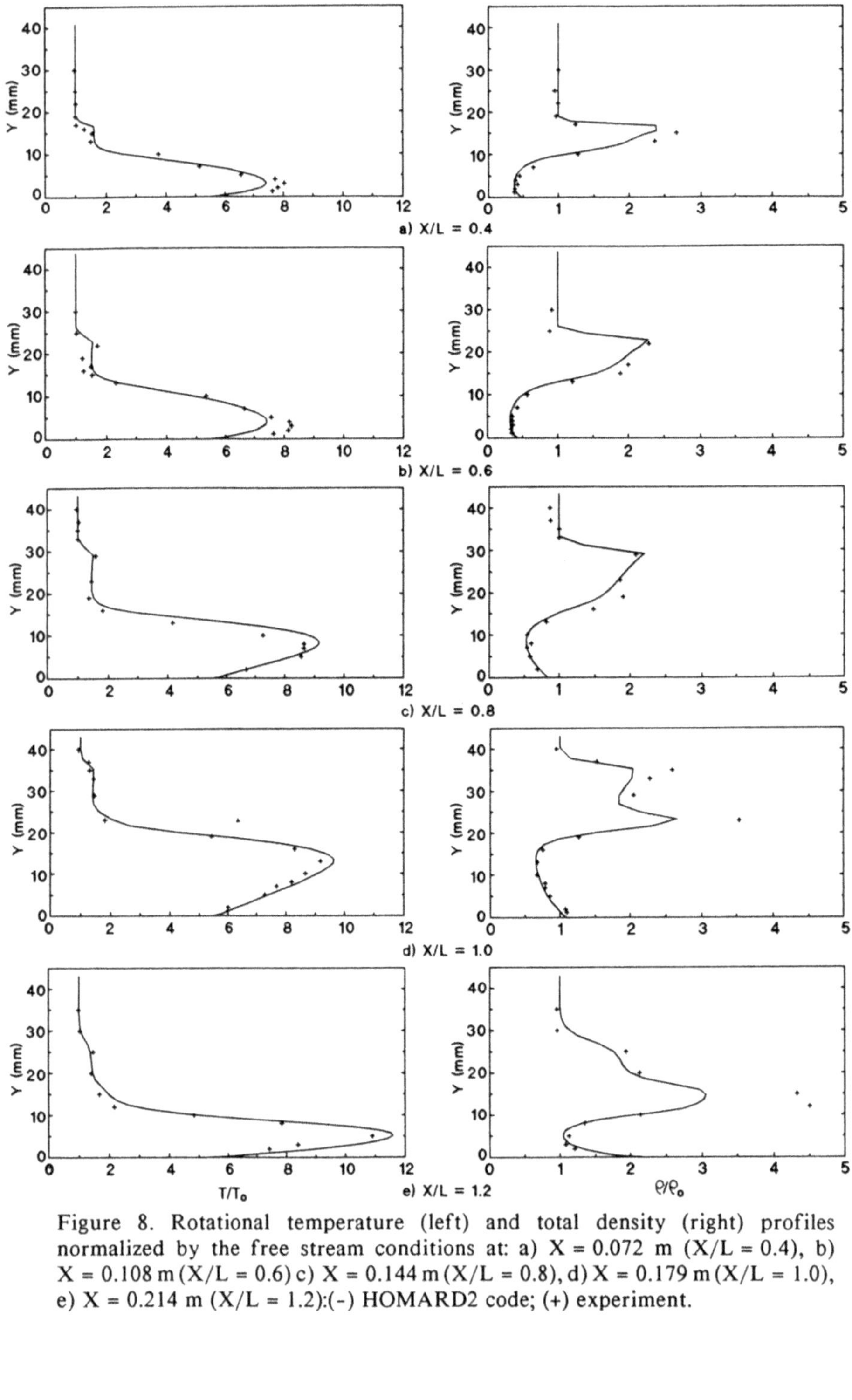

Figure 8. Rotational temperature (left) and total density (right) profiles normalized by the free stream conditions at: a) X = 0.072 m (X/L = 0.4), b) X = 0.108 m (X/L = 0.6) c) X = 0.144 m (X/L = 0.8), d) X = 0.179 m (X/L = 1.0), e) X = 0.214 m (X/L = 1.2):(-) HOMARD2 code; (+) experiment.

LASER-SPECTROSCOPIC MEASUREMENT TECHNIQUES FOR HYPERSONIC, TURBULENT WIND TUNNEL FLOWS

ROBERT L. MCKENZIE AND DOUGLAS G. FLETCHER
NASA Ames Research Center
Moffett Field, California, USA

1. Introduction

Until recently, aerodynamic measurements obtained away from a surface have been limited either to the use of intrusive probes to determine pressure, mass flux, and total temperature or to visualization methods for indications of the flow geometry. New laser-spectroscopic methods are now emerging which offer compelling opportunities to obtain simultaneous measurements of density, temperature, and their fluctuations, with unprecedented spatial and temporal resolution [1]. As a result, measurement techniques are now feasible which can be used in supersonic and hypersonic flows to characterize the effects of compressibility on turbulent behavior and to obtain quantitative descriptions of complex, three-dimensional flow fields.

Most of the laser-spectroscopic techniques incorporate laser-induced fluorescence (LIF) or Raman spectroscopy. The primary objective of this report is to describe the spectroscopic nature, present status, and capabilities of some LIF and Raman techniques which have been developed at NASA Ames Research Center. Particular emphasis has been centered on applications in hypersonic wind tunnels where measurements with more conventional probes are often inadequate or impractical. The requirements for laser-spectroscopic measurements in hypersonic flows are dictated by their intended aerodynamic research applications. By design, hypersonic flow fields over test articles are generally free of large stagnation regions. Consequently, in moderately heated facilities, they can have densities lower than 0.01 amagat and temperatures ranging from 50 K to 300 K. To provide sufficient sensitivity at the low temperatures, the spectroscopy must be based on individual rotational transitions in the interacting species. In addition, most applications are generally limited to chemically stable species which have been seeded into the flow or which occur naturally in air.

With these requirements in mind, two principal laser-spectroscopic approaches have been developed at NASA Ames Research Center for the measurement of temperature, density, and their fluctuations owing to turbulence in high-speed flows. Early work was centered on the use of near-ultraviolet lasers to induce fluorescence from trace concentrations of nitric oxide (NO) added to flows of nitrogen. More recently, advances in laser technology have allowed similar measurements to be made in unseeded air flows using vacuum-ultraviolet lasers to excite molecular oxygen.

A. Boutier (ed.), New Trends in Instrumentation for Hypersonic Research, 235–244.

2. Laser-Induced Fluorescence in Flows Containing Nitric Oxide

Prior to the availability of high-energy excimer lasers, a spectroscopic approach was developed which relied on the existing Nd:YAG-pumped, tunable dye laser technology. It was based a dual-frequency LIF technique in which two frequency-doubled dye lasers were each tuned to a near-ultraviolet, ro-vibronic transition in the Gamma Band spectrum of nitric oxide [2]. Pulse energies of 1 μJ could be achieved over the entire electronic band near 225 nm. The two independent, coincident LIF signals allowed density and temperature to be determined simultaneously. NO was added in concentrations of less than 100 ppm to a wind tunnel reservoir containing pure nitrogen. Because the NO fluorescence is quenched efficiently by oxygen and very little by nitrogen, the technique is most applicable in pure nitrogen flows.

The capabilities of the method were demonstrated in the turbulent boundary layer on the wall of a Mach-2 channel flow. Single-pulse measurements were obtained with a spatial resolution of less than 1 mm and uncertainties of less than 2%. A small blow-down wind tunnel was used which was designed to handle gas mixtures containing NO. Since NO is toxic in concentrations greater than 25 ppm, a test facility suitable for NO-LIF must incorporate all of the special features necessary to safely use toxic mixtures. Despite this inconvenience for general applications, the NO-LIF technique demonstrated the unique capabilities of laser-spectroscopic methods for aerodynamic measurements. Its success encouraged investigations of similar techniques which could be applied to unseeded air flows.

3. Raman Scattering and Laser-Induced Fluorescence in Air Flows

3.1. BACKGROUND

The development of high energy, vacuum-ultraviolet, ArF excimer lasers made feasible the use of LIF from oxygen molecules in air. Exploration of the concept was initiated in 1981, before such lasers were available commercially, with a study solicited by NASA-Ames and performed by Massey and Lemon [3]. They were able to construct a tunable, ArF laser and demonstrate its use to obtain O_2 LIF signals in room air. Their measurements indicated that the LIF was sufficiently intense and sensitive to temperature to be a viable measurement technique in unseeded air flows.

The results of Massey and Lemon and the subsequent commercial availability of high-energy ArF excimer lasers encouraged further development of the concept. However, unlike the two-laser system used for NO-LIF, the cost and complexity of excimer lasers made the use of two lasers unattractive. Alternatively, Gross and McKenzie [4] proposed that both the fluorescence emission and the Raman scattering from each pulse of one laser be used to provide the two independent signals required for simultaneous temperature and density measurements. Since then, Laufer et al. have characterized the spectroscopy of O_2-LIF [5] and demonstrated its use to measure temperatures down to 130 K in a cell at known pressures [6]. The development was continued by Fletcher who incorporated the Raman signal to obtain simultaneous measurements of temperatures and densities, without knowledge of any other thermodynamic parameters [7]. Most recently, he has demonstrated the use of the O_2-LIF/Raman technique to measure temperature, density, and their fluctuations in the turbulent boundary layer of the same Mach-2 channel flow which was studied previously using NO-LIF [8].

3.2. O_2-LIF/RAMAN SPECTROSCOPY

The O_2 LIF process is characterized by the energy-level diagram shown in Fig. 1. The tuning range of the ArF laser is centered at 193.3 nm with a bandwidth of approximately 0.5 nm. Within that band, absorbing transitions may be accessed in the $O_2\ B^3\Sigma_u^- \leftarrow X^3\Sigma_g^-$ Schumann-Runge band from v" = 0 to v' = 4,5. Hot-band vibrational transitions are also accessible, including v" = 1 to v' = 7,8; and v" = 2 to v' = 10,11,12. In each vibrational band, the rotational transition pairs, P(K") and R(K" + 2), are at nearly the same frequency. The pairs associated with K" = 15, 17, and 19 from v',v" = 4,0 have been the principle absorbing transitions used for LIF at room temperature and below. Below room temperature, the population of levels with v" > 0 is extremely small. However, the Frank-Condon factor for v',v" = 7,1 transitions is 200 times greater than for v',v" = 4,0 so that only a small population of the v" = 1 level is necessary to produce observable hot-band absorption. The effect starts to become significant at temperatures slightly above 300 K.

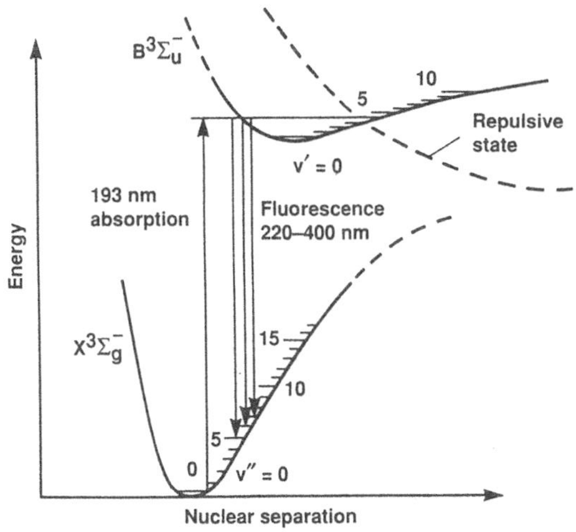

Figure 1. Energy-level diagram for O_2 laser-induced fluorescence by an ArF laser.

The $B^3\Sigma_u^-$ upper state of O_2 is crossed near the v' = 4 level by an array of repulsive states which are represented by a single curve in Fig. 1. Consequently, most molecules which are elevated to v' ≥ 4 by the LIF process predissociate within a few picoseconds. The molecules that do not predissociate must have a radiative lifetime which is equally short. Consequently, it is much less than the time between molecular collisions at most wind tunnel test conditions. Collisional quenching effects on the LIF signals may then be neglected, thereby greatly simplifying the signal analysis required to determine temperature and density.

For fluorescence originating from v' = 4, the Franck-Condon factors favor spontaneous transitions back to all $X^3\Sigma_g^-$ vibrational levels above v" = 4. The resulting fluorescence spectrum extends from 220 nm to greater than 400 nm. Although the total fluorescence signal is maximized if the entire band is observed, spectral interference from other laser-induced processes can limit the practical range of observation. In any case, at least half of the fluorescence energy is from wavelengths below 300 nm, which includes O_2 vibrational emission bands up to v',v" = 4,13.

The LIF emission provides a signal which is sensitive to both temperature and density. The Raman signal can be made sensitive only to density. Taken together, the two contributions to the laser-induced spectrum allow temperature and density to be determined from each laser pulse.

An experimental spectrum of laser-induced radiation from air at room temperature is shown in Fig. 2. It was obtained using an optical multi-channel analyzer (OMA) at the exit plane of a small spectrometer with a 0.5-mm entrance slit. Because the image of the laser beam lies with its axis orthogonal to the slit, the slit width also determines the length of the observed sample volume. In this example, the spectral width of each feature is dominated by the width of the slit image. In addition, the grating has been positioned so that the contribution from Rayleigh scattering below 196 nm does not illuminate the OMA photo-diode array. The Δv notation refers to the overall change in vibrational quantum number resulting from the entire Raman or LIF process. The features denoted by $\Delta v = 1,2$ are the Raman fundamental and first overtone transitions of O_2. The Raman $\Delta v = 1$ transition from N_2 falls between them. The strengths of the O_2 Raman scattering are resonantly enhanced at 193 nm, which significantly improves the viability of Raman measurements in air at low densities. At room temperature and below, the fluorescence underlying the O_2 $\Delta v = 1$ Raman transition is negligible because the v',v" = 4,1 Frank-Condon factor is so small. Hence, the $\Delta v = 1$ transitions of O_2 and N_2 may be assumed to result only from Q-branch Raman transitions for which all rotational components are at nearly the same frequency. The resulting, spectrally integrated signal is independent of temperature and is linearly proportional to the species density. The sum of O_2 and N_2 Raman contributions provide a maximum signal for determining the gas density. The O_2 $\Delta v = 2$ feature is not included in the sum to avoid the small influence of underlying fluorescence.

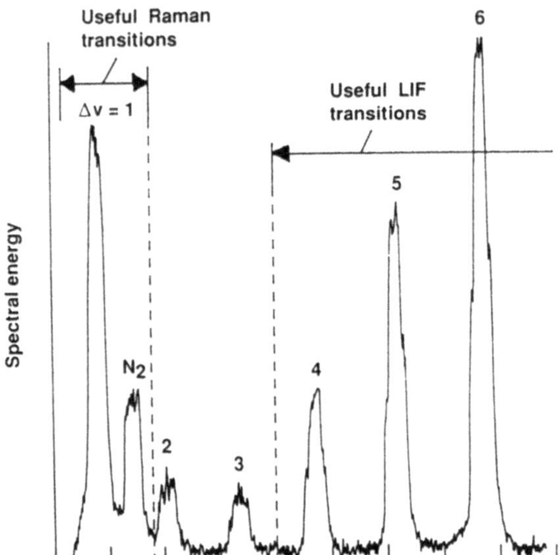

Figure 2. Emission spectrum from room air induced by an ArF laser with low fluence.

The LIF transitions at the wavelengths captured by the OMA diode array also appear in Fig. 2. The entire useful spectrum extends to $\Delta v = 13$ at 300 nm. The $\Delta v = 3$ transition is avoided because it contributes only a small fraction to the total fluorescence energy and includes a weak contribution from the second overtone of the resonance-enhanced O_2 Raman scattering.

Because the laser energy typically fluctuates ±5% from pulse to pulse, the Raman and LIF spectral energies for each pulse must be normalized by the corresponding laser energy. However, only the laser pulse energy can be measured accurately, but not its fluence (energy per unit area). Consequently, the observed radiative processes must be dominated by interactions for which the signal energies vary linearly with laser energy and are independent of the distribution of fluence within the beam cross section. The dominance of linear LIF processes depends both on the magnitude of laser fluence and on its wavelength. For example, Fig. 3 shows that as the laser fluence exceeds about 1.5 J/cm^2, both the Raman and LIF energy behave quadratically, depending on the transition excited. This behavior is the consequence of underlying multi-photon processes with excitation rates that increase non-linearly with laser fluence and become dominant at high values.

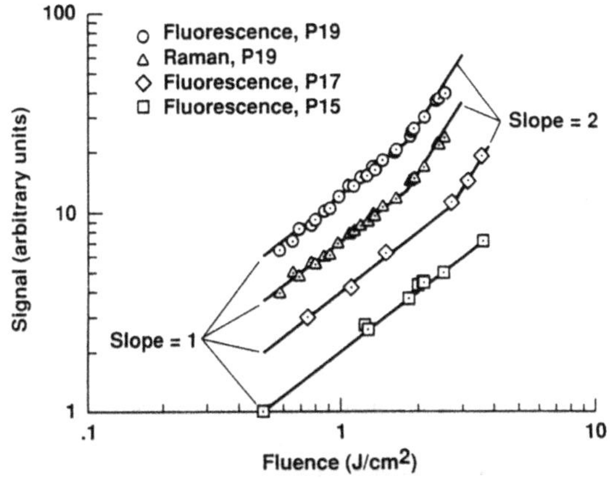

Figure 3. Dependence of total fluorescence and Raman signals on the spatially-averaged fluence of a narrowband ArF laser. (Note: The signal magnitudes for each case have been arbitrarily separated for clarity.)

The dependence of the non-linear processes on laser wavelength is demonstrated more explicitly in Fig. 4. It shows an excitation spectrum in which the total LIF energy is monitored while the laser is tuned through the absorbing O_2 transitions. The laser has been focused to achieve a high average fluence. The non-linear resonances, denoted by asterisks, do not appear when an unfocused beam of the same pulse energy is used. Most of them have been identified as resonances associated with the photoion, O_2^+. The resonant wavelengths of the non-linear processes influence the choice of O_2 transitions used to obtain LIF measurements. For example, the P19 and R21 transitions are both strongly affected. Interference underlying the Raman transitions also occurs at this laser wavelength, as shown by Fig. 3. The P17 transition can be affected similarly if the laser is not carefully tuned or sufficiently narrowband. The nearby R19 transition produces weaker LIF signals than those from P17 but lies farther from the non-linear resonance. Finally, the P15 transition is apparently free of nearby non-linear resonances for the range of fluence investigated.

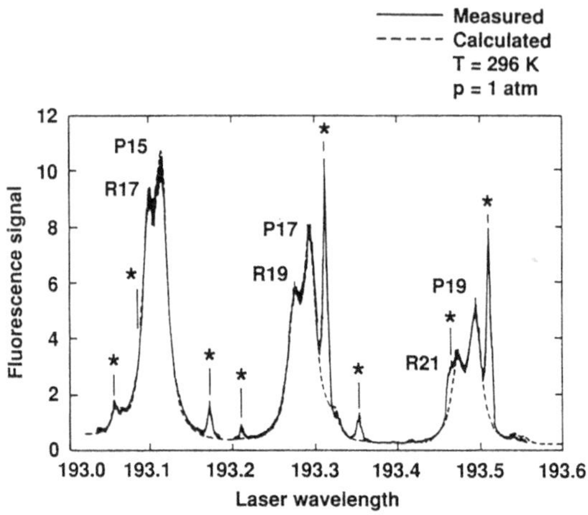

Figure 4. Oxygen excitation spectrum stimulated by a focused laser beam with an average fluence of about 1.5 J/cm^2. Multi-photon transitions are denoted by asterisks.

3.3. APPLICATION TO A MACH-2 TURBULENT BOUNDARY LAYER

Work is now in progress to demonstrate the use of the O_2-LIF/Raman technique for simultaneous measurements of temperatures, densities, and their fluctuations owing to supersonic turbulence [8]. The characteristics of the test flow have been described previously [2]. It includes a 12.1-mm thick, fully turbulent boundary layer that is adjoined by a relatively undisturbed flow of the same thickness.

In Fig. 5, the average temperature and density profiles obtained from O_2-LIF/Raman data are compared with distributions implied from pressure measurements using a pitot probe and a static pressure port in the wall below the sample volume. Spatial resolution of the spectroscopic measurements was about 1 mm. Agreement of comparable data is within ±3%. Since the flow is in thermochemical equilibrium and both state variables have been measured simultaneously, the O_2-LIF/Raman data may also be used in conjunction with the equation of state to determine the local pressure. The pressure distribution is nearly constant, as expected for a flat-wall boundary layer. These results are identical to those obtained previously using NO-LIF.

Figure 6 shows a comparison of the root-mean-square (rms) fluctuation amplitudes of temperature and density obtained from both spectroscopic methods and from hot-wire data. Again, the NO-LIF and O_2-LIF/Raman results are in agreement. They both show a region in the boundary layer near $Y/\delta = 0.6$ where the density fluctuations are exceptionally large. A histogram of the NO-LIF fluctuations displays a bimodal character which suggests that a weak wave was oscillating across the sample volume. Such density fluctuations also imply the presence of pressure fluctuations in that region. On the other hand, an analysis of the hot-wire signals to obtain density fluctuations requires the assumption that pressure fluctuations are negligible. Clearly, that is not the case in this flow and the effects of pressure fluctuations would not have been observed without the use of the LIF instrumentation. These comparisons demonstrate that the O_2-LIF/Raman technique can provide accurate measurements in air flows at the low temperatures and densities which are prevalent in high-speed wind tunnels.

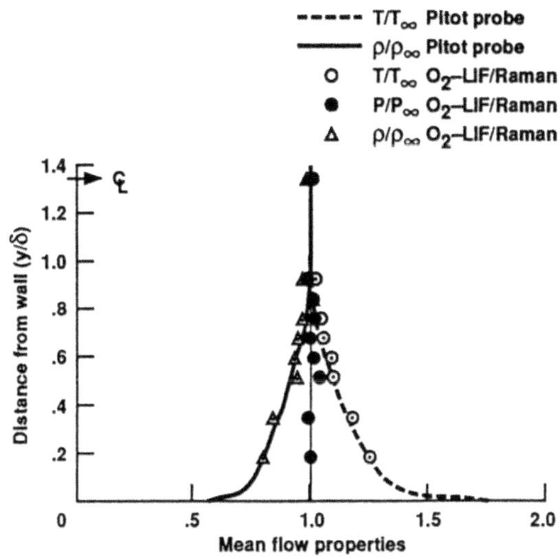

Figure 5. Time-averaged temperature, density, and pressure distributions in a turbulent boundary layer with a centerline Mach number $M_\infty = 2.06$, stream temperature $T_\infty = 158$ K, stream pressure $P_\infty = 0.35$ atm, and thickness $\delta = 12.1$ mm. Local values are normalized by their respective free stream values.

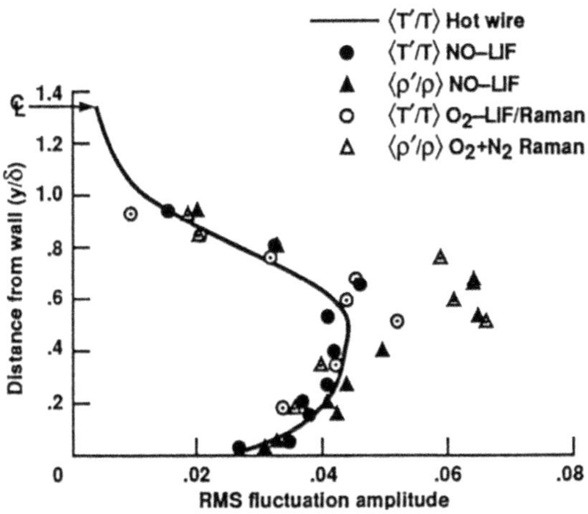

Figure 6. Comparisons of temperature and density rms-fluctuation amplitudes obtained with LIF techniques or implied by hot-wire anemometer data. Flow conditions are the same as in Fig. 5. For each variable, the fluctuation amplitude is normalized by the local time-averaged value.

4. Applications to Large-Scale Hypersonic Flows

A compelling application of the O_2-LIF/Raman technique is to flows in the NASA-Ames 3.5 ft. Hypersonic Wind Tunnel. Its test region is a open free-jet at Mach numbers of 5, 7,and 10, depending on the nozzle. The test environment serves as a relevant example for evaluations of the capabilities of O_2-LIF/Raman techniques.

LIF measurements have been demonstrated in which uncertainties greater than 1% may be characterized solely by the unavoidable effects of photon-statistical noise in the radiative signals [6]. In that case, the uncertainty or rms-noise/signal ratio for pulsed signals is simply equal to $[n_e]^{-1/2}$, where n_e is the average number of photo-electrons detected per pulse. This feature allows the corresponding uncertainties in temperature and density to be predicted in terms of the laser pulse energy and the efficiencies of the detector and collection optics. It may also be used to extend predictions based on the design parameters chosen used here to other values of pulse energy, f-number, etc.

For a given temperature, the weakest LIF signals are always obtained at the lowest densities. The density, $\rho = 0.02$ amagat, is representative of the free-stream density at Mach 7. To evaluate the capabilities of O_2-LIF/Raman measurements, the corresponding uncertainties in temperature measurement may be calculated using the most favorable set of optical parameters believed achievable and assuming that the optical system accepts fluorescence emission from all upper states with $\Delta v \leq 11$. In practice, more restrictive filtering of the emission spectrum may be required, depending on the experimental arrangement.

Among the significant parameters related to the performance of the collection optics system are its effective f-number and the transmission and reflection losses of its optical components. A system design based on f/2 optics, for which only 1.5% of the total radiative emission is collected, will still require a large device to match the focal lengths necessary in a large-scale facility. For example, in the Ames hypersonic facility, the optics can not be much closer to the flow centerline than 1 m, thereby requiring an f/2 system to have an aperture diameter of 50 cm. If a Cassegrain telescope is used, its two mirrors typically provide a combined reflectivity of 75% at 200 nm. In addition, the mirrors must be protected from heat and dust from the free-jet flow by several window panes of fused silica. Their combined transmission, including absorption and surface reflection losses, will be near 60%. Finally, isolation of the Raman spectra requires a monochromator with at least a grating, even if all other mirrors are eliminated. Grating efficiencies near 200 nm are typically less than 50%. Finally, losses at the monochromator entrance slit owing to imperfect imaging can exceed 10%. Thus, an overall transmission of no better than 20% should be expected. This value, combined with the collection efficiency of f/2 optics and a representative detector quantum efficiency of 12%, leads to an anticipated efficiency for conversion of emitted photons to detected photoelectrons of only 1/3000.

The laser pulse used for the calculations has been limited to 15 mJ, to stay within the range of fluence for linear behavior with a 1-mm spatial resolution at all laser wavelengths. However, linear behavior has been demonstrated for excitation of the P15 transition with pulse energies to 30 mJ, as Fig. 3 indicates.

The effect of the Raman signal uncertainty is also included in the temperature uncertainty calculation since the LIF signals are proportional to density. For the parameters given, the density measurements will contain a 5% rms-uncertainty owing to photon-statistical noise. However, density fluctuations owing to turbulence at hypersonic speeds are expected to be larger.

The variations in temperature measurement uncertainty are shown in Fig. 7 for narrowband laser wavelengths which are resonant with the absorbing transitions indicated. At the lowest temperatures, the P15 transition is superior and can be further enhanced in practice by using higher pulse energies. However, its useful range of temperatures is limited and superseded by the R19 transition above 140 K. The P17 transition is not attractive at low temperatures because it is most favorable in only a limited temperature range and susceptible to non-linear interference. Above 300 K, the sensitivity of any v',v" = 4,0 transition must be compared with that available from excitation of the hot-band v',v" = 7,1 transitions [6].

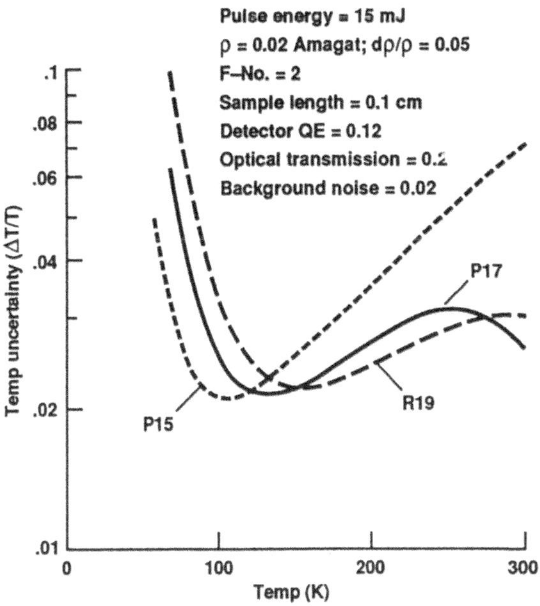

Figure 7. Predictions of temperature measurement uncertainties owing to photon-statistical noise for constant density and laser pulse energy.

5. Conclusions

These results demonstrate that laser-spectroscopic techniques can be used to obtain measurements of temperature, density, and their turbulent fluctuations and they show the feasibility of using a carefully tuned, narrowband ArF excimer laser and O_2-LIF/Raman spectroscopy to measure temperatures and densities at hypersonic wind tunnel conditions with uncertainties of less that 5%.

6. References

1. McKenzie, R. L. (1991) "Progress in Laser-Spectroscopic Techniques for Aerodynamic Measurements—An Overview," AIAA Paper 91-0059.

2. Gross, K. P., McKenzie, R. L., and Logan, P. (1987) "Measurements of Temperature, Density, Pressure, and Their Fluctuations in Supersonic Turbulence Using Laser-Induced Fluorescence," Exp. in Fluids, 5, pp. 372-380.

3. Massey, G. A. and Lemon, C. J. (1984) "Feasibility of Measuring Temperature and Density Fluctuations in Air Using Laser-Induced O_2 Fluorescence," IEEE Journ. Quantum Elect., QE-20, pp. 454-457.

4. Gross, K. P. and McKenzie, R. L. (1985) "Optical Diagnostics of Unseeded Turbulent Flows Using Ultraviolet Laser-Induced Processes in Air," Unpublished Research Proposal.

5. Laufer, G., McKenzie, R. L., and Huo, W. M. (1988) "Radiative Processes in Air Excited by an ArF Laser," Optics Lett., 13, pp. 99-101.

6. Laufer, G., McKenzie, R. L., and Fletcher, D. G. (1990) "Method for Measuring Temperatures and Densities in Hypersonic Wind Tunnel Air Flows Using Laser-Induced O_2 Fluorescence," Applied Optics, 29, pp. 4873-4883.

7. Fletcher, D. G. and McKenzie, R. L. (1991) "Simultaneous Measurements of Temperature and Density in Air Flows Using UV Laser Spectroscopy," AIAA Paper 91-0458.

8. Fletcher, D. G. (1992) "Measurements of Density, Temperature, and Their Fluctuations in Turbulent, Supersonic Flow Using UV Laser Spectroscopy," Unpublished presentation at the AIAA 30th Aerospace Sciences Meeting, Reno, NV.

Broad Band Excimer Laser Induced Fluorescence Applied To Hypersonic And Supersonic Air Flows.

F.DIONNET, J.MOUARD, D.PUECHBERTY.
CORIA CERTAM -- URA CNRS 230 -- INSA de ROUEN
B.P. 118
76134 Mont Saint Aignan-FRANCE.

Abstract :

The main objective of this work is to evaluate the potential of broad-band Excimer Laser Induced Fluorescence in high enthalpy air flow studies and particularly in ballistic tunnels. The ArF excimer laser (193 nm) is able to excite many rotational transitions of the nitrogen monoxide ε band and of the oxygen Schumann Runge band. LIF of NO was applied in a hypersonic ballistic tunnel to visualize the NO fluorescence in a projectiles wake. Then a related study of diatomic oxygen Laser Induced Predissociative Fluorescence demonstrates the capacity of the LIPF technique to measure temperatures in a flow without nitrogen monoxide.

I Introduction :

The Laser Induced Fluorescence (LIF) technique is well known for its application in the field of combustion, especially to provide concentration measurements of radicals such as OH, CH, CN or NO [1]. But this technique is also useful in high enthalpy supersonic and hypersonic air flow studies where nitrogen monoxide

A. Boutier (ed.), *New Trends in Instrumentation for Hypersonic Research*, 245–253.

is one of the major species. In addition excitation of the first electronic excited state of diatomic oxygen is now possible with the recent development of excimer lasers. Moreover, LIF is non-perturbating, instantaneous, local and species selective; thus this laser diagnostic technique is well suited for high enthalpy air flow studies.

II LIF Principle:

The principle of LIF can be summarized as the absorption of incident light followed by spontaneous emission. In a simplified two level model the population exchanges occur only between two energy levels; the excited level of energy E_2 and the fundamental level of energy E_1. These population exchanges between the two levels are accomplished by absorption, stimulated emission, spontaneous emission (also called fluorescence), electronic quenching and predissociation (depending on the excited species). The steady state fluorescence signal can be described by this equation:

$$S_f(N_t,T) = \frac{B_{12}\ U_v\ N_t\ F_{v''J''}}{(B_{12}+B_{21})\ U_v + A_{21} + Q_{21} + Q_p}\ A_{21}\ \frac{\Omega}{4\pi}\ V\ \varepsilon \qquad (1)$$

where $F_{v''J''}$ is the Boltzmann fraction which depends on temperature. N_t is the total number density (cm^{-3}), U_v is the spectral radiation energy density of the laser ($J\ cm^{-3}\ s$). B_{12} ($J^{-1}\ cm^3\ s^{-2}$), B_{21} ($J^{-1}\ cm^3\ s^{-2}$), and A_{21} (s^{-1}) are the Einstein coefficients respectively for the absorption, stimulated emission and fluorescence. V is the probe volume, Ω is the collection solid angle and ε is the transmittance of the optical setup. Q_{21} (s^{-1}) is the electronic quenching rate between the laser coupled levels E_1 and E_2, and Q_P (s^{-1}) the predissociation rate. The fluorescence signal depends on concentration and temperature and also on spectroscopic data (A_{21}, B_{12}, B_{21} and Q_P) and geometric data (Ω,V).The Einstein coefficients and the predissociation rate can be found in the literature. The collection solid angle and the probe volume can be measured. Therefore, LIF can be used for temperature or concentration measurements if the quenching rate Q_{21} is well known or negligible.

III LIF of nitrogen monoxide (NO):

A) Experimental setup:

There are many impulse facilities (i.e. wind tunnels, firing tunnels etc.) used to produce high enthalpy air flows. The ballistic tunnel, where a model flies through quiescent air, seems to be one of the most realistic impulse facilities for simulation of hypersonic atmospheric reentry [2].

Nitrogen monoxide (NO) LIF in a hypersonic flow was used in the L.R.B.A. (Laboratoire de Recherches Balistiques et Aérodynamiques - 27207 Vernon - France) hyperballistic tunnel to visualize the NO fluorescence of a projectiles wake. This tunnel is 1.2 m in diameter and approximately 100 m in length and was filled with dry air at a pressure of 64 mbar. The fluorescence is induced by a broad band ArF excimer laser Lambda Physik LPX 110 i cc. The laser emission spectra is centered around 193 nm with bandwidth of 0.7 nm (FWHM). The pulse duration is approximatively 17 ns which is nearly instantaneous relative to the flight time scale. A laser light sheet crosses the tunnel at right angles to the theoretical flight axis of the model and is stopped by a light trap. The fluorescence emission is collected perpendiculary to the laser sheet and above the tunnel by an intensified vidicon tube EG&G SIT 1254. The 200 ns intensified gate is triggered by a hyperfrequency detector located at 0.75 m before the measurement area. The fluorescence emission is collected through a colored filter (Schott BG24) between 200 nm and 300 nm to minimize scattered radiation from the medium. The measurement area is 30 x

30 mm^2 and is divided into 10000 elementary pixels. The real flight trajectory of the projectile is always different from the theoretical trajectory. Therefore it is necessary to record the real trajectory to situate the projectiles wake with respect to the measurement area. The model's trajectory is recorded by a set of photographic chambers distributed each 1.5 m along the tunnel. A set of sparks provides the flash lighting required to take pictures.

B) Results:

A two-dimensional nitrogen monoxide fluorescence image was obtained with this experimental setup in the wake of a hyperballistic axisymmetric projectile. The velocity of this object was approximately 4600 m/s. This image is taken at 66 mm measured from the projectiles base. The background noise was substracted from the actual image pixel by pixel. The noise was collected in quiescent air in the tunnel 20 s before the arrival of the ballistic object.

Three electronic bands of the nitrogen monoxide molecule are excited by the laser pulse:

the γ band $(A^2\Sigma\ (v'=3) \longleftarrow X^2\Pi\ (v''=0))$,
the ε band $(D^2\Sigma\ (v'=0) \longleftarrow X^2\Pi\ (v''=1))$,
and the β band $(B^2\Pi\ (v'=7) \longleftarrow X^2\Pi\ (v''=0))$.

A quantitative interpretation of the fluorescence image requires a relation between the fluorescence signal, the temperature and the concentration. Because of the large spectral width of the laser the monochromatic two-level model cannot be used. Therefore the synthetic absorption and fluorescence spectra must be calculated.

Because the absorption by rotational transitions occurs from fundamental levels with j" around 50.5, the γ band of NO excited by a broad-band ArF laser is not experimentally observed at 300 K [3]. Likewise, the $X^2\Pi$ (v"=1) levels of the ε band are not very populated at ambient temperature. Therefore at low temperature the absorption of the ArF laser is dominated by the β band. This band is also the strongest in the fluorescence spectrum if the pressure is low enough (P<0.1 mbar) because of its very elevated mean quenching rate. On the other hand, at higher temperature the levels $X^2\Pi$ (v"=1) and $X^2\Pi$ (v"=0; j"> 50.5) of the ε and γ bands are more populated, so the absorption is more intense. But quenching desexcitation (in particular self-quenching) is much more important for the γ band than for the ε band [4]. Consequently the γ band does not appear in NO fluorescence emission at elevated temperature. Therefore the modelling of NO fluorescence can be simplified into two parts depending on the temperature. First, the β electronic band is the only system which is important at low temperature (<500K) and very low pressure (<0.1mbar). Secondly, the ε electronic band is the predominant band at high temperature. But the desexcitation of β and ε bands is however very quenching dependent.

The fluorescence emission from the $B^2\Pi$ (v'=7) levels has been calculated and agrees very well with the results of Shibuya (Figure 1). However, in our model collisional quenching effects have been neglected which is a good assumption only at low pressure. But in our experimental conditions (P=64 mbar and T>2000K) the ε band predominates in the fluorescence emission. This band, like the β band, is depleted by quenching, but it is not possible to neglect these effects at the experimental pressure of 64 mbar. These quenching effects depend on pressure, temperature and collision partners and, as a consequence, they are very difficult to evaluate. In addition, because of the variation of the quenching rate, Q_{21}, with the location, relative concentration measurements cannot be determined. Therefore the obtained fluorescence image does not exactly represent the NO density distribution in the wake. Because of this inherent difficulty of

LIF the Laser-Induced Predissociative Fluorescence (LIPF), which is independent of quenching, has been developped in the last ten years [5].

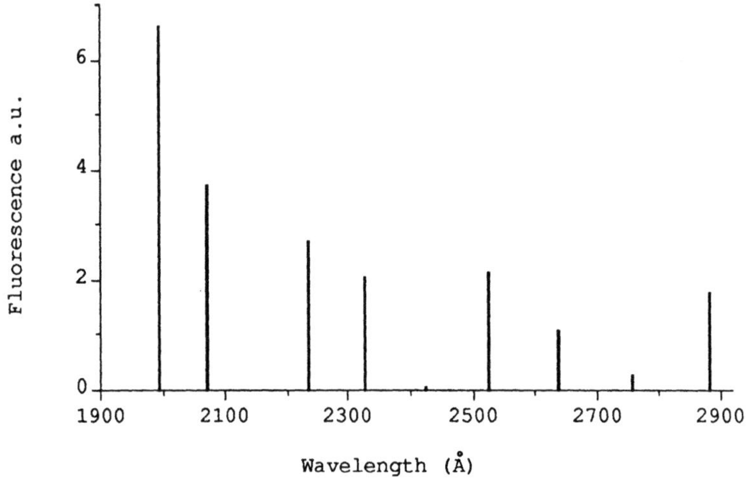

Fig. 1. Theoretical fluorescence vibrational spectrum of NO beta band, excited by a broad band ArF laser, at low presure and T = 300 K.

IV Potential of molecular oxygen broad band LIPF:

Thus, in the present study broad-band Laser-Induced Predissociative Fluorescence is examined for the diatomic oxygen molecule, which is a major species in dissociating air flow. Radiation from the same broad-band ArF excimer laser is used to excite several vibrational transitions of the Schumann Runge band ($B^3\Sigma$ <--- $X^3\Sigma$) of oxygen. These excited vibrational levels (v'=4, 7, 10, 14, 15) are predissociated. The desexcitation rate by predissociation Q_P (10^{12} s^{-1}) is larger than the electronic quenching Q_{21} ($\approx 10^9$ s^{-1}), so the fluorescence signal is independent of the electronic quenching in the following manner:

$$S_f(N_t, T) = \frac{B_{12}\ U_v\ N_t\ F_{v''J''}}{Q_p}\ A_{21}\ \frac{\Omega}{4\pi} V\ \varepsilon \qquad (2)$$

Similar to nitrogen monoxide, the two-level model cannot be used for a broad-band excitation source. Thus, the absorption and fluorescence synthetic spectra must be calculated to determine the relation between the fluorescence signal, the temperature and the concentration. The spectrally integrated fluorescence signal has been calculated between 200 nm and 300 nm as a function of temperature in the range from 500 K to 2500 K. These calculations show a strong dependence of the signal with the temperature (Figure 2), this evolution was already found by Lee et al.[6]. This was

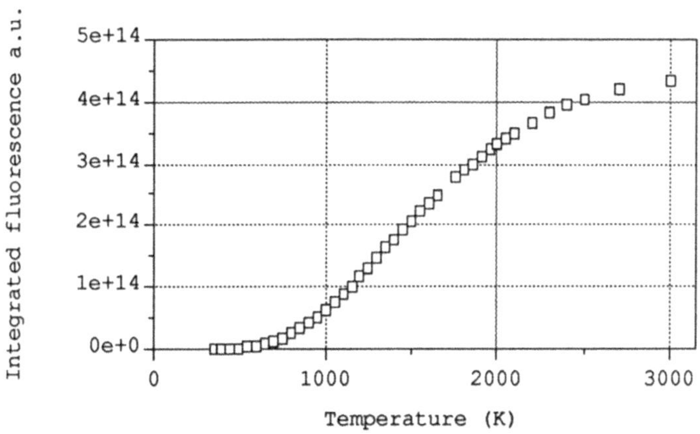

Fig. 2. Theoretical spectrally integrated fluorescence of molecular oxygen versus temperature.

confirmed in the present study by simultaneous measurement of the fluorescence signal and the temperature in a heated jet air flow (Figure 3). Thus the L.I.P.F. of diatomic oxygen is a preferred temperature measurement technique in the range 500K - 2500K. The predissociation of rotational excited levels enables one to neglect quenching effects but it has the major drawback of minimizing the fluorescence yield and thus of decreasing

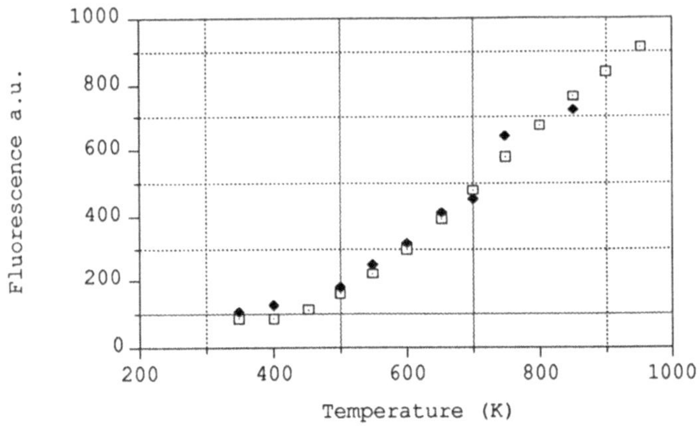

Fig. 3. □ Theoretical spectrally integrated fluorescence of molecular oxygen (transmittance of Schott BG 24 filter was included in the calculation).
◆ Measured spectral integrated fluorescence in a hot air jet.

dramatically the sensitivity of this temperature measurement technique. Therefore, the presence of nitrogen monoxide, which is more fluorescent than oxygen, is a major complication in measuring temperature by L.I.P.F. in a dissociated air wake flow. On the other hand, if the flow's enthalpy is not too high, the temperature is low enough to avoid formation of nitrogen monoxide, and measurements of temperature could be performed with LIPF of diatomic oxygen.

V Conclusion:

In conclusion classical LIF with a broad-band excimer laser is a suitable technique to obtain a single shot, two-dimensional, visualization of nitrogen monoxide in hypersonic air flows. In addition, when the temperature is low enough to avoid nitrogen monoxide formation, the LIPF technique of diatomic oxygen could be used for temperature measurements. In intermediate cases, where the NO fluorescence level is nearly equal to the oxygen fluorescence level, it becomes necessary to use a narrow-band laser [7].

References :

1. A.C. Eckbreth, A. Paul, A. Bonczyk, J.F. Verdieck (1979) "Combustion diagnostics by laser raman and fluorescence techniques" Prog. Energy Combust. Sci. 5, 253.
2. S.P. Sharma, C. Park (1990) "Survey of simulation and diagnostic techniques for hypersonic non-equilibrium flows" J. Thermophysics 4, 129.
3. K. Shibuya , F. Stuhl (1982) "Single vibronic emissions from NO $B^2\Pi$ (v'=7) and O_2 $B^3\Sigma$(v'=4) excited by 193 nm ArF laser" J. Chem. Phys 76, 1184.
4. H. Okabe (1978) "Photo-chemistry of small molecules" Willey Interscience Publication.
5. G.A. Massey, C.J. Lemon (1984) "Feasability of measuring temperature and density fluctuations in air using laser induced fluorescence" IEEE J. of Quantum Electronics 5, 454.
6. M.P.Lee, P.H. Paul, R.K. Hanson (1986) "Quantitative imaging of temperature fields in air using planar laser induced fluorescence of oxygen" Optics Letters 2, 75.
7. A.M. Wodtke, L. Huwel, H.Schluter, G. Meijer, H. Voges, P. Andresen (1988) "High-sensitivity detection of NO in a flame using a tunable ArF laser" Optics Letters 10, 910.

NONINTRUSIVE TEMPERATURE DETERMINATION BY LIF IN HYPERSONIC WIND TUNNELS

G. GUNDLACH and C. DANKERT
DLR - Institute for Experimental Fluid Mechanics
Bunsenstrasse 10
D-3400 Goettingen, Germany

ABSTRACT: LIF excitation spectra of nitric oxide have been measured in a hot supersonic freejet expansion in the vacuum wind tunnel (V2G) in Goettingen to determinate rotational temperatures. To interprete the experimental results, a computer program was developed that calculates the excitation spectra for temperatures up to 2000 K. By means of the LIF measurements in the free flow and in front of a model surface details of the heat transfer can be examined; this is relevant for the gas-surface interaction studies of reentry vehicles.

1. Introduction

The e^--beam-technique is well established in wind tunnel experiments for measuring temperatures, densities and velocities at densities up to 10^{15} *molecules/cm*^{3} [1], and CARS (Coherent Antistokes Raman Scattering) is used at higher densities, e.g. combustion processes [2,3]. The Laser-Induced Fluorescence (LIF) technique is on the way of becoming a standard tool for nonintrusive measurements in supersonic wind tunnel flow fields [4,5,6,7,8] in a wide density range. It can measure translational-, vibrational- and rotational temperatures by exciting selectively each molecular state at very low densities (10^{11} *molecules/cm*^{3}) as well as at high densities detecting predissociative molecular states (LIPF) [9].

In the Goettingen vacuum wind tunnels (V1G, V2G, V3G), LIF is used to investigate the influence of the internal degrees of freedom on the gas-surface interaction of supersonic flow vehicles at low densities of $10^{11} - 10^{16}$ *molecules/cm*^{3} and velocities of 1000 - 2000 m/s at Mach numbers of Ma = 7 - 20 . Due to the rapid expansion in Laval-nozzles and freejets there is a nonequilibrium of the internal temperatures in the flow [10]. The vibrational temperature freezes very rapidly near the stagnation temperature whereas the rotational temperature is nearly as low as the translational temperature. This causes the characteristic nonequilibrium-temperatures in V2G: $T_{trans} < T_{rot} << T_{vib} < T_0$ (300 - 2000 K). Up to now it is not possible to detect nitrogen with LIF, but nitric oxide can be excited with ultra violet lasers and detected with UV-sensitive photomultipliers or CCD-cameras [4,5]. For wind tunnel flow experiments a mixture of some percent of the probe gas NO in the carrier gas, normaly nitrogen or helium, is

A. Boutier (ed.), New Trends in Instrumentation for Hypersonic Research, 255–264.

used to detect laser-induced fluorescence. To identify the observed spectral lines, it is necessary to calculate the quantum mechanical rotational-vibrational transitions. By means of the calculated wavelengths and line intensities one can interprete the measured LIF spectra and determinate internal temperatures. Therefore a simple computer program has been developed for simulation of LIF excitation spectra of NO at temperatures up to 2000 K [11]. Within the tuning range of the ArF-Excimerlaser (192.9 - 193.8 nm), the γ (A-X), β (B-X), and ε (D-X)-bands of NO can be excited in dependence on the state population. These transitions can be calculated separately.

In this paper experimental spectra in the free flow and in front of a hot surface are presented and compared to the computational model.

2. Theory

2.1 FLUORESCENCE

When a single rotational vibrational line becomes excited by absorption of narrowband laser energy, the emitted signal after the spontaneous lifetime τ_K of the excited state is given by the total fluorescence intensity [12]

$$I = const \cdot I_L \cdot N'' \cdot q_{v'v''} \cdot S_{J'J''} \cdot \frac{1}{\tau_K} \; / \, (2J'' + 1) \tag{1}$$

with the transition probabilities $q_{v'v''}$ (Franck-Condon Factor) and $S_{J'J''}$ (Hoenl-London Factor). The fluorescence signal is a measure of the number density N'' of the absorbing molecular state. In thermal equilibrium it can be described by a Maxwell-Boltzmann distribution:

$$N'' = \frac{1}{Q} (2J'' + 1) \, e^{(-E_{vib}/kT_{vib})} e^{(-E_{rot}/kT_{rot})} \tag{2}$$

where $E_{vib,rot}$ and $T_{vib,rot}$ are the energies, respectively the temperatures of the internal states, and Q is the partition function that is given by the sum over all states. Within one vibrational band the rotational temperature can be obtained by a semi-logarithmic plot of the measured transition intensity versus the rotational energy:

$$\ln (I / S_{J'J''}) \sim -\frac{1}{k \, T_{rot}} \cdot E_{rot} \; . \tag{3}$$

For non-saturation spectroscopy the fluorescence must be normalized to the laserpower I_L.

2.2 TERM VALUES

The total term energy of a molecular state is given by the sum of the electronic, vibrational and rotational energies:

$$T = T_e + G(v) + F(J) \; . \tag{4}$$

For the vibrational term values the usual formula is

$$G(v) = \omega_e(v+\frac{1}{2}) - \omega_e x_e(v+\frac{1}{2})^2 + \omega_e y_e(v+\frac{1}{2})^3 \mp \ldots ; \tag{5}$$

ω_e is the wave number of the oscillation around the length x_e in the harmonic oscillator model. The higher terms consider the anharmonic correction, that is necessary for the description of a molecule.

2.3 THEORETICAL MODEL

To calculate the intensities the Franck-Condon factors of Nicholls [13], the Hoenl-London formulae of Bennett [14] and Earls [15] were used. The transition wavelengths were calculated by means of the following term value formulae of Herzberg [12] with constants of Huber/Herzberg [16], Hinz [17] and Amiot [18]. The electronic ground state of nitric oxide is a $^2\Pi$ state which is splitted due to the $\underline{L}-\underline{S}$ coupling in a $^2\Pi_{1/2}$ and a $^2\Pi_{3/2}$ state with the energy difference of $123cm^{-1}$. These states intermediate between Hund's coupling cases (a) and (b):

$$^2\Pi_{1/2}:\ F_1(J) = B_v\left[(J+\frac{1}{2})^2 - 1 - \frac{1}{2}\sqrt{4(J+\frac{1}{2})^2 + Y(Y-4)}\right] - D_v J^4 \tag{6}$$

$$^2\Pi_{3/2}:\ F_2(J) = B_v\left[(J+\frac{1}{2})^2 - 1 + \frac{1}{2}\sqrt{4(J+\frac{1}{2})^2 + Y(Y-4)}\right] - D_v (J+1)^4 \tag{7}$$

where B_v is the rotational constant and D_v takes into account the centrifugal force , i.e. the deviation from the rigid rotator model which is small in comparison to the rotational constant. $Y = A_v/B_v$ is the spin-orbit ($\underline{S}-\underline{L}$) interaction parameter, where A_v is a measure of the coupling strength.

The electronic $B^2\Pi_r$ state has the same configuration as the ground state and therefore the above formulae hold. The $^2\Sigma^+$ states A and D belong always to Hund's case (b); here the spin-rotation parameter must be considered. The term values are formed by

$$F_1(K) = B_v K(K+1) - D_v[K(K+1)]^2 + \frac{1}{2}\gamma K\ ,\quad K = J - \frac{1}{2} \tag{8}$$

$$F_2(K) = B_v K(K+1) - D_v[K(K+1)]^2 + \frac{1}{2}\gamma (K+1)\ ,\quad K = J + \frac{1}{2}\ . \tag{9}$$

The NO β-band ($B^2\Pi_r \leftarrow X^2\Pi_r$) consists of two clearly separated bands each of two branches: P_1, R_1 and P_2, R_2. The Q-branch and the satellite branches have no remarkable intensities [12]. At higher rotational energies the separation dissappears because the configuration changes to case (b).

In case of $^2\Sigma^+ \leftarrow {}^2\Pi_r$ transitions ($A, D \leftarrow X$) two subbands P_1, R_1, Q_1 and P_2, R_2, Q_2 can be observed - the Q-branch is the most intensive one. In any case Λ-type doubling can be neglected in all practical cases.

By means of a graphical user interface [11], diagrams of the spectra can be generated. The line broadening of the spectral transitions due to the Doppler effect can be considered by:

$$I(\nu) = I_o \cdot e^{-[c(\nu - \nu_o)/\nu_o \cdot v_w]^2} \tag{10}$$

with absorption frequency ν_o and the most probable velocity $v_w = \sqrt{(2kT)/m}$. Also the spectral resolution of the laser is included. This becomes important in the near of band heads where many transitions are very close together. At least this can change the intesities by overlapping excitation processes.

As an option the calculated intensities can be normalized by a Gaussian function, which is similar to the wavelength distribution of the ArF-laser emission (Eq.11).

3. Experiment

3.1 WIND TUNNEL

The vacuum wind tunnel provides low density flow at Mach numbers of Ma = 7 - 20. It works continuously, so that long measurement times are possible. Typically we have vibrational temperatures of 1000 K and rotational temperatures of 100 K. Usually no dissociation is observed. For the experiment presented here a new heater for temperatures up to 1100 K and stagnation pressures of 30 bars was used [19]. It runs with pure nitrogen as well as with many gas mixtures. **Fig.1** shows the experimental setup in V2G. The nozzle has a diameter of 0.7 mm and can be moved in flow direction relatively to the perpendicular incoming laser beam. The stagnation temperature of the gas can be controlled by a NiCr-Ni thermocouple mounted near the nozzle exit.Additionally the model set-up is sketched. The model is made of copper with a diameter of 5 mm, can be heated up to 1000 K and is also temperature controlled by thermocouple.

3.2 SEEDED BEAM TECHNIQUE

The standard freejet set-up using pure nitrogen or a NO/N_2 mixture is capable to deliver speeds up to 1200 m/s in nonequilibrium flow fields ($T_{rot} = 100\,K$, $T_{vib} = 1000\,K$) without dissociation. In order to speed up the flowing molecules up to 2000 m/s, the test gas (NO) can be mixed with a light gas (He) that accelerates also the heavier test molecules (seeded beam technique). Such a device was also tested using a mixture of 90 % He and 10 % NO. For future gas-surface interaction studies this apparatus will be redesigned for stagnation temperatures up to 2400 K (carbon heater) to enable flow velocities up to 4000 m/s in a continuous running wind tunnel.

3.3 LASER

The laser source is a tunable ArF-Excimerlaser Lambda Physik EMG 160 T MSC (192.9 - 193.8 nm). Narrowband tunable radiation (bandwidth 0.2 - 0.3 1/cm) is produced in the oscillator cavity by means of a diffraction grating. This radiation is amplified in a second cavity with unstable resonator optics (Cassegrain). The max. pulse energy is 100 mJ at a pulse length of 13 ns and a max. repetition rate of 250 Hz. The amplifier can generate broadband radiation of max. 50 mJ itself. If the amplifier is not completely locked by the oscillator the laser emission is a combination of narrow band and broad band radiation. The broad band part increases at both ends of the tuning range, so that there is a dip in the emission profile. This can be avoided by removing the Cassegrain optic while the output energy is reduced to approximately 30 % - 40 % of the initial value [20]. Additionally there are 3 strong absorption lines of the O_2 Schumann Runge $B^3\Sigma_u^-(v'=4) \leftarrow X^3\Sigma_g^-(v''=0)$-band within the tuning range. Therefore the beam path was purged by N_2 to reduce the oxygen. The wavelength distribution of the laser emission can be described by a Gaussian function [21]:

$$I_L(\nu) = I_L^o \, e^{\left(-\left[2(\nu-\nu_o)/\Delta\nu_{1/2}\right]^2 \ln 2\right)} \tag{11}$$

where ν_o is the middle frequency of the tuning range and $\Delta\nu_{1/2}$ is the half distance. During a measurement the pulse to pulse deviations are registrated by a UV-photodiode. For tuning the wavelength, the grating can be turned by a Oriel Motor Mike with encoder.

3.4 FLUORESCENCE DETECTION SYSTEM

Perpendicular to the laser-molecular beam plane the fluorescence is collected by f/14 suprasil optics (**Fig.2**). The photons are counted by a watercooled Thorn Emi 9789QB photomultiplier. The laser wavelength was blocked by a Schott BG24 glass filter. With gated integrator technique (SR250) the multiplier signals are sampled at a gate width of 50 ns and averaged over 10 pulses. The spectra can be saved and evaluated by the boxcar software SR270 from Stanford Research.

4. Results and Discussion

At first some static cell experiments were done to calibrate the LIF apparatus. **Fig.3** shows a static NO excitation spectrum at room temperature. The pressure was 19 Pa . It is compared with the computer calculation. The $B^2\Pi_r(v'=7) \leftarrow X^2\Pi_r(v''=0)$-band is the only one within the tuning range of the ArF-Excimerlaser. Therefore the spectrum is very simple and easy to identify. The $P_{1,2}$ and $R_{1,2}$-branches yield in line groups each of four transitions. Rotational quantum numbers from 23.5 to 31.5 are excited. In the computer diagram the Doppler broadening was included. The intensities are normalized to a Gaussian function (see Eq.11). The measurement was done without Cassegrain optic. The experimental resolution was $2\,cm^{-1}$. Considering the Gaussian distribution of the laser power over the tuning range, the temperature determination from the measurement has an accuracy of 5 % .

The spectrum changes drastically in the wind tunnel flow. **Fig.4** shows a measured spectrum in a He/10 % NO -freejet compared with the computer simulation. The stagnation conditions were $T_o = 1040\,K$, $p_o = 4$ bar . For the calculation the theoretical temperature values of the isentropic expansion $T_{vib} \approx 0.95\,T_o$ and $T_{rot} \approx T_{trans} \approx 100\,K$ were taken. At least the best agreement have been found for $T_{vib} = 1000\,K$ and $T_{rot} = 150\,K$.

Because of the low rotational temperature only rotational quantum numbers up to 22.5 are populated. Within the tuning range of the ArF laser this causes that only excitation of the

$A^2\Sigma^+(v'=5) \leftarrow X^2\Pi_r(v''=2)$-band and $D^2\Sigma^+(v'=0) \leftarrow X^2\Pi_r(v''=1)$-band are possible. There is no transition from the vibrational ground state. Very sensitive to the rotational temperature are the $D \leftarrow X$ transitions. This was observed from the first measurements in front of a hot model surface in N_2/20 % NO flow (**Fig. 5a,b**). Near the surface the $D \leftarrow X$ transition lines increase, indicating a higher rotational temperature due to gas surface interaction. To study details of this heat transfer mechanism [22] more quantitative measurements are necessary. Therefore the signal-to-noise ratio of the measured spectra must be improved. The measurement in Figs. 5a,b were measured with cassegrain optic. This should be removed although a lost in laser intensity is the result.It should be noted that the shown flow spectra were measured at total densities $< 10^{15}$ $1/cm^3$; but only approximately $5 \cdot 10^{11}$ $1/cm^3$ NO molecules could be really excited. Due to the low densities there are no quenching effects to be considered. Further on, each transition should be checked on saturation by laser radiation because this is critical for temperature determination.

References

[1] Cattolica, R.J., *Modern Developements in Electron-Beam Fluorescence.* RDG, 17th Int.Symp., Aachen ,Germany, 1990

[2] Kohse-Hoeinghaus K.; Meier, U.; Attal-Tretout, B., *Laser-Induced Fluorescence Study of OH in Flat Flames of 1 - 10 bar Compared with Resonance CARS Experiments.* Appl.Opt.29 (10), (1990) pp.1560-1569.

[3] Stricker, W.; Woyde, M., *CARS Temperature Measurements in High Pressure Flames.* Proceedings of the Joint Meeting of the German and Italian Section of the Combustion Institute, Ravelo, Italy, Sept. 11-14, 1989, Paper 1.1.

[4] Dankert, C.; Gundlach, G., *Laser-Induced Fluorescence Apparatus for Application in Rarefied Hypersonic Flow Fields. Part I: Static Cell Experiments with NO.* DLR-IB 222-91 A 11, Goettingen, Germany, 1991.

[5] Beck, W.H.; Dankert, C., Eitelberg, G.; Gundlach, G., *Preliminary Laser Induced Fluorescence Measurements in Several Facilities in Preparation for Application to Studies in the High Enthalpy Shock Tunnel Goettingen.* 30th Aerospace Sciences Meeting, AIAA 92-0143, Reno, NV, 1992.

[6] Niimi, T.; Fujimoto, T.; Shimizu, N., *Planar Measurements of Temperature in Rarefied Gas Flow by LIF Images.* RGD, 17th Int. Symp., Aachen, Germany, 1990.

[7] Hanson, R.K; Chang, A.T.; Seitzman, J.M.; Lee, M.P.; Panl, P.H.; Battles, B.E., *Laser-Induced Fluorescence Diagnostic for Supersonic Flows.* 28th Aerospace Sciences Meeting, (1990) Reno, Nevada, AIAA-90-0625.

[8] McMillin, B.K.; Palmer, J.L.; Hanson, R.K., *Two-Dimensional Temperature Measurements of Shock Tube Flows Using Planar Laser-Induced Fluorescence Imaging of Nitric Oxide.* AIAA 22nd Fluid Dynamics, Plasma Dynamics & Lasers Conference, (1991) Honolulu, Hawaii, AIAA-91-1670.

[9] Andresen, P., Bath, A., Groeger, W., Luelf, H.W.; Meijer, G.; ter Meulen, J.J., *Laser-Induced Fluorescence with Tunable Excimer Lasers as a Possible Method for Instantaneous Temperature Field Measurements at High Pressure: Checks with an Atmospheric Pressure Flame.* Appl.Opt. 27, pp.365-378 (1988).

[10] Dankert, C.; Bütefisch, K.-A., *Influence of Nozzle Geometry on Vibrational Relaxation in an Expansion Flow.* RGD, 9th Int. Symp., Göttingen, Germany, 1974.

[11] Gundlach, G.; Martinez-Schramm, J., *Calculation of Laser-Induced Fluorescence Excitation Spectra of Nitric Oxide for Application in Hypersonic Flow Fields.* DLR-IB to be published in 1992.

[12] Herzberg, G., *Molecular Spectra and MolecularSstructure, Vol.1: Spectra of Diatomic Molecules.* Princeton, NJ/USA, Van Nostrand 1950.

[13] Nicholls, R.W., *Franck-Condon Factors to High Vibrational Quantum Numbers IV: NO band system.* J.Res.Nat.Bur.Stand.-A 68A (5), (1964) pp.535-540.

[14] Bennett, R.J.M., *Hoenl-London Factors For Doublet Transitions in Diatomic Molecules.* Mon.Not.R.Astr.Soc. (1970) 147, pp.35-46.

[15] Earls, L.T., *Intensities in Transition in Diatomic Molecules.* Phys.Rev. 48 (1935) pp.423-424.

[16] Huber, K.P.; Herzberg, G., *Molecular Spectra and Molecular Structure, Vol 4: Constants of Diatomic Molecules.* Princeton, NJ/USA, Van Nostrand (1979).

[17] Hinz, A.; Wells, J.; Maki, A.G., *Heterodyne Frequency Measurement on the Nitric Oxide Funamental Band.* J.Mol.Spec. 119 (1986) pp.120-125.

[18] Amiot, C.; Verges, J., *Spin-Rotation Doubling in the NO Electronic States by Emission Fourier Transform Spectrscopy.* Chem. Phys. Lett. 66 (3), (1979) pp.570-573.

[19] Lehmkoester, K.; Legge, H., *A Heater for a Seeded Beam System.* DLR-IB 222-91 A 02, Göttingen, Germany (1991).

[20] Wodtke, A.M.; Huewel, L.; Schlueter, H.; Andresen, P., *Simple Way to Improve a Tunable Argon Fluoride Laser.* Rev.Sci.Instrum. 60 (4), (1989) pp.801-802.

[21] Shibuya, K.; Stuhl, F., *Fluorescence Lifetime and Collisional Quenching of the Predissociative NO* $B^2\Pi$ $(v' = 7)$ *State.* Chem.Phys. 79 (1983) pp.367-381.

[22] Legge, H., *Recovery Temperature Determination in Free Molecular Flow of a Polyatomic Gas.* Proceedings of the 14th Int. Symp. on Rarefied Gas Dynamics, Tsukuba Japan, 1984, pp.271-278.

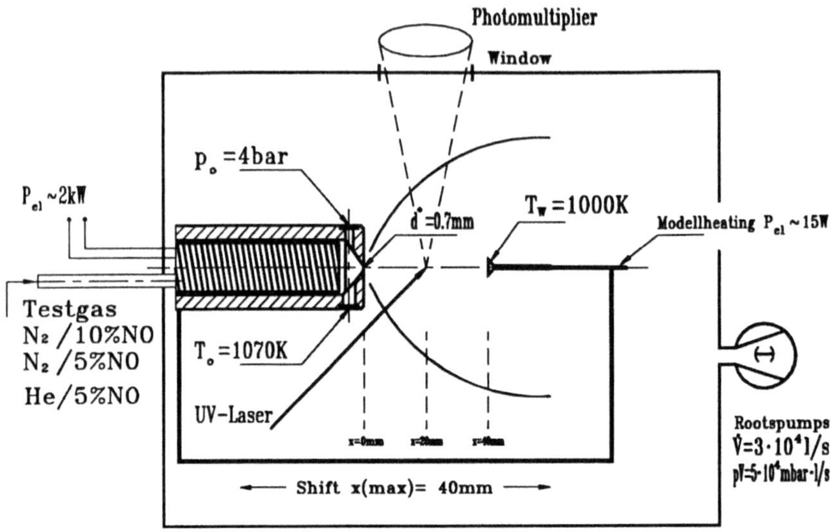

Fig. 1 Kanthal-heater with model in V2G.

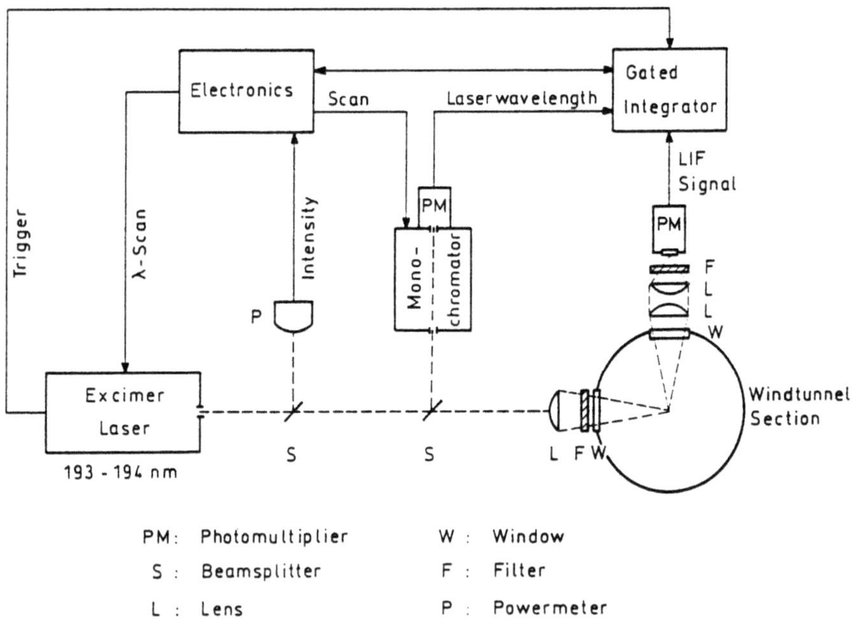

Fig. 2 Experimental set-up for LIF measurements.

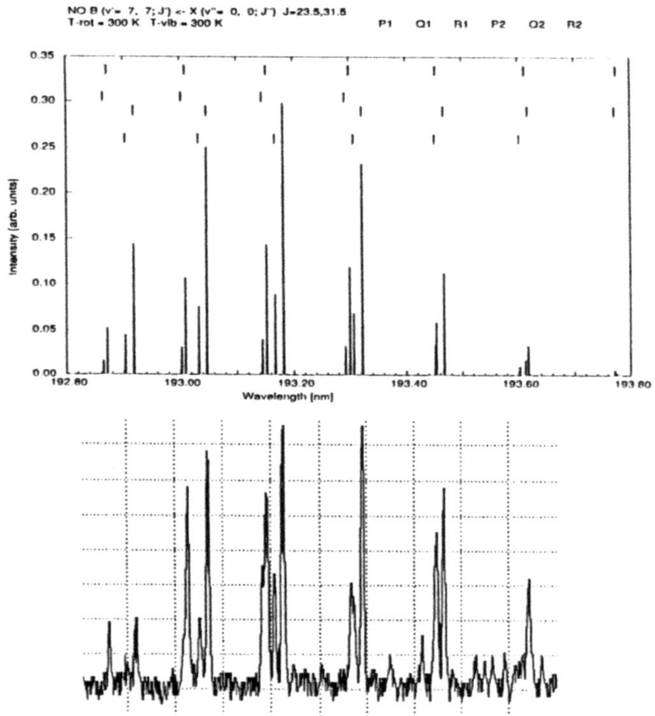

Fig. 3 Calibration measurement in comparison to calculated spectra at room temperature.

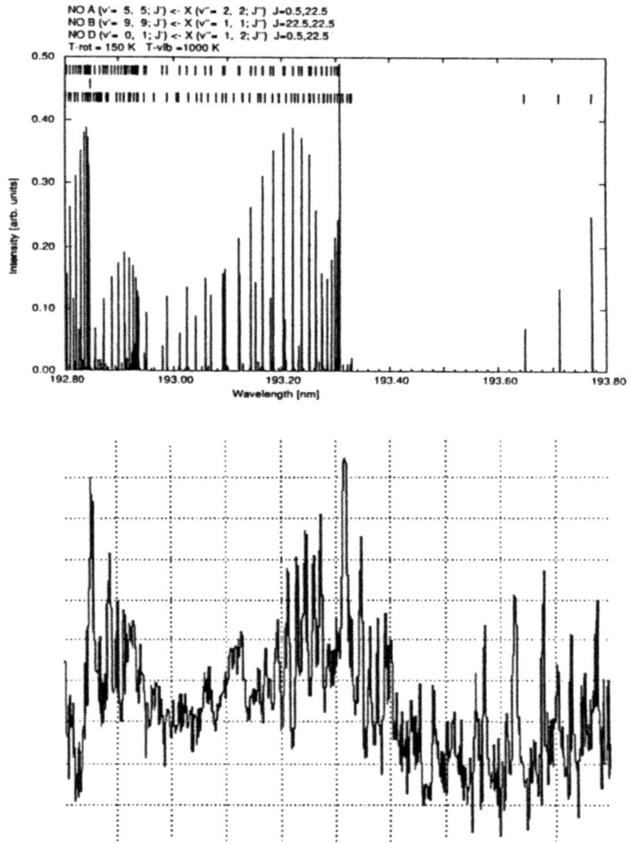

Fig. 4 He/NO 10 %-freejet spectrum at $T_o = 1040\,K$, $p_o = 4$ bar with computer calculation for $T_{vib} = 1000\,K$ and $T_{rot} = 150\,K$.

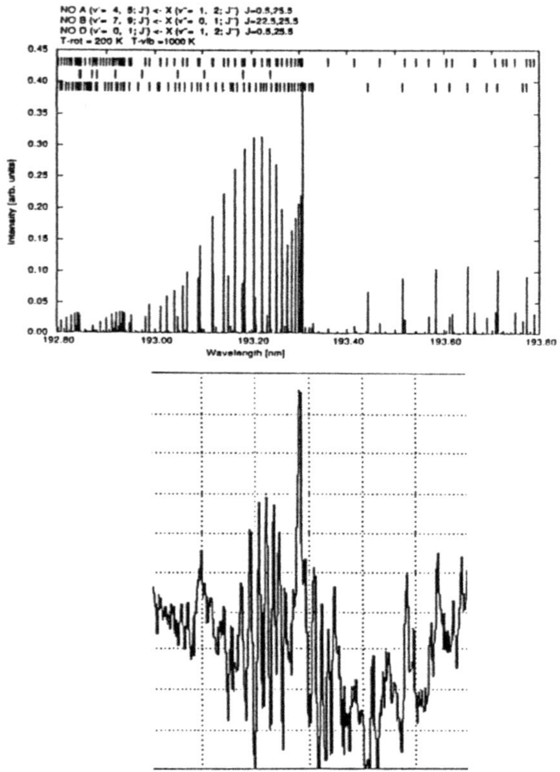

Fig. 5a N_2/NO 20 % freejet spectrum at $T_o = 1080\,K$, $p_o = 4$ bar with computer calculation for $T_{vib} = 1000\,K$ and $T_{rot} = 200\,K$.

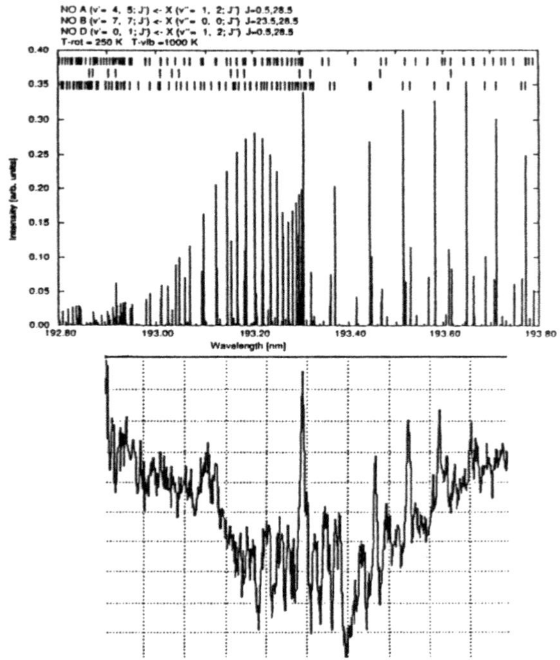

Fig. 5b N_2/NO 20 % spectrum in front of a hot ($T_w = 668\,K$) copper surface at $T_o = 1063\,K$, $p_o = 4$ bar, $x/d = 70$ with computer calculation for $T_{vib} = 1000\,K$ and $T_{rot} = 250\,K$.

RAPID PULSE ELECTRON BEAM FLUORESCENCE FOR FLOW FIELD DIAGNOSTICS

E. P. MUNTZ AND D. A. ERWIN
Department of Aerospace Engineering
University of Southern California
Los Angeles, California 90089-1191
USA

ABSTRACT. Recent developments in pseudosparks and related devices offer the possibility of high current (up to 1 ka peak current) short time (10 ns) pulses of 10 to 50 kV, collimated electron beams. These devices are compact and have observed repetition rates of 10^3 Hz. This paper discusses the application of fast, high current electron beam pulses to electron beam flow field diagnostics.

1 Background

Electron beam fluorescence (EBF) is a well established diagnostic technique for the study of hypersonic flows up to number densities around 10^{17} cm^{-3} and providing the flows are not strongly self-radiating. The chief limitation of EBF has been that the electrons scatter at high gas densities and that the fluorescence emission is difficult to intepret due to quenching in unknown quenching environments. Previously, Muntz, Kunc and Erwin (1987) and Wojcik et al. (1990) have suggested ways around both of these difficulties (Electron-Photon Fluorescence) by using a combination of an e-beam and laser photons. Cattolica (1989) has reported on some experiments relative to the EPF concept and Wojcik et al. (1990) on pulsed electron beam fluorescence (PEBF). After thinking about all of this for a year or two we have come to the opinion that a high current, pulsed electron beam offers the most interesting approach. High current beams are self focusing so that they can partly negate scattering at higher densities. They retain the chief advantage of EBF which is the simultaneous broad-band excitation of all the rotational and vibrational internal energy states of all flow species and they can be coupled with gated detection to avoid quenching related interpretation difficulties.

A. Boutier (ed.), New Trends in Instrumentation for Hypersonic Research, 265–273.

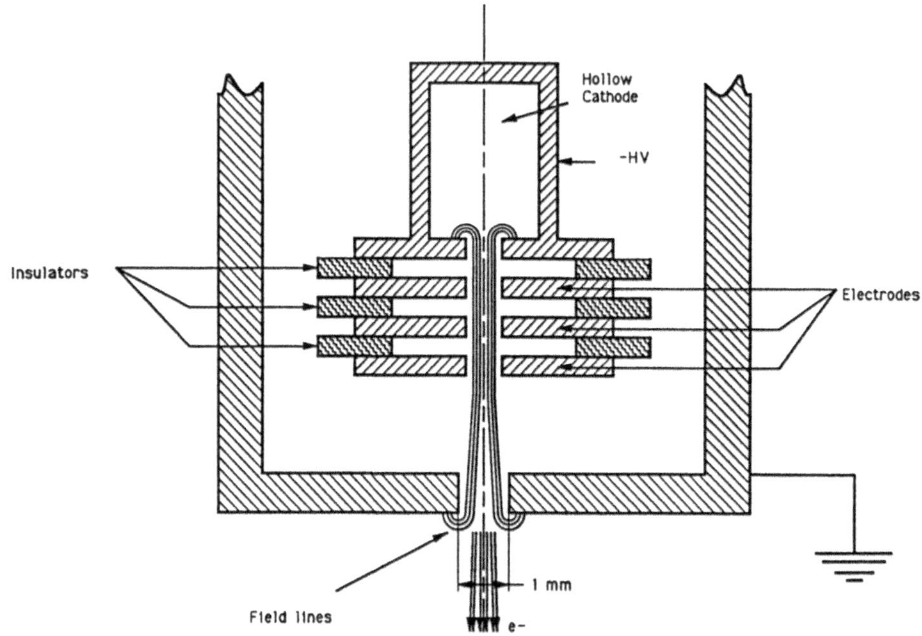

Figure 1: Hollow cathode pseudospark discharge after Frank (1989) and Benker (1989).

2 Pseudospark Devices

Pseudospark devices (Benker, 1989; Frank 1989) operate on the left hand side of the Paschen minimum. The discharge thus prefers long path lengths and an axial discharge generating energetic electrons is produced. A typical pseduospark electron beam generator (after Frank (1989)) is illustrated in Fig. 1 and features a hollow cathode.

Apparently related hollow anode c-w devices have been used for high current e-beam generation (Pham-Van-Diep, 1990 and references to others therein). A typical hollow anode device is shown in Fig. 2.

In both devices the discharge electrons are given enough ballistic energy to separate from the field lines at the device's exit and are observed to form a well collimated beam of electrons. The c-w hollow anode device produces beams that have energies close to the potential applied across the discharge (basically the cathode fall energy). The pulsed pseudospark discharge produces an electron beam with a broader energy range but with a mean between 0.5 and 0.7 times the breakdown potential (Benker, 1989). For present purposes only the pseudospark beam generator will be considered. Since the pressure (10–100 pa depending on gas, geometry and desired breakdown potential) in the pseudospark chamber is important care must be exercised for installations in gas dynamic facilities. One scheme is illustrated in Fig. 3 where the isolation chamber is maintained at a pressure at least a factor of three below either the test gas or the pseudospark chamber.

The pumping for the pressure isolation can be conventional or using cryogens for

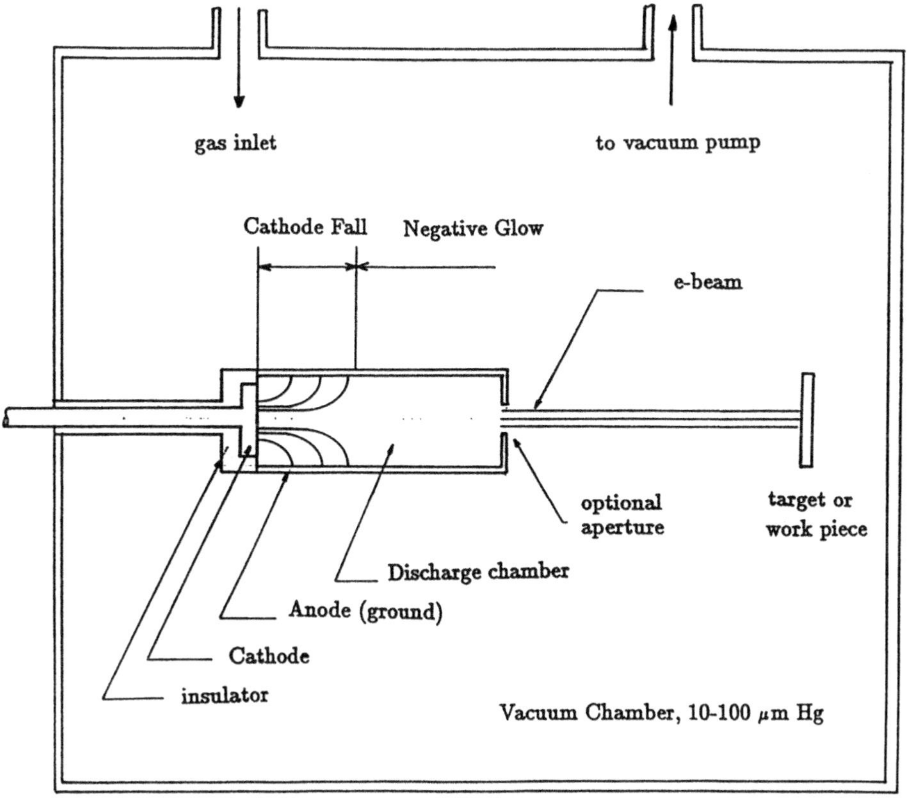

Figure 2: Hollow anode c-w electron beam generator.

portable applications such as flight experiments.

3 Analysis

Consider the fluorescence excited by a pulsed electron beam in a gas. For simplicity, a three-level system is assumed, the ground state (0) and two excited states (k, l) with the lower one (l) metastable. The details of molecular structure will be ignored but we do assume degeneracies g_k and g_l for the excited states. If $\Omega_{Q,k}$ is the quenching cross section for the state k and n_Q is the quenching species number density, a characteristic quenching time due to collisions is

$$\tau_{QC} = (\bar{C}\Omega_{Q,k}n_Q)^{-1} \tag{1}$$

where $\bar{C} = \sqrt{8kT/\pi m}$ is the mean molecular speed of the excited species of mass m, while k is Boltzmann's constant.

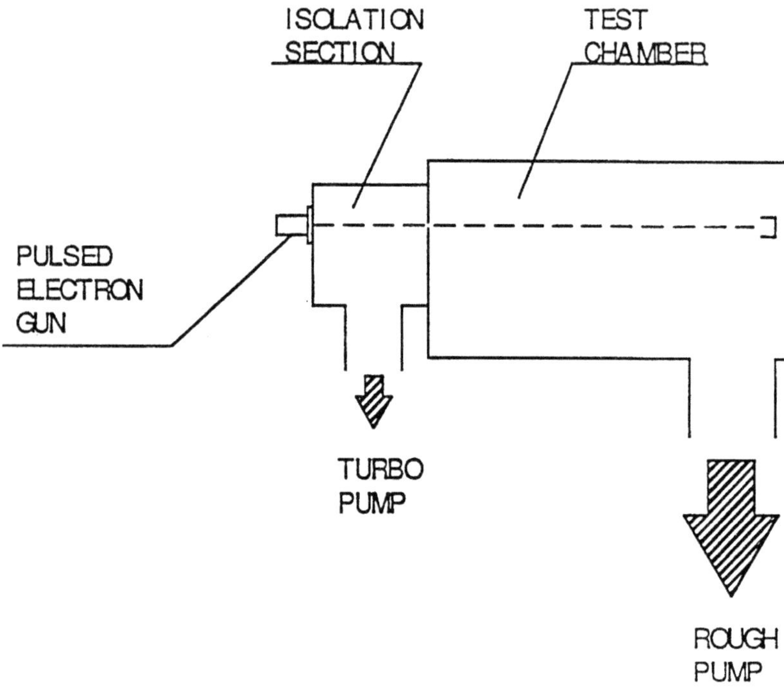

Figure 3: Pressure isolation for a pseudospark e-beam source.

For the pulsed electron beam, assume an instantaneous current I_B which is constant during the pulse. The instantaneous excitation rate per unit of distance along the beam is

$$\dot{N}_{0,k} = I_B \Omega_{0,k} n_g / 1.6 \times 10^{19} \tag{2}$$

where n_g is the number density of the gas specie of interest. If the beam has a cross sectional area A_B at any location, the excitation rate per unit volume is

$$\dot{n}_{0,k} = 6.25 \times 10^{18} I_B \Omega_{0,k} n_g / A_B. \tag{3}$$

The number of molecules per unit volume excited to level k in the e-beam pulse length τ_E is, from (3),

$$n_k = \dot{n}_{0,k} \tau_E = 6.25 \times 10^{18} (I_B \Omega_{0,k} n_g / A_B) \tau_E. \tag{4}$$

Taking into account spontaneous emission and quenching collisions, the time history of the excited state number density $n_k(t)$ is

$$n_k(t) = \frac{\dot{n}_{0,k}}{A_{k,l} + \bar{C}\Omega_{q,k} n_Q} \{1 - \exp[-(A_{k,l} + \bar{C}\Omega_{Q,k} n_Q)t]\}. \tag{5}$$

Note that for $t \to \infty$ (with $\tau_E \to \infty$ and $t < \tau_E$),

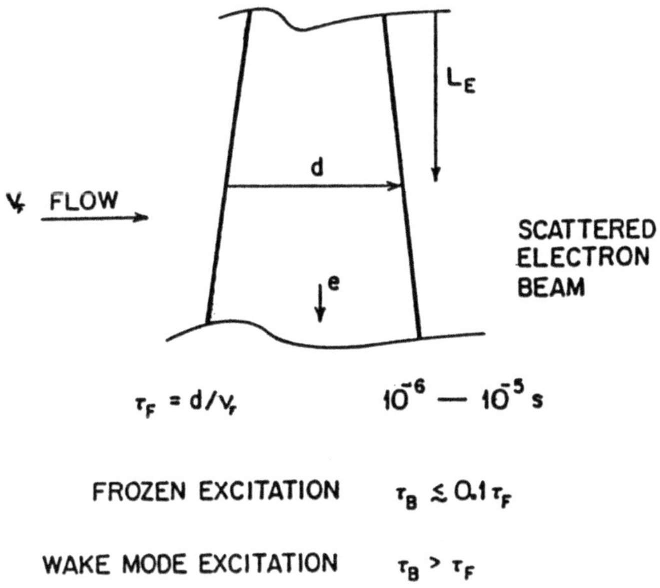

Figure 4: Electron beam scale times.

$$n_k = \frac{6.25 \times 10^{18} I_B \Omega_{0,k} n_g}{A_B[A_{k,l} + \bar{C}\Omega_{Q,k} n_Q]}. \tag{6}$$

If $\bar{C}\Omega_{Q,k}n_Q \gg A_{k,l}$ and $n_Q = n_g$ or n_Q varies linearly with n_g, we obtain the well-known result that $n_k(t \to \infty)$ is independent of n_g.

For very short times such that $(A_{k,l} + \bar{C}\Omega_{Q,k}n_Q)t \to 0$, eq. (5) gives

$$n_k(t) = 6.25 \times 10^{18} I_B \Omega_{0,k} n_g t / A_B. \tag{7}$$

The emitted power that can be observed as a result of the population $n_k(t)$ is

$$P_{k,t} = h\nu_{k,t} A_{k,l} n_k(t) \tag{8}$$

where $\nu_{k,l}$ is the wavenumber of the k, l transition and h is Planck's constant. Thus, for very short times on the scale of the characteristic "quenching" time, $\tau_Q = (A_{k,l} + \bar{C}\Omega_{Q,k}n_!)^{-1}$, the emission intensity will be directly proportional to the gas density n_g and the pulse length t (eq. (7)). The shape of the curve at longer times gives a direct measure of the quenching time τ_Q.

Consider the sketch in Fig. 4, where the question of a scale time for the electron beam pulse is addressed.

Assume that the flow velocity vector is normal to the electron beam. A conveneient characteristic time τ_F is obtained from the beam diameter d and the flow speed V_F : $\tau_F =$

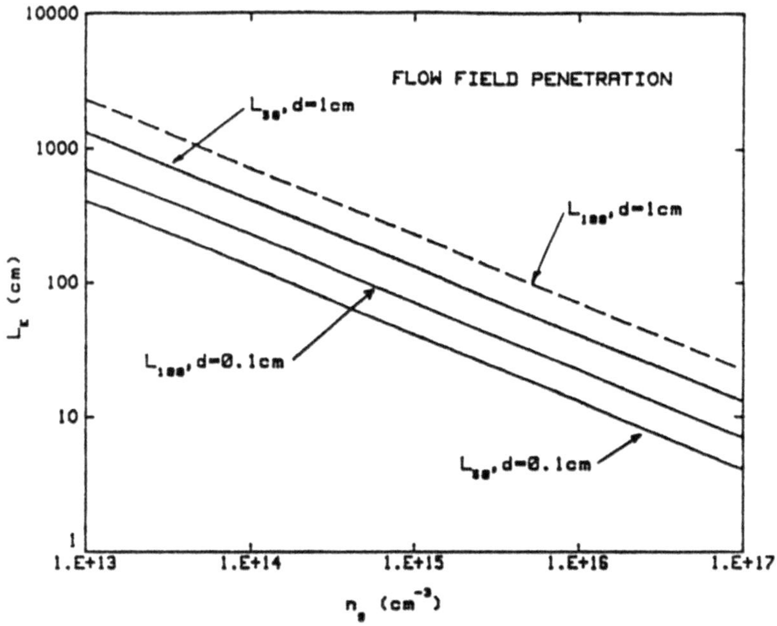

Figure 5: Distances L_E that 10 and 100 keV e-beam can travel in a gas before reaching a full width at half maximum of $d = 0.1$ and 1.0 cm.

d/V_F. Analysis of the excitation is most conveniently done for times much shorter or much longer than τ_F. This is because it is not always easy to know the exact current density profile of the beam. Typical values of τ_F will be from 10^{-6} to 10^{-5} sec. Thus, an electron beam pulse shorter then, say, 10^{-7} s will produce an excitation that is "frozen" in space. For times much longer than 10^{-5} s there will be continuous excitation wake downstream of the electron beam.

Electron beam spreading is important in relatively high density gases. Data from Muntz (1968) have been used to construct Fig. 5 which gives beam diameter for 50 and 100 keV electrons passing through uniform gases.

More recent data obtained by Cattolica (1989) are in essential agreement with the older data. For the pulsed high current electron beam of present interest both electrostatic focusing and magnetic focusing will likely be important. Muntz (1968) reviewed this some time ago for EBF applications concluding that generally only electrostatic focusing is important. The pulsed, high current beam will also be subject to magnetic focusing since typically it's perveance will be 10^{-4} compared to 10^{-10} or so for a 1 mA beam.

4 A Back of the Envelope Look at Pulsed Electron Beam Fluorescence

4.1 COUNTING STATISTICS

Passing a 10^3 amp beam of electrons through a flow field immediately raises questions about intrusive behavior. The thing to remember is that the pulse length will only be say 10 ns. As mentioned before 10 ns, even for a 5 km/s flow, is really very short. From eq. 4 the maximum local number density of excited molecules (i.e. ignoring spontaneous emission and quenching) after of beam pulse of τ_E duration is

$$n_k = 6.25 \times 10^{18} (I_B \Omega_{OK}/A_B) n_g \tau_E.$$

For $n_g = 3 \times 10^{17}$ cm^{-3}, $\tau_E = 10$ ns, an $I_B = 10^3 A$, $A_B = 5$ mm^2 and say $\Omega_{OK} = 10^{-18}$ cm^2; $n_k = 3 \times 10^{12}$ cm^{-3}. Assuming that there is a total spontaneous transition probability of 10^7 s^{-1} associated with the kth level, there will be emission during a 10 ns observation gate and from an emitting volume of 10^{-2} cm^3 of 3×10^8 photons. Allowing a factor 100 for the solid angle of the optics, a factor of two for detected quanta, a factor of 10 for distribution amongst vibrational levels and a factor of 30 for the rotational distribution leaves 3×10^4 quanta for the average emitted rotational line. Counting statistics are thus not a serious problem.

4.2 BACKGROUND EMISSION

It is of interest to compare the photon emission rate to the emission of photons by a black body. Consider 4000 Å, around where the fluorescence from nitrogen is strong. Assuming a spectrometer resolution of 0.1 Å a 7500 K black body produces about 25 times more detected photons from the surface of a 10^{-2} cm^3 volume than the pulsed e-beam fluorescence.

4.3 FLOW DISTURBANCES

For the case we have been discussing the maximum excited state density is 3×10^{12} cm^{-3}. If each excited state represents say 15 eV, the excitation energy density is 10^3 ergs cm^{-3} whereas the 300 K thermal energy is 6.3×10^3 ergs cm^{-3}. If all of the excitation energy were converted to thermal energy there would be about a 50 K temperature disturbance. However, because the beam pulse and observation times are much shorter (factor of about 30 in the example considered here) than the collisional quenching time little of this excitation energy will appear as an observed flow disturbance.

In terms of gross flow disturbances, the average energy added assuming a 10^3 Hz repetition rate for the e-beam is only 10^{-3} of the average thermal energy of the gas in the beam, assuming 300 K and a 10^5 cm/s flow velocity perpendicular to the beam. The local plasma density during the beam pulse will be quite high (10^{12} cm^{-3}) so there may be minor spectroscopic effects. These are not expected to be difficult to handle. Otherwise the spectral distribution of the emission intensity should be similar to c-w EBF.

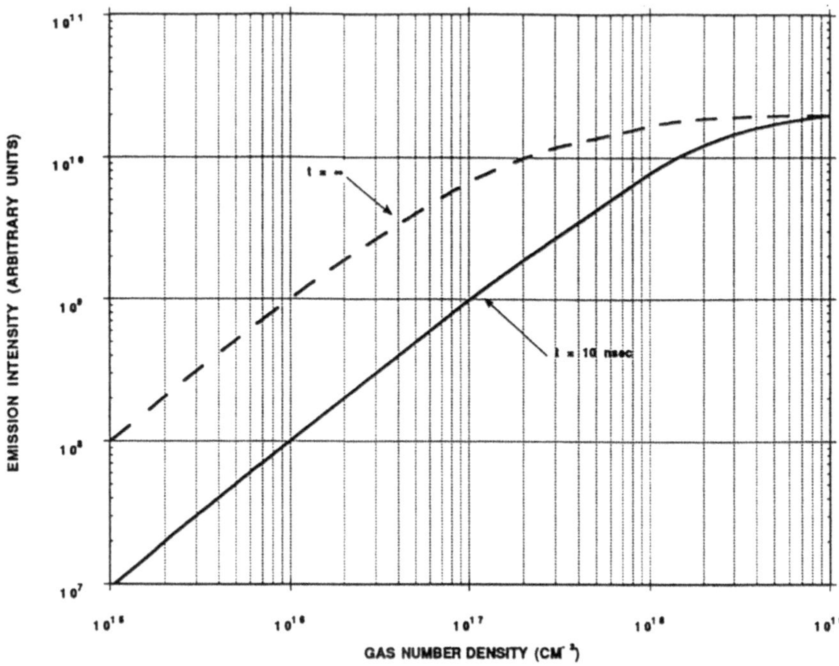

Figure 6: Emission intensity versus number density for c-w and 10 ns pulse electron beam with other parameters as discussed in the text.

4.4 QUENCHING AVOIDANCE

Typically $\bar{C}\Omega_{Q,k}n_Q t \simeq 5 \times 10^{-11} n_Q t$. For $t = 10$ ns and $A_{k,l} = 10^7$ s^{-1} a plot of $p_{k,l}$ from eqs. (8) and (5) with $(6.25 \times 10^{18} I_B \Omega_{0,k}/A_B) h\nu_{k,l} A_{k,l}$ arbitrarily set equal to one is shown in Fig. 6.

For this example $n_Q = n_g$. A comparison to the comparable c-w case is also presented in Fig. 6. Note that significant quenching begin for c-w operating at around 5×10^{16} whereas for the pulsed e-beam with gated detection the effects of quenching are delayed to about 10^{18} cm^{-3}. The price that is paid is of course evident in Fig. 6, an order of magnitude decrease in the observed light emission intensity at lower densities. As the beam pulse and observation time t are lengthened from 10 ns to ∞ the response curve will gradually transition for the lower to upper curves in Fig. 6 with all curves asymptotically to the raw saturation level at high densities.

If quenching is encountered at some flow condition, the pulsed beam offers the possibility for empirically determining the severity of the presumably unknown quenching. Various delay times after the e-beam pulse are assumed for gated observations of fluorescent intensity with the time of the e-beam pulse set at its average value. Extrapolation back to zero time, taking care to use delays reasonably larger than the gate width and the beam pulse provides an estimate of the zero quench fluorescence.

Since quenching cross section can be a function of local temperature and even rotational and vibrational levels, the ability to empirically determine the magnitude of any quenching correction is important.

4.5 ELECTROMAGNETIC INTERFERENCE

A fact of life attendant on any high current, rapidly pulsed e-beam is the accompanying electromagnetic interference with optical detectors. Great care needs to be taken to protect photodetector circuits as well as credit cards during pulsed e-beam operation.

5 Recommendation

The use of pseudospark generated, pulsed electron beam fluorescence as a hypersonic flow diagnostic appears attractive. The simultaneous generation of information on multiple species and quantum states is of particular interest for use in transient flow facilities. It is recommended that serious studies be initiated on pulsed e-beam fluorescence for application to hypersonic flow studies.

Acknowledgment

This work was supported by a NASA Hypersonic Research and Training grant NAGW-1061 and by AFOSR-88-0119.

References

E. P. Muntz, J. A. Kunc, D. A. Erwin, 1987, A Pulsed Electron-Photon Fluorescence Technique for Temperatures and Specie Concentration Measurement at Points in Relatively Dense, Unseeded Air Flow. AIAA-87-1526.

R. Wojcik, J. N. Schilling, D. A. Erwin, 1990, Rarefied Flow Diagnostics Using Pulsed High-Current Electron Beams. AIAA-90-1515.

K. Frank, J. Christensen, 1989, The Fundamentals of the Pseudospark and its Applications. Plasma Science, 17, 5, 748-753.

W. Benker, J. Christensen, K. Frank, H. Gundel, W. Hartmann, T. Redel, M. Steller, 1989, Generation of Intense Electron Beam by the Pseudospark Discharge. Plasma Science, 17, 5, 754-757.

E. P. Muntz, 1968, The Electron Beam Fluorescence Technique, AGARDograph 132.

G. Pham-Van-Diep, D. Weaver, D. DeWitt, J. Trelles, T. Randoph, E. P. Muntz, 1990, Properties of an Energetic O Atom Stream From a Gas Discharge Source. AIAA-90-1560.

R. J. Cattolica, 1991, Modern Developments in Electron-Beam Fluorescence, in Rarefied Gas Dynamics, ed. A. Beylich, VCH, Weinheim, 1552-1592.

ELECTRON BEAM FLUORESCENCE IMAGING FOR HYPERSONIC RESEARCH

Robert J. Cattolica
Department of Applied Mechanics and Engineering Sciences
University of California at San Diego
La Jolla, CA, 92093-0310, U. S. A.

ABSTRACT. The electron-beam fluorescence technique was developed as an experimental diagnostic for rarefied gas dynamic research over twenty years ago. In addition to extensive laboratory research, flight-research using electron-beam fluorescence has been successful in providing measurements of nitrogen rotational and vibrational temperature, and the density of nitrogen and oxygen in the atmosphere at altitudes above 80 km. Using advanced imaging detectors and improved electron beam technology, both flight research at altitudes down to 50 km and wind tunnel experiments at equivalent densities can be conducted to study hypersonic flows. In this paper we review previous flight research experience with the electron-beam technique and discuss the use of an imaging spectrograph for hypersonic flow field measurements.

1. Introduction

Renewed interest in hypersonic aerodynamics has been stimulated by the proposed development of new hypersonic space vehicles such as: the National Aerospace Plane (NASP) in the U.S., Sanger in West Germany, and the HERMES project of the European Space Agency. The design of these advanced vehicles will depend on computational fluid dynamics (CFD) codes to simulate hypersonic flight performance. To validate these CFD codes, new research facilities have been developed that require nonintrusive, optical diagnostic methods to provide measurements of the hypersonic flow field. The extension of wind-tunnel investigations to flight experiments on hypersonic research vehicles places additional demands on available optical diagnostic techniques.

Although laser-spectroscopic methods have evolved over the past ten years in a variety of fluid-mechanic applications, the electron-beam fluorescence technique offers a number of unique capabilities in the study of hypersonic fluid mechanics in both wind-tunnel and flight-research experiments. The evolution of electron-beam fluorescence in rarefied, hypersonic wind-tunnel research can be followed in the proceedings of Rarefied Gas Dynamic Symposia. In this review of electron-beam fluorescence for use in experimental hypersonic research, two principal topics are discussed: experience in the use of the technique in flight research and the influence of

A. Boutier (ed.), New Trends in Instrumentation for Hypersonic Research, 275–284.

advanced imaging technology on the extension of the technique to higher density hypersonic flows.

2. Electron-Beam Fluorescence as a Flight-Research Diagnostic

The use of electron-beam fluorescence as a flight-research diagnostic for hypersonic boundary-layer measurements was recommended [1] over twenty years ago. A more recent feasibility study [2] came to the same conclusion and recommended that in-flight hypersonic boundary-layer measurements could be made using electron-beam fluorescence spectroscopy of nitrogen. Modern developments in miniature electron-beam, optical-sensor, and electronic technology provide further support to these recommendations to perform hypersonic boundary-layer measurements using electron-beam fluorescence. To evaluate the potential success of such an experiment, a review of the results of previous flight research using the electron-beam fluorescence technique [3] is presented.

The application of electron-beam fluorescence to in-flight measurements was first demonstrated by Deleeuw and Davies from the University of Toronto [4,5] over twenty years ago. They developed a rocket-launched instrument package using a 2.5-keV, 5-ma electron beam with a two-channel detection system. This instrument was used on a sounding-rocket experiment to measured the nitrogen density and rotational temperature from 70 to 150 km from the electron-beam fluorescence of the (0,1) band of the B $^2\Sigma$ - X $^2\Sigma$ transition of N_2^+ at 428 nm. The nitrogen rotational temperature was determined from the ratio of portions of the (0,1) band using two filters with bandwidths chosen to give optimum sensitivity in the expected temperature range [6]. The fluorescence-based density measurements that were made were in reasonable agreement with an on-board ionization gauge and were consistent with the U. S. Standard Atmosphere from 100 to 140 km. The rotational-temperature measurements agreed with the U. S. Standard Atmosphere down to 110 km. Below 110 km, shock heating of the air during the descent portion of the flight increased the measured temperature. A subsequent flight instrument [7] with a five-channel fluorescence detection system was designed to detect the 2nd-positive system of N_2 at 414 nm. In-flight measurements of the N_2 and O_2 density were obtained with this instrument from 85 to 150 km in altitude Except for the oxygen fluorescence measurements above 115 km, which were possibly contaminated by auroral emission, both the O_2 and N_2 density measurements were in good agreement with the model atmosphere.

Using a similar instrument, O'Neil, et al [8] made measurements of the nitrogen vibrational temperature in the thermosphere from 90 to 170 km based on the ratio of the (1,2) band (423.6 nm) and the (0,1) band (427.8 nm) intensities of the first-negative system of N_2^+. Nitrogen concentration measurements from the (0,1) band agreed with an atmospheric density model from 145 to 175 km. At lower altitude (80 to 145 km) the measured nitrogen density was consistently lower than the model atmosphere. Analysis of the effect of the vehicle flow field on the fluorescence measurement location using a Monte Carlo calculation indicated that vehicle heating of the surrounding gas could have contributed to the observed lowering of the measured nitrogen density. The effect of vehicle perturbation of the flow field on in-flight rotational temperature measurements [4] of ambient nitrogen has also been analyzed by Haasz [9].

After the initial use of electron-beam fluorescence to measure atmospheric properties, subsequent flight experiments used higher beam accelerating voltage and high current to study electron scattering/propagation, as well as complementary optical emission. In these latter experiments no external beam current collection was used to prevent vehicle charging. Isrealson and Winckler [10] used a 40-keV, 70-ma beam at high altitude (140-210 km) to examine electron back scattering, and also monitored the 391.4 nm N_2^+ fluorescence to measure nitrogen density during the flight. Increasing the beam current to 800 ma, a 2.5-keV electron beam was flown by O'Neil, et al [11] in the altitude range from 80 to 120 km. The electron beam was operated without an external Faraday cup and the vehicle was charged to a maximum potential of ~ 30 volts. In their experiment, sufficient fluorescence signal was produced at an altitude of 92 km to record N_2^+ (391.4 nm) and $O(^1S)$ (557.7 nm) fluorescence with a ground-based spectrograph and two-color telephotometer. In a subsequent flight experiment by the same group [12], two 3-keV, 5-amp electron sources and a 1-keV, 1-amp electron source were flown. Operating for a period of 100 seconds at altitudes from 80 to 140 km, both ground-based and in-flight measurements of the fluorescence from N_2^+ (391.4 nm) and $O(^1S)$ (557.7 nm) were obtained.

In an experiment designed to examine the source of anomalously high electron-beam fluorescence from the (0,0) band of N_2^+ observed above 150 km, a 10-keV, 130-ma electron beam experiment was flown by Grandal, et al [13] from 95-220 km. Agreement with the expected atmospheric density variation was obtained below 140 km, but a constant, relatively high N_2^+ fluorescence signal was observed between 150 and 220 km and is as yet unexplained. Although electron-beam experiments have continued to be flown on flight experiments over the last decade, including the space shuttle [14], these have been used primarily for space-craft charging and plasma studies in the atmosphere, with no specific attention to measurement of the flow field surrounding the flight vehicle.

The previous in-flight experiments using electron-beam fluorescence as an optical diagnostic that have been summarized in the above discussion are listed in Table I. Electron-beam sources producing up to 5-amps current and accelerating voltages to 40 keV have been flown. Using single-point or area-averaged detection with simple optical detectors, the ambient N_2 density, rotational and vibrational temperature in the atmosphere have been measured from 80 to 200 km. Density measurements have been made both with and without an external Faraday cup on an arm extended from the flight vehicle to return current and prevent a build up of positive charge. Although vehicle charging from the electron beam was observed in the experiments without an external beam collector, the accumulated voltage on the flight instrument was not significant relative to the beam accelerating voltage.

In-flight electron-beam fluorescence measurements have been successfully demonstrated for applications above 100 km. For flight experiments below 100 km, the use of electron-beam fluorescence becomes more complicated. Unlike the high-altitude experiments listed in Table I, it will be necessary to isolate the electron-beam source from the ambient pressure to maintain a pressure below 10^{-4} torr to protect its cathode. This is an engineering problem that can be overcome.

Table I. Previous Flight Experience with Optical Detection of Electron-Beam Fluorescence.

Year	Researchers	E_V (keV)	I (ma)	Altitude (km)	Time (sec)	Measurement
1969	Deleeuw and Davies [4]	2.5	5	70-150	~100	$[N_2]$ and T_{rot} from (0,1) at 428 nm
1970	O'Neil, et al [8]	2.5	6.2	80-175	~100	$[N_2]$ and T_{vib} (1,2)/(0,1) bands
1971/ 1972	Haasz, et al [7]	2.5	4	85-175	~100	$[N_2]$and $[O_2]$
1974	Isrealson and Winckler [10]	40	70	140-210	~100	$[N_2]$ from $N_2^+(0,0)$ at 391.4 nm
1974	O'Neil, et al [11]	2.5	800	80-120	~180	space craft charging and $[N_2^+]$ & [O] from ground.
1975	O'Neil, et al [12]	1 (1) 3 (2)	1000 5000*2	110-135	~100	$[N_2^+]$ and [O] from ground base.
1976	Grandal, et al[13]	10	130	95-220	~280	$[N_2]$ from $N_2^+(0,0)$ at 391.4 nm
1982	Banks, et al [14]	1	100	250	9 hrs	space-craft charging and video of emission

Of more concern are the complications [15] to the underlying physics of the electron-beam fluorescence diagnostic for nitrogen gas densities greater than 10^{15} cm^{-3}. The details of the N_2^+ excitation and de-excitation process and beam-propagation effects become important at higher gas density. In particular, collisional quenching of the fluorescence signal will decrease the effective measured density and spreading of the electron beam will degrade the spatial resolution in higher density applications. Previously these issues have been avoided by working at lower pressure. For flight research applications, where it is necessary to collect data throughout the flight envelope, these effects can not be avoided.

Recent fluorescence imaging measurements of electron beam propagation [16] have been used to determine beam spread at flight altitudes from 48 to 71 kms over an equivalent pressure range from 0.80 to 0.05 torr. This fluorescence beam spread data was compared to previous beam spread based on measurements [17] of the dispersion of the primary electrons in the electron beam. Comparing beam spread as a function of gas target thickness (torr-cm), agreement with the previous primary electron data was obtained at a target thickness of 1.0 torr-cm. The fluorescence beam spread data extended to a target thickness of 6.25 torr-cm. The measured beam spread data was compared with a self-similar, multiple-scattering model [18] that over predicted the observed beam spread at the highest target thickness. A better model for predicting beam spread is necessary. Total fluorescence signal obtained by integrating the fluorescence intensity transverse to the beam axis was proportional to the nitrogen density down to an equivalent altitude of 48 km or a density of 0.8 torr at ambient temperature. Fluorescence quenching effects were not significant at this density/altitude and a quenching correction to the density measurement was not necessary.

3. Imaging Spectrograph for Hypersonic Flow Field Measurements

Very compact spectrographs suitable for hypersonic flight experiments can be designed using commercially available microchannel-plate image intensifiers and two-dimensional charge-coupled device (CCD) array detectors coupled by fiber-optic bundles. Since the entire spectrum is recorded in parallel, ICCD systems have a photometric advantage of more than two orders of magnitude over scanning systems covering the same spectral range. Using a imaging spectrograph, a wind tunnel/flight experiment to measure density and temperature profiles across the shock-boundary layer on a hypersonic vehicle or wind tunnel model can be designed based on the electron beam fluorescence technique. This experiment would use a collimated electron beam to produce a column of excited gas molecules in the boundary layer. The fluorescence from this excited column of gas is collected by an optical system and analyzed using an imaging spectrograph coupled to a two-dimensional ICCD detector. This spectrograph forms an image of the electron beam along one dimension of the CCD detector array, and disperses an emission spectrum along the other dimension of the array. Thus, the imaging spectrograph can both spectrally and spatially resolve the fluorescence generated by the electron beam.

Spatially resolved fluorescent spectra can be used to determine profiles of the density, rotational temperature, and vibrational temperature in the boundary layer. Density can be determined by measuring the total fluorescence as a function of position through the boundary layer. The fluorescent signal from the N_2^+ ion, for example, is linearly dependent on its density for number densities up to 10^{15} cm^{-3}. At higher number density, the relationship becomes nonlinear. The rotational temperature of N_2^+ can be determined by detailed spectral analysis of the B $^2\Sigma$ – X $^2\Sigma$ (0,0) band at 391.4 nm. This measurement requires a spectrograph with a resolution of 0.02 nm over a range of 1 nm. The vibrational temperature of N_2^+ can be determined by measuring the ratio of the fluorescence from the (0,0) band at 391.4 nm to the fluorescence from the (0,1) band at 427.8 nm. This measurement requires a spectrograph with a range of ~40 nm. The resolution of the spectrograph is not critical for the vibrational temperature measurement because the integrated intensity of each band is measured, not the detailed rotational structure of each band.

Since the rotational and vibrational temperatures in the boundary layer are probably not equal, both temperatures should be measured simultaneously. An imaging spectrograph with two gratings, one for rotational spectra and one for vibrational spectra, could be designed for this measurement. The basic layout of the instrument is shown in Fig. 1 One grating has the spectral resolution required for rotational temperature measurements. The second grating has the wide spectral range necessary for vibrational temperature measurements. The two gratings are tilted with respect to one another so that the imaged spectra (rotational and vibrational) are separated on the face of the ICCD sensor. Details of the design parameters for this imaging spectrograph can be found in a more complete feasibility study [2].

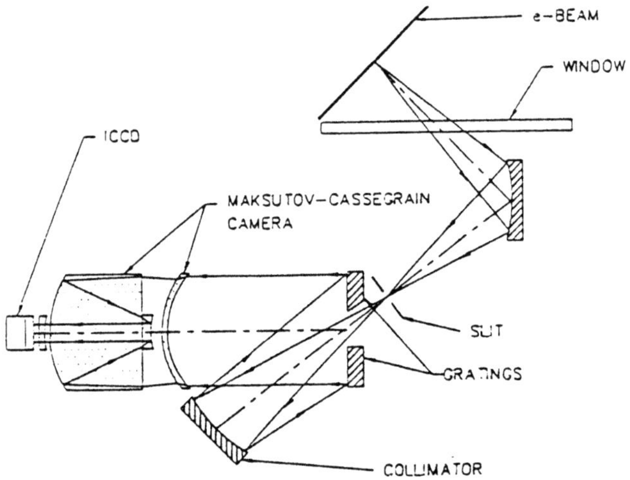

Figure 1. Imaging spectrograph for hypersonic shock-boundary layer experiment.

Since hypersonic flight research and wind tunnel experiments in transient facilities have restricted measurement time, an important issue in conducting such experiments is the minimum signal acquisition time to maintain a an acceptable signal-to-noise ratio in the measurement. The electron-beam fluorescence signal from the (v'=0,v"=0) band of the B $^2\Sigma$ – X $^2\Sigma$ transition from the nitrogen ion is directly proportional to the gas number density N_g. The photometric equation describing the detected signal P (photon counts/sec) from the electron beam fluorescence is

$$P = I\,E_{00}\,N_g\left(\frac{A_{00}}{1/\tau_{rad}+Q_0}\right)\frac{A_d\eta_d\eta_{os}}{16\,d\,e\left(f/\#\right)^2} \qquad (1)$$

where I is the beam current, E_{00} is the excitation cross section, A_{00} is the spontaneous emission rate for the transition, t_{rad} the radiative lifetime of the $B^2\Sigma(v'=0)$ state, Q_0 is the quenching rate, the detector area is A_d, η_d is the detector quantum efficiency, η_{osc} is the optical light collection efficiency, d is the beam diameter, e the charge of the electron, and (f/#) is the f-number of the light collection system. For an imaging spectrograph instrument for a flight research experiment we use appropriate values for the parameters in eq. (1) that are given in Table II. For density measurements pixel elements on the ICCD are summed across the rotational spectrum to give an effective detection array of 0.001 cm^2. The spatial resolution along the electron beam is 0.1 cm.

Table II. Parameter values for photometric equation to calculate electron beam fluorescence signal from the from $N_2^+(0,0)$ band of the B $^2\Sigma$ -X $^2\Sigma$ transition .

electron-beam current, I	10 mA
excitation cross section,	
E_{00} at 10 keV	1.500×10^{-18} cm^2
E_{00} at 20 keV	0.976×10^{-18} cm^2
E_{00} at 50 keV	0.510×10^{-18} cm^2
spontaneous emission rate, A_{00}	1.24×10^{7} sec^{-1}
radiative life time t_{rad}	66 nsec
detector area, A_d (1 mm x 0.1 mm)	0.001 cm^2
detector quantum efficiency, h_d	0.20
optical system efficiency, h_{os}	0.15
optical system, f/#	f/20
electron beam diameter, d	1 mm

Based on eq.(1) , the time required to collect sufficient signal from the electron beam fluorescence to maintain a signal-to-noise-ratio of 100 for a density measurement was calculated for different beam accelerating voltages (10 keV, 20 keV, and 50 keV) and the results plotted in Fig. 1 as a function of flight altitude. The number density for use in eq.(1) is determined from a simple atmospheric scale height model $n(h) = n_0 \exp(-h/7.8)$ where n_0 is the atmospheric number density at sea level and h is the altitude in units of km. For simplicity we have assumed that the gas temperature is constant as a function of altitude.

The three curves that are plotted in Fig. 1 show the increase in signal acquisition time as the electron beam energy is changed from 10, 20, and 50 keV. The longer signal acquisition time is the result of the decrease in the excitation cross section E_{00} with increasing beam voltage V_{eb}; $E_{00} \sim (V_{eb} / \ln V_{eb})^{-1}$. For all three electron beam energies the time required to measure the gas density with a signal-to-noise ratio of

100 is less than a 10 millisecond over the altitude range from 48 to 78 km. Under these conditions, operating the beam at 50 keV would provide the best spatial resolution without a significant increase in signal acquisition time. To measure the nitrogen rotational temperature from the electron-beam fluorescence with the equivalent signal-to-noise ratio, the signal acquisition time is increased by as much as factor of 100, since the rotational spectrum is dispersed over the pixel elements on the ICCD which can no longer be binned as was the case for the density measurement. Depending on the flight trajectory and time at altitude in an experiment, it may be necessary to use a beam energy lower than 50 keV in order to increase the excitation cross section and decrease the signal acquisition time required for the rotational temperature measurement. The necessary signal acquisition time can also be decreased by summing pixels in the spatial direction along the electron beam path. By increasing the spatial resolution to 1 mm along the electron beam through the boundary layer, the signal acquisition time can be reduced by one third.

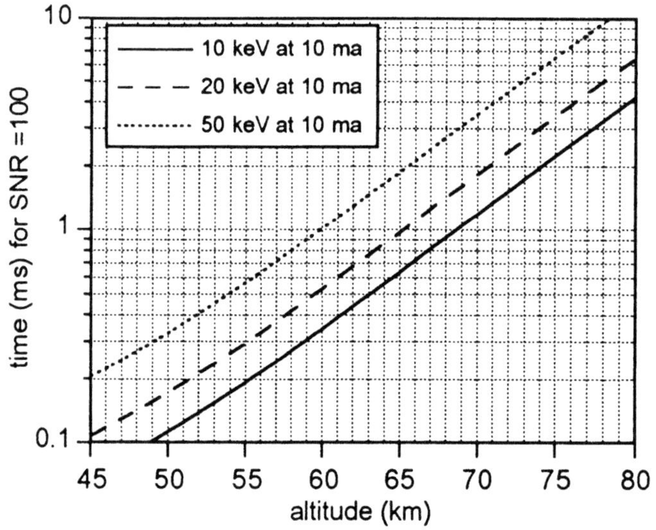

Figure 2. Signal integration time required to maintain a signal-to-noise ratio of 100 in the measurement of nitrogen density from the electron-beam fluorescence from $N_2^+(0,0)$ band (at 391.4nm) of the B $^2\Sigma$ -X $^2\Sigma$ transition as a function of the altitude for three electron beam energies: 10 keV, 20 keV, and 50 keV.

3. Conclusions

Previous experience with electron-beam fluorescence technique in flight research demonstrates that N_2 rotational and vibrational temperature, and density can be measured successfully at altitudes as low as 80 km. Although these experiments focused on the measurement of free-stream properties, with the simple addition of multi-element detector providing an imaging capability, it is possible to provide N_2 density and rotational and vibrational temperature measurements of the flow field through the hypersonic shock-boundary layer surrounding a hypersonic vehicle. These results should provide for a direct comparison with CFD modeling of hypersonic flows. In principle, flight research experiments down to an altitude of 50 km are possible before significant quenching of the fluorescence signal from the B $^2\Sigma$ -X $^2\Sigma$ transition of the N_2^+ ion that is used in the measurement of flow properties.

4. References

1. Camac, M., "Boundary Layer Measurements with an Electron Beam, " Avco Everett Research
2. Cattolica, R. J., Schmitt, R. L., Palmer, R. E., "Feasibility of Non-Intrusive Optical Diagnostic Measurements in Hypersonic Boundary Layers for Flight Experiments," AIAA 28th Aerospace Science Meeting, Paper No. AIAA-90-0627, Reno, Nevada, January 8-11,1990.
3. Muntz, E. P., "The Electron Beam fluorescence Technique," AGARDograph, 132, Von Karman Institute, Rhode-St.-Genese, Belgium, 1965.
4. DeLeeuw, J. H., and Davies, W. E. R., "Measurement of Temperature and Density in the Upper Atmosphere Using an Electron Beam," Canadian Journ. of Phys., 50, pp. 1044-1052, 1972.
5. DeLeeuw, J. H., and Davies, W. E. R., "Simultaneous Measurement of Temperature and Density by an Electron Beam Luminescence Technique," in *Small Rocket Instrumentation Techniques*, North-Holland, Publ. Comp.,Amsterdam, pp. 144-151, 1969.
6. Muntz, E.P., Abel, S. J., and Maguirre, B. L., IEEE Trans. Aerospace Supp., 21, p. 210, 1965.
7. Haasz, A. A., DeLeeuw, and Davies, W. E. R., "Thermophysics N_2 and O_2 concentration with Rocket-Borne Electron Beam Fluorescence Probes,"J. Geophys. Res., 81, pp. 2383-2391, 1976.
8. O'Neil, R. R., Pendleton, W. R., Jr., Hart, A. M., and Stair, Jr., A. T., "Vibrational Temperature and Molecular Density of Thermoshperic Nitrogen Measured by Rocket-Borne Electron Beam induced Luminescence," J. Geophys. Res., 79, pp. 1942-1958, 1974.
9. Haasz, A. A., "Effect of Reflected Molecules on Rotational Temperature Measurements with Rocket Fluorescences Probes," Canadian Journ. of Phys., 55, pp.370-377, 1977.

10. Israelson, G. and Winckler, J. R., "Measurements of 3914-A Light Production and Electron Scattering From Electron Beams Artificially Injected Into the Ionosphere," J. Geophys. Res., 80, pp. 3709-3712, 1975.
11. O'Neil, R. R., Shepherd, W. P., Reidy, W. P., Carpenter, J. W., Davis, T. N., Newell, D., Ulwick, J. C., and Stair, Jr., A. T., "Excede 2 Test, an Artificial Auroral Experiment: Ground Based Optical Measurements," J. Geophys. Res., 83, pp. 3281-3288, 1978.
12. O'Neil, R. R., Bein, F., Burt, D., Sandock, J. A., and Stair, Jr., A. T., "Summarized Results of the Artificial Auroral Experiment, Precede" J. Geophys. Res., 83, pp. 3273-3280, 1978.
13. Grandal, R., Thrane, E. V., Troim, J., "Polar 5-An Electron Accelerator Experiment within an Aurora.4. Measurements of the 391.4 nm Light Produced by an Artificial Electron Beam in the Upper Atmosphere," Planet. Space Sci., 28, pp. 309-319, 1980.
14. Banks, P. M., Raitt, W. J., White, A. B., Bush, R. I., and Williamson, P.R., "Results from the Vehicle Charging and Potential Experiment on STS- 3," J. Spacecraft, 24, pp. 138-149, 1987.
15. Rebrov, A. K., Karelov, N. V., Sukhinin, G. I., Sharafutdinov, R. G.,and Lengrand, J. C., "Electron Beam Diagnostics in Nitrogen: Secondary processes", in *Rarefied Gas Dynamics*, Part II. (S.Fisher ed.), AIAA Progress in Aeronautics and Astronautics , 74 (1980) 931-945.
16. Clapp, L. H., Twiss, R. G., and Cattolica, R. J.," Electron Beam Dispersion Measurements in Nitrogen Using Imaging of N_2^+ Fluorescence," *Rarefied Gas Dynamics*, VCH Verlagsessellschaft, Munich, Germany (1990).
17. Center, R. E., "Plural and Multiple Scattering of Fast Electrons in Gases", Phys. Fluids 13, No. 1 (1970) 79-88.
18. Lee, E. P.and Cooper, R. K., "General Envelope Equation for Cylindrically Symmetric Charged-Particle Beams," Particle Accelerators 7 (1976) 83.

FORCE MEASUREMENT IN HYPERVELOCITY IMPULSE FACILITIES

J.M. SIMMONS, W.J. DANIEL, D.J. MEE and S.L. TUTTLE
Department of Mechanical Engineering
The University of Queensland, Q 4072
Australia

ABSTRACT. A new technique is described and demonstrated for measuring drag with 100 μs rise time on non-lifting models in free piston shock tunnels. Extensions of the technique to measurement of lift and moment on a body and thrust from a scramjet nozzle are discussed.

1. Introduction

Hypervelocity impulse facilities such as free piston shock tunnels are capable of generating the high enthalpy flows associated with the external aerodynamics of spaceplanes at near orbital velocities and the internal aerodynamics of scramjet propulsion systems. Important model test parameters are drag, thrust, lift and moments but their measurement is complicated greatly by the very short test times of impulse facilities. A number of fast response force measurement techniques have been described in the literature (Sanderson and Simmons, 1991; Jessen and Grönig, 1989) but none is sufficiently fast to reach stress equilibrium with practically sized models in the 1 ms flows associated with free piston shock tunnels. In this paper we review and extend the novel approach of Sanderson and Simmons (1991) to direct measurement of drag and then describe preliminary designs of balances for measurement of thrust, lift and moments. The balances are based on interpretation of the transient stress waves propagating within the model and its support (sting).

2. Drag Measurement on a Short Non-Lifting Body

The first step in the development of the new force and moment balances was the successful demonstration of a balance for measuring drag on a 'short' cone at zero incidence (Sanderson and Simmons, 1991). With a short cone the number of stress wave reflections within the model during the test time is large so that the behaviour of the balance is relatively insensitive to discrete stress waves, a practical restriction that is removed in the next section.

The drag balance is shown diagrammatically in Figure 1(a). The non-lifting model, for example a cone at zero incidence, is attached to a 'sting' in the form of a slender elastic bar. The sting is suspended by several vertical threads so that it is free to move in the axial (free stream) direction. Its purpose is to provide a mass against which the model can react as it

A. Boutier (ed.), New Trends in Instrumentation for Hypersonic Research, 285–294.

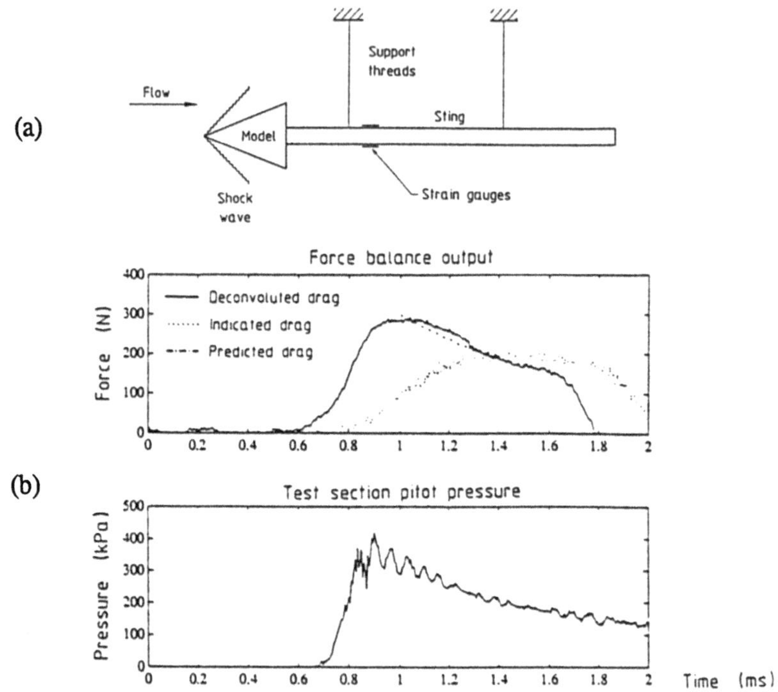

Figure 1. (a) Diagram of drag balance and (b) measured and predicted drag and measured Pitot pressure for short cone.

moves under the action of aerodynamic loading. The time-history of drag applied to the model can be determined from the output of strain gauges that respond to this reaction by way of the stress waves travelling along the sting.

The dynamic behaviour of the model/sting combination, treated as a time-invariant, causal, linear system with single input $u(t)$, single output $y(t)$ and unit impulse response $g(t)$ is described by the convolution integral,

$$y(t) = \int_0^t g(t-\tau)\, u(\tau) d\tau = g(t) * u(t) \tag{1}$$

The unit impulse response function $g(t)$ for the model/sting system is obtained numerically from a finite element analysis. The aerodynamic drag input $u(t)$ can then be obtained from $g(t)$ and measurement of the strain gauge output $y(t)$ by a numerical deconvolution process (performed in the frequency domain by Sanderson and Simmons, 1991). Inverse dynamic problems such as this are often poorly conditioned and difficulties arise through the amplification of measurement and numerical noise by the inversion procedure.

The short non-lifting body tested by Sanderson and Simmons (1991) was a solid aluminium cone with 15° semi-vertex angle and 200 mm length. The sting had a length of 2 m and was constructed from brass tubing (31.75 mm diameter and 1.63 mm wall thickness) to make its bending stiffness high. The sting was suspended horizontally in the test section of the T4 free piston shock tunnel at The University of Queensland and was enclosed in a close fitting shroud to shield it from the flow. The outputs from two strain gauges were sampled every 5 μs and averaged to cancel bending effects.

The unit impulse response of the model/sting system was obtained from the finite element computation by differentiation of the step response of the model/sting system. In this process two distributions of loading on the cone were assumed; (a) a uniform pressure distribution with negligible skin friction, and (b) a point load applied to the vertex. Loading distribution (a) is representative of that in the shock tunnel and loading distribution (b) enabled validation of the finite element computation by an experiment based on sudden removal of a known weight. An important feature of this drag balance with a short model is that the computed impulse response is typical of that of a first-order dynamic system and is time-limited, i.e. it decays almost to zero within the shock tunnel test time of about 1 ms.

Sanderson and Simmons (1991) report measurements in nitrogen of the drag on the cone at eight stagnation enthalpies from 2.6 to 33 MJ/kg. A representative measurement is reproduced in Figure 1(b) to show the effectiveness of the drag balance. The test section conditions were 3.96 MJ/kg stagnation enthalpy, 6.9 Mach number, 2640 m/s velocity, 6.64 kPa static pressure, 320 kPa Pitot pressure, 450 K static temperature and 0.0492 kg/m^3 density. It should be noted that the deconvoluted drag follows the falling test section Pitot pressure after about 100 μs from the peak of the pitot pressure. Because of its 400 μs mechanical time constant, the uncompensated balance is not able to follow the development of the flow over the model. However, the deconvoluted drag measurement follows the Pitot pressure until the arrival of the stress wave reflected from the free end of the sting. It also agrees well with drag predicted from the inviscid theory of Taylor and Maccoll (1932) for a perfect gas (frozen chemistry), attached conical shock and no base pressure.

3. Drag Measurement on a Long Non-Lifting Body

The design of the drag balance described in Section 2 is based upon achieving a large number of stress wave reflections within the model to make the balance insensitive to both the spatial and temporal distribution of loading. This was achieved by keeping the scale of the model small enough for the period of stress wave reflections within it to be an order of magnitude shorter than the tunnel test time. In this section the consequences of relaxing this model size restriction are considered and a model of larger scale is used to evaluate further the capabilities of the drag measurement technique. A solid aluminium cone of 5° semi-vertex angle and 425 mm length is attached to the sting described in Section 2.

3.1 NUMERICAL SIMULATIONS

A dynamic finite element program with internal material damping modelled as Rayleigh damping has been used to predict the response of the long cone to various applied loading distributions. In the test section of a shock tunnel a perturbation in flow conditions (e.g. a change in pressure) will not lead to an instantaneous, global change of loading on models of the length considered in this section. Instead, the effects of a small change in conditions at a reference location, say the leading edge of the model, will be propagated downstream at the speed of the flow plus the local speed of sound. It is important to understand the sensitivity of the drag balance to temporal and spatial uncertainties in the aerodynamic loading distribution.

The 'true' impulse response for the shock tunnel experiments is determined from a loading distribution in which pressure forces are applied uniformly in magnitude over the surface of the cone but in a temporally swept manner based on the speed of the flow and the local speed of sound. (Forces due to skin friction are neglected). The stress at the measurement station on the sting for a unit step load is plotted in Figure 2(a). The simulation is for a typical tunnel run in which the flow speed plus local sound speed is 4.3 kms^{-1}. The zero for the timescale is taken as the time at which the load is first applied at the vertex of the cone. The change in load at 1.1 ms is associated with the reflection of the stress waves from the free end of the sting.

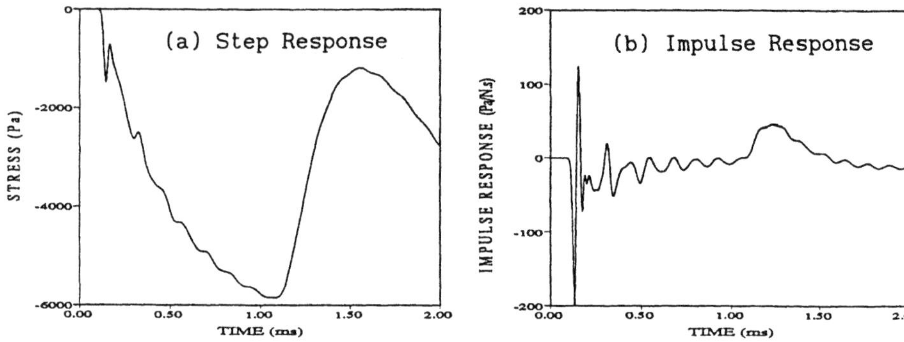

Figure 2. Step and impulse response for long cone.

The impulse response associated with this model and loading has been determined by differentiating the step response of Figure 2(a) with respect to time. The result is shown in Figure 2(b). The impulse response is not time-limited during the period of interest - the first one millisecond or so before the reflected waves arrive.

We now restrict consideration to the first millisecond; i.e. before the arrival of the stress waves reflected from the free end of the sting. In order to test the ability of the drag balance to indicate the load on the 5° cone model, three loading distributions have been applied in the finite element model. The first is the one described above and is assumed to be close to the manner in which loads would be applied in the shock tunnel. The second has the loads applied uniformly and instantaneously over the model surface. This will enables the sensitivity to temporal loading to be examined. The third has all the load applied at the vertex of the cone and assists in examination of the sensitivity to spatial loading. For each type of loading the impulse response was found by differentiating the computed step response with respect to time. The impulse responses are referred to as g_1, g_2 and g_3 respectively and are plotted in Figure 3. (Note that the vertical scales are different). The initial delays before the stress in the sting first changes are different for each loading case and have been removed to enable easier comparison of results.

The impulse responses all show an initial rise then an exponential decay. Superimposed on this are oscillations due to stress wave propagation and reflection within the model, the details of which are associated with the manner in which the loads are applied. From Figure 3 it can be seen that the form of the deviations from the exponential decay is similar for each loading case but the magnitudes are different. The more uniform load application of g_2 results in the magnitude of these oscillations being relatively small and the point load application of g_3 produces the largest amplitude deviations.

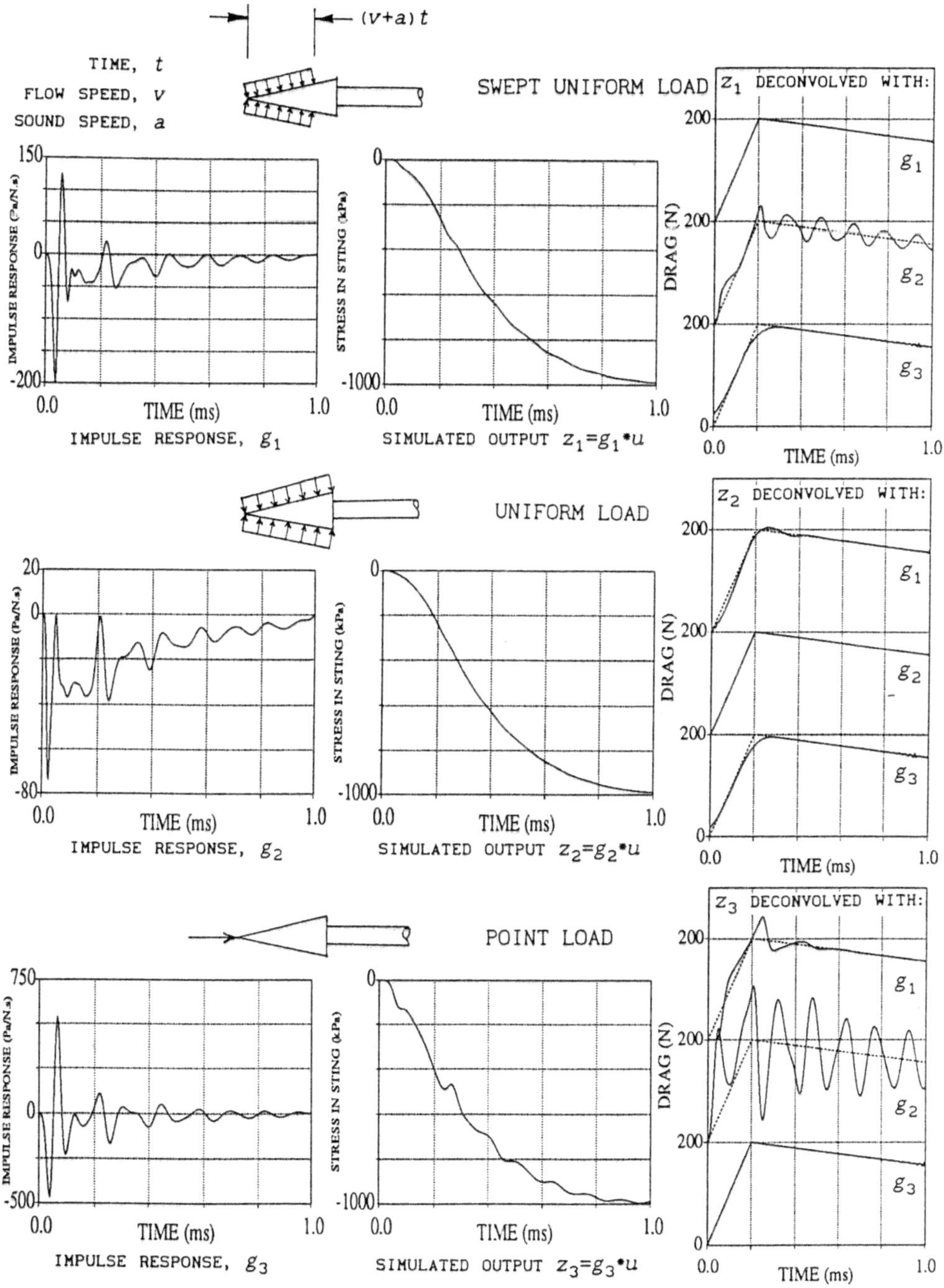

Figure 3. Numerical simulation of deconvolution with different impulse responses.

3.2 DECONVOLUTIONS USING DIFFERENT IMPULSE RESPONSES

As a test of the sensitivity of the deconvolution to the impulse response used, a temporal loading distribution, similar to that expected in the shock tunnel operating in undertailored mode, has been modelled. The applied drag, referred to as u, rises from 0 N to 200 N in 200 μs and then slowly decreases with time (see the dashed lines in deconvolutions in Figure 3). For each impulse response the stress which would be measured in the sting has been simulated by forward convolution of the impulse response with the applied drag, i.e. $(g_i * u)$.

Simulated outputs for the model loading u and impulse responses g_1, g_2 and g_3 are shown in Figure 3. It can be seen that the smaller the oscillations in the impulse response, the smoother is the simulated output. Each output has been deconvoluted with each impulse response to give an estimate of the input load. The algorithm used was functional minimisation with extended conjugate gradient (Prost and Goutte, 1984). Results are shown in Figure 3. As expected deconvolution of the simulated output with the impulse response used to generate that output reproduces the input almost exactly.

Noise on an output signal is known to be amplified in the deconvolution process. Similarly here the oscillations on the simulated outputs of Figure 3 lead to relatively large oscillations on the estimated loads when deconvolving with g_2. However, since the frequency of these oscillations is higher than those at which the drag load changes, digital filtering can be used to reproduce the applied drag quite successfully.

The oscillations seen in the impulse responses in Figure 3 are associated with propagation and reflections of stress waves within the model and will be of higher frequency for shorter models and lower frequency for longer models. The results presented in this section indicate that, for the scale of model being considered, the applied loading can be estimated well, even when the impulse response has been determined for an approximated spatial and temporal loading distribution.

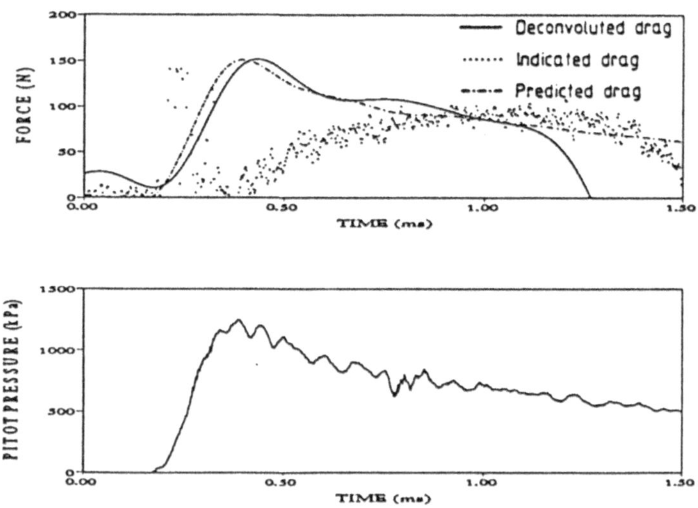

Figure 4. Measured and predicted drag and measured Pitot pressure for long cone.

3.3 EXPERIMENTAL RESULTS

Because of the smaller cross-sectional area of the 5° cone compared with the 15° cone, the level of drag is reduced and there is a consequent reduction in signal to noise ratio. Therefore more digital filtering was required in the deconvolution process. A typical result from the experiments of Tuttle (1990) is shown in Figure 4 where impulse response g_1 has been used in the deconvolution. The test section conditions with nitrogen as the test gas were 7.8 MJ/kg stagnation enthalpy, 5.9 Mach number, 3700 m/s flow speed, 21.0 kPa static pressure, 1060 kPa Pitot pressure, 970 K static temperature and 0.074 kg/m³ density. The predicted Newtonian drag is based on test section flow conditions with an allowance for skin friction and assuming zero base pressure. It can be seen that the drag balance performance is satisfactory for this longer model and lower signal level. Results obtained for deconvolution of experimental data with impulse response g_2 and g_3 confirmed the conclusions of Section 3.2

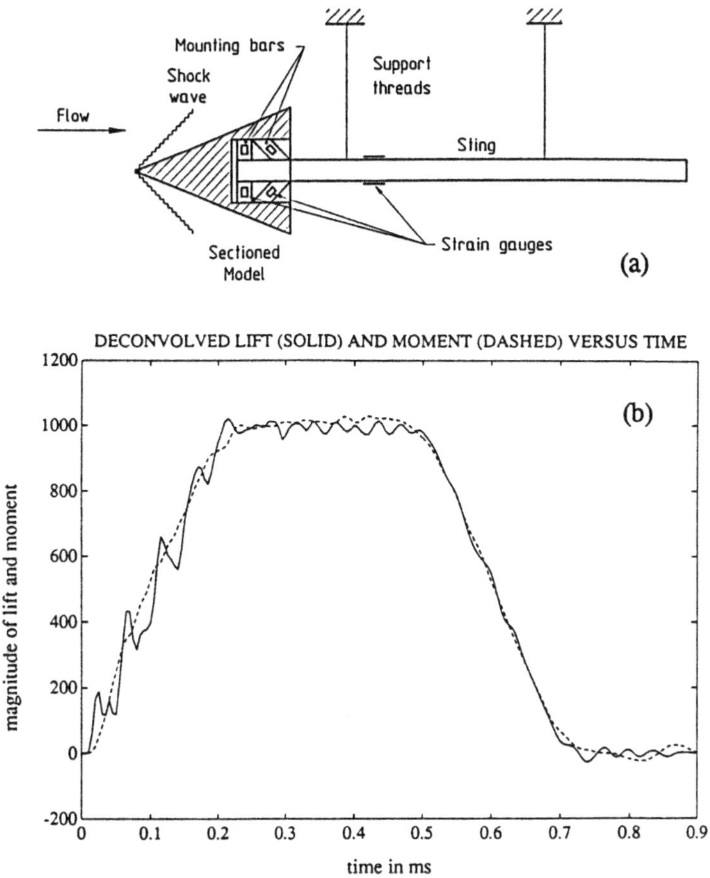

Figure 5. (a) Diagram of drag, lift and moment balance and (b) simulated lift and moment response to trapezoidal input loading.

4. Balance for Drag, Lift and Moment Measurement

In this section we present the design of a balance for measuring drag, lift and moment on the cone of Section 2 (15° semi-vertex angle) at incidence. In principle, average axial strain, transverse shear strain and bending strain could be measured in a single sting, such as that in Sections 2 and 3, to identify the drag, lift and moment components of aerodynamic loading, but in practice the bending strain will dominate the other two, presenting a very difficult measurement problem. A more subtle problem is a serious loss of high frequency information caused by the dispersion of the flexural stress waves and the domination by the low frequency modes of transverse vibration of the model on the sting. Deconvolution of the resulting signals has proven to be very difficult.

The above difficulties can be alleviated by measuring only axial strains in a sting with multiple elements. Even so, large amplitude, low frequency resonant effects characteristic of a second-order system are unavoidable, causing impulse response functions to be not time-limited and leading to loss of information in numerical processing of the smaller amplitude, high frequency information. A further complication is the inevitable coupling between at least one force component and the moment.

A variety of possible configurations has been evaluated using finite element models. At least three independent strain signals must be measured. The finite element models provide a 2 x 2 matrix of impulse responses to lift and moment loading. In the designs considered the drag component is decoupled and is not considered. It can be found by the techniques presented in Sections 2 and 3. Lift and moment were applied as distributed step loadings, all nodal loads being applied at the same time. Impulse responses were obtained by differentiation of the step responses. To simulate aerodynamic loading in the shock tunnel, a trapezoidal pulse of 700 μs duration was used for both the applied lift and moment, each of which were given arbitrary magnitudes of 1000 units.

The configuration shown in Figure 5(a) satisfies a number of design requirements. Only axial strains need be measured in each of four mounting bars which connect the model to a sting. These strains show time histories similar to those of strains in the sting itself, but do not lead to loss of high frequency information. The connection between the model and the sting is stiff enough to avoid problems of resonance of the model on the mounting bars but the bars have cross-sections that are small enough to produce measurable strain signals. The bars provide a sufficiently uniform strain field for errors due to gauge positioning to be insignificant.

For the configuration in Figure 5(a), Figure 5(b) shows the result of the numerical process of determining the impulse responses, obtaining the outputs from the trapezoidal inputs and deconvoluting the outputs with the impulse responses to recover the inputs. Both lift and moment are recovered with sufficient accuracy, allowing for some subsequent low-pass filtering, to confirm the effectiveness of the design as a preliminary to shock tunnel testing. The deconvolution algorithm used was 'error energy minimisation' (Prost and Goutte, 1984).

5. Scramjet Thrust Balance

The drag measurement technique described in Sections 2 and 3 is being extended to the measurement of the net thrust produced by a scramjet nozzle. Figure 6 shows the proposed configuration. A symmetrically diverging (11°), nozzle with parallel side walls and with two long stings attached, is suspended behind a scramjet combustor. The nozzle-sting system can move freely and the joint between nozzle and duct is such that there is no flow spillage and no transmission of combustor wall stresses to the nozzle. The production of positive thrust will move the nozzle closer to the combustor. The stings are shielded from the flow and the nozzle only sees the combustor flow.

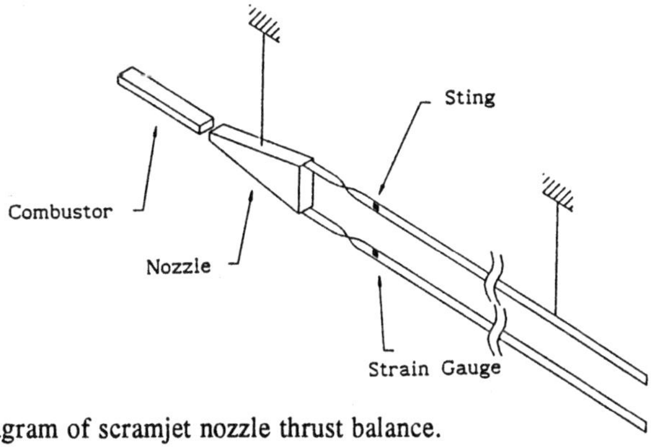

Figure 6. Diagram of scramjet nozzle thrust balance.

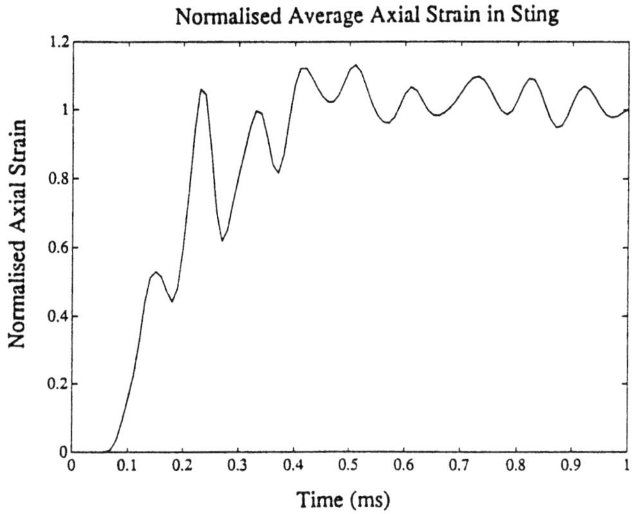

Figure 7. Numerical simulation of step response of thrust balance.

A rectangular (50 x 25 mm) scramjet combustor of approximately 300 mm length with central injection is proposed for the initial model. The effect of the impulsive pressure loading on the inner walls of the nozzle is to send stress waves of many orientations down the stings. Fortunately, the most coherent and readily identifiable of these is the axial tensile stress wave corresponding to the net axial thrust produced by the nozzle. The stings are rectangular in cross-section. The 11° angle of the nozzle means that the loading is almost transverse, yet the small axial component is sought. To increase bending stiffness and to minimise the impedance of the nozzle-sting joint, the sting is twisted through 90° just downstream from the nozzle. The axial strain is measured just aft of this twist. The twist eliminates shear strain and increases the overall rigidity of the system.

Figure 7 shows the axial strain in one sting due to a uniformly distributed step pressure load on all four nozzle walls and calculated using a finite element package. Structural damping was included. The drop after 1 ms is due to reflection from the free end of the sting. The oscillations on the strain signal are flexural vibrations of the nozzle walls caused by the nearly transverse loading and can be reduced by the use of stiffening ribs. Refinement of the finite element model spatially and temporally, the inclusion of material damping and low pass filtering should significantly decrease the noise level on the step response. The fast rise of the strain in the sting to the level of the step nozzle loading indicates that thrust measurement is feasible.

6. Conclusions

A drag balance has been developed and its use demonstrated in a free piston shock tunnel. The performance of the balance is shown to be insensitive to the assumed spatial and temporal loading distribution. Numerical simulations of stress wave propagation indicate that preliminary designs of balances for measuring lift and moment on a body and thrust from a scramjet nozzle are capable of performing in the short test time in shock tunnels.

7. Acknowledgments

The authors are grateful for the support received from the Australian Research Council under grant AE9032029 and the Queen Elizabeth II Fellowship Scheme (for Dr Mee) and from NASA under grant NAGW-674.

8. References

Jessen, C. and Grönig, H. (1989) 'A new principle for a short-duration six component balance', Experiments in Fluids, 8, 231-233.

Prost, R. and Goutte, R. (1984) 'Discrete constrained iterative deconvolution algorithms with optimized rate of convergence', Signal Processing, 7, 209-230.

Sanderson, S.R. and Simmons, J.M. (1991) 'Drag balance for hypervelocity impulse facilities', AIAA J.,29, 2185-2191.

Taylor, G.I. and Maccoll, J.W. (1932) 'The air pressure on a cone moving at high speed',Proc. Roy. Soc (London), Ser.A, 139, 278-297.

Tuttle, S.L. (1990) 'A drag measurement technique for hypervelocity impulse facilities', M.Eng.Sc. Thesis, Department of Mechanical Engineering, The University of Queensland, Queensland, Australia.

A SIX COMPONENT BALANCE FOR SHORT DURATION HYPERSONIC FACILITIES

C. JESSEN and H. GRÖNIG
Shock Wave Laboratory, RWTH Aachen
Templergraben 55
5100 Aachen
Germany

ABSTRACT: A new balance for six component forces measurements in short duration hypersonic facilities is presented. Construction and functioning of the system are described in detail. The third order calibration is explained and sample measurements on models in a four millisecond Mach 7.9 flow are given.

1. Measuring Principle

During the last three years a six component strain gauge balance for model-measurements in the Aachen shock tunnel TH 2 has been developed. The balance is characterized by a completely new geometry consisting of an inner plug, a connecting cross and an outer shell (see Figure 1). The strain gauges are positioned on all four sides of each arm of the cross. By a suitable wiring of the gauges the six components can be measured independently to a first approximation [1]. Main features of the balance are an outer diameter of 42 mm, a total length of 22 mm (without model mounting) and a lowest natural frequency of 2,4 kHz (without model).

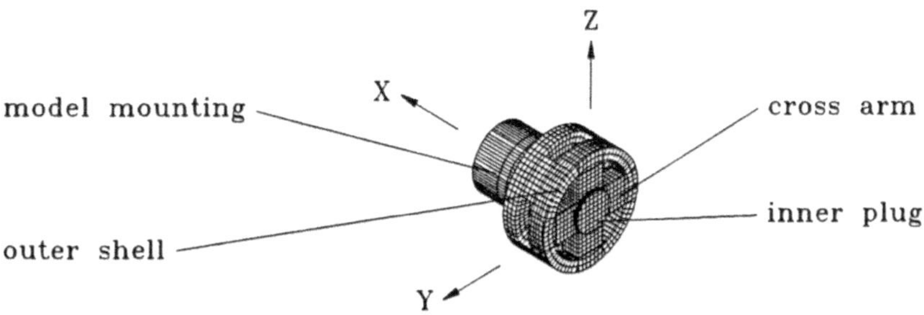

Figure 1. Finite element model of the balance.

A. Boutier (ed.), New Trends in Instrumentation for Hypersonic Research, 295–305.

The balance allows for measurement of all six force and moment components without acceleration compensation. The model, the balance and the support system have to meet certain requirements to make the measurement possible within the short testing time. In the following discussion the support system inside the test section is regarded as infinitely stiff. This means, that the dynamics of the model-balance system is not influenced by the sting. To illustrate the principle of the balance, a prototype constructed for the Aachen shock tunnel is discussed.

The basic idea of the balance is to connect the model to the outer ends of four cross arms. Forces on the model generate stresses and strains in these cross arms measured by strain gauges. The center of the cross is mounted to the model support in the test section. Due to the limitations in testing time, the balance has to have a fast response or - in other words - a high natural frequency in all components. In this context the term 'high' means, that at least three to five full cycles of the oscillation corresponding to a certain component of the balance fit into the useful testing time. For the Aachen shock tunnel it is required, that the lowest natural frequency of the balance together with the mounted model should not fall far below 1 kHz. The mass of the model significantly lowers the natural frequencies of the balance corresponding to the magnitude of the moments of inertia in the different components. Therefore it is desirable to obtain the highest possible natural frequency with the balance design. This leads to a stiff construction with small deformations and consequently low sensitivity to forces and moments in the strain gauge bridges. The geometry of the cross together with sensitive semiconductor strain gauges seem to be a good compromise between the two demands of stiffness and sensitivity.

2. Construction of the Balance

Studies on the finite element model of the balance [2] showed, that the lowest natural frequencies are determined by the geometry of the junction between the cross and the inner plug. The axis of the pitching oscillation is located in this cross section, so that the geometrical moment of inertia has to be maximized there.

Table 1 shows, how the balance components correspond with the natural frequencies. The disagreements between the calculations and the experiments are mainly due to the imperfect clamping of the inner plug in the experiments. It is difficult to assign a measured frequency to a certain component, but the lowest frequency could be clearly identified and is in reasonable agreement with the one that was calculated.

TABLE 1. Natural frequencies of the balance.

Component	Natural frequency calculated from finite element model	Natural frequency measured on the prototype balance
F_x: Drag	11.000 Hz	3.700 Hz
F_y: Side Force	6.000 Hz	~ 3.000 Hz
F_z: Lift	6.000 Hz	~ 3.000 Hz
M_x: Rolling Moment	15.000 Hz	5.300 Hz
M_y: Pitching Moment	2.700 Hz	2.300 Hz
M_z: Yawing Moment	2.700 Hz	2.300 Hz

The diameter of the outer shell has to be minimum with regard to the inner dimensions of the models. On the other hand, all forces are transmitted through the outer shell, so that it has to have sufficient strength. And finally the cross arms should be as long and flexible as possible to generate high strains for high outputs of the gauges.

Obviously a compromise has to be found. The prototype balance aimed at a 2.5 kHz lowest natural frequency and minimum outer dimensions. This led to a plug diameter of 18 mm, which also determines the width of the cross arms. To increase the sensitivity, the cross arms are conical towards the shell and the gauges are cemented to the surface at locations of maximum strains. The thickness of the outer shell influences the sensitivity of the F_y and F_z forces and the crosstalk of the components (interferences). The design aims at a certain force sensitivity, the interference-problem is regarded as secondary.

For the sake of small interferences, the cross arms have a quadratic cross section, so that only few measures have to be defined to fix the geometry of the balance, leading to a balanced sensitivity in the six components. The arrangement of the gauges on the cross and in the Wheatstone bridge circuit is illustrated schematically in Figure 2.

The prototype balance has been equipped with semiconductor strain gauges of high sensitivity (k = 140, R = 500 Ω) and small dimensions (0.38×1.27 mm^2).

The results that can be obtained with a given balance depend on the useful testing time, natural frequencies of the balance and moments of inertia of the model.

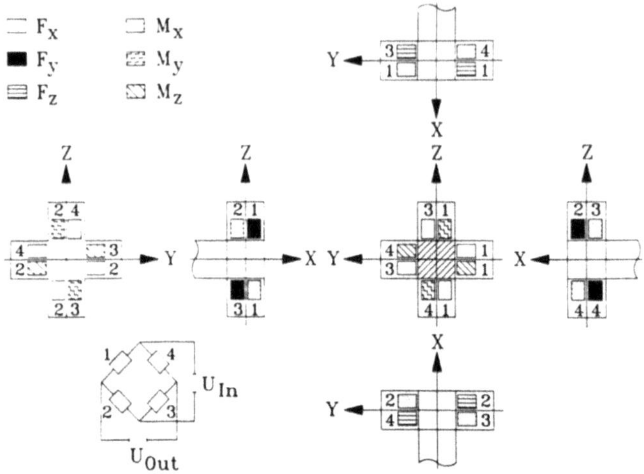

Figure 2. Arrangement of the strain gauges on the balance.

In a sort of 'trade off', sensitivity can be gained by accepting lower natural frequencies or building lighter models. Reducing the outer dimensions of the balance as much as possible allows for smaller and lighter models to be used. The model design is a complicated item in balance measurements that will not be discussed in this context. Only it should be mentioned that the choice of the material and the location of the center of gravity of the model with respect to the balance have significant influence on the overall performance.

The best performance is obtained, when the center of gravity of the model together with the balance is located at the critical cross section between cross and plug. The model itself must not have natural frequencies lower than the lowest frequency of the balance. As the aerodynamic loads can reach significant values, the model should be stiff to avoid aeroelastic effects that might not appear with the original object.

Whether or not the stationary aerodynamic loads fully develop during a millisecond flow around a complicated body at incidence is questionable [3]. Therefore one should remember, that balance design alone does not solve all the problems with forces measurement in a short duration flow.

The prototype balance, readily gauged, is shown in Figure 3. After the balance is gauged, the inner plug is mounted to the sting by means of thermal shrinking. The wires are led through a central hole in the sting, that is fixed to the model support in the test section. The shrinking itself is a delicate procedure, as the gauges may

Figure 3. Prototype balance before the sting is mounted.

not be heated above 100 °C while the sting has a temperature of 300 °C. The inner plug should only have a short overhang as otherwise the lowest natural frequency would decrease. In the case of the prototype balance, the free length of the plug is only three millimeters.

3. Calibration of the Balance and Data Reduction

Unlike classical strain gauge balances for wind tunnels, the construction introduced does not aim at accurate separation of the components. On the other hand - from the flight-mechanical point of view - it is clear that the pure forces and moments must be determined. Therefore it is necessary to carry out an accurate calibration especially in our case, where the interferences are known to be significant. As all signals are recorded digitally during the experiments, it is convenient to perform the interference calculations on a personal computer after the actual run.

Classical strain gauge balances give linear responses with a crosstalk typically below one percent. In this case a mathematical description of the balance behaviour by linear terms is sufficient, but a higher accuracy can be achieved by including square terms.

Unfortunately, a dependence including a square term cannot be described by a curve having point symmetry with respect to the origin of coordinates (that is expected due to the physical effects in a balance system). Therefore it seems necessary in our case of comparatively large interferences to use a more sophisticated description of the system taking into account also third order terms [4]. The relation between signal and load is then given by

$$S_i = R_{0i} + \sum_{j=1}^{6} a_{ij} Z_j + \sum_{j=1}^{6} \sum_{k=j}^{6} b_{ijk} Z_j Z_k + \sum_{j=1}^{6} C_{ij} Z_j^3 \qquad i = 1, 2 \ldots 6 \qquad (1)$$

with S_i - signal of the strain gauge bridge for component i,
R_{0i} - signal of bridge i without load,
Z_j - total load acting on component j,
a_{ij} - calibration factors for linear terms,
b_{ijk} - calibration factors for square terms,
c_{ij} - calibration factors for cubic terms.

In the static calibration process known loads Z_j are applied to the balance and the resulting signals S_i of all components are measured. Altogether 206 unknowns have to be determined. These are 36 a_{ij}, 36 c_{ij}, 126 b_{ijk} (symmetrical matrix), 6 R_{0i} and one for the load and moment of the calibration device, respectively. They are calculated from the calibration data by means of least square fit, which means that more calibrated load combinations lead to a higher accuracy. Four hundred individual load-cases are the minimum to obtain results from the algorithm. One thousand load-cases lead to good and 1.500 to very good results. All single components and all possible combinations of two force components have to be loaded.

The S_i are the measured and digitized quantities during an experiment. As the Z_j are the unknowns, the above nonlinear set of equations has to be solved for each set of six values S_i recorded at a certain instant. This is achieved using the Newtonian method of approximation for a set of nonlinear equations. After the calculation the pure forces and moments are obtained as function of testtime. Now a transformation of coordinates from the balance-fixed into the aerodynamic coordinate system is undertaken, resulting in the aerodynamic loads over testtime. The problem of obtaining the dimensionless aerodynamic factors is closely connected to the problem of determining the free stream conditions in the oncoming flow. This problem will not be discussed in this context, but it is worth noting, that deviations in dimensionless results obtained in different facilities on the same model do not only have their origin in inaccurate force measurement, but are also due to incomplete knowledge of the free stream conditions.

4. Experimental Results

Two different models were tested with the prototype balance in a M = 7.9 flow: a cone-cylinder and a 30° cone. The cone-cylinder was studied by Stevens [5] in 1950. He investigated small models (Ø 5.6 mm - 20 mm) in the test section of the Ames supersonic free flight wind tunnel. A Mach number range from 1 to 8 was covered by launching the models from different guns into the oncoming flow. To obtain the drag, the time-distance history was recorded through a four-station shadowgraph setup, so that velocity, deceleration and hence drag could be evaluated. A 60 mm diameter model of the same shape but with a rearward sting is presently investigated by Naumann [6] and a model with the same dimensions was tested with the balance described here. The flow about circular cones has been calculated by Jones [7] and drag measurements are presently conducted on a 30° cone 200 mm long with a completely new technique by Sanderson et al. [8], again a model with the same dimensions was tested with our balance. Figure 4 shows the results obtained with the cone-cylinder model at zero angle of attack.

Figure 4.
a) Forces on a 60° cone-cylinder model in a M_∞ = 7.9 flow at α = 0°.

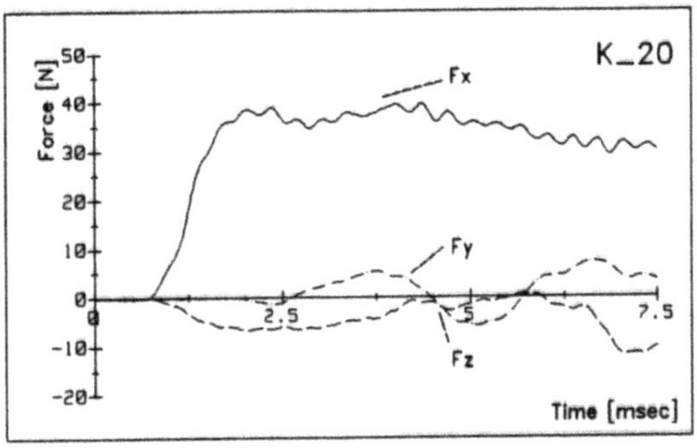

b) Moments on a 60° cone-cylinder model in a M_∞ = 7.9 flow at α = 0°.

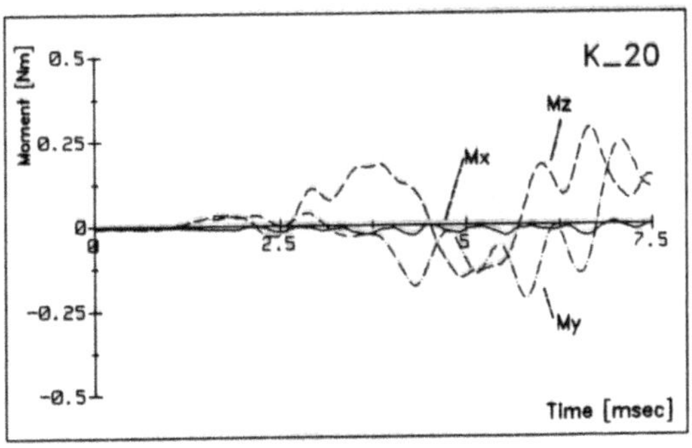

If the drag is made dimensionless with the dynamic pressure and the cross section area of the cylinder, a value of 0.556 is obtained, that agrees well with what Stevens has measured (Fig. 5).

Figure 5.
Variation of drag with Mach number for 60° cone-cylinder model from /5/, p. 1155 with own result at M = 7.9.

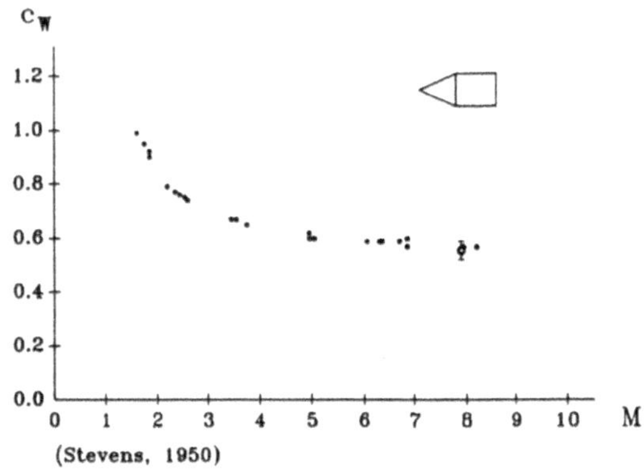

Figure 6 shows the results obtained with the cone model at zero angle of attack.

Figure 6.
a) Forces on a 30° cone model in a M_∞ = 7.9 flow at α = 0°.

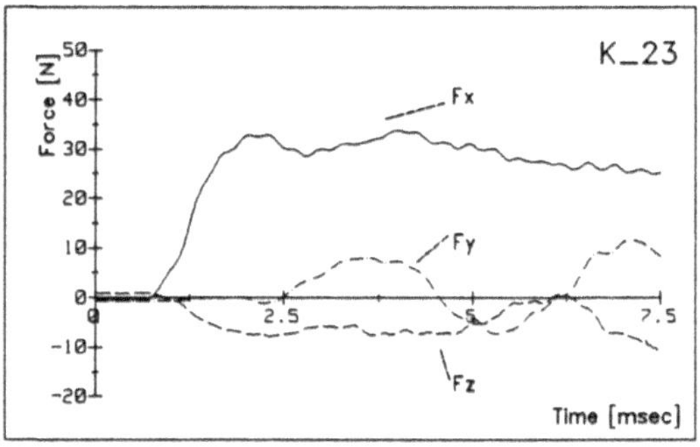

b) Moments on a 30° cone model in a M_∞ = 7.9 flow at α = 0°.

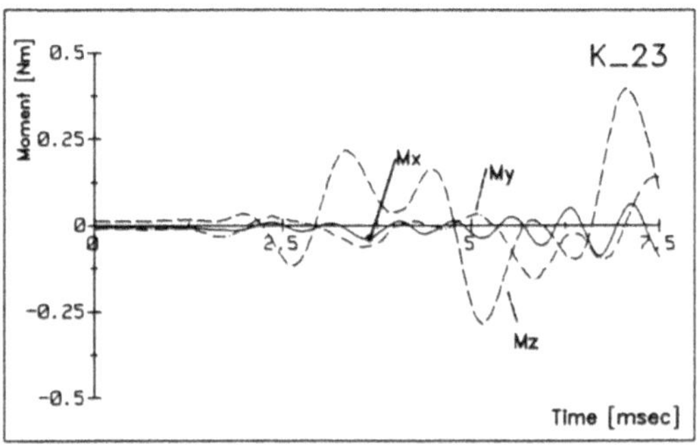

If the drag is made dimensionless with the dynamic pressure and the base area of the cone, a value of 0.155 for M = 7.9 is obtained. Jones gives a value of 0.1461 from his inviscid calculations for M = 8.

The main differences between our measurements and what Stevens measured or Jones calculated are the following: the measurements in the Aachen shock tunnel were performed with a conical nozzle and consequently in a conical flow with a negative pressure gradient along the model, and the model was fixed by a rearward sting of 36 mm diameter. For both effects corrections are possible [9], [10] and necessary, as the results obtained in a conical flow do differ from those in a parallel flow. All but the drag component should give zero output with zero angle of attack, for rotationally symmetrical models - which is obviously not the case.

The axial force F_x and the moment M_x already have satisfactory signal quality, which is remarkable because these two components usually make major problems with the classical balance design. The oscillations that occur could be easily filtered or eliminated by averaging, they originate from the natural frequencies of the corresponding balance components.

Unfortunately, the sting of the model support inside the test section performs oscillations with frequencies around 200 Hz, that produce forces and moments due to inertia. Oscillations with such low frequencies cannot be filtered without falsifying the signals originating from aerodynamic loads. To get rid of this problem, that is not directly connected with the balance itself, two solutions exist: either the model support in the test section is stiffened so that it's natural frequency rises considerably above 1 kHz, or the oscillations are compensated with the readings of accelerometers located close to the balance. As the oscillations do not occur during the first two milliseconds of the test flow, it is possible to take the results from this period without additional effort.

The negative lift-forces F_z occuring with both models are due to a negative angle of attack of -0.5°, that the model was accidently adjusted to.

5. Capabilities and Frontiers of Development

A prototype six component balance has been introduced, especially designed for the Aachen shock tunnel. The new geometry allows for balance measurements in the classical sense also during the millisecond flows of high enthalpy facilities. All six components are measured in one experiment and the aerodynamic loads are afterwords calculated using the data base of a third order static calibration. Measurements are possible over a wide range of forces and moments with a high resolution. Also can the balance endure the high overloads occurring between the actual test time and the breakdown of the flow without mechanical limit stops. As the balance is designed to operate in the subcritical region below the lowest natural frequency the model-balance-sting system has to be carefully matched to the available test time. Whether or not this matching is possible has to be decided for each individual case. For a balance with 50 mm outer diameter a measurement on a 0.6 kg model with a characteristic length of 0.3 m within 1 - 2 ms is possible; if the inner dimensions of the model do not allow for an installation of a 50 mm balance - it is not. On the other hand, if the model is lighter, a shorter test time is sufficient; or the other way round: if the test time is longer, the model may be heavier.

In conclusion of all aspects discussed it is obvious, that the system is suitable for a number of existing hypersonic facilities, but clearly has its limits where available test times are far below one millisecond and inner model diameters are much smaller than 40 mm. We believe that this area of application could be covered by balances in a hybrid technique, employing strain gauge and piezoelectric transducers. The aim of future work will be the compensation of sting oscillations, tests in the parallel flow of a contoured nozzle and the design of an advanced prototype, that will be calibrated with 1500 load cases.

Acknowledgements

Thanks are due to Mr. H. Schobben from the mechanical workshop for his excellent work in the manufacture of several prototype balances, to Dipl.-Ing. M. Lenartz, who performed the finite element calculations and patiently installed all the modifications in the geometry according to last knowledge, and to Dipl.-Ing. T. Schnittfeld, who successfully solved the problems occurring in the data reduction.

We would also like to thank the DLR Göttingen for their sponsoring and the extensive help with the calibration, and MBB Bremen for helping us with the calibration software.

References

[1] Jessen, C. and Grönig, H. (1989) 'A new principle for a short-duration six component balance', Experiments in Fluids 8, pp. 231 - 233.

[2] Lenartz, M. (1990) 'Auslegung einer Sechskomponentenwaage für den Stoßwellenkanal', Diplomarbeit, Stoßwellenlabor RWTH Aachen, Templergraben 55, 5100 Aachen.

[3] Mertens, J. and König, K. (1991) 'Short Time Force Measurement System', First European Symp. on Aerothermodynamics for Space Vehicles, ESTEC, Noordwijk, NL, 28 - 30 May 1991, esa SP - 318.

[4] Schnabl, F. (1987) 'Entwicklung eines Algorithmus zur Auswertung der Eichversuche an 6 - Komponenten - DMS - Waagen', Zeitschrift für Flugwissenschaft und Weltraumforschung 11, pp. 342 - 346.

[5] Stevens, V.I. (1950) 'Hypersonic Research Facilities at the Ames Aeronautical Laboratory', Journal of Applied Physics 21, pp. 1150 - 1155.

[6] Naumann, K.W., Ende, H., Mathieu, G. (1991) 'Technique for aerodynamic force measurement within milliseconds in shock tunnel', Shock Waves 1, pp. 223 - 232.

[7] Jones, D.J. (1969) 'Tables of Inviscid Supersonic Flow About Circular Cones at Incidence γ = 1.4', AGARDoghraph 137, Part 1, pp. 1 - 16.

[8] Sanderson, S.R., Simmons, J.M., Tuttle, S.L. (1991) 'A Drag Measurement Technique for Free Piston Shock Tunnels', 29th Aerospace Sciences Meeting, Reno, Nevada, AIAA - 91 - 0549.

[9] Pörtner, T. (1991) 'Berechnung von Korrekturgrößen für Kraft- und Momentenmessungen im Radialströmungsfeld hypersonischer Kegeldüsen', Interner Bericht: DLR Institut für experimentelle Strömungsmechanik, Abt. verdünnte/reale Gase, Bunsenstr. 10, 3400 Göttingen.

[10] Steinebach, D. (1990) 'Berechnung der Support-Interferenzen für die aerodynamischen Beiwerte eines Windkanal-Eichmodells', Forschungsbericht DLR - FB 90 - 26, DLR, Institut für theoretische Strömungsmechanik, Bunsenstr. 10, 3400 Göttingen.

MILLISECOND AERODYNAMIC FORCE MEASUREMENT TECHNIQUE FOR HIGH-ENTHALPY TEST FACILITIES

K. W. NAUMANN, H. ENDE, G. MATHIEU
French-German Research Institute Saint-Louis ISL
Rue de l'industrie 5
F-68301 Saint-Louis
France

ABSTRACT. For aerodynamic force measurement in short duration aerodynamic testing facilities we have developed a novel measurement technique. A fast-acting mounting support releases the model and grips it again after a free flight duration of some milliseconds. The model is equipped with accelerometers and may contain additional installations. Pitot pressure measurement and suitable data processing allow for direct straightforward time-dependent evaluation of the aerodynamic coefficients. This procedure is insensitive against nonlinearities or disturbances in the starting phase of the flow and compensates flow variations, if the flow is quasistationary and maintains a roughly uniform Mach number. The paper presents the measuring technique and, for example, experiments with a simple cone-cylinder model and with a body-wing model with built-in side-jets.

1. Introduction

1.1. GENERAL CONSIDERATIONS

Demands on hypersonic testing increase according to the complexity of future hypersonic vehicles: Air-breathing SCRAMjet engines, optimized lift/drag performance or good manoeverability entail an increase in number of relevant parameters and less tolerance in scaling. This holds particularly for the enthalpy of the test flow. Powerful high-enthalpy wind tunnels meet this demand, but only for milliseconds of measuring time.

At conditions like these conventional balances for measurement of aerodynamic force cannot be used, because measuring time is of the same order or less than the response time of the model/support/balance arrangement. The stress measured is produced as well by external aerodynamic load and inertia effects of model and support. Up to now 5 or more oscillation cycles during test time plus complicated procedures are needed to evaluate the aerodynamic force. Separation becomes more difficult, if the aerodynamic load history is not linear or unpredictible. This may occur particularly in the starting phase of

A. Boutier (ed.), New Trends in Instrumentation for Hypersonic Research, 307–316.

tests with models of complex stucture, or in general, if the models are equipped with installations to produce thrust or side-force.

On the other hand, force measurement is advantageous in the presence of strong pressure gradients, complex flow fields or different simultaneous effects of the same order of magnitude. In these cases, accurate deduction of surface force from pressure mapping is very expensive. Moreover, surface friction effects are not negligible for minimal-drag configurations (like waverider-like bodies, spaceplanes or SCRAMjet engines), and force measurement includes their contribution as well.

1.2. REQUIREMENTS FOR MILLISECOND AERODYNAMIC FORCE MEASUREMENTS AT ISL

In the ISL shock tunnel "B" (see Figure 1, with the flow conditions) we need aerodynamic force measurement mainly for investigations on:

- ▷ The force on a surface, produced by the interaction of side-jets and ambient hypersonic flow.
- ▷ The force on a surface resulting from external combustion in hypersonic flow.
- ▷ Hypersonic aerodynamics with real Mach- and Reynolds-number.

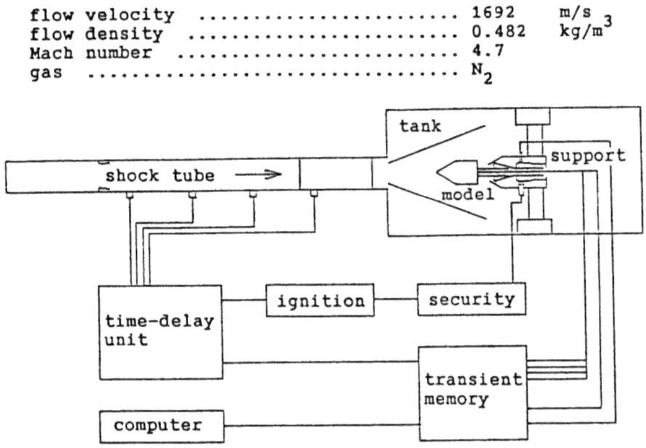

Figure 1. Sketch of ISL shock tunnel "B".

For these experiments it is important, that the interior of the models is kept sufficiently free to allow the installation of side-jets and on-board propellant supply. Moreover, budget considerations imposed the abandonment of mechanical insulation between support and tank. Electrically driven devices or magnetic suspension were discarded, because they produce electromagnetic disturbances in the measurement cascade. The complicated flow processes during the start of test flow and jets suggested a direct time-dependent measurement technique.

1.3. TECHNIQUES FOR AERODYNAMIC FORCE MEASUREMENT IN SHORT-DURATION TEST FACILITIES

Previous experiments with aerodynamic force measurement in short-duration facilities were mainly done by 3 methods (see Bernstein 1975 and Dayman 1966):

▷ Softly suspended models, equipped with accelerometers (Vidal 1956, Wittliff and Rudinger 1958, Reddy 1983). This technique allows direct measurement of force history. Usually the suspension springs need almost all space inside the model. Wire suspension (Sheeran and Duryea 1969) needs no space inside of the model, but is delicate.

▹ Free-flying models are the common technique in shock tunnels. A suitable release procedure (Pennelegion et al. 1967, Grauer-Carstensen 1971, Requardt and Kabelitz 1972) has to put them in the right place at the right time. Usually the motion of the very lightweight models is measured indirectly, with a series of photographs. This method of data registration allows only simple shapes, and the displacement-time history needs twice differentiation. Free-flying models with acceleration registration and telemetry could overcome these constraints, but to our knowledge have not yet been used.

▹ Rigid suspensions are designed to maximize the oscillation frequency of the model/support arrangement. State of the art is about 600 Hz (Stojanowski 1989) to 1 kHz (Jessen and Grönig 1991). Accordingly the minimum time needed is 5 milliseconds or more. This is too long for the ISL shock tunnel as well as for some other high-enthalpy facilities.

2. The Novel ISL-Technique

2.1. PRINCIPLE OF OPERATION

Our newly developed technique (for details see Naumann et al. 1991) is a combination of fixed and free flying model techniques. Its key feature is a special mounting support (Naumann et al. 1988). Figure 2 schematically shows the function cycle during an experiment. Initially the chuck of the mounting support is closed, fixing the model at its stem with its correct position and attitude. Just before test flow onset, it releases the model for the time of measurement, thus preventing parasitic tank and support oscillations from inflicting model motion. After the testing time it grips the model again before afterflow accelerates it too much. Model acceleration is measured by a set of small piezo accelerometers. Little displacement and small velocity during few milliseconds of testing flow allow the use of thin wires between model and recorder or control device. From acceleration and pitot pressure we calculate the aerodynamic coefficient history directly, without any integration or differentiation procedure. Thus slowly changing flow conditions can be tolerated as far as the demand of quasistationarity of the flowfield is fulfilled.

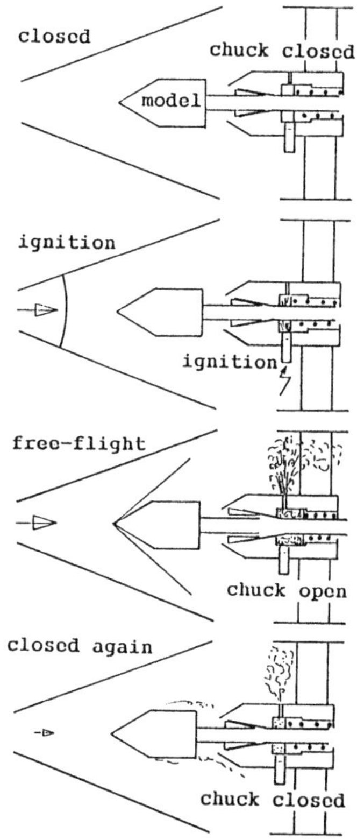

Figure 2. Illustration of the function cycle of the mounting support.

2.2. SUPPORT

Figure 3 shows mounting support Mk. 2 with side-jet model Mk. 1, retracted behind the test chamber for better visibility. For driving the opening mechanism we use pyrotechnics because of its simplicity and availability in ISL. The charge is 220 milligrams in total. The ignition current of 1 A at 90 V does not cause electromagnetic disturbances in the measuring cascade.

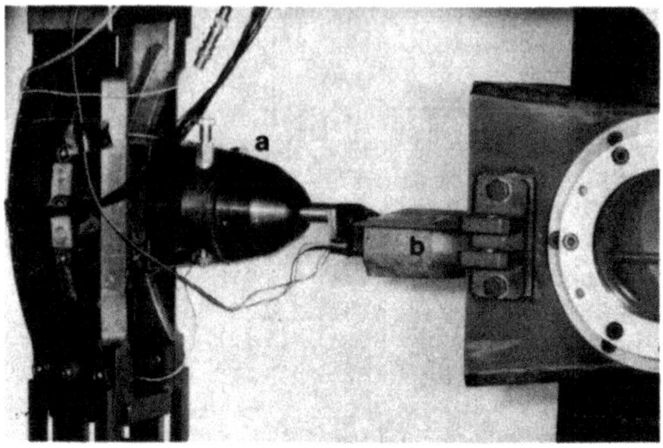

Figure 3. Installation of mounting support (a) and side-jet model (b).

2.3. MODEL CONSTRUCTION AND INSTRUMENTATION

The models may be of comparatively high cross-section area load, say 100 - 1000 kg/m^2, because dynamic pressure of roughly 1 MPa accelerates these models sufficiently. Accordingly we are free to use any material, mostly steel. Furthermore, thin umbilical wires cause no considerable restraint on model motion. The wires run through stem and support, well protected against the flow.

Bernstein (1975) discussed in depth various accelerometer arrangements. If centripetal accelerations are negligible, the minimum number needed is 3 pairs of accelerometers for 6 components of model motion. Mostly we use a greater number of accelerometers to have redundant information about the components of special interest. Up to now we use small quartz accelerometers PCB 309A with built-in amplifiers. They are mounted adhesively with resin on damping discs made from resin and micro-balloons, to attenuate the excitation of resonant oscillations of the accelerometers as result of shock loading. This occures when the incipient shock of the test flow hits the model, or at ignition of built-in pyrotechnical devices.

2.4. DATA RECORDING AND EVALUATION

Figure 1 shows the main equipment used for data registration, triggering of recorders and ignition. Any of the heat sensor signals can be used for triggering purposes, the maximum period of time being about 13 milliseconds.

With a Maurer transient memory we record 8192 data points with 500 kHz sampling frequency and 10 bit storage depth. This covers the whole relevant time of experiment for check of correct functioning of all parts involved. We use no filters in the measuring cascade, because they might suppress short overload peaks or oscillations mentioned above. For this reason we solely use numerical filtering procedures.

The attached HP-computer is sufficient for all data evaluation procedures and calculations. For any component "i" of model motion we take the respective acceleration and pitot pressure histories to calculate the corresponding aerodynamic coefficient directly. The model acceleration a_i typically depends on model area A, model mass m, the dynamic pressure of the flow $\rho_\infty \cdot W_\infty^2/2$ and the coefficient C_i. Using

$$a_i = (0.5 \cdot \rho_\infty \cdot W_\infty^2 \cdot C_i \cdot A)/m \tag{1}$$

and measured pitot pressure $\hat{p}_0$ with

$$0.5 \cdot \rho_\infty \cdot W_\infty^2 = (\hat{p}_0 - p)/C_{ps} \tag{2}$$

$$(\hat{p}_0 - p)/C_{ps} = \hat{p}_0 \cdot f(Ma, \kappa) = \hat{p}_0/C_{p0} \tag{3}$$

$$0.5 \cdot \rho_\infty \cdot W_\infty^2 = \hat{p}_0/C_{p0} \tag{4}$$

we get

$$C_i = (C_{p0} \cdot a_i \cdot m)/(A \cdot \hat{p}_0) \tag{5}$$

Accordingly the output of the accelerometers are typically normalized with the output of the pitot pressure transducer and then multiplied by $C_{p0} \cdot m/A$, which depends only on model properties, Mach number and κ. The calculation is straightforward and yields time-dependent coefficient histories. Moreover, this procedure to some extent compensates flow variations, if the flow is quasistationary and maintains a roughly uniform Mach number.

3. Examples of Measurements

In the following we present some examples of measurements to illustrate the procedures of measurement and data evaluation, and to give some idea about the capabilites of this technique.

3.1. VALIDATION EXPERIMENTS WITH A CIRCULAR CONE-CYLINDER

The experiments with the circular cone-cylinder shown in Figures 4 and 5 served as validation tests (for more details see Naumann et al. 1991). The centre of gravity of the model is adjusted in the centre of the interior body. 8 symmetrically monted accelerometers allow for recording of 5 acceleration components, three of them (X-, Z- and pitching accelerations) redundantly.

Fig. 6 shows a sample of complete signals of 8192 points from an experiment with pitch angele = -5°. The history of support combustion chamber pressure indicates correct function: 120 μs after ignition the model is set free. Model disturbance from the opening process is small and is attenuated by the test flow onset about 4 ms later. For the following time of test flow pitot pressure trace and acceleration signals (filtered with 3 kHz) show according histories. Irregular motions of the model indicate the end of free-flight, when the model either reaches the stop or is caught by the chuck.

Using the procedure described in chapter 2.4 we get the histories of the aerodynamic coefficients. Figures 7 and 8 show them for the 3 ms period of time of interest. Fig. 9 compares the drag coefficient from different experiments on the drag coefficient for shapes similar to that of our test model. We notice, that our results fit favourably.

3.2. EXPERIMENTS WITH A BODY-PLATE MODEL WITH SIDE-JETS

These experiments on side-jet control effectiveness at hypervelocity conditions mark the first operational use of the novel technique.

The model, which is yet shown in Figure 3, has three main parts:

▷ The body contains all necessary installations to produce the side-jets, i.

Figure 4. Circular cone-cylinder model.

diameter	59.2	mm
half-angle of the cone	30	deg.
length of the cylindrical part	70	mm
total length with sting	239.5	mm
mass with transducers	982	g
transverse moment of inertia I_Y	$1.943 \cdot 10^{-3}$	kgm^2

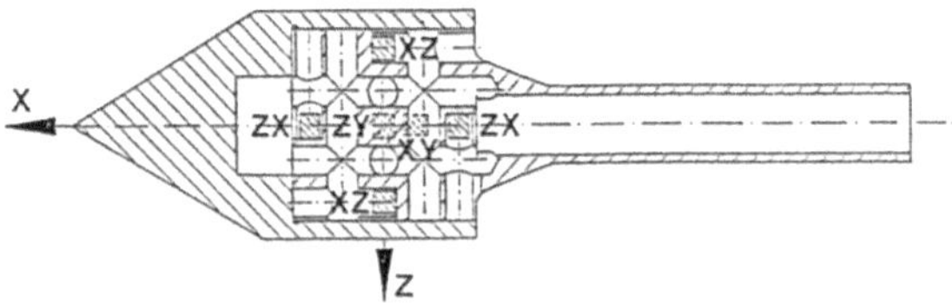

Figure 5. Sectional view of the cone-cylinder model, ik - designations of accelerometers indicate: i - direction of sensitive axis; k - axis of accelerometer location.

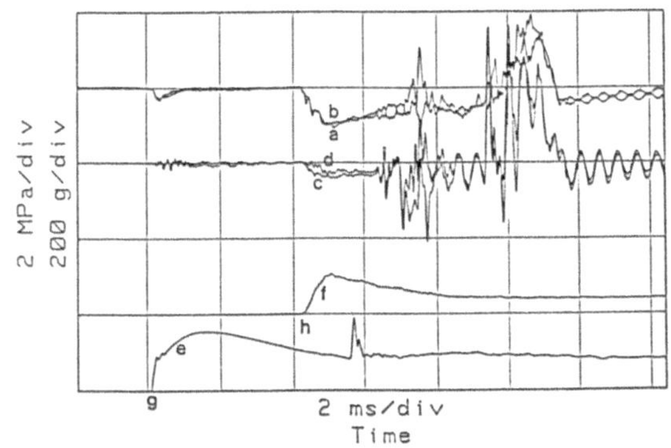

Figure 6. Signals from an experiment with the cone-cylinder model. a,b - accelerometers XY; c,d - accelerometers ZY; e - support combustion chamber pressure; f - pitot pressure; g - ignition; h - testing flow onset; i - model hits the stop.

e. two pyrotechical charges, a settling chamber and 1 to 3 parallel nozzles, which may have different shapes. The pyrotechnical charges of about 7 gram in total produce combustion gas with an internal pressure of roughly 50 MPa. Besides this, all accelerometers are mounted on the body, in excavations, as can be seen in Figure 10.

▷ The plate-wing, together with the upper side of the body, provides a flat surface, which is aligned with the oncoming ambient flow. It can be exchanged with other wings.

▷ At the stem the model is fixed before the test, and caught again, when the test flow is over.

The evaluation of the force on the surface of the model due to interaction of the side-jet and the oncoming flow must be carried out by distinct tests to measure the effects of side-jet alone and ambient flow alone. Figure 11 shows internal pressure and acceleration history of an experiment with one nozzle, without test flow. Figure 12 shows lift and drag coefficient histories with jet off, evaluated with the procedure of chapter 2.4.

Figure 13 shows, from a test with 2 jets and ambient flow, normal accelerations due to jet force a_{zj} (from p_{0j}), flow force $a_{z\infty}$

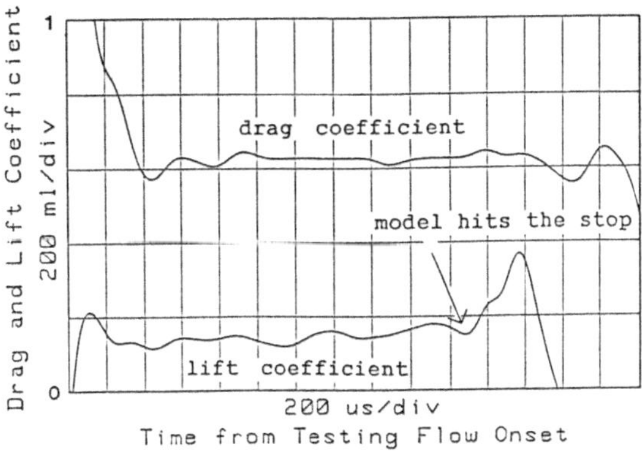

Figure 7. Aerodynamic coefficients of the cone-cylinder model. Pitch angle = -5°.

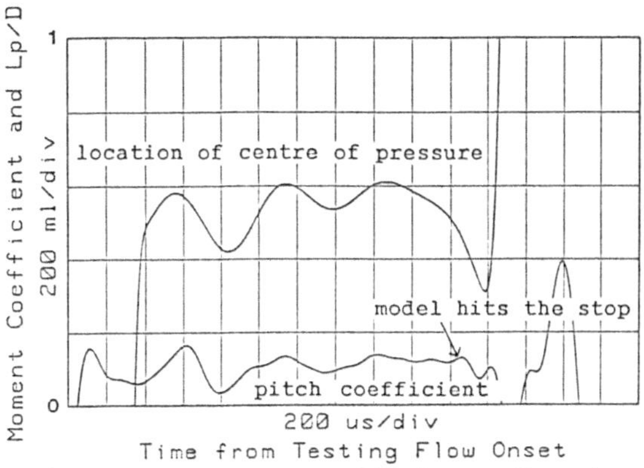

Figure 8. Moment coefficient (based on centre of gravity) and location of centre of pressure. Pitch angle = -5°.

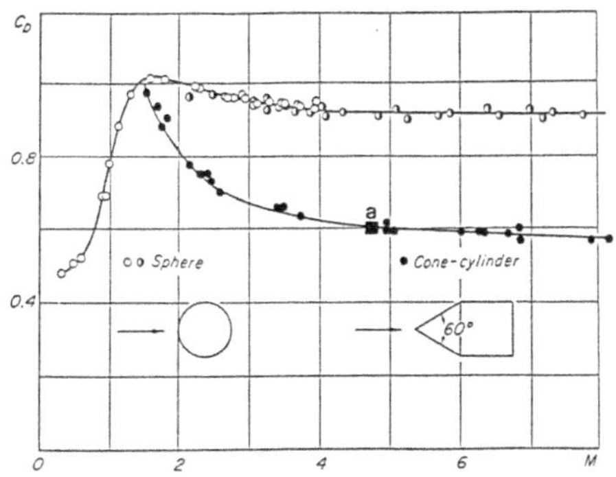

Figure 9. C_D without pitch from Chernyi (1961) and our experiments (a).

(from p_{pit}), and measured total acceleration a_{Zt}. Figure 14 shows the acceleration due to interaction of jet and ambient flow $a_{Zi} = a_{Zt} - (a_{Zj} + a_{Z\infty})$ together with a_{Zj} and $a_{Z\infty}$ and we notice, that in this case interaction force almost matches the jet thrust. The results reveal that our time of measurement is just sufficient to establish quasi-stationary flow conditions in this complicated case.

Figure 10. Body of the side-jet model.

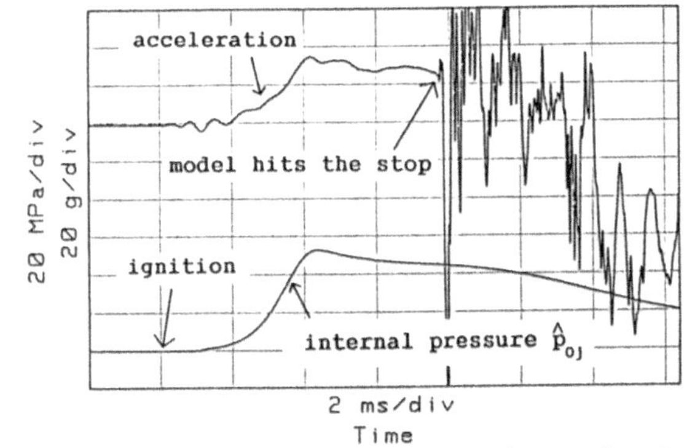

Figure 11. Lateral acceleration of the side-jet model due to jet thrust.

4. Results

We notice, that the experiments prove the sub-millisecond cabability of this technique even with a complex "body-wing" model with built-in jets. The limiting factor with regard to minimum time of measurement in this case is the complex flow structure, but not dynamic response of the model. Its size of about 140 mm length (without stem) and span is nearly the maximum possible size for our test chamber. We can expect, that larger models on the one hand will need longer time to establish quasi-stationary flow, on the other hand the dynamic response will be somewhat slower as well. In general it will be favourable to prefer simulation conditions with little characteristic length l and high gas density ρ_∞. This can readily be done at investigations on wave and friction phenomena (i. e. Mach-, Knudsen- and Reynolds - similarity) in equilibrium flow. For nonequilibrium flow

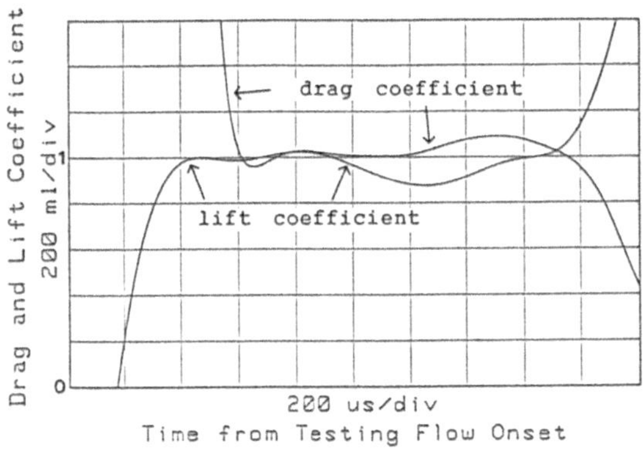

Figure 12. Lift and drag coefficient of the side-jet model with jets off.

and chemical reactions (Damköhler - similarity) as well as for fuel/air mixing processes the limitations will be more severe.

The force measurement technique presented can be used in any short-duration flow test facility without substantial and expensive modifications of this. Pyrotechnics for the opening mechanism is not obligatory, but just allows simpler design and operation. If pyrotechnics should be prohibited (e. g. at universities), a gas-driven or spring-driven opening mechanism could be used as well.

The straightforward time-dependent data evaluation procedure based on measured acceleration and pitot pressure proved to be very suitable for short-duration flow conditions. It is insensitive against nonlinearities or disturbances in the starting phase of the flow, which may result as well from aerodyamic effects, model dynamics and transducer behaviour. Up till now we cannot separate the respective contributions, and hence we are not able to quantify instationary aerodynamic force during the starting process of test flow and/or side jets.

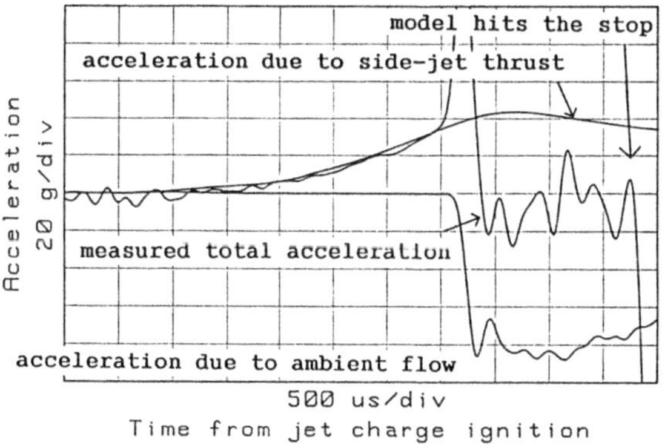

Figure 13. Lateral acceleration of the side-jet model equipped with 2 jets.

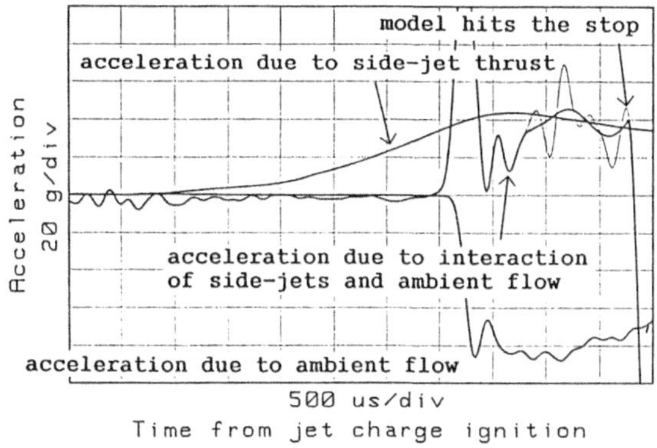

Figure 14. Interaction acceleration and acceleration due to jets and test flow. Lean line: cut-off frequency = 3 kHz; bold line where differing: 1.3 kHz.

The expense with respect to number and price of transducers and channels of transient memories does not exceed the order of equipment, which is usually available at well-equipped laboratories.

5. Conclusion

In addition to elaborating the experiments on interaction of pyrotechnically produced side-jets and ambient flow we plan to use this arodynamic force measuring technique for:

▷ Similar experiments with reservoir-supplied side-jets. Presently we are working on mechanisms, which are able to eject gases fast enough and in sufficient quantity.

▷ Experiments on lateral force created by external combustion. Smeets and Patz (1991) showed, that in this case the mixing length on models accomodated in our test chamber is sufficient. For these experiments we also have to develop devices to eject gases or liquids.

An other interesting topic is SCRamjet propulsion. If mixing of fuel and air and their combustion can be simulated in a short-duration high-enthalpy facility, with a model of sufficient dynamic properties, it is possible to build a model SCRAMjet engine with built-in fuel supply and ignition device. With such a model, thrust performance of the entire engine could be evaluated.

References

Bernstein, L. (1975) Force measurements in short-duration hypersonic facilities, AGARDograph 214.

Chernyi, C. G. (1061) Introduction to hypersonic flow, Academic Press, New York and London, p. 49.

Dayman, B. jr. (1966) Free-flight testing in high-speed wind tunnels, AGARDograph 113.

Grauer-Carstensen, H. (1971) 'Free-flight testing in a Ludwieg tube at high supersonic speeds' ICIASF '71 Record.

Jessen, C. and Grönig, H. (1991) 'The Aachen shock tunnel and its instrumentation' International Workshop on Hypersonic Facilities (Miyagi, Japan), Springer, to be published.

Naumann, K. W.; Ende, H.; Altmeyer,G. (1988) Eine Halterung für die 6-Komponenten - Kraftmessung im ISL Stoßrohr - Windkanal, ISL RT 525/88 (in german and french)

Naumann, K. W.; Ende, H.; Mathieu, G. (1991) 'Technique for aerodynamic force measurement in shock tunnel' Shock Waves 1:223-232.

Pennelegion, L.; Cash, R. F.; Shilling, M. J. (1967) Free-flight tests in the NPL 6 inch (15 cm) shock-tunnel of model HB-2 using multiple spark recording, ARC Current Paper CP 934.

Reddy, N. M. (1983) 'Aerodynamic force measurements in the IISc hypersonic shock tnnel' Proc. 14th Int. Symp. on Shock Waves and Shock Tubes, pp. 358-362.

Requardt, G. and Kabelitz, H.-P. (1972) Development and tests of the prototype of a free-flight system for the gun-tunnel, DFVLR-FB 72-14.

Sheeran, W. J. and Duryea, G. R. (1969) The application of the accelerometer force balance in short-duration testing, AIAA Paper 69-351.

Smeets, G. and Patz, G. (1991) Stoßrohrversuche an ebenen Plattenmodellen zur Außenverbrennung bei hohen Machzahlen, ISL-R 111/91.

Stojanowski, M. (1989) 'Mesures d'efforts six composantes en écoulement transitoire' AAAF 26ème Coll. d'Aerodynamique Appliquèe, Toulouse.

Vidal, R. J. (1956) Model instrumentation techniques for heat transfer and force measurements in a hypersonic shock tunnel, Cornell Aero Lab Rept. AD-917-A-1.

Wittliff, C. E.; Rudinger, G. (1958) Summary of instrumental development and aerodynamic research in a hypersonic shock tunnel, Cornell Aero Lab Rept. AD-917-A-2.

AERODYNAMIC FORCE MEASUREMENTS IN THE VKI LONGSHOT HYPERSONIC FACILITY

M. CARBONARO
von Karman Institute for Fluid Dynamics
Aeronautics/Aerospace Department
72, ch. de Waterloo
1640 Rhode-St-Genèse
Belgique

ABSTRACT. Measurements of aerodynamic forces and moments on reentry bodies are performed in the VKI Longshot hypersonic facility, using a sting-mounted strain gauge balance. Inertial corrections are applied to compensate the effect of inevitable model vibrations. This involves the measurement of model accelerations, using miniature sensors. Inertial corrections are optimized, taking into account accelerometers transverse sensitivity effects, time lags, signal filtering or smoothing, and tuning for the inertia of the vibrating masses.

1. Introduction

The principles of force measurements in short duration hypersonic facilities are thoroughly discussed in [1]. In such facilities, the test model is generally subjected to important vibrations, because of the impulsive nature of the aerodynamic forces applied to it. As a consequence, the direct measurement of such forces using a conventional, sting-mounted strain gauge balance is difficult, as the balance also measures the inertia forces of the vibrating model. These have therefore to be evaluated, and subtracted from the direct balance measurements, to yield the net aerodynamic forces. In order to evaluate the inertia forces, the model accelerations have to be measured, and the moments and products of inertia of the vibrating parts have to be known.

The techniques used at VKI to perform measurements of forces and moments in the Longshot hypersonic wind tunnel, that were developed in collaboration with Dassault Aviation, following lines suggested by Gaignebet [2], are here described.

2. The VKI Lonshot Hypersonic Wind Tunnel

This facility, fully described in [3] and sketched in Fig. 1, is a heavy piston gun tunnel, consisting of a 12.5 cm bore, 6 m long driver tube (initially pressurized with dry nitrogen gas to 300 bar at room temperature) and of a 7.5 cm bore, 27 m long driven tube (initially pressurized with dry nitrogen to about 1 bar at room temperature). These two parts are separated by a heavy piston with a mass of 1.8 to 9 kg, which is held at its initial position by an aluminum diaphragm. Rupture of this diaphragm releases the piston, which is shot down the driven tube, thereby

A. Boutier (ed.), New Trends in Instrumentation for Hypersonic Research, 317–325.

compressing the nitrogen gas in front of it into a 320 cm3 reservoir, where pressures as high as 4000 bar and temperatures up to 2500 °K may be attained. The compressed gas is trapped in the reservoir by the automatic closure of 48 poppet valves. The subsequent bursting of a secondary copper diaphragm located in the converging part of the contoured hypersonic nozzle allows the test gas to expand to Mach 14 into a 16 m3 test section, depressurized to a high vacuum (5 m Hg) before the test. The diameter of the nozzle exit is 42.7 cm, and that of the core of uniform Mach number is 24 cm. Due to the small reservoir volume, the usefull duration of a test is limited to about 10 to 15 ms, and the test conditions continually evolve with time.

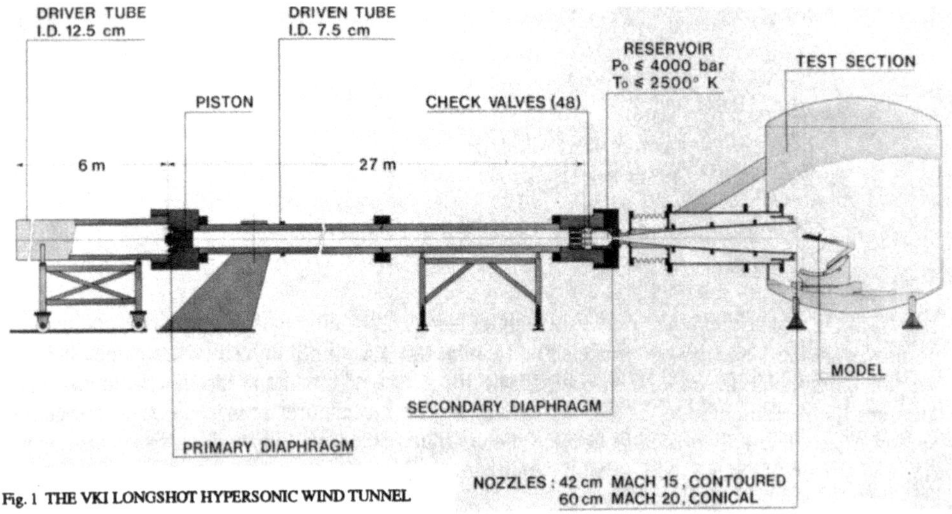

Fig. 1 THE VKI LONGSHOT HYPERSONIC WIND TUNNEL

To define the test conditions, the reservoir pressure and the pitot pressure in the test section are measured. The reservoir temperature is derived from a measurement of the heat transfer rate at the stagnation point of a spherical probe located in the test section [3].

The test model is sting-mounted in the test core. Two different stings are available and were used for the studies here described: a 1.2 m long one, supported on a circular sector mount, and a shorter one (0.5 m long), supported on the upper platform of a 5-degrees of freedom orientation mechanism.

Due to the short test duration, a fast, multichannel data acquisition system is required. Two of them were presently used: a unit, designed at VKI in 1980, the VKI High-Speed DAS, (48 channels with three multiplexers and three A/D, 12 bit converters operating at a frequency of 500 kHz), and a more recent commercially available unit, (64 channels, each one having its own filter/amplifier and its own A/D 16 bit converter operating at a frequency of 50 kHz per channel). The first unit is directly linked to a VAX 3500 computer, the second uses a PC-driven transient recorder to store up to 8192 data points per channel

3. Principle of Inertial Corrections

A model of a hypersonic vehicle, supported internally by a strain-gauge balance, in turn mounted on a sting, will experience strong vibrations when subjected impulsively to the gas stream of a facility like the Longshot. These vibrations are associated with the imperfect rigidity of the sting, to the necessary, though minimal, flexibility of the strain gauge balance, to the possible flexibility of some parts of the model (wings, winglets, stabiliser, flaps). They also are dependent on the mass, moments and products of inertia of the model. An action that seems obvious to reduce such vibrations, is to increase the model natural frequency, by decreasing model inertia (mass, moments and products) and by increasing rigidities (of the sting, the balance and the model) as much as possible. However this is in general not sufficient, and the strain gauge balance will provide outputs that show very large levels of fluctuations, with periods of the same order of magnitude as the test duration, preventing in most cases the evaluation of average values.

This indicates the need for application of inertial corrections, involving the measurement of model accelerations and the evaluation of the inertial forces to be subtracted from the measurements.

Alternatively, another solution would be to have a very flexible, or even a free mount of the model, which would start moving when subjected to the hypersonic stream, without any other constraint but its inertia. The measurement of model accelerations could then directly yield the forces and moments applied. This technique, which was also applied at VKI in the early times of the Longshot operation, on very light, simple conical models, was not applied to more complex models and will not be discussed here.

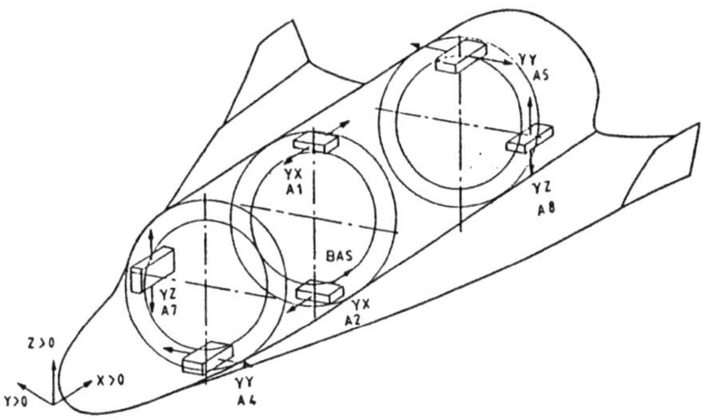

Fig. 2 POSITIONNING OF THE ACCELEROMETERS

In order to derive the inertial corrections as applied at VKI, a fundamental hypothesis is made, that the model is a rigid body, and that its vibrations result only from the elasticity of the balance and of its mount. A second simplifying assumption will be made, that the model vibrations are characterized by small rotation rates.

This allows to express, in a reference frame fixed to ground, the acceleration of a generic point of the model following Euler's rigid body dynamics,as:

$$\mathbf{a} = \mathbf{a_g} + \mathbf{R} \times (\mathbf{R} \times \mathbf{P}) + \mathbf{R'} \times \mathbf{P}$$

where $\mathbf{a_g}$ is the acceleration of the center of gravity, also expressed in an absolute reference frame, $\mathbf{P}$ and $\mathbf{R}$ are the point position vector and the instantaneous rotation rate vector, both expressed in a reference frame fixed to the moving body:

$$\mathbf{P} = X\,\mathbf{x} + Y\,\mathbf{y} + Z\,\mathbf{z}$$
$$\mathbf{R} = p\,\mathbf{x} + q\,\mathbf{y} + r\,\mathbf{z}$$

and p, q, and r are the rotation rates of the model around its body axes, with origin in the model center of gravity. In cartesian coordinates, the above expressions become

$$a_x = a_{g,x} - X(q^2 + r^2) + Y(pq - r') + Z(pr + q')$$
$$a_y = a_{g,y} - X(pq + r') - Y(p^2 + r^2) + Z(qr - p')$$
$$a_z = a_{g,z} - X(pr - q') + Y(qr + p') - Z(p^2 + q^2)$$

The purpose is to derive the three angular accelerations, p', q' and r' from measurements of linear accelerations (along x, y, or z direction) carried out at selected points on the model. This can be done, if six accelerometers are arranged in the model as shown in Fig. 2, such that accelerometers A1 and A2, A4 and A5, A7 and A8 have their measurement axis respectively aligned with x, y and z directions. The above expressions, after subtraction to eliminate the unknown accelerations of the center of gravity, lead to:

$$a1\text{-}a2 = -(X1\text{-}X2)(q^2+r^2) + (Y1\text{-}Y2)(pq\text{-}r') + (Z1\text{-}Z2)(pr+q')$$
$$a4\text{-}a5 = (X4\text{-}X5)(pq+r') - (Y4\text{-}Y5)(p^2+r^2) + (Z4\text{-}Z5)(qr\text{-}p')$$
$$a7\text{-}a8 = (X7\text{-}X8)(pr\text{-}q') + (Y7\text{-}Y8)(qr+p') - (Z7\text{-}Z8)(p^2+q^2)$$

which is a system of three equations in the six unknowns p, q, r, p', q', r'. It can be solved for the three unknown angular accelerations p', q' and r' only if the rotations rates p, q and r are sufficiently small, so that their squares and products can be neglected with respect to the angular accelerations. Alternatively three additional accelerometers (A3, A6, A9, respectively aligned along x, y, z axes) have to be added, to yield a more complex, nonlinear system of six equations in the six unknowns.

With the assumption of small rotation rates, the system becomes:

$$\begin{aligned}
&(Z1\text{-}Z2)\,q' - (Y1\text{-}Y2)\,r' = a1\text{-}a2\\
&-(Z4\text{-}Z5)\,p' \qquad + (X4\text{-}X5)\,r' = a4\text{-}a5\\
&(Y7\text{-}Y8)\,p' - (X7\text{-}X8)\,q' \qquad = a7\text{-}a8
\end{aligned}$$

and can be solved, provided the coordinates of the acceloremeters are such that the determinant of the system differs from zero. Once the angular accelerations are thus determined, the acceleration of the center of gravity of the model can be derived

$$a_{g,x} = a1 + Y1\,r' - Z1\,q'$$
$$a_{g,y} = a4 + Z4\,p' - X4\,r'$$
$$a_{g,z} = a7 + X7\,q' - Y7\,p'$$

Denoting by F and M the components, in a system of axes parallel to the body axes, of the forces and moments exerted by the balance on the model (the moments are here referred to the center of measurement of the balance), and by Fv and Mv the components of the forces and moments, in body axes, exerted by the gas stream on the model, the condition of dynamic equilibrium of the model leads to the following expressions of the inertial corrections

$$F_{v,x} = a_{g,x}\,m - F_x$$
$$F_{v,y} = a_{g,y}\,m - F_y$$
$$F_{v,z} = a_{g,z}\,m - F_z$$
$$M_{v,x} = A\,p' - F\,q' - E\,r' - M_x - (Y_c\,F_z - Z_c\,F_y)$$
$$M_{v,y} = -F\,p' + B\,q' - D\,r' - M_y - (Z_c\,F_x - X_c\,F_z)$$
$$M_{v,z} = -E\,p' - D\,q' + C\,r' - M_z - (X_c\,F_y - Y_c\,F_x)$$

where m is the mass of the vibrating body, A, B and C its moments of inertia and D, E and F its products of inertia. The quantities between brackets express the transport of the moments from the center of measurement of the balance, of coordinates Xc, Yc and Zc, to the center of gravity, at the origin of the body axes.

It is important to stress that the vibrating body, that must be considered to evaluate the mass, moments and products of inertia, is not the model alone, but it also includes "some" part of the balance, that vibrates with it. It is quite difficult to define precisely which part of the balance has to be considered, and therefore a convenient assumption has to be made. For the present studies the arbitrary, though reasonable, assumption that this part is the forward half of the balance, was made.

The net aerodynamic loads F_V and M_V thus obtained can then be expressed in a reference system with origin shifted to a conventional point (for instance at 40% of fuselage length, on the longitudinal body axis), and with axes rotated to become parallel to the wind direction, to get lift and drag. Side force and moments are left expressed in body axes.

4. Application of Inertial Corrections

Four series of tests were conducted at VKI, from 1986 to present, applying the described correction method. Specific programs and procedures are described here.

4.1. FEASIBILITY STUDY

In order to investigate the feasibility of the correction method a very simplified test case was first studied [4] in the Longshot, using an AMD-BA 1/210 scale model of the Orbiter.

This model was mounted on an internal, sting-type strain gauge balance (AMD-BA No.53.5) measuring only normal force and pitching moment. The model and half of the balance had a total mass of 316 g.

The need for correcting only two force components allowed to reduce to three the number of accelerometers. All three, denominated A7, A8 and A9, were measuring accelerations in the direction of the z body axis. A7 was placed on the model plane of symmetry, in a front bottom position (location of A4 in Fig.2), A8 and A9 were placed at the back, in a high position, symmetrically with respect to the position of A5 in Fig.2. This allows the calculation of the vertical acceleration of the center of gravity, and of the angular acceleration in pitch, again using rigid body and small rotations assumptions.

The accelerometers used were piezoresistive (Entran type EGAS-100-D), with range up to 100 g and with a -3 dB frequency response up to about 2000 Hz. Their calibration, with the complete data acquisition chain, was determined in steady state conditions, at 2 g, by turning upside-down the model with the accelerometers mounted in it.

During the Longshot tests, signals from the balance, from the accelerometers, and from a pressure transducer connected to a reference Pitot probe placed in the free-stream were recorded on the VKI High-Speed DAS at a 12.5 kHz sampling rate.

Sample results are presented: Fig. 3 shows a typical time trace of the free-stream dynamic pressure, indicating that maximum test duration, from first peak to beginning of the second rise (corresponding to jet closing) is 23 ms. Very high accelerations up to 90 g were recorded, and resulted in highly perturbed balance outputs, as shown in Fig. 4 for the normal force (even worse oscillations were present in the pitching moment traces). The large amplitude, low frequency perturbation, around 120 Hz, corresponds to the oscillation of the sting, while the higher frequency perturbation, around 400 to 600 Hz, of lower amplitude, corresponds to the balance vibration.

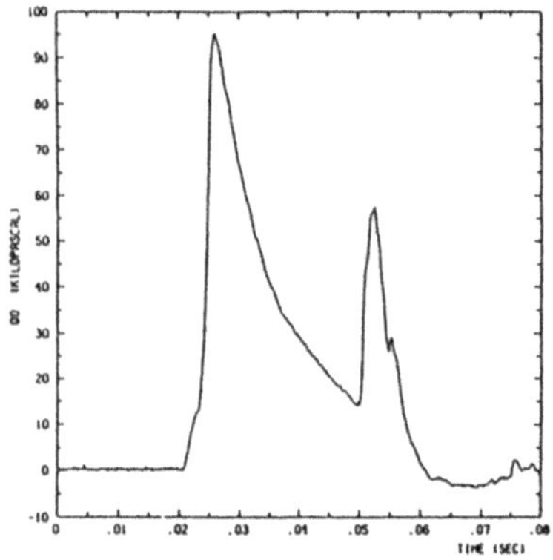

Fig. 3 LONGSHOT DYNAMIC PRESSURE

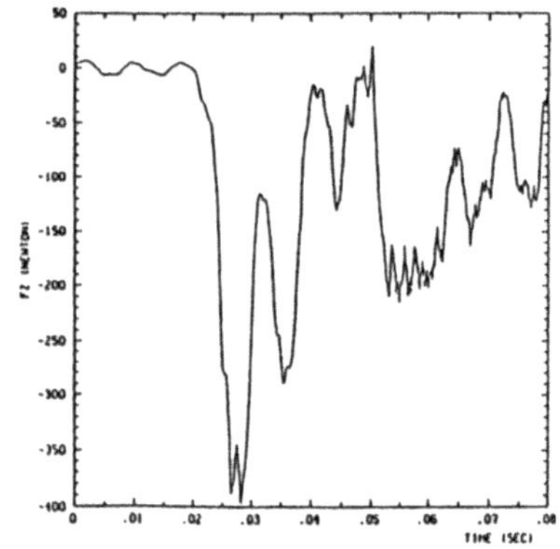

Fig. 4 NORMAL FORCE BALANCE OUTPUT

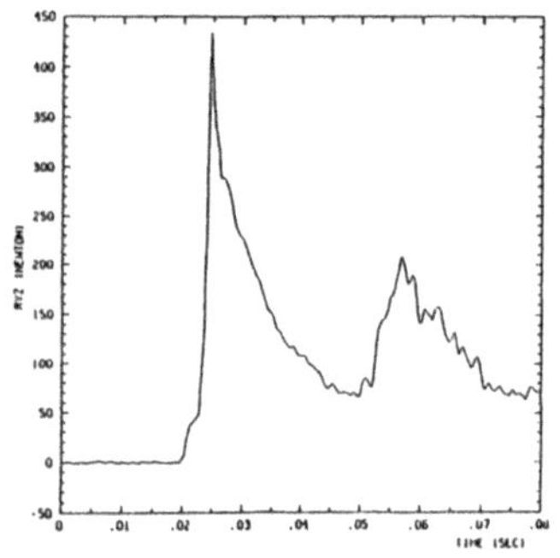

FIG. 5 NORMAL FORCE AFTER INERTIAL CORRECTION

FIG. 5 NORMAL FORCE AFTER INERTIAL CORRECTION

Optimized application of the inertial corrections resulted in the normal force trace of Fig. 5 which shows, as it should, a very similar trend to the dynamic pressure, with almost no residual oscillation, thus indicating the validity of the method. However this involved applying a forward shift in time to the accelerometer signals with respect to the balance signals.

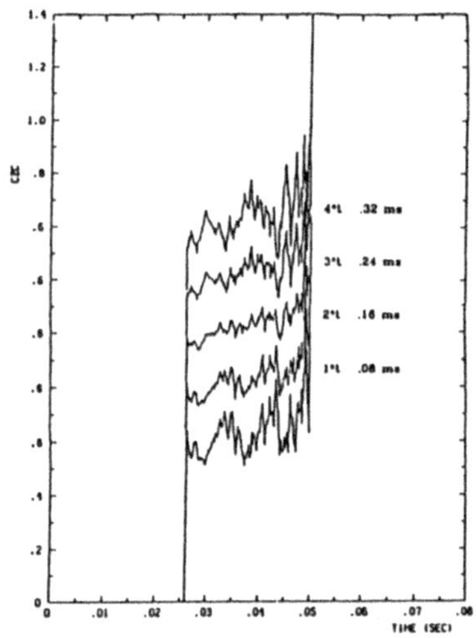

FIG. 6 EFFECT OF TIME SHIFT OF ACCELEROMETER SIGNALS

Indeed this was found necessary to compensate for the time lag introduced by the accelerometers, due to their frequency response, not as high as would have been necessary. Thus Fig. 6 shows the effect of various time shifts on the corrected normal force coefficient. Optimal time shift minimizes the residual fluctuation and corresponds to its 180° phase variation.

However Fig. 6 also shows the presence of a high frequency perturbation, not appearing in the balance traces, and therefore due to noise introduced by the accelerometers. In order to damp out these high frequency fluctuations, a simple second order numerical filter was also applied to the final data. Systematic tests were done at various cut-off frequencies, and the filter was found to be effective. It could eliminate, as shown in Fig. 7 the accelerometer noise and damp out the high frequency oscillations due to balance vibrations, but not those of lower frequency (120 Hz) due to the sting, as a too low frequency filter cut-off could cause large deviations because of the transient behaviour of the filter. The residual fluctuations in Fig. 7 disappeared with the optimal time shift defined in Fig. 6, and only a filtering to 600 Hz was needed to yield Fig. 5

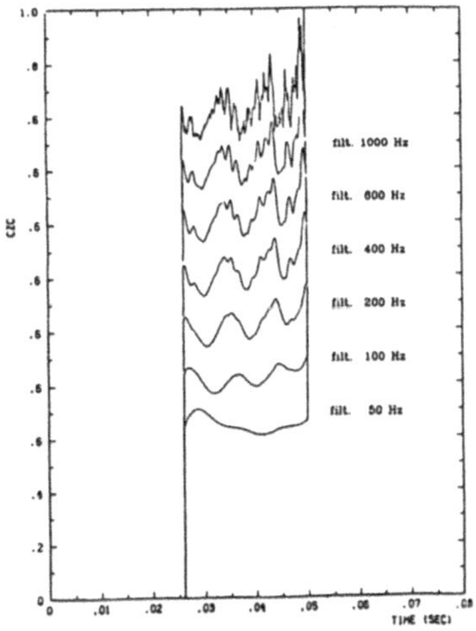

Fig. 7 EFFECT OF SIGNAL FILTERING

4.2. APPLICATION TO SIX-COMPONENTS FORCE MEASUREMENTS

Next step was the application of the method to a case where the six components of the aerodynamic force were to be measured. This was carried out [5] using a 1/60 scale Hermes model, manufactured at the NLR following AMD-BA specifications. The model was made out of magnesium, to yield a minimal mass (452 g). As its drawings were made using CATIA program, its moments and products of inertia were calculated. It was equipped with six Entran EGAS-100-D accelerometers, similar to those previously used, and positionned as shown in Fig. 2. As previously done, they were statically calibrated at 2 g by inverting the model position. Special care in avoiding touching the model had to be taken, as they were found to yield important zero shifts due to temperature effects.

The model was mounted on a semi-conductor strain gauge six-component balance (AMD-BA No. 71), specially designed to yield a high rigidity. The part of the balance that was assumed to vibrate with the model was again its fore half, having a mass of 421 g. Also the moments and products of inertia of this part of the balance were calculated.

Procedures were the same as previously, except for the data acquisition (the higher performance 50 kHz/channel, 16-bit conversion, Longshot dedicated commercial unit was used), and for a simple weight release test that was devised to easily check the the inertial correction procedure without the need of firing a Longshot test.

For this test the model with its accelerometers and its balance was mounted on the Longshot sting. A thin steel wire carrying a weight was attached to small hole drilled through one of the wings. Burning the wire releases the weight and results in a step change of the load applied to the model. Various combinations of loads are possible, by varying model orientation: normal force, rolling and pitching moment whith an horizontal model; or addition of axial force and yawing moment by pitching the model nose-up; or side force and yawing moment by rolling the model of 90°.

An example of weight release test result is given in Fig. 8 for the side force. The inertial correction transforms the irregular balance signal into the desired step, however residual

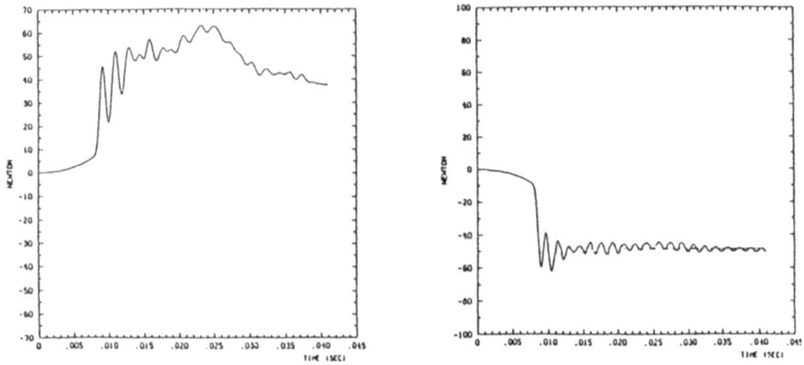

Fig. 8 SIDE FORCE DURING WEIGHT RELEASE TEST, BEFORE AND AFTER INERTIAL CORRECTION

fluctuations are present. A possible cause could be the arbitrary assumption that one-half of the balance is vibrating with the model. Therefore a systematic variation of the values of the mass and moments of inertia of the vibrating part was made in the data reduction program. Mass was varied from 70% to 150% of the nominal value. Increasing this mass generally caused a worsening of the corrections on axial and side forces, and an improvement for the normal force, with the low frequency oscillation due to the sting disappearing for a mass of 110% the nominal value. However the higher frequency oscillations never disappeared completely, and it was necessary to apply the numerical filtering technique described, or a numerical smoothing, to damp them out.

4.3. IMPROVEMENTS OF METHOD

The inertial correction method was further improved during following tests [6] carried out on another 1/55 scale model of Hermes, again manufactured at NLR according to specifications of AMD-BA. Model mass was 482 g. Another six-component balance, (VKI AR-89), specifically designed for these tests, but with design similar to the previous one, was used.

As the previous researches indicated the problems associated with the low frequency response of the piezoresistive accelerometers used, another type was selected. These were piezoelectric miniature accelerometers (Endevco model 2250AM1-10) with frequency response from 4 to 15000 Hz. To calibrate such accelerometers a static method is not suitable, and therefore a shaker with a reference accelerometer was used.

Another improvement made was the consideration of the sensitivity of each accelerometer to an acceleration perpendicular to its direction of normal use. This spurious transverse sensitivity is due to inaccuracies in locating the axes of the piezoelectric crystal. When the accelerometer is mounted on the model, another transverse sensitivity effect can arise from very small inaccuracy in positionning the casing of the accelerometer. This implies that each accelerometer signal could be a linear function of the three components of acceleration.

Two procedures were devised to compensate for this effect. The first was an iterative one: the formulae of section 3 were applied to get p',q' and r', in turn used to calculate the transverse accelerations at each accelerometer position, and these, multiplied by the corresponding value of transverse sensitivity were subtracted from each accelerometer output The procedure was applied iteratively. and a decrease in the RMS of the residual fluctuation in the corrected forces and moments was observed. However, after a limited number of iterations, the RMS increase again. Optimal number of iterations varied from 2 to 4, but was found to be dependent on the force component to be corrected. As this is an undesirable feature, another procedure was devised leading to a system of six equations in six unknowns (the three linear and the three angular accelerations of the model center of gravity). Indeed, each accelerometer output is a linear function of the three acceleration components at its location, which in turn are also linear functions of the six unknowns. Of course both procedures need the determination of the transverse sensitivities of the accelerometers when mounted in the model. This was achieved by mounting the complete Hermes model on the accelerometer calibration shaker. However, due to the rather important mass of the model, meaningful results were obtained at very low frequencies only, generally below 80 Hz. Procedure presently applied, which was found to be satisfactory, consists in mounting each accelerometer on a small trap-door which is exactly positionned in the model, and which can be removed from it, to be mounted on the calibration shaker. This allows the normal and transverse sensitivities to be determined at frequencies up to at least 1500 Hz. Application and evaluation of this procedure is presently under way

5. Conclusions

The techniques used at VKI to correct force measurements in the Longshot hypersonic tunnel were reviewed. Several modifications of the method were applied, to improve it. It was found important to use accelerometers with a high frequency response, and to take into account their transverse sensitivity effect. Low frequency oscillations in the results due to sting vibration are well corrected, while the higher frequency oscillations have to be damped by suitable numerical filtering or smoothing.

The effect of imperfect model rigidity, not considered in the present study, and is likely to cause oscillations at higher frequencies, remains to be studied.

References

[1] Bernstein L. (1975) 'Force measurements in short duration hypersonic facilities' AGARDograph 214.

[2] Gaignebet E., Hecknam, (1975) 'Nouvelle méthode d'acquisition de buffeting' 12ème Colloque d'Aérodynamique Appliquée, ENSMA-CERT, Poitiers.

[3] Simeonides G., (1990) 'The VKI hypersonic wind tunnels and associated measurement techniques ' VKI Technical Memorandum 46.

[4] Carbonaro M., Conniasselle R., (1987) 'Utilisation de la soufflerie Longshot pour des mesures de portance et de moment de tangage à Mach 15 sur une maquette d'Orbiter' VKI Contract Report 1987-26/AR.

[5] Vanmol D., Carbonaro M., (1989) 'Mesure d'efforts aérodynamiques à Mach 15 sur une maquette de l'avion Hermes ' VKI Contract Report 1989-27/AR.

[6] Vanmol D., Carbonaro M., (1990) 'Measurement of aerodynamic loads at Mach 14 on a Hermes model ' VKI Contract Report 1990-25/AR.

MEASUREMENT OF HYPERSONIC FLOWS

F. Kevin Owen
Complere Inc.
P.O. Box 1697
Palo Alto, CA 94302

ABSTRACT

Although extensive progress has been made in computational fluid mechanics, reliable flight vehicle designs and modifications still cannot be made without recourse to extensive wind tunnel testing. Future progress in the computation of hypersonic flow fields is restricted by the need for a reliable turbulence modeling data base which could be used for the development of empirical models for use in numerical codes. Currently, there are few compressible flow measurements which could be used for this purpose. In the paper, assessments will be made of the potential for laser velocimeter measurements of turbulent mean flow and fluctuations in hypersonic flow fields. The results of recent experiments conducted in two hypersonic wind tunnels will be presented.

INTRODUCTION

Current hypersonic flow field instrumentation is insufficient to meet present and future ground test requirements. Measurements are needed to establish the basic physical mechanisms and turbulence models which are required for the reliable prediction of transitional and turbulent hypersonic flow fields.

In recent years, experimental methods in lower speed regimes have made significant advances due primarily to the availability of high power lasers. Their introduction has enabled the field of laser velocimetry to expand from low speed, small scale, closely controlled laboratory applications to the measurement of compressible flows in large-scale wind tunnels (Ref. 1). The advent of the laser velocimeter allows us to measure velocity fluctuations directly in a linear, non-intrusive manner. Of particular value is the capability it offers to measure some of the compressible turbulent shear stresses, since this is an impractical task with hot wires (Ref. 2).

However, before laser velocimetry can be extended to hypersonic flow, some basic questions must be addressed. The primary question is that of particle size requirement for reliable response combined with adequate Mie scattering. Practical assessments must therefore be made of flow seeding capability and for the potential for laser velocimetry in hypersonic flows.

EXPERIMENTAL DETAILS

Laser velocimeter measurements have been made in two hypersonic test facilities. The first measurements were made in the Wright Laboratory, M=6 High Reynolds Number Wind Tunnel, which is an open jet, blow down facility described in Ref. 3. The

A. Boutier (ed.), New Trends in Instrumentation for Hypersonic Research, 327–336.

measurements were obtained in this facility on two model configurations, namely: on a zero pressure gradient smooth flat plate and in a pressure gradient flow imposed by the introduction of a 30 deg. ramp. The other laser velocimeter investigation was conducted in the NASA Ames 3.5 Ft. Hypersonic Wind Tunnel. In this facility, high-pressure air, which can be heated up to 3400°R in a pebble-bed heater, flows through an open jet test section to lower pressure spheres. The test model used in this study was a 10° cone-ogive-cylinder 79 inches long and 8 inches in diameter (Ref. 4). Measurements were also made across an oblique shock wave generated by the introduction of a 20 deg. flare installed 55 inches from the nose. Seed particle generators were designed and installed in each facility. The particles were injected into the plenum just upstream of the throat. The two component, forward scatter, fringe mode laser velocimeter systems, which were used for the flow field measurements, utilized the 4880 and 5145 Angstrom lines of an argon-ion laser. Further details of the optical, traverse control and data acquisition systems are given in Ref. 5.

TEST RESULTS

Mean boundary layer flow results obtained in the Ames facility (Fig. 1) are shown along with the mean profile measurements which were obtained from previous conventional probe measurements (Ref. 4.) The good agreement between the two measurement methods confirms the seed particle response for mean velocity measurements in the zero pressure gradient boundary layer.

The results of a more stringent test of the particle response and the laser velocimeter measurements are shown in Fig. 2 where the zero pressure gradient axial and vertical turbulence measurements are presented. These data show similarities in levels and trends to previous incompressible test results. The streamwise turbulence component has a pronounced maximum close to the wall whereas the vertical component, which is approximately half the axial value, is relatively flat in the wall region. These similarities are not altogether surprising since previous hot wire turbulence convection velocity measurements (Ref. 6) showed that the relative velocity between the disturbances and the local mean flow was always subsonic, which allows the turbulent bursts to propagate as they would in an incompressible flow. The axial component measurements are also compared with Klebanoff's incompressible results and previous hot wire hypersonic measurements in Fig. 3. There is reasonably good agreement between the hypersonic laser velocimeter and incompressible hot wire data when normalized by the wall friction velocity. This is in contrast to previous hot wire compressible flow results, reviewed in Ref. 4, which show a monotonic decrease with increasing Mach number. However, all these past hot wire results have been evaluated assuming zero pressure fluctuations which we would expect to become more important with increasing Mach number (Ref. 2). The turbulent velocity cross correlations are presented in Fig. 4, which shows the variation of the turbulent velocity correlation coefficient across the boundary layer. The maximum value of approximately -0.4 is in close agreement with incompressible shear layer observations.

Adverse pressure gradient effects on the mean and turbulent flow fields can be seen from measurements obtained at the same streamwise station on the flat plate model with and without the ramp (Fig. 5). These measurements, obtained at a station 0.3 boundary layer thicknesses ahead of the ramp, clearly show retardation of the flow due to the imposed adverse pressure gradient and a significant increase in turbulence level over a wide region. A comparison of the zero pressure gradient and ramp induced turbulence level profiles shows that the streamwise turbulent kinetic energy for the ramp flow is more than three times that for the flat-plate boundary layer. Turbulent mixing length scales, calculated using local RMS levels and mean flow gradients, are an order of magnitude larger, an indication of the large-scale, unsteady character of the flow field ahead of the interaction.

Turbulence levels based on local mean flow values exceed 30 percent in the wall region, so that significant hot wire measurement errors and flow interference would arise. At this high intensity, large--scale turbulence results in directional intermittency of up to 15 percent ahead of the time-averaged recirculation zone. Clearly, hot wire measurement errors associated with directional intermittency would be considerable in these regions.

The most stringent test of particle response was made by perturbing the flow and measuring the particle velocity variation across an oblique shock wave and shear layer generated by the introduction of a 20 deg. flare. Unfortunately, these attempts to determine particle response were complicated by the proximity of the shock to the shear layer on the flare and by shock boundary/layer interaction instabilities. The results of a scan taken 2 inches above the model surface are presented in Fig. 6 which shows the measured mean streamwise velocity and flow angularity distributions through the shock and shear layer region compared with conical flow theory and shadowgraph measurements of the shock location. The location of the measured mean velocity gradient is in good agreement with the shadowgraph shock location and the velocity change across the shock is comparable to conical flow predictions until the shear layer is encountered. The flow angularity measurements are consistent with conical flow predictions and the experimental flare angle. These comparisons indicate adequate particle response, since some of the velocity and flow angularity gradient discrepancies across the shock are probably caused by large-scale, time-dependent oscillations of the shock wave about its mean location. Indeed, attempts to measure particle response across a 30 deg. flare were unsuccessful as the increased tunnel blockage led to excessive flow field instabilities and extensive shock motions.

Shock motion has been confirmed by analyzing the velocity probability density distributions which are narrow in the freestream ahead of the shock where the turbulence level is low and wider in the more turbulent region within the shock layer. However, they are clearly bimodal in the region of the time-averaged shock location. These bimodal distributions are of most interest as they give a clear indication of particle response in hypersonic flow. The bimodal distributions shown in Fig. 7 are due to shock wave fluctuations around its mean location. Thus, if the particles follow the flow, the two bimodal peaks should be a measure of the velocity change across the shock. Since, when the instantaneous shock location is upstream of the focal volume, particles will register the lower velocity behind the shock and, when the focal volume is upstream of the instantaneous shock location, the higher freestream velocity will be recorded.

The shift from the dominant freestream peak ahead of the shock as the probe volume is traversed towards the plate, is a measure of the probability of shock passage through the focal volume. The location of the most symmetric bimodal distribution is the most likely time-average shock location. Thus, from these velocity probability density distributions we can determine the particle velocity change across the shock and estimate the mean shock location above the plate. These results compare well with theoretical velocity change predictions and optical observations of the mean shock location.

These measurements can also be used to assess seed particle response and dynamics in hypersonic flow. Using the measured velocity change and calculated transit time through the shock wave region, seed particle response characteristics can be calculated. These calculations show that the measured particle response is equivalent to that of a 0.4 micron, 1.0 s.g. sphere which undergoes a deceleration of almost seven million times the acceleration due to gravity i.e. -7DMO (1DMO =1million g). The size and deceleration of these particles is consistent with hypersonic modifications of the Stokes drag law and corresponds to a 95% response to velocity fluctuations of up to 10kHz..

Although these preliminary test results confirm the feasibility of obtaining flow field data, efficient laser velocimeter measurements of more complex flows will require prior flow field visualization to determine the regions of primary interest. With this in mind, minor modifications were made to the sending optics which enabled laser light to pass through a plane-cylindrical lens system. The lens assembly was designed so that a sheet of laser light was generated and optimized to provide vapor-screen flow visualization by illuminating liquefied vapor in the cross flow plane. The model selected was a generic hypersonic aircraft model which was tested at 15 deg. angle of attack. Fig. 8 shows the laser light sheet visualization of the cross flow plane 24 inches from the nose where the shock layer location and symmetric lee-side vortex structures are clearly visible. Streamwise scans of such flow fields will enable shock envelope and vortex trajectories to be determined and the characteristics of their interactions with control surfaces to be documented. Grid shot overlays will then enable an *a priori* determination of the extent and location of most useful laser velocimeter flow field documentation.

CONCLUDING REMARKS

Diagnostic tools are available to attempt the measurement of turbulent hypersonic flows, an area where comprehensive studies are lacking. Comparisons of new laser velocimeter turbulence measurements with previous hot wire results indicate that past data reduction assumptions can result in significant measurement errors in hypersonic flows. Extensive work is still needed to establish a reliable data base for turbulence modeling and to define the reliable ranges of hot wire and laser anemometer application.

ADDENDUM

Complere Inc. has recently made measurements in the Wright Laboratory 20-inch Hypersonic Wind Tunnel to determine particle response across an oblique shock wave generated by a flat plate model at -10 deg. incidence in a M=12.5 air flow. Details of the facility can be found in Ref. 7. The measurements, which were made above the plate, five inches from the leading edge, are shown in Figs. 9 and 10. In Fig. 9, the time of arrival information has been used to generate the first published laser velocimeter-time histories obtained in hypersonic flow. These data show the time-dependent nature of the velocity fluctuations in the region of the shock. Just above the shock, the flow is predominantly the freestream value with frequent bursts of retarded flow which occur when the probe volume is behind the shock. At the mean shock location, freestream and shock layer fluid velocity occurrences are equally probable, whereas, within the shock layer, large-scale, low frequency unsteadiness with occasional freestream velocity occurrences are evident. These time histories provide new insight into the unsteady nature of the flow.

The new probability density measurements shown in Fig. 10, clearly show the shift from the dominant freestream peak ahead of the shock as the probe volume is traversed towards the plate. This shift is again a measure of the probability of shock passage through the focal volume. The location of the most symmetric bimodal distribution being the most likely, time-average shock location. Thus, we can again determine the particle velocity change across the shock and estimate the mean shock location above the plate.

These results can be compared with theoretical predictions of shock strength and location and assessments made of particle response. If particle lag was significant, one would expect that the velocity change across the shock would be less than theoretical predictions. Surprisingly, the opposite result was obtained. That is, the measured velocity change across the shock was greater than the theoretical value. Also, the mean shock location was displaced further from the plate. These were both unexpected results. But, on reflection, the theoretical predictions only apply to sharp flat plate flow and, as shown in Fig. 9, a nominal nose bluntness of between 0.001 and 0.002 inch. could well account for these differences in shock strength and location. That is, nose bluntness will produce a stronger shock which will be displaced further from the body. This is indeed what the particle velocity measurements show.

These measurements have been used to further assess seed particle response and dynamics in hypersonic flow. Using the measured velocity change and calculated transit time through the shock wave region, seed particle response characteristics have been calculated. These calculations show that the measured particle response is equivalent to that of a 0.5 micron, 0.1 s.g. sphere which undergoes a deceleration of 15 million times the acceleration due to gravity i.e. a seed particle DMO number of -15. Although the seed particle to molecular DMO number ratio is less than 1000, these particles have a 95% response to velocity fluctuations up to 50kHz. and their characteristics are again consistent with hypersonic modifications to the Stokes drag law. The particle size is also consistent with our concept that the original seed material would break up into sub-micron particles at high stagnation pressures.

It is felt that these new test results are the most convincing evidence to date of particle response in hypersonic flow and emphasize the advantages of using low specific gravity seed materials. They clearly show that attempts to assess seed particle response must involve detailed studies of the velocity probability density distributions. Particle response measurements inferred from conventional time-averaged velocity measurements could well be flawed by their failure to account for the hidden, adverse effects of large-scale, time-dependent mean flow variations which, on closer examination, may well manifest themselves in the velocity probability density distributions.

REFERENCES

1. Owen, F. K., *Application of Laser Velocimetry to Unsteady Flows in Large Scale, High Speed Wind Tunnels*, ICIASF '83 Record, IEEE Publication 83CH1954-7, 1983.

2. Owen, F. K. and Fiore, A. W., *Turbulent Boundary Layer Measurement Techniques*, AFWAL-TR-86-3031, 1986.

3. Fiore, A. W. and Law, C. H. *Aerodynamic Calibration of the Aerospace Research Laboratories M=6 High Reynolds Number Facility*, ARL TR 75-0028, 1975

4. Owen, F. K. and Horstman, C. C., *On the Structure of Hypersonic Turbulent Boundary Layers*, J. Fluid Mech., Vol. 53, pt. 4, p. 611, 1972.

5. Owen, F. K., *Turbulence and Shear Stress Measurement in Hypersonic Flows* , Invited Paper 90-1394, AIAA 16th Aerodynamic Ground Testing Conference, June 1990.

6. Owen, F. K., Horstman, C. C. and Kussoy, M. I., *Mean and Fluctuating Flow Measurements of a Fully Developed, Non-adiabatic Hypersonic Boundary Layer*, J. Fluid Mech., Vol. 70, pt. 4, p. 393, 1975.

7. Gregorek, G. M. *Initial Calibrations and Performance of the ARL Twenty-Inch Hypersonic Wind Tunnel*, ARL 62-393, August 1962.

ACKNOWLEDGEMENT

The work described in this report is dedicated to my wife Dorothy after whom the DMO number is named. Financial support was provided in part by Wright Laboratory under contract number F33615-88-C-3014 and by NASA Ames Research Center under contract number NAS2-12853.

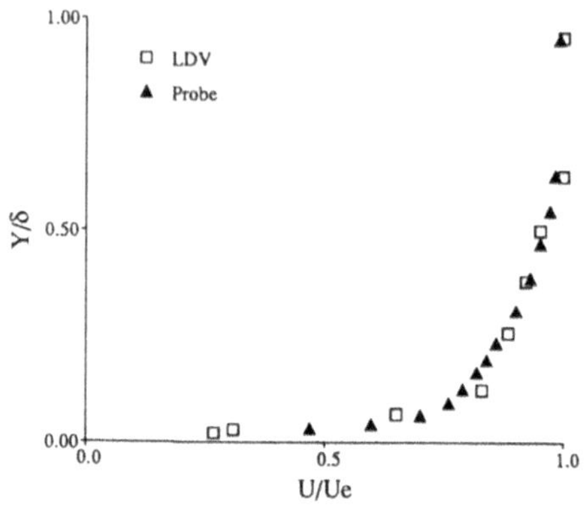

Figure 1 Comparison of Probe and Laser Velocimeter Data.

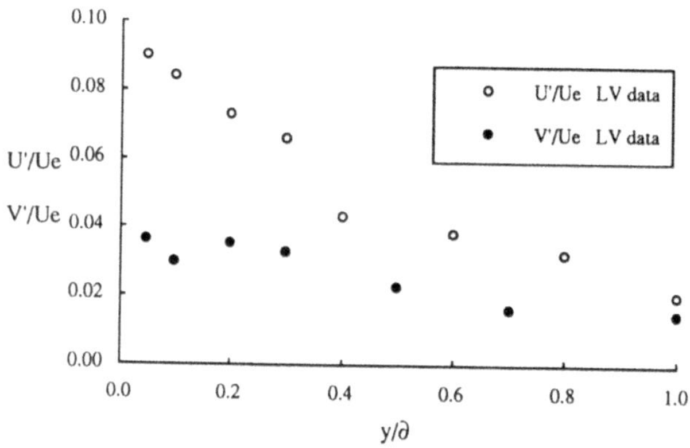

Figure 2 Velocity Fluctuations across the Zero Pressure Gradient Boundary Layer.

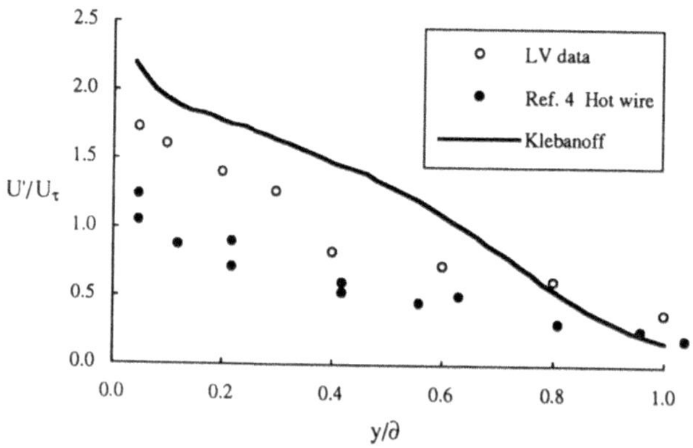

Figure 3 Comparison of Axial Velocity Fluctuations.

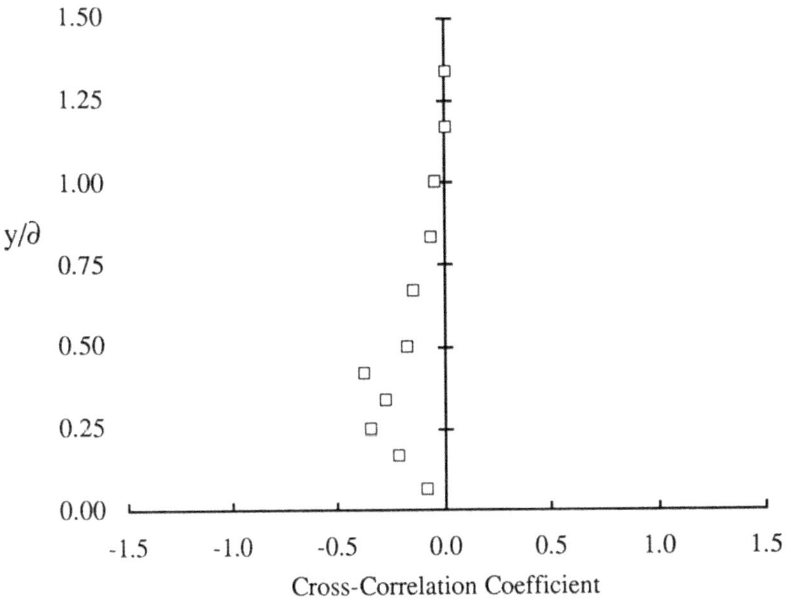

Figure 4 Turbulent Velocity Cross-Correlation Coefficient.

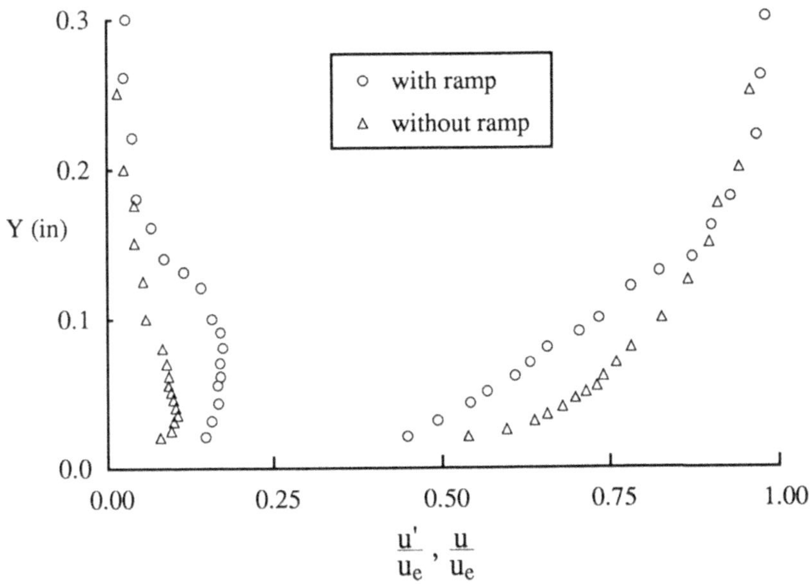

Figure 5 Mean and Turbulent Velocity Measurements in Zero and Adverse Pressure Gradient Flows.

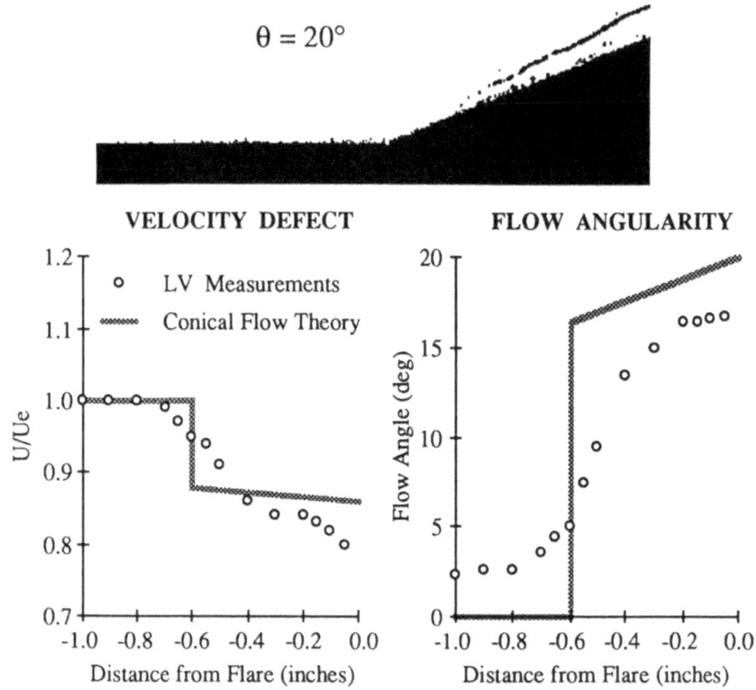

Figure 6 Laser Velocimeter Measurements
Across an Oblique Shock Wave.

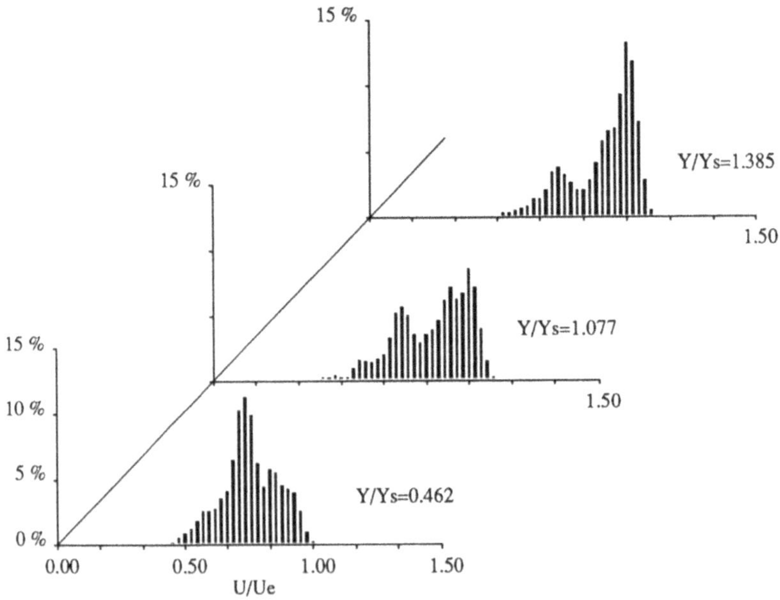

Figure 7 Particle Response in Hypersonic Flow.

Figure 8 Hypersonic Flow Visualization.

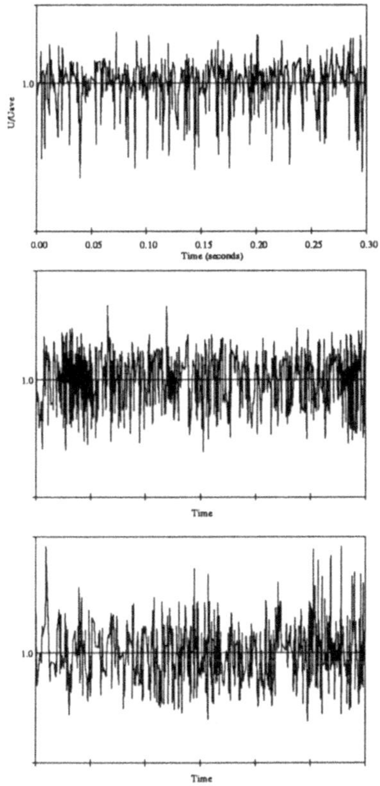

Figure 9 Laser Velocimeter Time Histories, M=12.5.

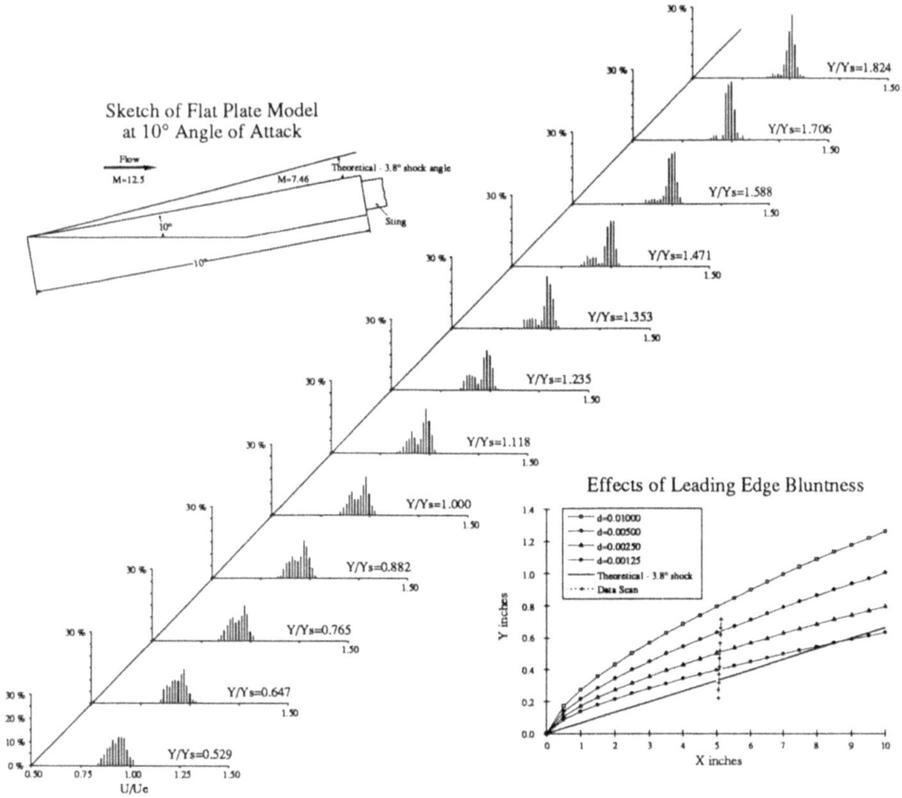

Figure 10 Particle Response in the Wright Laboratory 20-inch Hypersonic Wind Tunnel.

DEVELOPMENT AND ASSESSMENT OF AN LDV SYSTEM FOR MACH 6 FLOW FIELD MEASUREMENTS

William W. Hunter, Jr., Luther R. Gartrell, William M. Humphreys, Jr.,
David W. Witte and Cecil E. Nichols, Jr.
NASA, Langley Research Center
Hampton, Virginia 23665-5225
United States of America

ABSTRACT. A fluidized bed seeding system for the injection of Aluminum Oxide into a Mach 6 flow field is described. Laser transit anemometer tests verified the operational capability of the seeding system. Measurements across an oblique shock with a two-component laser Doppler velocimeter showed that the seed material spatially responded as 0.6 micrometer diameter particle. The estimated velocity vector bias error due to particle lag was 2.5% and 0.3% in angle and magnitude respectively at a point 8 millimeters behind the shock front in the uniform flow field of the shock layer.

1. Introduction

Supersonic and hypersonic flight vehicles and associated computational fluid dynamics codes need nonintrusive velocity measurements for design and performance verification. Mie scattering based laser velocimeter techniques have achieved a high level of success in atmospheric wind tunnels. Application of these techniques to supersonic facilities have also had a degree of success. The purpose of this work was to demonstrate the capability of Mie scattering laser velocimeter techniques to make useful velocity measurements in Langley's 20-inch Mach-6 Wind Tunnel. To achieve this goal two major tasks were pursued. The first task was to develop and demonstrate a particle seeding system that could reliably deliver particles into the wind tunnel's test section and model flow fields. The second task was to make in situ measurements of the nominal particle size in the facility under typical operating conditions. These measurements quantify the laser velocimeter performance capability.

2. Particle Seeding System

Successful application of Mie scattering based laser velocimetry to wind tunnels is ultimately determined by the properties of the radiation scattering particles. Selection of particle material is guided by the material's ability to respond to local flow field changes, the necessity to survive the facility's environment, and its radiation scattering properties. At the present, the particles typically used in the Langley Research Center atmospheric wind tunnels are polystyrene latex (PSL) spheres. This material is produced in large quantities

A. Boutier (ed.), New Trends in Instrumentation for Hypersonic Research, 337–346.

and is essentially monodispersed with nominal diameters of 0.8, 1.0, and 1.7 micrometers. The PSL particles, suspended in a 50% water-alcohol mixture, are injected into the tunnel flow through spray nozzles. The liquid carrier evaporates before entering the tunnel test section leaving only the particles to scatter the incident laser radiation. The relatively low PSL mass density, 1050 kg/m, negligible radiation absorption properties, and spherical shape provides a suitable scattering material.

Application of laser velocimeter techniques to supersonic and hypersonic wind tunnels requires a compromise in the selection of scattering seed particles because of the high temperatures often found in these facilities. Velocimeter measurements in the 20-inch Mach-6 Wind Tunnel require a material that can withstand temperatures up to 517°K.

Aluminum oxide, Al_2O_3, powder was selected for the seed material since it could survive the tunnel's high temperatures. This powder is commercially available in quantity as a very fine abrasive to produce smooth finishes on metals, composites, and ceramics. The supplier classifies the size of the material based on the crystal size [1]; 0.05, 0.3, and 1.0 micrometers. The crystal structure for the 0.3 and 1.0 micrometer powder is hexagonal; it is cubic for the 0.05 micrometer material. This powder comes in two forms, agglomerated and agglomerate free. Both types of 0.3 micrometer powder were used in the tests. The mass density of Al_2O_3 is approximately 3940 kg/m^3.

The particle seeding system consisted of four parts: 1) seed material, 2) fluidizing bed, 3) tunnel delivery plumbing, and 4) pneumatic controller. The Al_2O_3 powder was delivered into the tunnel's plenum chamber in a dry state. To accomplish this a fluidizing bed, figure 1, was made from a 305 millimeter length of 38 millimeter diameter schedule 80 stainless steel pipe fitted with a porous brass plug. The plug disperses the input air flow across its 25 millimeter diameter face. The dispersed air stream passing through the porous plug lifts the powered material and entrains it as it passes through the chamber. By sustaining a positive pressure across the fluidizing chamber a continuous flow of entrained

Figure 1. Exploded view of the 20-inch Mach 6 Wind Tunnel Seeding System fluidizing chamber.

particles is developed. Tests showed that approximately 6900 N/m^2 differential pressure across the bed provided an adequate flow rate.

The fluidized particles were delivered into the tunnel settling chamber through a 1 meter long, 25 millimeter diameter, tube inserted through a special fitting into an existing opening. The fitting provided passage for the tube into the high pressure settling chamber and permitted tube positioning with an attached stepping motor. This positioning capability allows location of the injected particles into streamlines that passed into the facility test section through a two dimensional expansion nozzle.

A pneumatic system, figure 2, controlled the flow of the Al_2O_3 powder. A bypass valve across the seeder prevents the ejection of seed material during tunnel start-up by equalizing the pressure on both sides of the chamber. The control valve joins an external 4.1×10^6 N/m^2 pressure supply to the seeder and regulates the positive flow pressure once the bypass valve was closed. Automatic control of the differential pressure across the seeder depended on the delta-p transmitter that determines the position of the control value.

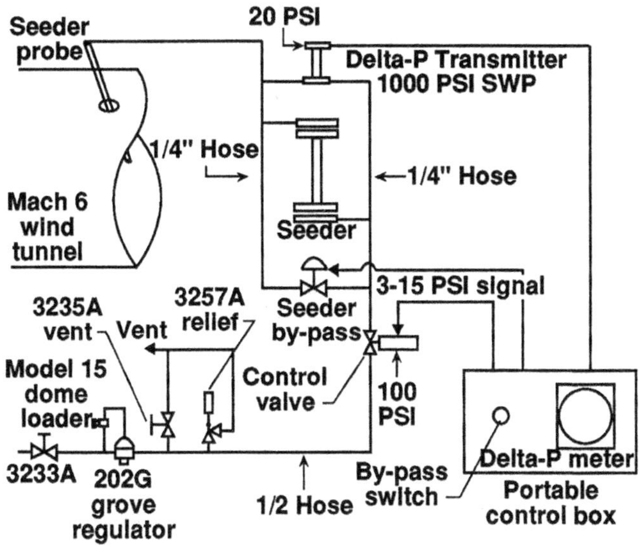

Figure 2. A schematic of the 20-inch Mach 6 Wind Tunnel Seeding System.

3. Laser Velocimeters

3.1 LASER TRANSIT ANEMOMETER

Laser Transit Anemometry (LTA) measures the transit times of seed particles entrained in the flow field as they cross the foci of a pair of parallel laser beams. The transit time of the ith individual particle, t_i, in conjunction with the known beam separation, s, provides a measurement of the velocity, v_i, of the particle in the plane normal to the optical axis of the system by

$$v_i = s/t_i.$$

Assuming that the particle faithfully tracks the flow field, v_i is a measure of the flow velocity in the measurement plane at the sample volume location. The beams are rotated about an axis that is equidistant from and parallel to the two beams. The ability to rotate the two beams allows for the determination of flow angularity in the measurement plane.

The optics package used for the test, mounted next to the tunnel, utilizes a circularly polarized argon-ion laser beam at a wavelength of 488.0 nanometers. The beam is sent through a Wollaston prism/lens assembly to split the beam into a pair of parallel beams. These parallel beams are then sent through a dove prism assembly mounted in a rotating ring controlled by a DC servo system to enable the two beams to be rotated precisely about a common optical axis. The beams pass through a final field lens assembly and are focused to form two Gaussian waists at the measurement location in the flow. For these tests the system utilized a focal length of 900 millimeters with a beam separation at the focal point of approximately 1230 micrometers.

The system operates in a 180 degree backscatter mode with the scattered light from particles crossing the sample volume being collected around the annulus of the transmission optics. This scattered light passes through the dove prism assembly and beam stop (which blocks the focused image of each beam) and enters fiber optic links connected to two separate photomultiplier detectors. Each photomultiplier sends its output to a filter-discriminator/pulse-center detector circuit which produces a TTL level pulse whenever the signal exceeds a preset threshold level. Thus, as particles pass through the sample volume, a series of pulses are produced on two channels, each one corresponding to a separate beam.

The method used to extract transit time information consists of computing the cross correlation, displayed in the form of a correlogram, between the input pulse streams occurring on the two data channels. The data acquisition procedure allows the determination of mean velocity, mean flow angle, and the turbulence intensity along this angle. The normal procedure consists of collecting correlograms between five and nine discrete angular positions preselected by the system operator. From this set of correlograms a measurement of the mean flow angle is made. Next, the beam orientation is set to the mean flow angle so that an additional correlogram containing the mean velocity and turbulence intensity information can be obtained.

3.2 LASER DOPPLER VELOCIMETER

Velocity measurements with the laser Doppler velocimeter (LDV) are based on the Doppler effect. Pairs of laser beams are crossed at their focal point to form a common region, the measurement volume. Individual particles moving through the measurement volume will scatter light at Doppler shifted frequencies that are proportional to the velocities of the moving particles. The velocities of these particles are given by

$$V = \lambda f_D / 2 \sin \frac{\theta}{2}$$

where f_D is the Doppler shift, λ is the wavelength of the laser beams, and θ is the cross beam angle.

The LDV system used in these tests is a conventional two-component system with a 5 watt (all lines) Argon laser, as shown in figure 3. The 488.0 nanometer and 514.5 nanometer laser wave lengths distinguish the two orthogonal velocity components. Transmitter optics are coupled to a rotational stage for precise orientation of the orthogonal components relative

Figure 3. The two-component Laser Doppler Velocimeter (LDV) installation in the 20-inch Mach 6 Wind Tunnel.

to a shock front angle. The focal length of the 83 millimeter diameter objective lens in the transmitter optical section is 750 millimeters. The transmitted orthogonal laser beam pairs are each separated 50 millimeters about the central optical axis forming a cross beam angle of 3.82°. The receiver is aligned 30° off-axis backscatter with respect to the transmit optics and uses a 750 millimeter focal length, 152 millimeter diameter objective lens. Using this arrangement, the spatial resolution is better than 500 micrometers. The received Doppler signals are separated into their corresponding velocity (frequency) components with optical color filters. A three degrees-of-freedom scan rig with a positioning accuracy of 1.0 micrometers supported the laser and LDV optics packages. The scan rig could be operated manually or by computer control. High-speed burst counters and a microcomputer processed the laser Doppler signals and data sets, respectively.

4. Test Facilities

4.1 20-INCH MACH 6 WIND TUNNEL

The Langley 20-Inch Mach 6 Wind Tunnel is a hypersonic blowdown facility with a maximum run time of approximately ten minutes. A schematic of this wind tunnel along with its operating range is given in figure 4. This facility possesses an injection system which allows the model to be inserted into the flow after the target tunnel conditions are established. A detailed description of this tunnel is provided in reference [2].

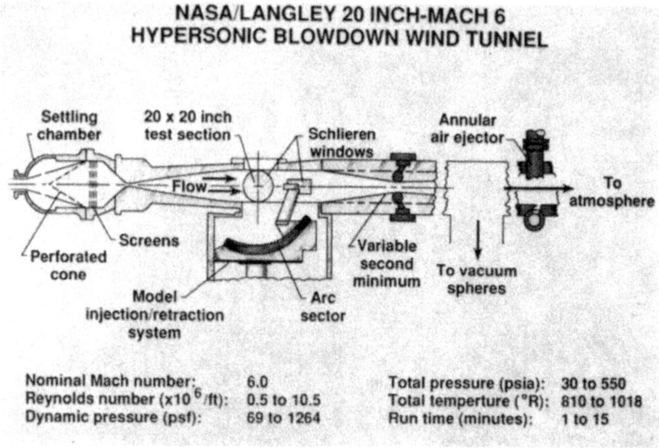

Figure 4. Schematic of the 20-inch Mach 6 Wind Tunnel.

4.2. MODEL

The model was a sting mounted 224 millimeter square flat plate made from stainless steel. The model was positioned so that it would be visible in the forward 432 millimeter diameter Schlieren window. When mounted at a nonzero angle of attack the flat plate is essentially a two-dimensional wedge producing a two-dimensional oblique shock with a uniform flow field in the shock layer. Testing was conducted at an angle of attack of 30° that generates a 40.8° oblique shock angle. The setting of the angle of attack was accomplished through the use of a prism embedded in the sting holder on the upper surface of the flat plate. A point source of light is reflected from a prism onto a calibrated angle of attack scale outside the tunnel. After model injection the angle of attack of the flat plate is increased until the reflected light reached the desired 30° mark on the calibrated scale.

5. Description of Tests

Two sets of tests were conducted. The first tests demonstrated the operational capability of the seeding system. For these tests an LTA system was utilized based on its ease of installation and proven ability to obtain measurements of Mach 4 supersonic flow fields [3,4]. Velocity surveys of the freestream were conducted along a line that was normal to and in a horizontal plane that passes through the tunnel centerline. The survey line extended from approximately 50 millimeters to 250 millimeters outboard of the tunnel centerline toward the tunnel side wall. An angle search was not performed since the freestream flow was assumed to be parallel with the test section walls. Thus, the beams were oriented to zero degrees and only a single correlogram collected at each measurement point. For this study two different tunnel supply conditions were used, 6.9×10^5 N/m^2 and 450°K and 2.1×10^6 N/m^2 and 464°K. The first was selected during initial tests of the seeding system and the second was used once the control setting of the seeding system was established.

The second set of tests measured the seed particles' velocity change across an oblique shock wave. The spatial relaxation distance of the particles across the shock was compared with theoretical predictions [5]. Based on this comparison an estimate of the in situ mean aerodynamic particle size was made. For these tests the two-component laser Doppler velocimeter (LDV) was selected. The LDV optical system alignment sensed two orthogonal velocity components, the velocity component parallel to the oblique shock front and the component normal to the shock front. Only the normal velocity component varies across the shock with the tangential velocity component unchanged. Therefore all the velocity spatial response information is contained in the measured normal velocity component. The results of the normal velocity measurements were compared with the theoretical predicted response.

For these tests, the Mach 6.0 tunnel was operated at a total temperature and pressure of 483°K and 2.4×10^6 N/m^2, respectively. The local flow conditions were:

Parameters	Front of Shock	Behind Shock
Temperature, °K	59	231
Pressure, N/m^2	2.2×10^5	3.9×10^6
Density, kg/m^3	1.3×10^{-1}	5.9×10^{-1}
Velocity, m/sec	920	737

After establishing flow, the model was injected into the test section and set to the selected angle-of-attack. Next the seeding system was activated and the seed delivery tube positioned to optimize the seed plume location in the flow field. Positioning was aided with a video camera/monitor to observe the light scattered by particles passing through the measurement volume. The two-components of velocity were measured along a path normal to the shock-wave at a position that was 178 millimeters from the leading edge and 51 millimeters outboard of the centerline of the model.

6. 20-Inch Mach 6 Tests Results

6.1 LASER TRANSIT ANEMOMETER RESULTS

The LTA tests showed that the seeding system could deliver particles across the required free stream region and deliver acceptable particle concentrations to achieve useful data rates, typically several hundred per second. Figure 5 shows the results of the surveys. The figure shows velocity measurements could be achieved over a broad range of the freestream, near the centerline to the edge of the freestream. The centerline region was not reached due a restriction in the scan rig placement for these tests.

6.2 LASER DOPPLER RESULTS

The initial tests in the LDV series used Al_2O_3 particles, classified as agglomerated 0.3 micrometers material. Measurements of the velocity component normal and across the oblique shock and comparison with theory [5] provided an estimate of an equivalent aerodynamic particle size. Results of these tests, figure 6, suggest 1.0 to 1.2 micrometers diameter particles. The measured velocity values approached the expected value of 133 m/sec, but 20 millimeters behind and normal to the shock the final value is still 8% high.

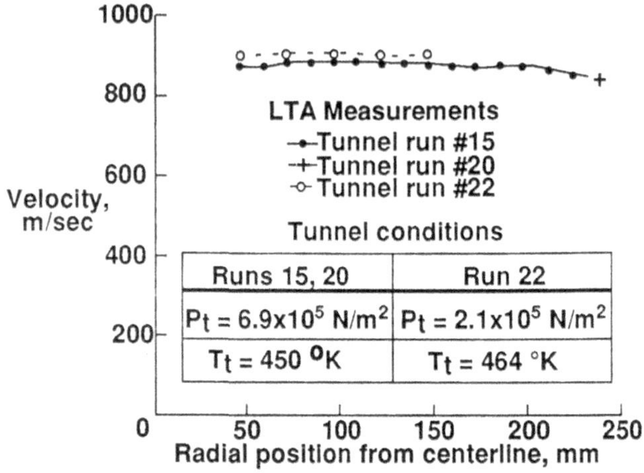

Figure 5. Results of Laser Transit Anemometer radial survey of the 20-inch Mach 6 Wind Tunnel test section.

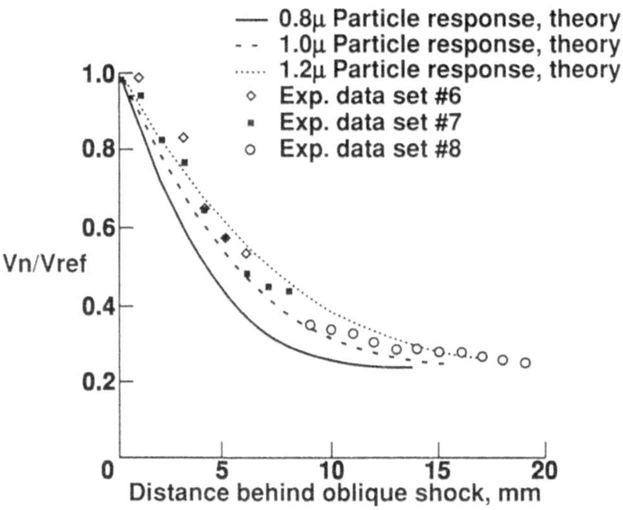

Figure 6. Comparison of LDV normal velocity component measurements using agglomerated 0.3 micrometer Al_2O_3, normalized to freestream reference velocity, with theoretical predictions for different particle sizes as a function of position behind a 40.8° oblique shock.

Further tests with similar but agglomerate free particles were conducted and the improved results are displayed in figure 7. This material is similarly classified as nominal 0.3 micrometers. Tests results with the new Al_2O_3 showed a significantly improved particle response and suggest 0.6 micrometers diameter particles. Now the measured velocity values converges to within 8% of the expected 133 m/sec value 8 millimeters behind the shock. And the values were within 5% of the 133 m/sec 12 millimeters behind the shock.

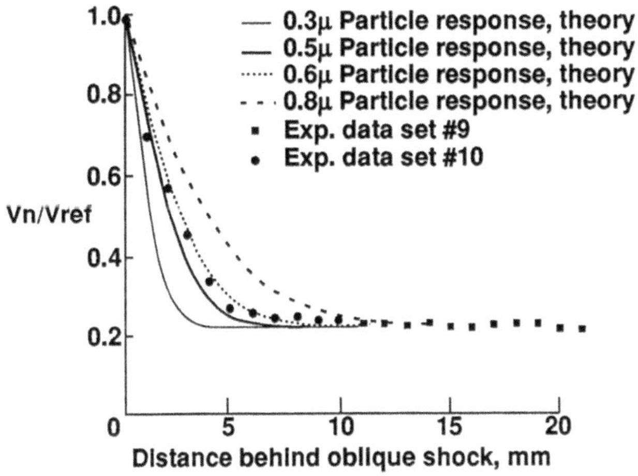

Figure 7. Comparison of LDV normal velocity component measurements using agglomerate free 0.3 micrometer Al_2O_3, normalized to freestream reference velocity, with theoretical predictions for different particle sizes as a function of position behind a 40.8° oblique shock.

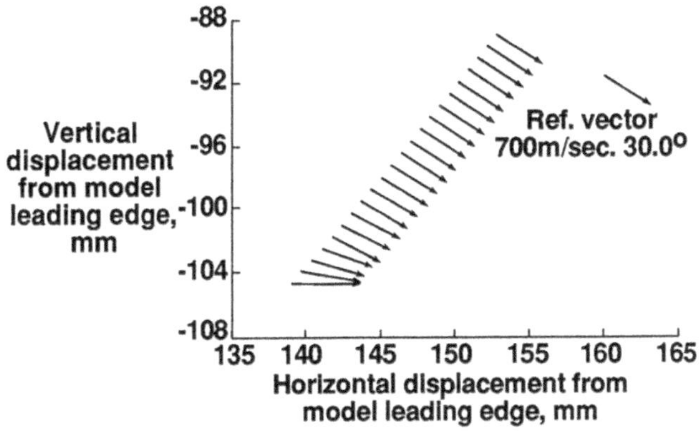

Figure 8. Velocity vector results of LDV evaluation tests with agglomerate free Al_2O_3, as function of vertical and horizontal displacement from the model leading edge.

Vector results of the tests series are shown in figure 8. It is important to note the relative velocity vector bias errors due to the particle lag with respect to the flow field. The error in the velocity vectors provides an overall measure of the impact of the particle lag induced error. It was noted above that an approximate +8% error in the normal velocity component existed at a point that is 8 millimeters behind the shock. This lag results in a +2.5% error in the vector angle and less than a +0.3% error in the magnitude. Similarly, at a point 12 millimeters behind the shock the +5% normal velocity component lag error results in a +1.5% and +0.015% errors in the vector angle and magnitude respectively. These relatively small inaccuracies occur because the tangential velocity components are free of particle lag induced errors.

7. Summary

A series of laser velocimeter tests with a laser transit anemometer showed that a fluidized seeding system could reliably deliver dry particles into a tunnel flow field through a high pressure and temperature settling chamber. Further tests with a laser Doppler velocimeter showed that the aluminum oxide particles used in these tests performed as nominal 0.6 micrometer diameter particles across an oblique shock generate by a flat plate at an angle-of-attack of 30 degrees.

Velocity vector results provide a quantitative assessment of the laser Doppler velocimeter system performances for these tests. At a point 8 millimeters behind the oblique shock a +2.5% error in the velocity vector angle and a +0.3% error in the magnitude existed. Similarly, at a point 12 millimeters behind the shock, an error of +1.5% and +0.015% errors existed in the vector angle and magnitude, respectively. The errors in the velocity vectors are mitigated by the lack of lag induced error in the tangential velocity component.

Finally, the results of this work provided an in situ measurement of the aerodynamic particle size in a hypersonic flow field. With this information, an estimate of the particle lag induced error can be estimated for a given flow condition. The results of this work provide a gauge with which to judge the applicability of Mie scattering based laser velocimetry techniques for a specific test situation.

8. References

1. Union Carbide Corporation, Speciality Powers, Specification Data Sheets, 1991.
2. Keyes, J. W.: Force Testing Manual for the Langley 20-Inch Mach 6 Tunnel. NASA Technical Memorandum 74206, 1977.
3. Humphreys, William M., Jr.; Hunter, William W., Jr.; Covell, Peter F.; and Nichols, Cecil E., Jr.: Laser Transit Anemometer Measurements on a Slender Cone in the Langley Unitary Plan Wind Tunnel. NASA Technical Memorandum 102775, 1990.
4. Hunter, W. W., Jr.; Ocheltree S. L.; and Russ C. E., Jr.: Laser Transit Anemometer and Pitot Probe Comparative Measurements in a Sharp Cone Boundary Layer at Mach 4. NASA Technical Memorandum 104101, 1991.
5. Nichols, R. H.: The Effect of Particle Dynamics On Turbulence Measurements With The Laser Doppler Velocimeter. Dissertation, The University of Tennessee, Knoxville, 1986.

LASER VELOCIMETRY APPLIED IN HYPERSONICS

A. Boutier, J. Lefèvre, F. Micheli

Office National d'Etudes et de Recherches Aérospatiales
BP 72, 92322 Châtillon Cedex, France

ABSTRACT. Difficulties arised by the extension of the application domain of laser velocimetry to very high velocity flows (more than 1000 m/s) are reviewed in this paper; the probe volume geometry has to be checked carefully so that the apparatus is sensitive to the smallest particles and the signal processor can operate along accurate conditions. The flow seeding remains the most important problem we successfully solved in most of the experiments performed at ONERA and described below, providing very confident results up to 1700 m/s. Velocities up to 3 900 m/s have also been measured.

1. Introduction

Laser velocimetry is currently applied in subsonic or transonic flows; as soon as the flow becomes supersonic, the particle drag across shock waves or generally speaking through significant local velocity gradients is the most important problem to be checked and solved. But up to now the use of laser velocimetry in hypersonic flows appeared as a real challenge.

At first we shall emphasize the use of an optimum optical set-up always chosen as a function of the signal processor characteristics; then we shall discuss the choice of aerosols, this problem having still to be investigated in details in spite of the good results already obtained. Finally we shall report experiments performed in four hypersonic facilities at Mach 10 and 12, i.e. in a 1400 - 1700 m/s velocity range, and finally at Mach 16 in a 2 500 - 4 000 m/s velocity rtange.

2. Specific Problems in High Velocity Flows

2.1. CHARACTERISTICS OF THE FLOW

Hypersonic flows present several difficulties for the application of laser velocimetry; among them we can point out the run duration generally too short, the high velocities involved leading to process high frequencies and the low density which does not facilitate the particles to reach the gas velocity.

Most of hypersonic flows are generated in blow down wind-tunnels: air at a high pressure (from a few bars up to some hundreds bars) comes from a reservoir where it is heated, crosses a throat of small dimensions (a few cm), expands across a nozzle (a few meters long) which creates the hypersonic flow inside the test chamber, and is usually sucked up into a vacuum tank.

When a flow is established a few minutes or a few seconds, to check all the classical problems, during first experiments done in a new facility, becomes very difficult to do because it is time consuming; moreover sometimes the repetition rate of the runs is very low: only a few runs per day. Nevertheless the most important following points must be looked at: flow seeding adjustment, optical alignment verification (stability of the optical set-up during the run, does a significant refraction effect exist?), displacement of the test section or of the model during the run, etc... Such validation procedures made in parallel with data acquisition are very long; during a run only a limited number of points is explored, and to build a velocity map of the flow it is necessary to be sure that the initial flow conditions can be repeated.

Hypersonic flows have typically Mach numbers higher than 5 or 7; but as a laser velocimeter measures

A. Boutier (ed.), New Trends in Instrumentation for Hypersonic Research, 347–360.

the velocity and not the Mach number, the problems may be completely different at the same Mach number depending upon the gas density and temperature (especially concerning the particle behaviour). It is why one may take advantage of the results already obtained at the high velocities herewith reported, but must carefully consider the flow conditions.

2.2. DIFFICULTIES IN LASER VELOCIMETRY

Most of the problems come from the particle behaviour; but in fact many cautions must be taken which all converge towards the purpose of improving the signal to noise ratio.

The geometry of the probe volume is a crucial point to discuss; the idea is to concentrate the maximum laser power inside the smallest area. Thus the comparison between classical fringe laser velocimeters and optical barrier velocimeters (two-spots or two-dashes LV) performances appears essential, keeping in mind that commercial two-spots apparatus are only designed in the backscatter mode for turbomachinery experiment, which must be modified into forward scatter set-ups to increase the signal to noise ratio on the smallest aerosols.

In the case of a fringe type laser velocimeter, reducing probe volume size contributes to signal increase, improves the spatial resolution (interesting feature in flow regions presenting important velocity gradients), but induces the processing of high frequencies. As old and new processors have their own limitations, we shall point out that presently the signal processor available for an experiment dictates the choice of the probe volume geometry.

Nevertheless the flow seeding remains the most important problem. "Light" and submicron particles are required; but due to the initial flow conditions in the settling chamber two other difficulties appear: they must be refractory (because temperatures involved are usually in the thousand K range) and the particle injector itself must be very strong because the pressure around the injection location during the run reaches often several bars (and sometimes some hundreds bars). Generally speaking, valves are necessary between the particle generator and the facility, because the injection point is located in a part of the circuit either linked to the vacuum tank or to the high pressure and high temperature reservoir.Therefore the particle injection procedure is generally very specific for each facility.

3. Optimization of Laser Velocimeter Parameters

3.1. SIGNAL TO NOISE RATIO

The general expression of the maximum signal to noise ratio issued from [1] is:

$$SNR_{max} = \frac{\sqrt{P}\ \alpha_r \sqrt{\frac{\pi}{4}}\ \sqrt{\sigma}}{2\sqrt{\frac{hf_0 . \delta f}{\eta}}\ \sqrt{S}}$$

- P laser power (watts)
- α_r total aperture cone angle of the collecting optics
- σ scattering cross section of a particle
- h Planck constant: 6.6×10^{-34} J.s
- f_0 frequency of laser light: 6×10^{14} Hz

δf bandwith of the signal processor
η quantum efficiency of a photomultiplier PM (0.1)
S lighted section in the probe volume

From this expression the following features clearly appear to optimize the detection of the weakest signals due to the smallest particles which follow the best the flow accelerations:

- P must be as high as possible, but one must not saturate the photodetector PM
- the collection solid angle must be as large as possible: usually at ONERA receiving optics are in the range F/5 - F/10
- new signal processors working in the frequency domain, compared to classical counters present from the point of view of instantaneous electonic bandwith a real advantage: for instance a gain of 10 on the SNR is generally obtained because in spectrum analysis an instantaneous bandwith of 100 KHz is currently achieved compared to a 10 MHz used bandwith with a counter
- the choice of a high quality PM tube is fundamental; it must have a quick rise time (to be sensitive to high frequency sinusoidal signals, i.e. in the 100 MHz range) and an optimum quantum efficiency in relation to the laser light frequency
- smaller is the probe volume, higher is the laser light intensity inside of it, and better is the SNR.

These general comments make appear that it is important to use in hypersonic flow investigations high power lasers, small probe volumes, new frequency domain processors, large aperture collecting optics used in the forward scattering mode and high quality PM tubes.

3.2. OPTICS AND SIGNAL PROCESSING

The first component to be checked is the argon laser itself; if it is obvious that a high power must be concentrated inside the smallest probe volume, it is very important to control that a TEMOO mode is really delivered by a powerful laser in order to get the light energy where it is theoretically expected. In fact anybody expects a gaussian profile of the laser beam inside the probe volume with a maximum intensity in the middle; as soon as the laser cavity is a little bit dirty for instance, the laser delivers a high power in the TEM01 mode, which is an annular beam with a "black hole" in the middle: this phenomenon must absolutely be avoided. In new argon laser technology, this phenomenon is easily controlled thanks to an intra-cavity adjustable aperture and to a permanent filtered air flow.

The probe volume size reduction is generally limited by the possibilities of the signal processor; for instance with counters a convenient accuracy (below 1%) is obtained if signal frequency does not exceed 50 MHz; new commercially available frequency domain processors can accept higher frequencies with a good resolution capability: presently DANTEC can reach 12O MHz with the BSA and AEROMETRICS 160 MHz with the DSA; TSI with the IFA 750 computing autocorrelation functions reaches 90 MHz. Nevertheless these apparatus claiming such high frequency possibilities must be really checked in real flow situations, because the very short transit times involved through small probe volumes in hypersonic flows generally reduce the optimum possibilities claimed by these apparatus (leading for instance to reduce the number of pixels taken into account in the recorded digitized signal for further processing).

The use of an optical barrier velocimeter appears really interesting because the SNR increase due to the probe volume cross section reduction easily reaches 10 or more. But such an apparatus must be developped and we shall present first tests performed in a Mach 12 flow at S4 MA and in a Mach 16 flow at F4. In these cases a two-dashes LV was operated in forward scatter mode: it allows the survey of flows more turbulent than the two-spot technique, very efficient but in laminar flow (but highly time consuming even in low turbulent flows).

In hypersonic flows it appears necessary to work with an optical set-up composed of a powerful argon laser (about 3 Watts per component), a receiving optics with a F/4 to F/10 aperture and PM tubes with a 10^6 gain (and a 2 ns rise time).

3.3. FLOW SEEDING

The particle seeding remains the most important problem. Submicron particles are necessary, and smaller they are, better they follow the flow fluctuations, but poorer is their scattering efficiency.

The smallest particle size detectable with a laser velocimeter (fringe or two-spots or two-dashes) is not well defined. One of the main difficulties often consists in dealing with the "natural" seeding of the flow (what is not easy to control!); if the "natural" flow seeding rate is not too high, the introduction of "known" particles generally leads to higher data acquisition rates on a more controlled (?) seeding. As particles are introduced upstream of the throat in order to reduce the flow disturbance, the high temperatures existing in these regions impose up to now to use refractory powders. The following table [2] gives some indications on the physical properties of some aerosols at high temperatures.

Aerosol	Density	Liquefaction Temperature (°C)	Vaporisation Temperature (°C)
SiO_2	2.2 to 2.6	1710	2230
TiO_2	4.26	1640	disappears!
Al_2O_3	3.5 to 4	2050	2977
ZrO_2	5.49	2700	4300

The liquefaction temperature must be considered for practical use, because beyond this limit the scattering efficiency drastically decreases.

In hypersonic flows high temperatures exist upstream of the throat, and due to the rapid flow expansion very low static temperatures (about 50 K) are found where velocity measurements must be done. Researches are necessary on particle chemistry in order to discover particles which vaporize (and not dissociate) in the hot flow part and could rapidly condense during the expansion at adequate sizes to follow the flow. This problem would have to be treated for each facility, because the temperatures, pressures and nozzle geometry are all different; usually air is dried in order to avoid ice formation, but one could take advantage of very tiny amounts of water condensation in some facilities, if it does not change the Mach number!

A last theoretical feature must be mentioned now, keeping in mind that it still has to be experimentally verified. After justified approximations in the general movement particle equations [3], the particle drag is usually described by the Stokes coefficient **k** :

$$k = \frac{18 \mu}{\rho_p D^2} = \frac{1}{T}$$

μ fluid viscosity
ρ_p volumic mass of the particle
D particle diameter
T time response of the particle

However in hypersonic flows, with low density and static temperature environments, corrections are proposed to the Stokes coefficient in order to extend its application range to flows where the Knudsen number is higher than 1. The modified time constant T becomes [4]:

$$T' = \frac{\rho_p \, D^2}{18 \; \mu} \left\{ 1 + \frac{C \; L}{D} \right\}$$

C is the Cunningham constant and L is the mean free path.

If this model is valid, at large values of Knudsen number, the particle response becomes proportional to D, instead of D^2. Hence it appears from [4] that big particles (5μm) of low density (0.1 g.cm^{-3}) could theoretically follow the flow as well as smaller ones (0.5 μm) of current density (1 g.cm^{-3}).

4. LV Application in Four ONERA Hypersonic Facilities

4.1. R5 CHALAIS-MEUDON (MACH 10)

This wind-tunnel is supplied by air at a pressure of 2.5 bars and a temperature of 1100 K; the flow expands to a free jet, 300 mm in diameter, with a theoretical speed of 1450 m/s, which has been confirmed by time of flight measurements performed on N_2 molecules thanks to the fluorescence induced by an electron beam. The gas density is about 4.10^{-4} kg.m^{-3}, thus very low.

The flow has been seeded with ZrO_2 particles issued from a cyclone generator; during the very short period available for these tests only these heavy particles have been tried; moreover they have not been dried, so that they may have created agglomerates of several μm.

The 1D fringe LV had the following characteristics:

laser power : 3 Watts
probe volume diameter : 200 μm
fringe spacing : 13.9 μm
receiving aperture : F/5 (Cassegrain telescope 200 mm in diameter looking at a 1 m distance)

Three signal processors were tried: an old HP 141T spectrum analyser (sequential analog analysis), a Dantec counter and a prototype of the DSA from Aerometrics. The results are summarized in the table below. These measurements were performed during several different runs lasting about 2 mn; the large discrepancies found are surely due to the lag of too big and too heavy refractory particles of ZrO_2. The counter results were so far from the two frequency analyser ones that we changed the optics in order to process lower frequencies, i.e. under 50 MHz. Then the counter indicated 1154 m/s in better agreement with the HP spectrum analyser giving 1170 m/s, however very far from the 1450 m/s expected.

	Theory	Dantec counter	HP spectrum analyser	DSA Aerometrics
f MHz	103.9	75	88	87.5 - 89.5
U m/s	1450	1047	1228	1221 - 1249
Tu %			7.7	6 - 4.8

These first feasibility tests were not quite satisfying; the counter must not be used at so high frequencies; frequency analysers (and especially the DSA which worked very well) provide consistent results, but the seeding problem must be carefully checked in order to reduce this unacceptable particle lag. We must notice that the very low density, combined with a short nozzle partially explains this important lag of 200 m/s.

4.2. R3 CHALAIS-MEUDON (MACH 10)

The initial conditions are: Pi = 120 bars,Ti = 1050 K; the Mach number is 9.95 and the free stream velocity is 1420 m/s.

The 1D fringe LV had the same characteristics as the previous one, except a larger probe volume (400 μm) and thus a larger fringe spacing (25 μm) leading to process frequencies in the 56 MHz range, which is an upper limit for an acceptable resolution of the Dantec counter used.

As the runs last a very short time (10 s), in order to prepare the LV adjustments a miniature wind-tunnel was built and installed inside the vacuum test chamber of R3 Ch; supplied with Helium at 10^5 Pa and 280 K, the Mach number is 3, but the velocity reaches 1500 m/s: moreover as the nozzle is only 150 mm long, this set-up constituted a severe test for the seeding. Among the many aerosols tried, only Aerosil (20 nm SiO_2 powder initially) provided 15 % of the velocity measurements within 2% of the computed velocity, and was consequently chosen for the real tests in R3 Ch.

The flow has been seeded with this Aerosil upstream the throat thanks to a cyclone generator working up to 150 bars; the pressure difference applied is typically about 5 bars in order to insure anyway at the generator input a pressure higher than the one of the flow.

A flat plate was settled in the test section at a 14.4° incidence, which creates an oblique shock inclined at 19.5° relative to the horizontal; therefore across the shock the Mach number decreases from 10 to 5.4 and the velocity from 1420 m/s to 1300 m/s [5]. The exploration line is represented on fig.1.

Statistics were established on 200 samples only in order to try to obtain several points (3 to 4) during one run; in order to suppress the effects of small flow variations from one run to another (or eventually during one run), all results were normalized by the velocity Vth deduced from the Mach number M and the initial temperature Ti, assuming that the gas remains perfect, thanks to the classical following formula where T is the static temperature:

$$Vth = \sqrt{\gamma RTM} \ with \ \gamma = 1.4 \quad R = 286 \ and \ T = Ti(1+0.2M^2)^{-1}$$

Results are drawn on fig.2. It appears that upstream the shock the measured velocities are higher than Vth by 1 % and present a low dispersion of 0.7 %. Let us notice that we can be confident in these results as the calibration of the LV is made thanks to a theodolite giving an accuracy in the one per thousand range [6]. Downstream the shock wave the velocity rapidly decreases but meets the computed velocity only 14 mm after the shock line: a significant particle lag appears, because in supersonic flows this distance is currently reduced to 2 mm; moreover the two last points are surely situated inside the boundary layer, which explains that the velocity does not stop decreasing.

4.3. S4 MODANE (MACH 12)

Air at a 125 bars pressure is heated up to 1500 K in an alumina pebble bed heater; when opening a quick acting valve the flow is established at Mach 12 for these experiments. Other nozzles are available to create Mach numbers of 10 and 6. About 16 sesonds after this opening the stagnation pressure is constant (125 bars) and the stagnation temperature slowly increases from 1250 K to 1350 K at the end of the run. The run lasts about 30 seconds until the pressure becomes too high in the receiving large vacuum spheres. The density was about 6.7 x 10^{-3} kg.m^{-3} and the expected velocity in the 1600 m/s range.

Two kinds of flow seeding have been tested. At first the natural seeding was used and provided low data acquisition rates (about 100 Hz): it is due to the alumina powder coming from the heater. Then we succeeded to inject Aerosil particles (as in the previous tests in Chalais-Meudon) thanks to the same robust injector and the data acquisition rates reached 10 kHz. The implement scheme of this injector is given on fig.3 which shows that many cautions must be taken when manipulating high pressures; moreover before the run the pressure at the injection point is nearly 0 (thus all the circuit of the injector must be isolated) and when the flow is established all the valves commanding the high pressure establishment must be remotely controlled for security reasons.

The 1D laser velocimeter installed had the following optical characteristics:

laser power : 3 Watts
probe volume : 600 μm
fringe spacing : 38.66 μm

same receiving optics as previously, unless that at the rear part of the Cassegrain telescope a box containing dichroïc plates makes possible to separate the scattered flux of the three more powerful lines of the argon laser (514.5; 488; 476.5 nm) and to lead them on three photomultipliers (fig.4); this enables to perform easily 1D, 2D or 3D measurements or to use the two dashes laser velocimeter (with a green start dash and a blue stop dash for instance).

The signal processing has been achieved with three different processors: a digital oscilloscope Lecroy 7200, a Dantec counter and the IFA 750 from TSI.

The results of runs 1014 and 1015 were obtained in the following conditions (fig.5 & 6 showing the mean velocity evolution during these two runs): natural seeding and signal processing with the Lecroy oscilloscope (only few data processed by the Dantec counter by the end of run 1015). The Lecroy apparatus worked along the following principle: the signal digitalization is operated on 50 successive particles, as soon as the signal level overcomes a fixed threshold; blocks of 50 successive particles are thus recorded so that the flow has been sampled quasi every 1.3 s with recording time intervals lasting about 500 ms. Then after the run, a FFT is automatically performed on each signal, providing instantaneous frequencies, and a mean frequency value is finally available for each block.

During run 1014, the LV measurements suddenly stopped because the laser ceased emitting due to a lack of water cooling. During run 1015, the Dantec counter measurements gave a 1651 m/s value in perfect agreement with the Lecroy analyser.

During run 1019 a pressure rake was moved along the flow axis, so that at the beginning of the run the probe volume was downstream the rake (thus in a wake during 16 s); then during 8 s the laser beams were stopped (no signal) and during the last 11 seconds of the run the LV probe volume was situated upstream of the rake. The flow has been successfully seeded with Aerosil during this run. All the data were obtained with the IFA 750 processor from TSI: statistics are made on 1024 samples (an histogram filtering does not eliminate more than 100 data and just slightly reduces the turbulence rate already low).

In the wake region, fig.7 shows the velocity evolution and that the Dantec counter measurements agree well with the TSI IFA 750. The two apparatus announce a quasi constant turbulence rate of 7 % which is a correct order of magnitude inside a wake. The measured mean velocity (around 1160 m/s) is about 75 % of the theoretical velocity of the incoming flow (around 1550 m/s), which is also a coherent result inside a wake.

At the end of the run, on fig.8 **the measured values are pretty high** (**1700 to 1708 m/s**): the quality and stability of these results obtained with the IFA 750 of TSI are very outstanding. The Dantec counter seems in this run to be sensitive to bigger particles which do not follow so accurately the flow. Another feature confirming the quality of the results issued from the TSI apparatus is the low value of the turbulence rate measured: between 0.7 and 1 % (instead of 2 or 3 % for the counter). This measured turbulence level coincides with the values usually found in subsonic or supersonic low turbulent flows with laser velocimetry, which are however always higher than the actual flow turbulence level, this phenomenon being due to the accuracy limitation of laser velocimetry (fringe stability, processor jitter and statistics uncertainties,etc).

Finally a two dashes velocimeter was installed using the same laser source (15 Watts argon laser) and the same receiving optics, but with the following probe volume geometry: 2 lines (one green and the other blue), 200 μm high and 10 μm thick, 402 μm set apart. These dashes are obtained by spectral dispersion of the lines of the argon laser by a prism and by astigmatism creation thanks to inclined plates. During run 1020 the measured values with this apparatus were 1692 m/s (in the same range as the previous ones), but in spite of the high quality signals viewed on the control oscilloscope a small number of measurements was performed due to the difficulties the electronic time analyser presented to work on too short pulses.

Data interpretation

At the beginning of runs (cf fig.5 & 6), for velocities lower than 1520 m/s, the stagnation temperature is not sufficient to insure that air liquefaction does not occur. So the most interesting results concern the second part of the runs when the stagnation temperature is high enough. In this part, the measured velocities exceed by 60 m/s (eventually 70 m/s in run 1019, fig.8) the velocities computed assuming the gas is perfect (cf.4.2). Therefore another theoretical approach has been undertaken.

From the initial pressure Pi and initial temperature Ti, the enthalpy Hi of the real gas at equilibrium is computed, using a Mollier diagram. The "real gas" velocity is then deduced from the equation $Hi = Cp\,T + 1/2\,U^2$ where T is the static temperature previously calculated with the perfect gas assumption. This calculation assumes equilibrium during the flow expansion through the nozzle, but does not take into account the fact that the vibrational temperature of N_2 molecules may be rapidly frozen after the throat; let us notice that a 10 K variation of T induces a relative variation of U less than 0.5%.

Then on fig.6 & 8 it clearly appears that this "real gas" theory provides a closer agreement between LV results and velocities computed from classical pressure and temperature measurements, these computed values being slightly higher than the LV values (about 0.6%).

Nevertheless in hypersonic flows, any data interpretation must be done with many cautions because several complex phenomena are involved, and only a measurement redundancy may confirm assumptions. For instance all previous calculations would have to be refined in order to take into account the vibrational energy of molecules or their vibrational relaxation due to small amounts of "pollutants", such as residual water vapour or our seeding.

4.4. F4 LE FAUGA-MAUZAC (MACH 16)

F4 is a hot-shot facility; air or N_2 is arc heated in a 10 l chamber where for the tests performed the reduced enthalpy reached 100 and the pressure 300 bars; each run lasts about 90 ms. It is not possible to seed the flow, so that laser velocimetry measurements are performed on unwanted natural pollution. The flow density at the nozzle exit is in the $4.10^{-4}kg.m^{-3}$ range; the velocity is about 4 000 m/s at the run beginning and decreases down to 2 500 m/s at the run end.

The fringe laser velocimeter (fringe spacing 39.7 μm) and the two dashes LV installed are similar to those described for S4 Modane tests. Signal processing is achieved only by FFT on signals individually recorded thanks to a Lecroy digital oscilloscope in order to evaluate signal quality.

The purpose of these tests was quite unusual: as presence of particles is highly unwanted, we tried at first to determine the seeding rate and secondly by measuring eventually velocities to get some order of magnitude of particle size due to the velocity lag. The main conclusions deduced from the data gathered during five runs are:

* during the first 10 ms of the runs, with a data acquisition rate of 1 to 3 kHz, one particle was measured at a **3 900 m/s** velocity (fig. 9), some others being at various velocities (3 000 - 2 400 m/s), but most of them were distributed inside the 1 700 $\pm$ 200 m/s range.
* during the main part of the run (10 to 90 ms), photomultipliers are saturated due to the presence of too many particles.
* after the useful part of the run, the flow still goes on with velocities in the 1 650 m/s range and a data rate of a few kHz (data obtained until 300 ms).

Fig. 10 shows the theoretical evolution of the velocity lag at the nozzle exit, as a function of the particle size, this being computed for different types of particles; the experimental particle lag is in the 1500 - 2500 m/s velocity range, which allows to think that the majority of copper or copper oxyde particles (with a density of 9) are between 15 and 30 μm in diameter.

As the facility is running the first evaluation tests, the results which have been obtained only describe what is going on presently, expecting of course lower pollution for the future.

5. Conclusion

The laser velocimetry application domain is greatly expanded after these tests towards very high velocities, because 3 900 m/s measurements have been obtained successfully. However these tests were first attempts and the particle lag across shock waves in these hypersonic flows must be still studied in more details: many other aerosols behaviour are planned to be checked.

These very encouraging results have been obtained thanks to the use of powerful TEM00 argon lasers in connection with good optics and principally the use of new signal processors which proved to be more efficient than the classical counters because of their better signal detection procedures and improved resolution at high frequencies.

Acknowledgments

The authors of this paper want to thank all the following people who greatly contributed by their actions to prove the feasibility of laser velocimetry measurements in hypersonic flows. M. BACHALO from Aerometrics and the French representant M. BAUCHE from Deltalab actively participated in the first tests in R5 Ch with the new DSA processor. All the results in R3 Ch are due to the permanent team work achieved in Chalais-Meudon by MM. d'HUMIERES, SOULEVANT, SOARES and PAPIRNYK.

In S4 Modane-Avrieux and F4 Le Fauga-Mauzac the mechanical stable implement of the laser velocimeters was done by M. BRET; the high velocities in S4 MA have been measured thanks to the very active contribution of M. ATTIACH from TSI France who operated with efficiency the new processor IFA 750; and we are finally indebted to M. SAGNIER who took into account the real gas effects to compute the velocity evolutions, allowing to validate by this way many concepts. Many thanks also to MM. FRANCOIS and BERTRAND for many helpful discussions about data interpretation.

References

[1] A. BOUTIER - Laser velocimetry - VKI Lecture Series on "Measurement Techniques for Hypersonic Flows" - Rhode Saint Genèse, May 28 - June 1, 1990 - ONERA TP 1990 - 58

[2] J. HAERTIG, W. FLECK - Anémommétrie laser : amélioration de la précision des mesures par élimination des erreurs dues à la présence de grosses particules. ISL Report n°10/74, 1974.

[3] G. TEDESCHI - Analyse théorique du comportement de particules relativement à la traversée d'une onde de choc - DEA Université Aix - Marseille II, 30 juin 1989.

[4] F.K. OWEN, W. CALARESE - Turbulence measurement in hypersonic flow. AGARD CP n° 428, Aerodynamics of Hypersonic lifting vehicles, Bristol (UK), 6-9 April 1987.

[5] C. d'HUMIERES, F. MICHELI, O. PAPIRNYK - Etude du comportement des aérosols pour la mesure en vélocimétrie laser. 2ème Congrès Francophone de Vélocimétrie laser - Meudon, 25-27 Sept. 1990.

[6] A. BOUTIER, J. LEFEVRE - Accurate calibration of 3D fringe laser velocimeters with a theodolite. 3rd Int. Symp. on appl. of laser anemometry to fluid mechanics. Lisbonne, 7-9 July 1986.

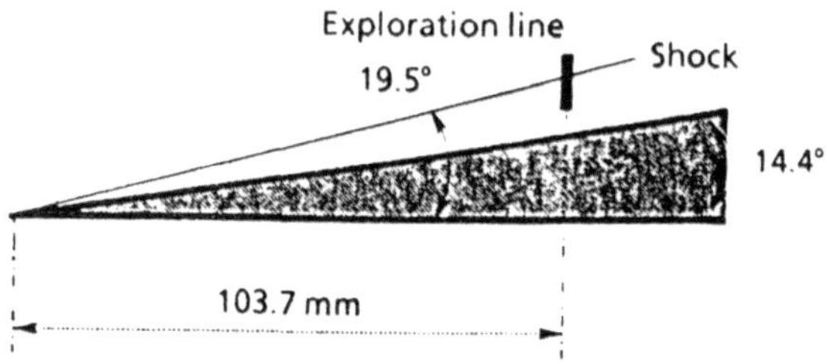

Fig.1 : Scheme of the flat plate and of the exploration line at R3 Ch.

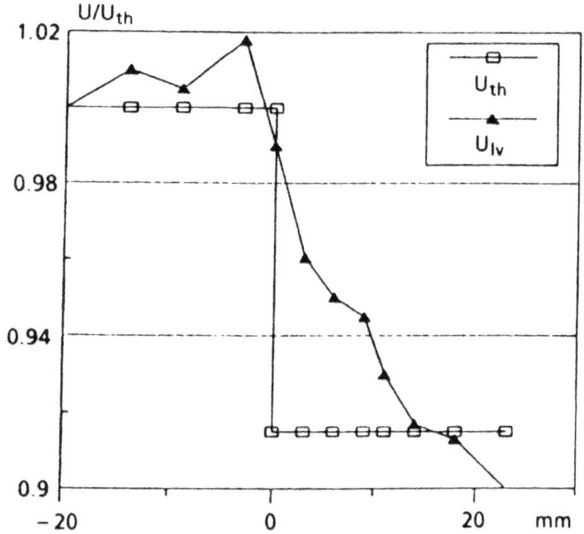

Fig.2 : Velocity profile across the shock at R3 Ch

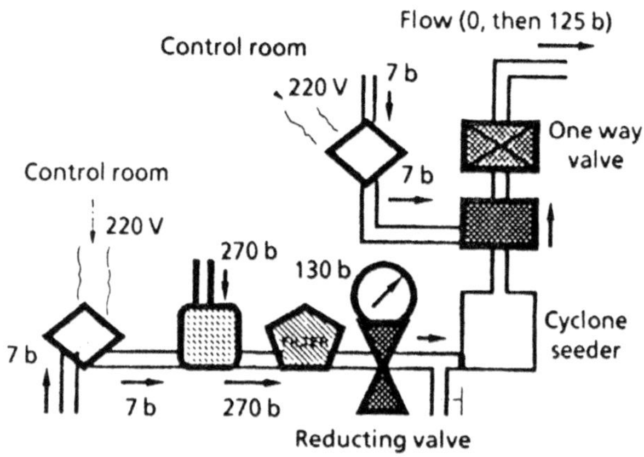

Fig.3 : Injection seeding device used at S4 Modane

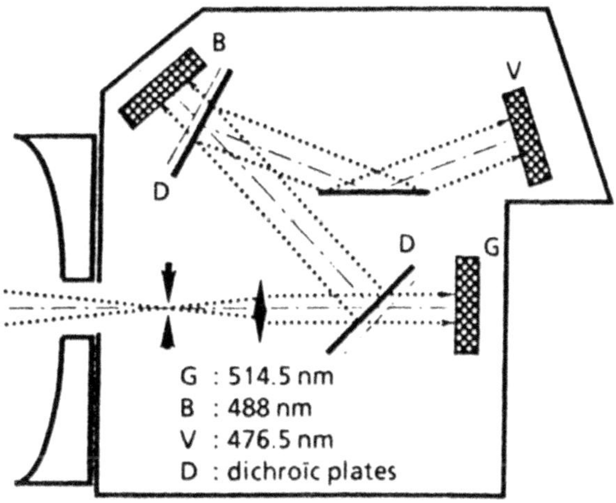

Fig.4 : Colour separation system at the rear part of the Cassegrain receiving optics.

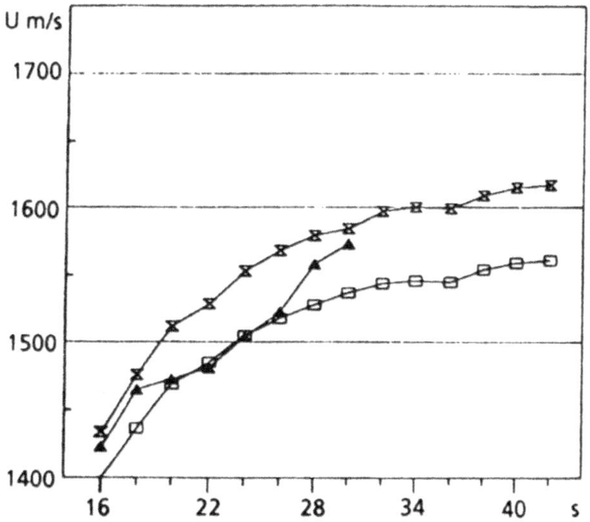

Fig.5 : Velocity evolutions during run 1014 at S4 Modane

perfect gas theory

real gas theory

LV measurements (LECROY)

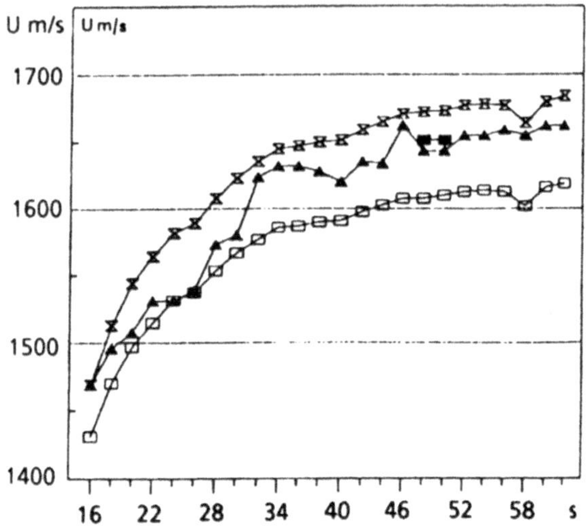

Fig.6 : Velocity evolutions during run 1015 at S4 Modane

perfect gas theory

real gas theory

LV measurements (LECROY)

LV measurements (DANTEC counter)

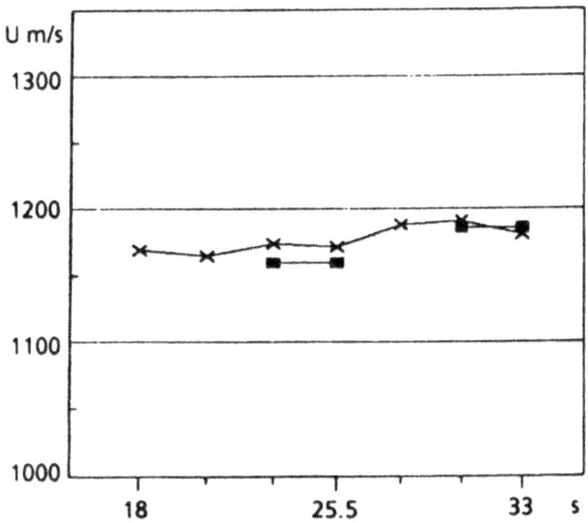

Fig.7 : LV measurement evolution inside the wake (run 1019 at S4 Modane)

IFA 750 from TSI

DANTEC counter

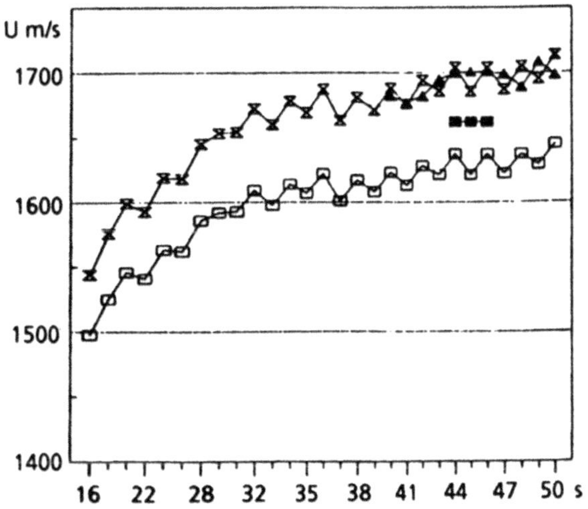

Fig.8 : Velocity evolutions during run 1019 at S4 Modane

—□— perfect gas theory

—⧗— real gas theory

—▲— LV measurements (IFA 750 of TSI)

—■— LV measurements (DANTEC counter)

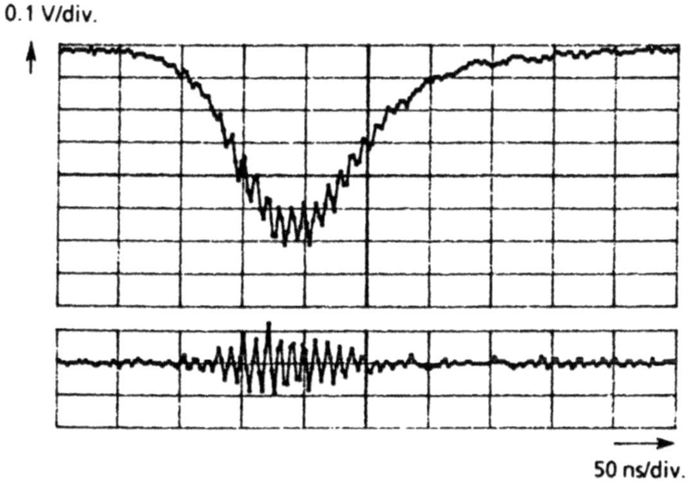

Fig.9 : Signal recorded by Lecroy digital oscilloscope for a 3 900 m/s particle

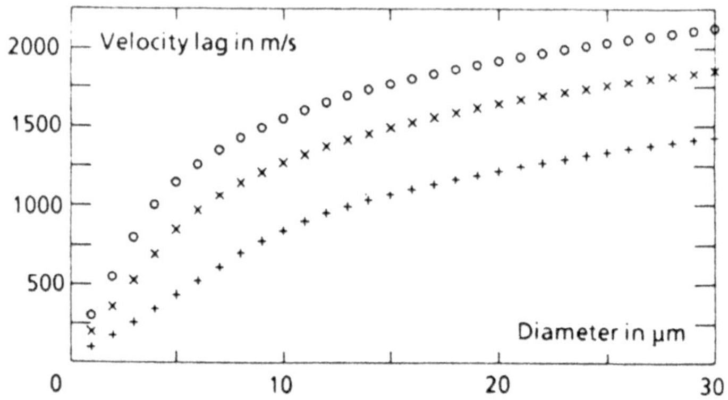

Fig.10 : Velocity lag evolution at F4 Mach 16 nozzle exit.

+	2 000 kg m^{-3} (teflon)
x	5 000 kg m^{-3} (ceramics)
o	9 000 kg m^{-3} (copper)

VELOCITY AND DENSITY DETERMINATION BY THE ELECTRON BEAM TECHNIQUE

E. HIRAI, K.A. BÜTEFISCH, C. DANKERT

DLR, Institute for Experimental Fluid Mechanics
Bunsenstraße 10, Göttingen, Germany

ABSTRACT: In aerothermodynamic gas flows the accurate knowledge of the flow velocity and the density is important as the Mach number is in the hypersonic range and relatively insensitive against real gas effects. This is true especially in the case of rarefied flows if the deviation of the continuum flow condition takes place, viscous effects have to be taken into account, and mixtures of gases have to be considered.
In order to get access to these quantities the electron beam technique has been developed for space related research in the hypersonic wind tunnel of the institute. This paper summarizes the main features of the technique concerning density and velocity determination, typical experimental set-ups, results of some investigations, and advantages and drawbacks of the technique. Finally, as a new possibility for flow diagnostics an electron photon fluorescence technique using metastable atoms or molecules is discussed.

1. INTRODUCTION

The electron beam fluorescence technique has been applied in the field of hypersonic rarefied gas dynamics at number densities from $10^{14} cm^{-3}$ to $10^{18} cm^{-3}$. Despite of many studies [1,2] for many years and considerable progress this technique remains still in the state of development and there are still many problems where further work is needed.

In 1967 the first application of the electron beam technique took place in the hypersonic low density wind tunnel V1G of the institute [3]. The stagnation temperature can be varied between 600 K and 2250 K at a stagnation pressure of 50 atmospheres absolute. The Mach number in the test section is about 20. Using nitrogen as test gas at stagnation temperatures lower than 1200 K condensation might occur. In a first approximation it was assumed that this should not be of great importance with regard to the main stream velocity.

Usually the flow velocity is not a quantity to be measured in a hypersonic wind tunnel if the composition of the gas is precisely known because the determination of the stagnation pressure, the stagnation temperature, and the Pitot pressure are sufficient to characterize the flow properties in the test section. But the accurate knowledge of the absolute flow velocity in such wind tunnel can provide

- at low stagnation temperatures the amount of condensation
- at high temperatures the accurate knowledge about the extent of relaxation of the excited vibrational energy levels of the molecules which can occur due to the rapid temperature drop during the expansion
- the effective specific heat c_p of unknown gases or gas mixtures.

Due to the high accuracy the electron beam technique is the only diagnostic technique to be sensitive enough to solve such problems. In the 1980s laser spectroscopic diagnostics have been evolved. According to the recent review of Cattolica [4] although progress has been made and electron photon fluorescence (EPF) as a new technique is quite noteworthy there seem to

A. Boutier (ed.), New Trends in Instrumentation for Hypersonic Research, 361–371.

be still some problems in the laser based spectroscopic diagnostics and still electron beam technique is of high interest.

The aim of this paper is to review the present understanding and problems related to velocity and density measurements and finally to discuss as a new possibility an electron photon fluorescence technique using metastable atoms or molecules.

2. PRINCIPLE OF MEASUREMENT

2.1 Density measurements

Various methods exist for the measurement of the number density with the aid of the electron beam technique. For monatomic gases as helium or argon usually the most intense lines of the emission spectrum are used. For molecular nitrogen the possibility to determine the intensity of a single rotational line, a single vibrational band or the total radiation exists. In general the intensity-density relationship obtained experimentally is of Volmer type, i.e. the intensity-density curve is nearly linear at moderate densities and shows saturation due to quenching when the density is further increased.
If quenching takes place an excited molecule loses energy due to a radiationless transition when it collides with another molecule. In order to determine the density of a flow field the relationship between intensity and density has to be found by calibration. The actual calibration curve is strongly dependent on experimental conditions and care must be taken in applying a calibration curve obtained under static conditions to determine the density in a flow. This is because besides the wanted fluorescence light production an electron beam interaction with the gas is associated with various phenomena that affect the calibration procedure in different ways.

Figure 1 shows a typical experimental set-up for density measurements with an electron beam in a hypersonic wind tunnel [5]. The beam is pulsed at a repetition rate of 5 KHz. This was done to increase the signal-to-noise ratio by means of the lock-in-amplifier technique. The emitted light passes a window and is focused by a lens outside the wind tunnel on a slit of a monochromator (0,75 m) and converted into an electrical signal by a photomultiplier. The width of the entrance slit is dependent on the optical resolution which is needed: for a single rotational line a resolution of 0.5 Å, and for a vibrational band a resolution of 5 Å is chosen.

A more qualitative result of the density distribution of a flow field is obtained if the electron beam is traversed along the main flow direction and the corresponding fluorescence light intensity is recorded by means of a camera [2,6]. Today video cameras are sensitive and fast enough to get access to the intensity distribution with a linear relationship with respect to the intensity. In Figure 2 the result of a photograph of the flow around a circular cylinder is shown. By applying image processing techniques the background intensity is removed and gray values in steps of 10 % are generated. Clearly the shock stand-off distance and the development of the shock is visible. If a more sophisticated image processing technique is applied the structure of the flow field behind the cylinder can be recognized. After the expansion immediately behind the cylinder a recompression takes place.

2.2 Velocity measurements

During the experiments when density measurements were performed, the generation of ions with long lifetime was observed. The idea came up to make use of them for velocity determination by recording their time-of-flight during their travel downstream among the surrounding neutral molecules along a known distance [7].

It has been found, that to measure the speed of supersonic flows in wind tunnels and free jet expansions, the new ion time-of-flight technique turned out to be a precise method [7,8]. Flow speeds from below 300 m/s up to 5000 m/s were measured without any severe difficulties and changes of the necessary equipment. The main advantage was that the measurement could be performed independently of significant changes of temperature and density. The density limits are extremely wide and vary in the range of $10^{12} - 10^{18}$ molecules/cm^3. The method has been proved in hypersonic nozzles, plumes of satellite thrusters, free jets and molecular beams. The investigated gases are H_2, He, Ar, N_2, Air, H_2O, CO_2, SF_6, CF_4, and $CClF_3$.

Today the possibility of replacing the electron beam by a tunable high power laser looks promising yielding the resonantly enhanced multi photon ionization (REMPI).
Figure 3 shows the experimental set-up for producing ionized molecules by the electron beam technique. The beam is pulsed for 1 μs at a repetition rate of 2-8 KHz. Downstream in the flow an ion collector acting like a Langmuir probe is positioned. From the time delay of the ion signal at the probe with respect to the corresponding electron beam pulse the time-of-flight of the ions is measured. When the probe is moved a well defined distance downstream or upstream the time-of-flight is measured again. From the known probe displacement Δs and the difference of the time of arrival of the ions Δt at the probe the flow speed can be calculated by:

$$u = \Delta s/\Delta t.$$

3. APPLICATION OF THE ELECTRON BEAM FOR VELOCITY MEASUREMENTS

3.1 Velocity measurement in pure gases

When applying the ion time-of-flight technique for velocity determination to pure gases in the case of the far field of an expansion flow there exists a very simple relation for the expected velocity. Depending on the specific heat c_p of the test gas the final velocity in the free jet follows the relation:

$$u = (2c_pT_0)^{1/2}$$

T_0 being the stagnation temperature.

An example of the high accuracy of this technique is shown in Figure 4. The measured velocity of a hypersonic nitrogen flow is plotted in comparison with theoretical predictions [7]. The lower curve depicts the result of the calculation with the assumption that the vibrational temperature is equal to the stagnation temperature T_0. The upper curve predicts the flow velocity assuming the vibrational state in equilibrium with the rotational and translational states. Most of the measured data neither lie on the upper nor on the lower curve representing the two limiting cases. But they do lie on the dashed line which was obtained by calculation if the vibrational temperature measured in the same experiment has been taken into account.

If other gases are considered the velocity can vary between u = 368 m/s for sulphur-hexafluoride SF_6 and u = 1765 m/s for helium under room temperature conditions. As shown in Figure 5 the measured velocities agree quite well with the prediction. Also for water vapor it was possible to determine the flow velocity. The measured flow velocities are sometimes higher than expected which can be explained by condensation of the gas during the expansion and the transfer of heat into kinetic energy [8,9].

The application of the method in the case of a H_2 free jet with velocities from 2800 m/s up to 4400 m/s represents the possibility of this method to be applied in flow fields with extremely high speeds. The highest value of 5000 m/s plotted in Figure 4 yields the result of flow sources available until 1991 and does not indicate any limitation of the ion time-of-flight technique [10].

For application in the new high enthalpy wind tunnels F4 at ONERA and HEG at DLR Göttingen this technique should be applicable. But probably the presence of free electrons in particular in the case of the arc-heated tunnel F4 has to be taken into account as the presence of more or less charged flow will influence the quality of the recorded ion signals. On the other hand at velocities of about 5000 m/s and a flight path of 100 mm the expected time-of-flight will be 20 μs and hence during the wind tunnel running time of 1 ms fifty individual measurements in sequence will be possible. If advantage is taken from the short pulse interval of 1 μs the number of measurements can be considerably increased and statistical means can be applied for data reduction. Due to the strong phase lock between successive ion signals and their corresponding starting pulse of the electron beam a reasonable signal-to-noise ratio may be expected even in the case a charged flow is present.

3.2 Measurements across a normal shock

For testing the method in a flow with a known velocity gradient a shock holder was chosen [11]. Figure 6 shows the result of the velocity measurement at a Mach number of about 24 across the shock which was derived from the time-of-flight measurement by graphical differentiation. The figure depicts in addition the corresponding density profile. Both velocity and density show the expected change along the Rankine-Hugoniot relations by a factor 6 known from continuum theory for high Mach numbers. But the positions of the shock transitions are very different and it is not possible to combine the velocity and the density variation to fulfil the law of mass conservation.

3.3 Streamline detection in thruster plumes

Following the initial assumption that the ions really follow the flow of the neutral molecules it should be possible to detect streamlines by the ion-tracing method with the pulsed electron beam [12]. For this purpose the electron beam was focussed at the nozzle exit (x=0). Keeping the electron beam at a fixed position, the ion-collector probe was shifted in steps in downstream positions and by moving the probe each time in crosswise direction y, the maximum height of the ion signal was searched. The corresponding coordinates within the flow field belonging to the obtained maximum of the ion signal was defined as a streamline point.

When the position of the electron beam y_E was kept at various positions along the radius perpendicular to the nozzle axis other streamlines could be recorded as shown in Figure 7 for

$$y_E / d^* = 0.8,\ 1.67,\ \text{and } 2.5.$$

The procedure worked well at stagnation pressures from 0.5 bar to 8 bar. It failed at lower pressures due to the dramatic loss of the ion signal height.

Figure 7 depicts the experimental results of streamline detection for two stagnation pressures: 0.5 bar plotted in the upper half and 8 bar in the lower half of the drawing. A considerable influence of the stagnation pressure on the initial spreading of the streamlines plume can be recognized: At the high Re-number referring to the higher stagnation pressure (lower half of the plot) the streamlines are more outwardly curved.

3.4 Velocity measurement in a free jet of gas mixtures

After having found the good results of flow velocity determination of pure expanding gas in free jets it was tried to apply the ion time-of-flight method to expanding gas mixtures.

Figure 8 shows the results of such a measurement in a free jet for a helium-argon mixture with a stagnation pressure of p_0 = 150 Torr. The limiting velocities agreed quite well with the prediction. For comparison, the computed change of the velocity as a function of the molar ratio of the two components is plotted in the figure. Up to $p/p_0 = 0.5$ the measured velocity exceeds the calculated value and for $p/p_0 > 0.5$ the measured values lie below the theoretical curve, [13]. The deviation between experiment and calculation can be explained by possible errors introduced when mixing the gases.

4. ELECTRON PHOTON FLUORESCENCE USING METASTABLE ATOMS AND MOLECULES

As indicated in the previous section, the electron beam technique is able to be applied for many practical problems. However, the electron beam technique seems not always to satisfy all diagnostic requirements. The question is whether there is any possiblity to improve the electron beam technique.

One of the big problems of electron beam technique is that spectral lines can not be excited selectively by the fast electrons. From this point of view laser-induced fluorescence (LIF) is much more promising. Unfortunately for nitrogen molecules which are of high interest and which need very high excitation energies commercial lasers which provides for short enough wave length are not available. Muntz et al. [14], proposed electron-photon flourescence technique (EPF), which combines electron beam technique and LIF for gas dynamics studies. The technique consists of electronic excitation of the $N_2^+(X^2\Sigma_g^+)$ state by an electron beam and subsequent selective excitation by laser light resulting in fluorescence in the $N_2^+(B^2\Sigma_u^+)$ ---> $N_2^+(X^2\Sigma_g^+)$ transition. Cattolica [4], however, pointed out that it can not be used for a rotational temperature measurement.

Several possible mechanism for ionization in a two component gas mixture which consists of target gas and inert gas are considered [15]:

1. direct ionization by the electron beam
2. Penning ionization via collisions between target molecule and a atom in a metastable state produced by electron impact
3. charge transfer reaction through collision of ionized atoms produced by electron impact with target molecules

Energy levels for ionization and metastable states of inert and some other important gas species are shown in Figure 9. Comparing the energy levels the process 2 seems to be very attractive. As indicated in some studies [1,6,16], the fact is well known that metastable atoms with a long life time which act as energy carriers enabling efficient reactions even at very low collision energy, extremely enhance fluorescence emission of the other species in gas mixture, for example in mixtures of N_2/He, Ar/He , CO_2/He, NO/N_2, N_2/Ar etc. This Penning ionization normally happens when the energy level of the metastable is nearly equal to or a little higher than the excited state of the target molecule. If in a gas mixture the energy level of the metastable atom is on the contrary lower than the excited state the lack of the energy difference between the metastable state and the exited state can be balanced by the light of a tunable laser. In this case, for an example in order to excite nitrogen molecules, the following process would be possible, Figure 10:

$$\mathrm{Ne} + e^- \text{ (electron beam) ---> } \mathrm{Ne}^* \text{ (metastable } 2^3P_1^0 \text{ or } 2^1S_0) + e^-$$

$$N_2(X^1\Sigma_g^+) + \mathrm{Ne}^* + h\nu \text{ (laser) ---> } N_2^+(B^2\Sigma_u^+) + \mathrm{Ne} + e^-$$

$$N_2^+(B^2\Sigma_u^+) \text{ ---> } N_2^+(X^2\Sigma_g^+) + h\nu \text{ (1st Negative System).}$$

Some similar processes were investigated in [17,18] and usually referred to as laser induced collisional energy transfer (LICET). This EPF process is completely different from that described by Muntz. This scheme is characterized by the important fact that no intermediate state is required. As a consequence the fluorescence of the 1st Negative System results directly from the condition of background gas $N_2(X^1\Sigma_g^+)$. In this process as an exciting source a tunable laser with the wave length 613.1 nm or 585.2 nm can be used.

From Figure 9 some other possibilities seem to be likely as well. For example the metastables $Kr(4^1S_0)$ together with $N_2(C^3\Pi_u)$ and the metastable $Xe(5^1S_0)$ together with $N_2(C^3\Pi_u)$. From the point of view that the mass ratio of the two gas components is desirable to be close to unity in order to eliminate the velocity slip between the two species the combination Ne/N_2 would be better than the others.

The above mentioned process is primary useful for density, species concentration, and temperature determination. In the case of density measurements the accuracy will considerably be improved since the accuracy of the rotational and vibrational temperature measurement is much better. In addition more detailed information about the population of each quantum state with respect to the ideal Boltzmann distribution will be available.

If velocity measurements are desired not only the previous described time-of-flight method but due to the selective excitation and small line width of the emitted light also the Doppler shift method [19] is applicable, since ring dye lasers provide for line width in the range of some KHz.

In this way using metastable atoms as an energy source for selective excitation in conjunction with photons of a laser many suggestions to much more improvement of the conventional electron beam technique are given. It is intended to prepare a fundamental experiment in order to confirm the above principle.

REFERENCES

[1] Muntz E.P., The electron beam fluorescence technique, AGARDograph 132, (1968).

[2] Bütefisch, K.A., Vennemann, D., The electron beam technique in hypersonic Rarefied Gas Dynamics.Progress in Aerospace Science Volume 15, p. 217, (1974)

[3] Wuest W., Koppenwallner G., Hefer G., Legge H., Der hypersonische Vakuumwindkanal der Aerodynamischen Versuchsanstalt Göttingen, Jahrbuch der DGLR, pp. 38, (1969)

[4] Cattolica R.J., Modern Developments in Electron-Beam Fluorescence, 17th Int. Symp. on Rarefied Gas Dynamics, p. 1581, (1990)

[5] Dankert C., Bütefisch K.A., Influence of Nozzle Geometry on Vibrational Relaxation in an Expansion Flow, 9th Int. Symp. on Rarefied Gas Dynamics, Vol. 1, p. 320, (1972)

[6] Mohamed, A., Electron Beam Probing of Blow-Down Hypersonic Flows, Ph. D. Thesis, University Paris Sud, (1991)

[7] Bütefisch, K.A., Absolute velocity determination in a hypersonic low-density flow. AIAA Journal, 1639, (1971)

[8] Baily, A.B., Cluster Intensing and Velocity Measurements in Condensed Flows, AEDC-TR-74-38, (1974)

[9] Dankert, C., Experimental Verification of Rocket Exhaust Plumes and Impingement Effects on Spacecraft Surfaces, Work Package 1: Plume Model Testing Part 2: Velocity Measurement Interner Bericht 222-84 A 47, DFVLR-AVA, Göttingen, (1984)

[10] Dankert, C., Velocity measurements in rarefied nozzle plumes. AIAA-88-2635, (1988)

[11] Bütefisch, K.A., Vennemann, D., Abslolute Velocity in a Rarefied Gas Flow by an Ion Time-of-Flight Technique, DLR-FB 73-115 (1973).

[12] Dankert, C. and Dettleff, G., Near-Field Expansion in Thruster Plumes, 17th Int. Symp. on Rarefied Gas Dynamics, p. 1003, (1990)

[13] Bütefisch, K.A., Velocity Determination in an Expansion Flow of Gases and Gas Mixtures in a Free Jet, Interner Bericht 252-76 H 10, DFVLR-AVA, Göttingen, (1976)

[14] Muntz, E.P., Kunc, J.A. and Erwin, D.A., A Pulsed Electron-Photon Fluorescence Technique for Temperature and Specie Concentration Measurement at Points in Relatively Dense, Unseeded Air Flows, AIAA 22nd Thermophysics Conference, AIAA-87-1526 (1987)

[15] Lester, M.I., Zegarski, B.R. and Miller, T.A., Laser-Induced Fluorescence Studies of Large and Small molecular Cation Produced by Using Electron Bombardment in Free Jet Expansion, J. Phys. Chem. 87, p. 5228, (1983)

[16] Piper, L.G., Cowles, L.M. and Rawlins, W.T., State-to-state excitation of NO ($A^2\Sigma^+$, $V = 0,1,2$) by $N_2(A^3\Sigma_u^+$, $V = 0,1,2)$, J. Chem. Phys. 85(6), p. 3369, (1986)

[17] Lukasik, J. and Wallace, S.C., Laser-Assisted Intermolecular Energy Transfer between Electronic States of Carbon Monoxide in the Vacuum Ultraviolet, Phys. Rev. Lett. 47, p. 240, (1981)

[18] Matera, M. and Mazzoni, M., Laser-induced collisional energy transfer: Experimental study of the spectral profile, Phys.Rev. A 41, p. 3766, (1990)

[19] Kawada, N., Kurita, K. and Hirai, E., Study on Velocity and Temperature Measurement in Free Jet Using Laser- Induced Fluorescence, Mitsubishi Heavy Ind. Ltd. Hiroshima R&D Center Report, 26A90050 (1990)

FIGURES

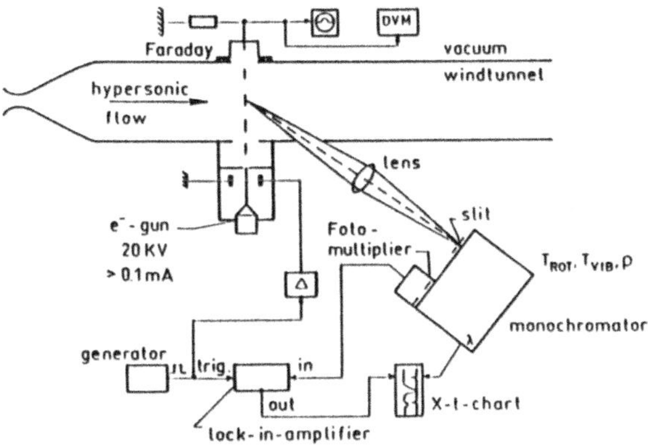

Fig. 1: Experimental set-up for density measurements by means of the electron beam technique

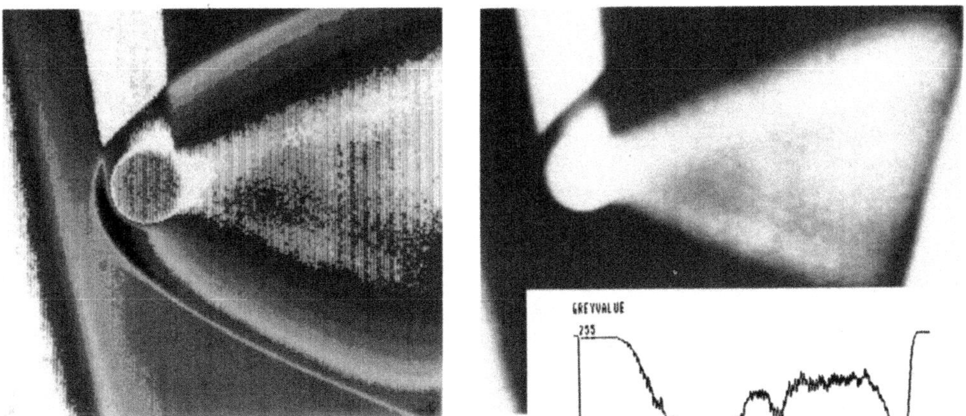

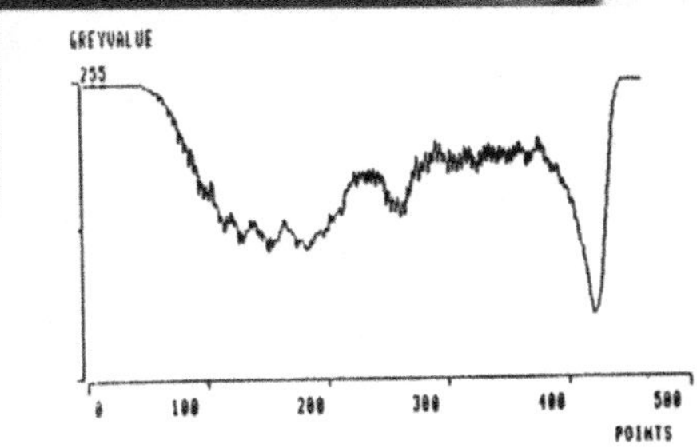

Fig. 2: Visualization of the flow field at M = 20 around a cicular cylinder by an electron beam swept in main flow direction

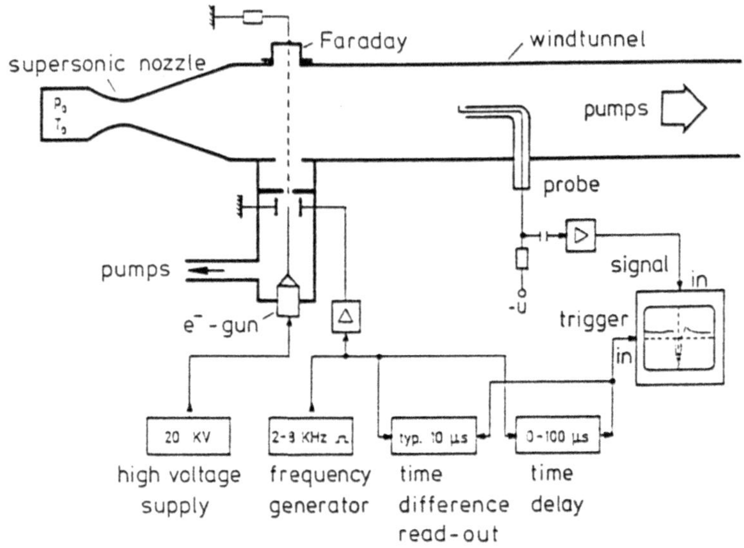

Fig. 3: Experimental set-up for velocity measurements by means of a pulsed electron beam

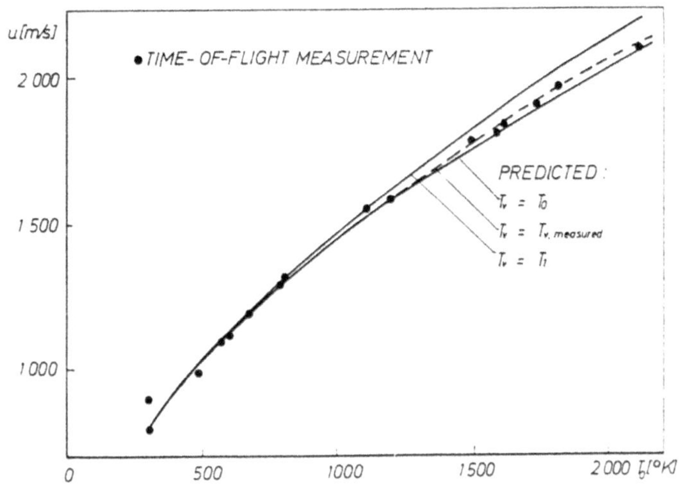

Fig. 4: Measured flow velocities as function of the stagnation temperature T_0

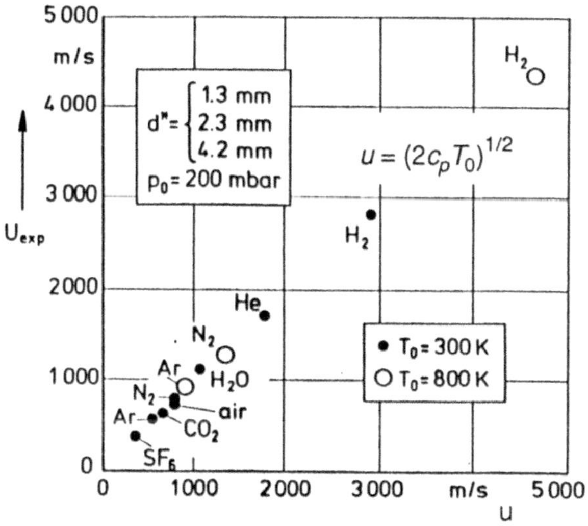

Fig. 5: Measured flow velocities for different atoms and molecules. d* = orifice diameter

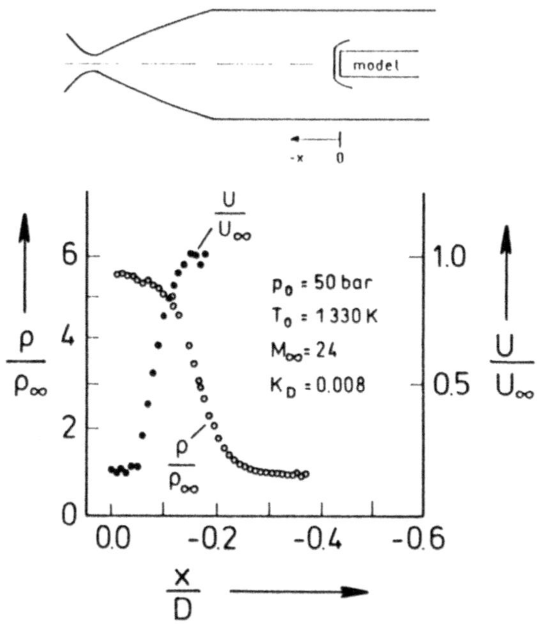

Fig. 6: Velocity measurements across a normal shock together with the corresponding density profile at M = 24

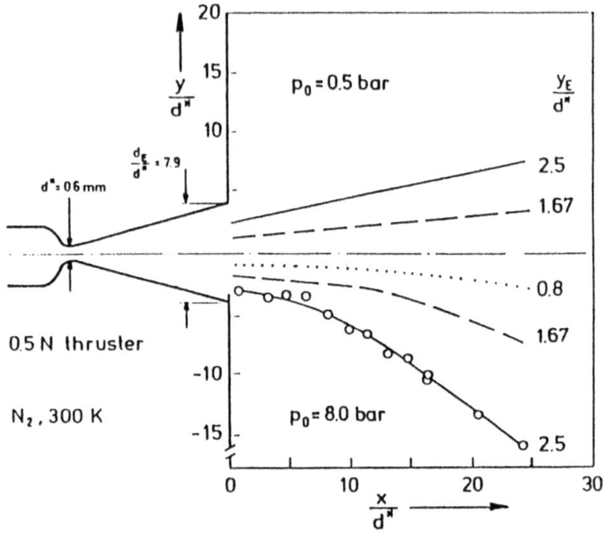

Fig. 7: Experimental results of streamline detection for two stagnation pressures: 0.5 bar plotted in the upper half and 8 bar in the lower half of the drawing

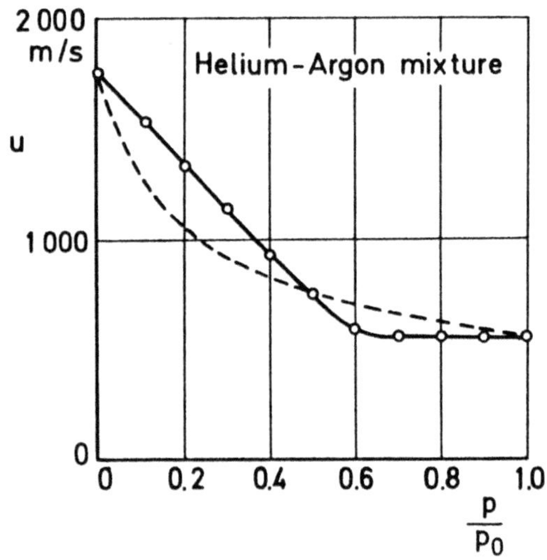

Fig. 8: Measurement of the flow velocity in a free jet for a helium-argon mixture with a stagnation pressure of p_0= 150 Torr.

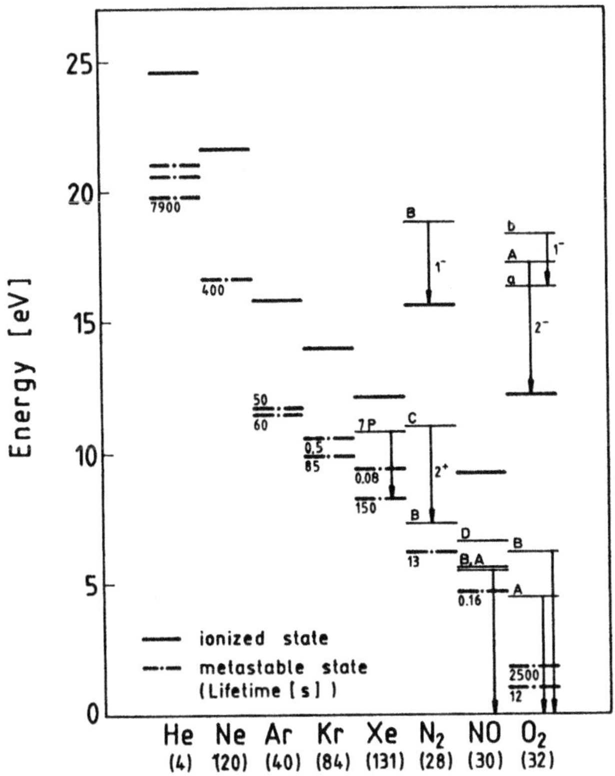

Fig. 9: Energy levels for ionization and metastable states of inert and some other important gas species

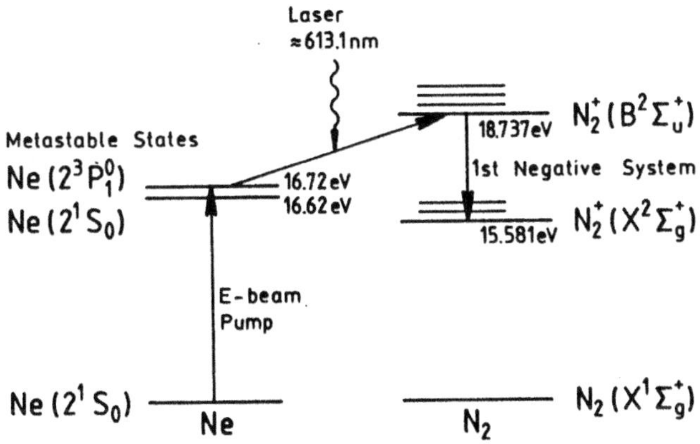

Fig. 10: Laser-induced collision energy transfer in Ne and N_2 mixture

ELECTRON BEAM VELOCIMETRY

Ajmal K. Mohamed
ONERA, B.P. 72, 92322 Châtillon Cedex, France

ABSTRACT. A pulsed electron beam has been used to tag a hypersonic flow with plasma columns. Non perturbative optical velocity measurement of the free flow is shown to be possible from spatial imaging of the movement of these luminous columns.

1. Introduction

In a flow of air or of pure nitrogen, the use of an energetic electron beam (typically of 25 keV energy) induces a complicated set of excitations in the gas all along the beam. Some of these excitations produce radiative emissions, mainly the first negative system (N_2^+1N) and the second positive system (N_22P) of molecular nitrogen in the 200-1000 nm wavelength range. These emissions are usually exploited to measure locally the density, the rotational and vibrational temperatures of N_2 all along the beam [1]. The excitation which is of interest here is ionization. This phenomenon can be exploited, by pulsing the electron beam, to induce columns of plasma which are convected by the flow. The plasma columns contain mainly N_2^+ ions and low energy secondary electrons e_s produced during the ionization process.

The secondary electrons are ejected with angular and energetic distributions which depend on the properties of the gas and the energy of the primary electrons e_p of the beam. Usually, the mean energy of the secondary electrons is around 1 or 2 eV. But, from Ishino's results [1], about 13% of the secondaries can have energy in excess of 20 eV, which is sufficient to excite molecules of the gas producing the same type of fluorescence (N_2^+1N and N_22P) than those from excitations by the primary electrons of the beam [2], with the difference that these secondary excitations can take place over a relatively long distance downstream the electron beam. This distance is the product of the flow velocity and the relaxation time of the secondary electrons. As these secondary electrons loose practically all their energy on their first collision with a molecule and can no more induce excitations on afterwards collisions, the relaxation time can be assimilated to the duration of the first collision. The movement of the plasma columns can be traced by the radiative emissions (afterglow) produced from the electron excitations. This can be done because the radiative lifetime of the excited molecules is very short ($\tau \approx 60$ ns [3]), which means that the fluorescence upon desexcitation is emitted practically at the point of excitation for flow velocity V less than 10000 m/s.

A. Boutier (ed.), New Trends in Instrumentation for Hypersonic Research, 373–380.

After tagging the flow with columns of plasma, velocity measurements can be done in two ways :

(i) by measuring the time of flight of the plasma columns between two choosen points through electrical detection of N_2^+ ions of the plasma with Langmuir probes or through optical detection of the afterglow accompanying the columns [4]
V = d/t where the distance d is known and the time interval t is measured.

(ii) by measuring the distance of flight of the luminous plasma columns during a known time interval
V = d/t where the time interval t is known and the distance d is measured.

The present article is restricted to the second type of measurement whose feasibility is shown from an experiment recently carried out at the ONERA R5 wind tunnel.

2. Plasma afterglow

The physical mechanism producing the afterglow is not completely understood up to now. From available spectroscopic data, the afterglow consists mainly of N_2^+1N emissions. But it is difficult to know the exact cause of these emissions which can be produced from two kinds of reactions : excitations of neutrals N_2 through collisions by energetic secondary electrons and excitations of the plasma N_2^+ ions through collisions by less energetic secondary electrons. Theoretical calculations [2] show that the first process is the main cause of the emissions up to 100 mm downstream the electron beam as shown on figure 1. This seems to be confirmed by photographic densitometry measurements which show that the afterglow, at a distance of about 100 mm downstream the electron beam, is 100 to 1000 times less intense than the emissions from the beam. The lack of complete knowledge of the exact phenomenon causing the afterglow is not really a drawback for velocity measurements in as much as we have sufficient light tracing the column movement.

3. Experimental setup

3.1 FLOW CHARACTERISTICS

The Mach number of the free air flow is 10, the density is 3.95×10^{-4} kg/m^3 (N_2 number density is $6{,}47.10^{15}$ molecules/cm^3), the temperature is 50 K and the theoretical velocity is 1450 m/s.

3.2 ELECTRON BEAM

The monoenergetic electron beam of 25 keV has a diameter of the order of 1 mm and a current flux intensity of 1.8 mA. As shown on figure 2, the electron gun is set to emit beam pulses of 2 μs every 10 μs. In a flow with a velocity of about 1500 m/s, these pulses induce plasma columns of a few millimiters in diameter which are convected downstream the electron beam. The columns are evenly spaced by a distance of 15 mm in so far as the velocity of the flow remains constant in the field of interest.

3.3 OPTICAL DETECTION

A circular flow field 150 mm in diameter and beginning 30 mm downstream the electron beam is observed with a Nikkor objective lens of 105 mm focal length and f/4.5 aperture (figure 2). The 30 mm shift is chosen to avoid direct the detection of emissions due to excitations by highly deviated primary electrons. The lens, having a transmission higher than 80% in the 200-800 nm wavelenth range, forms the image of the circular field on a 25 mm diameter high speed gated image intensifier (V3347U from Hamamatsu). The final image is recorded on a 400 ASA sensitive photographic plate immediately applied to the output surface of the intensifier in order to reduce optical coupling losses. To avoid high voltage problems in a low pressure medium, the whole detection system is placed outside the wind tunnel and observation is done at a working distance of 700 mm through a 100 mm diameter silica window.

The image intensifier has two purposes, the first one is to act as a high speed shutter in order to freeze the movement of the plasma columns. In our case the chosen 0.5 μs exposure time allows a displacement of 0.75 mm for the columns. The second purpose is to use the high gain (up to 9000) to amplify the low intensity light coming from the plasma columns.

4. Results

Figure 3 shows the picture obtained in the free flow. The contrast is low due to the very weak intensity of the afterglow and to the spreading of plasma columns. This spreading has different causes : the initial electron beam spreading, convection due to the flow of the N_2^+ ions at the moment of their creation with the pulsed beam and diffusion of the plasma components from the columns. Also, due to saturation problems, the intensifier has in fact a gain of unity and functions only as a high speed shutter (0.5 μs) to freeze the movement of the columns. To compensate the gain loss, a great number of images has to be accumulated on the same photographic plate. The shutter is set to open every 10 μs and synchronized with the electron beam pulses. The total duration of the flow ($\approx$ 70 s) is used to accumulate about 70s/10μs= 7x10^6 images in order to obtain the acceptable constrast of figures 3 and 4.

The velocity of the free flow can be deduced from figure 3 by measuring the distance in between two columns, the time of flight being given from the frequency of the electron beam pulses. In our case the measured velocity of 1480 m/s, with a 5% accuracy, is in good agreement with the theoretical value of 1450 m/s. The accuracy can be improved (to better than 0.5%) by determining more accurately the location of the maximum peak intensity in each column through image processing of the pictures.

Figure 4 shows the behaviour of the columns around a body (a sphere of 60 mm diameter) placed in the flow. The columns can be considered as isotemporal lines. Velocity downstream the shock wave cannot be deduced from the sole disposition of the columns as they do not give any information on the direction of the velocity. We have, nevertheless, a partial information which can be coupled with computer flow dynamics simulations to deduce the velocity field around a body.

5. Further developments

A more sensitive optical detector must be used to ameliorate the temporal resolution. Signal evaluation with now available high frequency pulsed intensified CCD cameras shows that the acquisition time to obtain valid images can be reduced to 10 μs. But the frequency of image acquisition will still be limited by the camera output which is usually at video frequency (25 Hz). The Langmuir probe option offers a better temporal resolution as its acquisition frequency is usually equal to the frequency of the pulsating electron beam.

Other types of radiative emissions can be used to trace the columns movement. For example, the chemiluminescent reactions of $N_2(A)$ metastables with NO producing the γ bands [5] or $N_2(A)$ with OH producing the A-X bands [6]. The metastables are produced by the electron beam and have a long radiative lifetime of 1.9 s. The real lifetime in presence of NO or OH will depend on the concentrations of these latter species. A too high concentration will destroy most of the metastasbles too closely to the electron beam. Traces of these species ($\approx 10^{10}$ molecules.cm^{-3}) in a flow of total density around 10^{16} molecules.cm^{-3} are sufficient to produce intense radiative emissions up to a few hundred millimiters downstream the electron beam. The emissions are about 1000 times more intense than the afterglow produced by the electron excitations in the plasma. Optical detection will be easier and the acquisition time can also be subsequently reduced.

6. Conclusion

The experiment carried out shows that the optical spatial imaging of luminous plasma columns induced by a pulsed electron beam can be a useful tool to measure the velocity of the free jet of a low density hypersonic flow. Compared to electrical Langmuir probe detection, the main advantages of this technique are that it performs non perturbative measurement and it gives a global view of the plasma columns in the flow.

7. Acknowledgement

This work was supported by DRET through contract BC N° 89-001-22/6.1.36.

8. References

[1] : MUNTZ E. P., "The Electron Beam Fluorescence Technique", AGARDograph 132, 1968

[2] : MOHAMED A. K., "Electron Beam Probing of Blow-Down Hypersonic Flows", Thesis, Université Paris Sud, 1991

[3] : SCHMORANZER H., HARTMETZ P., MARGER D., DUDDA J., "Lifetime Measurement of the $B^2\Sigma_u^+(v=0)$ State of $^{14}N_2^+$ by the Beam-Dye-Laser Method ", J. Phys. B : At. Mol. Opt. Phys., 22, p1761, 1989

[4] : BÜTEFISCH K. A., VENNEMANN D., "The Electron-Beam Technique in Hypersonic Rarefied Gas Dynamics", Progress in Aerospace Science, Vol 15, p217, Edited by D. KUCHEMANN, Pergamon Press Ltd, Oxford and New York, 1974

[5] : PIPER L. G., COWLES L. M., RAWLINS W. T., "State to State Excitation of NO ($A^2\Sigma^+$,v'=0,1,2) by $N_2(A^3\Sigma_u^+$,v'=0,1,2)", J.Chem.Phys., 85, p3369, 1986.

[6] : WATEGAONKAR S. J., SETSER D. W., "Excitation-Transfer Reactions from $N_2(A^3\Sigma^+{}_u)$ and $CO(a^3\pi)$ to OH", J. Phys. Chem., 94, p7200, 1990

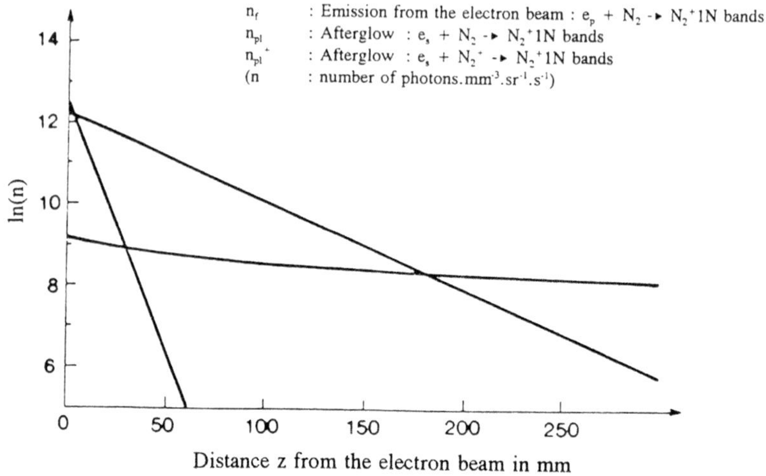

Figure 1
Afterglow intensity downstream the electron beam

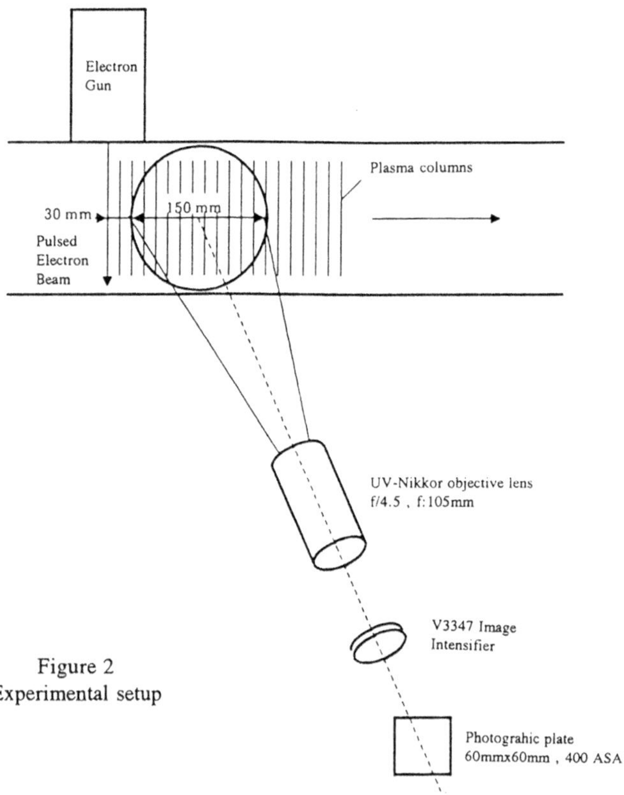

Figure 2
Experimental setup

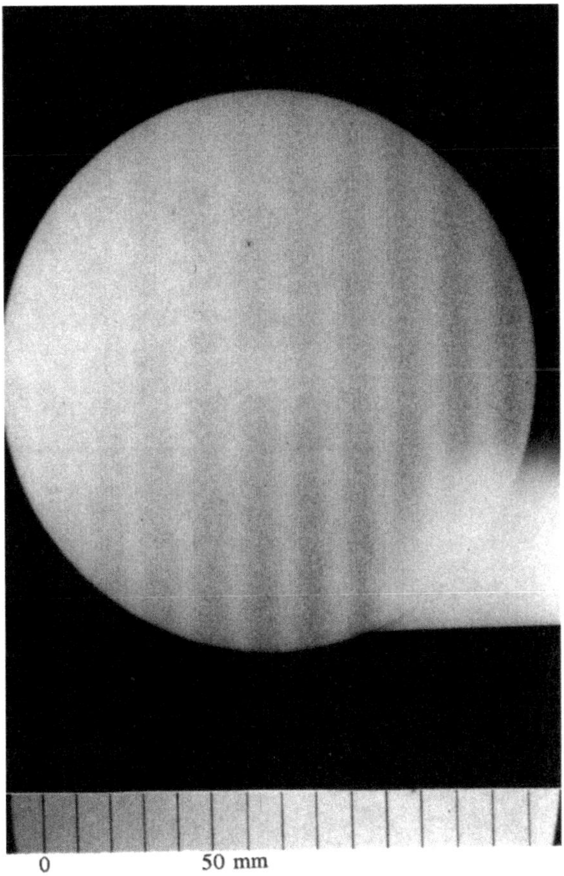

Figure 3
Plasma columns in the free flow

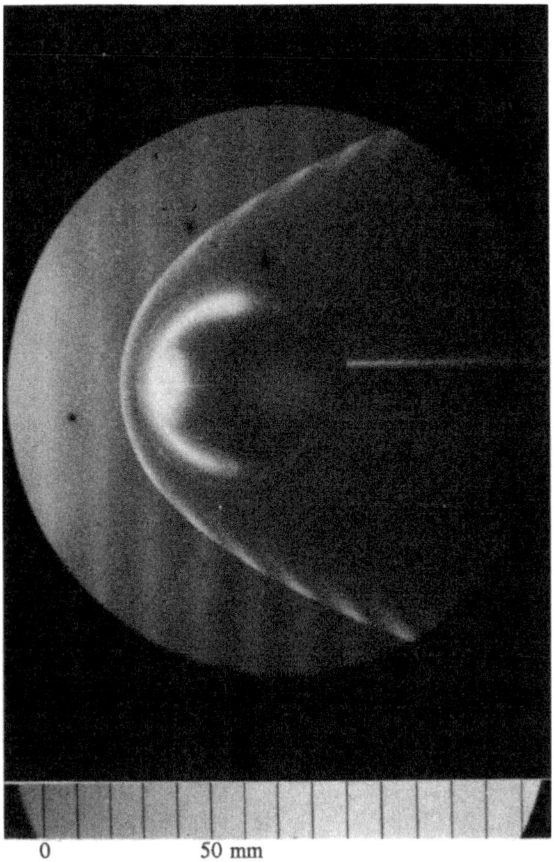

Figure 4
Plasma columns around a sphere ϕ 60 mm

PLANAR VELOCIMETRY IN HIGH-SPEED AERODYNAMIC AND PROPULSION FLOWFIELDS

J.C. McDANIEL, S.D. HOLLO AND K.G. KLAVUHN
Department of Mechanical and Aerospace Engineering
University of Virginia
Charlottesville, VA 22903, USA

1. Introduction

For analyses of low-speed aerodynamic and propulsion flowfields, engineers have traditionally relied on surface pressure measurements and in-stream pressure and velocity measurements made using physical probes. In high-speed applications, however, the use of physical probes is plagued by problems associated with interference effects (due primarily to shock waves in supersonic flows) and probe survivability (in high-temperature propulsion flows). The emergence of an increasing number of nonintrusive, optical measurement techniques in recent years is serving to alleviate these shortcomings in high-speed flow diagnostics. Laser-based techniques are becoming available for spatially and temporally-resolved measurements of the primitive thermodynamic variables and velocity components using a variety of elastic (i.e., Rayleigh and Mie) and inelastic (i.e., spontaneous and coherent Raman and fluorescence) scattering techniques [1].

In spite of all the recent effort in the field of optical diagnostics, there is still a serious need for quantitative velocity measurement techniques for use in high-speed flows. In regions of high-speed flows where the velocity gradients are not large, particle-based Mie scattering approaches may be used. These methods include pointwise laser-Doppler velocimetry (LDV), now employing semiconductor laser diodes for in-flight applications [2], and planar particle image velocimetry (PIV), with the recent reporting of holographic recording and optical correlation techniques for measurement of the out-of-plane velocity component [3]. In high-speed aerodynamic and propulsion flows of practical interest, however, the large particle inertia relative to that of the gas molecules causes particle lag and centrifugal effects [4]. Recent work [5] has shown that even in the presence of modest velocity gradients (a 10 degree turn in a Mach 2.5 flow), 0.6 micron diameter polystyrene latex particles have a relaxation distance of about 3 mm. Other recent work using PIV in an ideally-expanded Mach 2 jet [6] produced velocity profiles within 3% of the calculated isentropic values using 0.1 micron Al_2O_3 particles. Methods to correct for this particle velocity bias require full flowfield solutions [7] which are often not available.

Optical, non-particle-based, velocity measurement techniques fall into two major categories: tagging/time-of-flight methods and Doppler-shift methods. Miles [8] pioneered the RELIEF method of tagging a line of oxygen molecules via stimulated RAMAN scattering and then interrogating the tagged molecules via laser-induced fluorescence. Recent measurements of mean and root-mean-square velocity profiles in a low-speed circular jet were reported using RELIEF [9]. Krauss [10] developed a laser-ion time-of-flight method for velocity measurements in rotating gas centrifuges that used a focused laser to create ions and an ion probe for subsequent detection. Boedeker [11] has reported the use of an UV laser to tag flows containing water vapor, particularly high-speed propulsion flows. A KrF excimer laser generates a zone of enhanced OH concentration via the photolysis of water molecules and a second UV laser interrogates the generated OH via laser-induced fluorescence.

A. Boutier (ed.), New Trends in Instrumentation for Hypersonic Research, 381–390.

Doppler-shift methods measure the frequency shift of elastic or inelastic scattering from atoms or molecules in a flowing gas. Although shifted Rayleigh scattering is monitored using a molecular filter in one approach [12], most methods measure the frequency shift of the absorption by tuning a narrow linewidth laser across a resonant lineshape. A diode laser has recently been employed with a 10 kHz scanning rate for measuring the Doppler-shifted absorption by oxygen in a shock tube [13]. In order to add spatial resolution, measurement of the absorption frequency shift is made by monitoring the fluorescent emission as the laser wavelength is tuned. This method was demonstrated in high-speed nonreacting flows using sodium seeding in nitrogen [14], iodine seeding in air [15, 16] and NO seeding in argon [17]. The latter work utilized a rapid-tuning CW dye laser to measure velocity at a repetition rate of 4 kHz and also measure velocities with Doppler-shifted absorption from O_2 without seeding. The tuning method was extended to planar measurements using a CW argon ion laser to excite iodine in air [18]. A variation of this method in which the frequency of the narrow linewidth laser is fixed in the linear region of the absorption lineshape and changes in fluorescence signal are related to velocity-induced line shifts was demonstrated using iodine in nitrogen flows [19]. A different approach employed a broadband laser and related intensity changes to the shifted lineshape of NO seeded in nitrogen [20,21]. In reacting high-speed flows the laser tuning method has been used for velocity measurements of copper atoms [22] and atomic hydrogen [23]. Other optical velocimetry measurement techniques that have been demonstrated are based on the photothermal deflection effect [24], the phase shift of an induced grating [25] and Doppler shifted inverse Raman scattering [26].

The work reported herein is based on Doppler-shifted absorption with fluorescence monitoring using tunable, narrow linewidth lasers. A CW argon ion laser is used with iodine-seeded air in high-speed nonreacting flows, while a pulse-amplified CW dye laser is used with OH in reacting flows. These methods were developed specifically for steady high-speed flows which require extremely good spatial resolution due to their physical scale and three-dimensionality. These methods are applicable to flows in facilities which have run times of at least several minutes and provide optical access from at least one side.

2. Application to Nonreacting High-Speed Aerodynamic and Propulsion Flowfields

In steady, nonreacting flowfields with total temperatures below approximately 700 K, planar velocity measurements can be made using planar laser-induced iodine fluorescence (PLIIF). These techniques are based on the detection of the laser-induced fluorescence from a Doppler-shifted absorption line of iodine molecules seeded into the flow. Due to its high vapor pressure at room temperature (0.3 torr), iodine can be easily seeded into air flows and, with some care in facility design, can be safely used in large-scale wind tunnels. Iodine is readily excited by the powerful 514.5 nm line of an argon ion laser to produce strong visible fluorescence that can be recorded using an unintensified CCD array camera. The strong visible absorption and emission spectra of iodine and the lack of optical density problems with room-temperature seeding make the molecule uniquely well suited for seeding nonreacting air flows for planar velocity measurements.

Figure 1 illustrates the experimental setup for the PLIIF velocity measurements. A Spectra-Physics 171 argon-ion laser is operated on the 514.5 nm line to excite the strong P13, R15 absorption transitions in the (0,43) vibrational band of iodine. An intracavity air-spaced etalon is used to produce tunable single-mode, narrow bandwidth output. The relative frequency of the laser output is monitored using a scanning interferometer. An absolute frequency reference is provided by monitoring the fluorescence from a low-pressure iodine static cell. The laser beam is converted into a 0.1 mm thick sheet using a cylindrical lens and collimating spherical lens. A translation stage, controlled via a laboratory computer, is used to position the laser sheet in the flowfield. The resulting broadband planar fluorescence is imaged through a Nikon 50 mm f/1.8 lens onto a Photometrics CH210 liquid nitrogen-cooled, 512 X 512 element CCD camera. A multilayer reflective filter and a 560 nm longpass filter are used to block scattered laser light. An image intensified CCD camera is not required for this technique due to the intensity of the iodine fluorescence and the long integration times (up to 30 seconds) that can be used during these time-averaged measurements.

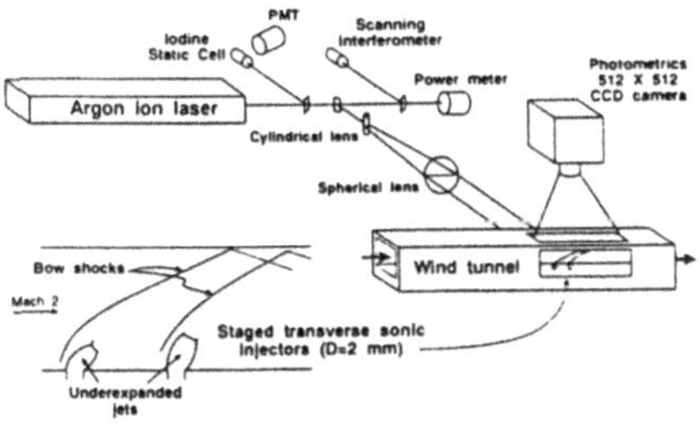

Figure 1. Experimental setup for PLIIF velocity measurements.

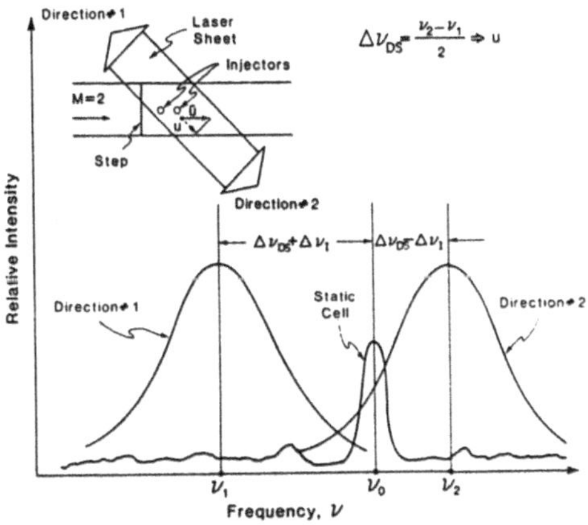

Figure 2. Velocity measurement using counter-propagating laser sheets to eliminate the collisional impact shift.

Cooling the CCD virtually eliminates the buildup of thermal dark current on the array during long exposures and allows images with extremely high signal-to-noise ratios to be collected. Digitizing the signal from the CCD to 14 bit resolution produces measurements that have a dynamic range exceeding four orders of magnitude.

Each planar velocity measurement consists of two linearly independent measurements of the total frequency shift of the iodine absorption spectrum. For each total shift measurement, the single-mode output of the laser is tuned in discrete steps over a 5 GHz range, from the center to the high frequency edge of the laser gain profile, by manually tilting the intracavity etalon. Since the primary iodine transitions are located about 2 GHz to the high-frequency side of the center of the argon ion gain profile, this tuning range is adequate to measure the maximum frequency shifts encountered in low-enthalpy supersonic flows. The axial mode spacing of the Spectra Physics 171 is 84.7 MHz; therefore, each scan consists of approximately 60 discrete frequencies. At each laser frequency, an image of the planar iodine fluorescence within the wind tunnel test section is recorded along with laser power and static cell fluorescence by a computer controlled data acquisition system. The series of images, properly normalized to account for the variation in laser power and iodine seeding fraction (monitored using a second argon ion laser operated broadband) during the scan, is a record of the iodine absorption spectrum at each point in the flowfield. The total shift of the absorption line, relative to the unshifted spectrum measured in the static cell, is determined at each point in the flowfield (i.e. each camera pixel) by a computer program that employs a Gaussian curve fitting routine to locate the linecenters of the shifted and unshifted transitions.

At pressures above 100 torr, a significant portion of the measured total frequency shift of the absorption line is due to the collisional impact shift of the iodine transition [16]. The collisional shift is dependent only on the local thermodynamic conditions and is always to a lower frequency for iodine molecules in air [16]. In order to determine the magnitude of the velocity-induced Doppler shift, the collisional shift must be removed from the measured total shift of the absorption line. This has been accomplished using two different methods. In the first method, the magnitude of the collisional shift is determined from measurements of the local pressure and temperature using a derived relationship that describes the thermodynamic scaling of the collisional shift [16,18]. This technique can be used in a wide variety of flowfields; however, its accuracy is inherently limited by the uncertainties in the temperature and pressure measurements. In the second method, the collisional impact shift is eliminated from the total shift measurements by using counter-propagating laser sheets [27,28]. As shown in Figure 2, the Doppler shift of the iodine absorption line at a point in the flow is equal in magnitude but opposite in direction for the

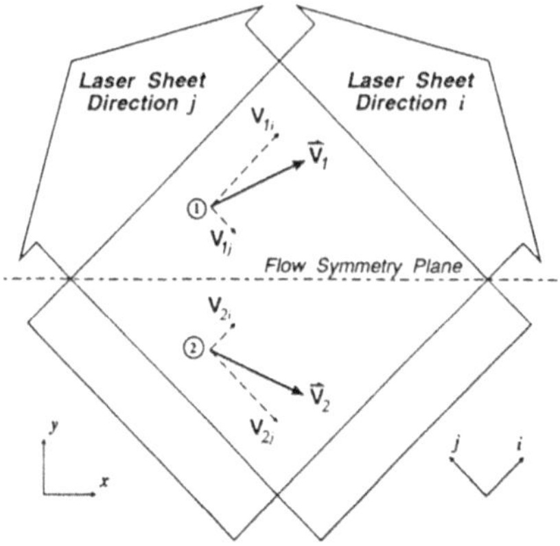

Figure 3. Simulated counter-propagating laser sheets in symmetric flowfields.

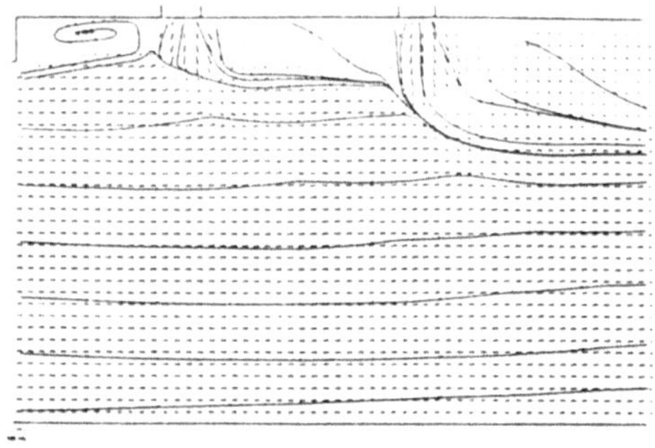

Figure 4. Velocity field measured in Mach 2 flow with staged transverse injection behind a rearward-facing step.

opposing laser sheet directions. The collisional shift, on the other hand, is insensitive to the direction of the laser sheet and is eliminated by taking the difference in the total frequency shifts measured using counter-propagating laser sheets. The result is equal to twice the Doppler shift. This method is considerably more accurate than the first due to the doubled sensitivity of the Doppler shift measurement; however, it requires optical access from both sides of the flowfield and a total of four total shift measurements for each planar velocity measurement. A modified version of this method can be used in symmetric flowfields to reduce the required number of total shift measurements to two. With this approach, the symmetry of the flowfield is exploited in order to simulate the use of counter-propagating laser sheets [29]. Corresponding points on either side of the symmetry plane have identical thermodynamic conditions and mirror image velocity components, as shown in Figure 3. Since the collisional shift is the same at both points, total frequency shift measurements from opposite sides of the symmetry plane can be used to eliminate the collisional impact shift in the same manner as measurements made using counter-propagating sheets. In Figure 3, for example, the total shift measured at point 2 using laser sheet direction ***i*** is equal to the total shift that would be measured at point 1 using a counter-propagating laser sheet, ***-j***. This approach retains much of the accuracy of the fully implemented counter-propagating sheets approach but requires only two total shift measurements and optical access from only one side of the flowfield.

Figure 4 shows the velocity field measured along the centerline of a Mach 2 test section configured with staged transverse sonic injection behind a rearward-facing step [18]. Due to the absence of symmetry in this plane and the lack of optical access from both sides of the flowfield, this measurement was made using the method in which the collisional impact shift is calculated from measurements of the local temperature and pressure. For clarity, only 2,000 of the 125,000 velocity vectors actually measured in this plane are shown in Figure 4. The lines shown on the vector plot represent the trajectory of a massless particle injected into the 2D velocity field (they are included as a visualization aid). Figure 5 is a higher-resolution closeup of the velocity field in the vicinity of the first jet. Note the high shear region between the edge of the recirculation zone in the base of the step and the high velocity core of the jet. Clearly this region of the flow could not be accurately measured by velocity techniques based on particle seeding or time-of-flight measurements.

Figure 6 shows the velocity field measured in the plane located 0.5 injector diameters above a flat plate configured with two staged circular injectors as shown in Figure 1 [30]. This flowfield is symmetric about the plane containing the injector centerlines; therefore, the measurement was made using the simulated counter-propagation method described above. For clarity, only 1/25th of the available data is presented in

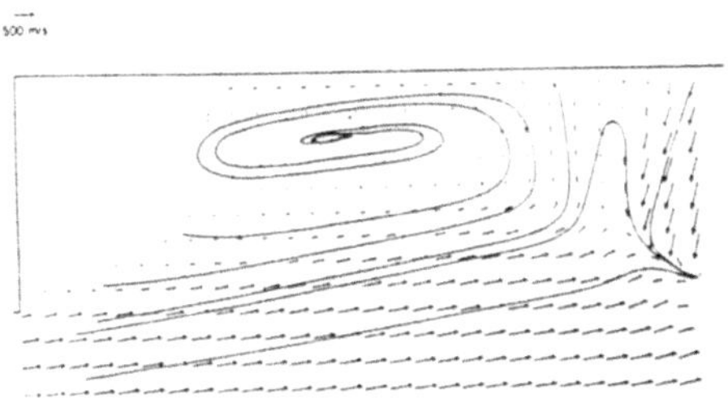

Figure 5. Closeup of velocity field between the step and the first injector.

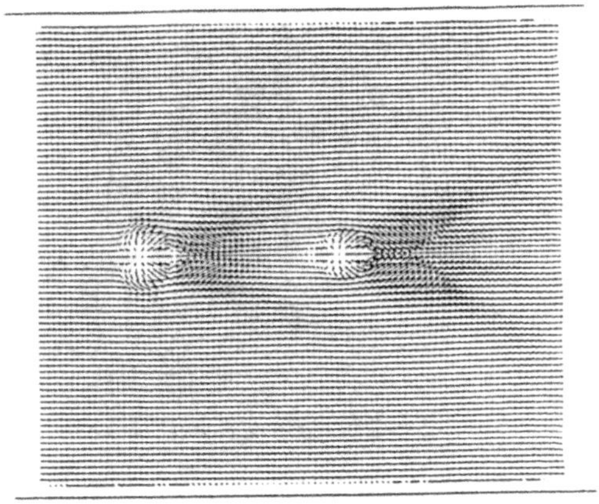

Figure 6. Velocity field measured using simulated counter propagating laser sheets.

the vector plot. The turning of the flow around the jets, the strong acceleration into the low-pressure regions in the wake of the two injectors and the generation of compression waves to turn the flow back parallel to the symmetry plane are readily apparent. With planar velocity measurement capability of this accuracy and spatial resolution, one can study many supersonic flow phenomena without the need to estimate particle lag effects.

3. Application to High-Speed Reacting Aerodynamic and Propulsion Flowfields

In steady high-temperature flowfields with chemical reaction, velocity measurements can be made using planar laser-induced OH fluorescence. The addition of iodine seed molecules into chemically reacting flows is no longer an option due to the chemical interactions between the seed material and the flow. However, the OH molecule is abundant in hydrogen-air flames and its (1,0) and (0,0) absorption bands of the $A^2\Sigma^+$-$X^2\Pi$ system in the 280-320 nm range are easily accessible with available UV laser sources.

The same approaches that were described for PLIIF can be used for OH PLIF provided that a sufficiently narrow linewidth, tunable, UV laser is available with enough peak power for planar imaging. At the present time, only a pulse-amplified ring dye laser system meets these requirements. Figure 7 illustrates such a system for OH PLIF velocity measurements. A Spectra-Physics 2030 argon-ion laser pumps a Spectra-Physics 380A ring dye laser, producing a 500 mW tunable CW beam at 566 nm with a linewidth of ~20 MHz. A fraction of the beam is chopped and sent to an iodine static cell and 2 GHz confocal interferometer to provide phase sensitive detection of absolute and relative laser frequency. The remainder of the beam is passed through an optical isolator and amplified by a Spectra-Physics DCR-3 Nd:YAG pumped Spectra-Physics PDA-1 pulsed dye amplifier to provide 2 mJ, 10 ns pulses at 566 nm. Simultaneous scans of a 1.2 GHz iodine absorption line with the CW and pulse-amplified beams revealed that no apparent broadening of the beam by the amplification process occurs. The pulse-amplified beam is frequency doubled in a Spectra-Physics WEX-1 wavelength extender, resulting in 0.2 mJ pulses at 283 nm with a linewidth of ~30MHz, which is less than 0.5% of typical OH linewidths in high-speed flames.

Pointwise measurements of velocity with the pulsed laser system using the counter-propagating beam approach have been demonstrated in the reacting underexpanded jet facility shown in Figure 8 [31]. The jet facility consists of a ceramic-lined combustion chamber which contains a stoichiometric hydrogen-air flame that is accelerated through an alumina tube into a low pressure chamber [32]. The 5 mm UV beam

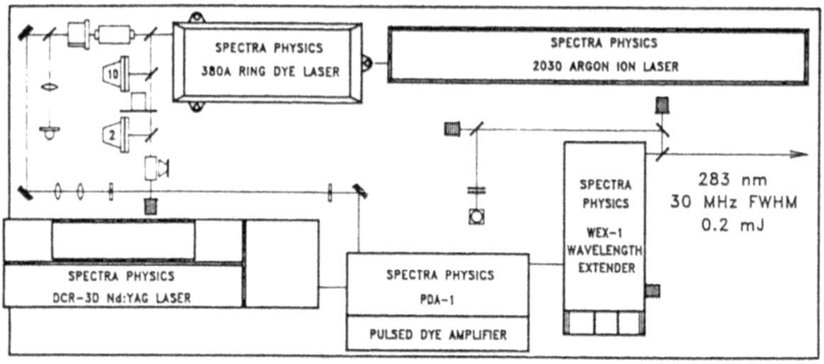

Figure 7. High-power, narrow linewidth, tunable UV laser system for OH LIF velocity measurements.

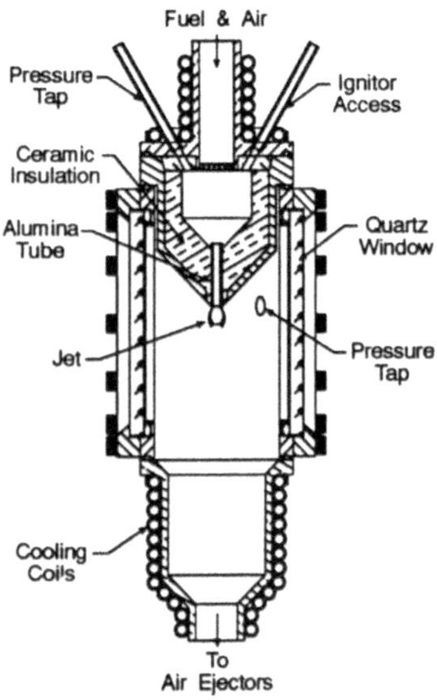

Figure 8. Reacting underexpanded jet facility.

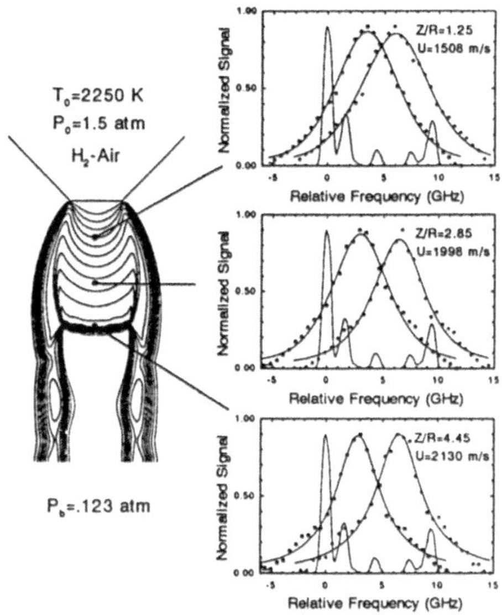

Figure 9. Doppler-shifted OH absorption spectra measured in reacting underexpanded jet.

was focussed with a f=150 mm lens on the jet centerline at an angle of 60° to the direction of flow. The laser was scanned through a 20 GHz region centered on the OH (1,0) $Q_1(5)$ absorption line. Fluorescence from the (0,0) band was collected by a photomultiplier through a 310 nm narrow bandpass filter and f/1.9 optics oriented at right angles to the beam. The fluorescence signal and laser power were averaged over ten pulses by boxcar gated integrators and fed along with the iodine static cell and interferometer signals to a computer for storage. Line center locations were determined by scaling the frequency axis to match the interferometer fringe spacing and fitting Gaussian and Voigt lineshapes to the iodine and OH signals, respectively.

Figure 9 shows the Doppler-shifted spectra from three areas of the jet. The increasing shift and narrowing lineshape are visible as the velocity increases and the pressure and temperature decrease from the nozzle down to the Mach disk. Figure 10 shows contours of the axial component of velocity calculated using an axisymmetric Navier-Stokes code with frozen chemistry where γ=1.25. The pointwise measurements are also shown along with the centerline profile from the calculation. Each data point represents an average over three scans. Preliminary analysis reveals a maximum deviation of 5% for a three scan average with a typical deviation of 3%. The results of the SPARK calculation shown here are also preliminary and exhibit a large error at the Mach disk which is evident when compared to a method of characteristics solution for the same jet conditions. This is a numerical error caused by insufficient grid resolution around the Mach disk region. The grid will be revised and reactions turned on for a future calculation that will include finite rate chemistry.

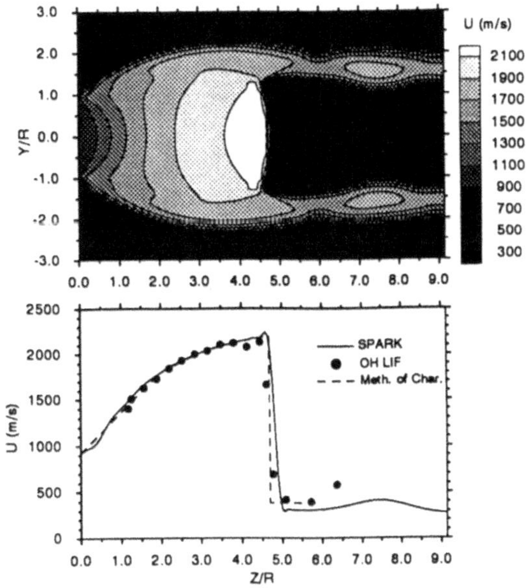

Figure 10. Comparison of pointwise OH LIF velocity measurements with axisymmetric Navier-Stokes calculation.

Planar velocity measurements have also been demonstrated in the reacting underexpanded jet [33]. For these measurements, the laser setup was similar to the one used for the pointwise measurements except the beam was expanded to a 1.5 cm wide sheet using a f=50 mm cylindrical lens and a f=150 mm spherical lens. The laser was tuned through a 25 GHz region centered on the unshifted OH peak location in order to capture the full profile at every point, regardless of the sign of the Doppler shift. Fluorescence was detected by a Princeton Instruments ICCD camera equipped with a 25 mm UV sensitive image intensifier that is fiber-optically coupled to a thermoelectrically cooled 576 x 384 pixel CCD array. The array is read with an A/D converter with 14 bit resolution. A Nikkor 105 mm f/4.5 UV lens with a 310 nm narrow bandpass filter was used to focus the image. A set of ten 100 x 125 pixel images, each representing an accumulation of ten laser pulses with a gate width of 50 ns, were averaged and stored at 0.625 GHz intervals. Figure 11 shows the Doppler-shifted fluorescence for a 9 GHz region in the center of the scan for the beam direction shown. Fluorescence from the right side of the jet peaks earlier in the scan as the laser is tuning up in frequency because the flow in this region is moving against the laser beam direction. The opposite is true for the flow on the left side of the jet. The decrease in signal from the nozzle to the Mach disk is due to the decrease in density and temperature. The sudden increase caused by the density and temperature rise after the Mach disk is also clearly shown. The program used to reduce the PLIIF velocity data is currently being translated to run on an IBM RS6000 to fit Voigt profiles to the OH lineshapes at each pixel in the array and calculate the shift relative to the iodine signal from the static cell.

An experiment is currently being planned for measuring velocities in a supersonic combustion wind tunnel using OH PLIF. Figure 12 illustrates the test section which models the flow inside the combustor of a scramjet engine. Air enters the test section at M=2 and T_0=1300 K and passes over 10°-compression swept ramps. Hydrogen fuel, injected from the base of the center ramp, mixes with the air and burns as the flow moves down the tunnel. The two side walls contain quartz windows to pass the UV laser beam and the third wall contains a quartz observation window. The beam orientations for measuring velocity within the first ten ramp heights from the base of the ramp are also shown. This region is directly affected by the injection geometry and is where most of the mixing occurs. Direction #1 is required for the areas of the measurement region where counter-propagation is not possible due to the presence of the ramp. By

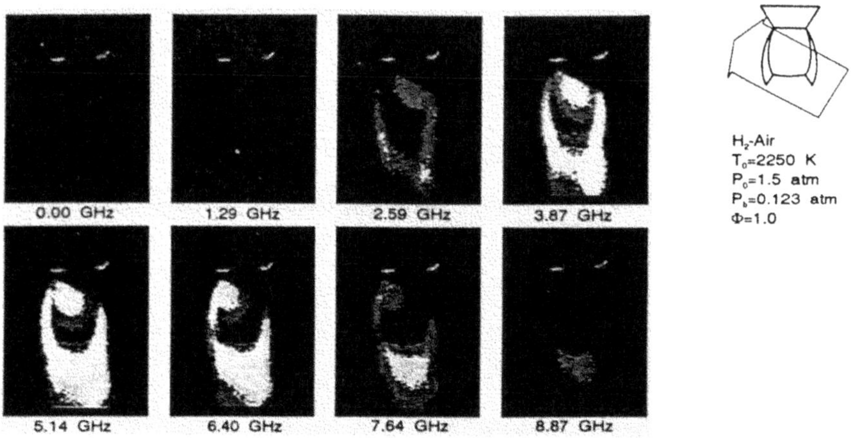

Figure 11. Doppler-shifted OH PLIF in a reacting underexpanded jet.

using a computer-controlled translation system, the camera and laser beam can be positioned throughout the flowfield to allow a complete survey of the measurement region with a single run of the supersonic combustion tunnel for each of the three required beam orientations.

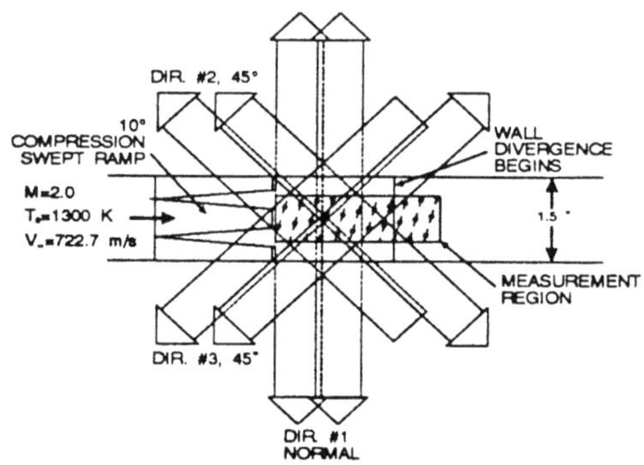

Figure 12. OH PLIF velocity imaging in a supersonic combustor.

4. Future Work

In addition to continuing the use of the two techniques described herein in studies of nonreacting and reacting high-speed flows, two new approaches are under investigation. The first uses a tunable, narrow linewidth KrF excimer laser to excite O_2 with a similar Doppler-shifted absorption measurement strategy. The second uses the third harmonic of the Nd:YAG laser to generate nitrogen ions with their subsequent detection by a second laser in a time-of-flight approach. For regions of the flowfields of interest where the velocities and velocity gradients are modest, particle image velocimetry (PIV) will also be employed. PIV or pointwise laser Doppler velocimetry (LDV) will also be used to obtain time-resolved velocity profiles, when this data is needed.

5. Conclusions

Techniques based on Doppler-shifted absorption with fluorescence monitoring using tunable, narrow linewidth lasers have been shown to provide high resolution velocity fields in steady, high-speed aerodynamic and propulsion flowfields. Planar laser-induced iodine fluorescence has been used effectively in nonreacting flows to study complex three-dimensional mixing processes in supersonic flowfields. Counter-propagating beam approaches have been described to eliminate non-Doppler impact shifts without a priori knowledge of the flowfield thermodynamic conditions. Another approach that simulates counter-propagating beams for symmetrical flows has also been described. Flowfield symmetry allows accurate velocity measurements where optical access is restricted to one side of the flow. Extensions of the PLIIF techniques are now being successfully applied to reacting high-speed flows. Velocity measurements using laser-induced fluorescence of naturally occuring OH molecules have been shown to provide accurate data in flows that are hostile to traditional seed species. An application of OH PLIF in a practical supersonic combustor has also been described.

6. References

1. McKenzie, R.L. (1991) "Progress in Laser-Spectroscopic Techniques for Aerodynamic Measurements - An Overview," AIAA Paper 91-0059.
2. Durst, F., Muller, A. and Naqwi, A. (1992) "Semiconductor Laser Doppler Anemometer for Applications in Aerodynamic Research," AIAA Journal 30, 1033-1038.
3. Coupland, J.M. and Halliwell, N.A. (1992) "Particle image velocimetry: three-dimensional fluid velocity measurements using holographic recording and optical correlation," Optics Letters 31, 1005-1007.
4. Maurice, M.S. (1991) "The Effect of Particle Dynamics on Laser Velocimetry Measurement Bias in Compressible, Vortical Flows," AIAA Paper 91-0292.
5. Chesnakas, C.J., Andrew, P.L. and Ng, W.F. (1991) "An LDV Evaluation of Particle Lag Prediction Techniques in Supersonic Flows," Yokohama International Gas Turbine Congress, Nov. 1991.
6. Lourenco, L. and Krothapali, A. (1991) "Velocity Measurement in a Supersonic Jet Using Planar Image Velocimetry," Bulletin of American Physical Society 36.
7. Maurice, M.S. (1992) "A Method to Quantify and Correct Particle Velocity Bias in Laser Velocimetry Measurements," AIAA Paper 92-0764.
8. Miles, R., Connor, J., Markovitz, E., Howard, P. and Roth, G. (1989) "Instantaneous supersonic velocity profiles in an underexpanded sonic air jet by oxygen tagging," Phys. Fluids A 1, 389-393.
9. Miles, R., Lempert, W., Zhang, B., Zhang, L., Zhou, D. and She, Z. (1992) "Fundamental Turbulence Measurements by RELIEF Flow Tagging," AIAA Paper 92-0007.
10. Krauss, R., Popp, C. and Scott, J. (1987) "Countercurrent Velocity Measurements in a Rotating Frame of Reference," Bulletin of the American Physical Society 32.
11. Boedeker, L. (1992) "Velocity Measurement in Rocket Exhaust and General Aerodynamic Flows by Photolysis of H_2O and Laser Induced Fluorescence of OH," AIAA Paper 92-0003.
12. Miles, R., Lempert, W. and Forkey (1991) "Instantaneous Velocity Fields and Background Suppression by Filtered Rayleigh Scattering," AIAA Paper 91-0357.
13. Philippe, L. and Hanson, R. (1992) "Sensitive Diode Laser Absorption Technique for Aerodynamic Measurements," AIAA Paper 92-0139.
14. Cheng, S., Zimmerman, M. and Miles, R. (1983) "Supersonic nitrogen flowfield measurements with the resonant Doppler velocimeter," Applied Physics Letters 43, 143-145.
15. McDaniel, J. (1983) "Quantitative Measurement of Density and Velocity in Compressible Flows Using Laser-Induced Iodine Fluorescence," AIAA Paper 83-0049.
16. Fletcher, D. and McDaniel, J. (1989) "Laser-Induced Fluorescence Technique for Quantitative Measurement in a Nonreacting Supersonic Combustor," AIAA Journal 27, 575-580.
17. DiRosa, M., Chang, A., Davidson, D. and Hanson, R. (1991) "CW Laser Strategies for Multi-Parameter Measurements of High Speed Flows Containing Either NO or O_2," AIAA 91-0359.
18. Hartfield, R., Hollo, S. and McDaniel, J. (1992) "A Unified Planar Measurement Technique for Compressible Flows Using Laser-Induced Iodine Fluorescence," AIAA Paper 92-0141.

19. Hiller, B. and Hanson, R. (1988) "Simultaneous planar measurements of velocity and pressure in gas flows using laser-induced fluorescence," Applied Optics 27, 33-48.
20. Paul, P., Lee, M., and Hanson, R. (1989) "Molecular velocity imaging of supersonic flows using pulsed planar laser-induced fluorescence of NO," Optics Letters 14, 417-419.
21. Palmer, J., McMillin, B. and Hanson, R. (1991) "Planar Laser-Induced Fluorescence Imaging of Underexpanded Free Jet Flow in a Shock Tunnel Facility," AIAA Paper 91-1687.
22. Marinelli, W, Kessler, W., Allen, M., Davis, S. and Arepalli, S. (1991) "Copper Atom Based Measurements of Velocity and Turbulence in Arc Jet Flows," AIAA Paper 91-0358.
23. Liebeskind, J. Hanson, R. and Cappelli, M. (1992) "Laser-Induced Fluorescence of Atomic Hydrogen in an Arcjet Thruster," AIAA Paper 92-0678.
24. Sell, J. and Cattolica, R. (1986) "Linear imaging of gas velocity using the photothermal deflection effect," Applied Optics 25, 1420-1428.
25. Trebino, R. and Hayden, C. (1990) "Velocity measurements using the phase shift of a laser-induced grating," Optics Letters 15, 1397-1399.
26. Exton, R., Hillard, M., Lempert, W., Covell, P. and Miller, D., "Molecular Flow Velocity Using Doppler Shifted Raman Spectroscopy," NASA Langley Research Center.
27. McDaniel, J. (1988) "Laser Methods for Nonintrusive Measurement of Supersonic Hydrogen-Air Combustion Flowfields," ICALEO 67, 209-218.
28. Westblom, U. and Alden, M. (1989) "Spatially resolved flow velocity measurements using laser-induced fluorescence from a pulsed laser," Optics Letters 14, 9-11.
29. Hollo, S., Hartfield, R. and McDaniel (1992) "Planar velocity measurements in symmetric flowfields using laser-induced iodine fluorescence," Submitted to Optics Letters.
30. Hollo, S., McDaniel, J. and Hartfield, R. (1992) "Characterization of Supersonic Mixing In a Nonreacting Mach 2 Combustor," AIAA Paper 92-0093.
31. Klavuhn, K., Gauba, G., and McDaniel, J. (1992) "High-Resolution OH LIF Velocity Measurement Technique for Supersonic Combustion Flowfields," AIAA paper 92-3422.
32. Klavuhn, K., and McDaniel, J. "Reacting Underexpanded Jet for the Calibration of Supersonic Combustion Diagnostics," In preparation.
33. Klavuhn, K., Gauba, G., Hollo, S. and McDaniel, J. "Planar Velocity Measurements in a Reacting Underexpanded Jet Using OH LIF," To be submitted to Optics Letters.

FILTERED RAYLEIGH AND RELIEF IMAGING OF VELOCITY, TEMPERATURE, AND DENSITY IN HYPERSONIC FLOWS FOR THE STUDY OF BOUNDARY LAYERS, SHOCK STRUCTURES, MIXING PHENOMENA, AND THE ACQUISITION OF IN-FLIGHT AIR DATA

R. MILES, W. LEMPERT, J. FORKEY, B. ZHANG, and D. ZHOU
Princeton University
Dept. of Mechanical & Aerospace Engineering
Room D-414 Engineering Quadrangle
Olden Street
Princeton, New Jersey 08544 U.S.A.

ABSTRACT. We describe here several new techniques which have potential for quantitative diagnostics in hypersonic flow fields. In particular, RELIEF flow tagging of oxygen and Rayleigh scattering are examined. Both have been implemented in a 200 mm x 200 mm high Reynolds number flow facility at Mach 3 and show great promise of being extended into the hypersonic range. Filtered Rayleigh Scattering is an especially attractive approach to hypersonic measurement in both ground-test facilities and on in-flight vehicles.

1. INTRODUCTION

Accurate and nonintrusive measurements of velocity, temperature, density, and species concentrations are needed for the study of hypersonic fluid dynamics. These measurements are very difficult to obtain due to the low density and characteristically short residence time associated with current ground-test facilities. There is, at the same time, a need to instrument hypersonic flight vehicles to provide air data for vehicle control and performance evaluation. While the use of seed gases or seed particles appears promising under some conditions [1,2], it is clear that the development of new techniques which do not require seeding is of prime importance. This is particularly true for measurements in large-scale test facilities which are difficult to seed, for measurements in unsteady flows where seeding uniformity and particle lag become important issues, and for in-flight measurements where seeding is impractical and ambient particle densities may be too low.

Over the past several years we have focused our research efforts on the development of nonintrusive diagnostic techniques which do not require seeding and can be applied to high-speed air flows [3,4,5,6]. Of particular interest for the hypersonic regime are the RELIEF flow tagging technique and Rayleigh scattering.

A. Boutier (ed.), *New Trends in Instrumentation for Hypersonic Research*, 391–398.

2. RELIEF FLOW TAGGING

For the RELIEF technique, the air is tagged by vibrationally exciting oxygen molecules and interrogating at a later time by laser-induced electronic fluorescence. This technique has been previously discussed [3] and is currently being implemented in a Mach 3 high Reynolds number facility at Princeton University. Lines can be instantaneously (10 nsecs) written into the flow field, and their displacement after an accurately measured time interval gives the flow velocity. Figure 1 shows a cross written into a Mach 2 supersonic jet after a 1 microsecond and a 12 microsecond time interval. The displacement of the cross gives an accurate measure of the flow velocity. Rotation of the cross and distortion of the lines can be used to determine vorticity and turbulence. To date, measurements have been made up to Mach 4. The extension of this capability into the hypersonic regime looks very promising, but will most likely require a shorter wavelength interrogation laser source in order to generate higher fluorescence intensities to compensate for low density.

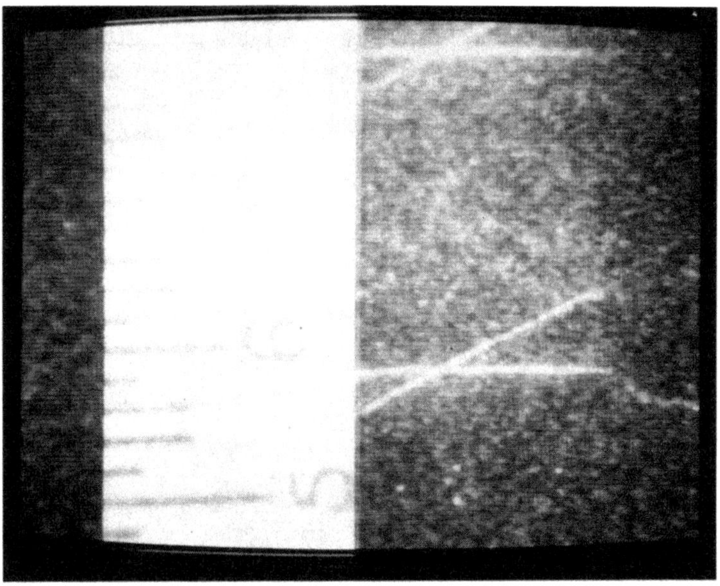

Figure 1. Cross written by RELIEF flow tagging into a vertical Mach 2 air jet after 1 μsec [bottom] and 12 μsec [top] delays. The ruler is calibrated in tenths of an inch and the measured flow velocity is 490.2 $\pm$ 2.2 m/sec.

3. RAYLEIGH SCATTERING

Rayleigh scattering is an alternative approach for observing hypersonic flow fields. Traditionally, the limitation of Rayleigh scattering has been background scattered light at the same frequency obscuring flow field data. We have approached this problem in two ways. The first is the development of ultraviolet Rayleigh imaging [4] which has permitted us to image high-speed flows, turbulent boundary layers, and shockwave/boundary layer interactions up to approximately Mach 3 [7]. The second is by filtering the Rayleigh scattering through a sharp cut-off optical filter [5].

3.1. Ultraviolet Rayleigh Imaging

Ultraviolet Rayleigh imaging takes advantage of the fact that the Rayleigh scattering cross section increases as the fourth power of the laser frequency. Therefore, by using ultraviolet laser sources, the Rayleigh scattering intensity is significantly enhanced. Simultaneously, background scattering from windows and walls is reduced due to low reflectivity of scattering surfaces in the ultraviolet. Thus, there is a dramatic increase in the signal-to-noise ratio. Imaging further enables us to discriminate against background noise sources by spatially isolating them. For example, scattering from large size particles in the flow field saturates several pixels in the image, but leaves the remainder of the picture unaffected. Figure 2 shows an instantaneous image of a shockwave/boundary layer interaction in a Mach 2.5 air flow where it is apparent that the curvature of the shock is affected by the turbulent structure. In this picture, the scattering is enhanced by the presence of a very low density ice fog caused by the condensation of residual water vapor or carbon dioxide in the flow. This condensation mechanism tends to highlight the turbulent boundary layer since the temperature of the flow field near the walls is warmer than in the free stream so condensation does not occur. Similar images with less contrast across the boundary layer have been taken with pure nitrogen where condensation is not a factor.

The laser source for Fig. 2 was an argon-fluoride excimer laser operating at 193 nm with a pulse energy of approximately 40 mJ. Similar lasers are commercially available with pulse energies approximately 10 times higher. This increased pulse energy will help offset the decreased scattering intensity due to the low density in hypersonic facilities. Further increase in signal occurs by sacrificing the two-dimensional imaging capability and doing simple line images. In that case, all of the laser energy is focused to a single line, enhancing the scattering per pixel by two or more orders of magnitude.

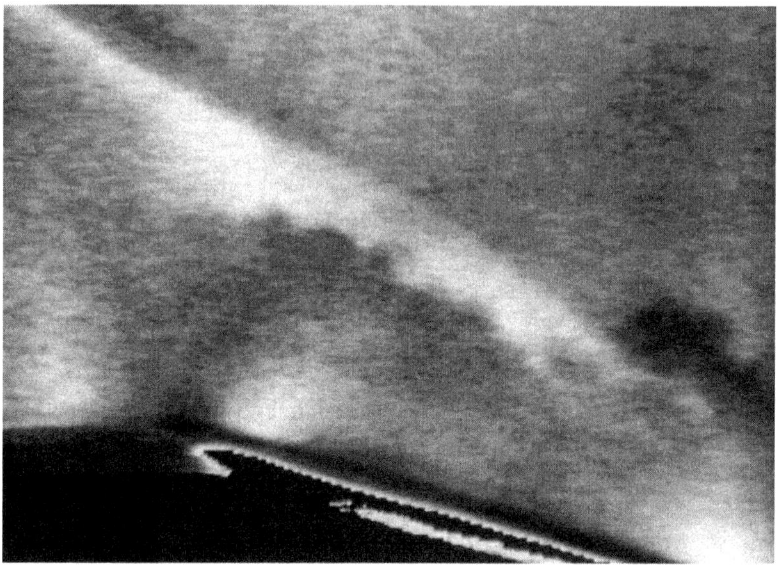

Figure 2. Instantaneous image of a shock/boundary layer interaction in a Mach 2.5 air flow. Flow direction is from right to left.

3.2. Filtered Rayleigh Scattering

An alternative way of eliminating background scattering is to use the newly developed Filtered Rayleigh Scattering approach [5]. In this case, a narrow linewidth laser is used so that the Doppler frequency shift due to the motion of the flow is larger than the laser linewidth. By passing the scattered light through a sharp cut-off molecular filter which is placed in front of the camera, background scattering from windows and walls and other non-moving objects can be eliminated, leaving only scattering from the high-speed flow field. This technique has been demonstrated in small-scale flow facilities up to Mach 2.5. Figures 3 and 4 show images in that Mach 2.5 facility with and without the optical filter.

These Filtered Rayleigh Scattering measurements have been taken using an injection-locked, frequency-doubled Nd:YAG laser together with an optically thick molecular iodine filter. Since the frequency-doubled Nd:YAG laser has an output at 532 nm, which falls in the green portion of the spectrum, all these measurements may be made through standard optical windows using conventional camera systems. The images shown here were taken with a laser energy of approximately 80 mJ. In order to extend this approach into thehypersonic regime, a higher power laser and line images may be required as discussed previously. An alternative approach is to apply the same optical filtering in the ultraviolet portion of the spectrum using a narrow linewidth ultraviolet laser source.

Figure 3. Rayleigh and background scattering from a Mach 2.5 vertical channel flow.

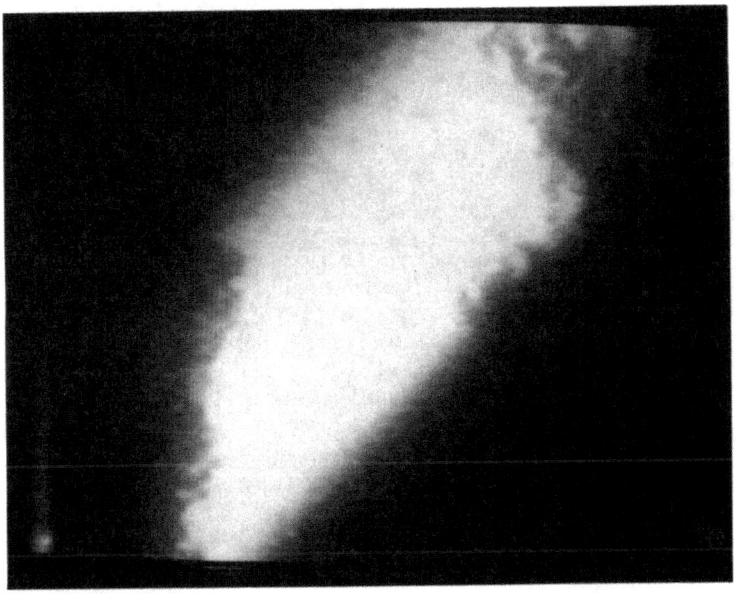

Figure 4. Filtered Rayleigh scattering from the same flow shown in Fig. 3.

With our current system we have been able to measure extinctions of background light in excess of 10^7. Thus, we expect that very good signal-to-noise ratios can be obtained using this approach. By tuning the laser source with respect to the edge of the optical filter, the velocity of the flow field can also be measured with Filtered Rayleigh Scattering. By measuring the rate at which the signal is attenuated by the filter as a function of the laser frequency, one can also determine the temperature of the flow field. As a consequence, this approach is a particularly interesting one to pursue for hypersonic flow diagnostics.

4. AIR DATA ACQUISITION

Filtered Rayleigh Scattering is also of interest for the measurement of in-flight air data. Using new high-power, diode-pumped laser sources, a compact device can be placed on a hypersonic vehicle and used to measure air speed, angle-of-attack, and slip angle, as well as air static temperature and density. In this configuration, a narrow linewidth laser beam is transmitted through an optical window and into the quiescent air in front of the vehicle. Backscattered Rayleigh light is collected and passed through the molecular filter into a detector. By accurately controlling the frequency of the transmitting laser source with respect to the molecular filter cut-off edge, the speed of the aircraft relative to the quiescent air can be determined. The magnitude of the signal is a measure of the air density and, as before, the slope of the filter cut-off as a function of laser frequency is a measure of the air temperature.

5. ADVANCED LASER DEVELOPMENTS

The successful implementation of these techniques is most heavily dependent upon the development of high-performance advanced laser sources. For the RELIEF technique, we have recently [8] developed a single laser tagging source which generates two overlapping output beams at the proper frequencies to vibrationally tag air. This technology appears to be scalable to higher powers and shorter pulse lengths in order to provide more efficient air tagging. Further development of the interrogation laser, particularly extending its frequency farther into the ultraviolet, will enhance the versatility of the RELIEF technique. Double-pulse and pulse burst laser sources can be used to generate time sequenced series of tagged lines or grids in order to follow dynamic processes.

For the Filtered Rayleigh Scattering, narrow linewidth, high power, frequency tunable laser sources are required. Fortunately, the limited tunability of the frequency-doubled Nd:YAG laser overlaps several absorption lines in iodine molecular vapor. Due to its high molecular weight and its low vapor pressure, iodine is an ideal gas for this application. A frequency-controllable, pulse burst laser of this variety will be particularly useful for capturing time-evolving high-speed structure and generating real-time measurements of flow velocity, temperature, and density. The extension of such a source into the ultraviolet will significantly enhance scattering intensities and may permit imaging in hypersonic facilities. The development of compact, high-power, narrow linewidth sources will be important for in-flight operations in order to minimize space, power, and weight requirements.

6. SUMMARY

With the new laser sources now available, there are numerous approaches to the acquisition of flow field parameters in the hypersonic regime. Flow tagging gives us the capability of measuring velocities to high accuracy and directly observing vorticity and turbulent structure. Other parameters such as Reynolds shear stress can be measured by writing grids into the flow. Rayleigh scattering mechanisms provide the capability of imaging flow field properties including density, temperature, and velocity. Shockwave/boundary layer images have already been taken in the Mach 3 regime and, with high-power ultraviolet sources, this capability can be expected in the hypersonic regime. Filtered Rayleigh Scattering will enhance the signal-to-noise ratio and provides the additional capability of extracting temperature and velocity from Rayleigh scattered light.

REFERENCES

1. M. Zimmermann and R.B. Miles, "Hypersonic-Helium-Flow-Field Measurements with the Resonant Doppler Velocimeter," Appl. Phys. Lett. 37, (1980), p. 885.

2. F. Owen, "An Assessment of Laser Velocimetry Potential in Hypersonic Flow," AIAA Paper #AIAA-92-0010, AIAA 30th Aerospace Sciences Meeting, Reno, Nevada, January 6-9, 1992.

3. R.B. Miles, J.J. Connors, E.C. Markovitz, P.J. Howard, and G.J. Roth, "Instantaneous Profiles and Turbulence Statistics of Supersonic Free Shear Layers by Raman Excitation plus Laser-Induced Electronic Fluorescence (RELIEF) Velocity Tagging of Oxygen," Experiments in Fluids 8, (1989), p. 17.

4. R. Miles and W. Lempert, "Two-Dimensional Measurement of Density, Velocity, and Temperature in Turbulent High-Speed Air Flows by UV Rayleigh Scattering," Applied Physics B 51 (1990), pp. 1-7.

5. R.B. Miles, W.R. Lempert, and J. Forkey, "Instantaneous Velocity Fields and Background Suppression by Filtered Rayleigh Scattering," AIAA Paper #AIAA-91-0357, AIAA 29th Aerospace Sciences Meeting, Reno, Nevada, January 7-10, 1991.

6. R.B. Miles, J.J. Connors, P.J. Howard, E.C. Markovitz, and G.J. Roth, "Proposed Single-Pulse Two-Dimensional Temperature and Density Measurements of Oxygen and Air," Optics Letters 13, 3, (1988), p. 195.

7. M. Smith, A. Smits, and R. Miles, "Compressible Boundary-Layer Density Cross Sections by UV Rayleigh Scattering," Optics Letters 8 (17), (September 1, 1989), p. 916.

8. W.R. Lempert, B. Zhang, R.B. Miles, and G. Diskin, "Simplifications of the RELIEF Flow Tagging System for Laboratory Use," AIAA Paper #AIAA-91-0356, AIAA 29th Aerospace Sciences Meeting, Reno, Nevada, January 7-10, 1991.

2D VELOCITY AND TEMPERATURE MEASUREMENTS IN HIGH SPEED FLOWS BASED ON SPECTRALLY RESOLVED RAYLEIGH SCATTERING

Richard G. Seasholtz
Optical Measurement Systems Branch
NASA Lewis Research Center
Cleveland, Ohio 44135
U.S.A.

ABSTRACT. The use of molecular Rayleigh scattering for measurements of gas velocity and temperature is evaluated. Molecular scattering avoids problems associated with the seeding required by conventional laser anemometry and particle image velocimetry. The technique considered herein is based on the measurement of the spectrum of the scattered light. Planar imaging of Rayleigh scattering using a laser light sheet is evaluated for conditions at 30km altitude (typical hypersonic flow conditions). The Cramer–Rao lower bounds for velocity and temperature measurement uncertainties are calculated for an ideal optical spectrum analyzer and for a planar mirror Fabry–Perot interferometer used in a static, imaging mode. With this technique, a single image of the Rayleigh scattered light from clean flows can be analyzed to obtain temperature and one component of velocity. Experimental results are presented for planar velocity measurements in a Mach 1.3 air jet.

1. Introduction

Laser diagnostics based on molecular scattering are receiving increased attention for use in aerospace test facilities. These methods can provide information directly related to the parameters of the molecular state of the gas including temperature, density, and bulk velocity. The elimination of the need for seeding is an important advantage of molecular scattering methods compared to conventional laser anemometry. Of the various molecular scattering techniques, Rayleigh scattering is the simplest and least expensive to implement. Rayleigh scattering has chiefly been used to obtain gas density by measuring the total Rayleigh scattered power. More information, however, is available from the spectrum of the Rayleigh scattered light.

The Rayleigh scattering spectrum is directly related to the molecular velocity distribution function, which is a function of the thermodynamic parameters (temperature, density, pressure), composition, mean velocity, and turbulence parameters. Light scattered at a single scattering angle provides information related to one component of the molecular velocities. Thus one component of the mean velocity can be determined (from the shift of the spectral peak). The width of the spectrum is related to the gas temperature and turbulence. For low turbulence, gas temperature can be obtained; or, for a known temperature, turbulence intensity can be obtained.

Extraction of the parameters of interest (temperature and velocity) from the spectrum can be accomplished using a sharp cutoff atomic or molecular absorption

A. Boutier (ed.), New Trends in Instrumentation for Hypersonic Research, 399–408.

filter [Shimizu et al.(1983); Miles et al.(1991)], or with a high resolution spectrometer, such as a Fabry-Perot or Michelson interferometer. Point measurements of Rayleigh scattering spectra with scanning Fabry-Perot interferometers have been used to measure temperature in a hydrogen-air flame [Cattolica et al.(1976); Pitz et al.(1976)] and velocity and temperature in the exhaust of a small hydrogen-oxygen rocket [Seasholtz et al.(1991)]. Rayleigh scattering has been proposed for atmospheric temperature measurements [Schwiesow and Lading(1981); Rees et al.(1984)]. The application of single pass and multi pass scanning Fabry-Perot interferometers for high speed velocity point measurements has been previously analyzed [Seasholtz (1991)]. Imaging velocity measurements of seeded flows using a Michelson interferometer have also been reported by Seiler and Jäger (1983). In this paper we examine an imaging technique to measure velocity and temperature using a planar mirror Fabry-Perot. The technique is an extension of 1D temperatures and number density measurements in a heated air jet [Lock et al.(1992)] and is similar to Fabry-Perot imaging techniques for measurement of instantaneous velocity of solid surfaces [Gidon and Behar (1986)], temperature and velocity in the upper atmosphere [Sivjee et al.(1980)], and solar corona measurements [Chandrasekhar et al.(1981)].

We first analyze a hypothetical planar imaging experiment where the measurement region is illuminated with a laser light sheet, and the Rayleigh scattered light is detected with a 2D array detector. It is assumed that the medium under study is a single component gas with no turbulence and contains no particles. The Cramer-Rao lower bounds for velocity and temperature measurement errors are calculated. We first calculate the error bounds for an ideal spectrum analyzer, which gives a basis with which to compare the performance of spectral measurements made with real instruments. The error bounds are then calculated for a planar mirror Fabry-Perot interferometer used in the static imaging mode. A specific example typical of hypersonic flow environments is discussed.

The use of 2D Rayleigh imaging for velocity measurements is demonstrated with an experiment designed to measure the axial velocity component in a Mach 1.3 supersonic free jet.

2. Theory

In this section we obtain theoretical lower bounds for uncertainty in velocity and temperature measurements for two cases where the scattered light is detected with a 2D array detector. The first case assumes that an ideal instrument is used to measure the spectrum. The second case assumes that the spectrum is measured with a Fabry-Perot interferometer operated in a static, imaging mode (this is the technique described in the Experiment section). Only errors resulting from the Poisson statistics of the detected light are considered. Thus these results represent lower bounds for measurement errors.

2.1. 2D RAYLEIGH IMAGING DIAGNOSTIC

Consider a planar imaging experiment where a laser sheet uniformly illuminates a region in the object plane corresponding to the image of an array detector having N_r rows (measured perpendicular to the beam propagation direction) and N_C columns of pixels. Let the detector pixel size be $L_x \times L_x$ (square pixels) and let the system have a magnification M (i.e., the measurement plane has dimensions $(N_C L_x \times N_r L_x)/M$).

The expected number of detected photons per pixel can be written [Eckbreth(1988)]

$$\langle NPP \rangle = \frac{E_l}{N_r} \frac{\epsilon n L_x \lambda_0 \Omega}{hcM} \left[\frac{d\sigma}{d\Omega}\right] \tag{1}$$

where E_l is the laser energy, λ_0 is the laser wavelength, $(d\sigma/d\Omega)$ is the differential Rayleigh scattering cross section, n is the molecular number density, Ω is the solid angle of the collected light, h is Planck's constant, c is the velocity of light, and ϵ is the overall detection efficiency (including detector quantum efficiency and system losses).

As an example, we consider air at 30km altitude (T = 230K, p=0.0118atm, $n=3.83\times10^{23}$ m^{-3}), λ_0=532nm, E_l=1J, L_x=23μm, N_r=384, $M\Omega$=0.05 (f/4 at M=1), ϵ=0.05, and $(d\sigma/d\Omega)=6.1\times10^{-32}$ m^2/sr. For this case, equation 1 gives $\langle NPP \rangle$=9.4 detected photons per pixel.

2.2. SPECTRUM OF RAYLEIGH SCATTERED LIGHT

For a single component, low density gas with a Maxwellian velocity distribution, the normalized spectrum of the scattered light is [Fabelinskii (1968)]

$$S(f)\,df = \frac{2\sqrt{\pi}}{aK}\, e^{-[2\pi(f-f_o)-\mathbf{K}\cdot\mathbf{V}]^2/a^2K^2} df \tag{2}$$

where $a = (2\kappa T/m)^{1/2}$ (m = molecular mass; κ = Boltzman's constant). The spectral peak is shifted from the laser frequency $f_0=c/\lambda_0$ by $\mathbf{K}\cdot\mathbf{V}/2\pi$, where $\mathbf{V}$ is the mean velocity of the gas, and $\mathbf{K}=\mathbf{k}_s-\mathbf{k}_o$ ($\mathbf{k}_s$ and $\mathbf{k}_o$ being the wave vectors of the scattered and incident light). The magnitude of K is a function of the scattering angle θ_s and the wavelength λ

$$K = |\mathbf{K}| = (4\pi/\lambda)\sin(\theta_s/2) \tag{3}$$

The observed Rayleigh scattering spectrum can be expressed as a set of counts

$$\langle n_j \rangle = G_R\, S(f_j)\, \Delta f \tag{4}$$

which are the expected number of counts in frequency interval f_j to $f_j + \Delta f$; G_R is the expected total number of counts. The spectrum (eq. 4) is thus a function of a set of three unknown parameters that can be expressed as the vector

$$\boldsymbol{\alpha} = [G_R, V_K, T] \tag{5}$$

where V_K is the velocity component along $\mathbf{K}$ (i.e., $V_K=\mathbf{K}\cdot\mathbf{V}/K$).

2.3. CRAMER-RA0 LOWER BOUND

The Cramer-Rao lower bound [Whalen(1971)] for the variance of the estimate of the parameter α_i is given by

$$V(\hat{\alpha}_i) = [\Gamma^{-1}]_{ii} \tag{6}$$

where Γ is the Fisher information matrix given (for Poisson statistics) by [Lading and Jensen(1980)]

$$\Gamma_{ij} = \sum_q \frac{1}{\langle n_q \rangle} \frac{\partial \langle n_q \rangle}{\partial \alpha_i} \frac{\partial \langle n_q \rangle}{\partial \alpha_j} \tag{7}$$

If the parameters are uncorrelated, the estimate of the variance of α_i is simply the inverse of the ith diagonal element of the Fisher matrix elements given by

$$V(\hat{\alpha}_i) = \frac{1}{\Gamma_{ii}} \tag{8}$$

2.3.1. *Ideal Spectrum Analyzer.* If we assume that the parameters are uncorrelated, we can easily evaluate the Fisher matrix elements and use equation 8 to get the Cramer-Rao bounds. Expressed as standard deviations $\sigma(\alpha_i)= [V(\alpha_i)]^{1/2}$, the lower bounds for measurement uncertainties for velocity and temperature are

$$\sigma(V_K) = \frac{a}{(2G_R)^{1/2}}, \qquad \sigma(T) = \frac{2T}{(2G_R)^{1/2}} \tag{9}$$

We now consider a subregion where it is assumed the parameters do not vary. The total number of photoelectron counts due to Rayleigh scattering detected from this subregion is then $G_R = \langle NPP \rangle \times$ number of pixels. The size of the subregion determines the spatial resolution of the measurements For example, assuming a 20×20 array of pixels (which corresponds to 0.5mm×0.5mm) and $\langle NPP \rangle$=9.4 (as given in the above example), we have G_R=3760 counts, T=230K, and a=370m/s. Thus the uncertainties in velocity and temperature are $\sigma(V_K)$=4m/s and $\sigma(T)$=5K. Note that the velocity uncertainty is proportional to the square root of temperature.

2.3.2. *Fabry-Perot Interferometer.* The above error estimates were based on the use of an ideal instrument to measure the spectrum. In this section we perform the error estimation based on measurement of the intensity distribution of light imaged through a planar mirror Fabry-Perot interferometer. The expected number of detected photoelectrons $\langle NDP \rangle$ for the q_{th} pixel is

$$\langle NDP_q \rangle = \langle NPP_q \rangle \int\int S(f) I_t(f, \theta_r) df \, dA_{pixel} \tag{10}$$

where the Fabry-Perot transmission function [Vaughan(1989)] is

$$I_t(\psi) = \frac{1}{1 + F \sin^2(\psi/2)} \tag{11}$$

with ψ being the phase delay of the light for one pass through the interferometer given by

$$\psi = \frac{4\pi\mu df \cos\theta_r}{c} \tag{12}$$

and $F = (2N_R/\pi)^2$, where N_R is the reflective finesse. In this equation, f is the frequency of the light, μ is the refractive index of the medium in the Fabry-Perot cavity (taken as unity here), d is the Fabry-Perot mirror spacing, and θ_r is the angle between the ray and the optic axis. For simplicity, we neglected spectral broadening due to the finite size of the aperture of the collection optics.

We now select, for this example, a particular Fabry-Perot configuration (mirror spacing d=15mm; finesse N_R=20) and numerically evaluate the Fisher matrix elements

$$\Gamma_{ij} = \sum_q \frac{1}{\langle NDP_q \rangle} \frac{\partial \langle NDP_q \rangle}{\partial \alpha_i} \frac{\partial \langle NDP_q \rangle}{\partial \alpha_j} \tag{13}$$

TABLE 1. Lower bounds for uncertainty in velocity and temperature measurements.

Fringe number	Fringe radius	$\sigma(V_K)$, m/s	$\sigma(T)$, K
1	38 pixels	14	44
2	69	19	49
3	90	22	61
4	106	24	71
5	121	25	68
11	185	20	81
19	286	24	83
ideal		4	5

where the sum is over the same 20×20 pixel subregion used above. The results of this calculation are given in Table 1 for several locations of the subregion (denoted by fringe radii). The velocity uncertainty is about 20m/s (5 times that achievable with ideal spectrum analyzer) and the temperature uncertainty is about 40-80K (10 times that of ideal). In practice, the assumption of constant finesse will not be valid; vignetting reduces the effective number of reflections in the interferometer cavity, thereby decreasing the finesse at larger fringe radii. This means that the finesse must be measured, as is done in the experiment described later in this paper.

2.4. OTHER CONSIDERATIONS

The example presented above considered an idealized measurement situation. There are always other factors that degrade measurement accuracy. For instance, if the gas is made up of an unknown mixture of component gases, the spectrum cannot be uniquely related to the temperature. If the flow is highly turbulent, and the measurement is obtained over a duration much greater than the turbulence time scale, the spectrum is broadened with both temperature and turbulence contributing to the spectral width. Thus, if one of these is known, the other can be determined from the spectral width, but both cannot be simultaneously determined. The Gaussian spectrum used in the example occurs only when the molecular mean-free-path is greater than the wavelength $\Lambda=2\pi/K$ involved in the scattering process (i.e., $y\equiv\Lambda/mfp<<1$). For values of $y > 1$, collective effects in the molecular velocities become important and must be taken into account in evaluating the spectrum shape. For $y>>1$, relatively simple continuum models can be used [Clark (1975)]; the spectrum in this case splits into three peaks and is referred to as Brillouin scattering. For the transition region, $y\simeq1$, the spectral calculations are more involved and kinetic theory models, such as the Tenti S6 model [Tenti et al.(1974)] must be used. In general, ambient conditions fall into the transition region of $y\simeq1$, while the lower densities typical of hypersonic flows have $y<<1$.

Particles in the flow will strongly scatter. In some cases where the particle number density is low, the images can be processed to remove the obvious particles. If the particle number density is so large that the particles images overlap and

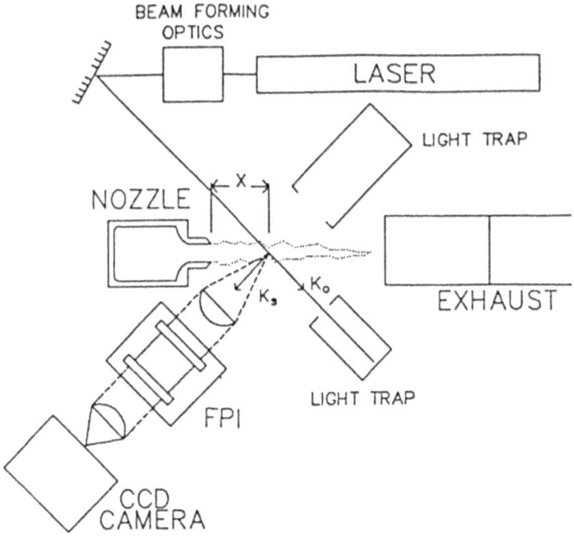

Figure 1. Experimental setup for measurement of velocity in Mach 1.3 free jet using imaging Fabry-Perot interferometer.

cannot be removed, the spectral shape is changed from that of pure molecular scattering and temperature measurements cannot be obtained. It is important to note, however, that particles do not prevent the measurement of velocity (provided there is negligible particle lag) since the spectral peak due to particles is at the same frequency as the peak of the molecularly scattered light.

3. Experiment to Map Velocity in Supersonic Free Jet

3.1. EXPERIMENTAL SETUP

To demonstrate the use of Rayleigh scattering for measuring velocity, we conducted the experiment shown in figure 1. The Mach 1.3 supersonic free jet (9.9mm exit diameter) was mounted on a linear translation stage so it could be moved along its axis. The beam from an argon-ion laser(2.5W, 488nm) was formed into a sheet (about 10mm high by 0.2mm thick) using a combination of cylindrical and spherical lenses. The light sheet crossed the flow at 45°. The receiving optics were aligned at 45° to the flow direction and 90° to the propagation direction of the laser beam. Thus the **K** vector was aligned with the flow, and the axial velocity component was measured. A 250mm focal length f/3.6 lens was used to collect and collimate the scattered light. The collimated light was passed through a planar mirror Fabry-Perot interferometer (70mm diameter aperture and 14.96mm mirror spacing) and focused onto the CCD array with a 225mm focal length lens. The CCD array had 384 rows by 576 columns of 23μm x 23μm square pixels (giving a field size in the flow of 9.8mm×14.7mm) and was cooled to about -50° C to reduce dark current. The CCD pixel data was digitized with 12 bit resolution and transferred to an 80386 PC via an IEEE-488 interface for storage and analysis.

The initial step in the measurement process was to measure the instrument function [$I_t(\psi)$, for $f=f_0$]. A diffusely scattering target was placed in the field of view and illuminated with laser light scattered from a rotating diffuse target (the

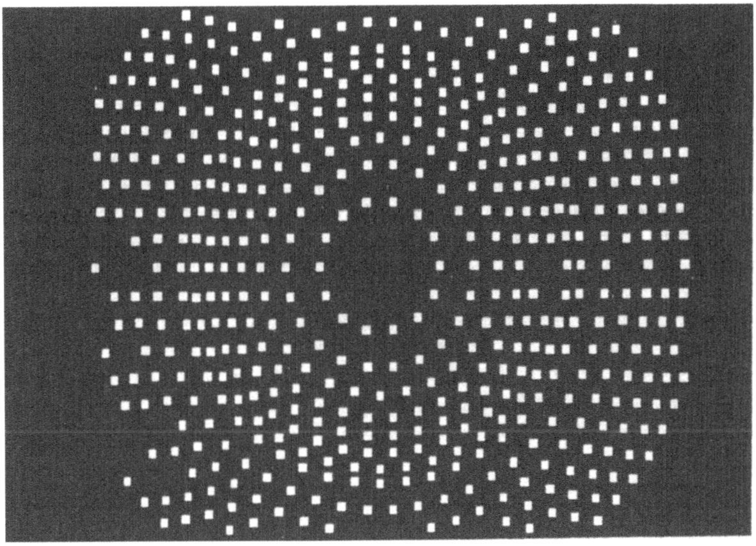

Figure 2. Pattern of subregions used for data analysis (5×5 pixel subregions shown; actual size of regions used was 20×20 pixels).

rotating target was used to reduce speckle in the recorded image. The Rayleigh scattered light from the flow was then recorded. Exposure times were 0.1s for the instrument function and 1.0s for the Rayleigh scattering.

Although the air supplied to the nozzle was filtered, it was not feasible to supply filtered air to the air surrounding the jet. Particles in the ambient air were thus mixed into the flow, and temperature measurements were not realizable (even though the technique described by Kay and Sadler (1991) was used to reduce the effect of particles).

3.2. DATA REDUCTION

The data reduction was done in two steps. First, the instrument function image was analyzed to determine the center of the circular fringe pattern, the finesse, and the phase corresponding to the laser frequency. Because the effective finesse and phase varied over the image plane, it was necessary to determine them as a function of location in the image (typically, the finesse varied from about 20 at the center to about 4 at the edges of the image). About 500 20×20 pixel subregions were used. An integral number were positioned around the each of the inner fringes and around every other outer fringe (so the subregions did not overlap). A typical pattern of subregions is shown in figure 2. A 4th order, 2D polynomial (15 parameters) was then fit to these values of phase and finesse. The coefficients for the polynomial fits were saved for the second step (analyzing the image of the Rayleigh scattered light).

In the second step, the image of the Rayleigh scattered light in each subregion was fit to a model function given by equation 10 using the S6 model of Tenti et al.(1974) with the gas properties of air. The velocity data obtained at each subregion were fit to a 2D 4th order polynomial, which was used to generate contour plots of constant axial velocity.

3.3. RESULTS

Figure 3 shows constant axial velocity contours for 5 axial locations along the jet axis (X=20,30,50,70, and 90 mm). Recall that these planes are oriented at 45° to the jet axis. At the locations closer to the nozzle exit, the maximum velocity is in good agreement with the 423m/s velocity calculated from the measured total pressure and temperature (assuming isentropic flow). As the measurement plane moves downstream, the maximum velocity decreases as the core flow mixes with ambient air. The contours are elliptically shaped because of the orientation of the measurement plane. We also can see that the center of the jet is shifting to the right as it moves downstream. This is probably caused by a slight misalignment of the jet axis.

The uncertainty in the velocity estimates due to photon arrival rate statistics is quite small, less than 10m/s. Errors due other sources most certainly outweigh these statistical errors. One potential error source is the drift in the Fabry-Perot mirror spacing between recording the instrument function and the Rayleigh data. The r.m.s. deviation of the contours from the individual subregions data was about 10-15m/s.

The data reduction requires rather lengthy processing, chiefly because of the numerical integration in equation 10. The time per iteration for each subregion is about 15s using a 80386/33MHz/Weitek 3167. Thus an image with 500 subregions and an average of 5 iterations, requires about 10hr to process. No attempt was made to optimize the code at this time, so it should be possible to substantially reduce the data reduction time.

4. Concluding Remarks

The results of this work show the feasibility of using measurements of the spectrum of molecular Rayleigh scattering as a diagnostic suitable for many high speed flow experiments. The technique is capable, under certain restrictions, of mapping gas density, velocity, and temperature.

The experimental work reported here used a CW argon-ion laser, which results in time-averaged measurements. Use of a pulsed laser and gated detection would offer several advantages. Measurements based on a image from a single laser pulse would give 2D maps of the gas parameters at an instant of time. The effect of background light, such as flame luminosity, could be greatly reduced. And, particle images, if not too dense, should be easier to identify and remove from the flow field image.

A large number density of particles in the flow generally will prevent temperature and density measurements. It is important to note, however, that velocity can be obtained with or without particles, provided the particles follow the flow. Care should be taken to prevent spurious laser light from reaching the receiving optics. Even though this light can be accounted for in the data reduction, it will degrade the signal-to-noise ratio and measurement accuracy. This problem is particularly severe in internal flows where it is very difficult to eliminate all spuriously scattered light. The technique described by Miles et al. (1991) for blocking light at the laser frequency should be useful for high velocity flows where the frequency of the Rayleigh scattered light is shifted well away from the laser frequency.

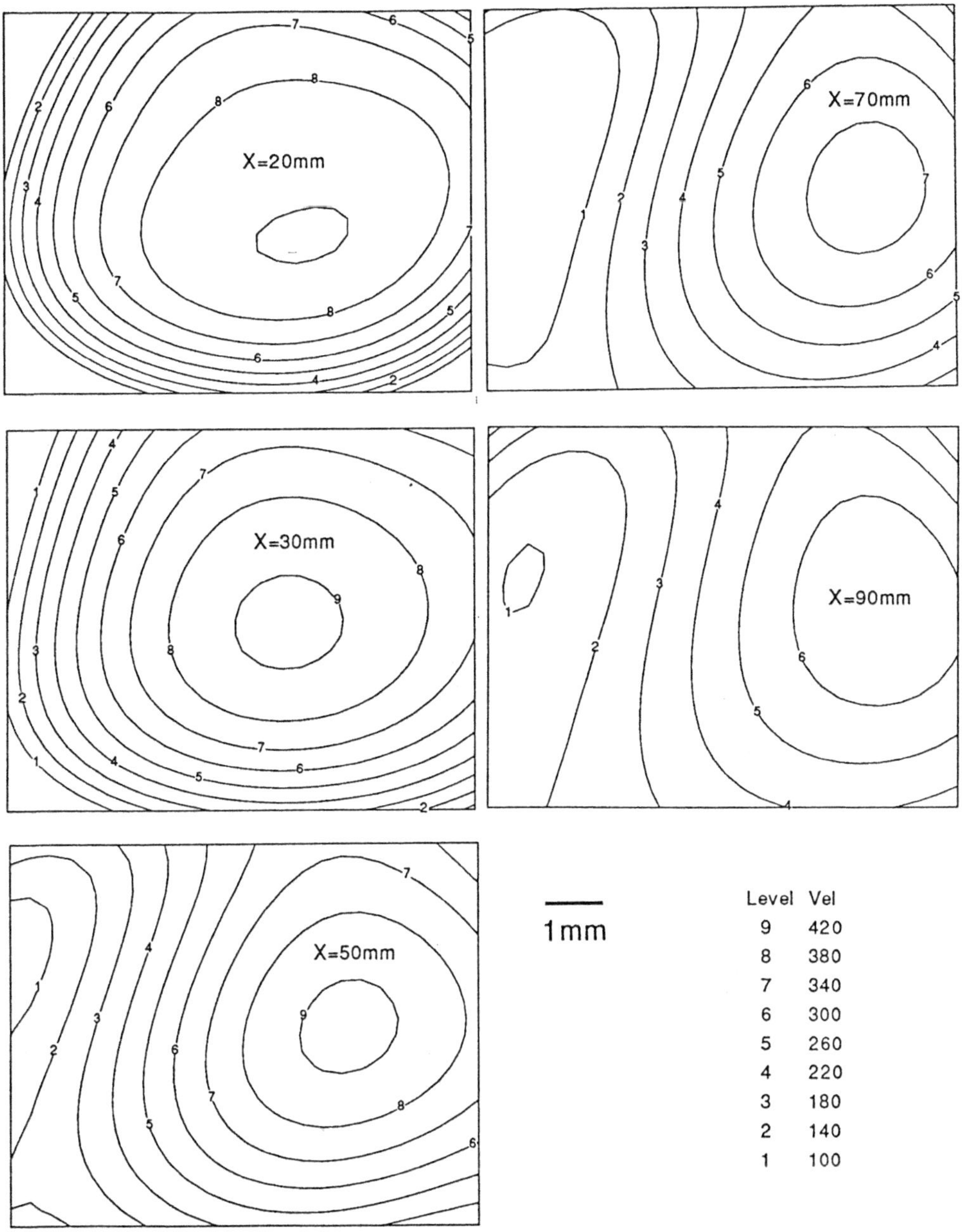

Figure 3. Measured constant axial velocity contours. Distances from nozzle exit are X=20,30,50,70,90mm. Isentropic velocity = 422m/s.

REFERENCES

Cattolica, R., Robben, F. and Talbot, L. (1976) 'The interpretation of the spectral structure of Rayleigh scattered light from combustion gases', AIAA 14th Aerospace Sciences Meeting, Washington, AIAA paper 76-31.

Chandrasekhar, T., Desai, J.N. and Angreji, P.D. (1981) 'Temperatures and differential line-of-sight velocities observed in the solar corona during a total solar eclipse', Appl. Opt. 20, 2172-2173.

Clark, N.A. (1975) 'Inelastic light scattering from density fluctuations in dilute gases. The kinetic-hydrodynamic transition in monoatomic gas', Phys. Rev. A 12, 232-244.

Eckbreth, A.C. (1988) Laser Diagnostics for Combustion, Temperature, and Species, Abacus Press, Tunbridge Wells, Kent, 367.

Fabelinskii, I.L. (1968) Molecular scattering of light, Plenum Press, New York, 263.

Gidon, S. and Behar G. (1986) 'Instantaneous velocity field measurements: application to shock wave studies', Appl. Opt. 25, 1429-1433.

Kay, L. and Sadler, D.A. (1991) 'A method for processing CCD images to remove cosmic rays and other randomly positioned spurious events - theory and experiment', Measurement Science and Technology 2, 532-535.

Lading, L., Jensen, A.S. (1980) 'Estimating the spectral width of a narrowband optical signal', Appl. Opt. 19, 2750-2756.

Lock, J.L., Seasholtz, R.G., and John, W.T. (1992) 'Using Rayleigh-Brillouin scattering to determine one-dimensional temperature and number density profiles of a gas flow field', Appl. Opt. 31, (to be published).

Miles, R.B., Lempert, W.R. and Forkey, J. (1991) 'Instantaneous velocity fields and background suppression by filtered Rayleigh scattering', AIAA 29th Aerospace Sciences Meeting', Reno, AIAA paper 91-0357.

Pitz, R.W. et al.(1976) 'Temperature and density in a hydrogen-air flame from Rayleigh scattering', Comb. and Flame 27, 313-320.

Rees, D. et al.(1984) 'The Doppler imaging system: initial observations of the auroral thermosphere', Planet. Space Sci. 32, 273-285.

Schwiesow, R.L. and Lading, L. (1981) 'Temperature profiling by Rayleigh-scattering lidar', Appl. Opt. 20, 1972-1979.

Seasholtz, R.G., Zupanc, F.J. and Schneider, S.J. (1991) 'Spectrally Resolved Rayleigh Scattering Diagnostic for Hydrogen-Oxygen Rocket Plume Studies', AIAA 29th Aerospace Sciences Meeting', Reno, AIAA paper 91-0462.

Seasholtz, R.G. (1991) 'High-speed laser anemometry based on spectrally resolved Rayleigh scattering', Fourth International Conference on Laser Anemometry, Cleveland.

Seiler, F. and Jager, W. (1983) 'Flow visualization with Doppler-pictures', Tenth International Congress on Instrumentation in Aerospace Simulation Facilities, Saint Louis.

Shimizu, H., Lee, S.A., and She, C.Y. (1983) 'High spectral resolution lidar system with atomic blocking filters for measuring atmospheric parameters', Appl. Opt. 22, 1373-1381.

Sivjee, G.G., Hallinan, T.J., and Swenson, G.R. (1980) 'Fabry-Perot interferometer imaging system for thermospheric temperature and wind measurements', Appl. Opt. 19, 2206-2209.

Tenti, G., Boley, C.D. and Desai, R.C. (1974) 'On the kinetic model description of Rayleigh-Brillouin scattering from molecular gases', Can. J. Phys. 52, 285-290.

Vaughan, J.M. (1989) The Fabry-Perot Interferometer, History, Theory, Practice and Applications, Adam Hilger, Bristol, Chapter 3.

Whalen, A.D. (1971) Detection of signals in noise, Academic Press, New York 324-331.

TEMPERATURE AND HEAT FLUX MEASUREMENTS - CHALLENGES FOR HIGH TEMPERATURE AEROSPACE APPLICATION

Richard D. Neumann
University of Dayton
Research Institute
300 College Park
Dayton, OH 45469-0111, USA

Thesis Statement

THE FUNDAMENTAL PROBLEM IS
THAT YOU AND I WOULD LIKE TO HEAR
BROAD, SWEEPING UNIVERSAL TRUTHS
ABOUT THIS SUBJECT, THERMAL SENSORS,
BUT IT IS DOMINATED BY THE
DETAILS OF THE PARTICULAR APPLICATION

These details involve the materials system complexity;
the characteristics of those materials
in which measurements are made;
the presence, if any, of sub-surface cooling
and the comparative use of
"add-on" or "build-in" thermal sensors which
infer heating rates by either understanding
the thermal capacity of a material or
the rate at which heat conducts through that material

INTRODUCTION AND BACKGROUND

The measurement of high temperatures and the inference of heat transfer data is not strictly a problem of either the high temperatures involved or the level of the heating rates to be measured at those high temperatures. It is a problem of duration during which measurements are made and the nature of the materials in which the measurements are made. Thermal measurement techniques for each application must respect and work with the unique features of that application.

In the past 30 years, high heat flux has successfully been measured on a number of high temperature flight test systems. Among these programs were (1) the NASA "FIRE" program, (2) the NASA "RE-ENTRY F" program and the (3) the NASA SPACE SHUTTLE program. The table below lists the peak measured

A. Boutier (ed.), New Trends in Instrumentation for Hypersonic Research, 409–436.

heating rates on these vehicles, the flight duration of the experiment as well as the peak wall temperature achieved on the sensor.

Flight Vehicle	Peak Heating Rate (Btu/Ft^2Sec)	Peak Measured Temperature ($^\circ F$)	Test Duration (Sec)
Space Shuttle Forebody	16.2	2015 $^\circ F$	1200
Space Shuttle Flap	26.9	2311 $^\circ F$	1200
FIRE	1100.0	2440 $^\circ F$	25
Re-entry F	545		3.2

Each of these was a unique experiment with quite different test goals. As a result, many of the flight-measurement features differed in detail from experiment to experiment.

The Space Shuttle, a lifting body, generated data for the longest duration with the substantial surface wall temperature. The other experiments, while measuring higher heating rates at comparable surface temperatures, obtained these data during far shorter test durations.

Matthews et al [1] discussing the need for high temperature heat flux gages stated that...

> " ...reliable heat flux gages are currently limited to relatively low operating temperatures and gages that can operate at temperatures of 2000^oF and above are required to measure the heat delivered to a structure during ground and flight testing."

in spite of this, flight tests have generated reliable heat flux data at surface temperatures substantially greater than 2000^oF. Admittedly, in each case the thermal instrumentation was well integrated into the thermal protection system in a manner which allows the efficient inference of heat transfer from temperature measurements. The basic thermal instrumentation on the Space Shuttle, for example, was a single platinum thermocouple placed near the surface of the tile, as shown in figure 1, with the surface temperature and the heat flux inferred from this thermocouple through the use of a thermal model. In both FIRE and RE-ENTRY "F" in-depth temperature plug gages, shown in figures 2 and 3, were used to define heat transfer. This instrumentation initially inferred heating through the temperature drop between adjacent thermocouples in the plug although post test analyses suggested that for the short test durations, multiple gages placed in depth were not required and a single thermocouple defining the heat capacity of the surface material would suffice.

Why would a quote in 1991 argue with successes of the past 30 years? The problem is NOT that heat transfer cannot be deduced at high operating temperatures but that either heat transfer cannot be deduced at high temperatures for more general applications or the problem lies in the definition of the term "heat flux gage". Each of the cases just discussed were carefully crafted from the materials viewpoint, from the test duration viewpoint and from the instrumentation viewpoint to facilitate measurements.

Matthews continues by saying that:

> "... (Heat flux) gage design emphasis is placed on COMPATIBILITY ISSUES associated with integrating the gage within the structure...and minimizing the gage's influence on the surrounding material response, particularly in actively-cooled structures"

This statement is a key. For each example case, the instrumentation and materials in which it was placed were thermally compatible. Compatibility issues define whether a commercially-available add-on thermal sensor or a build-in thermal sensor should be used in a specific application and whether accurate heat flux can be deduced at all. The corollary, is that NOT ALL STRUCTURES CAN BE INSTRUMENTED TO DEDUCE HEAT TRANSFER today. That is the basis of Matthew's comments and a substantial challenge for the measurement community. A more correct interpretation of Matthew's statements, which I might offer, would be...

> Significant deficiencies exist in deducing heat transfer from temperature measurements on ARBITRARY MATERIALS and DURING ARBITRARY TEST PERIODS and these problems are made worse by the selection of "add-on" rather than "design-in" thermal sensors. Examples of successful high temperature, high heat flux measurements are the result of careful, integrated designs where the instrument, the flight vehicle and the test acquisition phase of the flight have been properly selected to achieve successful data.

Several concepts have been briefly introduced and these concepts require further discussion.

THERMAL GAGES AND HEAT FLUX GAGES: AN OPERATIONAL DEFINITION

In this paper, thermal gages are instruments that measure temperature. They include thermocouples, resistance thermometers and fiber-optic based temperature measurements. Heat flux gages are self-contained instruments that use thermal measurements, with a thermal model, to infer causal heat flux.

ADD-ON VS DESIGN-IN HEAT FLUX GAGES

Heat flux is deduced through thermal models which interpret the measured temperature on or in the structure in terms of the heating that caused those measured temperatures. In so doing, the process must not alter the thermal character of the surface being measured. Heat flux gages are highly localized embodiments of those thermal models; concepts committed to hardware. A cross-section of heat flux gages for low and high temperature applications are defined in the tables below.

Dimensions of Low Temperature Heat Flux Sensors

Heat Flux Sensor Type	Typical Diameter (Ins.)
Thin Film resistance Thermometer *	0.0030 x 0.200
Schmidt-Boelter Gage	0.250
Coax Gage *	0.0150
Plated Thermocouple Gage *	0.008 x 0.008

* Requires proper materials to form a valid thermal model

Dimensions of High Temperature Heat Flux Sensors

Heat Flux Sensor Type	Typical Diameter (Ins.)
Water-cooled Gardon Gage *	0.0625
High Temperature Gardon Gage *	0.250
Vatell Gage	0.100 x 0.125
In-depth Thermocouple Gage *	0.125

* No surface thermocouple measurement

It is possible that heat transfer can be deduced through add-on, commercial heat flux sensors but it is also possible that heat flux must be deduced by distributing thermal sensors within the structure of the hardware being measured and using that hardware as the "heat flux gage" itself. In the Space Shuttle example the TPS material had thermal sensors imbedded in it and the material with the imbedded thermal sensors became the heat transfer gage. For the FIRE and RE-ENTRY "F" examples either the beryllium structure could be thought of as having thermal sensors imbedded in it or the thermocouple plug could be thought of as a thermally transparent add-on heat flux gage.

There were also add-on heat flux gages used on the Space Shuttle in
the external tank [2] and added to the orbiter heat shield [3]. These examples are discussed in appendix A. In both cases, the metallic add-on heat flux gages were poorly integrated into the insulative TPS material of the Space Shuttle and the resulting heat transfer "measurements" of an unacceptable quality.

Thermally incompatible heat flux sensors present two very different problems which must be addressed: 1) the sub-mold-line transfer of heat between the parent structure and the sensor which may destroy the "assumed" thermal model and 2) surface boundary layer distortion of CONVECTIVE flow over the structure housing the heat transfer gage. This distortion is caused by the step function change in the wall boundary condition with no corresponding change at the "edge" of the boundary layer. Since the surface is heated by the gradient of the static temperature distribution through the boundary layer, the sensed heat flux is sensitive to wall temperature disturbances. This is a FLOW FIELD-INDUCED phenomena present in a convective flow which has a developed boundary layer but not in a radiative calibration rig or on the stagnation point of model placed in a convective flow. The aerodynamic problems of temperature mismatch were discussed both by Carnahan [4] and Praharaj [5]. Both of these problems can be addressed through well-defined numerical techniques which correctly model the flow. The material mismatch problem is found in all high temperature or long time applications of thermal instrumentation.

Both adding-on thermal instruments or designing-in those instruments are potentially acceptable techniques within limitations defined by the specifics of the experiment. The details of the specific application may dictate one technique over the other. In either case, the basis of heat flux measurement must be respected. Clearly, in the case of the add-on gages for the Space Shuttle application, the gages were not well thermally integrated into the TPS system and the measured results were, as a consequence, not representative of the structure with no instrument installed.

Thermal Gage Location Within the Structure:

The inference of heat transfer requires the precise placement of thermal gages with the structure to be measured. Large gradients of temperature are present within the structure and imprecisely located thermal sensors reflect inaccuracy in the inferred heating rates. Knowledge of thermal sensor location is far more precise in add-on heat gages than in design-in heat gages. The instrument vendor precisely locates these sensors and calibrates the heat gage to assure they are properly located. This is not the case for design-in heat sensors for which thermal sensors are placed within a structure either during or after the fabrication process. Examples abound of poor data generated as a result of thermal gages imprecisely located within the structure or shifting within that structure during the process of test. One such example is, again, the Space Shuttle instrumentation shown in figure 1. In this case, the platinum thermocouple was placed within the soft, insulative tiles during the fabrication process without strict quality controls on the installation process [6]. The result was some ambiguity concerning the actual in-depth placement of the thermocouple wire during the installation and later during the flight-to-flight refurbishment of the tiles containing the thermal gage. Hodge initially identified the problem which is most acute in short duration flight maneuvers and defined a very sophisticated software tool to understand it [7]. Similar studies were conducted by Jones et al [8]. Figure 4. from their paper indicates that the accurate location of thermocouples was required to match heating rate data deduced from thermocouples with heat transfer data deduced from calorimeter gages. In-situ calibration of test hardware containing design-in heat gages is extremely important as well as the collateral use of parameter estimation techniques to generalize that calibration information. The paper by Kipp and Eiswith [9] proposed in-situ calibration of instrumentation that is as true today as it was a decade ago.

THERMAL CAPACITY VS CONDUCTION RATE HEAT FLUX INSTRUMENTS

Although aerodynamic heating is numerically computed from the slope of the static temperature in the boundary layer near the wall boundary condition, aerodynamic heating is experimentally inferred though its influence on the structure as shown in figure 5. The numeriscist worries only about fluids assuming a perfectly responding structure but the experimentalist worries about both questions of materials response as well as questions of the boundary layer fluid.

As noted earlier by example, successful instruments can be based either on the premise of capturing the heat pulse within the structure or determining the rate of heat transfer through the structure. The Shuttle, because of its insulative tile thermal protection system, essentially captured the heat pulse within the tiles whereas the FIRE and RE-ENTRY "F" programs were instrumented with gages capable of either capturing the heat pulse or measuring the rate of heat transfer through the structure. Heat capacity gages are indirect measures of heat transfer requiring a sometimes complicated thermal model to interpret measured temperature in terms of causal heating rates. Heat conduction rate gages are direct reading gages creating a temperature-difference signal proportional to causal heat transfer rates. This is, of course, an idealized description of the problem assuming only a direct relationship between causal convective heating to the surface and conductive dissipation of that heating within the structure; a cLassic low temperature application.

At high temperatures, the causal convective heating may be dissipated either through internal dissipation, ideally one-dimensional conduction into the skin of the flight vehicle or re-radiated back to space (or other structural elements if you are really unlucky). As the surface temperature of the structure increases, more of the heat radiates (as the 4th power of the wall temperature) and correspondingly less conducts inward. At some point during the flight operation of an uncooled surface, almost all of the heat is re-radiated away from the surface. In the limit, rate gages define zero heat transfer THROUGH the surface although substantial surface heating TO the surface is still occurring. In this limiting case, the sole indicae of heat flux is through a surface temperature measurement that is used to infer heating through calibration of the emissivity of the surface material. On the Space Shuttle, this situation occurred rapidly as shown in figure 6.

Rate gages, while direct reading, may not always measure the total heat load to the surface. All high temperature rate gages must measure not only the rate of heat transfer through the structure but also the absolute level of temperature at the surface. Some do not and others accomplish this only through calibration. Measuring both rate and level requires additional channels of information compounding the difficulty of the measurement process. Finally, heat capacity gages require a larger volume of structure for dissipation as the test duration increases and/or the thermal conductivity of the structural material increases. It is not always possible to capture the heat pulse within a structure and for those cases where it is not, rate gages may be a better choice.

SUB-SURFACE COOLING

Actively cooled materials are feasible and of increasing interest. While cooling reduces the material surface temperature (and the heat re-radiated to space from an instrumentation perspective), cooling also provides an additional heat loss mechanism. Heat capacity measurements, in this environment, require that the heat loss to the coolant be measured. Direct reading heat-rate measurements applied in this environment require that all possible heat paths be measured and that direct measurements be made within the thickness of the panel being cooled and within the portion of that cooled panel at which the temperature gradient (due to conduction) is linear. These thickness dimensions can be very small, challenging the design and integration of thermal instruments.

WHAT IS MEASURED AND WHAT IS CALIBRATED OUT

Re-radiation:

Not all the elements of aerodynamic heating can be measured with either thermal capacity or heat rate sensors. One element that requires calibration is surface re-radiation. The heating due to radiation follows the equation

$$\dot{q}_{Radiation} = \epsilon \sigma T_{Wall}^{4}$$

Evaluating radiation from the wall is inherently inaccurate resulting in an error of +/- 10% or more in heating rate. The emittance measurement has experimental scatter and the surface temperature measurement converted to a digital signal has experimental scatter whose magnitude depends on the type of thermal sensor used and the quality of the data train (from sensor to digital output). Further, the measured temperature is raised to the 4th power compounding the error and finally, the measured temperature may not be the actual surface temperature but a sub-surface measurement that must be related to surface conditions either numerically or through calibration techniques.

Chemical Energy:

Another increment of heating that normally requires calibration is that associated with incomplete recombination of dissociated boundary layer flow. Static temperature increases through the shock system about the vehicle and the deceleration of the flow in the boundary layer may dissociate the air. Low density flows, which limit energy transfer in the gas, may lock in the stagnation region chemical activity that sweeps over the surface. If this occurs, the catalytic character of the surface material and that of the thermal gage placed in that material may recombine the dissociated flow. This catalytic character must be calibrated. Upstream history effects are important as well as the local conditions at the instrument in question. Calibration of the surface catalysticity is possible to accomplish but surface aging as the flight surface is repeatedly flown is also important.

Our flight experience with catalytic surfaces is through the Space Shuttle flights. The non-catalytic nature of the Shuttle tile coating (a glassy material) was noted to reduce the measured heating rate up to 60% of a fully catalytic wall [10]. It has been noted that the sudden release of chemical energy over a catalytic gage placed in a non-catalytic surface results in a temperature rise greater than the equilibrium value that would be present for an entirely catalytic surface and gage [11]. Evidence of aging phenomena was also noted and reported by Jones [12] who noted a 20% increase in heating from flight 2 to flight 5 of the Space Shuttle due to either catalysis or emissivity changes.

This situation argues for in-situ calibration of thermal sensors on flight vehicles; yet another measurement challenge. In-situ calibration is more important for thermal sensors which are designed-in to the structure rather than added-on to the structure.

THERMAL MODELS AND THEIR IMPORTANCE:

Heat transfer is inferred by strategically placed temperature measurements through the use of a thermal model of the temperature dissipation. We implicitly use thermal models while we tend to forget the approximations that limit their application. Whether the causal relationship is seemingly trivial, a 1D conductive flow through the structure defining the flow within a so-called heat gage or whether it is complicated, full 3D time varying dissipation through a built-up structure, the process is conceptually the same. This paper will not discuss thermal models and their limitations but even a cursory review of the literature demonstrates the problems which are continuing today.

THE SIMILARITIES AND DIFFERENCES BETWEEN GROUND TEST AND FLIGHT APPLICATIONS

While this paper deals with flight measurements. The reader might deduce that either high temperature measurements are not made on the ground or that the instrumentation and techniques for ground-based measurements are solved or the same as for flight. None of this is the case.

Ground test applications cover a spectrum of thermal and heat transfer needs and goals. Aeromechanic testing is generally at low temperatures because of experiment design. High temperature ground test goals are structural and propulsion development. While the structural testing uses materials and structural concepts in common with flight goals, the propulsion testing may present ground test unique instrumentation problems. Propulsion tests focus on defining propulsion efficiency without the added cost and complexity of using flight-weight structural materials. This is accomplished by fabricating the model structure out of heat-sink copper. The challenge is to "measure" the highly non-uniform surface heat transfer within highly-conductive copper. A particular concern is the measurement of localized heating peaks within copper structures. These peaks create lateral heat conduction paths that may well invalidate the installed instrumentation.

Another application of ground-test experimentation has been often proposed is the need for a pre-flight operational validation of flight instrumentation in ground test facilities. While, on the surface, this appears to be useful, the practical aspects of the problem for high temperature instrumentation make such a test of questionable technical value. Some of these technical issues involve:

1. Defining the radiation environment between the "model" surface and the test-peculiar surroundings.

2. Incorporating sufficient structure about the instrument to correctly simulate the flight structure at flight-simulated test conditions.

3. Generating a clean, simulating test flow without the presence of test-induced contaminants.

Other technical solutions - primarily analysis techniques- should first be investigated and discarded as inappropriate.

The Complexity of the Materials System:

In past examples the instrumented materials system was homogeneous and thick enough to capture the thermal pulse for the duration of the flight experiment. The instrumentation of such systems presents no significant technological instrumentation problems. Instrumentation problems arise when systems goals of weight-efficiency conflict with the test goals to measure temperature and the infer heat flux on those materials. To illustrate, Figure 7 from a paper by Grallert and Keller [13] shows the unit weight of various TPS materials as a function of the wall temperature of that material. Representative metallic TPS candidates are lighter than the ceramic tiles of the Space Shuttle program as the surface temperature increases. The metallic concepts are superior at high surface temperatures but the challenge is to install thermal sensors within or onto these "...thin metallic foils..." and to infer heat transfer rates from those thermal measurements.

Figure 8 shows both the stiffened heat shield (the ceramic shingle design) that is being considered for the Hermes system and the metallic multiwell concept. The material of the structures as well as the thickness of the typical surface layer are shown. Recalling the dimensions of thermal sensors previously presented, installing any thermal sensor is difficult enough but inferring heat flux through thermal sensor measurements while not altering the thermal character of the surface being measured is certainly a challenge for high temperature aerospace application.

Apart from the thicknesses of the materials, the materials selected conduct heat across the section and radiate that heat both inward and back to space. Finally, the material fabrication process, particularly for coated molybdenum will make bonding of gages to coated materials impossible.

Seven advantages are noted for these advanced material systems which are indicative of advanced materials thinking; instrumentation installation is not one of them. Complex material systems are certainly a challenge to instrument if they can be instrumented at all.

CHALLENGES FOR HIGH TEMPERATURE AEROSPACE APPLICATIONS

I see several contemporary challenges in the development of measurement technology. These challenges will now be discussed in order.

CHALLENGE 1:

TO CAPTURE THE CHARACTER AND LOCALIZED PEAK VALUES WITHIN HIGHLY NON-UNIFORM HEATING REGIONS

The characteristics of highly non-uniform heating regions are (1) localized peaks in the imposed heating to the surface, (2) incomplete knowledge of the location of heating peaks due to real gas effects and (3) high thermal gradients along the surface of the material driven by the gradients in imposed heating.

Both ground test and flight test measurements are concerned first with identifying the level of the peak and secondarily defining the structure of the interaction. Either fields of individual point sensors are applied to define peaks by capturing the heating pulse whose location is guided by our incomplete knowledge of the distribution or survey testing is applied to map the entire interaction process in sufficient detail. Error in the measurement process are not random but a bias which produces lower data than actually present. These errors are due to: (1) the size of the heat sensor substantially larger than the spike in heating being measured, (2) the thermal model relating measured temperature to the imposed heating is incomplete and ignores characteristic temperature gradients along the surface or (3) the measurement span of the temperature sensor is inadequate to the measurement (significant for survey testing techniques).

The challenges that I observe are to: (1) develop a compatible system between the types and locations of physical sensors (thermal or heat transfer sensors) and the thermal model that uses these data to define a heat transfer distribution. In the case of highly peaked distributions, the resulting thermal model may be applied to a single sensor or to a sensor field as a group. The need is to define temperature gradients along the surface as well as through the material system. (2) to create thermal and heat transfer sensors that are properly scaled to the characteristic dimensions of the peaked interaction which implies very small sensors as well as tightly packed thermal sensors. (3) to efficiently manage the volume of data extracted from a field of either isolated sensors or temperature maps produced by various survey techniques and (4) to broaden the temperature span of survey techniques. The last two items will be discussed in later sections of this paper.

CHALLENGE 2:

TO MANAGE LARGE VOLUMES OF THERMAL INSTRUMENTATION IN ORDER TO EFFICIENTLY DERIVE CRITICAL INFORMATION

On the ground and in flight substantial amounts of raw data are generated that must be managed in acquisition, storage and manipulation. Both the volume and complexity of test data are increasing today. Single test runs can acquire over 100 million data elements. These large volumes of data must be reduced in an efficient but complete manner. This challenge considers the efficient transformation of data into information.

Ground Test:

Survey sensing techniques now produce digital data that both increase the effective numbers of sensed points and multiply the volume of data to be manipulated. Each "frame" of data can deliver 262 thousand data values and frames of data must be taken rapidly and sequentially to accurately map the test surface. Video refresh rate of 30 to 60 frames per second are commercially available and the test duration of 1 to 10 seconds are typical. Each run then can generate as many as 100 million data values. Admittedly, a small percentage of these values merit full reduction and much can be gleaned from even a partial reduction of these data but the information contained in these data may require a rather sophisticated reduction for complex interacting flows. Consider, for example, the evaluation of conduction losses along the surface caused by highly non-uniform heating of less than perfect insulators; the required information is captured in the data but the computational effort to reduce the data is not trivial. Numerical analyses by Dorignac and Vulliemre [14] demonstrate that these problems exist even for ideal test surfaces and must be considered in the data reduction and analysis.

Initial, zero order, data reduction of ideal insulative surfaces approximates the actual data reduction equation will a 1D closed form relationship. Using this technique, each data run requires about 4 minutes of computation on an Intel 80386 based machine. Balageas [15], who has written extensively on survey techniques at ONERA, points out that:

> " The snare to be avoided, and it is not a small one, is to keep from being swamped by the flood of data generated by this technique. Processing methods will have to be developed that are thrifty in computational time and memory space, but sophisticated enough that what is of interest can be distinguished from what is secondary or even useless, with user-friendly graphic postprocessing".

Flight Test:

Managing large volumes of high-frequency flight data with reasonable downlinks requires on-board information processing. This was observed over a decade ago by Galleher [13] who stated that

> "...as flight performance measurements become more demanding...more and more on-board processing and data compression techniques will have to be devised

> that are acceptable to both system designers and experimenters..."

This paper quote referred to the flight test of BMO's **A**dvanced **M**aneuvering **R**eentry **V**ehicle, AMARV, on which 60, in-depth thermocouple plugs (similar to those on the FIRE and RE-ENTRY F programs were used. In each of these plugs 3 to 4 high-temperature Tungsten 5% Rhenium/ Tungsten 26% Rhenium thermocouples were placed.

CHALLENGE 3:

TO ACCOMMODATE THERMAL SENSORS INTO PRACTICAL FLIGHT STRUCTURES

Our history is rooted in the thermocouple measurement of temperatures within blocks of material. Our future requires the use of whisker-like sensors to define heat transfer within shim stock. Wind tunnel heat transfer, in the past, was inferred using "thin skin models" having backface mounted thermocouples. Today we can directly measure temperature gradients within that thin skin.

The challenge is to develop more highly integrated sensors into increasingly complex and non-uniform structures without disturbing the natural heat paths of these material systems. The challenge is to transform measurements from add-on to designed-in and, in so doing, to approach the concept of a smart skin.

CHALLENGE 4:

TO BROADEN THE CAPABILITIES OF THERMAL SURVEY TECHNIQUES TO REPLACE DISCRETE GAGES IN FLIGHT AND ON THE GROUND

Thermal survey techniques have been used for 30 years to observe heat patterns on the ground and in flight. Initial wind tunnel applications of temperature paint were made in the mid 1960's. Flight test examples are documented on the X-7A [16] and the X-15 flight programs (figure 9) where shock interactions and boundary layer transition were observed. The reasons for those survey techniques of the 1960's still exist and although the quality and sophistication of techniques have improved, there are several challenges which remain.

Survey techniques emcompass older irreversible temperature

sensitive paints and newer reversible surface coatings such as thermographic phosphors and infrared (IR) measurements. The newer survey techniques generate large amounts of data that must be acquired, stored and reduced. A challenge is the ability to handle those data efficiently. On the positive side, the effective gage density s increased by a factor of 500 and all that data is "recorded" even though only a fraction may initially be reduced.

Recall that these survey techniques measure surface temperature and not heat flux. Heat flux is derived by a thermal model which relates measured surface temperatures to the causal aerodynamic heating. Ideal thermal models are designed on the basis of one-dimensional dissipation of aerodynamic heating with the structure. More general applications of this technique will require multi-dimensional inverse analysis which is more computationally intensive.

GROUND TEST APPLICATIONS:

There are several un-resolved needs for survey testing in ground tests facilities. Two examples are: (1) the need to define the complex boundary layer transition front on 3D models during both aerothermodynamic studies and the longer duration studies of overall forces and moments. (2) the need to more-fully understand the complexities of internal flows dominated by shock interactions. The challenges are to broaden the range of temperature response, eliminate false signals and manage the volume of derived test data.

The range of heat transfer measurable with survey techniques is limited. These techniques are only partially useful in hypersonic facilities. The limitations of the techniques can be observed in the measurement of shock interaction regions where high gradients and an order of magnitude variation occurs between the peak heating and the undisturbed heating level. Survey techniques with limited band-width bias the heating in interaction regions to lower peak heating than are actually present. [17]

Transition Front Measurements:

While the accuracy of CFD techniques at high Reynolds numbers critically depends on defining the transition front as an experimental input and while that front is difficult to predict for highly three-dimensional flows, few, if any experimental studies map transition fronts as an input to CFD validation. Most force and moment studies measure only gross vehicle forces and never provide the instrumentation to state whether the boundary layer is laminar, turbulent or, most likely, some of each. Matthews et al [1] observes that:

> "...unfortunately, the ability to predict the boundary layer state accurately - laminar versus turbulent - continues to elude the aerodynamicist".

The fundamental technology exists to observe the state of the boundary layer. The challenge is to implement that technology in the required production test facilities.

Internal, Shock Interaction Flows:

There are internal flows dominated by shock interaction processes. In none of these can the thermal effects be adequately defined through any practical number of discrete thermal sensors. Survey techniques are required to satisfactorily measure the thermal loads. This poses the challenge of placing the survey sensor within the restricted passage being measured and the challenge of managing the difficult, technique-peculiar errors involved in these measurements.

FLIGHT APPLICATIONS:

The most striking, modern application of survey techniques applied to flight vehicles is/was the use of and infrared (IR) sensor installed on the vertical tail of the Shuttle vehicle, Columbia. The overall features of this technique, its placement on the Shuttle and the views from the infrared camera are shown in figure 10.

Originally conceived in the 1970's and conceptually simple in design, this experiment has been difficult to install in the orbiter and successfully use to generate complete data. The specific technical difficulties encountered have been defined as following:

- -A protective plug over the windows that wouldn't jettison when required
- -Inadequate cooling of the windows as heating levels increased during re-entry into the sensible atmosphere.
- -Erratic operation of the camera scanning mechanism.
- -Massive amounts of raw data that require efficient reduction

Figure 11 demonstrates the high temperatures that occur on the observation windows; temperatures that can only be numerically deduced from an extrapolation of inner pane measurements through the use of thermal modelling techniques; a supplementary challenge in thermal measurements.

Apart from these technical issues, there is a more difficult and fundamental challenge: the interaction between the inevitable "tweaking" of a complex experiment such as this and the operational imperatives of a schedule-driven, multi-goal flight vehicle such as the Space Shuttle.

The problems are not fundamental in nature but true engineering problems of creating a flight test measurement system that works.

CHALLENGE 5:

TO PROVIDE SUPPORTING INSTRUMENTATION CONDUITS WHICH CONNECT THE MEASUREMENT POINTS TO THE THERMALLY CONTROLLED DATA ACQUISITION SYSTEM

Data acquisition requires hardware from the surface measurement point back to an environmentally-stabalized location on the vehicle where data transmission or on-board recording can be accomplished.

A significant challenge in making flight measurements is in implementing this chain of hardware. The material systems of the flight vehicle as well as the duration of the flight are important considerations. The Space Shuttle, by virtue of its cold structure concept, attained a benign thermal environment within inches of the point of measurement but other possible flight configurations may not. Hot structures would create a far more difficult thermal instrumentation situation.

The X-20A (Dyna Soar), a hot structure flight vehicle concept, required 3700 feet of 1800 deg F wire and connectors to connect the 750 sensors to be placed on each flight test system. The wire of that day consisted of inconel tubing 0.090 ins OD, magnesium oxide electrical insulation inside the sheath and two electrical connectors.

These high temperature conduits can include simple thermocouple leads, regulated power lines and, possibly, fiber-optic lines. Hellbaum [18] proposed "...platinum films laid down on a substrate of alumina..." to define transition for flights at lower Mach numbers. Platinum films are powered resistance thermometers that require a constant current flow through the measurement film. Similarly, high temperature microphones and photodiodes were suggested for this function. Each of these gage types is electrically-powered and the difficulty of delivering that power increases as the temperature of these conduits increases.

In the 1990's these lines may include fiber-optic bundles connected to surface temperature sensors or fiber optic conduits to interrogate or observe the thermal characteristics of a remotely located surface of interest.

Apart from conduits there is also a need for high temperature connectors to facilitate instrument replacement as well as in-line amplifiers to boost signal strength, regulators to provide precise power to surface measurements and even cooled lens systems to direct the images remotely sensed.

All of these problems are accentuated if flight duration increases and/or thermal conductors are selected as material systems. The Space Shuttle was one model of TPS and not the unique technology demonstration for future systems which must be instrumented.

CHALLENGE 6:

TO DEVELOP A CLASS OF "VEHICLE TENDING" THERMAL SENSORS TO ASSURE THE INTEGRITY OF FLIGHT VEHICLES IN AN EFFICIENT MANNER

There have been consistent thermal problems over the past 30 years caused by uncontrolled internal flow due to leakage of boundary layer air through the flight structure. Flight configurations are not the homogeneous structures but mechanically built-up of many separate elements held firmly in place with Sauerizen (R) or similar indispensable materials. They present many possible internal flow paths which are driven by large hypersonic pressure differences. McBride, 1983 [19] and 1986 [20], termed this flow "sneak flow" as he outlined Space Shuttle experience. He concluded (1) sneak flow is important for any reusable thermal protection system (TPS) large or complex enough to require interfaces and (2) it is difficult to make quantitative predictions of sneak flow effects. Because the problem is severe and the analysis is complex; because every hypersonic flight system has demonstrated sub-surface heating and the best "sensor",to date, is a discolored surface, the challenge is to create a new class of vehicle tending thermal sensors.

Vehicle-tending thermal "gages" are a new class of sensors which do not produce a point measurement of either temperature or heating rate but develop a sense of leakage through regions which, ideally, would have none. These developmental sensors monitor the health of joints and gaps determining the severity of imperfect seals.

The X-15 nose wheel door created a gap that allowed hot, boundary layer air to enter the wheel well and destroy the instrument lines located there. Figure 12 is a photograph of that situation. The X-15 only flew at Mach 5. Figure 13 from McBride indicates the Space Shuttle penetrations including doors, gaps and coves that all could produce potentially serious sub-mold-line flows. These regions are caused by the aero-thermo-elastic effects of hypersonic flight which cannot be simulated in ground test. They are driven by pressure differences across the produced gap and require an understanding and modeling of the sub-surface flow paths.

McBride, 1983 [19] observed that "...penetration thermal instrumentation (on the Space Shuttle) was only adequate. More sensors were required at difficult-to-predict environment locations. Available DFI (developmental flight instrumentation) should have been more concentrated. MORE EXTENSIVE USE OF PASSIVE TEMPERATURE-SENSITIVE DEVICES COULD HAVE BEEN MADE." Of the approximately 6000 recorded measurements (about 2000 of them temperature) available on the orbital flight test orbiter, 627 were dedicated to TPS elements BUT ONLY 90 OF THOSE ARE STRICTLY RELATED TO PENETRATIONS. Table I of AIAA 83-1486 outlines in greater detail the TPS penetrations and their instrumentation.

These regions are characterized by a large seam on the vehicle which could leak at any location on that line or in that region. While point sensors could be placed in such regions, the question is where to place them and what kinds of data are required from them to define leakage.

What we require is an overall impression of whether leakage occurred and whether that leakage was significant. The challenge is to develop a sensors that achieve these goals rather than to measure temperature. The challenge is to produce better coverage with fewer sensor assignments.

One conceptual recommendation is to use ablative (or phase change) overcoat surrounding a distributed sensor. The sensing surface would create a signal proportional to the amount of the overcoat removed as a result of the temperature exceeding a defined threshold level. Perhaps an ablative coating could be applied over a fiber optic bundle through which light is being transmitted and received. In a sense, this is another application of a smart skin. This technique was also discussed by Measures, 1989 [21].

CONCLUSIONS:

The conclusions I draw from this material are:

1. The discussion of high temperature, high heat flux

measurements is contingent upon the details of the experiment proposed. Those details are (1) the type, complexity and dimensions of the material system (2) the duration of the flight experiment and (3) the selection of "add-on" or "design-in" heat gages for those measurements.

2. The technical challenges in those heat flux measurements are fundamentally two: (1) the construction of ever smaller physical thermal sensors and (2) the development of efficient inverse thermal models that relate those thermal measurements to causal heat transfer.

3. Six areas of technical challenge have been postulated. These treat the heat flux measurement problem in a broader context; the delivery of information on thermal questions to a customer. They are intended to start a discussion concerning thermal measurement technology.

Finally, a conclusion implicit in this review paper is that thermal instrumentation is an enabling technology that makes possible both flight test and ground test programs. Instrumentation should not be an after-thought of a larger systems-related program or relegated to the catalog purchase of "proven" devices (proven on the last flight vehicle). If considered early and funded adequately, instrumentaton can enable tests otherwise impossible and/or reduce the cost of even routine tests substantially. More and more temperature measurements must be designed into the developing structural component and must be considered as an integral part of that development process.

Appendix A
Add-On Thermal Gages Applied to the Space Shuttle Hardware

A. Space Shuttle External Tank

Commercially available Schmidt-Boelter and "pill-type" heat flux sensors were installed in the insulative foam surface of the external tank. These metallic gages present a non-uniform surface temperature to the boundary layer; the gages being relatively cold and the surrounding insulator being hotter. Praharaj [1], reviewing the experience, noted that severe temperature mismatch was present producing "...a large measurement error in a convective flux environment...". "...The underprediction of these island measurements was 100% or more in the peak heating region...". He further notes that these effects can be "...successfully factored out of the flight data in undisturbed regions..." but that "...temperature mismatch effects in the interference regions are not dealt with in the existing literature and consequently were not factored out of the flight measurements...". Praharaj further notes that "...the choice of sensors for future space vehicles must consider this effect (temperature mismatch) and efforts must be made to reduce the temperature mismatch effects on the measurements...". Finally, Praharaj notes that "...temperature along with heat flux should be measured so that one can be derived from the other. This would help eliminate erroneous readings in a much easier fashion..."

The analysis of undisturbed temperature mismatch used by Praharaj was due to Westkaemper [22] following the expression:

$$\frac{\bar{h}(W,L)}{h(W,0)} = F(L/W)\,\frac{(T_{w_1}-T_0)}{(T_{w_2}-T_0)} + H(L/W)\,\frac{(T_{w_2}-T_{w_1})}{(T_{w_2}-T_0)}$$

where: $F(L/W) = \frac{5}{4}\,\frac{[1-(L/W)^{0.8}]}{(1-L/W)}$

and: $H(L/W) = \frac{5}{4}\,\frac{(L/W)^{0.8}}{(1-L/W)}\,[(W/L)^{0.9}-1]^{8/9}$

where: L is the running boundary layer distance to the gage
W-L is the gage width dimension
T_{w_1} is the wall temperature of the surrounding material
T_{w_2} is the gage surface temperature
T_0 is the recovery temperature

B. Space Shuttle Orbiter

Little has been written concerning the use of Schmidt-Boelter gages on the orbiter. Figure A-1 is a sketch of the installation. The sketch pre-dates the flight and may not represent actual flight hardware. Figure 4 of this paper shows data from these gages and indicates that problems exist with these gages relative to imbedded thermocouples to which they are compared.

C. General Comments:

Hornbaker and Rall [23] presented an excellent review article on this phenomenon as on many aspects of thermal instrumentation. The reader would do well to review this and the several other review papers which were written by these authors.

APPENDIX B

A REVIEW OF THERMOSENSING ELEMENTS AND THEIR USE IN AEROSPACE APPLICATIONS

The standard thermo-sensing elements used to measure temperatures were/are wire thermocouples. There are a series of these thermocouple material combinations available with which to measure temperatures at different temperature levels. Table II from Moffat [24] shows representative material pairs:

Material Designation	Temperature Limit, deg R	Output, mv/ 100 deg F
Chromel-Alumel	2290	2.20
Platinel	2650	2.20
Platinum-Rhodium	3730	0.43
Tungsten-Rhenium	4630	0.76

Note that Tungsten/Rhenium thermocouples must be placed in an inert atmosphere. "The main problems were centered around embrittlement of the Tungsten leg (of the thermocouple) and oxidation".
Tungsten thermocouples have been successfully used in high temperature flight tests by applying sheathed configurations.

High temperature gages also suffer from a progressive de-calibration (of the thermocouple) with time due either to changes in the composition of the material or changes associated with grain growth and the annealing out of the residual cold work from fabrication".

The progressive oxidation problems are eliminated through sheathing the thermocouple in a protective material. Very small sheathed thermocouples are currently available with outside diameters as small as 0.008 ins (0.2mm). One concern with thermocouple sheathing is understanding precisely the location of the thermocouple junction within the sheath. This problem, annoying for the measurement of temperature, is a critical deficiency when sheathed thermocouples are a part of a built-in heat transfer measurement system.

As small as these devices have become, the general rule of thumb is that the thermocouple assembly should have an outer diameter roughly 20% of the thickness of the material into which it is placed to avoid excessive conduction down the thermocouple wires. That would place the minimum material thickness at 0.040 inches or greater; far thicker than anticipated applications shown by Grallert and Keller.

This problem can be circumvented by: (1) avoiding the problem by using a non-conductive thermo-sensor material (such as a fiber-optic-based thermal sensor) or (2) integrate the sensor into the thermal analysis of the material; either directly through an inverse technique containing the actual structure elements are approximately through correction factors developed to account for conduction losses down the wire.

As the temperature to be measured increases, the selection of thermocouple materials decrease as well as the output sensitivity of available thermocouple materials.

Newer thermocouple configurations are plated rather than wire. The output sensitivity of these plated thermocouples are about half that of the corresponding wires. Techniques are, in principal, available to plate extremely small and thin single and multiple thermocouples on "selected" substrate. Plated thermocouples have been studied by several groups: the Vatell heat flux gage is one attempt to use this technology in a fabricated heat gage. The major challenge is broadening the domain of applicability of these plated thermo-sensors.

REFERENCES

1. AIAA Paper 91-1441
2. AIAA Paper 84-1750
3. ISA Paper "The Development and testing of Development Flight Instrumentation for the Space Shuttle Thermal Protection System", ISBN 87664-407-8
4. Aerotherm Corp. Report 75-135, 1975
5. NASA CR 179074
6. NASA CP 2283
7. AIAA Paper 83-0482
8. AIAA Paper 84-1763
9. ASME Paper 81-ENAs-13
10. AIAA Paper 84-0224
11. AIAA Paper 82-0944
12. AIAA Paper 87-1508
13. AIAA Jour. Aircraft, V 28, Nr. 6, pp 410-416
14. Rech. Aerosp. 1991-1, pp 67-71 (English Version)
15. Rech. Aerosp. 1991-4, pp 51-68 (English Version)
16. WADD TR 60-567, p 239
17. ASD TR 92-xxx (to be published)
18. AIAA Paper 88-4651A
19. AIAA Paper 83-1486
20. AIAA Paper 86-1261
21. "Smart Skins with Nerves of Glass", Progress in Aerospace Sciences, V 26, pp 289-351
22. Jour. Aero. Sci, 11/61, pp 907-908
23. ISA Preprint 8.3.63
24. Moffat, R.J., International Heat Transfer Conference, 1982

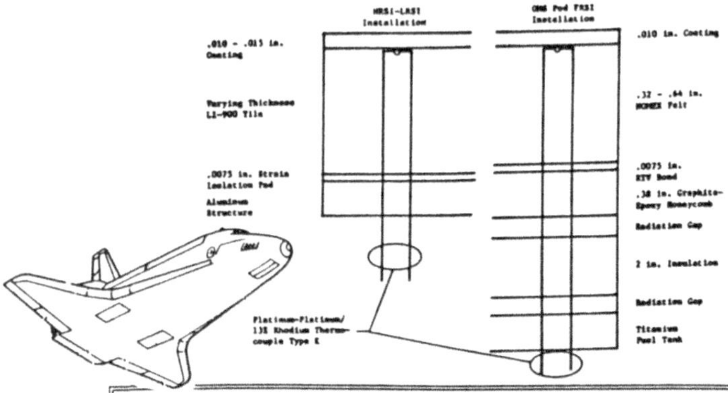

Figure 1, Space Shuttle Thermal Instrumentation

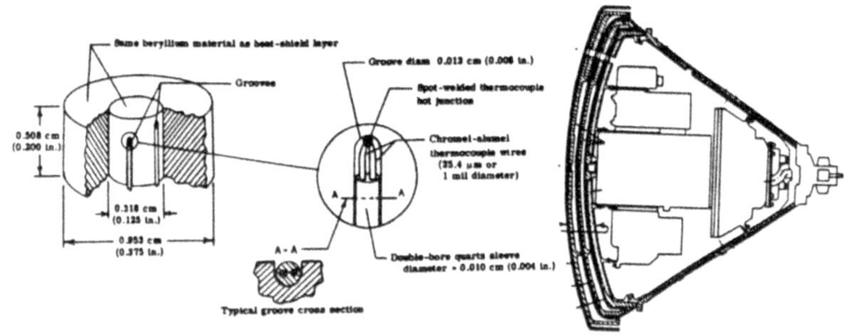

Figure 2, NASA FIRE Project Thermal Instrumentation

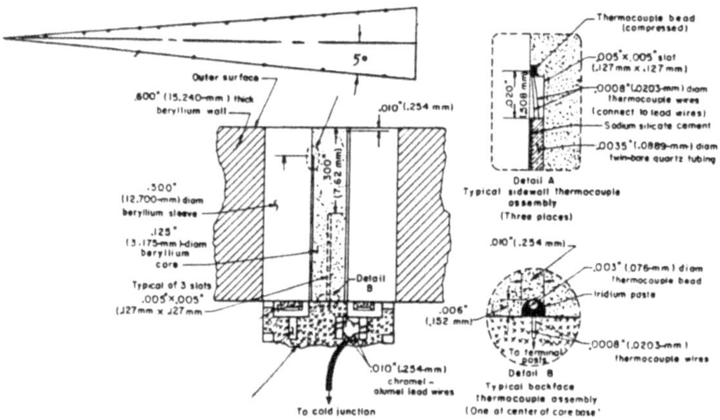

Figure 3, NASA Re-entry "F" Project Thermal Instrumentation

Gross difference is due to uncertain depth of thermo-couple below the outer mold line.

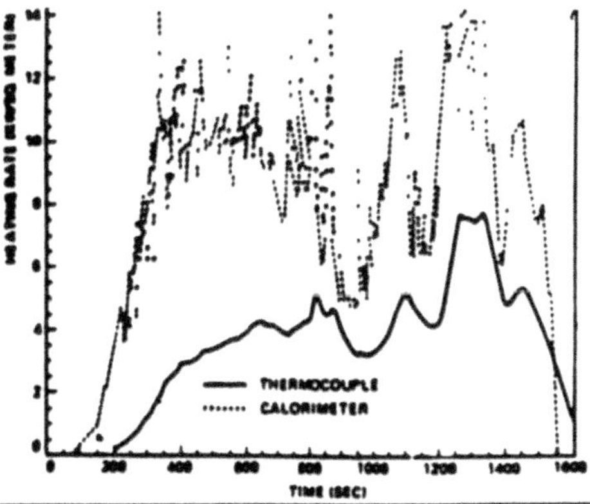

Figure 4, Difference Between Calorimeter and Thermocouple-Based Heat Transfer Inference

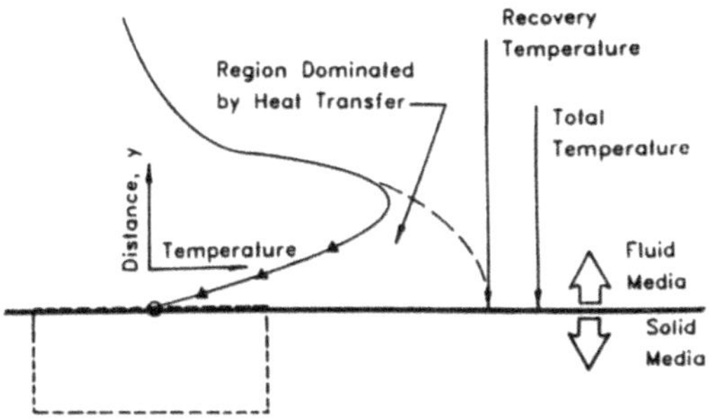

Low temperature heat flux relates the convective input with conductive dissipation within the structure

High temperature heat flux relates the convective input with conductive dissipation within the structure and radiation relief to space

Figure 5, The Process of Heat Transfer Rate Inference

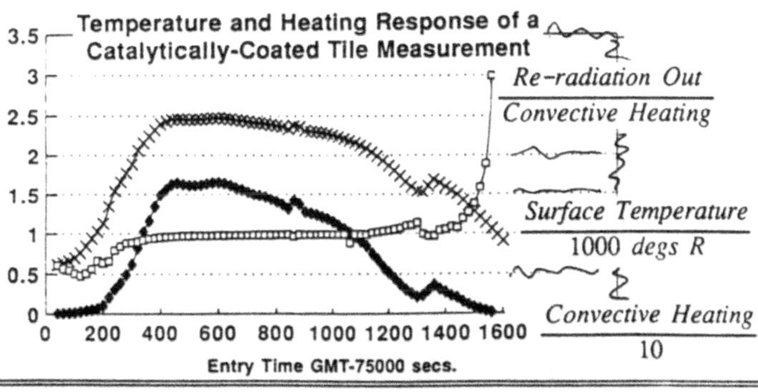

Figure 6, Temperature and Heating Response of a Catalytically- Coated Tile Measurement on the Space Shuttle

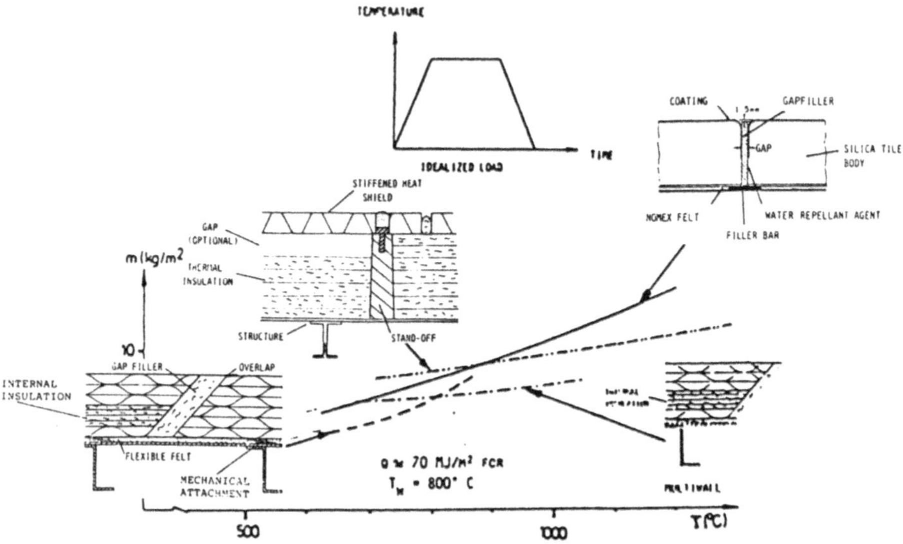

Figure 7, Trend of TPS Mass Per Area vs Temperature for Representative TPS Candidates from Gallert and Keller

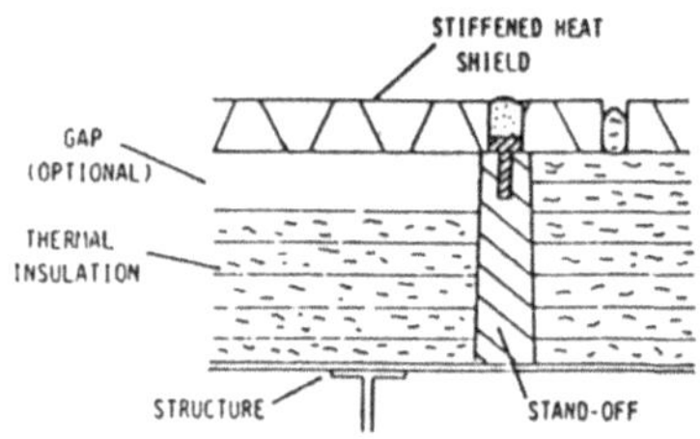

Stifferend heat shield is a rigid ceramic shingle of carbon fibers in a silicon carbide matrix or silicon carbide fibers in a silicon carbide matrix. Shield surfaces are, typically 0.020 ins thick.

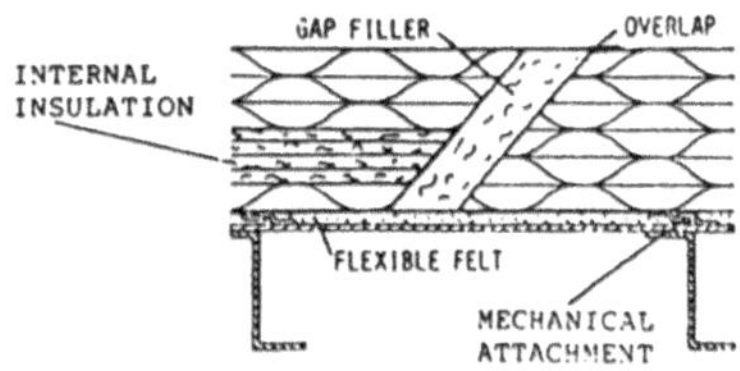

Metallic multiwall concept is either nickel or cobalt based alloy for use up to 1000 degs C and molybdenium based alloy for use up to 1300 degrees C. Typical thickness is 0.002 to 0.004 ins.

Figure 8, Details of the Scale of Proposed TPS Systems from Grallert and Keller

Figure 9, Early Use of Temperature Sensitive Paint on the X-15

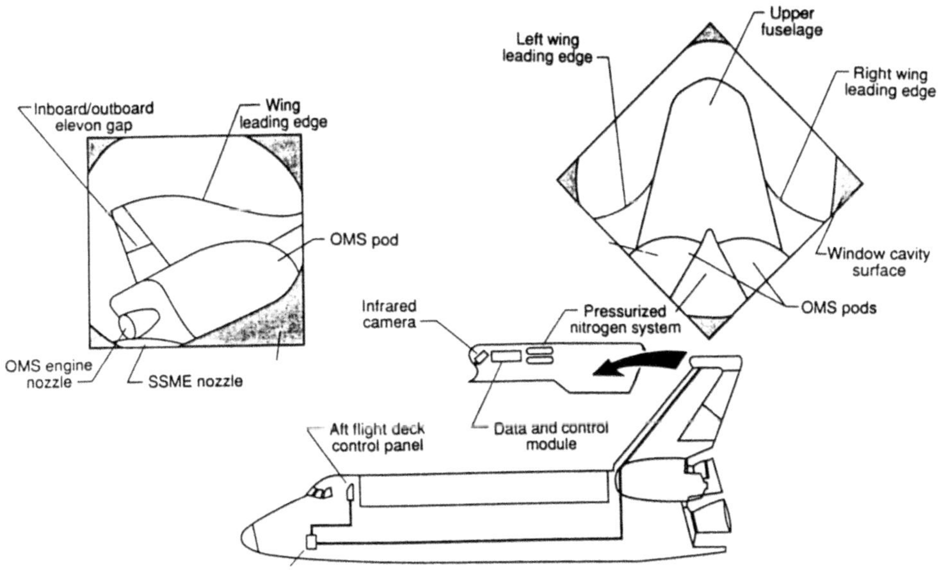

Figure 10, Overall View of the IR Flight Experiment on the Space Shuttle Flight Vehicle, Columbia

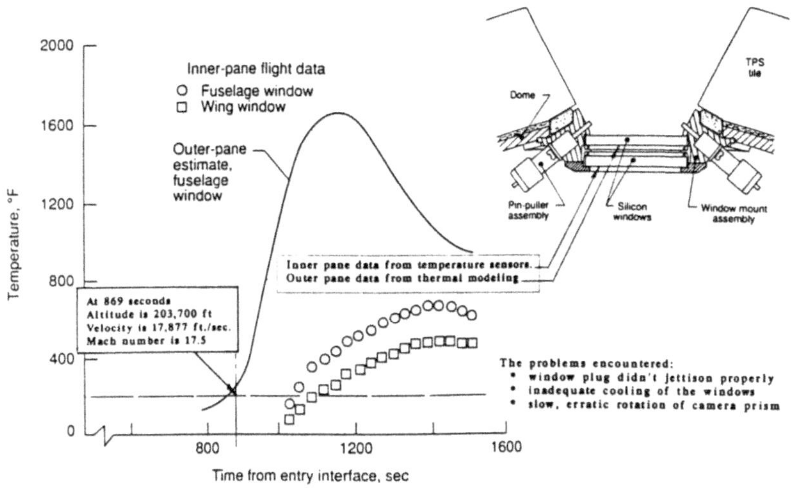

Figure 11, Discussion of the Problems in the IR Flight Experiment

Figure 12, Example of "Sneak Flow" in the X-15 Nose Wheel Cavity

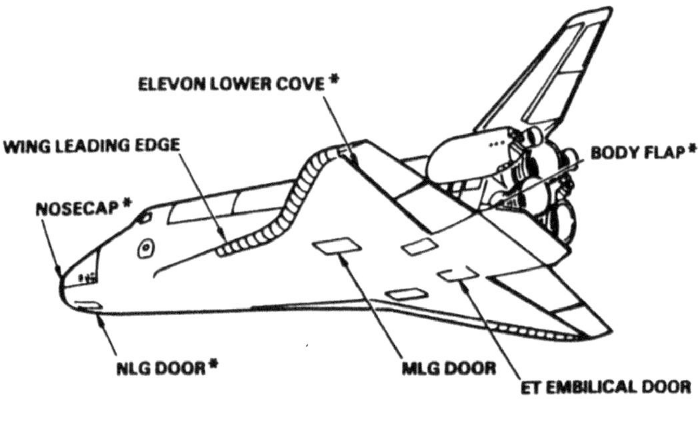

Figure 13, Penetrations on the Space Shuttle Orbiter

HEAT TRANSFER MEASUREMENT TECHNIQUES USED OR IN DEVELOPMENT AT ONERA/CHALAIS-MEUDON

H. CONSIGNY, Y. LE SANT, F. BOUVIER, P. BAUMANN, M.C. MERIENNE
Experimental Aerodynamics Division
Office National d'Etudes et de Recherches Aérospatiales (ONERA)
92320 Châtillon - France

ABSTRACT. This paper presents briefly the various techniques currently used in the ONERA/Chalais-Meudon conventional hypersonic blow-down wind tunnels for measuring heat transfer rates. A special attention has been given to the advantages, limitations and difficulties associated with each of these techniques as well as to the related improvement and development work.

1. Introduction

The determination of a heat transfer rate has always been considered as a difficult task since this physical quantity can only be measured through its direct consequence, i.e. a temperature change. There exist several types of heat transfer gages corresponding to various principles and technologies but the fact that most hypersonic facilities have a maximum running time of the order of a few seconds makes that steady state gages have to be eliminated to the benefit of gages especially designed for transient thermal phenomena (i.e. governed by the heat conduction equation with stationary or varying boundary conditions). Despite this limitation, a wide choice still subsists and it is clear that each laboratory has to select a set of techniques in relation with its own facility characteristics and the particular problems it has to face with.

At ONERA, in the early sixties, the thin film and the thin skin techniques were introduced for single point measurements while the thermosensitive paint technique was developped for semi-quantitative thermal mapping purposes, [1]. In fact, it now turns out that the last two techniques were extensively used while the thin film technique all things considered, did not receive much attention and development effort.

Whithin the framework of the Ariane and Hermes new context, it was decided in 1987 ([2]), to re-introduce the thin film technique, to apply the surface thermocouple concept (coax-gages), to improve the thermosensitive paint method and to evaluate a "new" survey technique based on the use of luminescent compounds. The purpose of this paper is to describe the work undertaken these last few years at the Experimental Aerodynamics Division on these various topics. The infrared thermography technique will not be mentionned in this paper since the development work in this area is mainly conducted by the Physics and Systems departments.

2. Single point surface temperature techniques

The determination of heat transfer rates in wind tunnels from the measurement of model surface temperature changes has been known for a long time ; the main interest of this method is to allow

A. Boutier (ed.), New Trends in Instrumentation for Hypersonic Research, 437–447.

the study of extremely fast phenomena and for this reason it was extensively used in very short (10 ms, eg. shock tubes) or short (500 ms, eg. isentropic compression tubes) duration facilities. The possibility of using such a technique in facilities having a running time of several seconds like the ONERA R2Ch/R3Ch/R5Ch/S4MA blow-down wind tunnels was found attractive since the combination of a fast response time and a relatively long test duration would make it possible to obtain data for varying conditions (like for example model pitch sweeps) and consequently to improve greatly data productivity. The perspective of collecting results for different wall temperatures distributions as encountered during a single long test was also interesting.

2.1. PRINCIPLE AND REALISATION

The original technique (summerized in ***Fig. 1***) is based on the very simple theory of the one-dimensional flow of heat into a plane "semi-infinite" homogeneous solid having constant thermal properties. The relevant theory can be found in several references (e.g. [3]) and will not be recalled here. The model surface temperature rise may be measured by several ways but the most common gages are platinum thin-films deposited on insulating substrates or surface thermocouples installed in metallic models.

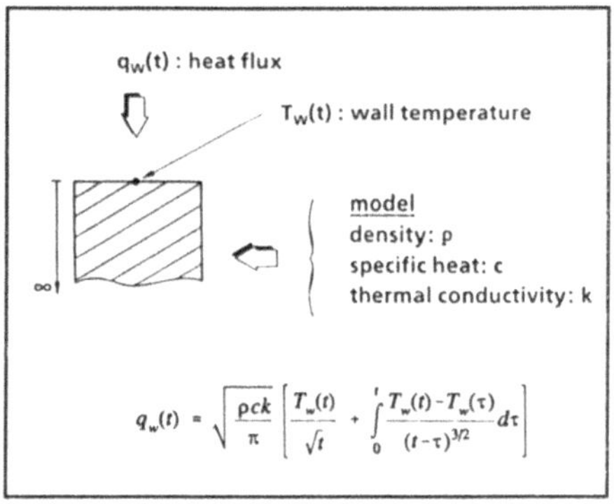

Fig. 1 : Surface temperature technique : basic principle -

The thin films used at the Experimental Aerodynamics Division are "hand made ". This rather classical method proved in fact to be efficient since it does not require sophisticated pieces of equipment and can be set up by people close to the wind tunnel which has the advantage to reduce equipment delays. An example of a simple model made from Macor® machinable ceramic is shown ***Fig. 2***. Thin film gages have a high sensitivity and a response time of several kHz, [2]. Moreover, the relatively low value of diffusivity of Macor makes them particularly suitable for measurements on thin and/or curved model parts as well as to reduce lateral conduction effects. The main disadvantage of this technique remains in the difficulty of machining complicated models and in the fact that the gages are likely to be destroyed in very severe environments although the systematic use of a very thin silica coating may limitate this problem.

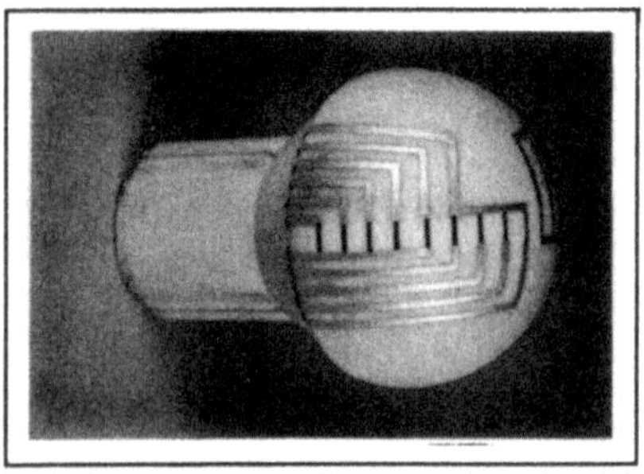

Fig. 2 : Macor sphere equipped with platinum thin-films -

The second kind of gages used at ONERA is shown schematicaly in ***Fig. 3***. The choice of the 17-4 PH stainless steel is justified by the fact that the thermal product $(\rho ck)^{1/2}$ of the three materials involved are close to each other so that the model wall can be, at least in a first approximation, considered as homogeneous (this point will be discussed later). As shown in ***Fig. 3-a***, a second chromel wire may also be installed in the constantan jacket in order to measure the gage back-face temperature ; this arrangement is necessary only for thermal properties calibrations (see below) since for actual wind tunnel tests, a nul heat flux boundary condition on the back-face is sufficient to derive the heat flux at the front-face from its temperature rise measurement. In order to increase the number of measurement points on a given model and also for reasons that will be given in section 2.4., the arrangement shown in ***Fig. 3-b*** which corresponds to a constantan jacket installed in a constantan model, is recommended. Coax-gages are very robust and don't need any specific temperature calibration. This technique is considered as very reliable but its main disadvantage lies in its relatively poor sensitivity. A picture of an Hermes model tested in S4MA and equipped with approximately 200 gages of this type is shown in ***Fig. 4***.

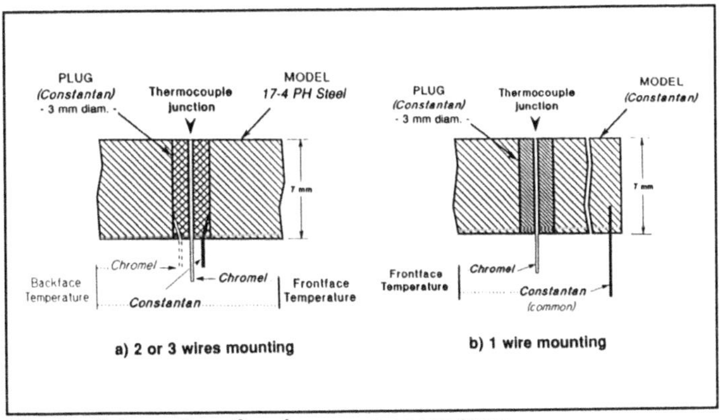

Fig. 3 : Schematics of various coax-gages types -

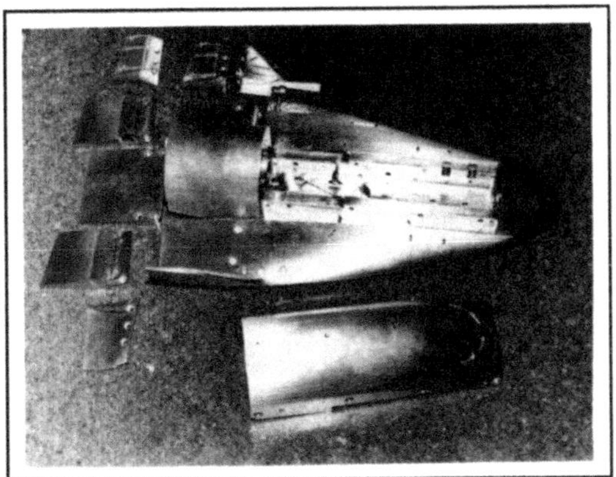

Fig. 4 : Hermes model equipped with approximately 200 coax-gages -

2.2. DATA REDUCTION PROCEDURES

The data reduction procedure associated with the basic semi-infinite gage is rather straightforward. Unfortunately, in most cases, significant departures from this ideal situation are encountered especially for "long" running times. Finite thickness and curvature effects as well as lateral conduction effects are likely to be important on metallic models equipped with coax-gages (which moreover violate the homogeneity assumption) and the high temperature usually experienced by insulating Macor® ceramic substrates raises the problem of the effect of temperature on the thermal properties (see section 2.3). In the case of an assumed constant heat flux, simple analytical correction formula can be derived to take into account thickness, curvature and varying thermal propertie effects ([4]), as long as these effects are supposed to remain "small".

The technique which consists in cumulating approximative corrections to cope with real situations is questionable and the need for a more general data reduction procedure is obvious. This new procedure (see ***Fig. 5***) described in details in [5] and [6], is based on the solution on the 1-D non-linear conduction equation by a finite volume method and can deal either with plane, cylindrical or spherical surfaces. This routine, which is now in use for standard wind tunnel data reduction, is also a valuable tool for the instrumentation designer (gage behaviour modelization and analysis, thermal property calibrations cheks, etc...).

Unfortunately such a data reduction procedure does not solve all problems and the example presented in ***Fig. 6*** gives a strong evidence of the magnitude of the lateral conduction effects which are still to be accounted for. In fact, wall curvature leads always to flow pressure gradient and hence to large heat transfer spatial variations so that, in some conditions, it appears unrealistic to try to improve the data quality by taking into account only one of these two effects. In some cases, compensating effects are such that the simplest data reduction gives the best results.

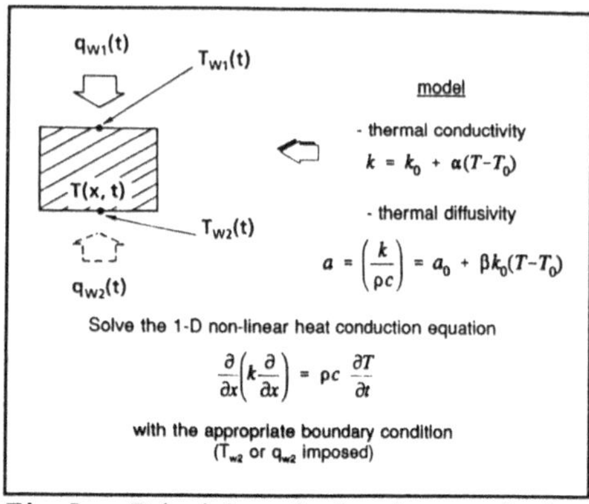

Fig. 5 : Principle of the improved data reduction procedure -

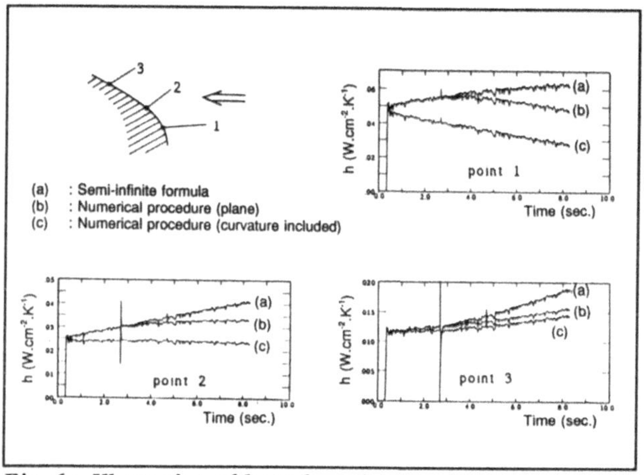

Fig. 6 : Illustration of lateral conduction effects -

The consideration of lateral conduction effects would require, in the most general case, the use of a 3-D conduction code which is far beyond the present capabilities of every day testing. It is shown in [6] and [7] that a local approach can be derived. However, it must be pointed out that with such a method, the availability and the quality of detailed boundary conditions is of a crucial importance ; in the case of discrete intrumentation, this problem is particularly critical.

2.3. CALIBRATION OF THERMOPHYSICAL PROPERTIES

It is essential to apply a reliable calibration technique of the substrate or gage thermophysical properties since the accuracy on these data influences directly the accuracy on the final heat flux results.

For the simple semi-infinite model, it is sufficient to calibrate the so-called thermal product $(\rho ck)^{1/2}$. This can be easily done by exposing such a gage to a known heat flux step using a well calibrated source and by recording the front surface temperature time history. Identification of measured and calculated temperature rises gives $(\rho ck)^{1/2}$.

When the more elaborate data reduction procedure of ***Fig. 5*** is used, one has to determine both the thermal diffusivity (a) and the thermal conductivity (k). If constant a and k are expected, a well known method consists in considering the one-dimensional flow of heat into a perfectly insulated slab of thickness l exposed to a constant heat flux q at its front surface. It can be easily demonstrated that the knowledge of l, q and the measurement of the front and back surface temperatures rises gives a and k.

The case of varying thermal properties is rather more complicated to deal with, since the non-linear heat conduction equation has to be considered. To solve this problem, Maulard ([4]) suggested to use a perturbation method to solve analytically the conduction problem in the slab as considered before. With the assumption of a linear variation of a and k with temperature (see ***Fig. 5***) it was further demonstrated by Maulard that for a Fourier number larger than 0.5, both front and back temperature rises are parabolic in terms of time and their difference varies linearly with time. Identification of measured temperature histories with theoretical ones allows to determine a_0, k_0, α and β.

Calibration tests based on this method were conducted for both coax-gages and thin films gages. Some results reported in [8], as well as unpublished data, show that there still remains an uncertainty on the effect of temperature on properties of Macor®. As mentioned in [9], an effect of the firing temperature may also be suspected. For this reason, it is recommended to run a calibration on every new material batch.

2.4. EXAMPLE OF GAGE NUMERICAL SIMULATION

Modern numerical techniques applied to conduction problems can be of a great help to aerothermal instrumentation specialists in order to understand the behaviour of heat transfer gages, to analyse any departure from an ideal situation and also to quantify the accuracy of the data reduction procedures.

The example presented in this section is related to the study of a coax-gage using a 2-D finite element conduction code. The problem consisted in analysing the influence of thermophysical properties in the data reduction procedure used for wind tunnel testing. As shown in ***Fig. 7***, a simplified coax-gage, supposed to be made from an homogeneous constantan jacket was represented in his 17-4 PH steel environment (it can be considered in a first approximation that the thin chromel wire does not greatly contribute to the conduction process). A constant heat transfer rate was applied to the upper surface during a simulated 10 sec. test and the computed temperature rises were obtained at any nodal point of the gage with the 2-D code. The calculation was performed with the known diffusivity and conductivity values of constantan and 17-4 PH steel. The temperature elevation at the point where the thermocouple junction is supposed to take place, was then processed through the 1-D data reduction procedure as if it was measured during an actual wind tunnel test. The comparison between the computed wall heat tranfer rate and the applied one, allows to evaluate the errors due to the values of thermophysical data that were introduced in the 1D routine.

The results of this analysis which are presented in ***Fig. 7***, show that the use of constantan properties is the most appropriate choice at the very beginning of the test while steel properties would be more suitable when long test durations are considered. The data that are effectively used at the Chalais-Meudon R2Ch/R3Ch wind tunnels lead to a satisfactory compromise since the maximum error is less than 5% over the test duration (10 sec). This analysis shows however that the hypothesis of an homogeneous material for this type of gage leads to a small but significant error. For this reason, the technique which consists in building the gage and the model from the same material (constantan) has to be preferred as indicated in section 2.1. Similar calculations were conducted to discuss the influence of the precise geometry of the back-face of the gage, see [6].

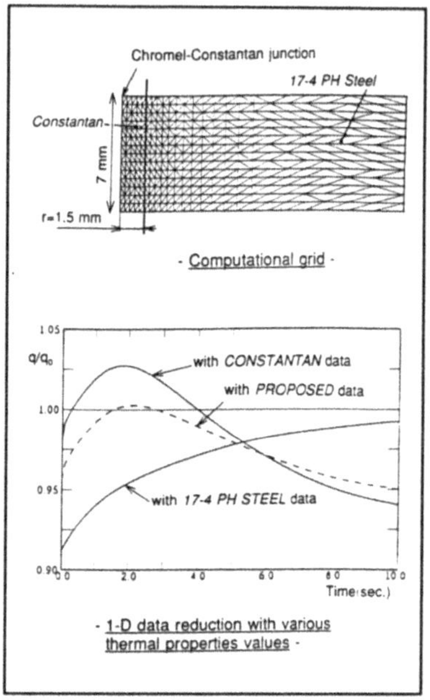

Fig. 7 : Example of heat transfer gage numerical simulation -

2.5. NON ISOTHERMAL WALL EFFECTS

Non-isothermal wall effects are a direct consequence of non uniform heat transfer distributions. They always appear in wind tunnel testing, especially when long running times are encountered. This problem is obviously enhanced when non-homogeneous walls are considered: this might be the case when, for various well founded reasons, different materials are used to built the model or when thermophysical properties of the gage and of its immediate surroundings are significantly different.

It is quite easy to demonstrate that the wall temperature level, at least in the range corresponding to most situations encountered in conventional hypersonic wind-tunnel testing has a rather small effect on the heat transfer coefficient.

The more difficult case of arbitrary non uniform wall temperature distributions has to be examined using a boundary layer code. The example presented in ***Fig. 8*** compares the laminar heat transfer distribution computed on a steel-ceramic flat plate at a Mach number close to 5 for different wall temperature distributions representative of what would be encountered in a real 10 sec. test. It is demonstrated that a significant thermal perturbation is introduced by the ceramic insert. Similar simulations conducted on metallic model equipped with Macor® gages or inversely on insulating models equipped with coax-gages lead to the conclusion that such configurations have not to be recommended from a fluid mechanics point of view, even if for the instrumentation specialist it appears to be possible to cope with such exotic situations from a pure thermal and data reduction point of view.

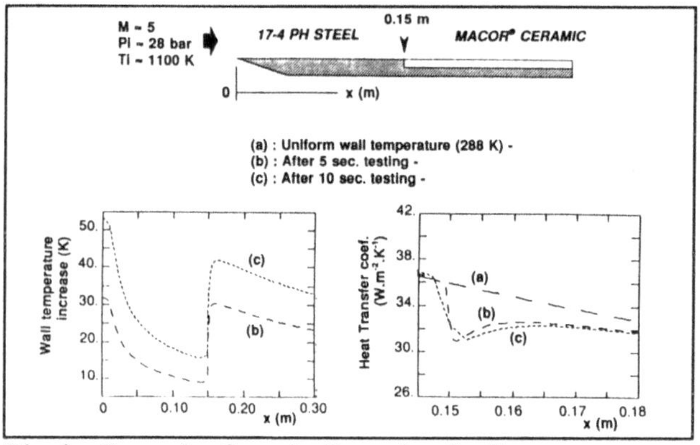

Fig. 8 : Example of non isothermal wall effects -

3. Thermosensitive paint technique

3.1. PRINCIPLE AND LIMITATIONS

This "semi-quantitative" survey technique has been used for a long time in the R2Ch/R3Ch hypersonic wind tunnels, [10]. It is based on the fact that thermosensitive paints produce color changes at a certain temperature level (or more precisely when a certain quantity of heat is absorbed). This technique could be considered, as a surface temperature technique but the early development stages proved that it is more appropriate to use a direct calibration procedure, by comparing the color change time histories on the model tested, to what is obtained on particular geometries like spheres (see ***Fig. 9***).

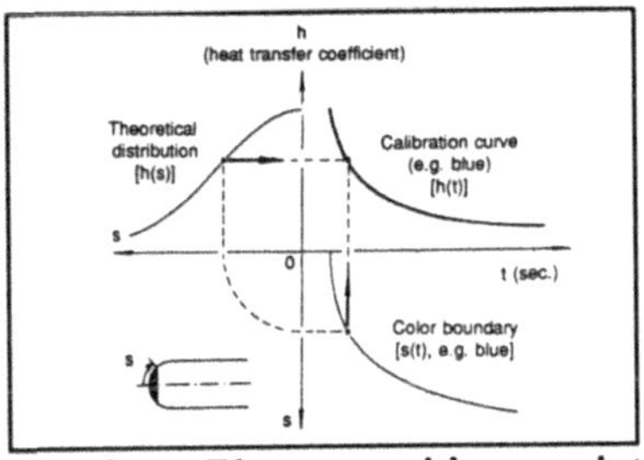

Fig. 9 : Thermosensitive paint technique : basic data reduction procedure -

Usually, the method is applied on model casted from a very low diffusivity material (Norcoat®) so that in most cases, the three available color change lines can be observed (green-blue around 100°C, blue-yellow around 180°C, yellow brown around 250°C) ; moreover, a semi-infinite behaviour can be guaranteed for a 10 sec. run as soon as the wall thickness is larger than approximately 2.5 mm.

This very particular technique generates satisfactory results although some justified criticisms may be addressed to its accuracy. Its main interest lies in the obtention of detailed iso-heat transfer coefficient lines and in the accurate determination of hot spots as proved by the example presented in ***Fig. 10***. An important limitation, mainly due to the use of Norcoat®, is the difficulty to obtain an accurate representation of model very thin parts. A second difficulty is associated with the data reduction which appears to be lengthy and particularly tedious.

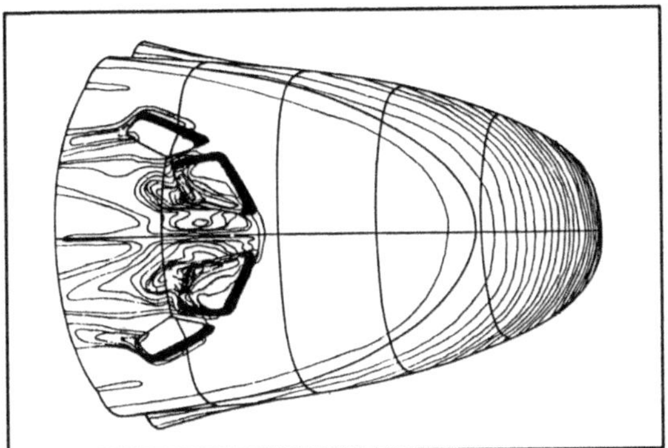

Fig. 10 : Example of iso-heat transfer coefficient mapping -

3.2. TECHNIQUE IMPROVEMENT FOR MODEL THIN PARTS

Thermosensitive paints can be used as a surface temperature technique on model machined thin parts provided that an appropriate data reduction procedure is applied. In order to limitate thickness effects, it is however recommended to use a low diffusivity material and for this reason Macor® seems to be well adapted. The color changes of the paint that will be used for the test are firstly temperature calibrated, using a Macor® sphere according to the simple procedure presented in ***Fig. 11***, then the data reduction method summarized in ***Fig. 12*** is applied to derive the heat-transfer coefficients on both model sides. When necessary, one of the main limitations of the method can then be easily overcome.

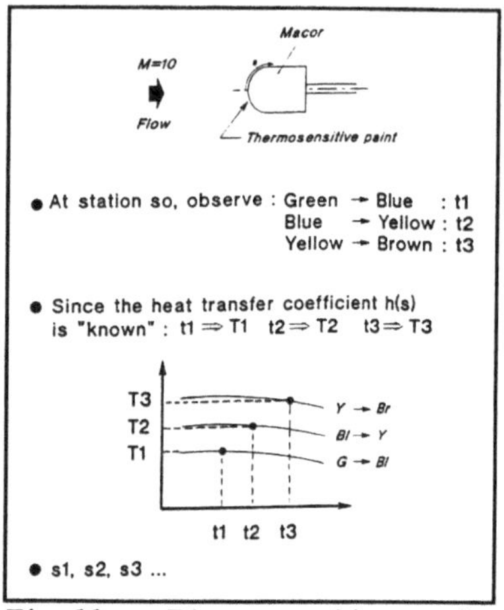

Fig. 11 : Thermosensitive paint technique : color changes calibration

3.3. THE USE OF DIGITAL IMAGE PROCESSING

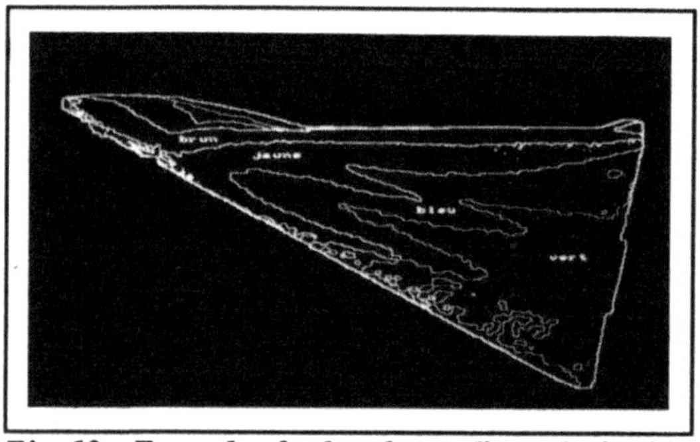

Fig. 13 : Example of color change lines obtained by digital image processing -

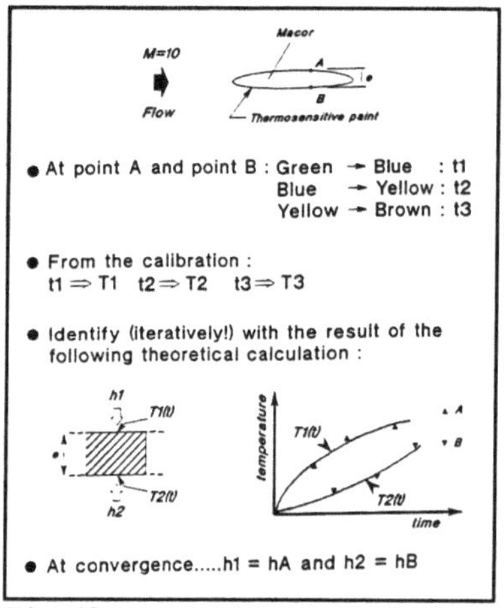

Fig. 12 : Thermosensitive paint technique : data reduction procedure for model thin parts -

In order to reduce the data reduction work, the possibility of using Digital Image Processing techniques has been carefully examined in collaboration with other departments of ONERA. At the very beginning of the study, it turned out that the rather straightforward operation for an human eye which consists in finding the location of the boundary between two colors, raised quite unexpected difficulties for a 3-colors image processor. Thus, most standard methods used for edge detection did not lead to reliable results for this particular problem, except the multispectral analysis and more especially segmentation of 2D-histograms ([11]) which proved to be very promising, as shown by the example presented in ***Fig. 13***. The design of a procedure that could be applied to a complete video sequence including the calibration phase, is now under development but difficulties associated with the image background quality, the spatial and temporal lighting defaults (shadows and bloomings), the model attitude change at the tunnel start-up have now to be resolved.

4. Photoluminescence technique

The temperature sensitivity of phosphors compounds has been known for a long time and as far as aerothermal wind tunnel problems are considered, single point wall temperature measurements were reported as soon as 1953 (a complete litterature review on this subject can be found in [12]). The coming of video imaging technology, of fast computers and of efficient digital image processing algorithms have given a new interest to this technique which was recently re-applied, [14] and [15].

4.1. PRINCIPLE AND PRESENT APPLICATION BASES

The so-called luminescent coating is a composite of a thermostable binder and a crystalline phosphor painted onto the model. When stimulated with U.V. light, it emits a visible radiation whose properties depend (among other factors) on the pigment temperature, see ***Fig. 14***.

There are a great variety of thermographic phosphors and several ways to extract pertinent information that is related to the temperature data, [15]; the method presently set up at ONERA employs a rare earth oxysulfides doped with the Europium ion. The main advantages of such compounds are listed in [16].

The method of luminescence analysis itself, uses a bichromatic optical measurement of the visible spectrum emitted by the phosphor : it is based on the ratio of the intensities measured for two particular bandwidths which is called the "Color parameter" and more exactly on the "Reduced Inverse Color Parameter" (CRI), which is the ratio of the color parameter at the current temperature to the color parameter at some reference temperature (e.g. the room temperature).

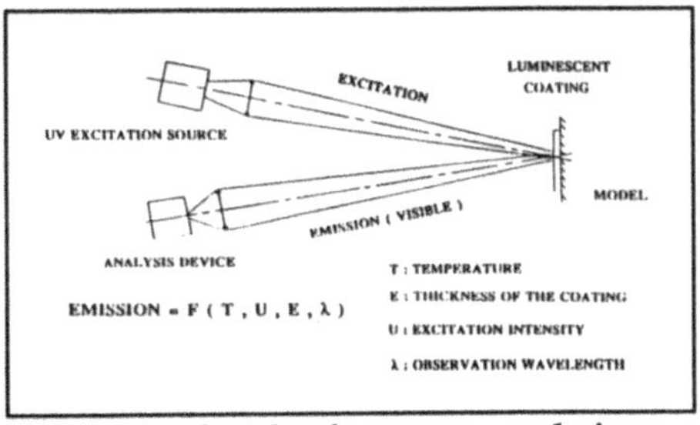

Fig. 14 : Photoluminescence technique : basic principle -

4.2. BASIC INVESTIGATIONS

At the very beginning of the study, it appeared essential to carefully investigate the influence of various parameters outside the wind tunnel in order to evaluate the potential interest of this technique and to check that no critical limitation would appear later in wind tunnel environment.

The first concern was to design and build the single point photothermometer schematically shown in ***Fig. 15***, which consists in two optical measurement channels, each one associated with a removable optical filter adapted to the emission spectrum of the luminescent compound being studied and with a photomultiplier tube.

This basic piece of equipment was extensively used to study the phosphor formulation as well as the influence of various parameters like UV excitation intensity, coating thickness, view angle, local pressure, maximum admissible temperature etc...Different optical filtering (interferential-narrow transmission- and broadband) were also investigated. A detailed analysis of this rather complete set of experiments is available in [12] and [16].

4.3. PRELIMINARY WIND-TUNNEL RESULTS

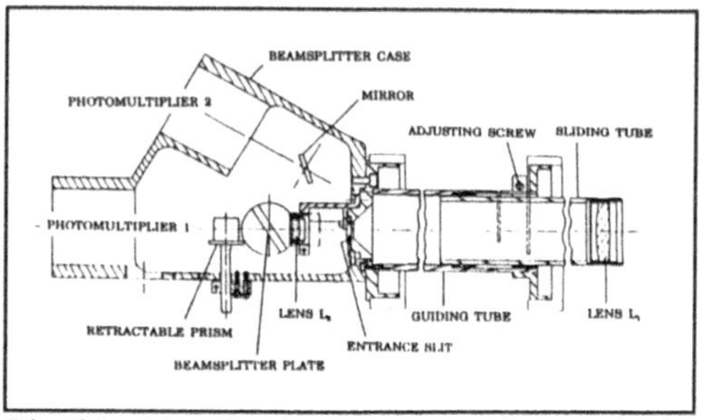

Fig. 15 : Photoluminescence technique : single-point photothermometer -

After investigation of the method's capabilities, preliminary wind-tunnel tests were conducted in the R2Ch M=6 and R3Ch M=10 blow-down facilities on simple model shapes. The three materials mentioned in this paper (Norcoat®, 17-4PH steel and Macor®) were tested and in the two later cases, thin films or coax-gages were installed underneath the luminescent coating whose thickness was varied between 200 µm and 20 µm approximately.

From the data analysis reported in [12], one must point out the following main observations:

- a thick (~200 µm) coating applied onto metallic models leads to unacceptable response times (of the order of several seconds). Significant faster response times were obtained with thin (~20 µm) coatings, the counterpart being in these cases, a rather poor signal to noise ratio.
- quite satisfactory results were obtained on Macor® substrate as illustrated in ***Fig. 16***. It has also to be observed that a good quantitative comparison is achieved with thin film data although difficulties were sometimes encountered to precisely localize the measurement point.

These results are considered as very encouraging and the effort is now devoted to develop an automatic processing software of video images that would definitively demonstrate the interest of the method.

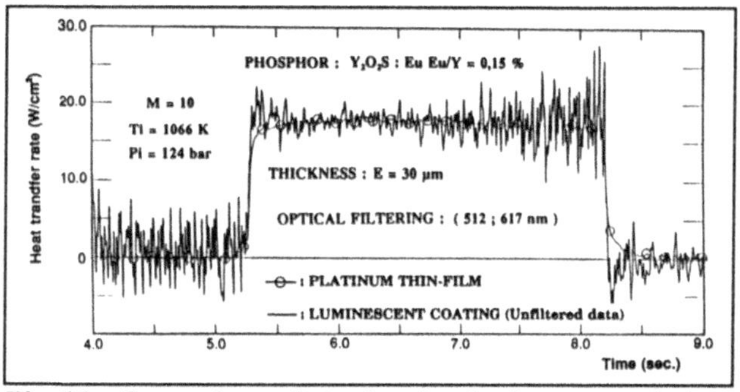

Fig. 16 : Comparison of single-point luminescent technique and thin film technique (MACOR® model) -

5. Concluding remarks

This paper has briefly presented the results of the work undertaken at ONERA these last few years to improve the heat transfer measurement techniques used in ONERA conventional hypersonic wind tunnels. The main observations concerning the techniques discussed are the following :

- Single point surface temperature gages like platinum thin-films or coax-gages provide accurate results but care must be taken of undesirable effects which could lead to large errors if they are not correctly and completely accounted for.
- The effort to develop a quantitative and reliable survey technique must be pursued. The luminescence technique appears to be a very promising candidate.

- Computational techniques are valuable tools for instrumentation design as well as for data reduction improvement. Asociated to Digital Image Processing, they will probably lead in the near future to great progress in thermal flux measurements on models set in hypersonic flows.

Aknowledgments : The work reported in section 4. was partly supported by DRET (French Ministry of Defence).

References :

[1] CHEVALLIER J.P. and TAILLET J. : "Récents progrès dans les techniques de mesure en hypersonique", Progress in Aerospace Sciences, n°12 - ONERA TP n°1055, 1972.
[2] CONSIGNY H., MENTRE V., BETREMIEUX A. : "Methodes de mesures de flux de convection dans les souffleries R2 et R3 de Chalais-Meudon - Mise en oeuvre actuelle et développements envisagés", Colloque AAAF 1987 - ONERA TP n°1987-85, 1987.
[3] SCHULTZ D.L. and JONES T.V. : "Heat Transfer Measurements in Short Duration Hypersonic Facilities", AGARD-AG-165, 1973.
[4] MAULARD J. : "Etude du fluxmètre à température superficielle pour mesures dans les tubes à chocs", ONERA NT n°139, 1969.
[5] MENTRE-LE SANT V. : "Amélioration des méthodes de mesure de flux par la technique des températures superficielles", Thèse de Doctorat, Université Pierre et Marie Curie Paris VI, 1988.
[6] CONSIGNY H. and LE SANT Y. : "Détermination des échauffements cinétiques sur maquettes en soufflerie par la méthode des températures superficielles", ONERA TP n°1990-104.
[7] SIMEONIDES G. : "The VKI hypersonic wind tunnels and associated measurement techniques", VKI TM n° 46, November 1990.
[8] MENTRE V. and CONSIGNY H. : "An Improved Data Reduction Technique for Heat Transfer Measurements using Surface Thermocouples or Thin Films", 12th ICIASF Meeting, Williamsburg - June 1987 - ONERA TP n°1987-74, 1987.
[9] MILLER C.G. : "Comparison of Thin-Films Resistance Heat Transfer Gages with Thin-Skin Transient Calorimeter Gages in Conventional Hypersonic Wind Tunnels", NASA TM n°83197, 1981.
[10] PONTEZIERE J. and BETREMIEUX A. : "Méthode de mesure des flux thermiques pariétaux par peintures thermosensibles", ONERA TP n° 470, 1967.
[11] GIRARD M., BOHL J.C., BAUMANN P. : "Etudes préliminaires pour l'automatisation du dépouillement des mesures de flux thermiques par peintures thermosensibles", Colloque AAAF 1989 - ONERA TP n°1989-196.
[12] BAUMANN P. : "Thermographie par photoluminescence en soufflerie hypersonique à rafale", Thèse de Doctorat de l'Université de Caen, Mars 1992.
[13] BUCK G.M. : "Applications of Optical Fluorescence Thermography Imaging System", Langley Aerospace Test Highlights, 1989.
[14] NUTT K.W. : "Thermographic Phosphor Technique", in VKI L.S. n° 1990-05, 1990.
[15] LEROUX J.P. : "Mesure de température par revêtements photoluminescents. Les Techniques de l'ingénieur, R2790, 1989.
[16] BAUMANN P. : " Mesure ponctuelle de température de surface en soufflerie hypersonique par photoluminescence", Congrès OPTO Mars 1991 - ONERA TP n°1991-94.
[17] NEUMANN R.D. : "Aerothermal Instrumentation", in AGARD Short Course on Hypersonic Aerothermodynamics, VKI, May 1988.

AEROTHERMODYNAMIC MEASUREMENT TECHNIQUES EMPLOYED IN NASA LANGLEY HYPERSONIC FACILITIES

CHARLES G. MILLER
NASA Langley Research Center
Experimental Hypersonics Branch
16 Victory Street, MS 408
Hampton, VA 23665-5225 USA

ABSTRACT. Measurement techniques routinely used at the NASA Langley Research Center to obtain qualitative and quantitative global and discrete values of heat transfer rate to the surface of models tested in conventional-type hypersonic wind tunnels and an impulse facility (NASA Hypulse at the General Applied Science Laboratory) are presented. The application of thin-film resistance gages for discrete measurements in conventional and impulse facilities is discussed along with the relative-intensity two-color thermographic phosphor technique, which has "revolutionized" experimental aerothermodynamics at Langley, and the absolute intensity infrared emission technique. Experimental aerothermodynamics at Langley over the last decade is reviewed, present capabilities discussed, and a look into the future is made.

1. Preface

The intent of the following introduction is to provide background information on experimental aerothermodynamics at the NASA Langley Research Center (LaRC). Descriptions of the nine LaRC hypersonic wind tunnels (referred to as the Hypersonic Facilities Complex (HFC)) used for aerodynamic/aerothermodynamic/fluid dynamic testing are provided by Miller and Smith, 1986 and Miller, 1990. Representing about half of the conventional-type (as opposed to impulse) hypersonic wind tunnels in the USA, these facilities collectively provide a range of Mach number from 6 to 22, unit Reynolds number from 0.03 to 40 million per ft, and most importantly for blunt configurations, normal shock density ratio from 4 to 12. Four different test gases (air, nitrogen, helium, and CF_4) are used and total temperatures range from ambient (as low as 480°R) for the helium tunnels to 3500°R for the nitrogen tunnel. With one exception, these blowdown-to-vacuum facilities are relatively small, economical to operate, possess run times in excess of 15 seconds, have model injection/retraction systems, and are ideally suited for basic fluid dynamic research and fast-paced aerodynamic/aerothermodynamic parametric studies for proposed aerospace vehicles. A number of transient aerothermodynamic measurement techniques have been used at LaRC since the late 1950's to accommodate a broad spectrum of studies and the requirements/limitations of the various hypersonic wind tunnels. (Excellent reviews of aerothermodynamic measurement techniques are provided by Schultz and Jones, 1973 and by Neumann, 1989.) Those techniques currently utilized at LaRC are addressed herein; while not new techniques (e.g., each is discussed by Schultz and Jones, 1973), they have been refined considerably.

2. Introduction

The demise of interest in hypersonics in the USA during the 1970's and well into the 1980's naturally corresponded to a demise in aerothermodynamic testing. Aerothermodynamicists moved into other disciplines during this period and the capability to perform aerothermodynamic studies

A. Boutier (ed.), New Trends in Instrumentation for Hypersonic Research, 449–459.

nearly vanished at the Langley Research Center. Renewed interest in hypersonics within NASA in the mid 1980's, initially due to the Aeroassist Flight Experiment (AFE) followed by such programs as the National Aero-Space Plane (NASP), Pathfinder High Energy Aero-Braking (HEAB) concepts, Assured Crew Return Vehicle (ACRV) (formerly Crew Emergency Return Vehicle (CERV)), Personnel Launch System (PLS) and Advanced Manned Launch System (AMLS), ushered in an urgent need to activate and upgrade hypersonic facilities, restore the capability to fabricate precision models, implement instrumentation and data acquisition system upgrades and most importantly, reestablish the expertise to perform hypersonic aerodynamic and aerothermodynamic testing.

The primary technique used to measure discrete values of heat transfer rate on models tested in Langley hypersonic wind tunnels in the 1960's and 1970's was the thin skin transient calorimeter technique (e.g., Miller, 1981). This technique generally involved the fabrication of metallic models having thin skin sections to which thermocouple leads were attached to the back surface of the skin. Great care was required to provide measurement of heat transfer rates to an accuracy better than 10 percent. Thermal mappings were obtained via the phase change paint technique developed at Langley by Jones and Hunt, 1964, 1966. This technique involves coating (spraying) the surface of a model, generally made of a highly filled epoxy casting compound (Emerson and Cuming, Inc. brilliant red or black stycast 2762FT), with a paint that changes from an opaque solid to a clear liquid at a known temperature. Melt lines representing isotherms are formed upon injection of the coated model into the flow and estimates of the heat transfer rate can be made using the thermophysical properties of the model and the time required to reach the phase change temperature. Although a qualitative technique, measurements of heat transfer rate to an uncertainty of 15 to 20 percent are possible. Primary disadvantages of this technique are: (1) at any instance in time, the only information available is the location of the isotherm; that is, this technique does not provide global information as the surface temperature in regions of melted or unmelted paint are unknown; (2) the paint melting process appears to be instantaneous for regions of the model having nearly the same value of heat transfer rate; and (3) the model requires cleaning and respraying after each run. The phase change paint technique played a significant role in aerothermodynamic studies at Langley for over two decades.

Although there were only a few researchers at Langley performing aerothermodynamic studies in the late 1970's and early 1980's, significant changes occurred in methodology nevertheless. For example, in 1981 a family of biconics were tested in the Langley Expansion Tube, an impulse facility generating flow velocities between 17000 and 23000 ft/sec and having a run time of approximately 300 msec, whereby heat transfer distributions were measured via thin film resistance gages (Miller, et al, 1985). Each model contained five contoured Macor (a machinable glass ceramic;trademark of Corning Glass Incorporated) substrates and each substrate contained eight palladium thin film gages sputtered onto the surface. Because the substrates were relatively thick (0.2 inch) and wide (0.2 inch), Miller attempted tests of these same models in the Langley 31-Inch Mach 10 Tunnel in the early 1980's. Subjecting the models to an exposure to the free stream flow of approximately one second, highly accurate heat transfer distributions were measured. This first attempt to use a relatively large number of thin film gages on a model in a conventional (as opposed to impulse) hypersonic wind tunnel at Langley was deemed to be highly successful. Because of the many advantages thin film gages offer over the thin skin technique, the thin film technique became the measurement method of choice at Langley for benchmark aerothermodynamic studies. It was in this time frame that significant advances were being made in the development and application of computational fluid dynamics (CFD) at Langley for hypersonic flows. Of the measurements made in hypersonic wind tunnels, computationalists sought heat transfer measurements for the purpose of computer code calibration. (Codes might predict measured surface pressures and shock shapes to within a few percent, yet miss the corresponding heat transfer rate by perhaps 25 to 35 percent.) Advances were made in the application of the thin film gage technique to conventional hypersonic wind tunnel testing in the early to mid 1980's. The capability to fabricate high fidelity (aerolines machined to within $\pm$0.002 inch tolerance) models entirely of Macor, which were heavily instrumented (first with up to 120 thin film gages and later with up to 300 gages), was developed; also developed were the signal conditioning and data acquisition systems required to handle these number of gages. Such models are quite challenging to construct, requiring considerable time to machine, drill for leads, polish, instrument and calibrate (typically require 6 to 18 months), are costly and are fragile, thereby requiring careful handling. Nevertheless, this technique is Langley's standard for benchmark heat transfer studies for reasons to be discussed in a subsequent section entitled Measurement Techniques.

The major hypersonic program in the USA is NASP. When the Langley Research Center became involved with this fast-paced, classified program, the techniques readily available for the aerothermodynamic assessment of various configurations were thin skin, thin film and phase change paint. A number of thin-film models were constructed for the NASP program and the phase change paint was widely utilized in the early days of the program; no thin-skin models were fabricated for tests at Langley. The fast pace of the NASP program clearly demonstrated the need for a quantitative aerothermodynamic measurement technique for which models could be constructed rapidly and at relatively low cost, yet provide detailed spatial information.

The goal was to develop a global (i.e., optical), quantitative heat transfer measurement technique that would be applicable to all hypersonic tunnels at Langley. Techniques considered were the phase change and color change paints, infrared emission, thermographic phosphors and encapsulated liquid crystals, all based on advanced imaging systems. Experimental aerothermodynamics at Langley was essentially revolutionized when Buck 1988, 1991 successfully demonstrated the two-color, relative intensity thermographic phosphor technique. Following a number of successful laboratory experiments, Buck applied his technique in the 31-Inch Mach 10, 20-Inch Mach 6 CF_4 and 22-Inch Mach 20 Helium Tunnels. The results of these evaluations were extremely encouraging and lead to continued development and refinement (Merski, 1991). An extremely important ingredient to the successful implementation of the thermographic phosphor technique was the development of fabrication techniques for rapid production of high-fidelity low-cost models (discussed subsequently).

Approximately half the tests recently performed in the hypersonic wind tunnels at Langley managed by the Space Directorate are aerothermodynamic studies, and the majority of these have been with the phosphor thermography system. Simultaneous development and implementation of the absolute intensity infrared (IR) emission technique by Daryabeigi, 1992 has provided friendly competition to the phosphor thermography system. The IR system has the advantage of not requiring a surface coating, thus models having extremely smooth surfaces may be tested, nor illumination of the surface; and the system presently possesses a wider temperature range than the phosphor system and may be applied to ceramic or metallic models. Although a major contributor to Langley aerothermodynamic studies, the absolute intensity IR technique, like the relative intensity phosphor technique, is still considered qualitative. The phase change paint technique has not been used for the past two-to-three years.

Presently, the basic philosophy at Langley is to design and fabricate models for one type of measurement; for example, models involving discrete surface measurements are generally dedicated to either pressure or to heat transfer measurements so to optimize the information obtained. However, both the phosphor and infrared thermography techniques have provided the opportunity to examine the feasibility of simultaneous aerodynamic and aerothermodynamic measurements via ceramic force and moment models and ceramic pressure models containing on-board electronically scanned pressure (ESP) transducer modules.

These optical, global techniques have provided the aerothermodynamicists with powerful tools for the rapid, relatively inexpensive assessment of aerothermodynamic characteristics of proposed aerospace vehicle concepts. They have also served to narrow the advantages of large hypersonic wind tunnels, since each pixel of the video camera may be viewed as a sensor and a larger model does not necessarily provide a decided advantage from a data information viewpoint. This is not to imply that larger models do not provide advantages, particularly in higher values of length Reynolds number, improved fidelity for thin leading edges, and thicker material to improve the assumption of one-dimensional conduction as used to reduce measured surface-temperature time histories to values of heat transfer rate (e.g., Miller, 1981).

Routinely used measurement techniques at Langley providing qualitative and quantitative heat transfer distributions will be discussed in some detail in the following section.

3. Measurement Techniques

3.1. THIN FILM RESISTANCE GAGES

All thin-film gages used to date in Langley hypersonic wind tunnels have been manufactured in-house using technology developed for the Langley Expansion Tube in the 1970's (Miller,1981). The substrate is generally Macor, although quartz is also used, and sized for flow exposure times of about 1 sec; that is, the width and thickness are selected so the substrate behaves as a semi-infinite slab for 1 sec. The sensing element may be sputtered onto the optically polished surface of the substrate, vapor deposited or painted. Until recently, most elements were sputtered palladium, applied in a serpentine pattern 0.04 by 0.05 inch and about 1000 to 1500 Å thick; the nominal resistance is 120 ohms. The pads where the sensing element is connected to the two leads are made relatively thick in an effort to make the resistance of these pads negligible. An aluminum oxide or silicon dioxide overlayer, approximately 3000 to 5000 Å thick, is sputtered or vapor deposited over the sensing element and connector pads primarily to enhance gage durability for conventional hypersonic wind tunnel testing and to minimize shorting in the ionized flow of hypervelocity facilities. Connections between the pads and wires are made using a conducting epoxy. Gages are calibrated (to determine the change in resistance with temperature) in a well-stirred oil bath for temperatures from 535° to 685°R which covers the range of surface temperatures experienced in most tests; when required, this calibration can be extended using an oven at higher temperatures. Details of the construction of sputtered thin film gages and their calibration are provided by Miller, 1981.

Because of increased requirements for thin-film gage fabrication in recent years primarily due to the NASP program and to the activation of the NASA Hypulse facility (formerly the Langley Expansion Tube) at the General Applied Science Laboratory (GASL), heavy demands were placed on in-house resources, particularly the sputtering equipment. To help alleviate the log jam that occurred, other fabrication techniques were employed, namely vaporization and painting. Presently, about as many gages are painted with platinum paint and fired in a ventilated oven as are sputtered. Sputtering is the preferred method for two-dimensional surfaces and because it was believed to have less influence on the thermal properties of the substrate than the paint, which penetrates and alters the properties of the substrate. Such an alteration of the substrate properties near the surface may be of less significance for tests in conventional wind tunnels as opposed to impulse facilities because of the increased thermal depth. To account for this effect, painted and sputtered gages are also calibrated using a water-to-air comparison technique to deduce substrate thermal properties.

Models containing thin film gages are either fabricated from stainless steel and slots machined in the surface to accept contoured Macor or quartz substrates, or fabricated entirely from Macor. The all-Macor models sometimes contain instrumented Macor inserts and/or the gages are placed directly on the model surface. The all-Macor approach is preferred because it eliminates the uncertainty associated with discontinuities in model material and their thermal properties (e.g., Schultz and Jones, 1973). There are distinct disadvantages to all-Macor models, however; they are difficult to machine, hence generally expensive, may be easily broken, and are difficult to polish to an optical finish.

Langley hypersonic wind tunnels are equipped with 128 or 256 channels of signal conditioning (NEFF Instrument Corporation system 620/series 300) for thin film gages and may be augmented with an additional 128 channels via a portable system. Each channel has a differential amplifier and supplies the gage with a constant current not to exceed 5 mA so to avoid the effects of ohmic heating. These channels are interfaced to a 15-bit, analog-to-digital (A/D) data acquisition system (DAS) having programmable gains and filters per channel (NEFF system 620/series 600) and controlled via a Hewlett Packard (HP) 9000 series model 375 computer networked to a central HP model 750 computer. Although capable of higher rates, this system typically interrogates each channel at a rate of 50 samples/sec. The recent addition of a portable 20-channel NEFF system 490 having 12 bit analog-to-digital conversion (ADC) and 1 MHz programmable amplifier/filter with 2 megaword RAM will provide the capability to utilize the fast response of thin film gages to examine unsteady surface flow phenomena. Heat transfer rates are determined from the measured temperature-time histories using the numerical method discussed by Miller, 1981.

One significant disadvantage of thin-film gages is they are susceptible to damage or modified properties when impacted by solid particles carried in the flow. (The introduction of thin-film gages is the primary reason that in-line filters have been, or soon will be, installed in nearly all the hypersonic wind tunnels at Langley.) On the positive side, the thin film technique provides the model surface temperature at all times, including prior-to and during injection of the model into the flow, hence time histories of the heat transfer coefficient. This technique has high sensitivity, accurately measuring surface temperature changes of less than 2°R such as may occur on the base or leeward surface of a model, and extremely fast response (typically, less than 1 msec) if so desired. Gages may be placed close (typically 0.1 inch) to one another to provide detailed distributions, and their locations on the surface are accurately known. Most importantly, of the measurement techniques utilized at Langley, the thin film technique provides the greatest accuracy (within 5 to 8 percent) for the heat transfer rate.

As standard test procedure, the conditions for the first run in a test series with thin-film models are repeated as the final run. Data from the two runs are compared prior to removal of the model from the facility. If repeatability is within acceptable limits (generally 1 to 2 percent), this implies the gages were stable throughout the test series. If repeatability is not acceptable, additional runs may be required to examine the effect of changes in the gages which may occur due to particle impact or to annealing of the gage element. In most cases the thin film gages are recalibrated in terms of resistance vs. temperature after the last test. Gages for which the change in thermal coefficient of resistance is less than one percent between pre-test and post-test calibrations are used with a high level of confidence.

3.2 THIN-SKIN TRANSIENT CALORIMETER GAGES

Until the early 1980's, most heat transfer models tested at Langley used the conventional thin-skin transient calorimetry technique. The rate of heat storage in the model skin is inferred from thermocouples attached to the inside surface of the skin. The thin skin is generally continuous, as opposed to slug inserts, and usually of type 347 or 17-4 stainless steel for which the variation in specific heat with temperature is accurately known. The temperature of the inside surface is generally measured using chromel-alumel or iron-constantan thermocouples since most facilities can accommodate a relatively large number (256) of either of this type of thermocouple. Sometimes thermocouple wire as small as No. 36 is used to minimize conduction along the wire, but because this wire size is difficult to work with, No. 30 wire is more often used. The two wires for each thermocouple are welded to the inside surface at precise locations corresponding to the desired location on the outer (wetted) surface, and are usually spaced no closer together than about 0.030 in. Because the uncertainty in the heat transfer rate determined with this technique is directly proportional to the uncertainty in the skin thickness, the inside surface of the skin is worked to a smooth finish prior to wire installation and the thickness measured at 5 locations around each set of wires to provide an accurate "average" value. Typical skin thickness is 0.015 to 0.030 in.

A transient is obtained by rapidly injecting the model into the test flow from a shielded position, where it was maintained isothermally at or below room temperature. A slide wire or switches are used to provide an indication of the model location with time during the injection. This information is used to provide an initial estimate of zero time corresponding to the instant the model shock layer is fully immersed in the test core of the tunnel during injection. The measured temperature-time history is used, along with the thermal properties of the skin and the skin thickness, to infer the heat transfer rate and corresponding value of the heat transfer coefficient. Zero time is then determined by reducing the data to distributions of the heat transfer coefficient for a number of potential zero times about the initial guess (typically 5 values). The final value for zero time is obtained by observing the convergence of these data with time. Because of the difference in surface area between the inner and outer surfaces of most thin-skin models, a geometric correction factor must be applied to the measured skin thickness. This correction factor is, naturally, small for models having a large radius of curvature. Additional information on this technique is provided by Miller, 1981.

The primary advantages of thin skin models compared to thin film models are they are durable and faster and less expensive to fabricate and instrument. Relatively few thin skin models have been tested in Langley hypersonic wind tunnels over the past 6 years and these have been blunt configurations (e.g., the baseline AFE configuration, planetary entry probe concepts, and representative low lift-to-drag aerobrake configurations).

3.3 THERMOGRAPHIC PHOSPHORS

The relative-intensity two-color thermographic phosphor technique (Buck, 1988, 1991; Merski, 1991) has become the most widely used heat transfer measurement technique at Langley. The model, generally ceramic, is coated with thermographic phosphors. When illuminated with ultraviolet light, electrons are excited and emit visible radiation during their subsequent relaxation to lower energy levels. The probability that the relaxation process emits visible radiation is temperature dependent and the amount and the spectral distribution may be used to determine local temperatures. Phosphor materials are selected for two-color emission band spectrum that matches front-end filters on conventional color video cameras. A three-chip co-site sampling CCD (Charge Coupled Device) camera, which has true color separation in the red, green, and blue (RGB) bands and spatially congruent detection arrays, is used. Early applications of this technique utilized blue-green thermographic phosphors, having secondary and primary emission bands at 450 nm and 520 nm, respectively; these phosphors were irradiated with 100 watt mercury arc lamps with broadband ultraviolet filters centered at 365 nm. Two materials were used, one with a temperature sensitivity range of 520°R to 710°R and the other with a range of 560°R to 810°R. Currently, a phosphor mixture is used consisting of both a broadband thermographic phosphor and a narrowband rare-Earth phosphor. For this mixture, the green and red camera filters are used and a temperature sensitivity range of 480°R to 860°R is obtained. A measured intensity ratio is used to obtain quantitative temperature information using digital image processing. Software for the original system was developed at the Oak Ridge National Laboratory under an interagency agreement.

Application of the laboratory developed prototype phosphor thermography system in various hypersonic wind tunnels was extremely encouraging. An improved system was designed (Merski, 1991), manufactured, delivered to Langley and is undergoing acceptance tests at this writing. A major advantage of the new system is that it possesses the hardware and software for target recognition capability. The model geometry is entered into the acquisition computer and 16 points are tracked simultaneously during the injection process and the testing period. The software, which is proprietary, handles spatial transformations and treats image interpolations using a frequency mapping. The new system will acquire images at a minimum of 30 frames per second and the vast amount of data will be stored on a studio video recorder; this data may be read from the recorder and digitized at real-time rates. A calibration procedure that includes the response of each individual camera pixel is included. The capability to detect and record three phosphor wavelengths is included so that the temperature range might be increased. Provision for a second camera, synchronized with the first camera, is provided to allow for three-dimensional heat transfer analysis. The data reduction software developed by Merski and incorporated into this new system will provide near quantitative, global heat transfer (as opposed to temperature) distributions on models immediately following a tunnel test.

Models used with the thermographic phosphor technique should have low thermal conductivity, small thermal deformation, smooth nonporous surface, uniform and well-defined thermophysical properties and be able to withstand repeated and prolong testing in all hypersonic facilities. Part of the refinement of the phosphor technique has been the development of a ceramic slip casting technique for the rapid fabrication of inexpensive heat transfer models (Buck, 1991; Merski, 1991). A high-purity silica ceramic material capable of withstanding temperatures up to 2500°R was selected which satisfied the previously mentioned criteria. Presently, all models used with the phosphor technique are made in-house. Models up to 30 inches in length and models having thin components, such as wings and fins, and sharp leading edges (diameter less than 0.005) have been fabricated and tested. The capability of thin ceramic sections such as wings to withstand the pressure loads generated in Langley hypersonic wind tunnels has been demonstrated repeatedly.

Various techniques have been tested for applying phosphor coatings and presently a silica ceramic binder is used. Coatings, typically 0.05 inch thick, are deposited using a compressed air spraying technique. These coatings survive high temperature cycling without altering the binder optical properties. The primary disadvantage of the coatings has been the induced surface roughness they produce, although considerable progress has been made to provide smoother coatings. At this stage of development, the fabrication and the coating of models is more art than science.

3.4 INFRARED EMISSION

The surface temperature of a model may be inferred from observations of its radiation at infrared wavelengths. Infrared imaging has been used in hypersonic wind tunnels for over two decades (e.g., Schultz and Jones, 1973; Neumann, 1989), but suffered from such problems as poor image resolution, low data acquisition rate and the inability to accurately provide absolute temperature measurements. Because of recent advances providing higher image resolution, standard video format data acquisition, radiometric (absolute temperature) measurements and computer based digitized image processors, the infrared technique has received renewed interest within the hypersonic testing community at Langley (Daryabeigi, 1992). Presently, two commercially available infrared imaging systems are used in the Langley HFC; both are single detector imagers employing optical-mechanical scanning to provide two-dimensional temperature distributions for the 8 to 12 mm band pass of the electromagnetic spectrum. One system has a frame frequency of 6.25 Hz and the other is 30 Hz; the 30 Hz camera provides 380 samples per line and 200 lines per frame. Standard video format is used. Stored information is manipulated and then transferred to a digital image processor. The precision of these infrared imaging systems for providing steady-state temperature measurements has been recently evaluated in the laboratory for temperatures from 400°R to 1390°R and found to be within 3 percent (Daryabeigi, 1992). Wind tunnel test sections have been equipped with zinc selenide windows capable of transmitting 8 to 12mm wavelengths and with suitable anti-reflection coating. Because of losses in the optical path (e.g., window and mirror attenuation and absorption/emission of the intervening medium which may consist of gases with unsymmetrical structures such as H_2O, CO_2, etc.), it is necessary to calibrate the system with the model in the test section. Models made of insulating material are preferred for many reasons, as discussed previously, one reason being the higher emissivity in the infrared spectrum provides the camera with a stronger signal. Presently, the surface emissivity is measured as a function of wavelength at ambient temperatures only; future plans call for the measurement of emissivity over a range of temperature. The overall sensitivity depends on the viewing angle. In his recent paper, Daryabeigi addresses transmission losses in infrared transmitting windows and the optical resolution and spatial frequency response of imagers, compares digital with analog recording, determines the directional variation of the emittance of model materials, and develops algorithms for correcting infrared data for model curvature.

A third infrared imaging system has recently been delivered to Langley for use in the HFC. This system is an Inframetrics 760 having a stirling cycle cooler, 8-12 mm spectral response, 30 Hz framing rate with 380 samples per line and 200 lines per frame, 40 deg horizontal by 30 deg vertical field of view (FOV) and 8-bit digitizer to be upgraded to 12-bit in the near future. It is expected to be heavily utilized in the Mach 6 and 10 air tunnels in which a major portion of aerothermodynamic testing at Langley is performed. It should be noted that the infrared technique is not applicable to the 20-Inch Mach 6 CF_4 tunnel because of emission absorption by this heavy gas.

Work is in progress within Langley's Instrument Research Division to develop this technique to a quantitative capability for complex models; that is, to develop the capability to correct for emittance variation with temperature and angular direction. Heating distributions have been measured with this technique on hemisphere models at Mach 6 and 10 in air for which the heating levels and distributions may be accurately predicted. In general, the agreement of measurement with prediction was within 5 percent over most of the hemisphere. This technique has been used to detect complex heating patterns, including boundary layer transition to turbulence, on force and moment, pressure and heat transfer models, and to examine plumes doped with freon; it is an extremely valuable tool for Langley aerothermodynamicists and fluid dynamicists.

4. Sample Applications

The first all-Macor thin film model tested at Langley in a conventional-type hypersonic wind tunnel was the AFE baseline configuration (tested in 1987 and reported by Micol, 1989). (The first all-Macor thin film model was a 2.5-inch-diameter sphere cone tested in the Langley Expansion Tube in the late 1970's.) The model diameter was 3.7 inches and approximately 80 thin film gages were installed (sputtered) onto only one-half the blunt forebody surface. These glass-ceramic models were machined via a numerical cutting machine and hand worked to a tolerance of 0.002 inch for the outer aerolines prior to polishing to an optical finish. Inputs for model aerolines were obtained directly from the geometric description developed for CFD computer codes, thereby insuring the same geometry was tested in the wind tunnels as used in the computer codes. This is standard procedure for hypersonic CFD code calibration studies performed at Langley. Companion AFE thin film models fabricated from stainless steel and slotted to accept contoured Macor substrates containing both sputtered palladium and painted platinum thin film gages were also made. These models, having a 1.5 inch or 2.5 inch diameter, were tested in the NASA Hypulse facility in air at a velocity of 18,000 ft/sec and in helium at 22,000 ft/sec; again, the run time was only 300 msec. They were also tested in the Mach 6 CF_4 and Mach 10 air tunnels to provide direct comparisons with the results obtained in the real-gas environment provided by Hypulse. All Langley thin film models are designed and fabricated such that they may be tested in impulse facilities having run times in microseconds or conventional hypersonic wind tunnels whereby the model is exposed to the flow for a few seconds.

Representative time histories of the heat transfer coefficient as measured on the all Macor AFE model in the 31-Inch Mach 10 Tunnel are presented by Micol, 1989. Time histories over the nose and cone section of the forebody were very well behaved, with the heat transfer coefficient remaining constant to within a percent or two during a 1 to 2 second period of exposure to the flow. On the shoulder of the AFE, the assumption of one-dimensional conduction was violated as indicated by an increase in the coefficient with time. This time history was shown by Micol so to illustrate a short coming for transient aerothermodynamic measurement techniques that utilize the assumption of one-dimensional heat conduction to infer heat transfer rates on surfaces having small radius of curvature. For this case, extrapolations of the coefficient-time histories to zero time provided what were believed to be reasonable estimates of heating on the shoulder of the AFE. A sample of the first comparison of predicted heating distributions for the AFE to measured distributions is shown in Fig. 1. With the exception of the nose region, the Langley developed LAURA code, a Navier Stokes solver, predicted the heating distribution quite well (within the experimental uncertainty). Causes for the differences observed in the nose region have since been resolved via modifications to the code and refinements in the grid.

An all Macor thin film model of a modified shuttle orbiter was fabricated about the same time as the all Macor AFE models (Micol, 1991). A stainless steel orbiter model was also fabricated; both the Macor and the stainless steel models were slotted in the spanwise direction to accept contoured Macor substrates having thin film gages sputtered 0.15 inch apart; each model contained approximately 150 thin film gages. As with the AFE models, these orbiter models were machined via a numerical cutting machine using inputs from a geometric description in a CFD computer code. Although the steel model was constructed primarily to serve as a backup in the event the Macor model was broken, these models provided the opportunity to examine whether discontinuities in the surface thermal properties, such as occurs at a Macor-steel junction, provide a first order influence on the inferred heat transfer rates. Comparisons of heat transfer distributions measured on the two models at Mach 6 in air revealed excellent agreement for relatively low Reynolds numbers corresponding to laminar boundary layers, thereby indicating no appreciable influence of surface property discontinuities for this case.

The last example of the application of thin film gages is for an impulse facility. A relatively heavily instrumented (145 thin film gages) flat plate (4 inches wide, 10 inches long) with fin(s) model was recently tested in NASA Hypulse to examine shock/laminar boundary layer interaction phenomena at hypersonic-hypervelocity conditions. Preliminary analysis of the data revealed excellent run-to-run repeatability (within 5 percent) and agreement between prediction and measurement for the flow over a clean plate was within the experimental uncertainty (5 to 6 percent). Because the data are preliminary and unpublished, sample time histories for the 300 msec test period and reduced data are not presented herein. However, as is well recognized, time histories measured in impulse facilities reveal valuable information as to the performance of the facility as well as the instrumentation and data acquisition system but their interpretations are often quite subjective (Miller, et al, 1985). For this and other reasons, this writer believes it is essential that time histories be presented in reports to assist in establishing the creditability of the results.

Two examples of the application of the phosphor thermography system will be discussed. The first example demonstrates how this technique can provide aerothermodynamic information quickly and inexpensively (both relative terms). This technique was used to assess the aerothermodynamic characteristics of a proposed inflatable ballute device attached to the solid rocket motor of the AFE; the purpose of this ballute was to aerodynamically control the trajectory of the motor upon jettison during entry into the Earth's atmosphere. Beginning only with a rough sketch of this configuration, construction of ceramic models was initiated and tests were performed in the Mach 10 tunnel just a little over three weeks later. Thus, the AFE community was provided information within one month of the request that revealed the ballute concept as tested may not be a viable candidate. (This configuration generated a region of very high heating on the ballute surface because of a shock/shock interaction that occurred at incidence.)

The last example presented herein illustrates how the qualitative phosphor thermography technique nicely augments quantitative discrete measurements via the thin film technique. The global heating obtained on the windward surface of the modified shuttle orbiter configuration at Mach 10 and a = 20° (Micol, 1991) revealed the so-called streak-heating phenomena due to the fuselage-wing junction. A spanwise heating distribution inferred from this global information at an axial station 75 percent of the length is compared to measurements made with thin film gages in Fig. 2. The agreement between these two measurement techniques is, in general, quite good.

5. Future Capabilities and Projections

The future of experimental aerothermodynamics will, naturally, depend on a number of factors. Most importantly will be the viability of hypersonics; that is, customers for aerothermodynamic information. Another primary factor will be the impact of rapid advances in CFD. An idealized projection of experimental aerothermodynamics at Langley for the next decade is made in this section. This projection is idealized since it is based on the assumption that hypersonics within the Agency remains viable and that experiment and prediction complement, as opposed to dominate, one another; that is, remain equal partners.

To provide a starting point, a recently performed study is used to illustrate the pace, accuracy and expense associated with a high priority experimental program. This study involved the hypersonic aerodynamic and aerothermodynamic assessment of the first NASP configuration to be developed by the consortium. Within 6 months of receiving the aerolines at Langley, a number of high fidelity aerodynamic and aerothermodynamic models were fabricated, including models for parametric configuration buildup studies, testing was performed at Mach 6 and 10 in air, and the data reduced, analyzed and disseminated. This study was performed entirely in-house and cost was driven primarily by model fabrication. In the interest of time, aerolines were read directly to numerical cutting machines thereby avoiding the time and cost of model design. The relative intensity thermography phosphor technique played a major role in reducing the time required for the aerothermodynamic phase of the study, yet provided global mapping information for a wide range of geometric and flow parameters, and reducing the cost of model fabrication.

Optical, video based infrared emission and relative intensity phosphor thermography systems will provide the vast majority of aerothermodynamic information from conventional-type hypersonic wind tunnels in the 1990's. These systems, taking advantage of advances in data imaging and acquisition and in computer and mass storage technology, will provide quantitative, global three-dimensional heating distributions via the simultaneous utilization of multiple cameras or a mirror system. (Daryabeigi recently demonstrated the feasibility of simultaneously obtaining thermal mappings on the windward and leeward surfaces of a model). Designers of aerospace vehicles will be provided with global pressure (e.g., DeMeis, 1992) and heat transfer mappings measured simultaneously. The same capabilities provided to conventional-type hypersonic wind tunnels via these thermography techniques will be expanded to impulse facilities having test times of only a few milliseconds. High speed, high resolution video cameras coupled to advances in fast response phosphorescent materials will make the phosphor technique applicable to impulse facilities. Discrete aerothermodynamic measurement techniques in conventional-type facilities will decline in usage as the optical thermography systems improve. Ceramic models will be the norm instead of the exception and procedures for fabricating hi-fidelity models will continue to improve. Most importantly for the discrete and global techniques, particularly for conventional-type facilities, will be the utilization of two and three dimensional heat conduction solvers. This will negate the restrictions associated with the assumption of one-dimensional conduction. These new data reduction tools will explicitly account for the temperature dependence of model material properties. Until such time as optical systems capable of quantitative measurements become available for impulse facilities, thin film gages will continue to be refined. Miniaturization will be achieved such that detailed circumferential (resolution to within 2 deg) heating distributions may be measured on cylinders as small as 0.2 inch diameter. These gages will be capable of providing accurate values of heat transfer rate even for severe shock/shock interaction phenomena (Aslam, et al, 1992). Overlayers will be developed that provide protection to the element in the presence of plasmas without degradation to gage response, and allow the construction of durable thin film gages with catalytic surfaces.

Emphasis in the 1990's will be on the refinement of existing aerothermodynamic measurement techniques and their application for calibration of CFD codes and the design of thermal protection systems for aerospace vehicle concepts. Other techniques, such as thin film thermopiles, ultra-high temperature thin films, etc., will be studied and developed in laboratories and may provide attractive capabilities for ground based and for flight testing.

6. Acknowledgments

The author extends his sincere appreciation to Kamran Daryabeigi for development/ application of the infrared emission technique; to Gregory Buck and Ronald Merski for development/application of the relative intensity technique; and to Thomas Horvath for refinement/application of the thin film technique. These experimentalists provided the vast majority of the diagnostic information presented in this overview of aerothermodynamics at Langley.

7. References

Aslam, M., Masood, A., Fredicks, R. J., and Tamor, M. A. (1992) "Thin Film Diamond Sensor Array for Harsh Aerospace Environments," SPIE Proceedings 1694-20.

Buck, G. M. (1988) "An Imaging System for Quantitative Surface Temperature Mapping Using Two-Color Thermographic Phosphors," 34th International Instrumentation Symposium, Albuquerque, NM, May 2-5, 1988.

Buck, G. M. (1991) "Surface Temperature/Heat Transfer Measurement Using a Quantitative Phosphor Thermography System," AIAA 91-0064.

Daryabeigi, K. (1992) "Global Surface Temperature/Heat Transfer Measurements Using Infrared Imaging," AIAA 92-3959.

DeMeis, R. "Paint Under Pressure," Aerospace America, pp. 40-41, March 1992.

Jones, R. A., and Hunt, J. L. (1964) "Use of Temperature-Sensitive Coatings for Obtaining Quantitative Aerodynamic Heat-Transfer Data," AIAA J. vol. 2, no. 7, July 1964.

Jones, R. A., and Hunt, J. L. (1966) "Use of Fusible Temperature Indicators for Obtaining Quantitative Aerodynamic Heat-Transfer Data," NASA TR R-230.

Merski, N. R. (1991) "A Relative-Intensity Two-Color Phosphor Thermography System," NASA TM--104123.

Micol, J. R. (1989) "Experimental and Predicted Pressure and Heating Distributions for an Aeroassist Flight Experiment Vehicle in Air at Mach 10," AIAA 89-1731.

Micol, J. R. (1991) "Aerothermodynamic Measurement and Prediction for a Modified Orbiter at Mach 6 and 10 in Air," AIAA 91 -1436.

Miller, C. G. (1981) "Comparison of Thin-Film Resistance Heat-Transfer Gages with Thin-Skin Transient Calorimeter Gages in Conventional Hypersonic Wind Tunnels," NASA TM-83197.

Miller, C. G., Micol, J. R., and Gnoffo, P. A. (1985) "Laminar Heat-Transfer Distributions on Biconics at Incidence in Hypersonic-Hypervelocity Flows," NASA TP 2213.

Miller, C. G. (1985) "Experimental and Predicted Heating Distributions for Biconics at Incidence in Air at Mach 10," NASA TP 2334.

Miller, C. G., and Smith, F. M. (1986) "Langley Hypersonic Facilities Complex - Description and Application," AIAA 86-0741-CP.

Miller, C. G. (1990) "Langley Hypersonic Aerodynamic/Aerothermodynamic Testing Capabilities - Present and Future," AIAA 90-1376.

Neumann, R. D. (1989) "Aerothermodynamic Instrumentation," AGARD R-761, pp. 4-1 to 4-40.

Schultz, D. L., and Jones, T. V. (1973) "Heat-Transfer Measurements in Short-Duration Hypersonic Facilities," AGARD AG-163, February 1973.

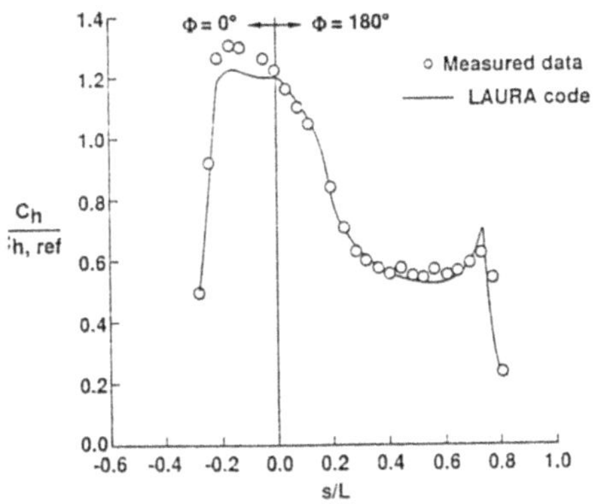

Fig. 1 Measured and predicted heat transfer distributions for AFE. (All-Macor thin-film model, M=10, α = -5°; from Micol, 1989)

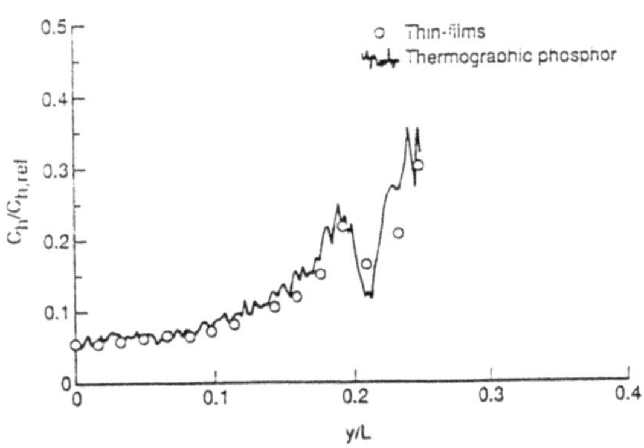

Fig. 2 Spanwise heat transfer distributions for modified orbiter measured via thin-film gages and phosphor thermocouple. (M=10, α = 20° x/L = 0.75; from Micol, 1991)

AEROTHERMAL TEST METHODOLOGY AND TECHNIQUES FOR THE DEVELOPMENT OF HYPERSONIC VEHICLES

R. K. Matthews
Calspan Corporation/AEDC Operations
Arnold Engineering Development Center
Arnold AFB, TN 37389

Abstract. The development of hypersonic vehicles requires a large volume of high-quality heat-transfer data which will be primarily used to validate codes. In addition, tests must be conducted to demonstrate the survivability of materials/structures in the extreme thermal environment of hypersonic flight. The methodology relating these tests and the techniques available to the experimentalist need to be clearly understood. This paper provides a brief survey of the methodology, typical facilities, and the aerothermal test techniques that have been used in the U.S. to develop hypersonic vehicles.

1. Introduction

The next generation of lifting hypersonic vehicles poses many technical challenges. Despite their exposure to severe aeroheating and aerodynamic loading, components should be reusable, light, and maintainable. Vehicle surfaces will typically be exposed to very high heating and, therefore, the design of certain components such as leading edges, control surfaces, protuberances, and shock interference regions is very complicated. Joints and seals must protect underlying cooling passages and/or storage tanks from severe environments without degradation. Unfortunately, the ability to predict the boundary-layer state accurately—laminar versus turbulent—continues to elude the aerodynamicist. A successful response to the challenges of component design lies in the proper integration of engineering analysis and experimentation to verify the engineering design of the vehicle.

The key role of analysis cannot be overemphasized. The design of the U.S. space shuttle required an enormous amount of wind tunnel testing and, despite the advances in Computational Fluid Dynamics (CFD), the coming generation of hypersonic vehicles will probably also require extensive testing. However, CFD will clearly play a much more significant role and will be used to guide test programs and to extrapolate data to flight. More fundamentally, the objective of current and future test programs will be to validate codes whereas, in the past, the experiments emphasized the generation of a massive database. Code validation experiments typically require high-quality surface and flow-field measurements.

*The research reported herein was performed by the Arnold Engineering Development Center (AEDC), Air Force Systems Command. Work and analysis for this research were done by personnel of Calspan Corporation/AEDC Division, operating contractor for the AEDC aerospace flight dynamics facilities. Further reproduction is authorized to satisfy needs of the U. S. Government.

A. Boutier (ed.), New Trends in Instrumentation for Hypersonic Research, 461–470.

2. Methodology

A common approach used to solve aerothermal issues is based on combining analysis with experimentation. It is imperative that analytical techniques be used to plan the test and to analyze the final data. The two fundamental phases (see Fig. 1) in the development process are:

1. defining the flight thermal environment
2. demonstrating hardware survivability

In defining the thermal environment, the versatility of analytical tools in combined with the experimentally measured heating distributions. These data, obtained on scaled models in simulated flow environments, are used to verify the accuracy of the analytic tools. The important simulation parameters are Mach number and Reynolds number.

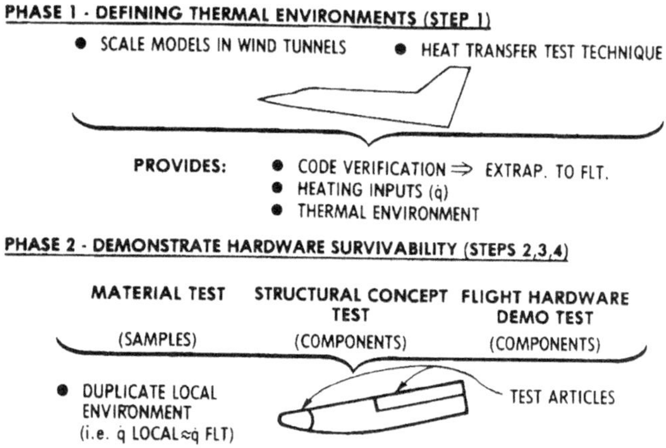

Fig. 1. Methodology for aerothermal structures/materials development.

The second major phase in the development of structural components is "demonstration of hardware survivability." The general approach is to use analysis tools to design tests that simulate the flight environment, and then to expose flight components to this environment in ground test facilities. Material selection is perhaps the most difficult and the most important step. Determining thermophysical properties and characterization of materials requires many hours of laboratory experiments. Despite these efforts, it is often difficult to predict material failure modes in a flight environment. Material survivability can be a function of many variables. Wind tunnels and arc-heated facilities often use the wedge testing technique[1] to produce local flow environments that simulate flight. The primary test results are test article condition after the run (i.e., survivability). The simulated flight environment is characterized by surface temperature, heating rate ($\dot{q}$), local surface pressure (p), and shear (τ). Of course, run time (or exposure time) is also a key test variable.

The test article may range from a simple insulated material panel to a complex structure with liquid hydrogen cooling passages. If the test article incorporates backside cooling, it is important to simulate both the aeroheating ($\dot{q}_{in}$) and the heat removed by the coolant ($\dot{q}_{out}$). Therefore, the mass flow, heat capacity, temperature, and viscosity of the coolant must be duplicated in the ground test.

3. Typical Facilities

Two of the primary facilities used to generate the shuttle database[2] were Tunnels B and C of Arnold Engineering Development Center (AEDC). These continuous flow, hypersonic tunnels played a dominant role in defining the shuttle thermal environment by providing a large quantity of high-quality data in a perfect gas environment.

The Aerothermal Wind Tunnel C, one of three wind tunnels in the AEDC von Kármán Facility (VKF), has also been used extensively to evaluate materials. Tunnel C (Fig. 2) is a continuous-flow supersonic/hypersonic wind tunnel which has three interchangeable, axisymmetric contoured nozzles with 25-in. and 50-in.-diam nozzle exits. The nominal tunnel operating conditions are also shown in the figure. The relatively new Mach 8, 25-in. diam nozzle provides a unique high Reynolds number ($7.8 \times 10^6 ft^{-1}$), high-enthalpy, perfect gas wind tunnel capability. Hypersonic Wind Tunnel B is very similar to Tunnel C and provides for testing of large models (50-in. diam test section) at Mach numbers of 6 and 8. Both Tunnels B and C are equipped with a model injection system which allows removal of the test article from the test section while the tunnel remains in operation.

The model injection system and the continuous operation are the key elements that provide the high productivity of these tunnels. For example, heat-transfer tests have been conducted with a run rate of a model injection every 2 min. Of course, Reynolds number and model changes interrupt this rate, but it is common to obtain over 100 runs in one 8-hour shift.

M	P_{0MAX} PSIA	T_0, °R	EXIT DIAM., IN.	RUN TIME,	RUNS/HR
4	180 100	1,660 1,900	25	CONTINUOUS*	5 - 30
8	1,900	1,900	25		↓
10	1,900	1,950	50		

*(WITH MODEL INJECTION SYS)

Fig. 2. Tunnel C operating conditons.

The primary advantages of these large continuous flow facilities when compared to impulse or shock tunnels are:

- higher productivity (by a factor of 10 to 100 depending on the specific test matrix)
- wider variety of test techniques to address specific problems
- higher quality data; well-defined calibrated free stream; repeat data easy to obtain (over 25 years of experience).

The disadvantages of these facilities are that total test costs are high and, in general, they are for perfect gas development testing versus real gas research.

Three arc heaters are available at AEDC for performing hardware survivability tests. The H1 and HR test units have nozzle exit diameters of up to 4 in. and exhaust to atmospheric static pressure providing a free-jet test rhombus. With a chamber pressure of 100 to 120 atm, the heating rates and pressures of reentry or ascent can be properly simulated. The new H2 test unit (Fig. 3) has recently been calibrated and is documented in Ref. 3. This facility has nozzle exit diameters of 9, 24, and 42 in. All three test units have a rotating model injection system allowing from 5 to 8 test samples to be evaluated during one run. Run times vary from 45 sec for the "hottest" condition in H1 to 10 min for the less severe conditions in H2. Instrumentation and measurement requirements can be accomplished with the existing thermocouple, pressure, pyrometers, radiometers, high-speed cameras, real-time laser recession system, and online recording and computer systems.

4. Test Techniques

The test techniques available to measure aero-heating are listed in Fig. 4 along with a reference which illustrates the use of the technique. Thermal mapping techniques provide a comprehensive look at the entire model and are often used to identify the location of high heating rates (e.g., shock interaction). However, the uncertainty of thermal mapping data is of the order of ± 15 percent; whereas, the discrete measurement techniques can produce ± 6-percent data which are more reasonable for code validation test.

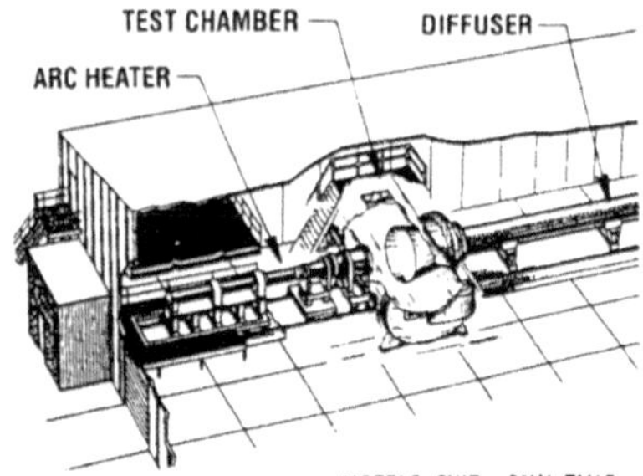

M	$P_{O_{MAX}}$, PSIA	T_O, °R	NOZZLE EXIT DIAM, IN.	RUN TIME, MIN	RUNS/WK
4 TO 5	500 - 1,000	~8,000	9	2 TO 20	1 TO 5
6 TO 8	↓	↓	24	↓	↓
7 TO 8.5			42		

Fig. 3.Heat-H2 test unit.

	ADVANTAGES	DISADVANTAGES	REFERENCE
THERMAL MAPPING			
• PHASE-CHANGE PAINT	VIVID ILLUSTRATION OF HOT SPOTS HIGH SPATIAL RESOLUTION	MUST REAPPLY PAINT, DATA PRESENTATION CAN BE CONFUSING	4
• IR SCANNING CAMERA	COMPUTER-GENERATED PLOTS AND COLOR MAPS, AND NONINTRUSIVE	SPATIAL RESOLUTION	5
• THERMOGRAPHIC PHOSPHOR	COMPLETE MODEL, GOOD SPATIAL RESOLUTION	MODEL PREPARATION AND DATA PRESENTATION	6
DISCRETE MEASUREMENTS			
• THIN-SKIN	HIGH QUALITY DATA, DENSE SPACING	EXPENSIVE MODEL FAB, CONDUCTION EFFECTS	7
• COAX GAGE	EASY TO INSTALL, CONTOURABLE, DURABLE	LOW OUTPUT, SHORT TEST TIMES	8
• SCHMIDT-BOELTER GAGE	HIGH OUTPUT, SLIGHTLY CONTOURABLE, VERY DURABLE	FAB AND CALIBRATION TIME REQUIRED	9
• GARDON GAGES (HI TEMP, LOW TEMP)	YEARS OF EXPERIENCE, FAST RESPONSE	GAGE ATTRITION RATE, NOT CONTOURABLE	10
• THIN-FILM	DENSE SPACING, FAST RESPONSE, CAN BE USED ON SMALL RADII	RELATIVELY DIFFICULT INSTALLATION, MATERIAL CRACKING	11

Fig. 4.Test techniques available for measurement of heat transfer.

5. Thermal Mapping

5.1 PHASE CHANGE PAINT TECHNIQUE

The Phase Change Paint technique of measuring the heat transfer to a model surface was developed by Jones and Hunt.[12] This technique assumes that the model wall temperature response is similar to that of a semi-infinite slab subjected to an instantaneous and constant heat-transfer coefficient.

Prior to each run, the model is cleaned and cooled with alcohol and then spray painted with a paint that melts at a specific temperature. The model is installed on the model injection mechanism at the desired test attitude, and the model initial temperature is measured. The model is then injected into the airstream for approximately 25 sec; during this time the model surface temperature rise produces isotherm melt lines. The progression of the melt lines is photographed with 70 mm sequenced cameras operating at one frame per second. An example of phase-change paint data compared to thin-skin data is presented in Fig. 5.

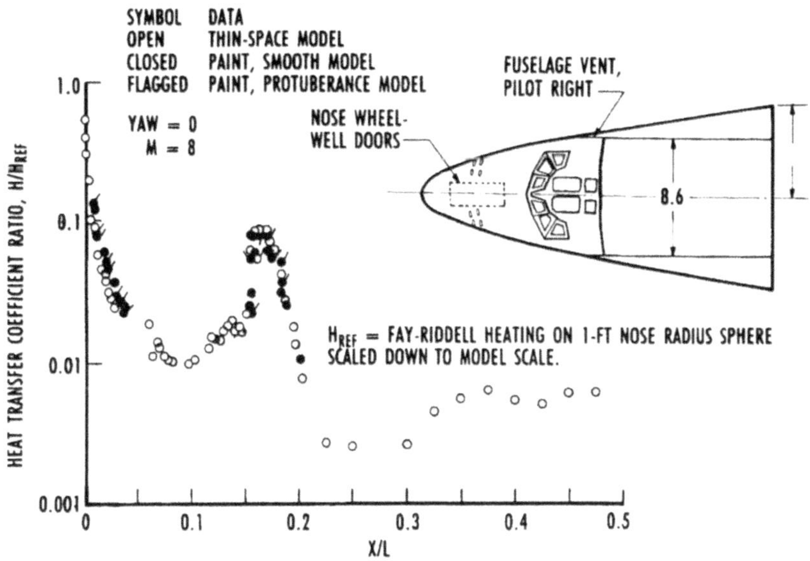

Fig. 5. Leeward centerline heat-transfer distributions at $Re_{\infty L} = 8.6 \times 10^6$, $\alpha = 30$ deg.

5.2 THERMOGRAPHIC PHOSPHOR PAINT

Thermographic phosphor paints are sensitive to temperature and, as temperature increases, their luminosity decreases. The luminosity of these paints is typically excited by ultra-violet or laser light and as the temperature patterns develop, photographs and/or video recordings are made. Thermographic phosphor paints have been used both in wind tunnel testing and in turbine engine tests. In the wind tunnel these paints measure surface temperature patterns which are used to infer heating rates. In turbine engine tests the phosphor paints have been applied to turbine blades to measure temperature.

In the 30-year history of the evolution of this test technique, clearly the most significant advances have been made in the "data handling" area. In the early days film reading, data reduction, and data presentation was done manually requiring many, many manhours. Today, computers do the entire job and display color pictures and plotted data within seconds after a run.

6. Discrete Measurements

6.1 THIN SKIN

The thin-skin technique has been used for many years and remains one of the most accurate and reliable methods available. The reduction of thin-skin temperature data to coefficient form normally involves only the calorimeter heat balance for the thin skin. Thermal radiation and heat conduction effects on the thin-skin element are neglected and the skin temperature response is assumed to be due to convective heating alone. A complete discussion of the thin skin test technique as applied to a particular test situation (including presentation and analysis of the resultant data) is contained in Ref. 13, and an example of the data quality that can be obtained is presented in Fig. 6.

6.2 COAX GAGE

The coax gage is simply a surface thermocouple which is comprised of an insulated Chromel® wire fixed concentrically within a Constantan jacket. The thermocouple junction is formed at the sensing surface by blending the two materials together with a file. This filing process provides one of the primary advantages of the coax gage; that is, it can be contoured to conform exactly to the model surface. The extremely fast response of these gages (50 μsec to a step input) makes possible their application to continuously moving models or moving components of models.

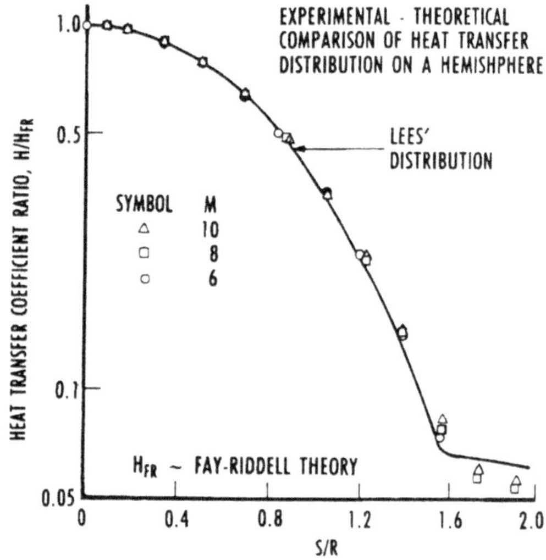

Fig. 6. Illustration of thin-skin data quality.

The sketch in Fig. 7b illustrates the installation of coax gages in the body flap of the Orbiter model which was used to evaluate the continuous-sweep concept. The results presented in Fig. 7c are typical of data obtained during testing of the Orbiter body flap and illustrate the excellent agreement obtained between sweep measurements and discrete (fixed-angle) measurements. These data are from the same gage operated in both a discrete (fixed point) and sweep mode. Such results indicate that reliable heat-transfer data can be obtained if certain precautions are followed. In particular, the data acquisition rate, the model sweep rate, and the method of data filtering (or averaging) used to reduce data scatter must be coordinated to meet test data quality requirements. Also, care must be taken to avoid violating the assumed semi-infinite solid thermal response of the model wall. This implies that sweep data should be obtained only during moderate (less than 5 sec) aeroheating exposures. For further details regarding test results and analysis of the advantages and restrictions of this technique, see Ref. 8.

6.3 SCHMIDT-BOELTER GAGE

AEDC has made operational a heat-flux transducer, the Schmidt-Boelter gage, which alleviates many deficiencies found in the thermopile Gardon gage discussed in Ref 10. Schmidt-Boelter gages have seen considerably wider usage in recent years for heat-transfer measurements in continuous-flow wind tunnels and flight test applications. This is primarily due to the attractive operating characteristics common to this type sensor. These include (1) excellent durability, (2) good sensitivity, (3) self-generating output signal directly proportional to incident heat flux, (4) continuous service temperature of approximately 700°F, and (5) semicontourability.

Experimental calibrations of Schmidt-Boelter gages are performed using a radiant heat source. These experimental procedures enable a calibration scale factor (CSF) to be obtained for each gage allowing heating rates to be calculated. AEDC has recently developed a high-

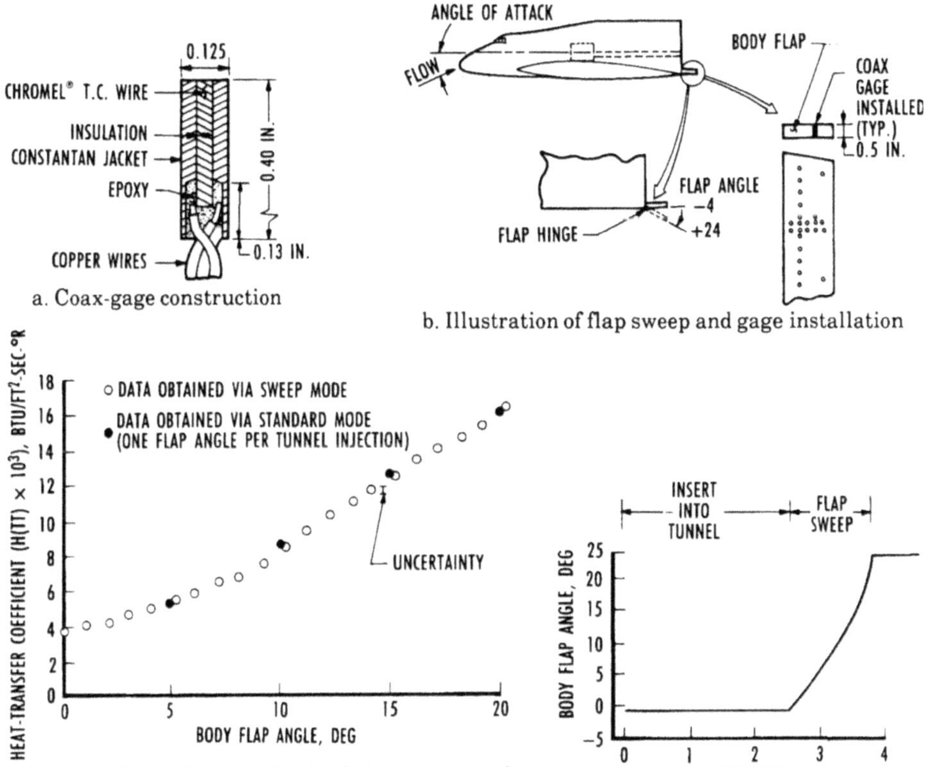

a. Coax-gage construction

b. Illustration of flap sweep and gage installation

c. Comparison of data obtained in sweep and standard modes

d. Body flap sweep profile

Fig. 7. Continuous sweep coax-gage data.

heat flux gage calibrator (Fig. 8). Through efficient collection of the radiant energy from a high-intensity xenon arc lamp, a lab calibration apparatus (Fig. 8a) has been developed to deliver up to 1,500 Btu/ft^2-sec. Emphasis has been placed on uniformity (Fig. 8b), repeatability, and continuous monitoring of the heat source. Continued development is underway to calibrate a gage at elevated temperatures to account for gage material property variations with temperature.

6.4 THIN FILM

The principles of operation and applications of the thin-flim gage are widely documented, however, their use in continuous flow wind tunnels has been restricted by film survivability problems. Reference 11 describes a technique of installing the films on a machinable glass ceramic (MACOR) and the measurement of Shuttle wing leading edge heating which has proven to be very successful. This technique also utilized a material heat conduction model to improve data quality.

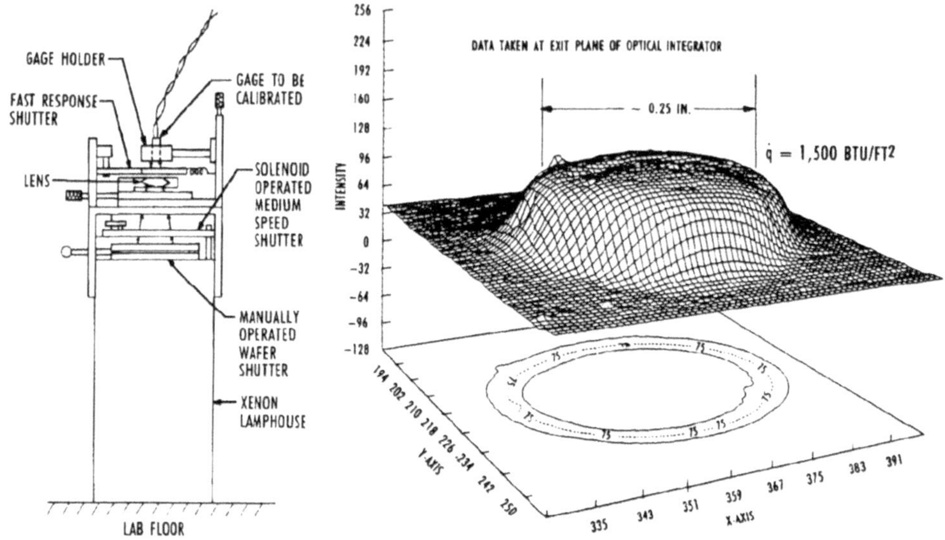

a. Apparatus b. Heat flux iniformity

Fig. 8. Capability for high heat flux calibration.

7. Materials/Structures

In general, materials/structures testing can be grouped into the four categories listed in Fig. 9, and discussed below.

	ADVANTAGES	DISADVANTAGES	REFERENCE
• SCREENING TEST	RELATIVELY SIMPLE, MANY TEST ARTICLES RUN IN SHORT TIME	ONLY RELATIVE RANKING BETWEEN TEST ARTICLES	14
• CHARACTERIZATION TEST	ALLOWS EVALUATION AT SELECTED FLIGHT CONDITIONS	REQUIRES TESTING OVER A RANGE OF SEVERAL PARAMETERS, $\dot{q}$, T_W, SHEAR	15
• COMPONENT SURVIVABILITY	PROVIDES "YES" OR "NO" ANSWER FOR SIMULATED FLIGHT CONDITION	NO QUANTITATIVE DATA FOR "OTHER" FLIGHT CONDITIONS	16
• COMPONENT THERMAL RESPONSE	CAN PROVIDE TEMPERATURES FOR VARIETY OF SIMULATED FLIGHT CONDITIONS	DATA ONLY VALID FOR SPECIFIC COMPONENT TESTED	17

Fig. 9. Categories of aerothermal materials/structures test techniques.

Screening tests make relative comparisons among many candidate test articles in a constant (or repeatable) test environment. It is desirable to fabricate several identical test articles to investigate the repeatability of failure modes.

Characterization tests are the inverse of screening tests in that one test article design is exposed to a variety of test conditions. For example, the test variables may be temperature and pressure, and each is held constant while the other is varied over the range of interest. The data can be developed into an algorithm that characterizes the test article.
Component survivability tests basically answer the question, "Will component X survive environment Y?" The disadvantage of this type test is that no quantitative data are produced for "other" flight conditions.

Component thermal response tests consist of a fully instrumented test article exposed to a wide variety of conditions so that a relatively sophisticated math model can be developed. The math model can be used to predict internal component temperatures for a variety of flight profiles. The disadvantage of this technique is that it is only valid for the specific component that is used to generate the database. If the component design changes after the test, it may be necessary to rerun the entire test for the new design.

8. Concluding Remarks

The complexities of hypersonic vehicles challenge the experimentalist to develop new facilities and innovative test techniques. The variety of test techniques available to the experimentalist continues to grow, and the importance of his selection is compounded by rising test costs. There are many factors that influence the selection of a test technique, but clearly the number one consideration must be the *test objective*. A precisely defined test objective, coupled with comprehensive pretest planning, are essential for a successful test program.

Over the years AEDC has developed a comprehensive inventory of test techniques for both defining thermal environments ($\dot{q}$-data) and for the demonstration of hardware survivability (materials/structures testing). Some of the specific recent accomplishments include:

- calibration of the new H2 arc-heated wind tunnel
- high Reynolds number operation with a 25-in.-diam Mach 8 nozzle in Tunnel C
- fabrication and development of a high heat flux calibrator ($\dot{q} \approx$ 1,500 Btu/ft^2-sec).

9. References

1. Matthews, R. K. and Stallings, D. W. "Materials Testing in the VKF Continuous Flow Wind Tunnels." AIAA 9th Aerodynamic Testing Conference, Arlington, TX, June 7-9, 1976.

2. Fehrman, A. H. and Masek, R. V. "Study of Uncertainties of Predicting Space Shuttle Thermal Environment." McDonnell Douglas, St. Louis, Report MDC-E 0639, June 1972.

3. Davis, L. M. and Carver, D. B. "Initial Calibration of the HEAT-H2 Arc-Heated Wind Tunnel." AEDC-TR-91-16 (AD-A245072), January 1992.

4. Matthews, R. K. and Gilley, G. E. "Reduction of Photographic Heat-Transfer Rate Data at AEDC." AEDC-TR-73-90 (AD-762928), June 1973.

5. Boylan, D. E., Carver, D. B., Stallings, D. W., and Trimmer, L. L. "Measurement and Mapping of Aerodynamic Heating Using a Remote Infrared Scanning Camera in Continuous Flow Wind Tunnels." AIAA 10th Aerodynamic Testing Conference, San Diego, CA, April 19-21, 1978.

6. Trimmer, L. L., Matthews, R. K., and Buchanan, T. D. "Measurement of Aerodynamic Heat Rates at the von Karman Gas Dynamics Facility." International Congress on Instrumentation in Aerospace Simulation Facilities, September 1973.

7. Carter, L. D. and Kaul, C. E. "Heat Transfer Tests on the Rockwell International Space Shuttle Orbiter With and Without Simulated Protuberances." AEDC-TR-75-20 (AD-A012876), July 1975.

8. Wannenwetsch, G. D. and Martindale, W. R. "Roughness and Wall Temperature Effects on Boundary Layer Transition on a 0.0175-Scale Space Shuttle Orbiter Model Tested at Mach Number 8." AEDC-TR-77-19 (AD-A038895), April 1977.

9. Matthews, R. K., Nutt, K. W., Wannenwetsch, G. D., Kidd, C. T., and Boudreau, A. H. "Developments in Aerothermal Test Techniques at the AEDC Supersonic/Hypersonic Wind Tunnels." Vol. 103, *AIAA Progress in Astronautics and Aeronautics*, 1986.

10. Stallings, D. W., Matthews, R. K., and Jenke, L. M. "Recent Developments in Aerothermodynamic Test Techniques at the AEDC von Karman Gas Dynamics Facility." International Congress on Instrumentation in Aerospace Simulation Facilities, September 1979.

11. Wannenwetsch, G. D., Ticatch, L. A., Kidd, C. T. and Arterbury, R. L. "Measurement of Wing-Leading-Edge Heating Rates on Wind Tunnel Models Using the Thin-Film Technique." AIAA Paper 85-0972, June 1985.

12. Jones, Robert A., and Hunt, James L. "Use of Fusible Temperature Indicators for Obtaining Quantitative Aerodynamic Heat-Transfer Data." NASA-TR-R-230, February 1966.

13. Matthews, R. K. and Eaves, Jr., R. H. "Comparison of Theoretical and Experimental Pressure and Heat-Transfer Distributions on Three Blunt Nosed Cylinders in Hypersonic Flow." AEDC-TR-67-148 (AD-819547), September 1967.

14. Ievalts, J. O. and Matthews, R. K. "Aerothermal Evaluation of High-Temperature Structural Materials for Use in High Speed Missile Design," AEDC-TR-79-38 (AD-A080023), January 1980.

15. Matthews, R. K., and Harper, D. C. "Aerothermal Test of the Space Shuttle External Tank Insulating Material." AEDC-TR-75-94 (AD-A017497), November 1975.

16. Knox, E. C. "Thermal Response and Reusability Testing of Advanced Flexible Reusable Surface Insulation and Ceramic RSI Samples at Surface Temperature to 1200°F." AEDC-TR-79-62 (AD-A097711), April 1981.

17. Wannenwetsch, G. D. and Matthews, R. K. "Prediction Techniques Used in the M505A3E2 Fuze Auto-Ignition Investigation." AIAA-81-1041, June 1981.

MEASUREMENTS ON MODELS FOR HYPERSONIC REAL GAS CONDITIONS

H. OLIVIER, M. VETTER, C. JESSEN and H. GRÖNIG
Shock Wave Laboratory, University of Technology Aachen
Templergraben 55, 5100 Aachen
FRG

ABSTRACT: Pressure and particularly heat transfer measurements on the surface of a model in the test section flow of a shock tunnel are discussed. The heat transfer gauges include thin films, thermocouples and calorimeters. A new calibration technique for heat transfer gauges is introduced which consists of a fast dipping method. The method of Fay and Riddell has been used in several laboratories to deduce the stagnation temperature from the measured heat flow to a sphere yielding, in general, too high values. Using the velocity gradient at the stagnation point introduced by Stokes one obtains more realistic stagnation temperatures.

1. Introduction

Shock tunnels are ground test facilities which provide high enthalpy flows encountered in hypervelocity flight. Different driver techniques yield flow conditions with specific stagnation enthalpies between 2 MJ/kg up to about 30 MJ/kg [1]. Shock tunnels are well described in the literature regarding the conventional and free-piston driver technique [2]. A gaseous detonation driver seems to present a promising alternative to the free-piston driver owing to the absence of fast moving masses (up to about 1750 kg and 300 m/s) [3, 4] yielding, however, similar stagnation enthalpies. Since all shock tunnels are short-duration facilities special measuring techniques have to be applied [5, 6]. Classical methods have been used to obtain the pressure and heat flux distributions on models in the conventional Aachen shock tunnel TH 2. Mostly commercial pressure gauges (Kulite, Endevco) are used, they are mounted in the model wall behind a small cavity connected by a hole of 0.7 to 1 mm and 3 - 6 mm length to the outer surface. Figure 1 gives an example of

A. Boutier (ed.), New Trends in Instrumentation for Hypersonic Research, 471–480.

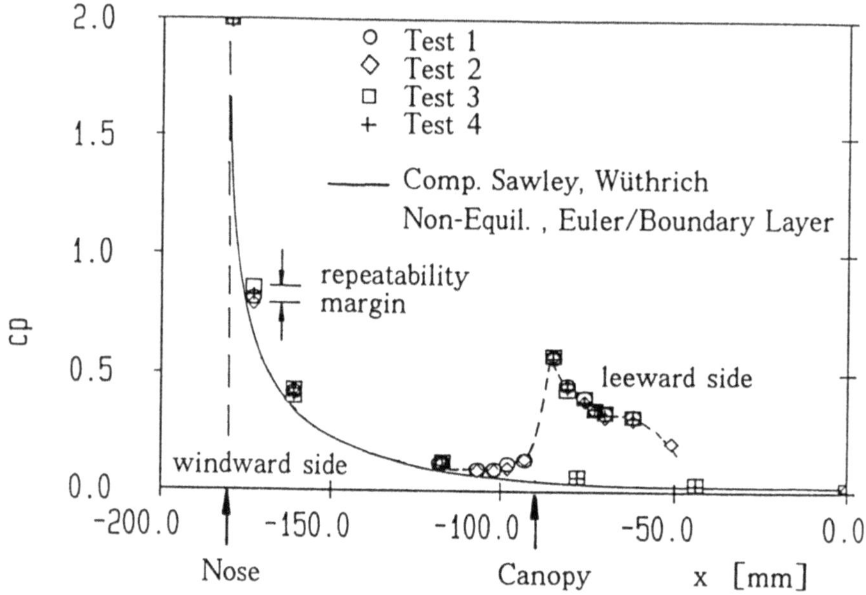

Figure 1. c_p versus x-coordinate for the plane of symmetry of a double ellipsoid [7, 8].

pressure measurements in the plane of symmetry of a double ellipsoid. Here the results of four tests are shown to demonstrate the reproducibility of the shock tunnel [7]. For the windward side a comparison with calculations are given [8]. Section 2 is devoted in more detail to heat transfer measurements applying thin film gauges, thermocouples and calorimeter gauges. Since these gauges, in general, measure the temperature rise during the test time the deduction of the heat flux requires an accurate knowledge of the thermal data of the gauge material [6]. There are several methods described in the literature to calibrate heat transfer gauges [5, 6]. In Section 2.2. we introduce a new one which to our best knowledge has not yet been used to calibrate heat transfer gauges for shock tunnel testing [9]. In Section 3 we briefly discuss the deduction of the free stream values in the test section from the measured pitot pressure and heat flux to a sphere [10]. This is done using the theory of Fay and Riddell [11]. As noted by other authors, too, this method renders stagnation temperatures higher than the nozzle settling temperature. One obtains lower temperatures if one introduces Stokes' velocity gradient at the stagnation point behind the bow shock [12] instead of taking the value from Newtonian theory as proposed by Lees [13] and further used by Fay and Riddell.

2. Heat Transfer Measurements

From the time history of the surface temperature one can deduce the heat flux to the model. Besides infrared techniques, liquid cristals or paints, the 'classical' method of measuring local temperatures delivers heat fluxes with a good accuracy. The probes used belong essentially to the following three types: thin film resistance thermometers, thermocouples and calorimeter gauges. Thin film gauges and thermocouples have a time response of a few microseconds while that of calorimeter gauges is longer owing to their larger heat capacity. The latter both have advantages in regions of heat fluxes higher than 300 W/cm^2 where thin film gauges reach their performance limit. Since these gauges have been used for many years there exists a large literature on them [6]. In the following we will present some new approaches of the three types of gauges mentioned.

2.1. THIN FILM GAUGES

A thin film heat flux gauge consists of a thin metal resistor mounted on an electrically insulating material, the substrate, as shown in Fig. 2 a. The thermal conductivity of the substrate should be much less than that of the film. Therefore quartz, hard glass (Pyrex) or machinable glass (Macor) usually serve as substrates. Sintered ceramics (ZrO_2) yield thin film gauges having high thermal sensitivity and mechanical strength. Since the making of these gauges is described in detail [14] here only a few remarks are in order. Two leads of platinum thermocouple wire are sintered into the ceramic material at temperatures of 1560 °C. The surface of the substrate is ground and polished to almost optical quality as can be seen from Fig. 2 b. A platinum or nickel film is deposited on the finished surface by means of vacuum evaporation through a mask of photo resistant lacquer. An additional

Figure 2. a) Scheme of a thin film gauge. b) Photograph of thin film gauges on graph paper.

protective and/or insulating layer of e.g. germanium may be evaporated on the metal film.

In the experiments the temperature history of the thin film resistance depends on the surface heat flux $\dot{q}_s$ in the following way

$$\dot{q}_s(t) = \sqrt{\frac{\rho c k}{\pi}} \left[\frac{T_s(t)}{t^{1/2}} + \frac{1}{2} \int_0^t \frac{T_s(t) - T_s(\tau)}{(t - \tau)^{3/2}} d\tau \right] \tag{1}$$

provided the one-dimensional theory may be applied and the surface measuring film has negligible effect on the heat conduction process [6]. ρ, c, k are density, specific heat and heat conductivity of the substrate, T_s its surface temperature. The voltage drop of the thin film resistance is related to the temperature by

$$\Delta U = \alpha R_0 I_0 [T_s(t) - T_s(0)] \tag{2}$$

where α denotes the temperature coefficient of the resistor, R_0 is the inital resistance of the film and I_0 the constant current through the film. One sees from (1) and (2) that the values $\sqrt{\rho c k}$ and α are needed to evaluate the heat flux.

Two types of calibration are necessary: a static calibration to determine the temperature dependence and a dynamic one to obtain the data to calculate the heat flux. Figure 3 shows the results of the static calibration. Static temperatures above 200 °C are critical because the thermal expansion of platinum is larger than that of the ceramic substrate. For short periods the sensors can withstand much higher temperatures, e.g. 750 °C for 90 seconds. The calibration data are approximated by a linear regression. In this example the correlation coefficient of 0.9995 proves the expected linearity from which the temperature coefficient α = 0.104 Ω/K is found.

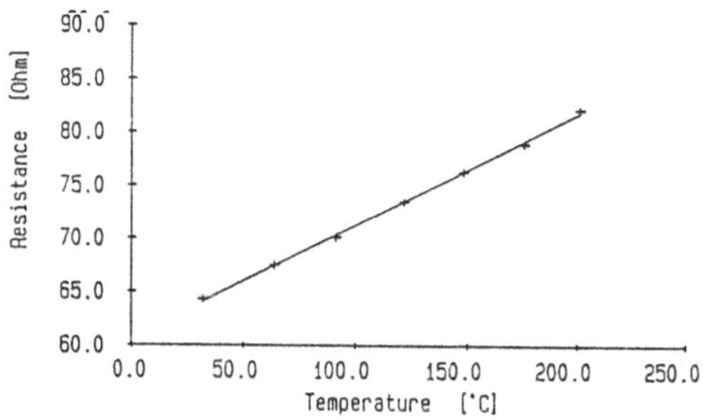

Figure 3. Static calibration points with a straight regression line.

Figure 4 shows the results of the $\sqrt{\rho c k}$ calibration. A rectangular voltage pulse heats the film, which is integrated in a Wheatstone bridge circuit, such that its response should ideally follow a square-root function of time. Figure 4 b indicates good agreement of the experimental data because the curve has a correlation coefficient of 0.999 after square-root transformation and linear regression. Deviations of the correlation coefficient to lower values indicate that the substrate is not perfect, and the gauge has to be ground again. The non-linear portions of the straight lines are excluded from the calibration, because they indicate a nonperfect bridge compensation.

Figure 4.

a) Dynamic calibration: 20.6 V voltage pulse and the response of thin film in air and glycerine in mV

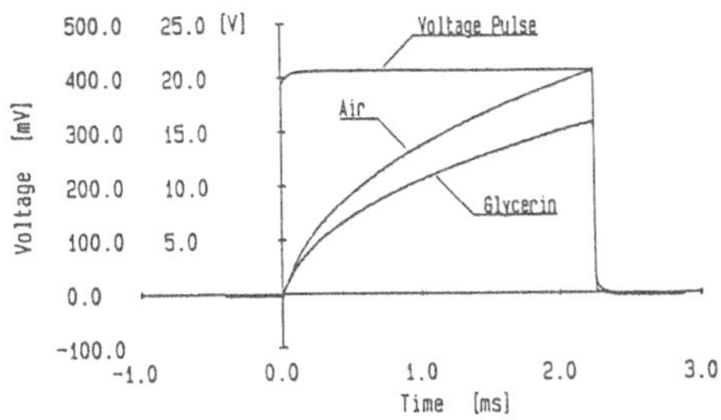

b) Thin film voltages versus square-root of time.

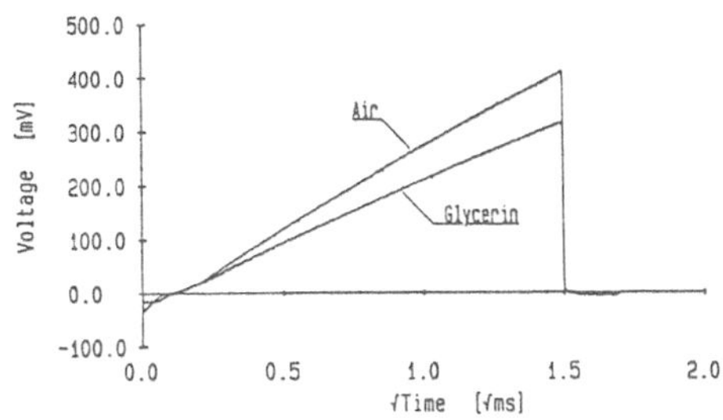

Although the values given above are valid only for an individual sensor they can be regarded as representative. About two hundred sensors have been calibrated so far individually and their calibration data scatter around ten percent. According to the proposal of Skinner [15] the calibration is carried out in air and glycerine to evade the problem and error source of measuring the area of the sensitive film. The average temperature dependence was found to be about 0.1 Ω/K and the average value of $\sqrt{\rho c k}$ was about 0.32 $J/Kcm^2s^{1/2}$. Typical results are given in Fig. 5 together with the response of thermocouples. Figure 5 shows the responses of a thin film gauge and a thermocouple mounted on the upper side and the leading edge of the wing of an ELAC model, respectively. Both probes display an almost parabolic temperature variation which yields nearly constant heat fluxes of about 15 and 100 W/cm^2, respectively. The test time with constant flow properties extends from about 3 to 6 ms. Thin film gauges in the winglet of that model measure heat fluxes as high as 250 W/cm^2.

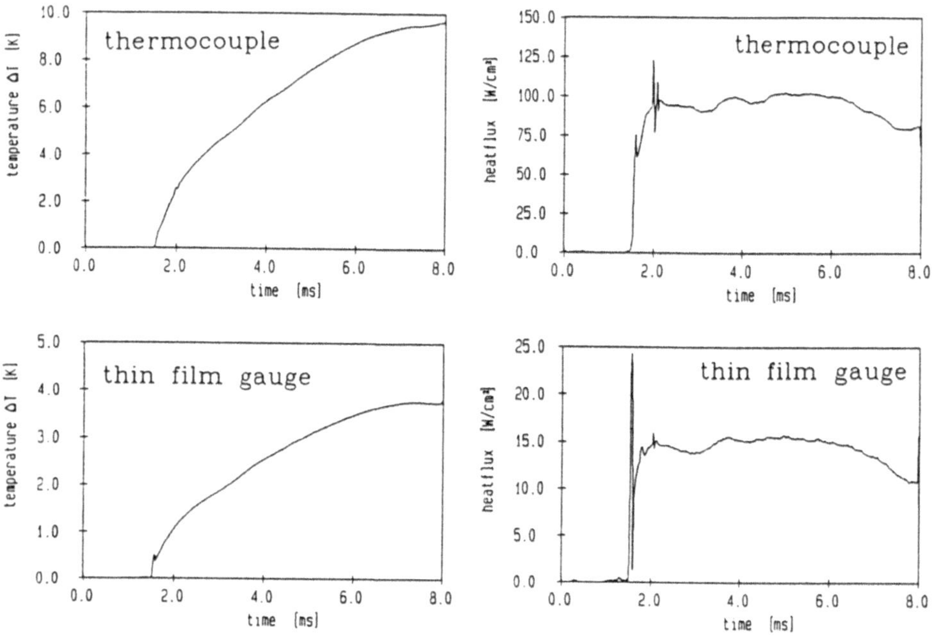

Figure 5. Temperature and heat flux history of a thermocouple and a thin film gauge. ELAC model $\alpha = -5$; $M_\infty = 7.3$; $p_0 = 190$ b; $T_0 = 2700$ K. Locations: TC in the leading edge of the wing, TFG on upper side.

2.2. THERMOCOUPLES

For heat fluxes higher than about 300 W/cm^2 we used coaxial thermocouples (chromel/constantan) commercially available (Medtherm Corporation). One thermocouple element (a tube, 1.55 mm outer diameter) is swaged over the second element (a wire) with an insulation 10 μm thick between them. The thermocouple junction is formed by filing. The response time of these gauges is about one microsecond. The behaviour of this type of thermocouple and the deduction of the heat flux are the same as for the thin film gauges because the layer forming the thermocouple junction is of the order of micrometers and less.

For the calibration of the thermocouples a new method was applied by which also thin film gauges can be recalibrated. This method is based on the familiar heat diffusion problem which arises when two semi-infinite bodies of different temperatures are suddenly brought into ideal contact. The solution of the problem yields a constant contact temperature T_m which depends on the two initial temperatures T_{10} and T_{20} of the two bodies and their $(\rho c k)$-values according to the following relation

$$\frac{T_m - T_{20}}{T_{10} - T_m} = \frac{(\rho c k)_1^{1/2}}{(\rho c k)_2^{1/2}} \quad . \tag{3}$$

To determine the $(\rho c k)$-value of the thermocouple one has to know that value of the other, in this case glycerine, the initial temperatures and has to measure the contact temperature by the thermocouple itself. The sudden contact of the probe to be calibrated with the fluid can be accomplished by fast dipping the probe into the fluid making sure, that the heat exchange is predominantly by conduction [9].

Figure 6 shows the voltage rise of a thermocouple after it has been dipped into a glycerine bath of about 90 °C. The voltage of the thermocouple rises to a corresponding constant temperature T_m within about 200 μs. This indicates that the method is reliable and takes the same period as the test time of the shock tunnel. The $(\rho c k)$-value of 8.22 $J/Kcm^2 s^{1/2}$ obtained for the chromel/constantan element agrees very well with the ONERA value of 8.43 $J/Kcm^2 s^{1/2}$.

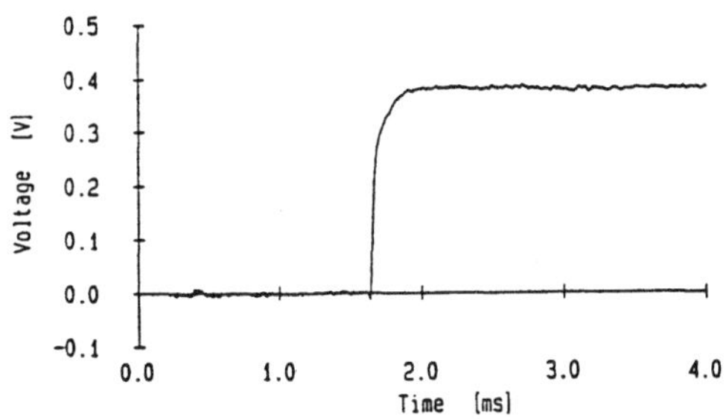

Figure 6. Calibration of a thermocouple: voltage rise of the thermocouple to a corresponding constant temperature T_m within 0.2 ms.

2.3. CALORIMETER GAUGES

The calorimeter gauges which are considered here belong to the so-called slug type [6]; they measure the local heat flux like thin film gauges or thermocouples. They usually consist of a small disk of copper or silver of 2 to 3 mm in diameter and about 0.15 - 0.2 mm thickness. They are mounted in an insert of low thermal conductivity, and the temperature of the slug is measured at the rear side usually by a thermocouple. If it is assumed that any heat losses from the rear side of the calorimeter can be neglected the measured heat flux to the surface may be written as

$$q_s = \rho c k \frac{dT}{dt} \quad . \tag{4}$$

Figure 7 shows the response of a copper slug calorimeter with a copper-constantan thermocouple in a model at nozzle stagnation conditions of 3800 K and 360 bar. (The copper slug calorimeter were manufactured by the Mechanics Institute, Academia Sinica, Beijing, and provided by Prof. H.-r. Yu, which is gratefully acknowledged.) The voltage of the thermocouple shows the expected nearly linear dependence on time. Taking the time derivative of the curve of Fig. 7 a one obtains the heat flux according to (4) and shown in Fig. 7 b. The dashed lines indicate the mean value of the heat flux during the test time between about 2.5. and 4.5 ms.

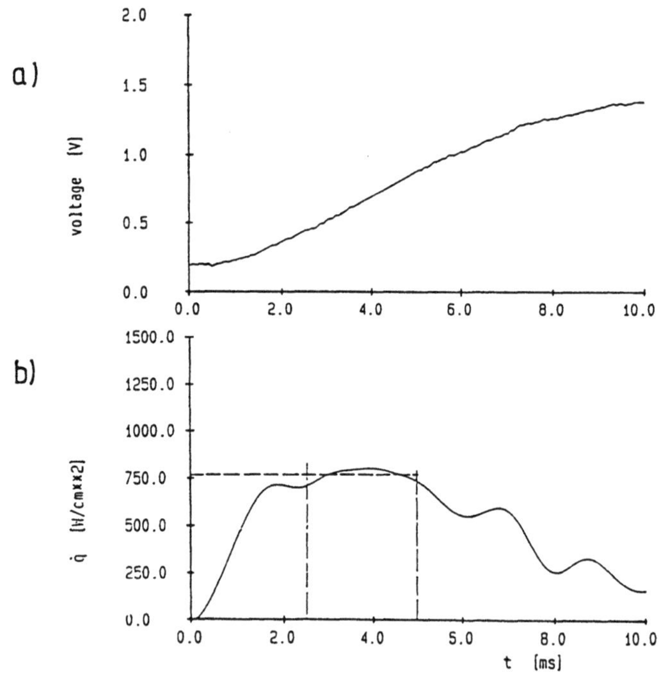

Figure 7. a) Voltage of a copper slug calorimeter in a model at condition III, $\alpha = 40^\circ$ ($T_0 = 3800$ K, $p_0 = 360$ bar; $M = 6.9$; $u_\infty = 3050\ ms^{-1}$). b) Deduced heat flux from a, dashed lines indicate mean value and duration of test time.

Advantages may be seen in the very smooth traces of the thermocouple voltage, however, fast variations of the heat flux may sometimes not be recorded. Under certain conditions one even takes advantage of the time dependence of the thin film gauge signals as this yields additional information. Because the sensitivity of the thermocouples are only about one sixtieth of the thin film gauges miniaturized preamplifiers may be used in the model to feed the usually long cables to the recording system to improve the signal to noise ratio.

3. Special Applications, Evaluation and Outlook

In some short-duration wind tunnels spheres of a few centimeters in diameter are used to measure the surface temperature by means of thermocouples [1]. The aim is to deduce the free stream conditions in the test section from the measured values of surface temperature and pitot pressure. One has to use this indirect method because the nozzle stagnation conditions are even more difficult to measure particularly under high enthalpy conditions ($T_0 > 5000$ K, $p_0 > 1000$ bar). By an iteration the free stream conditions are determined from the measured surface temperature, pitot pressure and a third property, e.g., the static pressure which influence the iteration less [10]. This procedure makes use of the heat flux relation of Fay and Riddell [11], and one generally notices that the stagnation temperature deduced by means of this relation is too high. An essential variable in this relation is the velocity gradient at the stagnation point behind the bow shock of the sphere. Olivier [16] succeeded to improve this method when he introduced the velocity gradient according to Stokes [12] instead of the Newtonian value which was taken by Lees [13] and subsequently by Fay and Riddell [11]. The velocity gradient according to Stokes depends on the stand-off distance of the shock which is obviously missing in the Newtonian approach. The stagnation temperatures determined with the corrected method are now 10 to 12% lower than the previous ones and in good agreement with the expected nozzle stagnation temperatures [16].

In the advent of more and more sophisticated measuring techniques (like LIF or CARS) which are developed to measure free stream properties in high enthalpy tunnels, the 'classical' methods are still valuable tools. In particular, improvements have been made in developing thin film thermometers regarding their robustness and their ease to make; improvements have also been made in the calibration of thin film gauges and thermocouples and finally in the deduction of the properties of the free stream in the test section of a high enthalpy shock tunnel.

References

[1] Grönig, H. (1992) 'Shock tube application: High enthalpy European wind tunnels'. In: K. Takayama (ed.), Proceedings of the 16th Int. Symp. on Shock Waves, Springer Verlag Berlin, in press.

[2] Stalker, R.J. (1987) 'Shock tunnels for real gas hypersonics'. In: AGARD CONFERENCE PROCEEDINGS: Aerodynamics of Hypersonic Lifting Vehicles, AGARD - CP - 428, November 1987, paper 4.

[3] Morrison, W.R.B., Stalker, R.J., and Duffin, J. (1990) 'New generation of free-piston shock tunnels'. In: Kim, Y.W. (ed.), Current Topics in Shock Waves, Americ. Inst. of Phys., New York, pp. 582 – 587.

[4] Yu, H-r., Esser, B., Lenartz, M., and Grönig, H. (1992) 'Gaseous detonation driver for a shock tunnel'. To be submitted to Shock Waves.

[5] Emrich, R.J. (ed.) (1981) 'Fluid Dynamics'. In: L. Marton and C. Marton (eds.), Methods of Experimental Physics, Academic Press, New York, Chapters 5 and 7.

[6] Schultz, D.L. and Jones, T.V. (1973) 'Heat transfer measurements in short-duration hypersonic facilities', AGARDograph No 165, Febr. 1973.

[7] Vetter, M., Olivier, H., and Grönig, H. (1991) 'Flow over double ellipsoid and sphere - experimental results'. Workshop on Hypersonic Flows for Reentry Problems, Part II, INRIA, GAMNI-SMAI, Antibes, France, Vol. 4, pp. 1 - 3, Vol. 7, pp. 25 - 35.

[8] Sawley, M.L. and Wüthrich, S. (1991) 'Non-equilibrium hypersonic flow simulations using a coupled Euler/boundary layer method'. In: Aerothermodynamics for Space Vehicles, Proceedings of the First European Symposium, ESA SP-318, ESA Publications Div., ESTEC, Noordwijk, The Netherlands, pp. 387 – 392.

[9] Neumann, N. (1992) 'Kalibrierung von Oberflächenthermometern zur Wärmestrommessung', Student's thesis, RWTH Aachen.

[10] Olivier, H., Vetter, M., and Grönig, H. (1991) 'High enthalpy testing in the Aachen shock tunnel TH 2'. In: Aerothermodynamics for Space Vehicles, Proceedings of the First European Symposium, ESA SP-318, ESA Publications Div., ESTEC, Noordwijk, The Netherlands, pp. 377 – 384.

[11] Fay, J.R. and Riddell, F.R. (1958) 'Theory of stagnation point heat transfer in dissociated air', JAS 25, pp. 73 - 85.

[12] Stokes, G.G. (1880) 'Mathematical and Physical Papers', Vol. 1, Cambridge University Press, pp. 38 - 41.

[13] Lees, L. (1956) 'Laminar heat transfer over blunt-nosed bodies at hypersonic flight speeds', Jet Propulsion, Vol. 26, pp. 259 - 269.

[14] Jessen, C. and Grönig, H. (1991) 'A new method for manufacture of thin film heat flux gauges', Shock Waves 1, pp. 161 - 164.

[15] Skinner, G.T. (1962) 'A new method of calibrating thin film gauge backing materials', Cornell Aeron. Lab., Buffalo, New York, Rept. CAL – 106.

[16] Olivier, H. (1992) 'An improved method to determine the free stream conditions in hypersonic facilities'. To be submitted to Shock Waves.

THIN GRADIENT HEAT FLUXMETERS DEVELOPED AT ONERA

P. Kayser, M. Rodrigues, J.C. Godefroy
Office National d'Etudes et de Recherches Aérospatiales
BP 72, 92322 Châtillon Cedex, France

SUMMARY. To meet the experimental needs of aircraft engine manufacturers, ONERA has undertaken research and development of thin gradient heat fluxmeters for measuring steady or quasi-steady fluxes. Two types of transducers are presented, the first one is a simple gradient fluxmeter usable up to 230°C, sticked on aerodynamical profiles, the second one is a multijunction gradient fluxmeter usable up to 1000°C, directly deposited on profiles or turbine blades. The principle, the design and the main characteristics of the sensors are described, as well as the techniques used for their calibration.

1. Introduction

To meet the experimental research needs of aircraft engine manufacturers, wind channel and calibration facility's users, ONERA has undertaken research and development of thin film transducers with the support of the DRET[1]. The planned applications concern surface temperature, heat fluxes, mechanical stress and blade tip clearance measurements.

Thin film transducers offer the capability of being implemented without machining of the profile. They don't disturb the flow and quite fully preserve the heat exchanges.

As regards heat fluxes, ONERA has developed two types of transducers for measuring steady or quasi-steady fluxes (convection or conduction-type exchanges):

- a simple gradient fluxmeter usable up to 230°C which can be sticked on aerodynamical profiles or blades,

- a multijunction gradient fluxmeter usable up to 1000°C, devoted to turbine hot stages and made of superimposed thin films directly deposited on profile or turbine blade.

The purpose of this paper is to present the two types of transducers, their principle, design and main characteristics.

2. Moderate temperature range-thin film fluxmeters

2.1 Principle

The transducers make use of the thermoelectric effect appearing at the junction-point between two partially superimposed metals deposited as thin layers on electric insulating substrate.

For the sensitive elements, thermocouples have been preferred to resistive films because the thermoelectric voltage does not depend on the conductor resistance.

[1]DRET: French Directorate for Armament, Research, Studies and Techniques.

A. Boutier (ed.), New Trends in Instrumentation for Hypersonic Research, 481–492.

The heat flux passing through the tranducer can be calculated by the Fourier relation:

$$\phi = \frac{\lambda}{e}(T_S - T_d) \qquad (1)$$

in which ϕ is the flux expressed in Wm^{-2} , λ the heat conductivity of the thermal insulating layer (Kapton film of figure 1) expressed in $Wm^{-1}K^{-1}$, e the thickness of this layer expressed in meters, and $T_s - T_d$ the difference between the temperatures measured on opposite sides of the thermal insulating layer by the deposited thermocouples (figure 1).

2.2. Fabrication and description of the transducer

The technological processes for fabricating the fluxmeters are the following ones:

- ultrasonic cleaning of the Kapton polymide films in an alkaline bath;
- a Kapton film (M1) (figure 2) receives, on both front and back sides, partially superimposed metal linings of nickel and copper (thickness 0,2 μm) forming the two elements of the thermocouples. The metallic layers are deposited by evaporation under vacuum, through metallic rigid masks. A protective gold layer (thickness 1,2 μm) is deposited on the connection pads of the copper conductors;
- films M2 and M3, also in Kapton, are glued to M1 by hot pressing using an epoxy glue. A 0,2 μm thick nickel layer (gard metal deposit) is evaporated on to the surface of the M2 film;
- after thermoelectric calibration, the assembled transducer is installed, also by glueing and pressing, using a mould, to the wall on which the measurements are to be made;
- copper wires are hard-soldered on nickel and copper (covered with gold) layers, using tin-lead soldering.

It is important to note that several transducers may be obtained simultaneously on a same substrate, and thus enable the spatial analysis, along a profile chord for instance, of the thermal phenomenon to be observed.

2.3. Metrological characteristics

2.3.1. Sensitivity of deposited thermocouples. The surface temperature T_s obtained with these transducers is related to the voltage E_s by the approximate relationship:

$$E_s = a\,T_s + \frac{1}{2}\,b\,T_s^2 + \ldots \qquad (2)$$

The sensitivity is the Seebeck coefficient, or thermoelectric power of the thermocouple which expresses this sensitivity. It does not only depend on the nature of the metals used, bu also on their stape; for a thin-film thermocouple it is usually lower than that obtained with the same metals used as solid wires. Lastly, it varies slightly from one production batch to another.

Table 1 summarizes the characteristics of the thin-film thermocouples realized at ONERA within the framework of the study on thin-film transducers and shows, for comparison, the characteristics obtained with thermocouples made of solid elements.

table 1 - Compared characteristics of Ni/Cu thermocouples

Ni/Cu thermocouples / Characteristics		Solidwires	Thin-film
Coefficients	a, μVK^{-1}	21.89	15
	b, μVK^{-1}	$1.54.10^{-2}$	$3.18.10^{-2}$
Slope of characteristic, linearized between 0 and 100, μVK^{-1}		23.2	17.6
Dispersion of a, percent		-	± 10

This peculiarity of thin-film deposited thermocouples is expressed by two practical consequences:

- the sensitivity of every thin-film thermometer and fluxmeter will have to be determined by calibration,

- junction points between thin films and solid wires may be either maintained at the reference temperature of the circuit (0°C) or at a different temperature, provided that a corrective term be applied to the indicated voltage.

2.3.2. Influence of the thermal characteristics of the thermal insulating layer. We saw that the precision of the flux measurement is related to the knowledge of the thermal insulating layer conductivity λ and thickness e. But these values vary with temperature in a poorly known way, and it is only by calibration that systematic errors can be avoided.

2.4. Calibration of heat fluxmeters

Two calibration benches are used:

- the first one, of conductive type, is used for the determination of the thermoelectrical sensitivity of each thermocouple,

- the second one allows the global calibration, by means of a laser beam, of the fluxmeters already installed to the wall on which the measurements are to be made.

2.4.1. Temperature calibration. The principle of the device is shown in figure 3. The fluxmeter is pressed between two heating systems. The usefull area is 200 x 30 mm^2 . By varying the temperature T_1 , a family of characteristics $\Delta E_s = f(T_1)$ is determined.

The conduction losses are reduced by placing the set up in a vacuum cell.

2.4.2. Heat flux calibration. The principle of the bench is shown in figure 4. A 100 Watts YAG laser (wavelength = 1,06 μm) is associated to an optical device able to generate a 7 x 7 mm^2 energy homogeneous beam. The system insures a maximum 200 $kW.m^{-2}$ calibrated heat flux. The light/thermal energy conversion is obtained by using a black paint deposited onto the surface of the sample: 98% of the energy is absorbed for wavelengthes less than 3 μm, up to 150°C.

The heat fluxes used to calibrate the moderate temperature range-fluxmeters are in the range 3 to 20 $kW.m^{-2}$. The accuracy of the flux value is about 10%, due to fluctuations generated by the optical system and the laser source.

2.5. Applications of thin-film fluxmeters

For instance, different multisensor fluxmeters (12 to 28 sensors) have been developed for the following applications:

- Minos 900 fixed blades to be tested in wind channel,

- turbine compressor's disk to be tested according to different working conditions (gas flow, rotation speed, carter to disk clearance) (figure 5),

- models simulating iced helicopter rotor blades,

- model simulating space shuttle profile to be tested in a low gas pressure gust wind channel.

3. High temperature range-thin film fluxmeters

Such sensors, studied to meet the needs of experimental research on the hot stages of aircraft engines usable up to 1000°C, require the development of refractory insulating layers on nickel base superalloy as well as the use of noble metals for sensitive elements.

Of the various techniques for manufacturing the superimposed thin films of these transducers, RF diode cathodic sputtering has been selected. This technique produces the desired result with a minimum investment cost. It also avoids any further operations off the sputtering facility, such as oxidation of the primer film.

3.1. Principle of multijunction fluxmeters

As shown on figure 6a this type of transducer includes a first coat of NiCoCrAlY, designed to improve the relief of the superalloy surface and to ensure (with the yttrium) the adherence of the alumina layer deposited on top (1). The fluxmeter itself consists of:

- an even number of Platinel type thermocouples deposited in a single film and connected in serie, half of which are placed on top of a calorimetric film of zirconia (figure 6b),

- tiny bars of zirconia that acts as the thermal resistance of the transducer,

- a protective film of alumina or a set of three optically absorbent films, for measuring added radiative fluxes.

The advantage of such a transducer is that it is more sensitive than simple gradient fluxmeters (the sensitivity is proportional to the number of junction pairs) and does not contain a thermocouple inserted between two layers of oxides which endows it with a better mechanical strength at high temperature.

However, this structure does introduce a slight error in the evaluation of the fluxes, and also exhibits surface roughness. Nevertheless, the simplicity of construction can in most cases justify the choice of this design.

Other multijunction fluxmeters (2) (figure 7) include a flat upper layer of insulating material and give an exact measurement because, due to the presence of the surface film made of this insulating material, it places identical thermal resistances at the level of the different junctions. This puts us in a configuration identical to that of a fluxmeter with superimposed thermocouples. This multijunction fluxmeter (figure 6), on the other hand, gives an exact measurement only if:

1) the flux is constant in the area occupied by a given pair of thermocouples,
2) the temperatures at A and B (figure 6b) are identical,
3) the bars of zirconia do not perturb the fluid flow along the wall.

The first condition is generally met. As far as the second condition is concerned, a model has shown (figure 8) that the systematic error introduced is about 0.1%.

Finally, for the third condition, the 10 μm thickness of the zirconia bars is very much less than that of the thinnest boundary layers (about 100 μm) and consequently the flow is disturbed very little.

3.2. Manufacture and characteristics of the fluxmeter films

The films are deposited by RF diode cathode sputtering. The conditions for depositing the NiCoCrAlY, alumina and Platinel thermocouples have already been published (1).

The nominal compositions by weight of the targets used are the follows:

NiCoCrAlY : Ni ; 23 Co ; 21 Cr ; 8,5 Al ; 1,5 Y

Platinel

- positive element: 55 Pd ; 31 Pt ; 14 Au
- negative element: 65 Au ; 35 Pd

Yttria stab. Zirconia: 93 ZrO_2 ; 7 Y_2O_3

As the emf delivered by the Platinel thermocouples is about 42 μVK^{-1}, with $\lambda = 2\ Wm^{-1}K^{-1}$ and $e = 10^{-5}$ m, the minimum measurable flux is about 150 Wm^{-2}.

As far as the sensitivity of the fluxmeter with Platinel is concerned the transducer can deliver 0,7 $\mu VkW^{-1}m^2$.

3.2.1. NiCoCrAlY coating. As stated in section 3.1., this coating promotes adherence of the insulating layer of alumina. Studies have already shown (3 to 5) the importance of yttrium in this adherence. This primer layer is stable when deposited hot (700°C), if the structure evolves later, the any variations in the volume due to structural change will be smaller, and thereby cause less disturbance.

3.2.2. Insulating layer of alumina. Obtaining this layer has raised the greatest development difficulties (1). Firstly, the film is expected to withstand the effects of the differential expansion existing between itself and the superalloy, without spalling or cracking, from room temperature up to 1000°C (expansion coefficient $\simeq 8.10^{-6}K^{-1}$ for Al_2O_3 and $\simeq 16.10^{-6}K^{-1}$ for the supperalloy). The solution is to suitably heat the substrate (800°C) during deposition in order to prestress the alumina in compression at room temperature. When it reheats in use, this prestress will compensate the differential expansion effects.

Secondly, the problem of adherence must be examined. It has been shown (6, 7) that a film containing given compressive stresses and determined adherence may have a maximum thickness beyond which the film spalls. As concerns the necessary electrical insulation, this fact led us to determine a thickness of 8 μm for this alumina layer (1).

3.2.3. Yttria partially stabilizied zirconia films. The problem linked to differential expansion and adherence are similar to those of the alumina layer but the deposition parameters are different. The results obtained so far, with simultaneous heating and biasing of the substrate, appear to bear out those of other studies already published (8).

3.2.4. Platinel thermocouples. The electrical output signal of a gradient fluxmeter comes from a differential temperature measurement. By its principle, this signal will therefore be one of small amplitude. This fact made it necessary to look for a thermocouple that would have a high sensitivity. Of existing high-temperature thermocouples, the Platinel (Engelhard registered trade name) type was selected (9).

As is true for any thermocouple deposited in thin films, the thermoelectric sensitivity obtained is less than the sensitivity of bulk thermocouples (42 μVK^{-1} , here). The best deposition conditions were obtained by using a pumping station withe a cryogenic pump, and fitting a Meissner liquid nitrogen trap into the sputtering chamber (1). The use of high temperature necessarily entails annealing of the deposited thermocouples, and this yields a sensitivity of as much as 97% of the bulk value.

The main difficulties in the technological development of these deposited thermocouples, were the adherence of the negative thermoelement and a possible oxidation of the metallization at high temperature.

The adherence of the negative thermoelement could be improved by applying a sublayer consisting of the alloy AuPd itself, sputtered in an argon atmosphere enhanced with 10% oxygen.

Finally, the thermal aging of the deposited Platinel thermocouples may cause an oxidation of the palladium in the alloys. So it is necessary to deposit a protective film of alumina, about 0.2 μm thick, to protect the metal coatings from oxidation in contact with air at high temperature.

3.3 Results

Flat laboratory models are under characterization. The life time of zirconia bars deposited on alumina layers has been evaluated. These bars can withstand more than 3 x 6 hours at 1000°C under argon. Thermal aging that has already been done on flat fluxmeters remains to be evaluated. As stated in section 3.2.4. deposited thermocouples have a sensitivity of 97% of the bulk value (42 μVK^{-1}) after annealing. This leads use to assume that the sensitivity of these fluxmeters will conform with what was expected (section 3.2).

4. Conclusion

Similar studies of moderate and high temperature range-thin film transducers that have been conducted in the laboratories of various countries show importance of developing such technologies, in particular for experimental research needs of aircraft engine manufacturers.

Moderate temperature range thin film-fluxmeters are available for use in quasi-static mode. The extension of the use of such transducers can be envisaged in the low temperature domain (cryogenic wind channel).

On the other hand, the present course of study to apply the deposition technologies that have been developed for high temperature thin film fluxmeters to real turbomachine blades is under way.

References

[1] - Godefroy J.C., Gageant C., François D. and Portat M. (1987), J. of Vac. Sci. and Tech., 5, n° 5, p. 2917-2923.

[2] - Van Dorth A.C., Van Der Graff F. and Steenvoorden G.K. (1983), Sensors and Actuators, 4, p. 323-331.

[3] - Smegill J.G. (1987), Materials Science and Engineering, 87, p. 261-265.

[4] - Whittle D.P. and Stringer J. (1980), Phil. Trans. R. Soc. Lond. A295, p. 309-329.

[5] - Wood G.C. and Stott F.H., Proceedings of High Temperature Corrosion, March 2-6, 1981, San Diego.

[6] - Thornton J.A. and Hoffman D.W. (1989), Thin Solid Films, 171, p. 5-31.

[7] - Schneider K. and Grünling H.W. (1983), Thin Solid Films, 107, p. 395-416.

[8] - Knoll R.W. and Bradley E.R. (1984), Microstructure and phase composition of sputter-deposited zirconia Yttria films. Mat. Res. Soc. Symposium Proc. Vol. 30, Elesevier Science Publishing Co.

[9] - Manual on the use of thermocouples in temperature measurement, A.S.T.M., STP 470 A, 1974.

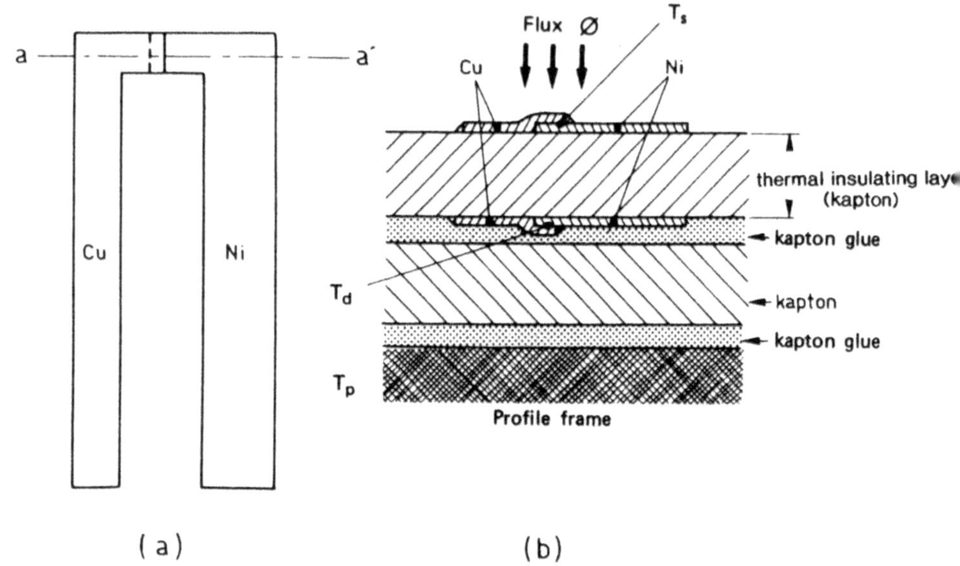

Figure 1. Principle of a simple gradient heat fluxmeter
a) top view
b) cross-section along aa'.

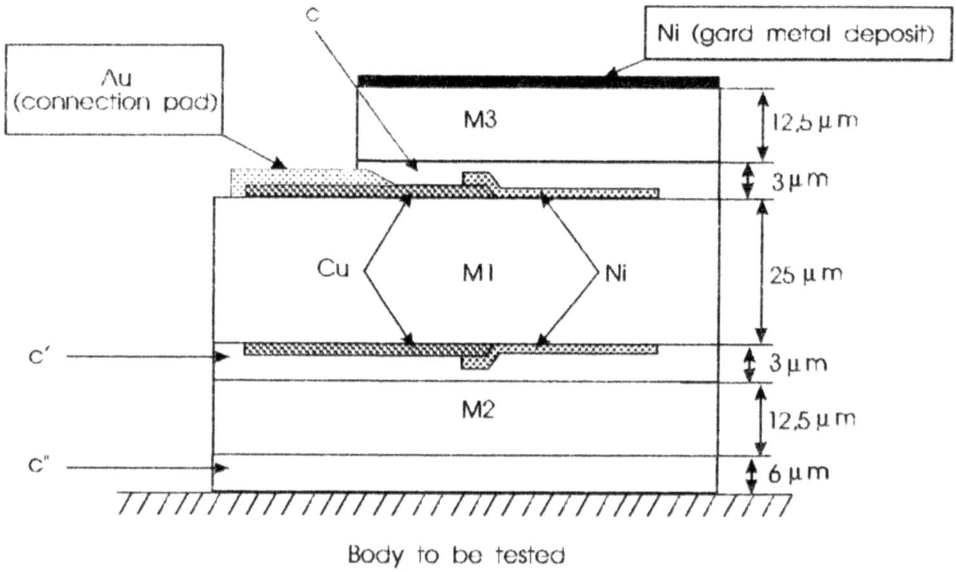

Figure 2. Schematic representation of the moderate temperature range-thin film fluxmete
(cross-section)
M1, M2, M3 : kapton films
C, C', C" : kapton glue.

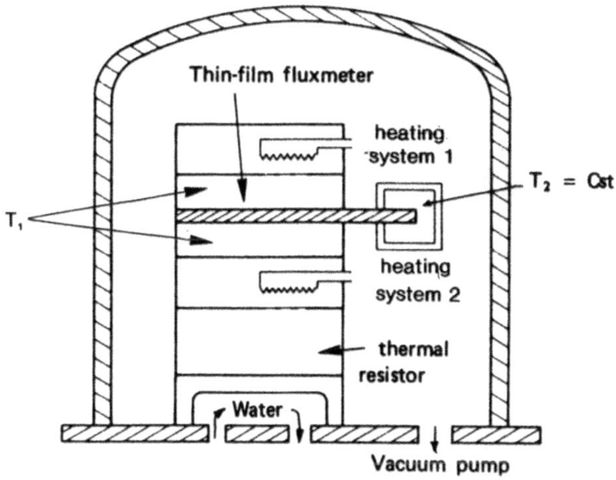

Figure 3. Principle of the temperature calibration device for moderate temperature range-film fluxmeters.

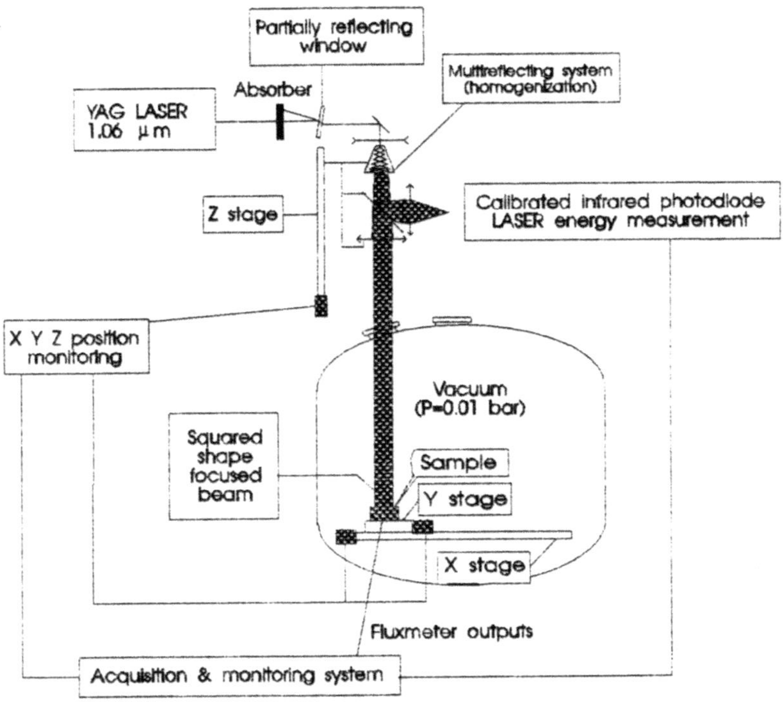

Figure 4. Principle of the heat flux calibration device for moderate temperature range-thin film fluxmeters.

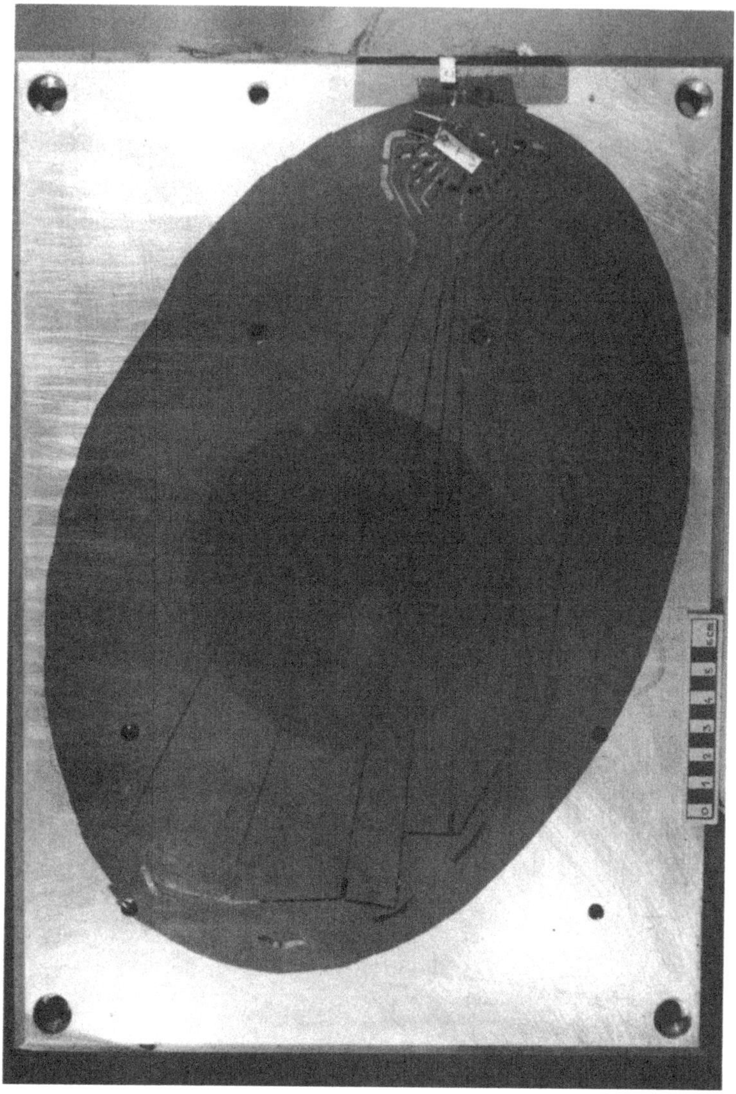

Figure 5. Thin film sensor (6 fluxmeters).

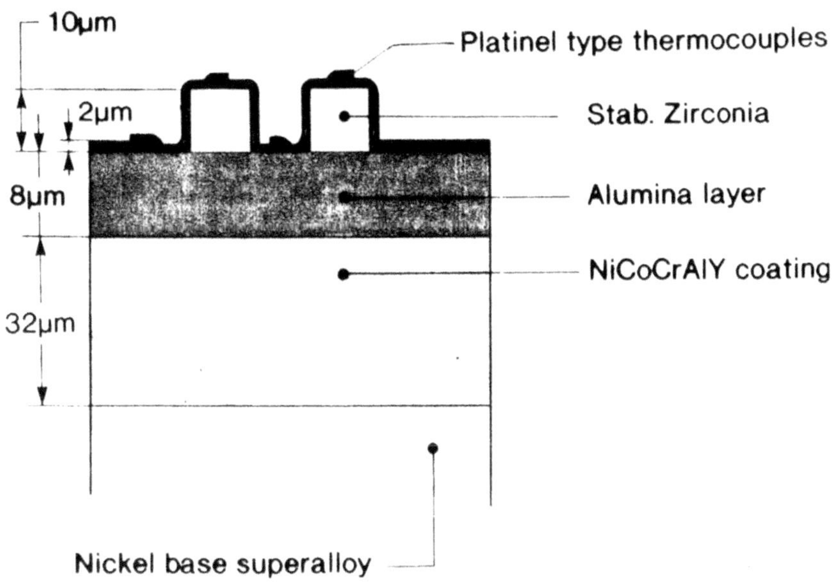

Figure 6a. Principle of multijunction heat fluxmeter (cross-section).

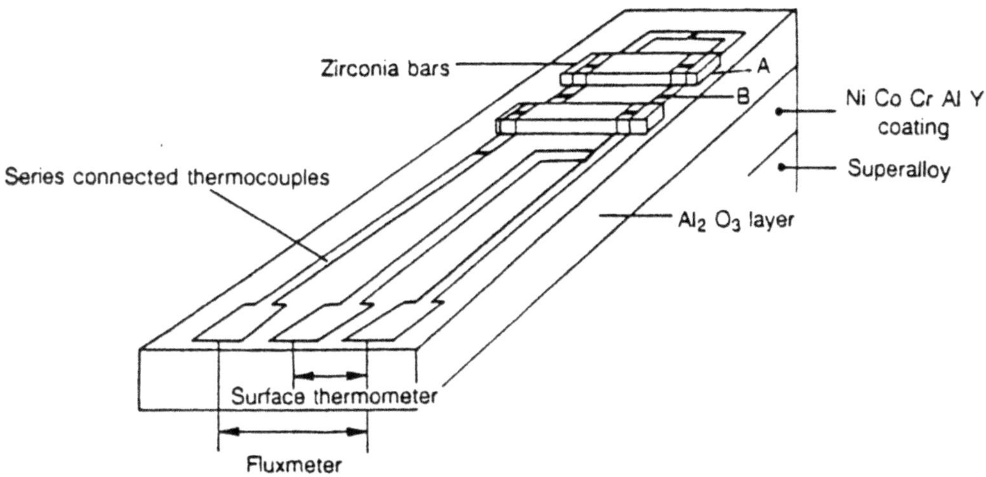

Figure 6b. Flat laboratory model representation of multijunction heat fluxmeter.

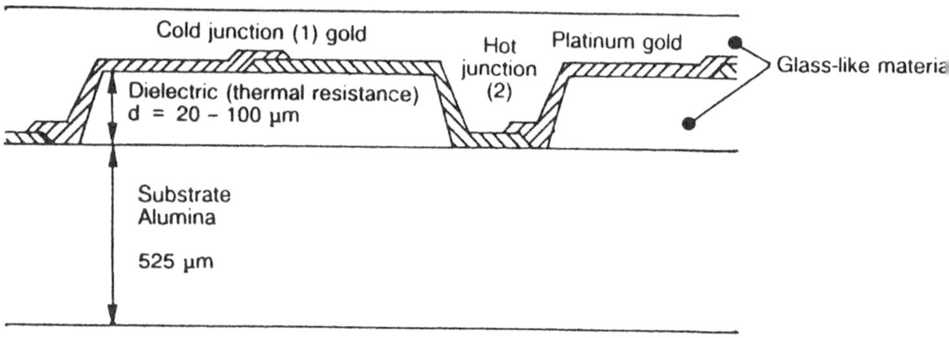

Figure 7. Side view of the heat fluxmeter studied by A.C. Van Dorth et al.

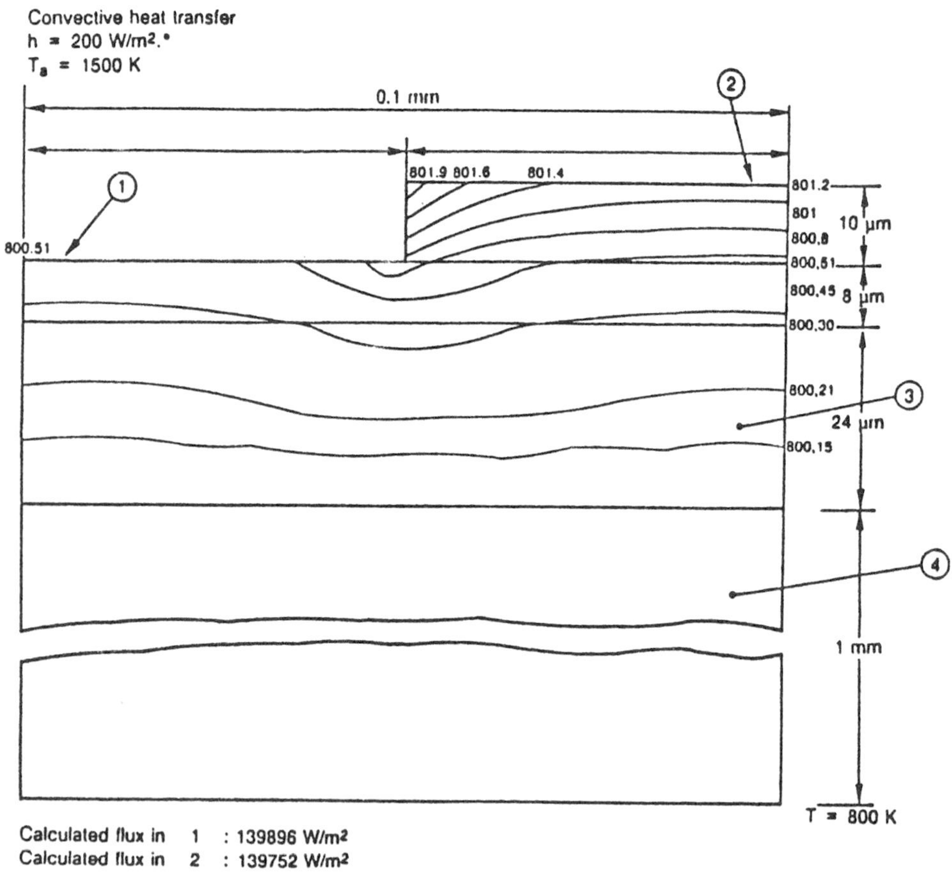

Figure 8. Numerical simulation of the isotherms in a transducer element, limited by symmetry on the left at junction n° 1, and on the right at junction n° 2.

HYPERSONIC AERODYNAMICS RESEARCH WITH AN INFRARED IMAGING SYSTEM

G.M. Carlomagno, L. de Luca, G. Cardone
University of Naples, DETEC
P.le Tecchio 80 - 80125 Naples
Italy

ABSTRACT. Applications of an Infrared Scanning Radiometer referring to flow visualizations and heat transfer measurements performed in a blowdown hypersonic wind tunnel on delta wing and ellipsoids models are discussed. Tests are carried out within the European Community space program *Hermes* to develop the first european space shuttle. Comparisons of IR data with oil film flow visualizations, thermocouple measurements and numerical results are presented.

1. Introduction

In the last decade the good results attained by means of a widespread use of Infrared Scanning Radiometer (IRSR) in convective heat transfer problems have proved IRSR to be one of the best tools to overcome several limitations of standard heat flux sensors, both from the measurement and the visualization points of view.

In effect, the need to solve the problems concerned with high-speed flight, related, for example, to the design of modern re-entry shuttles, recently produced a renewed interest about the aerothermodynamics of space vehicles. Therefore, parallel to the development of more sophisticated numerical codes, an accurate and comprehensive wind tunnel testing, including a more detailed measurement of thermal loads, is required.

Both qualitative and quantitative measurement methods are necessary. The former (e.g. surface flow visualization by oil films or temperature sensitive coatings) yield generally an overall view of the most critical regions, the latter (thermocouples, wall calorimeters, thin films) provide accurate information to validate numerical codes. Standard quantitative techniques typically give zero-dimensional measurements, i.e. each measurement is taken at a single point over the model surface. Surface flow visualization techniques can yield a two-dimensional surface mapping of the flow field, but they may affect the flow boundary conditions near the solid wall.

The essential features of an IR imaging system are: it is truly non-intrusive; it allows for a fully two-dimensional mapping of the surface to be tested; the video signal output may be treated by digital image

A. Boutier (ed.), New Trends in Instrumentation for Hypersonic Research, 493–502.

processing [1,2]. Accordingly, it can be successfully applied to both *thin-skin* and *thin-film* techniques [3-7].

As mentioned in Ref. [8] the exploitation of a computerized quantitative IR thermography needs the solution of several problems mainly concerned with: an accurate characterization of the IR imaging system performance; the calibration of the IR camera; the use of external additional optics and/or mirrors, e.g. to increase the spatial resolution; the determination of the surface emissivity; the correct geometrical identification of the measured points; the design of the optical access window including the choice of the most appropriate IR materials.

In this paper significant results referring to flow visualizations and heat transfer measurements performed with an IRSR in a blowdown hypersonic wind tunnel on delta wing and ellipsoids models are discussed. Present tests are carried out within the European Community space program *Hermes* to develop the first european space shuttle.

2. Experimental rig and procedure

Experimental tests are carried out in the hypersonic blowdown tunnel H210 of CEAT, Poitiers, having a test section diameter of 210mm. The main operating conditions are: stagnation temperature T_o = 800°K, stagnation pressure 4.5MPa < p_o < 10MPa. Unit free stream Reynolds number Re_m is changed mainly by varying the stagnation pressure and ranges from 1.2×10^7/m to 1.9×10^7/m. An appropriate nozzle is employed to obtain a nominal free stream Mach number M = 8.15. In effect Mach number slightly depends on the stagnation pressure (due to the variation of the nozzle boundary layer thickness) and ranges from M = 8 for p_o = 4.5MPa to M = 8.2 for p_o = 10MPa [9].

After starting of the tunnel, the model (initially resting in a remote position at room temperature) is vertically injected into the stream. Injection time is about 0.15s and testing duration is of the order of a few seconds. The time rise of the surface temperature of the model is measured by viewing the latter with the IR camera; a time sequence of thermograms is generally recorded during the test run. It has to be pointed out that thermograms themselves are used to obtain the surface flows behaviour. Comparisons are also made with other flow visualizations carried out by means of temperature sensitive paints and oil films.

2.1 DELTA WING WITH FLAP MODEL

The tested model of delta wing with flap (or delta wing/ramp configuration) has a swept angle of 70°, a ramp angle of 15° and a bottom wedge angle of 11° [4]. Two versions of this model are used for the tests. The first one, which is made of crosslinked elastomer (Rhodorsil RTV 147), is used for tests carried out by means of temperature sensitive paints and infrared measurements on the ramp in order to visualize Goertler vortices. The second version of the delta wing model, which is used to measure heat fluxes by means of the thin-skin technique, has an upper surface made of a 0.5 mm thick SS plate. In order to enhance the detection of the thermal image the model surface is coated with a black paint. All tests are performed by placing the model at zero incidence for three different Reynolds numbers

Re(based on free stream conditions and on the wing length L = 126.9 mm) namely: 1.6x106, 2.0×10^6 and 2.4×10^6. In this paper IR data will be mainly discussed. Comparisons of IR measurements with data obtained with the other techniques have been discussed in Refs. [4,5].

The thermographic system employed for the delta wing study is basically the AGEMA Thermovision 782 connected, via TIC 8000 A/D Converter Board, to an IBM/AT computer. An application software is developed, which can perform: noise reduction by numerical filtering; computation of temperatures; geometrical identification of isotherms; image restoration; heat transfer correlations. In particular, the software yields the dimensionless heat transfer coefficient expressed in terms of Stanton number that is defined as

$$St = q / [\rho\, c\, V\, (T_o - T_w)] \tag{1}$$

where ρ, V, T_o and c are respectively the mass density, the velocity, the stagnation temperature and the specific heat at constant pressure of the free stream. The history of the measured wall temperature T_w is related to the wall heat flux q entering the delta wing surface by the classic relationship of the *thin-skin* sensor.

In order to detect the formation of Goertler vortices on the ramp downstream of the delta wing, the AGEMA 880 camera is used which has a higher spatial resolution. In this case the heat transfer coefficient is inferred by means of the semi-infinite wall thermal sensor (thin film); it is also necessary to include into the dedicated software an image restoration algorithm, which takes into account both imaging and sampling as well as lateral thermal conduction degradations [10,11]. Image restoration is performed by means of an FFT inverse filter algorithm:

$$f(x,y) = \mathcal{F}^{-1}\left\{G(\nu,\mu)/H(\nu,\mu)\right\} \tag{2}$$

where $f(x,y)$ is the restored image as a function of the spatial coordinates x and y. G is the Fourier Transform of the recorded (degraded) image, while H is referred to as a generalized *System Transfer Function*. Both G and H are functions of the spatial frequencies ν and μ. Imaging and sampling as well as lateral thermal conduction effects may be accounted for by means of the so-called *Average Sampled System Modulation Transfer Function* (*ASSMTF*) and *Temperature Amplitude Transfer Function* (*TATF*), respectively. Then, by assuming the two effects to be cascaded, the relationship $H = ASSMTF \times TATF$ is employed in eq. (2).

2.2 ELLIPSOIDAL MODELS

A simple and a double ellipsoid (this last roughly simulating the fore part of the Hermes shuttle shape), are studied. The geometry of the double ellipsoid is indicated in Ref. [4]; the parameters of the simple ellipsoid correspond to those of the lower part of the double one. Models made of different materials are used for both bodies. A first version of the simple ellipsoid is a stainless steel thin wall model (0.5*mm* thick) which is used to measure heat fluxes by means of thermocouples with the *thin-skin* technique [7]. A second version of this body is made of solid NORCOAT 4000 to measure heat fluxes by means of IRSR with the semi-infinite wall (*thin-film*) technique. Thermographic data are also obtained on a solid NORCOAT simple ellipsoid model.

NORCOAT· 4000 is a relatively low thermal conductivity material, which is made of silica hollow microspheres drowned in a silicon elastomer base mixture. Being the NORCOAT normal emissivity coefficient sufficiently high (the measured value in the wavelength range of interest is .93), there is no need to coat it with a (thermally) black paint in order to enhance IR radiation detection. Indeed, the variation of the emissivity coefficient as a function of the view angle has been measured in some ad hoc tests. It has been found that for angles of view up to 60° the emissivity remains practically constant and equal to .93, while for view angles greater than 60° it rapidly falls down.

The models are generally tested at Mach number equal to 8.15, unit Reynolds number of 1.66×10^7/m and angles of attack α equal to 25°, 30°, 35°. For α = 30° some tests are also carried out to detect induced transition.

For the heat transfer measurements, because of the short measuring time, the thermal penetration depth can be generally considered small as compared to the local radius of curvature almost everywhere on the model surface: therefore, from this point of view,the 1D semi-infinite wall thermal model can be applied. The time history of the measured wall temperature T_w is correlated to the net exchanged heat flux $q = q_c - q_r$ which represents the difference between the convective heat flux and the radiative one.

The AGEMA 880 IR system is used to test ellipsoidal models. Details about IR system and experimental set-up are given in Ref. [7].

3. Results

3.1 DELTA WING

A typical thermogram (i.e. temperature map which may be also interpreted as surface flow field visualizations) obtained over the delta wing/flap configuration by means of IRSR is reported in Fig.1. Generally thermograms are taken for two different levels of thermal range. In fact, since the order of magnitude of the temperature on the wing is smaller than the one on the ramp,in order to improve accuracy, for each testing condition two thermal image sequences are recorded by setting two different levels of the scanner thermal range. The white zones at the wing leading edge and over the ramp indicate temperatures above the full-scale value. Three distinct regions are visible in the temperature map of Fig. 1. Within the two outer regions the flow appears essentially two-dimensional: the monotonic windwise decreasing of temperature (and then of aerodynamic heating) uncovers the typical laminar boundary layer development. In the two corners upstream of the ramp, where temperature presents local minima, the flow is separated. The interpretation of the temperature field in the inner region is rather complex. Separation seems to occur also in a region around the symmetry axis. This region reduces and almost disappears near the ramp where temperature slightly increases in the windwise direction on the wing centerline. The above interpretation well agree with the results of oil film visualizations [12].

The relief representation of the surface distribution of the Stanton number, as defined by eq.(1), over the delta wing for a Reynolds number of $2.4x10^6$ is depicted in Fig. 2. The plot is obtained by matching the two image sequences including details about the wing and the ramp. X and Y coordinates are streamwise and spanwise respectively, S is the ramp span. The locus of maxima St corresponds to the reattachment line. Data referring to wing and ramp edges are cut off because of the different thickness of the model there. For this Reynolds number value, as shown also by oil film visualization, the two separation regions in the two corners upstream of the ramp are reduced in size.

St spanwise distributions on the ramp, close to the reattachment line (X/L = 1.15) are reported in Fig. 3 for the three tested Reynolds numbers. Note the marked relative minimum on the symmetry axis for the lowest Reynolds number. Open symbols refer to data obtained with thermocouples.

A particular attention is also paid to visualize the presence of a transversal instability of the boundary layer over the ramp downstream of the delta wing model, that leads to the formation of counter-rotating vortices in the reattaching flow, referred to as Goertler vortices. Such vortices originate from a boundary layer instability due to the curvature of the dividing streamline that joins the separation and reattachment points.

In order to achieve the highest possible spatial resolution (the expected vortices wavelength is of a few millimeters), proper extension rings are used in conjunction with the standard 7° IR lens. Furthermore, to avoid the model to be scanned along the vertical direction, which has a spatial resolution lower than for the horizontal one, the IR camera is 90° rotated around its roll axis. In the above mentioned conditions the AGEMA 880 system works at a spatial frequency for which the corresponding ASSMTF is about 25%. Being in this case the field of view reduced to about $21x21mm^2$ with the use of long extension rings, the thermal image is well sampled, i.e. the system is only imaging-dependent. Hence, the corrections to be applied to the coarse data are those of removing blur due to the imaging and tangential thermal conduction effects.

The thermogram of Fig. 4 shows the restored temperature map relative to the small square frame viewed on the ramp downstream of the delta wing model. Testing conditions are: M = 8.15, $Re/m = 1.6x10^6$, $\alpha = 0°$.Since the flow comes from the bottom of the thermogram, the presence of the spanwise vortices is made evident by the horizontal periodic variation of the temperature. The average wavelength of the vortices present on the ramp results to be of about 2mm. The entity of the corrections due to the image restoration is shown by the plot of Fig. 5 where a typical Stanton number spanwise profile is reported. The dashed line represents the coarse data, the dotted line the ones restored from imaging degradation effects, the continuous line the final result which takes into account also the lateral conduction degradation. As may be seen, the corrections range from 10% to 15%. The average level of Stanton number as measured in the present tests well agrees with that typically reported by other authors [12].

3.3 ELLIPSOIDAL BODIES

The thermograms of Fig. 6 shows the temperature map for the 25° case referring to the aerothermodynamic heating over the simple ellipsoid leeward side. On both sides of the symmetry axis two clear wide regions (delimited by the outer curved black lines and by the inner ones almost parallel to the longitudinal axis) are evident. In these regions the flow is separated. Reattachment of the flow coming from model sides occurs in the central region comprised between such pockets [7]. For α = 30° and α = 35° the flow configuration is qualitatively the same; however, the separation line moves upstream as the angle of attack increases.

The thermograms of Fig. 7 reveals the windward side temperature map for an angle of attack α = 30°. In this case the model is placed in the tunnel in such a way that its nose goes away from the viewing lateral window. As a consequence, the edge zones toward the model tip, having a very high curvature, are scanned with a view angle which can become very high; hence, the actual NORCOAT emissivity coefficient falls to very low values there and the model size appears to be reduced. The continuous streamwise decrease of temperature from the nose, which is evidenced by the succession of bands of different gray level, has to be correlated to the laminar boundary layer development over the model.

In order to show the IRSR capability to detect also induced transition, a surface defect (a few tenths of millimeter in depth) is present on the model. Such a defect induces turbulence whose presence has to be correlated to the (hot) clear "wake" which is evident in the thermogram.

The thermogram of Fig. 8 shows the side view of the double ellipsoid for α = 30°, obtained by setting a relatively high scanner measuring temperature range. As a consequence, thermogram gives detailed information about the aerodynamic heating mainly on the lower ellipsoid, due to the higher temperature attained there. The temperature field recorded on the cockpit appears in this case substantially flat, except at its fore part in correspondence of flow reattachment. The dark region extending in the direction of the principal axis of the ellipsoids corresponds to the cockpit-lower ellipsoid junction, where the model seems colder mainly because of the directional emissivity effect.

Heat transfer distributions over the windward side obtained by IRSR are generally validated by analogous data obtained from thermocouple measurements with the thin wall stainless steel (simple ellipsoid) model and from 3D computational solutions [13,14]. For α = 30°, Fig. 9 reports such a validation as far as the streamwise Stanton distribution along the direction Y = 0 is concerned. Star symbols represent thermocouples *thin-skin* data, while the continuous line refers to the IR *thin-film* data. Note that in this latter case data are cut off in the nose region for $X \leq -41$ mm because they became meaningless there; in fact, because of the low viewing angle, the directional emissivity coefficient is undetermined near the model nose. As already mentioned, the X scale is corrected with respect to the draft image to account for the model geometry and incidence.

With reference to Fig. 9 it has to be pointed out that IR and thermocouples data are taken in the very same flow conditions; nevertheless, in the latter case, being the stainless steel thermal conductivity relatively high, the measured wall temperature becomes only slightly higher than the initial one at which model is injected into test section (i.e.

about 10°C). On the contrary, in the case of NORCOAT, due to its relatively low thermal conductivity, the surface temperature rises up to a few hundreds degrees centigrade. Therefore, in order to compare IR and thermocouples heat transfer data, it is necessary to make use of the standard definition of Stanton number:

$$St' = q/[\rho c V(T_{aw} - T_w)] \tag{4}$$

which is based on the local adiabatic wall temperature of the stream, T_{aw}. This last is computed by means of the relationship:

$$T_{aw} = T_o \left[\frac{T_e}{T_o} (1-r) + r \right] \tag{5}$$

where r is the recovery factor and T_e is the local temperature at the boundary layer edge. To reduce data of Fig. 10 it is assumed $r = Pr^{1/2}$ (laminar flow) and, in order to evaluate T_e/T_o, local Mach number is drawn from the inviscid 3D numerical solution of Ref. [15].

The agreement between *thin-film* and *thin-skin* data seems quite satisfactory. The discrepancy about the location of *St'* minimum has been discussed in Ref [7].

As far as the St distribution over the leeward side of the double ellipsoid is concerned, this shows a characteristic horseshoe shaped separation line on the cockpit, the size of which reduces as the angle of attack is increasing. The St streamwise distributions in the symmetry plane are reported in Ref. [7] for the three tested angles of attack. In Fig. 10 the St spanwise distributions along the line X = 0 are reported. For $\alpha = 25°$ and $\alpha = 30°$ two peaks (symmetric with respect to ellipsoidal axis) are present which correspond to the intersection of the above mentioned horseshoe shaped line with the line X = 0. Peaks seem to reduce with increasing angle of attack. For $\alpha = 35°$ one peak only (located on the symmetry axis) is present; in fact, at the highest tested angle of attack the horseshoe line practically disappears and the secondary separation on the cockpit is strongly moved upstream.

4. Conclusions

A computerized IR imaging system is employed to visualize the flow field and to measure the aerodynamic heating over models in the blowdown hypersonic wind tunnel H210 of CEAT (Poitiers).Experimental tests are performed on a delta wing/ramp configuration and on simple and double ellipsoid models at Mach number of 8.15 and various unit Reynolds number as well as angles of attack . The *thin-skin* and the *thin-film* techniques are used to reduce temperature data into heat transfer coefficient (Stanton number).

Quantitative *St* distributions are compared with data obtained on stainless steel thin wall models, where temperature is measured by thermocouples and computational solutions. For both quantitative and qualitative results, IR data well agree with the ones reported in the literature.

5. ACKNOWLEDGMENTS

This work has been sponsored by Centro Italiano Ricerche Aerospaziali (CIRA) and Avions Marcel Dassault-Breguet Aviation within the European HERMES Research Project.

6. REFERENCES

1. Wendt, J.F. (1989), Infrared Thermography. 2nd Joint Europe/US Short Course on Hypersonics, Colorado Springs.
2. Carlomagno, G.M.; de Luca, L. (1989), Infrared Thermography in Heat Transfer. Chapter 32, Handbook of Flow Visualization, (Ed. Yang, W.J.), 531-553, Hemisphere.
3. Simeonides, G.; Van Lierde, P.; Van der Stichele, S; Capriotti, D.; Wendt, J.F. (1989), Infrared Thermography in Blowdown and Intermittent Hypersonic Facilities. AIAA paper 89-0042.
4. Carlomagno, G.M.; de Luca, L.; Alziary de Roquefort, T. (1989), Heat Transfer Measurements with an Infrared Camera in Hypersonic Flow. In Computers and Experiments in Fluid Flow, (Eds. Carlomagno, G.M. and Brebbia, C.A.), Comp. Mech.- Springer Verlag, 467-476.
5. Carlomagno, G. M.; de Luca, L.; and Alziary de Roquefort, T. (1991), Mapping and Measurement of Aerodynamic Heating by means of IR Thermography, in Multi-Phase Flow and Heat Transfer (Eds. Xue-Jun Chen et al.), Vol. 2, 1316-1325, Hemisphere.
6. Balageas, D.L.;Boscher, D.M.;Deom, A.A. et Fournier, J. (1991), Application de la termographie infraruge passive et stimulee a la measure des flux termiques en soufflerie, to be published on la Recherche Aerospatiale.
7. de Luca, L; Cardone, G.; Carlomagno G.M.; Aymer, D. and Alziary, T. (1992), Flow Visualization and Heat Transfer Measurement in Hypersonic Wind Tunnel, Experimental Heat Transfer, 5, 65-79.
8. de Luca, L. (1991), Computerized IR Thermography for Convective Heat Transfer Measurements, in Computational Methods and Experimental Measurements V, (Eds. Sousa A., Brebbia C.A. and Carlomagno G.M.), 347-358, CMP/Elsevier.
9. Alziary de Roquefort, T. (1985), Les souffleries hypersoniques du centre d'etudes aerodynamiques et thermiques de Poitiers. 22eme colloque d'aerodynamique appliquee, Lille.
10. de Luca, L. and Cardone, G. (1991), MTF Cascade Model for a Sampled IR Imaging System, Applied Optics, 30, 1659-1664.
11. de Luca, L. and Cardone, G. (1992), Experimental Analysis of Goertler Vortices in Hypersonic Wedge Flow, Thermosense XIV SPIE Meeting, Orlando.
12. Gashard, G: Da Costa, J.L.; Aymer, D., and Alziary, T., (1987), Interaction onde de choc-couche limite en ecoulement hypersonique tridimensionnel transitionnel, CEAT Rpt. 1 RDMF 86, Poitiers.
13. Argyris, J.; Doltsinis, I.St.; (1990), Hypersonic Viscous Flow past Double Ellipse and Double Ellipsoid - Numerical Results. Workshop on

Hypersonic Flows for Re-entry Problems, Antibes, Tome 4 - part I, INRIA GAMNI-SMAI, 11-30.

14. Reidelbauch S., (1990), Numerical Simulation of Laminar Hypersonic Flow past a Double Ellipsoid, Workshop on Hypersonic Flows for Re-entry Problems, Antibes, Tome 4 - part I, INRIA GAMNI-SMAI, 410-428.
15. Radespiel, R.; Herrmann, U.; and Longo, J.M.A., (1990),Problem 6: Flow Over a Double Ellipsoid (Test Case 6.1.7). Workshop on Hypersonic Flows for Re-entry Problems, Antibes, Tome 4 - part I, INRIA GAMNI-SMAI, 363-379.

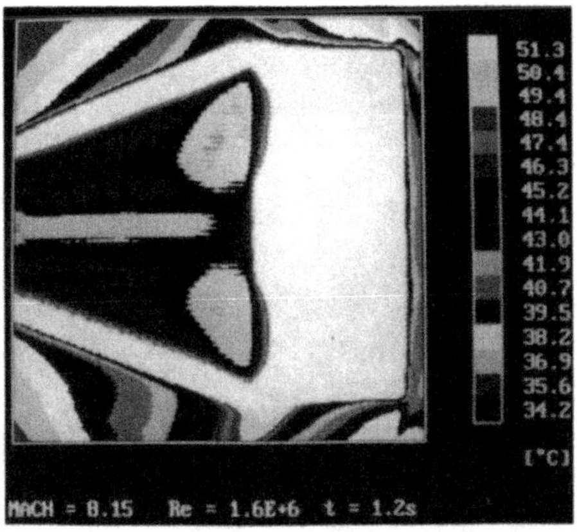

Fig. 1

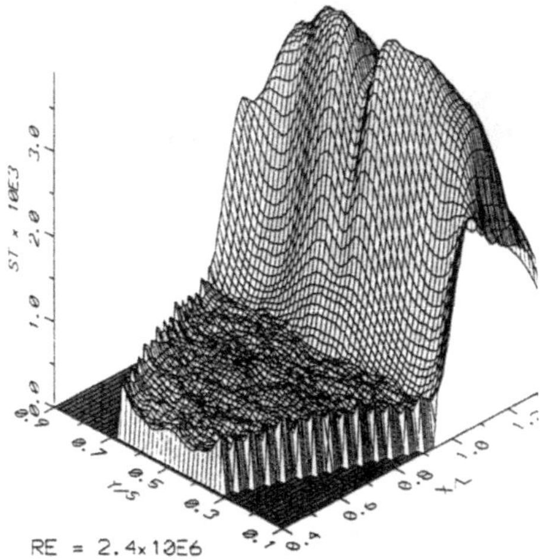

Fig. 2

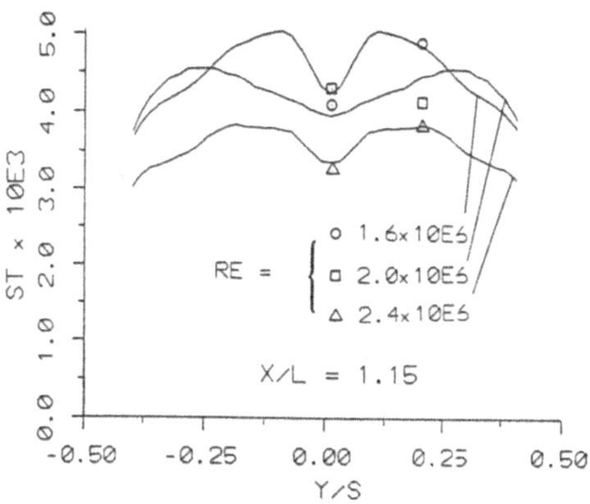

Fig. 3

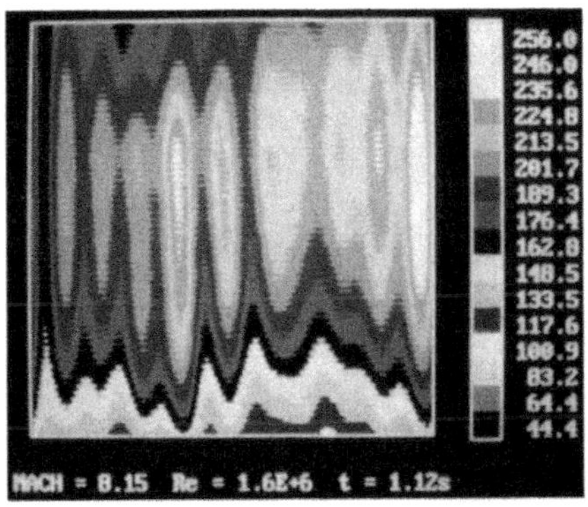

Fig. 4

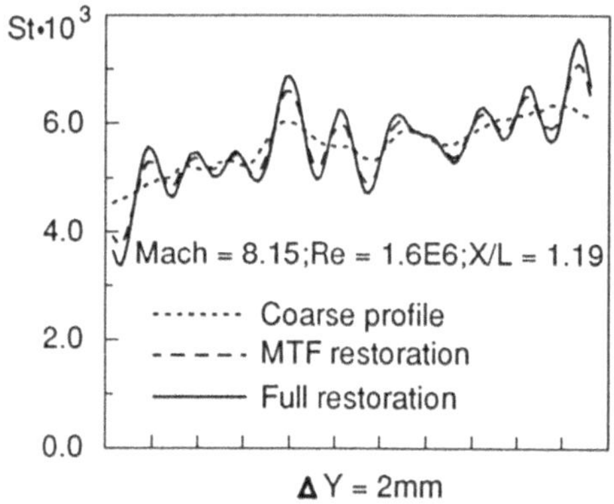

Fig. 5

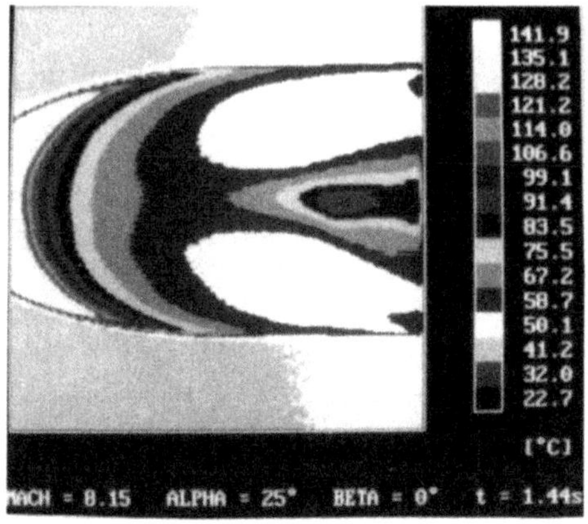

Fig. 6

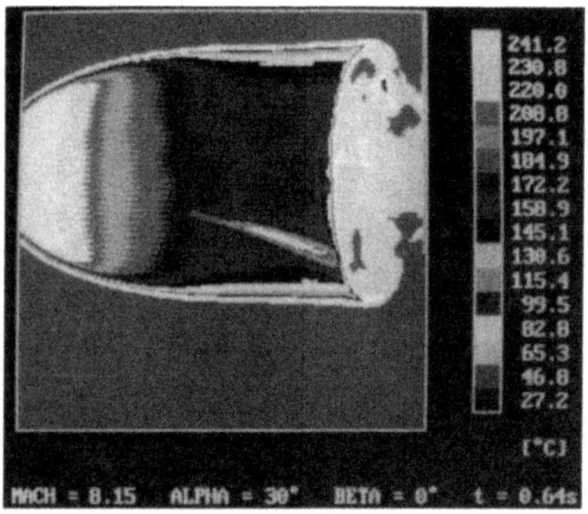

Fig. 7

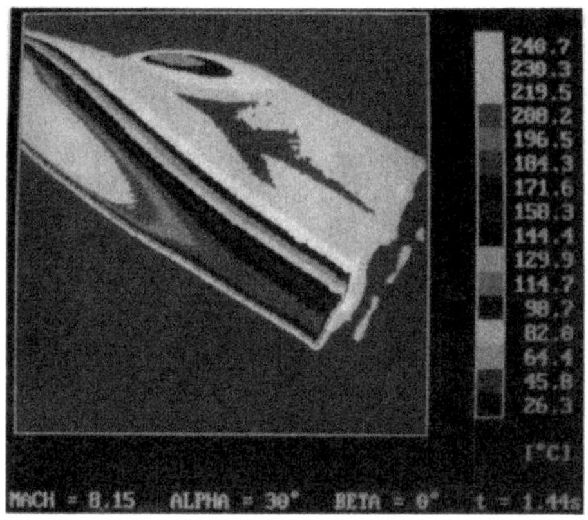

Fig. 8

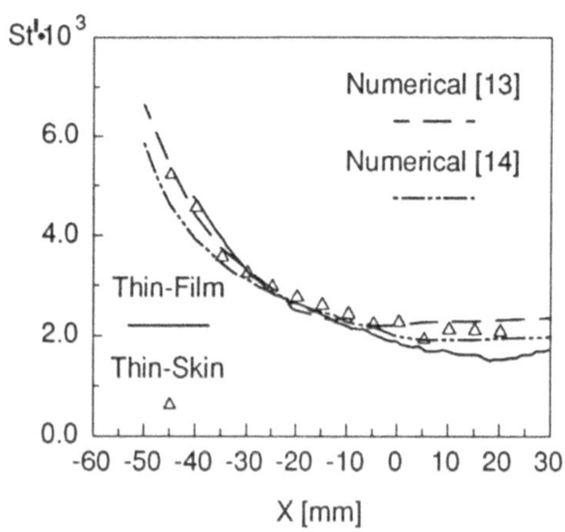

Fig. 9

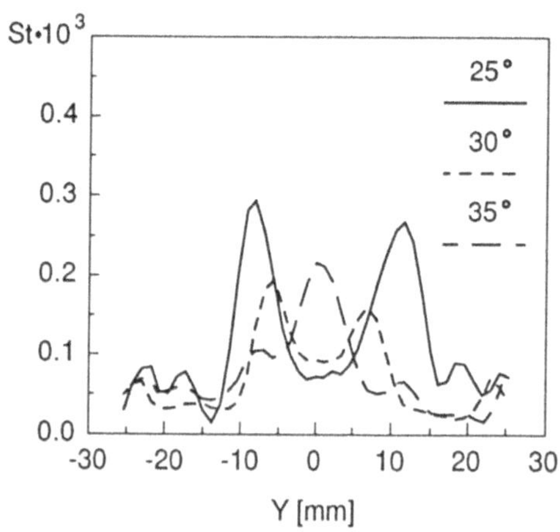

Fig. 10

HEAT TRANSFER MEASUREMENTS USING INFRA-RED THERMOGRAPHY IN RAREFIED FLOWS.

M F WESTBY
Aerodynamics Department
Defence Research Agency
Farnborough
Hampshire
GU14 6TD
Great Britain

ABSTRACT.

Tests have been carried out in the Royal Aerospace Establishment Low Density Wind Tunnel to determine the heat transfer rate on a flat plate and a series of sharp slender cones. The tests were carried out in the transitional rarefied flow regime. An infra-red scanning camera system was used to determine surface temperatures and hence heat transfer rate. Data were compared with two analytical theories and one existing set of data. Results show that the IR system is at least as good at determining heat transfer rates as a thermocouple technique and has better spatial resolution. The technique is also sensitive enough to detect changes in the state of the flow over the surface of models.

1 Introduction

In continuum flow CFD can be used with some confidence for the prediction of heating rates on complex bodies. However, there are currently no theoretical or computational methods which can be used with total confidence for the prediction of aerodynamic heating in rarefied flows. This leaves experimental studies as the prime method for obtaining such data. It is important therefore, that an experimental technique is available which is able to measure heat transfer rates reliably and accurately in rarefied flows.

A measurement technique which appears to have the necessary attributes is infra-red thermography, using a scanning radiometer. IR systems are commercially available and have been used with considerable success in various types of hypersonic wind tunnel over a number of years. The main advantage of the system is that it is completely non-intrusive, consisting of a passive, remote detector. Also, data can be obtained at any point within the field of view of the scanner (depending on the spatial resolution of the scanner and the orientation of the surface being observed).

Two test programmes have been carried out in the Royal Aerospace Establishment Low Density Tunnel (LDT) related to aerodynamic heating. Both of these have involved the use of infra-red thermography to measure the rate of change of surface temperature on thin walled bodies. An initial test[1] established that IR thermography showed promise as a useful technique for determining heat transfer rates in the LDT. It was

A. Boutier (ed.), New Trends in Instrumentation for Hypersonic Research, 503–512.

therefore decided, as a second stage, to test some very simple bodies for which analytical predictions of surface heat transfer rates existed.

2 Experimental Details

2.1 DESCRIPTION OF TEST FACILITY.

The RAE LDT[2] is a continuously operating facility which uses pure nitrogen as its test gas. It uses a graphite resistance heater to heat the test gas which is then expanded through the nozzle into an open-jet working section. The tunnel is connected to a set of mechanical vacuum pumps which have a sufficiently high pumping speed to sustain continuous running. The facility uses a contoured nozzle which generates a uniform flow with a Mach number of 9.84. The tests described in this paper were carried out at three calibrated test conditions. Reservoir pressure ranged from $1.791\times10^5 Nm^{-2}$ to $2.055\times10^5 Nm^{-2}$ and reservoir temperature was varied between 1170K and 1660K. The unit freestream Reynolds number range was $6.3\times10^4 m^{-1}$ to $1.24\times10^5 m^{-1}$.

2.2 DESCRIPTION OF MODELS USED.

Two types of model were constructed. The first type was a flat plate model 200mm long by 75mm wide. The surface upon which measurements were to be made was 0.5mm thick stainless steel with the leading edge ground sharp. The stainless steel plate was supported on an aluminium frame with the underside of the steel plate shielded from the free stream. At no point where measurements were made was the steel in contact with its supporting frame.

A series of slender sharp cones were also tested with half angles of 4°, 6°, 7°, 8° and 10°. These models were made of nickel using an electroforming technique, whereby a thin layer of nickel is deposited on a steel mandrel, which is removed at the end of the process. The cone models all had a nominal wall thickness of 0.4mm although, due to the manufacturing technique, there were slight differences between the five models. It should be also noted that the cone models were originally designed to be used as force testing models, where wall thickness is unimportant. Therefore, the wall thickness may not be constant along the length of the models.

All of the models were sprayed with a high temperature matt black paint. This gave the models an emissivity in excess of 0.9.

2.3 TESTS CARRIED OUT.

The flat plate model was mounted so that tests could be carried out with the plate at angles of incidence between 0° and 30°. Measurements were made with the Infra-Red scanner over the length of the plate at three freestream test conditions and at angles of incidence of 0°, 5°, 10°, 15°, 20°, 25° and 30°. Oblique shock relationships were used to determine the flow conditions over the plate, downstream of the leading edge shock wave. As readings were only made along the centre-line of the model, it was assumed that data were obtained where the flow was two-dimensional.

All tests carried out using the cone models were at zero incidence where a sharp cone is the axisymmetric equivalent of a flat plate in truly two-dimensional flow. Again, three freestream test conditions were used and oblique shock wave relations were used to calculate the flow conditions at the surface of the cone. Data were gathered along one meridian of each cone model at each test condition.

All of the tests carried out were repeated at least once to check the repeatability of the results. In some cases, especially where the flat plate was set at the larger angles of incidence, more repeat runs were needed with the scanner system set up differently to ensure that good readings had been made both near to the leading edge of the model and at the rear.

2.4 HEAT TRANSFER INSTRUMENTATION.

Surface heat transfer rates were measured during these tests by using an infra-red scanning camera system (an AGA 780), with its associated hardware and software. The system consisting of a camera, a main control unit (chassis) and a digital tape recorder (DTR). The scanner measures the intensity of electro-magnetic radiation in the 3 to 5μm waveband emitted by the body under test. This intensity is related to the surface temperature of the body by the Stefan-Boltzmann law. Thus, the temperature of a body can be determined, provided it is a black body emitter and that the system is calibrated accurately overall.

Calibration was carried out inside the working section of the tunnel to avoid the necessity of accurately determining the emissivity of the models and the transmissivity of the working section IR transparent window. A calibrated thermocouple was attached to the back of the surface of one of the models and the IR scanner was pointed at the front of the surface in the same position as the thermocouple. The entire model was then heated up using a diffuse hot-air blower until a stable temperature was reached. Measurements of surface temperature were then made using both the thermocouple and the IR scanner. This procedure was repeated at several different temperatures covering the whole range encountered during the tests. The IR system was then calibrated against the thermocouple output.

During runs, the output signal from the scanner was amplified and conditioned by signal conditioners contained in the main chassis. The DTR recorded images processed by the main chassis at a rate of one every second. After the completion of a run, the processed images were downloaded from the DTR into a microcomputer and further processed by a software package which produced a series of temperature maps of the model in the wind tunnel test section.

2.5 EXPERIMENTAL PROCEDURE.

The Low Density Tunnel uses a graphite resistance heater to produce the high stagnation temperature needed to avoid liquefaction of the working fluid (pure Nitrogen) in the test section. This results in a "warm-up" period of approximately 1 to 2 minutes (depending on freestream conditions) at the start of each run before the flow is correctly established with the reservoir conditions stabilised at the correct values.

It is important that the surface of the model under test is at a low and reasonably uniform temperature at the start of the measurement period. Therefore, the model to be tested was mounted on an injection rig mounted in the roof of the working section. This rig held the model outside the high-temperature free jet until stable flow conditions had been obtained and useful measurements could be made. The rig consists of a pivoted arm which can swing a model from outside the free jet to the centre of the core flow in under 0.1 second. This is fast enough for it to be assumed that heating of the model commences uniformly over the whole surface of the model when it is injected.

Once a stable flow had been established with the correct stagnation conditions, and just before the model was injected into the flow, the DTR was started. Up to 20 images were captured and recorded by the DTR at a rate of 1 per second. The scanner system was set up so as to record temperatures between approximately 20°C and 90°C. This allowed accurate derivation of heat transfer rates (see section 2.6) without excessive temperature gradients developing along the length of a model.

A run was terminated when a substantial part of the surface of the model under observation had exceeded the upper temperature limit of the scanner system (*ie* greater than 90°). At this point the tunnel was shut down, the DTR was stopped and the model was cooled down before the start of the next run.

2.6 DERIVATION OF HEAT TRANSFER RATES.

To obtain surface heat flux data, a series of thermal images were obtained of the model under test during the course of a run. Data was acquired as soon as possible after the model had been injected into the flow at the start of the run. Spot temperature measurements were obtained at several points on the surface of the model after the data from each run had been downloaded into the microcomputer and processed by the software. Readings were taken from each image, providing a time-history of the surface temperature of the model these points. From this time-history, the rate of change of temperature, dT_w/dt, could be calculated.

A local value of heat transfer rate per unit area ($\dot{q}$) could then be deduced from the value of dT_w/dt and the thermal properties of the model wall material, using the relationship

$$\dot{q} = \rho_w c_w y \left(\frac{dT_w}{dt} \right) \quad (kWm^{-2}) \tag{1}$$

where ρ_w is the density of the wall material, c_w is the specific heat capacity of the wall material and y is the thickness of the wall.

Two basic assumptions are implicit in this relationship. First, it has to be assumed that there is no temperature gradient across the thickness of the wall, *ie* that each element of the surface acts as a calorimeter. Secondly, the rate of heat conduction within the wall has to be small compared to the rate of convective heat transfer from the flow to the wall.

The stagnation temperature of the flow is an order of magnitude higher than the model wall temperature throughout the run. Thus, if the two previous assumptions are proved to be reasonable, the temperature rise throughout the course of the run should be linear with time at any point on the surface of the model.

The first assumption (zero temperature gradient through the thickness of the wall) was reasonable as the wall thickness was less than 0.5mm on each model and the gas pressure and density on the back face of the wall was very low in all cases. The initial heat pulse should have reached the inner surface of the wall material in less than 1 second. Thereafter, as there was no thermal path away from the back surface of the wall, there should have been no temperature gradient across its thickness.

The second assumption (neglecting internal heat conduction) is reasonable if temperature gradients along the surface of the models are small. This allows each element of the surface to be treated in isolation, with no reference being made to the surrounding elements of the surface. This was ensured by keeping the run time sufficiently short that the temperature of the wall was not raised to a point where internal heat conduction caused by temperature gradients became significant. To check that errors were not being caused by conduction, plots of temperature versus time were made. If heat conduction had been significant, dT_w/dt would have slowly increased as energy was transferred along the surface of the model by conduction. This was not detected in any of the plots of temperature versus time and so it could be assumed that the effect of internal heat conduction was negligible.

3 Experimental Results

3.1 PRESENTATION OF RESULTS.

Results are presented in Figs 1 to 9. Fig 1 shows the increase of temperature with time on the flat plate at an angle of incidence of 0°. It can be seen that temperature increases linearly with time, justifying the assumption that heat conduction within the wall of the model can be neglected. Fig 2 shows the surface heat transfer rate per unit area, $\dot{q}$, against distance from the leading edge of the plate. These results are derived from those given in Fig 1.

In order to plot the results meaningfully, the flow conditions behind the bow shock wave had to be calculated at each angle of incidence (or cone half angle) and at each of the freestream flow conditions.

To determine the flow over the plate, oblique shock relations were used to determine Mach number, static pressure, temperature and density behind the leading edge. Charts and tables[3] were used to calculate the flow conditions over the surfaces of the cones. The heat transfer rates per unit area were then non-dimensionalised into the form of Stanton number, where

$$St = \frac{\dot{q}}{\rho_2 U_2 C_p (T_o - T_w)} \tag{2}$$

Stanton number data from the cone tests were divided by $4\sqrt{3}$ (see ref 4) in order that it could be compared with the flat plate data. Note that hereafter, all Stanton numbers quoted for the cone tests are those values which have been divided by $4\sqrt{3}$.

Unit Reynolds number downstream of the oblique shock wave was calculated using the Sutherland formula to evaluate the values of viscosity, where

$$\mu_2 = \frac{1.4296\times10^{-6}\, T_2^{3/2}}{T_2 + 117.0} \qquad (kgms^{-1}) \tag{3}$$

Data from one run (that illustrated in Figs 1 and 2) are also shown in Fig 3. Stanton number is plotted against Reynolds number, based on conditions downstream of the oblique shock wave and distance from the leading edge of the plate. Figs 4 and 5 plot Stanton number against Reynolds number for all of the data from the tests on the flat plates and cones respectively.

When plotting heat transfer data obtained in rarefied flow, it is conventional to plot $M_2^3 St$ against $\bar{\chi}$, where

$$\bar{\chi} = M_2^3 \sqrt{\frac{C^*}{Re_{2,x}}} \tag{4}$$

$$C^* = \frac{\mu^* T_2}{\mu_2 T^*} \tag{5}$$

and

$$T^* = T_o \left[\frac{1}{6} + \frac{\frac{1}{2} T_w}{T_o} \right] \tag{6}$$

Data from all of the tests carried out on the flat plate model are plotted in Fig 6. Results for the sharp cones are plotted in Fig 7.

3.2 COMPARISON WITH THEORY.

Data were compared with two theories, viscous strong interaction theory[5], where

$$M_2^3 St = 0.0788 \left[1 + \frac{2.5 T_w}{T_o} \right]^{\frac{1}{2}} \bar{\chi}^{3/2} \tag{7}$$

and a reference enthalpy method[4], where

$$M_2^3 St = \frac{0.332}{\sqrt{Re_{2,x}}} \left[\frac{T_2}{T^*} \frac{\mu^*}{\mu_2} \right] (Pr^*)^{-2/3} \tag{8}$$

It can be seen from Fig 8 that the magnitude of all of the heating data from the flat plate tests is higher than the highest of the predictions by a factor of almost 2. However, the slope of the line of data is similar to that of the reference enthalpy method at values of $\bar{\chi}$ less than about 6. A point to note is that at a value of $\bar{\chi} \approx 6$, the slope of the flat plate data line increases to become close to that of the viscous strong interaction method. This shows that the IR system is sufficiently sensitive to detect differences in the state of the flow over the surface of the body in rarefied flow conditions (*ie* from merged or inviscid flow).

The data from the cone tests is in better agreement with the predictions but does not show the clear trends exhibited by the flat plate data. The reason for this is unclear but a possible explanation is that if the angle between the surface under investigation and the scanner is too acute, the temperature reading from the system is lower than the actual surface temperature. At the nose of the pointed cones it is difficult to be sure that the readings are being taken exactly along the meridian which is normal to the axis of the camera. Any small deviations from this line will result in readings which are erroneously low. This could well be the source of the deviation from the trend seen in the cone results at values of $\bar{\chi}$ greater than about 6. All of these data were acquired near to the noses of the cone models.

3.3 COMPARISON WITH EXISTING DATA.

Fig 9 compares the current data with that obtained by Hendry[6] using thermocouples attached to the back of a flat plate. It can be seen that over the range of $\bar{\chi}$ from around 2 to 30 the data coincide, within the bounds of experimental error. This shows that the IR thermographic technique can provide results at least as good as a more traditional method of measuring surface temperature. Also, the IR method has the advantage of resolution - it can produce temperature maps of whole surfaces. By comparison, a model instrumented with a relatively small number of thermocouples mounted in its surface can only produce discrete data points which may not adequately show important flow features.

4 Conclusions.

Tests carried out in the RAE Low Density Tunnel have shown that a commercially available Infra Red scanner system is capable of measuring surface heat transfer rates accurately and repeatably in rarefied flows. It has advantages over other measurement techniques which use discrete transducers and sensors mounted on the surface of the model under test. Model manufacture is much less complex and the positions at which data can be obtained are not constrained by the position of fixed sensors.

5 List of Symbols

C^*	Chapman Rubesin constant
C_p	specific heat capacity of Nitrogen at constant pressure
c_w	specific heat capacity of model wall material ($Jkg^{-1}K^{-1}$)
M	Mach number
Pr	Prandtl number
$\dot{q}$	heat transfer rate per unit area (kWm^{-2})
Re	Reynolds number
St	Stanton number
T	temperature (K)
t	time (s)
U	flow velocity
x	distance from the leading edge (m)
y	thickness of model wall (m)
μ	viscosity ($kgms^{-1}$)
ρ	density (kgm^{-3})
$\bar{\chi}$	viscous rarefaction parameter

Subscripts and superscripts

o	reservoir conditions
w	conditions at the model wall
x	based on distance behind the leading edge
2	conditions behind the bow shock wave
*	evaluated at the intermediate temperature given using equation (6)

6 References

1 M F Westby — Heat transfer rates on a blunted cone in rarefied hypersonic flow using infra-red thermography. Royal Aerospace Establishment Technical Memorandum Aero 2186 (1990).

2 M F Westby — The Royal Aerospace Establishment Low Density Wind Tunnel. Defence Research Agency (Aerospace Division) Technical Memorandum Aero 2277 (1991).

3 - — Equations, tables, and charts for compressible flow. NACA Report 1135 (1953).

4 L F Crabtree, R L Dommett, J G Woodley — Estimation of heat transfer to flat plates, cones and blunt bodies. Royal Aircraft Establishment Technical Report 65137 (1965).

5 J E Wallace, A F Burke — An experimental study of surface and flow field effects in hypersonic low density flow over a flat plate. Cornell Aeronautical Laboratory (1964).

6 J D Hendry — Local leading edge heat transfer on a flat plate in rarefied hypersonic flow. Institute of Engineering Research, University of California, Berkeley Report No AS-63-8 (1963).

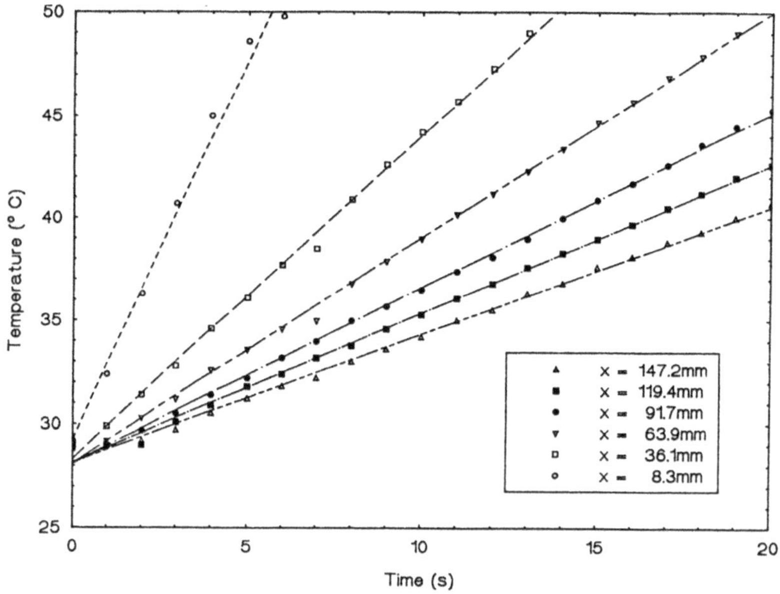

Figure 1. Variation of surface temperature with time at six locations on the surface of the flat plate. Plate at zero incidence. $Re_{2,x} = 1.24 \times 10^5\ m^{-1}$

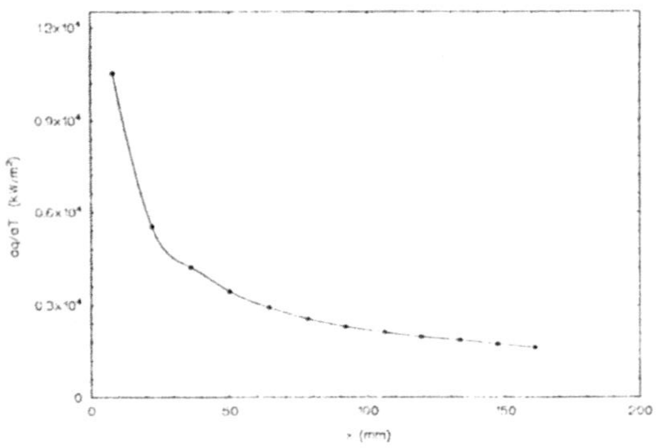

Figure 2. Variation of $\dot{q}$ with distance along plate. Data for $Re_{2,x} = 1.24 \times 10^5$ m^{-1}.

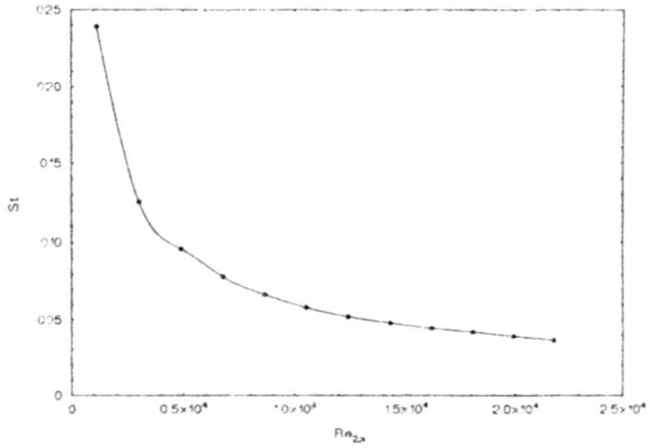

Figure 3. Variation of Stanton number with Reynolds number along plate. Data for $Re_{2,x} = 1.24 \times 10^5$ m^{-1}.

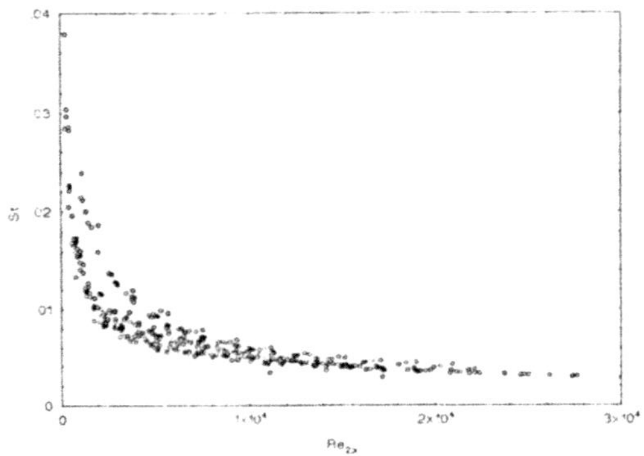

Figure 4. Variation of Stanton number with Reynolds number. Data for all plate tests.

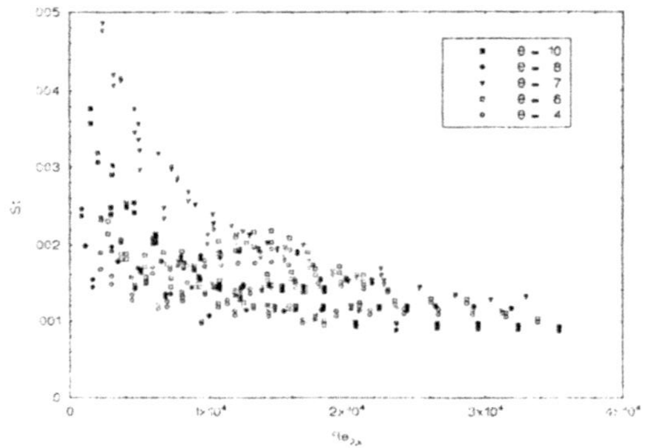

Figure 5. Variation of Stanton number with Reynolds number. Data for all cone tests.

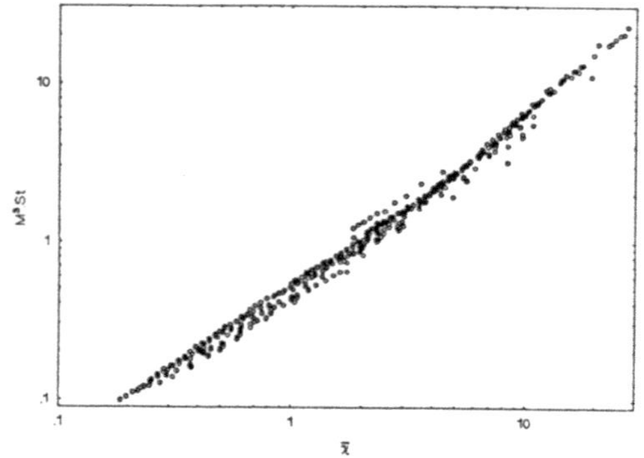

Figure 6. Variation of M^3St with $\bar{\chi}$. Data for all plate tests.

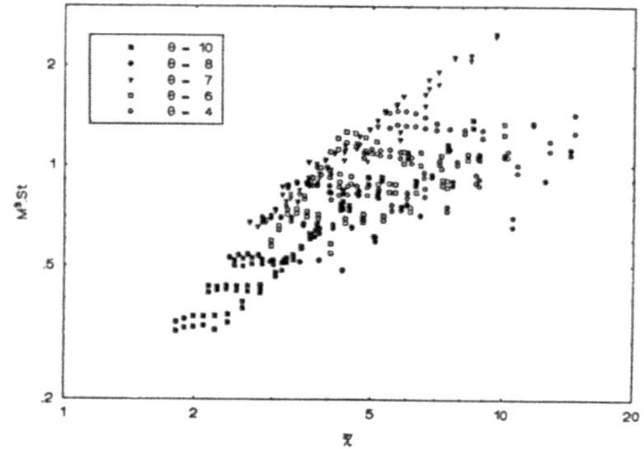

Figure 7. Variation of M^3St with $\bar{\chi}$. Data for all cone tests.

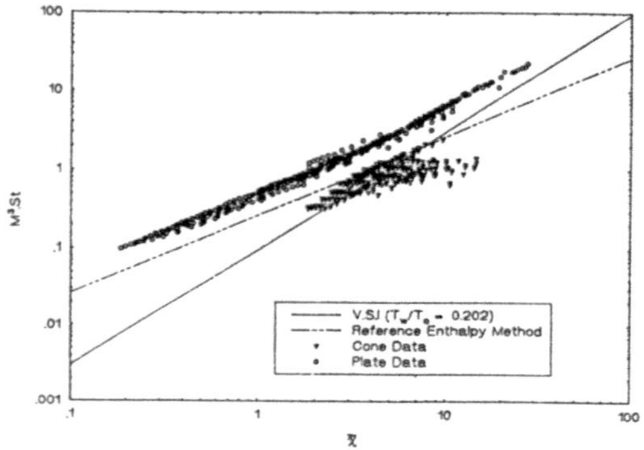

Figure 8. Comparison of all experimental data with Viscous Strong Interaction theory and a Reference Enthalpy method.

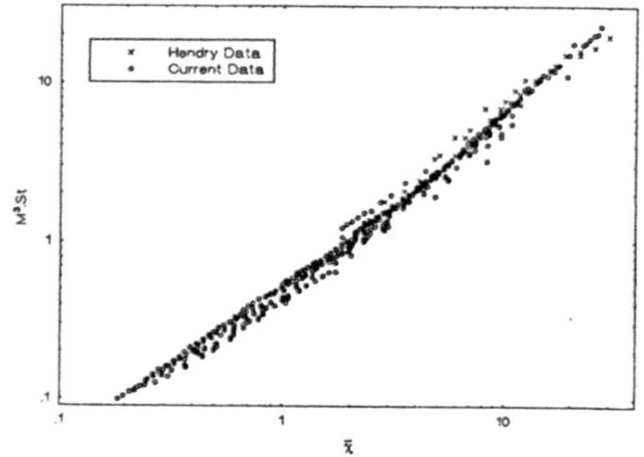

Figure 9. Comparison of data form plate tests with previous experimental data (Hendry[6]).

INFRARED THERMOGRAPHY FOR HOT-SHOT WIND TUNNEL

S. Barbé, D. Boscher, J. Deschamps, G. Gauffre, A. Girard
Office National d'Etudes et de Recherches Aérospatiales
BP 72, 92322 Châtillon Cedex, France

SUMMARY. Two high frame rate devices are described:
- a 1D scanning camera with a rate of 2500 lines per second,
- a staring infrared camera of 64 × 64 pixels with a rate of 400 images per second.
The first tests at F4 are planned in spring 1992.

Introduction

Infrared thermography is a technique commonly used for deriving heat transfer coefficients from model thermal mapping in wind tunnels, from transonic range to hypersonic.

In the case of high enthalpy F4 facility, any commercially available infrared camera, working at 50 images per second is strictly irrelevant because of the short duration of a single run, varying between 20 and 100 ms.

Two high frame devices are considered to be used and shortly described here below:

- a line-scanning camera with a rate of 2500 lines per second,
- a staring camera, using a HgCdTe array of 64 × 64 detectors, with a frame rate of 400 images per second.

1. High frame line-scanner

The experiment for the line scanner is simple (Fig. 1). A commercial infrared camera (Agema 880 LWB with a HgCdTe detector, measuring in the wavelength range 8 - 12 μm) observes the model through a ZnSe 210 mm i.D. window. The angle between the wind tunnel axis and the viewing direction is about 74°, in order to see the stagnation point. The vertical sweeping of the camera is blocked, permitting line scanning with a rate of 2500 lines per second. With a view angle for the camera of seven degrees and a distance between the camera and the model of about 2 meters, the field of view is of the order of 20 cm, which allows measurements with a spatial resolution of about 2 mm. The temperature resolution of such a camera is 0.1°C, for a black body like model, and the measurement range is typically from ambient temperature to 1000°C.

The video signal is sent to an analogical numeric converter, 12 bits, 2 MHz, placed on an acquisition board developed for the Agema infrared cameras by Cedip, which numerizes 250 points from a line. This board is installed into a micro computer Compaq 386/25. The

A. Boutier (ed.), New Trends in Instrumentation for Hypersonic Research, 513–516.

computer itself is able to process the data to calculate temperatures and convective heat fluxes.

The numerized levels are converted in video voltage, then in surface temperatures, taking into account calibration curves for the camera and local emissivity of the model. To calculate this emissivity, a superposition of an infrared image, recorded by a first experiment with the total field (the camera scanning in both the directions), with a surface mesh is made. The mesh projection is adjusted on the infrared image by translation and rotations of the mesh. The adjustment allows us first to know the angle between the normal to the surface and the viewing direction and then calculate local emissivity, second to determine the point where the temperature is calculated on the model. From temperature evolution, convective fluxes are calculated, using a 1D semi-infinite model and taking into account radiative heat losses.

2. Staring infrared camera at high frame rate

2.1. DESCRIPTION

It is based on the use of a HgCdTe focal plane array of 64 × 64 detectors, from CENG-LIR, cooled at 77°K, and sensitive in the 3 - 5 μm spectral range.

Specially designed for its use at F4 facility, it is divided in two units:

- the camera unit (Fig. 2) located inside the vacuum test chamber. It includes:
  - the array of detectors associated with its cryogenic device at liquid nitrogen temperature,
  - two commutable lenses (F = 25 mm and F = 70 mm, with F/2 aperture) for "wide" and "narrow" field of view,
  - the electronic set up delivering a DC 12 bits/2 MHz output signal. The clock frequency used is 26 MHz. This unit is airtight and pressurized near normal pressure;

- the computing unit, placed in the monitoring room of the wind tunnel includes a MOTOROLA 32 bits computer, a high speed recorder, for acquisition and treatment of the data.

Several specific numerical codes have been developed in association with the camera. One of them is specially devoted to the correction of the spatial inhomogeneities in the sensitivity of the 4096 detectors.

The experimental arrangement is shown figure 3.

2.2. PERFORMANCES MEASURED IN LABORATORY

- Frame rate: 400 images per second
- Integration time adjustable between 39 and 2500 microsecond
- Acquisition time: 500 ms
- Storage capacity: 200 images
- Thermal resolution: 0.1 K around 320 K
- Thermal range around 300 K
  - 20 K for 600 μs as integration time,
  - 250 K for 39 μs as integration time.
- Linear field of view (FOV) and Spatial Resolution (SR):

. Focal length F = 25 mm	FOV: 300 × 300 mm	SR: 5 × 5 mm
. Focal length F = 70 mm	FOV: 100 × 100 mm	SR: 2 × 2 mm

A sequence of 36 successive images is shown figure 4. The thermal contrast between the blades of the rotating disk and the background is 4 K around 300 K.

The camera will be installed in the F4 wind tunnel during spring 1992.

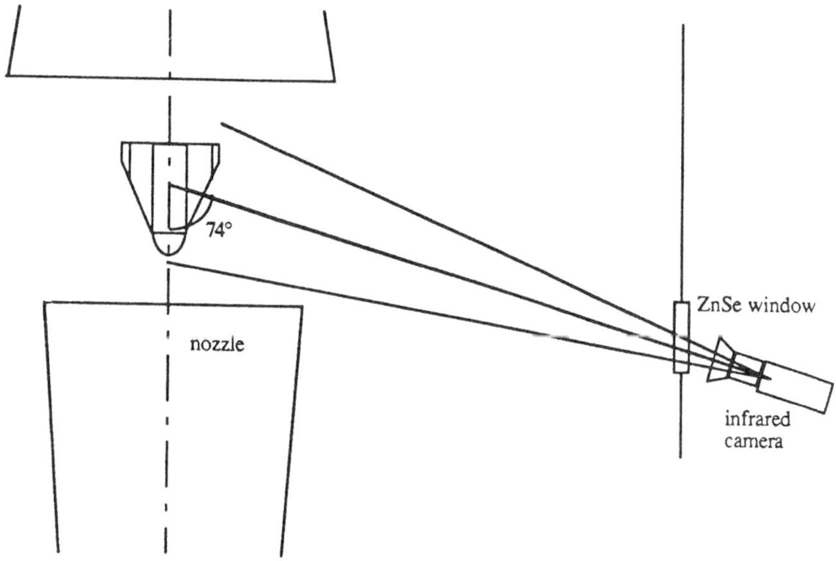

Figure 1 - Experimental arrangement for the line scanner thermography.

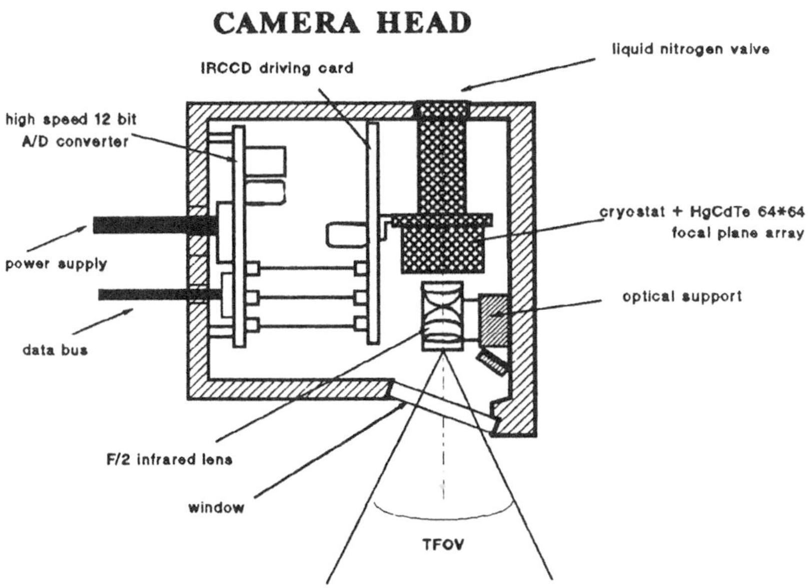

Figure 2 - Camera head.

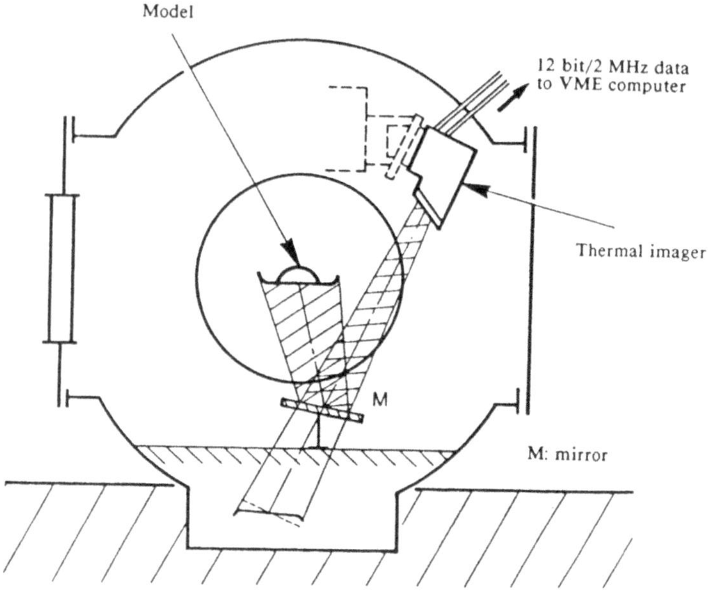

Figure 3 - Implementation of the camera in F4 wind tunnel.

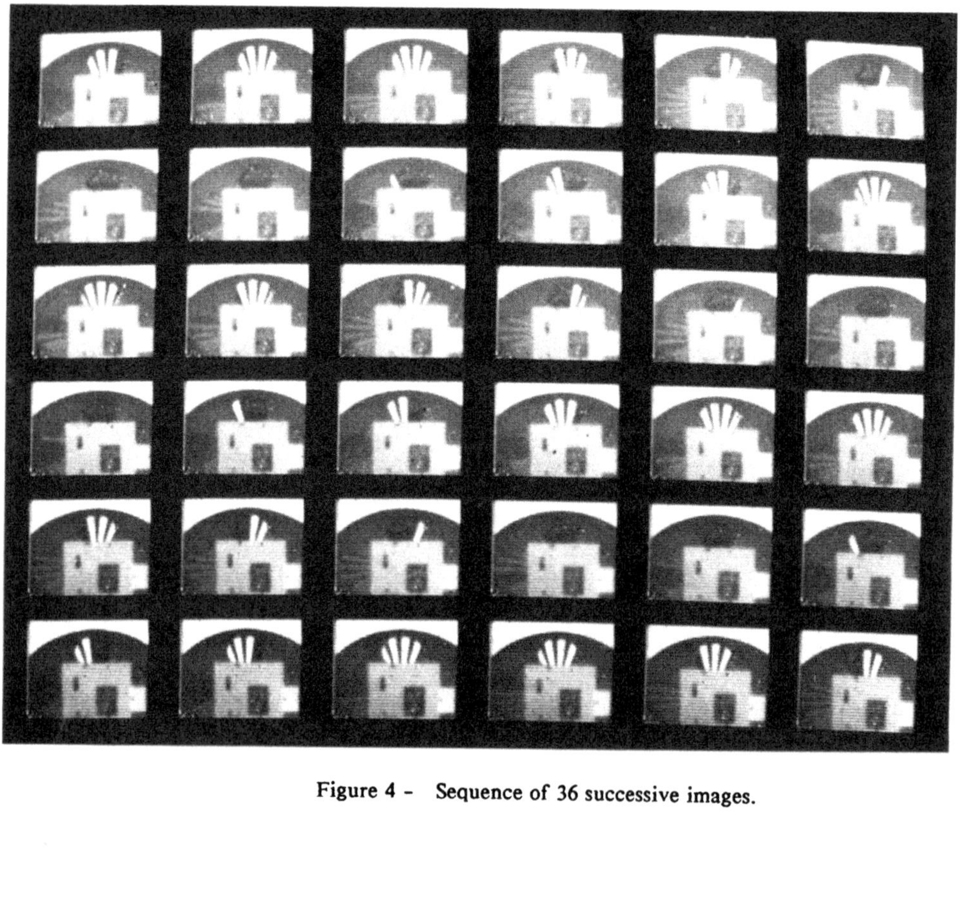

Figure 4 - Sequence of 36 successive images.

Experiments on the catalytic efficiency of heat protection materials in the arc heated wind tunnel (LBK) in Cologne.

A. Gülhan, V. Jörres and K. Kindler
DLR, WT - WK KP

Summary

The arc heated wind tunnel LBK of DLR is used to investigate the response of heat protection material to high enthalpy flows. The experimental study is performed using flat plates. The temperature distribution on a thermal protection material and the heat flux rate of a catalytic wall are used to investigate the effect of noncatalycity on the temperature field of the flat plate. The model consists of two parallel parts of a flat plate. The cooled copper surface represents the reference part to measure wall pressure and heat flux rates. Using the results of these measurements the same energetic conditions for the different noncatalytic surface materials can be set in different tests. The surface temperatures are measured by pyrometers. The development of the surface temperature of Al_2O_3 gives an indication for the catalytic reaction on the flow field.

1. Introduction.

The reentry process of space vehicles takes place in thermodynamic nonequilibrium flow in the vicinity of the body. In contrast to the real conditions the incident high enthalpy flow in front of the shock in wind tunnels is in nonequilibrium. Therefore the experimental results are transferable to the real aerodynamical field with some restrictions. But with respect to the conditions near the surface the validity of the results for heat protection and gas-surface interaction is not affected.

The interpretation of flow conditions in the wind tunnels is supported by equivalent theoretical analysis based on a realistic thermodynamical nonequilibrium model. Using these theoretical models, which treat the flow field quasi-onedimensionally, numerical codes are developed. This assumption does not restrict the interpretation of the flow field generated in the wind tunnel. The comparison of numerical and experimental results is valuable for classifying the experimental research with respect to the physical environment.

2. The arc heated wind tunnel of DLR.

The facility used for the investigation of heat protection material is the arc heated wind tunnel LBK of DLR in Cologne. The main parts of the facility are described in many reports [1, 2, 3, 4, 5]. The essential feature of several arc heated wind tunnels

A. Boutier (ed.), New Trends in Instrumentation for Hypersonic Research, 517–529.

is the testing time of more than one hour. Therefore the surface conditions can achieve the values equivalent to free flight conditions during simulation in the test facility.

The main parts of an arc heated wind tunnel are:

- the arc heater and the power supply
- the test chamber with expansion nozzle, free jet and diffuser
- the heat exchanger
- the supply equipment

The electrical power-supply system of the LBK is based on a thyristor-controlled rectifier consisting of four units of 354 kW each (600A, 590V), which can be switched alternatively to a parallel or series connection. The current is controlled during the test while the effective voltage depends on the resistance of the arc heater.

An arc heater of the hollow-electrode concept (Huels-type) has been applied for the experiments. Using the arc heater one should try to enlarge the voltage as much as possible, especially in the case of high mass flow rates. There is a reason for using high voltages with respect to flow quality. Contamination of the gas flow by burning off of electrodes cannot be completely avoided. This effect increases with current strength and is dependent on the total pressure. A value of contamination less than 10 ppm is reached for total pressures of 10 bar in the LBK.

The LBK is operable with two test legs (L2K and L3K). The L2K, used for the reported experiments, has a cylindrical test chamber with a diameter of 2.6m and with a length of 2m transversal to the flow direction. Tubes of different length can be flanged on the test chamber for changing the distance between the nozzle exit and diffuser entrance. The conical nozzle has an angular aperture of 12°. For the reported experiments a nozzle with a throat diameter of 29 mm and an exit diameter of 200mm is used.

The models and probes are mounted on a multiple holder, which can be moved in the range of 1000mm transversal to the flow direction and 1500mm in flow direction. The leading edge of the used flat plate model is positioned 100mm behind the nozzle exit.

The test chamber of L2K can be disconnected from the pumping system by a gate valve of a diameter of 1m. Downstream of this valve a diffuser is installed consisting of several cylindrical segments and one part for reducing the cross section diameter to 0.91m. On the axis of the diffuser a cylinder with a conical nose is mounted. By shifting of this central body of 0.5m diameter the area ratio of the diffuser can be changed continuously.

The heat exchanger is installed downstream of the diffuser to cool down the heated gas to a temperature equivalent for the intake of the pumping system. The heat exchanger is composed of a staggered bank of pipes with five rows.

The vacuum pumping system of the LBK consists of three Roots blowers (R107, R87, WGK12000), one rotary plunger vacuum pump S720 and one liquid seal pump (Elmo).

The supply equipment for the gas mass flow rate control system can be used at gas mass flow rates up to 300 g/s. It consists of a pressure adjustment system and a mass flow rate control system. The gas mass flow rate is measured by an instrument based upon the Coriolis effect.

3. Theoretical Model

The calculations of a quasi-one-dimensional flow are performed through a nozzle of specified geometry. The reservoir conditions are calculated from data on the settling chamber pressure in front of the nozzle and the total mass flow rate. Both values are easily measured and are used to calculate total temperature and total enthalpy based on equilibrium flow solution from the reservoir to the throat.

The gas composition of the computed flow consists of up to 20 chemical species. Species are atoms, atomic ions, diatomic molecules and molecular ions and electrons. For calculations of thermochemical equilibrium the concentrations of the species are expressed in terms of the independent elements (N, O, e). Using this technique the number of equations which has to be solved simultaneously to determine the equilibrium species mole fractions from the species number is reduced to the number of concerned elements.

The flow calculations through a nozzle are performed by the Non-equilibrium Arc Tunnel Analysis code developed by the Avco Systems Division [6].

Actual flows in arc heated wind tunnels show some radial non-uniformity. In some flow fields the non-uniformity is sufficiently moderate that the quasi-one-dimensional description of NATA can be considered to be roughly valid and useful. In other test conditions, the non-uniformity is so severe that the usefulness of an analysis based on radially uniform flow appears doubtful.

The actual flow in an arc-heated wind tunnel is a non-equilibrium expansion. Due to the rash expansion the thermodynamical equilibrium cannot be achieved downstream the nozzle throat. The equilibrium and frozen flow are only limiting cases where the reaction rates are infinitely high or zero, respectively. In some circumstances the equilibrium or frozen flow calculations may approximate the flow in some parts of the nozzle.

The accuracy of the transport properties as viscosity and translational thermal conductivity appears to be governed in almost all cases by the accuracy of the data on atomic and molecular cross sections used in the calculations of Chapman-Enskog. The collision cross sections are known only within about 10 to 20%.

NATA contains an approximate calculation of laminar boundary layer development on the nozzle wall. The choice of the method was guided by the requirement that solutions including boundary layer should not consume more computer time than the basic inviscid solutions. The chosen technique is an integral method developed by Cohen and Reshotko [7], which uses curvefits to certain boundary layer properties which are based on similar solutions for a Prandtl number of unity and a viscosity proportional to the absolute temperature. In this case the procedure is independent on the quotient of the square of the free velocity u_e and the total enthalpy h_o, the hypersonic parameter $\sigma = u_e^2/2h_o$. But the curvefits of Cohen and Reshotko are replaced in NATA by the similar solutions carried out by Dewey and Gross [8] which include the dependence upon Prandtl number, the viscosity-temperature exponent ω ($\mu \propto T^{\omega}$) and the hypersonic parameter σ.

Because the energy equation is not included in the Cohen-Reshotko correlation method, the heat flux is calculated from the shear stress using a Reynolds analogy factor based on similar solutions. Thus the accuracy should be comparable with that of the momentum thickness.

The NATA code provides calculations of the heat flux and pressure on models inserted into the free-stream flow at specific locations. There two general types of models are treated: the stagnation point of an axisymmetric body or a two-dimensional probe, and the flat surface of a blunt wedge.

3.1 Stagnation point calculations.

In the stagnation point case heat flux and shock standoff distance are calculated for both the flat faced and hemispherical axisymmetric models. The calculations are carried out assuming either equilibrium flow behind the shock or a frozen flow with species mole fractions behind the shock equal to those ahead of it. The NATA calculation of the stagnation-point heat transfer to models is based on a correlation formula which incorporates results of analyses by Fay and Riddell [9]. This formula correlates measurements of stagnation-point heat transfer in air within the accuracy of about $\pm$ 20% at temperatures up to the onset of ionization effects. Only for the frozen boundary layer a modification of the Fay and Riddell formula has been made to permit the inclusion of the effects of the surface catalytic efficiency. This modification in the NATA code uses a modification suggested by Goulard [10]. Although Goulard assumes the catalytic factor to be dependent on the surface recombination rate constant and the stagnation point velocity gradient, NATA treats the catalytic factor as an input constant. Therefore the product of heat flux rate and the square root of body radius, $q\sqrt{R_b}$, becomes independent of body radius in NATA. That is not the case in Goulards modification due to the dependence on the stagnation point velocity gradient.

3.2 Wedge flow calculations.

The quantities calculated by NATA on a blunt wedge are the heat flux, the surface pressure, the displacement thickness of the boundary layer, and the shock ordinate. The positions on the wedge where these calculations are carried out are under input control. The wedge is assumed to be inserted into a uniform flow with the conditions at the location of the leading edge without a variation of the free stream conditions with distance along the wedge.

The wedge models to be tested have blunt leading edges to avoid the high heat transfer there. Chung et al. [11] have formulated an analytical treatment of the combined effects of bluntness and boundary layer displacement on the hypersonic flow over a wedge. The Cheng theory is modified by Kemp [12] to be valid for a realistic size of the specific heat ratio of the gas. Therefore the modification of Kemp is used in NATA. This modification brought the theoretical prediction of the heat transfer and the surface pressure into much better agreement with experimental data than the data corresponding to the Cheng theory can do.

NATA uses additional modifications of the Cheng-Kemp results. The used constants are modified by increasing the influence of the ratio of the surface pressure to the free stream pressure.

4. Experimental Setup.

The experiments are carried out in the L2K test leg of the arc heated wind tunnel LBK of DLR in Cologne [4]. The test conditions are defined by the total conditions upstream of the nozzle throat and by the profile measurements of Pitot pressure and heat flux rate in different test section areas. After the definition of the test conditions the flat plate is moved into the test section for measuring the surface pressure distribution and the heat flux rate at three distinct points of the surface. During the flat plate tests the temperature field of different heat protection materials is measured by pyrometers.

4.1 Probes

The Pitot pressure is measured by a cooled probe with an outer diameter of 12 mm and an orifice diameter of 3 mm. Due to this design the response time is low enough to pick up the Pitot pressure profile continuously by shifting the probe through the flow field.

In high energy flow fields the measurement of the socalled cold wall heat flux is of main interest especially in characterizing thermal protection systems. The scheme of the cross section of the sensing element of the probe used during the experiments of this report is same as shown for the flat plate in **Fig. 1**. The temperature development of the slug is measured at two points, at the front surface and in one cross section in front of the rear surface. Using a special data reduction procedure [13] the time dependent heat flux can be evaluated.

4.2 The flat plate model

The flat plate model has a leading edge radius of 2.5 mm to produce a detached shock layer as small as possible without generating an unacceptable high heat transfer at the plate. On the bottom side of the model the thickness behind the leading edge increases to about 60 mm to allow enough space for installations of the measuring equipment. The main dimensions of the model are 225mm x 130mm x 60mm (L*B*H). It consists of an upper part to contain the measurement technique and a bottom part to protect the inner part of the model against the flow field. Both parts consist of nonscaling, stainless, and chemically inert austenitic VA-steel.

To enlarge the heat conduction and to intensify the catalycity of the surface the upper part is covered with a copper surface layer of 3mm thickness. The upper surface of the flat plate is arranged into two parallel parts side by side: in one part three bores are arranged for the incorporation of the copper cylinders to measure the heat flux, and on the other part a cutout with an opening of 144mm x 57mm x 3mm is provided to carry the different flat plates of heat protection material. Ten holes are distributed over the whole surface of the flat plate for the measurement of the surface pressure. The scheme of the arrangement of the slugs for measuring the heat flux at the flat plate is shown in Fig. 1. The center lines of the cylinders are positioned at 80mm, 120mm, and 160mm behind the leading edge of the flat plate.

5. Results

5.1 Comparison of calculated and measured results of the flow field.

The flow field conditions are calculated from the measured gas mass flow rate, the total pressure and the nozzle throat area by using the NATA code [6]. In **Fig. 2** we have plotted the development of the species concentrations of main interest for the settling chamber temperatures less than 6000°K. It is shown that immedeately after the nozzle throat the species mole fractions of N_2, O_2, O and NO are nearly constant for nonequilibrium flow. That means frozen flow is established in the expansion part of the nozzle. The development of the gas dynamic properties temperature, pressure, density and Mach number are calculated using the boundary layer displacement thickness based on the integral method by Cohen and Reshotko. In **Fig. 3** the development of the static pressure along the nozzle is plotted and the test chamber pressure is signed which has been measured during the tests. For the free expansion from the nozzle into the large test chamber the pressure of the test chamber not necessarily has to be equal the static pressure of the nozzle flow But the difference is indicative for the incompatibility of the nozzle expansion ratio and the characteristic of the pumping system for the used gas mass flow rate. A comparison of the Pitot pressure measurement and the calculation of the stagnation point pressure gives some information about the usefulness of the assumptions of the calculations. Measured values inbetween 20 and 23 mbar (**Fig. 4**) correspond poorly with the calculated value of 18 mbar. That means the boundary layer displacement thickness is underestimated in the calculation. Independent of this effect is the result that the development of all data is nearly the same for the non-equilibrium and the frozen solution.

The heat flux rate measured at the stagnation point of a copper slug (**Fig. 5**) is in disagreement with the calculated value due to the effect that the method of Cohen and Reshotko calculates the heat flux rate using the Reynolds analogy factor.

5.2 Comparison of calculated and measured results at the flat plate.

An essential intention of the flat plate experiments has been the critical examination of the gas-surface reaction of the heat protection material. The comparison of the experimental results with an analytical model would be helpful for the appraisal of the effect of the surface material.

In the experiments the surface pressure distribution, the surface heat flux rate at three positions and the surface temperature distribution on the heat protection material have been measured. In general the comparison of pressures seems to be nearly independent of the nonequilibrium state of the gas flow. Therefore it can be used for proving the adequacy of the boundary conditions of the computation. The comparison of the measured surface pressure distribution with the calculated values is used for testing the applicability of the boundary conditions in the computation to the experimental situation.

In the discussed case the test chamber pressure is much less (by a factor three) than the calculated static pressure at the leading edge of the flat plate. On the other side taking the Pitot pressure at the leading edge equivalent to the measured value on the centerline of the flow field, the calculated distribution of the surface pressure at the flat plate is higher than the measured distribution (**Fig. 6a**). The deviations of the calculated static and Pitot pressure from the measured values of the flow are indicative to an underestimated boundary layer displacement thickness. In **Fig. 6b** the comparison of calculated and measured wall pressure distribution is performed for identical static pressure in the flow in front of the flat plate. The agreement is satisfying for the data of the flat plate in a conical flow field computed by the NATA code. Necessarily the estimation of the boundary layer should be adapted to the wind tunnel conditions.

But the comparison of the calculated and measured heat flux rates at the surface provides strong deviations (**Fig. 6c**). Some reasons for the deviation are known. In the described case the deviation due to the Reynolds analogy is increased by the fact that the calculation of the heat flux rate at the surface of the flat plate is carried out for frozen flow and the copper slugs are representing catalytic surface conditions. Unfortunately the decreasing slope of the axial heat flux distribution could not be observed in the experimental data. This fact cannot be explained either due to the reasons mentioned above for the poor agreement.

5.3 Surface temperature at the heat protection material.

While the applicability of the NATA-code for LBK-test conditions has been proved comparing the calculated pressures and heat flux rates with measured values, temperature measurement on the surface of thermal protection material is used to get some information about the catalytic effect of the material itself.

The surface temperature of the heat protection material is measured by a pyrometer, which is shifted line by line over the flat plate. The investigated material has been Al_2O_3. The emission coefficient is tested by comparing the data of a thermocouple with a one color pyrometer in an oven at temperatures up to 550 C. The emission coefficient detected by this procedure has been 0.85 independent of the used temperature range. The accuracy of the temperature data measured with the pyrometer is better than .5% compared with thermocouple measurements for temperatures above 200 C. In **Fig. 7** the isothermes of the pyrometer measurements are plotted. At the corner at $x/L = 1.0$ and $y/B = 1.0$ the cooling effect of the copper part of the flat plate can be seen. In the same way it is observed that the area inbetween $0.2 < y/B < 0.4$ is affected by the cooled copper boundaries slightly. Normally in conical flow fields the temperature distribution in flow direction gets its maximum in the front part of the plate and a continous decreasing trend follows subsequently. This behaviour is not be seen in the **Fig. 8**. The unusually broad maximum temperature region is only explainable by the effect of recombination of atomic oxygen at the surface of the flat plate or in the boundary layer near the flat plate. The conductivity of the thermal protection material cannot be the reason for that figure as it can be seen from the development of the values at the border of the flat plate. As a certain catalytic material copper has been tested at the same flow conditions and shows a peak maximum at x/L = .5 at much lower temperature values than the thermal protection material.

6. Conclusion

The arc heated wind tunnel LBK has been used for an experimental study of the thermal protection material. The experiments have been carried out with a flat plate model consisting of two parallel parts, one for measuring the surface heat flux rates at three points and the other for carrying the heat protection material. For the first tests Al_2O_3-material has been used. The temperature distribution on the surface indicates a catalytic behaviour of this material.

The comparison of the wind tunnel data agrees poorly with the numerical results of the NATA code, especially with respect to the heat flux data measured at the flat plate as expected.

Further tests with other heat protection materials will give the possibility to classify them with respect to their catalytic efficiency.

7. Bibliography

[1] Fiebig, M., Neuberger, W. *Der Plasmawindkanal PK2 des DFVLR-Instituts für Angewandte Gasdynamik. Auslegung und Beschreibung der Anlage*. DLR-FB 70-35 (Aug. 1970)

[2] Fiebig, M. et al. *Die Plasmawindkanalanlage des DFVLR-Instituts für Angewandte Gasdynamik*. DLR-Mitt 71-29 (Dez. 1971)

[3] Fiebig, M., Kindler, K., Papanikas, D. G. *Hochleistungsbrenner und Expansionsdüsen zur Untersuchung von Hochenthalpieströmungen*. DLR-Mitt 75-06 (Feb. 1975)

[4] Kindler, K. *Arc heated wind tunnel facilities*. IB-39113-87-C-24 (Dec. 1987)

[5] Kindler, K. *Der lichtbogenbeheizte Windkanal*. Sonderbeilage der DLR-Nachrichten Heft 66, S. 18 (Febr. 1992)

[6] Bade, W. L., Yos, J. M. *The NATA code - Theory and analysis* NASA CR-2547 (June 1975)

[7] Cohen, C. B., Reshotko, E. *The compressible laminar boundary layer with heat transfer and arbitrary pressure gradient*. NACA Rept. 1294 (1956)

[8] Dewey, C. F., Jr., Gross, J. F. *Exact similar solutions of the laminar boundary layer equations*. Rep. RM-5089-ARPA, RAND Corp., (1967)

[9] Fay, J. A., Riddell, F. R. *Theory of stagnation point Heat transfer in dissociated air*. J. Aero. Sci., vol. 25, no. 2, (Feb. 1958), pp.73-83, 121

[10] Dorrance, W. H. *Viscous hypersonic flow*. McGraw-Hill Book Co. (1962), pp.92,93

[11] Cheng, H.K., Hall, J.G., Golian, T.C., Hertzberg, A. *Boundary-layer displacement and leading-edge bluntness effects in high-temperature hypersonic flow*. J. Aerospace Sci., vol. 28, no. 5 (May 1961), pp. 353-381

[12] Kemp, J.H., Jr. *Hypersonic viscous interaction on sharp and blunt inclined plates*. AIAA J., vol. 7, no. 7 (July 1969), pp. 1280-1289

[13] Kindler, K. *Ermittlung einer zeitlich veränderlichen Wärmebelastung auf einen endlichen Zylinder*. Vortrag 72-112, 5. Jahrestagung der DGLR in Berlin, (Okt. 1972)

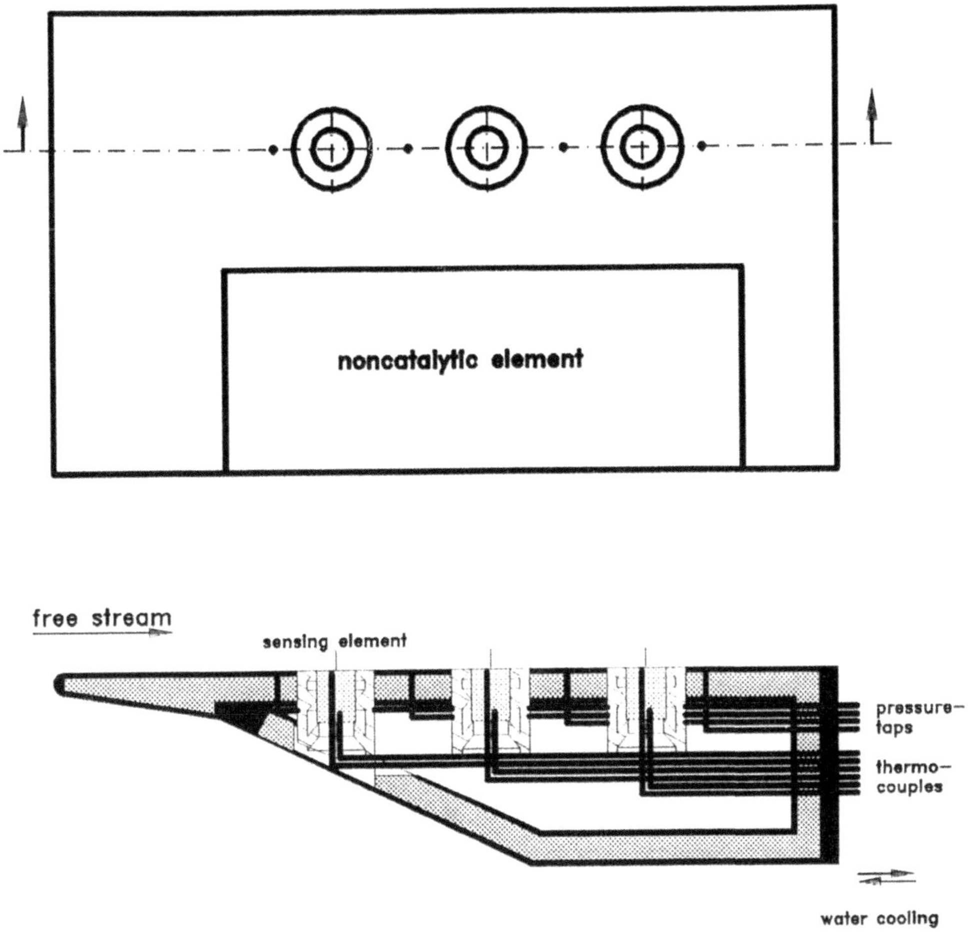

Fig. 1: Scheme of the flat plate model

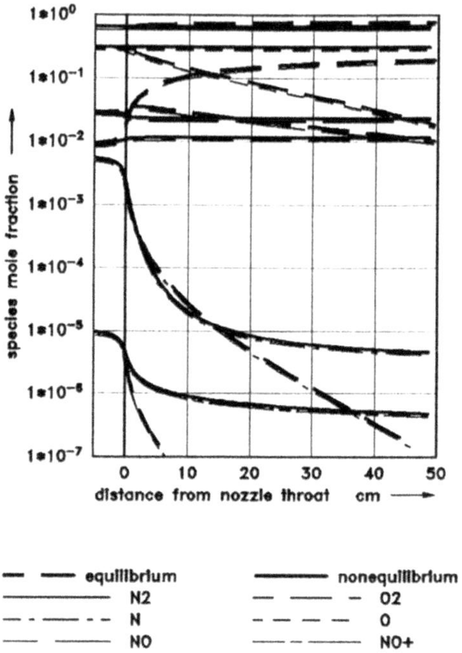

$H_o = 8.84 \frac{MJ}{kg}$ $p_o = 0.850$ bar $T_o = 4381.$ K $\rho_o = 0.055 \frac{kg}{m^3}$

Fig. 2: Development of the species concentrations.

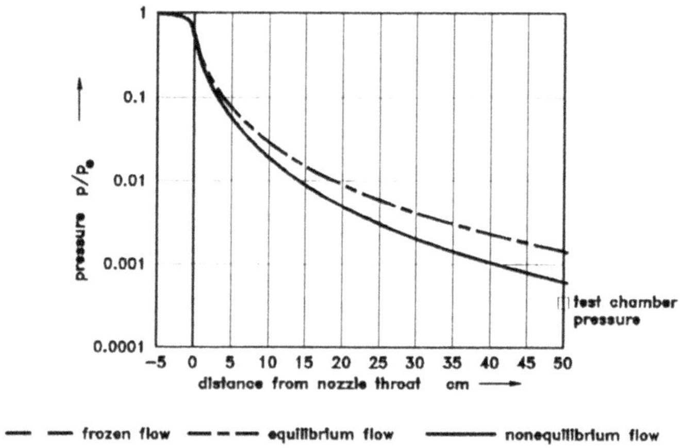

$H_o = 8.84 \frac{MJ}{kg}$ $p_o = 0.850$ bar $T_o = 4381.$ K $\rho_o = 0.055 \frac{kg}{m^3}$

Fig. 3: Development of the pressure in the nozzle.

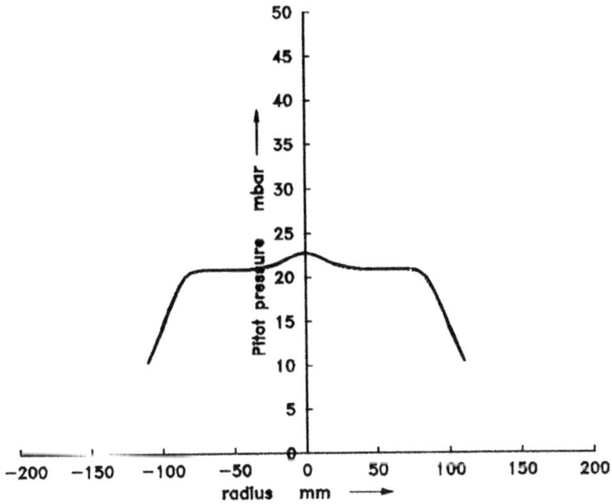

$p_o = 850$ mbar, $\dot{m} = 30 \frac{g}{s}$, $H_o = 8.8 \frac{MJ}{kg}$

Fig. 4: Pitot pressure profile 200 mm behind the nozzle exit.

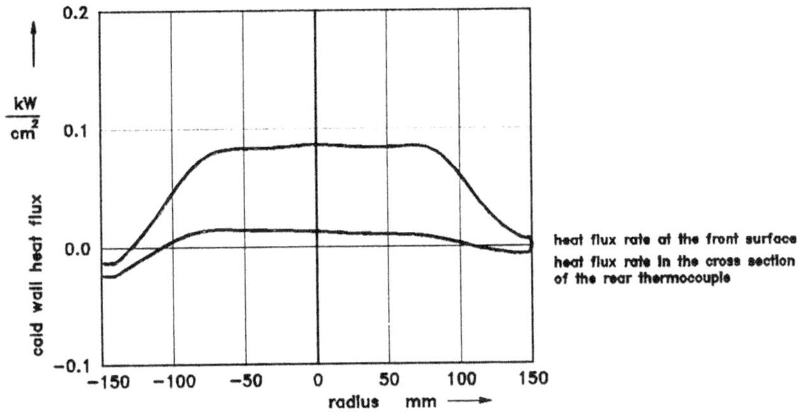

$p_o = 850$ mbar, $\dot{m} = 30 \frac{g}{s}$, $H_o = 8.8 \frac{MJ}{kg}$

Fig. 5: Stagnation point heat transfer 200 mm behind the nozzle exit.

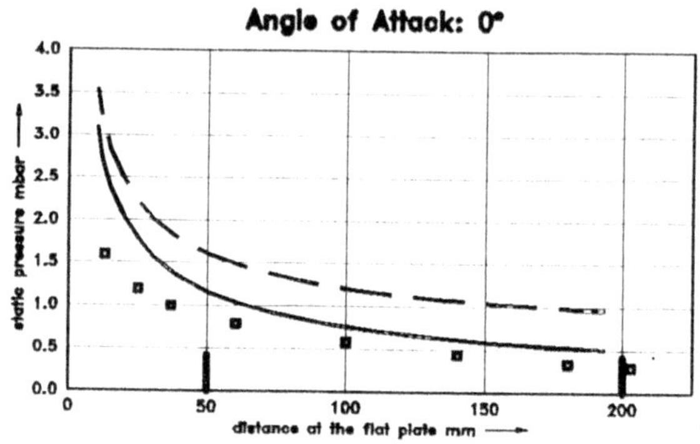

a) Wall pressure compared for equal Pitot pressure at the leading edge

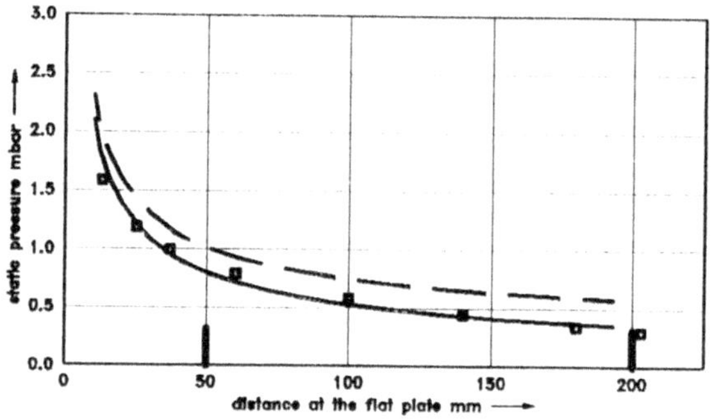

b) Wall pressure compared for equal static pressure at the leading edge

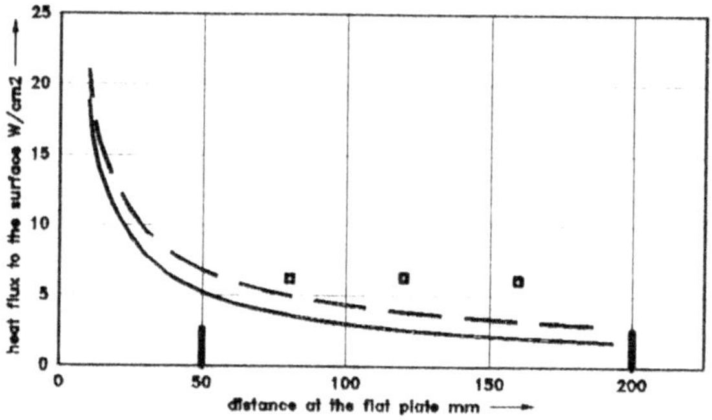

c) Surface heat flux rate

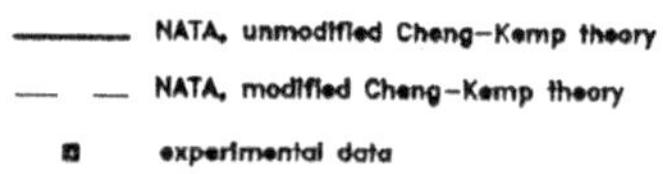

Fig. 6: Experimental data at the flat plate compared with data calculated by the NATA code

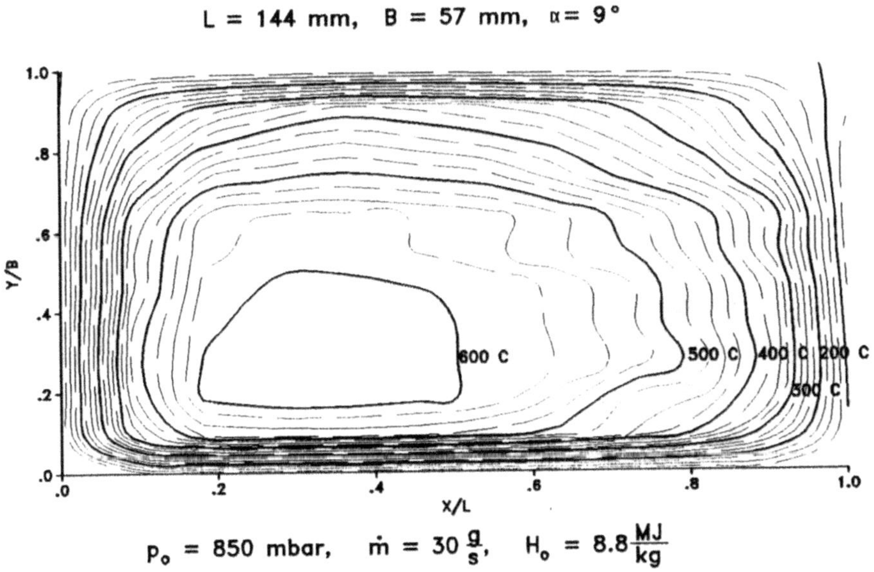

Fig. 7: Isothermes on flat plate of Al_2O_3 150 mm behind the noz zle exit

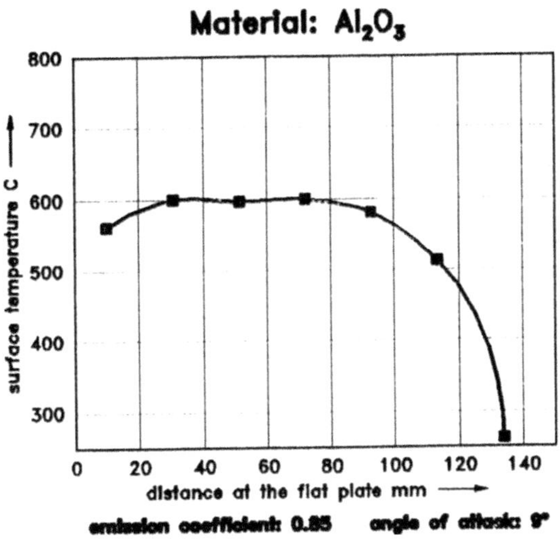

Fig. 8: Temperature profile on the flat plate at y/B = 0.4

INFRARED THERMOGRAPHY CHARACTERIZATION OF GÖRTLER VORTEX TYPE PATTERNS IN HYPERSONIC FLOWS

D. Boscher°, B. Baudouy°, A. Déom°, M.- C. Coët *, J. Délery * and D. Balageas°
°Thermophysics Division, * Aerodynamics Department
ONERA, BP 72 F-92322 Châtillon-Cedex, France

ABSTRACT. Infrared thermography is now currently used in wind tunnel for acquisition of heat transfer maps on aerodynamic surfaces. One of the advantages of the technique lies in the possibility of interpreting rather complicated configurations and heating patterns. Among them, a characteristic configuration producing spatially modulated aerodynamic heating in the spanwise direction is obtained, in hypersonic, on a two-dimensional compression ramp following a flat plate. These structures are frequently attributed to the formation of Görtler vortices which would be caused by the concave curvature of the flow in the reattachment region downstream of the hinge line. Such structures have received considerable attention because the configuration is representative of the control surfaces on hypersonic vehicles. For this particular configuration, infrared thermography was successfully used, allowing a quantitative analysis of the heat transfer modulation. In particular, parameters like the relative amplitude of the heat transfer coefficient (h) modulation, amplitude of the local recovery temperature (T_r) modulation, spatial frequencies of theses modulations were obtained from the time-resolved thermography of the model surface. It appears that the two modulations (h and T_r) are in phase opposition, fact until now unexplained.

1. Introduction

For many years, infrared thermography has been in common use in wind tunnels for localizing areas where the transition between laminar and turbulent regimes takes place [Gartenberg, 1991, Gauffre, 1988]. Progresses achieved in the technology of infrared scanners permit now the determination with a good precision of transient wall-temperature distribution on models in wind tunnels. The combination of this technique with heat transfer analysis and modelling allows to determine quantitatively the distribution of heat transfer coefficients [Balageas et al., 1990, 1991a, 1991b]. This allows a really quantitative mapping of the heat transfer rate distribution on models, specially for super and hypersonic flows for which high heating of models are easily obtained.

We shall present here results obtained using this technique, concerning the aerothermal analysis of the phenomena which take place in the reattachment zone of a ramp, in hypersonic.

2. Interaction phenomena on a hypersonic ramp

The deflection produced by a control surface on a hypersonic vehicle is at the origin of a shock wave leading to a strong interaction with the boundary layer which develops on the vehicle. If this

A. Boutier (ed.), New Trends in Instrumentation for Hypersonic Research, 531–544.

deflection goes beyond a limit value, the intensity of the shock can become such that it induces separation of the boundary layer. Then a rather complex flow establishes itself in the vicinity of the control surface [Délery, 1989]. The boundary layer separates at a point S which can be located well upstream of the hinge line. Downstream of S, a shear layer, well detached from the wall, develops. In most situations, this shear layer reattaches on the control surface at the reattachment point R.

The separation phenomena leads to very typical wall pressure and wall heat transfer distributions whose features depend on the nature - laminar or turbulent - of the incoming boundary layer. Thus, at separation a first pressure rise occurs, the pressure tending to a plateau value if the separated zone is sufficiently extended. A second pressure rise, much more ample that the first one, takes place during the reattachment process. More important for practical applications, is the distribution of heat transfer in the interaction region. Considering a laminar boundary layer, separation itself at S is accompanied by a small decrease in heat transfer. This decrease is followed by a sharp increase in the reattachment region, the heat transfer value reaching a peak at a short distance downstream of the reattachement point R. For a turbulent initial boundary layer, a slight rise in heat transfer occurs at separation and the peak at the reattachment can be more than 10 times higher than for a laminar boundary layer.

The high heat transfer values then existing on the control surface are due to the impact on the wall of the separated shear layer. This phenomenon leads to extremely efficient convective heat transfer process producing heating effects which can be well in excess of those exisintg at the blunted nose of the vehicle. Thus, prediction of the heat transfer distribution on a control surface, or on any part of the vehicle where separation is likely to occur, is of crucial importance for the proper sizing of thermal protections.

Basic experiments on this problem are classicaly performed by considering a simplified geometry most often constituted by a flat plate - with sharp or rounded leading edge - followed by a rectilinear ramp representing the control surface. This two dimensional arrangement simpifies data interpretation and flow simulation by numerical codes.

3. Infrared thermographic setup and software

A long-wave monodetector camera (AGEMA 782) has been used for the present tests. For data acquisition and storage, real time digitizing was chosen. This solution offers very good signal-to-noise ratios. The digitizing hardware is capable of supporting tests of long duration (which is not the case in the present experiments for which duration is between 5 to 7 s) and is easily transportable. The system, called PTR-9010®, was developed under ANVAR grant, in cooperation with the CEDIP Company and under ONERA licence [Potet et al., 1989]. It is based on a Compaq 386 microcomputer carrying a high-speed access 680-Mbyte formatted hard disk and a safeguard unit (WORM of 2x325 Mbyte). Two specific boards were developed to digitize and store the images in real time, and these were integrated into the microcomputer. The central RAM can hold all of the images (6.25 four-raster images per second, each containing 256 lines of 250 points digitized in 12 bits). The recordable time is then, limited to a few seconds. For longer duration tests, the data is stored in real time on the disk, with two possibilities: either one out of two images with 256 points per line or all of the images but with only 125 points per line.

Software tools exist to generate a synthetic overall view of the test, but also to obtain fine analyses of certain sequences and print out clear, readable documents. The program is user-friendly, accessible by menu, and is able to sort out rapidly what is of interest for analysis. In particular, the software is able to provide the following functions : (i) dynamic viewing of the test by presentation of all

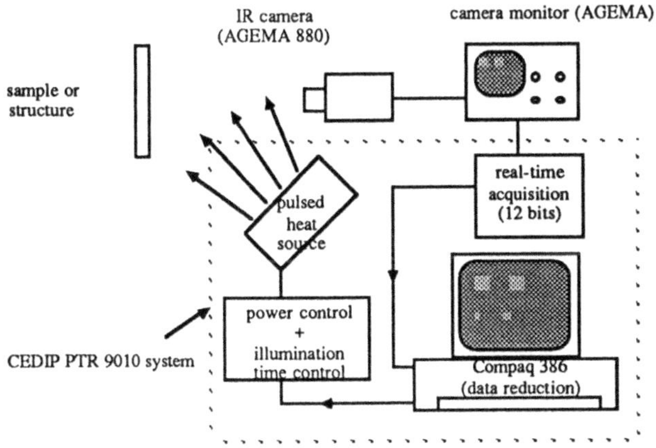

Fig. 1 - Schematic diagram of the Cedip-Onera PTR 9010® System used for both passive and active infrared thermography in ONERA wind tunnels.

of the rasters in store, in rapid succession, (ii) display of several rasters at the same time, for easier comparison and to detect variations, (iii) filtering/smoothing if necessary, (iv) identification of model elements in the infrared image, (v) converting radiance images to temperature images, (vi) converting temperature images to convective flux or heat transfer coefficient images, and (vii) extraction of time-varying thermograms at particular points on the model.

Furthermore, the system also offers the possibility of performing active thermography. Thus the microcomputer can control a unit of radiant energy power directed on the model. A schematic diagram of the system is given in Fig. 1.

4. Calculation of convective heat fluxes ans heat transfer coefficients

From the time variation of the model surface temperature distribution, convective heat fluxes and heat transfer coefficients have to be deduced. The advantage of the thermographic technique is mainly the possibility of elaborating 2-D or 3-D inverse methods, once the whole surface distribution is known. In fact, we shall consider here 1-D problems, each pixel being considered as thermally independent of the others. The 2-D and 3-D refinements are always possible but they are time consuming and not necessary in the present case.

The models here considered being poor conductors of heat, the thermal modelling is based on the semi-infinite medium assumption. The flux, at every moment can be deduced from the thermogram $T(t)$ by the following integral formula (see for example [H. Consigny et al., 1987]):

$$\phi(t) = b/\sqrt{\pi}\,[(T(t) - T_o)/\sqrt{t} + 0{,}5 \int_0^t [T(t) - T(\tau)]/(t-\tau)^{3/2}\,d\tau] \quad (1)$$

In this expression, ϕ is the density of the heat flux at the surface and b the effusivity of the material the surface is made of ($b = \sqrt{k}\sqrt{c}$, k being the heat conductivity and c the heat capacity per unit volume). If the recovery temperature, $Tr(t)$, is known, the heat-transfer coefficient h is then found from the flux and from the thermogram by the expression:

$$h = \phi(t) / [Tr(t) - T(t)]. \qquad (2)$$

If the recovery temperature is unknown and varies slowly, the heat-transfer coefficient can be found by taking the derivative of the flux:

$$h = - \partial\phi / \partial T, \qquad (3)$$

and the recovery temperature can be found by the relation: $Tr(t) = T(t) + \phi(t)/h$.

5. Test arrangement

The tests were performed in the hypersonic wind tunnel R3Ch of the ONERA's Chalais-Meudon Center. All results were obtained with the following flow conditions:

- Upstream Mach number : 10,
- Stagnation pressure : 125 bars,
- Total temperature : 1040 K,
- Reynolds number for the caracteristical length of the model (0,25 m from the leading edge to the hinge line) : Re= 2.1 10^6.

Fig. 2 presents the schematic diagram of the set-up. The model, a flat plate with a metallic sharp leading edge followed by a 15° angle ramp, is made of Norcoat 4000 ®, a material currently used for thermosensitive paint tests.

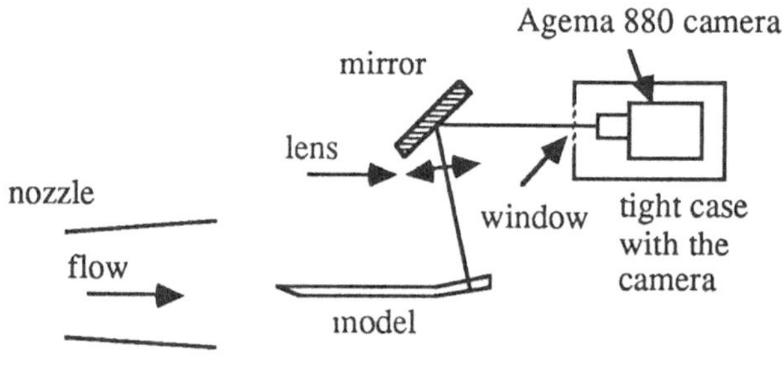

Figure 2 - Test configuration in ONERA's R3Ch wind tunnel for heat transfer measurements on a ramp at M = 10, p_i = 125 bars, T_i = 1040 K and Re_L = 2.1 10^{-6} (L = 25 cm, being the distance between the leading edge and the beginning of the ramp, whose angle is β = 15°).

The camera, with a 7°x7° or 12°x12° objective, is placed in a pressurized vessel and an extra lens can be positionned between the camera and the model to obtain a smaller field of view, when a better spatial resolution is needed. Fig. 3 presents the temperature distribution on the ramp, showing clearly a structure made of streaks that is sometimes related to Görtler vortices. The interest of the smaller field is obvious.

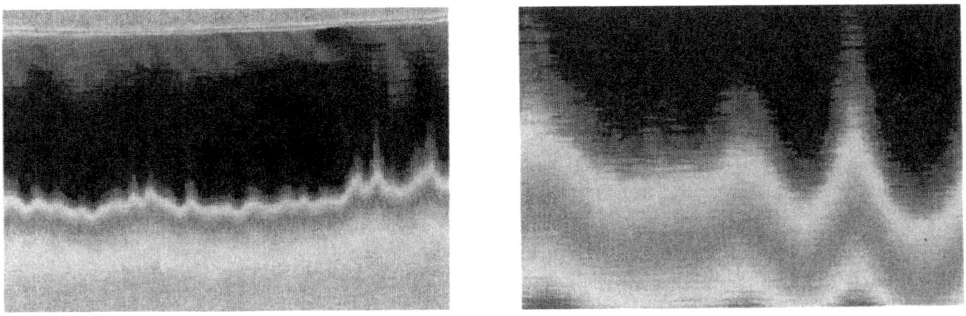

Figure 3 - Temperature (arbitrary unit) distribution on the ramp at a given time during the run. *Left:* large field of view; *Right:* small field of view (using the extra lens). The wind is coming from the lower side to the upper side.

6. Data reduction

The quality of the measurements appears on Fig. 4, where the temperature-time history of a given pixel is shown. The signal-to-noise ratio is good enough for there to be little noise in the heat flux-time history.

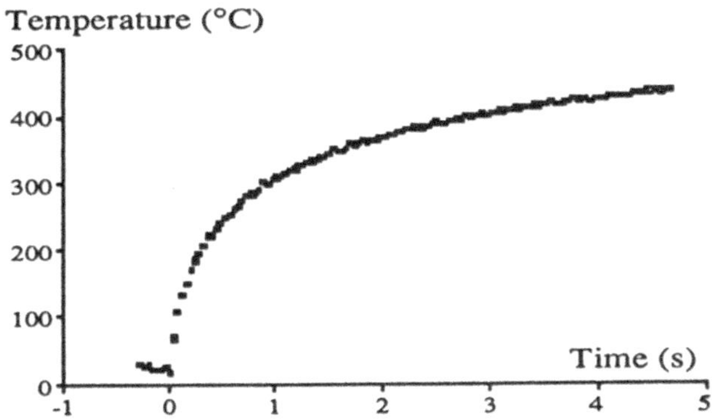

Fig. 4 - Temperature-time history of a given pixel extracted from the series of temperature images (at abscissa x = 35 mm from the beginning of the ramp).

Using the formula (1), from the evolution of the wall temperature (see Fig. 5), the heat flux entering the wall, ϕ_{cond}, can be deduced. As shown in Fig. 5 it is necessary to take into account the radiation flux emitted by the wall, ϕ_{rad}, to deduce the convective heat flux ϕ_{conv}:

$$\phi_{conv}(t) = \phi_{cond}(t) + \phi_{rad}(t)$$

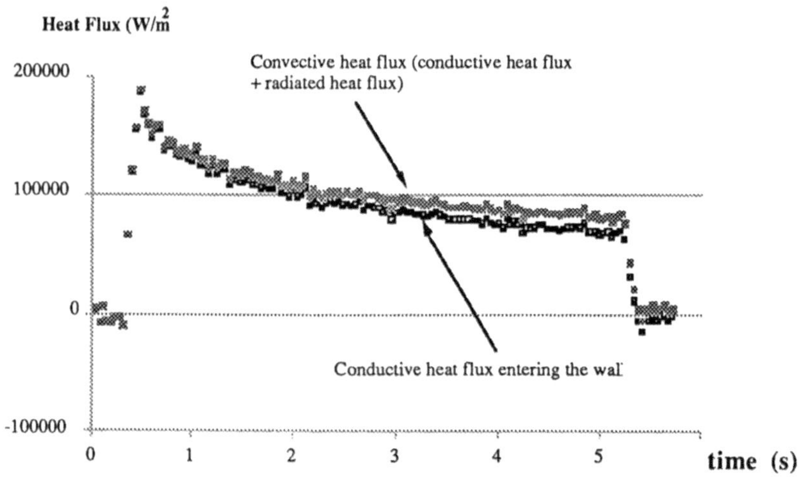

Fig. 5 - Time evolution of the convective heat flux of a given pixel, calculated from the temperature-time history extracted from the series of temperature images (at abscissa x = 35 mm from the beginning of the ramp).

Thanks to the low conductivity of the material used for the present model (Norcoat ® 4000 material), the wall temperature during the test reaches values not too far from the recovery temperature. Thus, it is possible to deduce the heat transfer coefficient without considering the recovery temperature. Fig. 6 presents the convective heat flux for a given pixel as a function of the local wall temperature. The linear variation allows the identification of this coefficient (slope of the curve) and of the true local recovery temperature (abscissa at the origin).

The same operation performed for each pixel yields the heat transfer coefficient image (left) and local recovery temperature image (right) of Fig. 7.

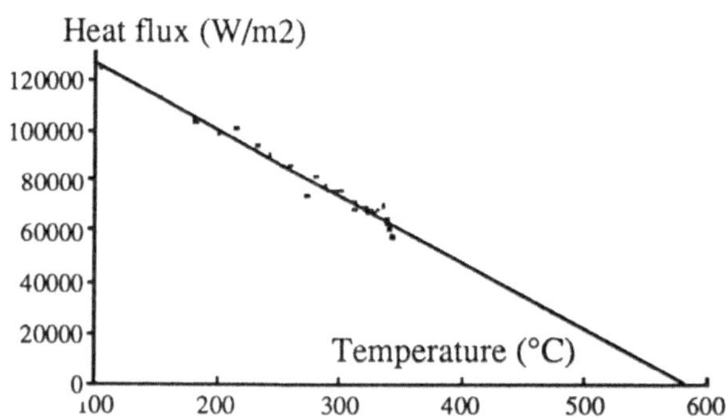

Figure 6 - Measured convective heat flux as a function of the local wall temperature for a given pixel at x=35mm.

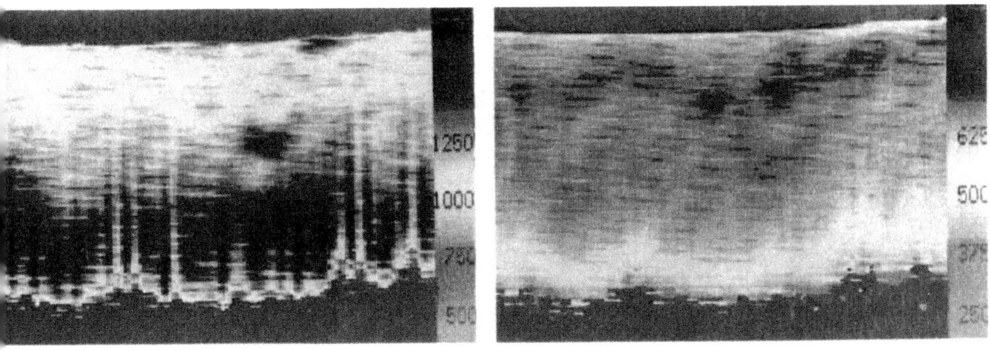

Figure 7 - Heat transfer coefficient ($Wm^{-2}K^{-1}$x 5) distribution on the ramp deduced from the slope of the flux-versus-wall-temperature curves *(left)*. - Local recovery temperature (°C) distribution on the ramp deduced from the abscissa at the origin of the flux-versus-wall-temperature curves *(right)*.

7. Analysis of the heat transfer coefficient and local recovery temperature distribution

Let us analyze the heat transfer coefficient and local recovery temperature distribution as given by the two images of Fig. 7 .

7.1. LONGITUDINAL DISTRIBUTIONS

From the two images of Fig. 7 we can extract longitudinal distributions of h and Tr. For instance, at the lateral location z=185 mm, near of the right end of the model, we obtain the heat transfer coefficient distribution presented in Fig. 8. The origin of x abscissa is the hinge line.

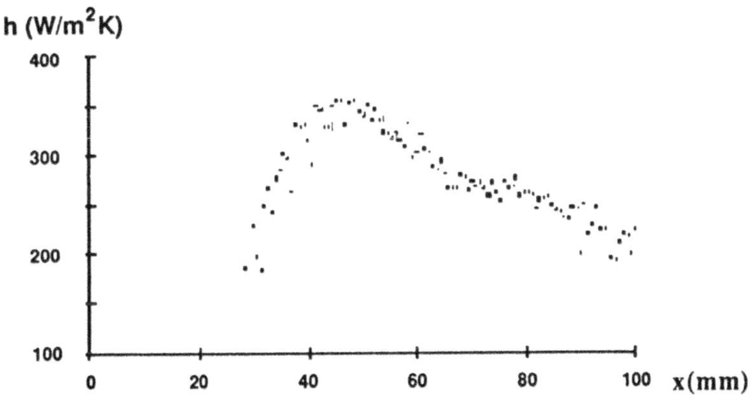

Figure 8 - Longitudinal distribution of heat transfer coefficient in the reattachment zone at z=185 mm.

From the heat transfer coefficient, and using the free stream conditions (density, velocity and theoretical recovery temperature) the Stanton number distribution is obtained (see Fig. 9).

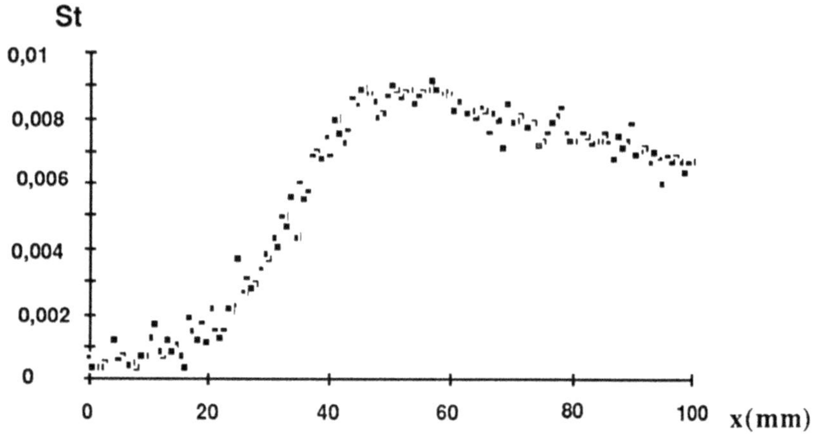

Figure 9 - Longitudinal distribution of Stanton number in the reattachment zone at location z=185 mm.

From the h distribution which is in relatively good agreement with previously published results from thermocouple measurements [Délery and Coet, 1990], the transitional nature of the flow is confirmed (the sharp increase of the heat transfer in the reattachment zone being much higher than for a fully laminar interaction). In some region, like for z=140 mm, there is a second maximum in the heat transfer distribution, phenomenon probably related to a transition effect (see Fig. 10).

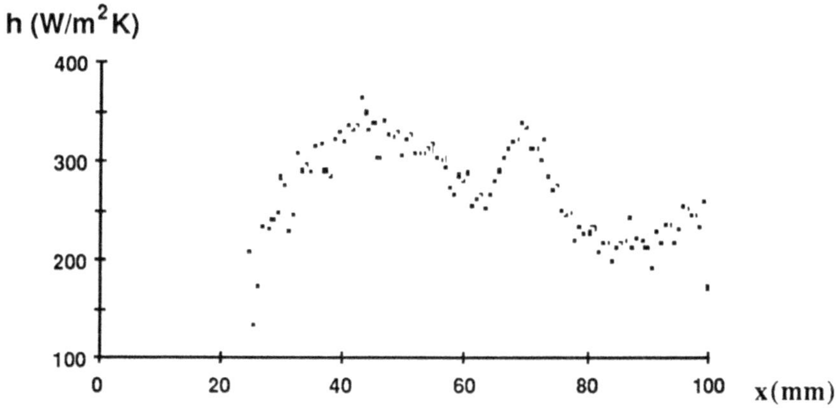

Figure 10 - Longitudinal distribution of heat transfer coefficient in the reattachment zone at z=120 mm.

Similar longitudinal distributions are obtained for the local recovery temperature. They present slighter variations at the beginning of the ramp and a plateau at the rear part (see Fig. 11). At the beginning of the ramp, there is a large difference between the theoretical recovery temperature calculated for the flat plate and the experimental one, which is much lower than predicted.

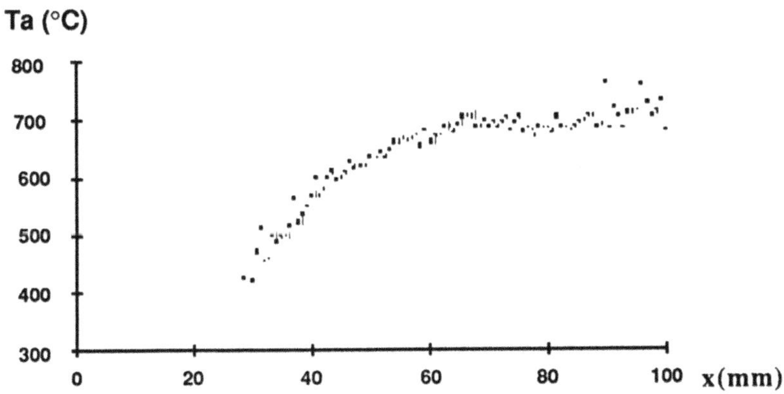

Figure11 - Longitudinal distribution of local recovery temperature in the reattachment zone at z=185 mm.

7.2 SPANWISE DISTRIBUTIONS

The lateral distributions of h and Tr, taken from the two images of Fig. 7, present modulation as expected. This is shown in Fig. 12 and 13 where the lateral distributions of these two parameters are presented for the location x=48 mm from the hinge line, where the convective heat flux is maximum.

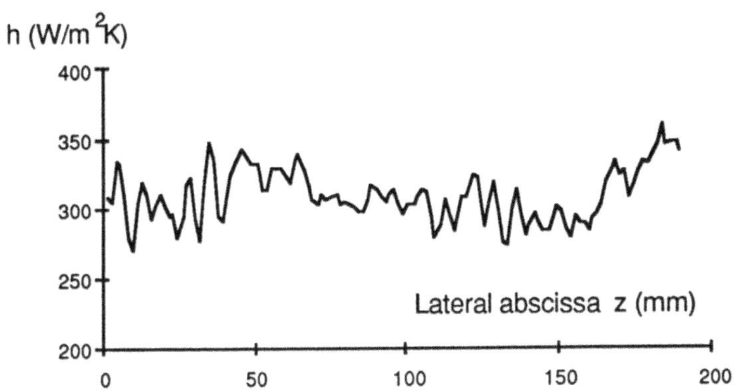

Figure 12 - Spanwise distribution of local heat transfer coefficient in the reattachment zone at x = 48 mm.

This modulation is not regular and it appears that several spatial frequencies are co-existing. This fact is quantitatively demonstrated by operating, line by line, a Fourier transform of the thermographic images. The so-obtained spectral distributions are presented in Fig. 14, for both the large and narrow fields of view. On Fig. 14 left, which corresponds to the large-field of view, the spectral analysis reveals various space periods from the full field to some mm. A maximum of amplitude is obtained in particular for a space period of 7.3 mm, which may correspond to the vortices qualitatively seen on the images of Fig. 7. This particular spatial period is identified with better accuracy and signal to noise ratio in Fig. 14 (right) issued from the small field image. The striation typical period is 6.7 mm, value close to that identified in the large field image.

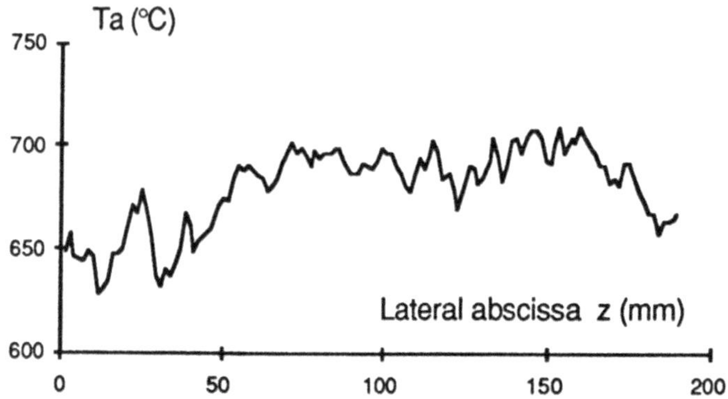

Figure 13 - Lateral distribution of local recovery temperature in the reattachment zone at x = 48 mm.

In fact, one can suppose that the interaction analyzed is disturbed by the irregular transition occurring on the model, which can be related to the variability of the surface state and the surface geometry of the sharp leading edge. An attempt of an experimental demonstration of such relation was recently made

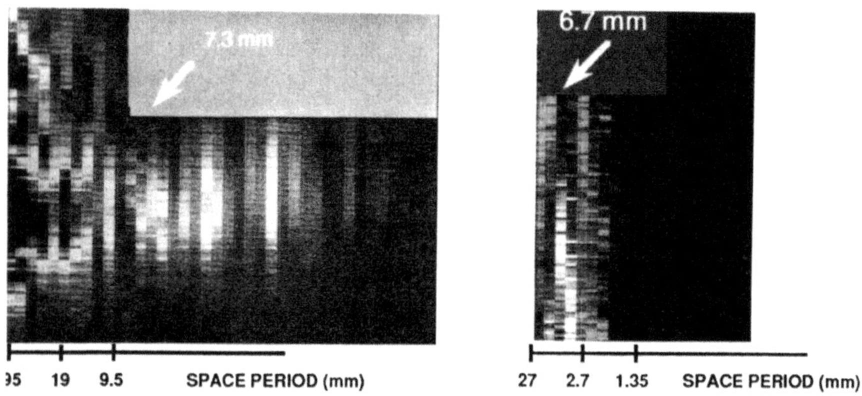

Figure 14 - Line-by-line Fourier transform of the thermographic images showing the spectral distribution of space-frequencies present in the lateral modulation of the flow on the flap. *Left* : large-field image ; *Right* : small-field image.

by Simeonides and al. [Simeonides et al., 1991]. Thus the observed field seems to be the superimposition of this phenomenon and of a more regular pattern centered on the 6.7 mm spatial period. This critical influence of the leading edge is also evidenced by the comparison of the spanwise distributions of heat transfer coefficient obtained with two models of identical geometry, in the same flow conditions, at R3Ch, but resulting from thermocouple temperature measurements in one case (curve 2) and from infrared thermography (curve 1) -see Fig. 15. The same modulation pattern are observed, but the large space periods are not correlated. Furthermore , in the case of the second series of tests , with the infrared camera ,

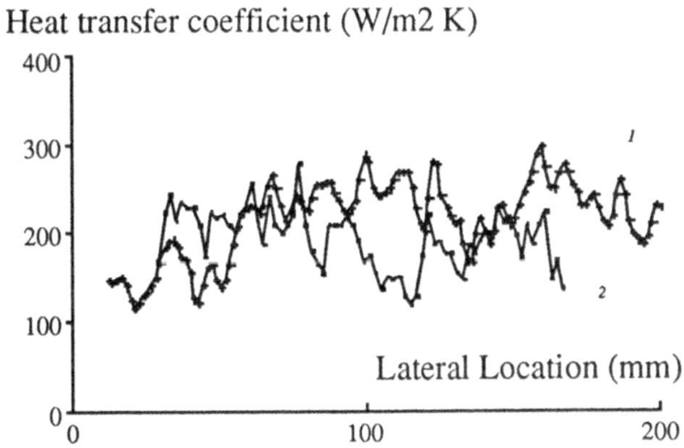

Fig. 15 - Spanwise distribution of heat transfer coefficient in the reattachment zone, at x = 35 mm. Comparison between thermocouple measurements [Chanetz et al., 1991] on a metallic model (*curve 2)* and present experiments on an insulating model (*curve 1).*

the large space period modulations varied erratically during the very first runs. The repolishing of the leading edge had also an influence.

Now, if we consider both lateral modulations (h and Tr), an interesting fact appears : these two modulations are in phase opposition, i.e. a maximum of Tr occurs when a minimum of h is obtained, and reciprocally. This is demonstrated in Fig. 16, where the following normalized parameters, plotted as functions of the lateral abscissa, are compared : (h-240)/120 and (740-Tr)/90. With this normalization, the two distributions are identical. This is particularly true when considering small spatial periods (smaller than 50 mm on this graph). For the large periods, which could correspond to the transition phenomena, a unique normalization does not lead to so good an agreement.

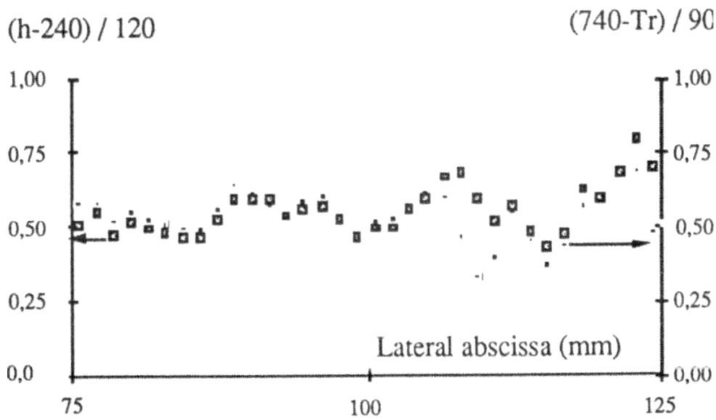

Figure 16 - Demonstration of the phase opposition of *h* and *Tr* lateral modulations, using normalizations. The *h* and *Tr* spanwise distributions are those presented in Fig. 12 and 13 related to x = 48 mm.

Let us come back to the local recovery temperature as identified by the previously described method. The existence and physical meaning of such a space-variable recovery temperature is questionable. Nevertheless, it is interesting to notice that the lateral modulation of the h coefficient is more regular when determined by the second method which makes no assumption on the recovery temperature value, than determined using the theoretical constant value (Fig. 17). Such a fact seems to indicate that this recovery temperature has a really physical significance. The theoretical constant recovery temperature and the variable one used for this identifications are presented on Fig. 18.

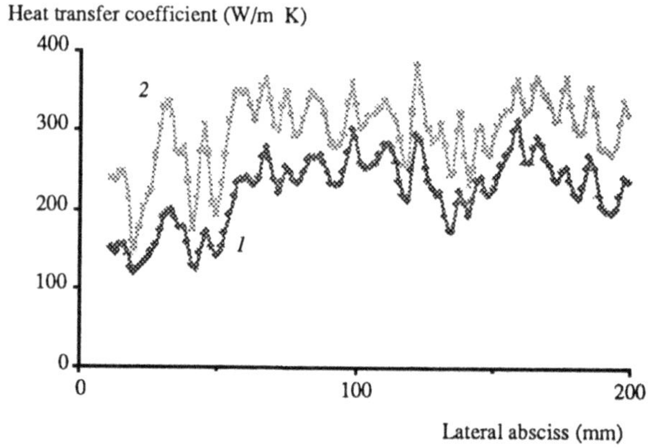

Fig. 17 - Spanwise distribution of heat transfer coefficient at x = 35 mm: comparison between the derivation assuming a uniform theoretical recovery temperature (*curve 1*), and the second procedure using the flux vs wall-temperature curves (*curve 2*).

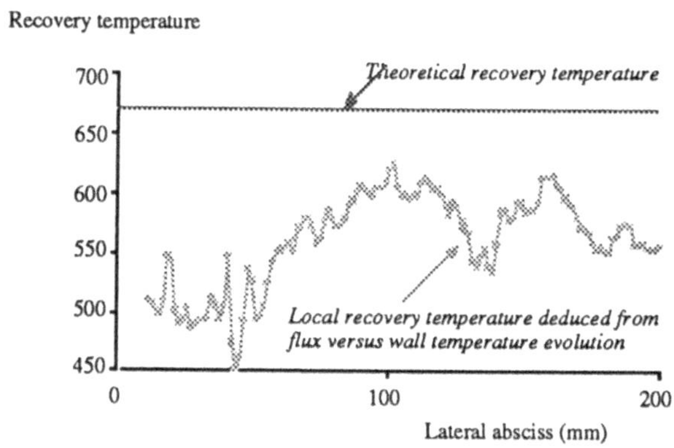

Fig. 18 - Lateral distribution of recovery temperature at x = 35 mm and comparison to the theoretical recovery temperature calculated for the flat plate.

7. CONCLUSION

The infrared thermography, now currently used as a heat transfer mapping method for super and hypersonic wind tunnel tests was used for analyzing the reattachment zone on a flap at Mach 10. The method, thanks to the tremendous amount of informations delivered, allowed to show several interesting phenomena until now not displayed.

The main results of this short series of tests are the followings :

i) in the present case, the spanwise modulation of the heat transfer on the ramp could be due to the conjunction of two phenomena : the laminar to turbulent transition and the occurrence of vortices in the reattachment zone, such three-dimensional structures being always observed in nominally two dimensional reattachments.

ii) the transition induces modulations characterized by erratic large space periods and large amplitudes. The space periods, which seems related to the surface and geometric quality of the leading edge, range from some tens of mm to the full width of the model.

iii) the modulations induced by the vortices are more regular, and characterized by smaller space periods, one of which seeming predominant and equal to 6.7 mm in our configuration. The amplitudes of these modulations are rather limited and near of ten per cent of the total heat transfer.

iv) the heat flux modulation is due to concomitant modulation of the heat transfer coefficient and of the local recovery temperature. On a large part of the ramp, in particular between the hinge line and the maximum heat transfer coefficient region, this last parameter is much smaller than the theoretical recovery temperature.

v) The two modulations, heat transfer coefficient and recovery temperature, are in phase opposition, i.e. a maximum of heat transfer coefficient occurs when a minimum of recovery temperature is obtained, and reciprocally. The physical explanation of this phenomena remains to be found, in relation with the vortices presence. If we can suppose that the maximum of heat transfer is located in downwash regions (see Fig 19), the variation of local recovery temperature is not so obvious.

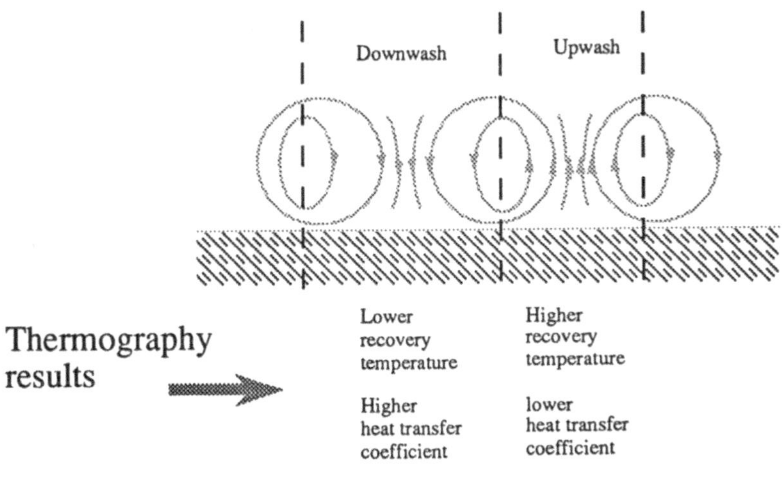

Fig. 19 - A possible configuration of the vortices and the wall heating spanwise modulation.

References

Balageas D.L.,Boscher D.M., and Déom A.A. (1990), 'Measurement of convective heat transfer coefficients on a wind-tunnel model by passive and stimulated infrared thermography', in Infrared Technology XVI, SPIE Proc. vol. 1341, 339-357.

Balageas D., Boscher D., Déom A., and Fournier J. (1991a), 'Measurement of convective heat-transfer coefficients in wind tunnels using passive and stimulated infrared thermography', La Recherche Aérospatiale, 1991-4, 51-72.

D. Balageas, D. Boscher, A. Déom, and G. Gardette (1991b), 'Heat transfer measurements in ONERA supersonic and hypersonic wind tunnels using passive and active infrared thermography', in Proceedings of the First European Symposium on Aerothermodynamics for Space vehicles, ESA SP-318, pp. 235-240.

Chanetz B., Coet M.C. and Delery J. (1991), 'Etude expérimentale des variations transversales de flux de chaleur dans un recollement sur un dièdre en écoulement hypersonique', Rapport Technique ONERA , no. 15/4362 AN.

Consigny H., Mentré V., and Bétrémieux A. (1987), 'Méthodes de mesure de flux de convection dans les souffleries R2 et R3 de Chalais-Meudon. Mise en œuvre actuelle et développements envisagés', Colloque d'Aérodynamique Appliquée AAAF, Poitiers, 26-28 oct. 1987, TP ONERA n°1987-185.

Délery J. (1989), 'Shock/shock and shock wave/boundary layer interactions in hypersonic flows', Special Course on Aerothermodynamics of Hypersonic Vehicles, AGARD Report n°761.

Délery J., Coet M.-C. (1990), 'Experimentation on shock wave/boundary layer interactions produced by two-dimensional ramps and three-dimensional obstacles', Workshop on Hypersonic Flows for Reentry Problems, Antibes, January 22-26.

Gartenberg E., and Roberts S. Jr. (1991), 'Twenty-five years of aerodynamic research with infrared imaging', in Thermosense XIII, SPIE Proc. vol. 1467, pp. 338-353.

Gauffre G. (1988), 'Détection de la transition laminaire turbulent par thermographie infrarouge', La Recherche Aérospatiale, 1988-2, 11-22.

Potet P., Garçon L., Balageas D., Boscher D., and Déom A. (1989), 'Appareillage industriel de contrôles non destructifs par thermographie infrarouge stimulée', Journée Etude Soc. Fr. Thermiciens, Paris 18 janv. 1989.

Simeonides G., Vermeulen J.P., Boerrigter H.L., Wendt J.F. (1991), 'Quantitative heat transfer measurements in hypersonic wind tunnels by means of infrared thermography', Von Karman Institute preprint 1991-29/AR.

INTRODUCTION TO SESSION 7: RECOMMENDATIONS FOR FUTURE RESEARCH IN HYPERSONIC INSTRUMENTATION

S. L. OCHELTREE
National Aeronautics and Space Administration
Langley Research Center
Hampton, Virginia 23665-5225 USA

ABSTRACT: An overview of the NATO Advanced Research Workshop Session 7 is presented. It describes the process followed to obtain a group consensus on the main technical recommendations for each of the five technical sessions of the Workshop and presents the general conclusions and recommendations for future research agreed upon by the workshop participants.

1. Introduction

Session 7, the last session of the NATO Advanced Research Workshop, was organized to provide an open forum for the workshop participants to reach a group consensus on the main conclusions of the workshop and to formulate a set of recommendations for future research in hypersonic instrumentation. The format of this last session was to have the five lead rapporteurs for each of the earlier five technical sessions present their draft summary reports and recommendations to the group. Each draft report was prepared by the three rapporteurs who led each of the earlier session group discussions and was based on both the technical papers presented in their sessions and the group discussions that followed. Following each draft report presentation to the workshop group in Session 7, there was a group discussion period where the final consensus position for the workshop participants was agreed upon. The session papers that follow are reports on the final conclusions and recommendations that resulted from the process just described.

The last part of Session 7 provided for a group discussion that led to the development of a set of general conclusions and recommendations for the Workshop as a whole. These latter overview comments were meant to capture the broader technical issues, concerns, and recommendations that were expressed by the group participants throughout the five technical sessions. There were eight general recommendations formulated and agreed upon by the workshop participants in this last session.

2. General Conclusions and Recommendations

The eight general conclusions and recommendations produced at the Workshop were:

A. Boutier (ed.), New Trends in Instrumentation for Hypersonic Research, 545–548.

2.1. Advanced measurement technology is critical and enabling to the development of hypersonic flight vehicles and requires high level management attention and support. The group identified four classes of enabling hypersonic research experimental data that require advanced measurement technology development and instrumentation. These were:

2.1.1. Aerospace vehicle design data obtained from wind tunnel and propulsion test facilities. An advanced measurement capability in these test facilities is needed to acquire the fundamental experimental test data base to support vehicle design in the flight regimes that cannot be obtained through existing engineering or Computational Fluid Dynamics (CFD) predictive capabilities.

2.1.2. New test facility free stream characterization and calibration. The complete free stream flow field characterization of both new and an existing facilities is the essential first step to defining the "input conditions" that will provide a high quality hypersonic experimental data base for all future engineering design and CFD code validation testing. Hypersonic test facilities place the most extreme performance demands on advances in new instrumentation that must survive and perform measurements in their highly hostile environments, e.g., high temperature; vibration; micro- to millisecond test times; chemically-reacting, high-speed flows; and simultaneous, global multi-parameter data sets for both model surface and disturbance-free gas flow properties.

2.1.3. Fluid physics and CFD code development and validation. All future advances in the improved understanding of hypersonic fluid physics and CFD code development and validation must rely on a high quality experimental data base obtained either from ground-based test facilities or future flight test data. The development of new measurement technology is enabling to acquire this essential experimental data base.

2.1.4. Flight test data. Even with the most advanced and sophisticated ground test facilities, there are still significant flight test envelope conditions that cannot be adequately simulated in ground test facilities. Flight testing under extreme flight envelope conditions places the most stringent demands on instrumentation. These flight test constraints limit the domain of advanced measurement technologies to those that can survive the hostile flight environment and yet, can provide accurate measurements with the additional constraints of limited size, weight, power consumption, etc.

2.2. The diversity of hypersonic test requirements will demand the development and application of a broad spectrum of advanced measurement techniques and sensors. A common theme agreed upon throughout the workshop deliberations was that multiple solutions must be addressed to satisfy many of the advanced hypersonic measurement requirements because of the wide variety of experimental test conditions and requirements. Further, the high risk nature of the hypersonic advanced measurement technology R&D program will require pursuit of multiple approaches to the most critical measurement requirements in order to reduce risk in the experimental test program.

2.3. It is essential that a close collaboration be established between the research data user, test facility engineer, and the instrument developer beginning with experiment definition and continuing through final demonstration. The workshop group cited the importance of a thorough scientific definition of the experimental data requirements, experimental test conditions and measurement accuracies as a necessary first step to successful hypersonics experiments. The instrument developer is a critical part of this initial experiment definition process. This is particularly true in hypersonic testing, since it is often found that the present state of the measurement technology will not support the test requirements. With this realization, and involvement of the instrument developer, an alternative experiment may be designed which is achievable or else an aggressive measurement technology development program must be defined and begun. The workshop participants felt that this early dialogue -- which involves the data user, test engineer and instrument developer -- and continues throughout the course of the experimental test program cannot be over emphasized.

2.4. Major extensions in the state of the art are needed in many areas (e.g., nonintrusive gas diagnostics, high temperature sensors) and will require a long-term (5-7 years) R&D investment to mature. As indicated earlier, major advances in measurement technology are needed to meet the needs of the hypersonic ground and flight test research program. Further, new fundamental scientific discoveries and technologies outside of the aerospace community must be aggressively monitored and considered for development and application to meet the new measurement technology requirements.

2.5. CFD code experimental validation will place the highest demands on facility data accuracy and on instrumentation with well-validated, low measurement uncertainties that must be supported. It was pointed out repeatedly during the workshop that the most demanding measurement requirements are dictated by CFD code validation tests which require well-calibrated test facility free stream input conditions, high-accuracy mean model surface conditions, and global multi-parameter gas flow field data with the highest spatial resolution. Further, to develop and test turbulence CFD models, these measurement requirements must be extended to include fluctuating quantities with wide measurement bandwidth--up to 100 kHz, turbulence power spectra and cross correlation functions of measured values. These sophisticated requirements combined with experimental uncertainties in the 1 to 10% range require significant advances in measurement technology to achieve.

2.6. A balanced near-term/long-term investment strategy is required that will address the following:

2.6.1. Extend mature technologies that have been successfully used in the past but which need to be reactivated or investigated for extension beyond the experimental test domain of their prior use. Examples cited in the workshop include electron beam fluorescence (former class) and laser Doppler velocimetry (latter class).

2.6.2. Exploit new emerging technologies in other disciplines. The workshop participants felt strongly that it was only through the adaption and extension of the latest emerging technologies -- e.g., lasers, fiber optics, electro-optic devices, laser spectroscopy, high-temperature sensor concepts and materials -- that the required quantum advances in hypersonic measurement technology can be achieved.

2.6.3. Develop advanced measurement and sensor concepts at university, industry, and federal laboratories. This element of the investment strategy, which was strongly endorsed by the workshop group, cites the essential managed coordination between the university, industry, and government laboratory research staffs to draw upon the strengths of each element in a synergistic way. Long-term support for university labs classed as "Centers of Excellence" and seed funding to small innovative private companies must be the cornerstones of this investment strategy.

2.7. New calibration techniques and facilities need to be identified and developed to satisfy measurement uncertainty requirements under both ground and flight test conditions. The issue of calibration techniques and facilities for new instrumentation technology was identified in the workshop as a significant issue that needs to be addressed along with new measurement techniques and sensors. The development of special calibration techniques and facilities for extreme test conditions -- e.g., high heat flux levels and high temperature -- can be as costly, time consuming, and technically challenging as the development of the new sensor or measurement technique itself and needs to be included as an integral part of the measurement technology R&D program.

2.8. Flight test measurement requirements should be identified and appropriate flight sensor development programs supported early in the program. Although the main thrust of this workshop addressed hypersonic ground facility instrumentation, the subject of extending new measurement techniques and sensors to flight testing was discussed. It was agreed that flight sensors are an evolutionary extension of those developed for ground testing which can be miniaturized and hardened for flight. As a consequence, potential flight sensor technologies should be identified and support begun early in the ground test development program so that there is a well-planned transition from ground test to flight.

3. Summary

In summary, Session 7 of the NATO Advanced Research Workshop provided a forum for the workshop participants to develop a final consensus on the main conclusions of the workshop and to formulate a set of recommendations for future research in hypersonic instrumentation. The general conclusions and recommendations of the group are presented above while the final reports of the lead rapporteurs for each of the technical sessions are presented in the following papers.

FLOW VISUALIZATION AND SPECTROSCOPY IN HYPERSONIC FLOWS: A REVIEW

NATO Advanced Research Workshop
New Trends in Instrumentation
for Hypersonic Research
ONERA La Fauga-Mauzac
France
April 27-May 1, 1992

JAMES TROLINGER
MetroLaser, 18006 Skypark Circle 108
Irvine, CA 92714

GEORG EITELBERG
DLR, 10 Bunsenstrasse
3400 - Göttingen, RFA

and

MARC RAPUC
AMD-BA, 78 Quai Marcel Dassault
92 214 - St. Cloud
France

ABSTRACT. This paper is based upon a session of the NATO Advanced Research Workshop, *New Trends in Instrumentation for Hypersonic Research* held at the ONERA La Fauga Facility in France during the week of April 27, 1992. The discussion includes some of the frontiers of the technology of flow visualization and spectroscopy as well as a discussion of the current development needs and trends. Included in the discussion are optical integrated measurements such as resonance absorption, schlieren, interferometry, and holographic methods. The discussion shows that while the technology is mature in a broad sense, a significant number of new development areas exist such as resonant holography and phase shifting holographic interferometry. The maturity of the technology makes it immediately applicable to many problems and the untapped potential offers considerable room for improvement of existing capability. The methods which are described can be used in harsh environments and have the potential for becoming flight test diagnostics for the measurement of temperature, density, constituency, and velocity.

1. Introduction

When an optical wave front traverses a gaseous medium, the wave front can be modulated in phase, direction, and intensity by the medium. The analysis of the integrated effect of the medium upon the wave can characterize the medium if enough information is recorded. At the simplest level, some aspects of the otherwise invisible flow field can be made visible through various optical processing procedures. At the highest level, some aspects of the flow can be accurately quantified. The fundamental principals underlying optical flow visualization techniques described herein are the following:

1. The refractive index changes in fluid mediums of varying density and/or constituency which means the wave fronts of optical waves propagating through these mediums are phase shifted

A. Boutier (ed.), New Trends in Instrumentation for Hypersonic Research, 549–561.

and deflected. Gas density can be derived from refractive index if the constituency is known. Refraction can be measured directly with schlieren systems. Phase shift is usually measured with interferometry.

2. When the exciting wavelength is near the resonance of a gas, the complex refractive index changes radically such that absorption and refraction properties are strongly affected. The analysis of absorption and refraction at resonance can be done by either tuning the wavelength of a narrow band illuminating diagnostic beam through the resonances of the medium or by comparing the input and output spectra of a broad band illuminating beam.

Flow visualization as well as spectroscopic methods are information intensive diagnostics and have been limited in practical application for many years by the capability of affordable image processing hardware and software. In recent years, the availability of small powerful computers is fast removing such limitations. Also, new lasers are simplifying the application and interpretation of such methods. Miniaturization hardware such as fiber optics, diode lasers, and small electronics hardware are opening the way to the development of small, rugged, lightweight instrumentation. New methods such as phase shifting interferometry are improving the resolution and accuracy of interferometry. Holography is expanding the capability of both interferometry and spectroscopy by providing a convenient method of storing wave front information in a form which makes its processing simpler. So even though each of the methods under discussion here is already widely used, the field is extremely dynamic and the capabilities of these methods will be extended in future years as the supporting technologies are advanced.

These methods can provide quantitative density measurements of two dimensional and axisymmetrical fields since the integrated measurement can be solved to provide point information. When flow fields become more spatially complex, the method becomes semi quantitative. In some situations, the measurement of integrated data over a sufficient number of angles of view can lead to three dimensional analysis through the process of tomography (tomography is the determination of a spatial distribution when only the projections through the distribution are given).

In the study of aero-optical phenomena, the desired measurement is the total integrated affect on the wave front, so these methods are finding broad application in this area.

2. Advantages

This class of instrumentation possesses a number of advantages over most of the more recently emerging optical diagnostics such as laser induced fluorescence and Raman techniques. The instruments are essentially in a high state of maturity, some of them having been in routine use for many years. This makes their incorporation into hypersonic testing more straightforward since many test facilities are already equipped with the required optical access. The results are often easier to communicate to the end user since aerodynamacists are already familiar with their use and interpretation. The methods currently in use are relatively simple, employing well known recording materials and data reduction procedures. They are readily adaptable to electronic recording and data reduction methods which have evolved in recent years. The expansion potential is extremely promising, which means that the methods already in use can be retrofitted with advancements that will enhance the system capability. For example, facilities which are already equipped with schlieren systems can be upgraded to incorporate new holography methods. Holography methods can add interferometry capability to such a facility. Facilities already using holography will be expandable to incorporate phase shifting interferometry, on-line recording and reduction, and resonant holographic methods.

3. Emerging Technology

A considerable effort is being made to extend the methods under discussion for use in hypersonics research. Hypersonic testing requires diagnostics at lower densities and higher velocities than the methods easily reach. Also, high temperatures result in dissociation and ionization with a resulting loss of knowledge of the gas state and constituency. The state of the gas can become an important factor in the simulation, so its measurement can become a major issue for modelers and for those desiring to simulate flight conditions.

In some instances, pulsed facilities are the only facilities that can simulate flight conditions. Accommodating short run times is another challenge being accepted by diagnostics technique developers. The environment in which such tests must be carried out offers still another challenge. For example, simply getting light in and out of the test facility without introducing spurious information on the light wave is not a simple task.

The following examples of emerging technology were presented in the workshop, illustrating some of the approaches being taken to deal with the problems of hypersonic testing.

3.1 Schlieren Methods

A good example of the refinement of a mature technology for application in hypersonic research was presented by Bize et al[1]. They have designed, constructed and tested a high sensitivity schlieren system for the ONERA La Fauga F4 Hotshot facility. This facility has been designed to cover the HERMES trajectory through gas densities ranging from 10^{-6} to 10^{-2} kg/m^3. The minimum detectable optical path difference is given by

$$\Delta s_{\min} = KL\Delta\rho_{\min}$$

Where K is the Dale/Gladstone constant, which for air is 2.27 x 10^{-4} (Kg/m^3)$^{-1}$, and L is the length of the gas flow. This requires the ability to measure an optical path length change from 2.3×10^{-2} to 2.3 micrometers per meter of path length or in terms of wavelength, a range from about $\lambda/20$ to 5λ per meter of optical path.

To extend the sensitivity to its maximum, the authors have employed long focal length (5 meter) mirrors (50 cm. diameter) polished to an accuracy of 0.1 arc seconds. Three major factors limit the achievable (in practice) sensitivity of schlieren systems: 1) vibration, 2) air current perturbations outside of the facility, and 3) distortions and noise caused by the windows of the facility. These have been minimized in the current design by 1) mounting the knife edge on a rapidly controllable mount with feedback to eliminate the effects of vibration on the knife edge, and vibrationally isolating the entire system, 2) placing the entire optical system in a vacuum chamber, and 3) removing the windows entirely, since the optics are all inside the vacuum of the test facility. Also, by using the phase contrast method instead of a normal knife edge, another factor of 10 is gained in sensitivity (the phase contrast method employs a phase shifting step instead of a knife edge). With these refinements the system can resolve a refraction as small as 0.01 arc seconds and can measure a step change in optical path of less than $\lambda/500$. This schlieren system is quite possibly the most sensitive schlieren system of its size in use for aerodynamics.

As an additional refinement, the authors plan to employ laser illumination to allow the high level of ambient light near the stagnation region of the models to be removed from the visualization.

3.2 Spectroscopic Methods

Bize et al[1] also describe a method that was employed for extending the use of emission spectroscopy to the La Fauga F4 facility by employing fiber optics to collect the light inside the facility and to conduct it to a high resolution grating spectrometer some distance away. Measurements were made of Cu, Fe, OH, O, and NH UV bands from 300 nm to 330 nm to evaluate various parameters in the flow. Temperature measurement is one of the primary objectives. This diagnostic also provides an evaluation of measures taken to reduce the metallic pollution of the flow.

Limbaugh[2] and McGregor[3] similarly described methods for adapting highly refined emission/absorption spectroscopic procedures to extremely harsh test conditions, specifically in the testing of propulsion systems. By employing a heated plate as an IR (1-3 Micrometers) source and measuring the transmission through the flow field, Limbaugh can deduce the temperature and species concentration in the exhaust of a propulsion system. A band model approach to solving the radiative transfer problem provides quantitative determination of these properties when the integral representing the measured illumination can be solved. Solution of the integral is straightforward only for symmetrical distributions of the field. By staying in the short wavelength band, Limbaugh shows how quartz fibers can be used to conduct the light in and out of a test facility to ruggedize the instrument. Primary candidates for measurement are the 4.3 μm band of CO_2 and the 2.7 μm band of H_2O. Work closer to the 1 μm bands would allow use of uncooled silicon or germanium detectors, but the method has not yet been extended to this regime to date. NO, OH, CO, and NO_x are also candidate molecules.

McGregor described the requirements and potential for applying resonance absorption methods in hypersonic flow using non laser illuminators. Typically, a source is produced with a discharge tube containing the species under study and the receiver is a high resolution spectrometer (**Figure 1**). The instrument can provide a measure of gas density and temperature if the integral representing the measurement can be solved and if broadening effects caused by Doppler shift or other effects can be incorporated. Considerable work is still needed in the physics of the process to make the method accurate. Ultimately, if rugged, low cost lasers in the correct wavelength become available, the method could be simplified. **Figure 2** shows a candidate design for an in-flight diagnostic using the technique. Fiber optics conducts the light in and out of the rectangular combustor.

Smeets[4] reviewed a variety of interferometers specifically developed to view extremely weak phase objects (optical path length changes less than $\lambda/500$). Of special significance is the fact that these instruments were developed and used routinely over 20 years ago, and now they should be easily adaptable to meet some of the newly emerging requirements in present day hypersonic flow diagnostics. A Wollaston prism differential interferometer shown in **Figure 3** provides a capability to quantify optical path length changes as small as $10^{-5}\ \lambda$. Using the above formula, this leads to a minimum detectable density change of about 2×10^{-8} Kg/m^3 per meter of optical path which covers essentially the full range of ground testing for most hypervelocity aircraft. The method biases the interferometer into its most sensitive region and reduces optical noise by canceling out light which is not contributing to the measurement of changing refractive index. When the image can be divided into small regions probed by individual rays which are then sensed electronically, the frequency response of the interferometer is virtually unlimited, making it an excellent candidate for modern day hypervelocity testing in pulsed facilities. The device has actually been used to map the density profiles surrounding a ballistic projectile as it passed through the interferometer.

Variations of the Wollaston prism interferometer have been employed for different problems. One variation represents one of the precursors to modern optical image processing for enhancing the sensitivity of flow visualization. One interferogram is produced when the object is present using one polarization orientation, and a second exposure is made without the phase object being present with the other polarization. This procedure makes the second recording complementary to the first in intensity with respect to the fixed illumination. Adding the two recordings effectively smoothes out the constant information, enhancing the signal-to-noise ratio of the temporally changing information. A recording of a sound wave of phase strength less than $\lambda/500$, using the interferometer, vividly demonstrated the high sensitivity made possible through image processing and setting the interferometer in its most sensitive mode. This type of interferometer can be adapted directly into almost any conventional schlieren system. The chief drawback is the difficulty in quantifying the data, limiting present use of the method to flow visualization. Smeets states that the method can be characterized by a typical sensitivity of 10^{-7} Amagat/meter (An Amagat is the density of the gas normalized to its value at STP.)

Instrument sensitivity enhancement in flow visualization must always be accompanied by procedures to remove or account for the variations in the flow field outside of the region of interest, which can dominate the measurement. This is especially true in hypersonic testing where the test region can be relatively small. Removing the test chamber windows and the ambient air itself, as described in the schlieren system above, is one way of achieving this. In Smeets interferometer, this is accomplished essentially by subtracting off such effects in real time. Techniques described below offer still other ways to abate ambient effects.

3.3 Holographic Interferometry

Four presentations were given describing advanced holography developments and applications[5,6,7,8] representing what is probably the most dynamic area in flow diagnostics.

Since holography provides a technique to store optical wave fronts for later retrieval, holographic flow visualization and interferometry would appear to be an obvious application of holography. The modulated wave front can be stored in a hologram for later retrieval, at which time it can be reconstructed for processing with any of the conventional optical flow visualization procedures including schlieren, shadowgraph, moiré, and many types of interferometry; so the single recording can lead to a multitude of representations of the flow .

The primary features of holographic flow visualization are the following:

1. Information-carrying wave fronts of arbitrary complexity can be stored in holograms for later retrieval.
2. The holograms can be recorded in nanosecond or shorter time frames.
3. Data is stored in a narrow wavelength range.
4. Holograms can be recorded at high frequency (megahertz).
5. Many holograms can be superimposed on a single recording.
6. Data is stored in a format that is readily accessible to optical processing procedures.

Storing entire wave fronts allows a number of extremely important and unique direct and quantitative comparisons, additions, subtractions, correlation and filtering of wave fronts such as the following:

1. Wave fronts that existed at different times.
2. Extremely aberrated and complicated wave fronts that differ by a small amount.
3. Wave fronts of different wavelengths.
4. Multiple wave fronts existing simultaneously.

Table 1 summarizes some of the advanced holographic techniques that support flow diagnostics.

Improved data reduction hardware and software may be the single most important new development in flow visualization holography. In recent years, great strides have been made in improving the efficiency and use of holography data. The application of modern image analysis and enhancement procedures such as logarithmic detection, normalization, subtracting off the non uniform background, floating thresholds, bandpass filtering and others have greatly simplified automated data reduction. Only through the routine processing and use of the additional data made possible by holography can the methods compete with more established and accepted techniques which may offer less, but which require little data processing before use.

Phase shift interferometry is a relatively new interferometric procedure for producing a continuous phase map from a series of three or more interferograms of an object field. At any point in an interferogram, the intensity is given by

$$I = I_0\left(1+\gamma\cos[\phi+\theta]\right)$$

Where I_0 is the background (D.C.) intensity, γ is the fringe contrast, ϕ is the desired phase and θ is an adjustable constant phase shift over the interferogram. By measuring intensity for three different values of θ , three equations are produced which can then be solved for ϕ. The interferogram can be sampled by a detector array during the phase shifting operation and a phase map is produced almost instantaneously (with a 2π ambiguity). The features and benefits of using this procedure include using more of the data in the interferogram, filling in data between the fringes in the interferogram, removal of optical noise, improvement of sensitivity, and rendering the process more easily automated. This represents a significant improvement over more conventional methods of locating fringe centers.

Spring et al[6] describe a system and an application in the NSWC hypervelocity facilities which represent various edges of the state of the art in applied holographic flow diagnostics. The system employs double plate (sandwich) holography in which a reference wave front is recorded on one hologram, the test field wave front is recorded on a second hologram, then the two wave fronts are superimposed by placing the two holograms together and reconstructing both wave fronts simultaneously. The phase shift operation for phase shift interferometry can be implemented with the hologram "sandwich" in various ways, and was achieved in this system by moving one of the plates relative to the other during reconstruction. Automated image interferogram reduction using this process is one of the significant features of the system allowing large amounts of data to be quantified. Image processing systems and software which now are relatively inexpensive would have been prohibitive in cost for most users only a few years ago. The study illustrates the use of holographic interferometry to evaluate CFD codes. Comparisons of the interferometry results with CFD clearly illustrate where the codes fail to predict the flow about models in hypersonic flows. **Figure 4** shows such a comparison where the CFD solution predicted by a three dimensional Euler code breaks down in the turbulent boundary layer. Some of the weaknesses of the technique are illustrated in this figure which underlined the problem of refraction near strong density gradients, which makes the data analysis difficult near the shock wave.

Eitelberg et al[5] reported on the development of an advanced on-line holocamera for the HEG shock tunnel at DLR. The system employs a thermo-plastic recording device, which allows holograms to be recorded, developed electronically in place, and reconstructed within seconds after recording. The system also employs multiplex holography using two separate reference waves. When the two wave fronts are recorded with two different reference waves, they can be independently reconstructed. Consequently, a constant phase shift can be added independently to either of the two waves during reconstruction, providing a capability to use phase shift interferometry.

Surget and Dunet described holographic interferometry systems designed for use in two ONERA facilities, F4 at La Fauga and F5 at Chalais-Meudon. The density in these facilities ranges from about 4×10^{-3} to 6×10^{-5} kg/m^3. The change in optical path length for a typical model is much less than one wavelength. The authors' approach to sensitivity increase is to employ a multiple pass interferometer. Although this has been done before, previous methods for recording multipassed beams have led to almost intolerable noise caused by the difficulty in recording only the desired beams which make n passes through the facility. When a light beam passes back and forth through a test facility, some method must be devised to select only that beam which has passed through the facility the desired number of transits. This can be achieved holographically by path matching a relatively short coherence length reference wave with the object wave that has made the desired number of passes. The remaining light exposes the hologram but does not interfere with the reference wave. This method, called coherence filtering, works reasonably well for two passes, but results in lower quality data as the number of passes increases. A time filtering method was developed by these authors that actually gates the recording, electro-optically shuttering out all light except that which traverses the facility by the desired number of times. **Figure 5** illustrates the system capability of recording the flow field over a wedge in the F5 facility. The interference fringes are almost straight in the single pass case, whereas the three pass case shows the improved sensitivity. This configuration leads to quantifiable interferometry data with a sensitivity gain of three. The time filtering method could increase sensitivity by orders of magnitude. Once again, methods for eliminating ambient effects would almost certainly be required to maintain the useful sensitivity.

Trolinger et al[8] describe a series of experiments that demonstrate the use of resonant holographic interferometry spectroscopy (RHIS) in detecting OH in a flame. RHIS is a new type of flow diagnostics technique that combines the power of holography with the species selectivity of spectroscopy to provide three dimensional images of the density of selected species in complex flows. The technique is particularly suitable to study high speed mixing processes, to measure minor species in the presence of major species, and to observe major species in a heterogeneous low pressure environment. Both experiments and modeling are being conducted to establish the feasibility of RHIS in measuring the concentration of water vapor, oxygen, and OH in high speed propulsion applications. In addition to the species concentration, the resonant holographic technique has the potential of providing temperature, pressure, and flow velocity.

The potential benefits of producing holographic recordings at a wavelength near a resonance of one of the constituent gases or a seed gas, have been recognized for many years, but hardware and technology have only recently proven adequate for practical application. At resonance, the refractive index varies drastically resulting in a potential improvement in sensitivity with the attendant capability to apply interferometry at much lower pressure. Also, the measurement can be made selective to spectroscopically diagnose a specific constituent. The resulting data can lead to quantitative measurement of concentration, temperature, pressure, and velocity, producing an extremely powerful diagnostics capability which extends far beyond conventional holographic interferometry.

Species Concentration can be determined by producing a hologram of the field of interest with a beam of wavelength close to a spectral line, superimposed upon a hologram at a slightly different wavelength (see **Figure 6**), either far from the line or symmetrically separated from line center, and simultaneously with the first hologram. When these two wave fronts are interfered by reconstruction with a single beam of light, the resulting interferogram characterizes only the resonant species, since the effects of all other species are contained on both wave fronts. Sensitivities as high as ten parts per million (for NO, OH, and iodine) are detectable in principle. The authors also cited other studies[9] that conceptually compared RHIS with Raman and laser induced fluorescence, for measuring **concentration, temperature, pressure,** and **velocity**, which reached the conclusion that RHIS actually excels in several ways. Perhaps the greatest advantage lies in the fact that RHIS does not depend upon a knowledge of quenching and other radiative decay paths, which is probably the greatest of the practical problems limiting the accuracy of LIF.

4. Conclusions

Flow diagnostics and spectroscopy fill a vital need in hypersonics research. Mature technology already allows the global visualization of a flow field, clearly depicting properties of boundary layers, transition, and shock wave interactions. The spectroscopic methods make the measurement of constituents such as NO, OH, H_2, O, and O_2 possible, while also providing a mechanism for measuring temperature. The more advanced spectroscopic measurements can also measure velocity. Flow visualization and spectroscopy methods provide candidates for assisting in vehicle design, facility calibration, modeling, and flight testing.

Recommendations for required future efforts emerging from this session of the workshop include the following:

1. Characterize the methods from hypersonics prospective.
2. Additional comparisons with computation and other methods needed to check accuracy.
3. Insure that modelers, facility developers, and testers know capabilities and potential of the methods.
4. Refine methods for hypersonics application. Specific areas requiring attention are sensitivity, quantification, and adaptation to facilities, models, and aircraft.
5. Prepare some of the methods for flight applications.

TABLE 1

Advanced Holographic Techniques Supporting Flow Diagnostics

TECHNIQUE	FEATURES
1. Phase shift and heterodyne interferometry:	Improves sensitivity of interferometry and simplifies data reduction.
2. Multiplex holography:	Allows simultaneous storage and independent reconstruction of different types of wave front information by using different reference waves. Provides a method to implement phase shift and heterodyne interferometry, optical subtraction and correlation of two image conditions.
3. Sandwich holography:	Allows different wave fronts to be stored on separate recordings, reconstructed separately, and directly compared optically during reconstruction. Further simplifies phase shift and heterodyne interferometry, optical subtraction and correlation of two image conditions.
4. Multiple wavelength recording:	Allows additional information about the subject to be stored (i.e. different variables that affect the refractive index).
5. Particle image displacement velocimetry:	Provides three dimensional velocity fields and enhances flow diagnostics.
6. On-line systems - Thermoplastic and TV holography.	Provides quicker access to data.
7. Fiber optic holography:	Improves optical access, simplifies path matching, and ruggedizes the optical system.
8. Resonant Interferometry:	Allows species identification and increases sensitivity.
9. Holographic optical elements:	Reduces weight and volume and simplifies optics.
10. Automated Data Reduction:	Makes the handling of large quantities of data that are available in the hologram tractable.

5. References

1. "Schlieren Device and Spectroscopic Measurements in F4" D. Bize, G. Dunet, M. Philbert, A. Roblin, J. Surget, Proceedings of the NATO Advanced Research Workshop, New Trends in Instrumentation for Hypersonic Research, Held at ONERA La Fauga, 27 April-1 May, (1992).

2. "Considerations for the Application of the Infrared Emission Absorption Technique to Hypersonic Propulsion Diagnostics", C.C. Limbaugh, Proceedings of the NATO Advanced Research Workshop, New Trends in Instrumentation for Hypersonic Research, Held at ONERA La Fauga, 27 April- 1 May, (1992).

3. "Molecular Electronic Resonance State Techniques" W.K. McGregor, Proceedings of the NATO Advanced Research Workshop, New Trends in Instrumentation for Hypersonic Research, Held at ONERA La Fauga, 27 April- 1 May, (1992).

4. "Interferometry in Rarefied Gas Flows", G. Smeets, Proceedings of the NATO Advanced Research Workshop, New Trends in Instrumentation for Hypersonic Research, Held at ONERA La Fauga, 27 April- 1 May, (1992).

5. "Holographic Interferometry on the High Enthalpy Shock Tunnel in Gottingen (HEG), G. Eitelberg, B. Fleck, T. J. McIntyre, Proceedings of the NATO Advanced Research Workshop, New Trends in Instrumentation for Hypersonic Research, Held at ONERA La Fauga, 27 April- 1 May, (1992).

6. "The Use of Holographic Interferometry for Flow Diagnostics", W. C. Spring III, W. J. Yanta, K. Gross, and C. A. Lopez, Proceedings of the NATO Advanced Research Workshop, New Trends in Instrumentation for Hypersonic Research, Held at ONERA La Fauga, 27 April- 1 May, (1992).

7. "Multipass Holographic Interferometer for the High Enthalpy Hypersonic Wind Tunnel F4", J. Surget and G. Dunnet, Proceedings of the NATO Advanced Research Workshop, New Trends in Instrumentation for Hypersonic Research, Held at ONERA La Fauga, 27 April- 1 May, (1992).

8. "Resonant Holographic Interferometry; A Multipoint, Multiparameter Diagnostics Tool for Hypersonic Flow", J.D. Trolinger, R. K. Hanson, B. Yip, and B. Battles, Proceedings of the NATO Advanced Research Workshop, New Trends in Instrumentation for Hypersonic Research, Held at ONERA La Fauga, 27 April- 1 May, (1992).

9. "Flow Diagnostics By Resonant Holographic Interferometry," Sirota, J.M., and Christiansen, W.H., AIAA Paper 90-1550, AIAA 21st Fluid Dynamics, Plasma Dynamics and Lasers Conference, Seattle, WA, June, 1990.

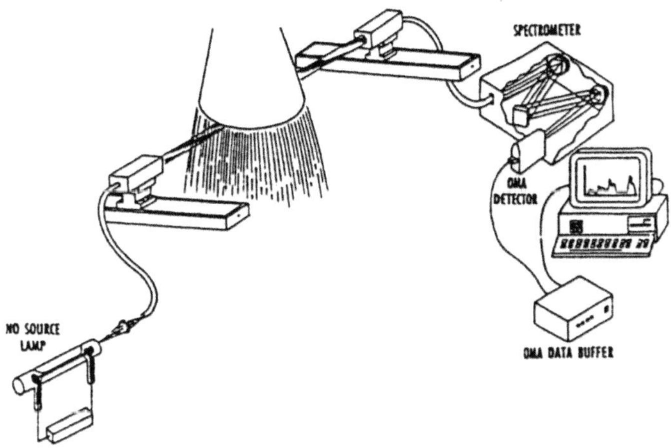

Figure 1. Illustration of the NO Resonance Absorption Measurement System.

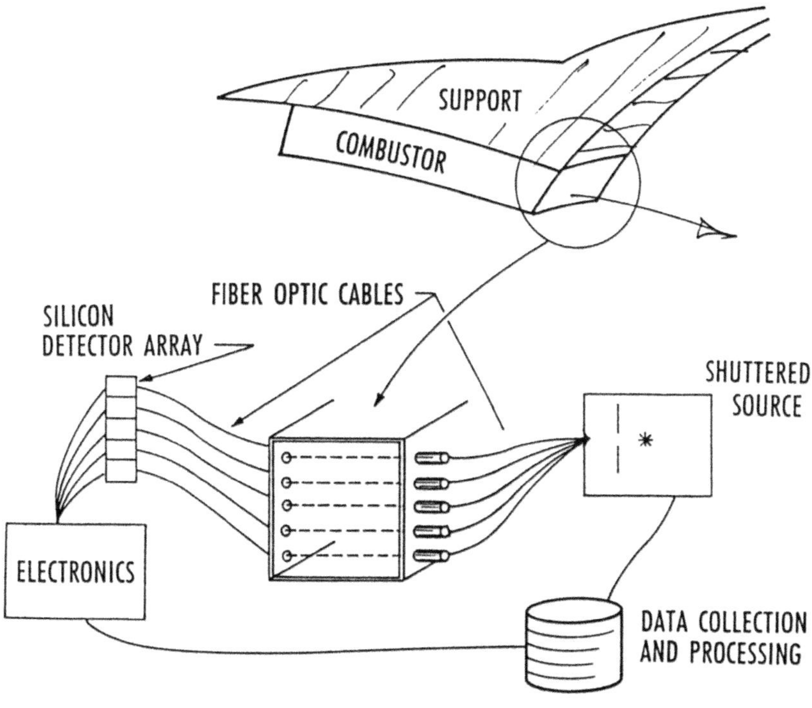

Figure 2. E/A Concept for Combustor Diagnostics

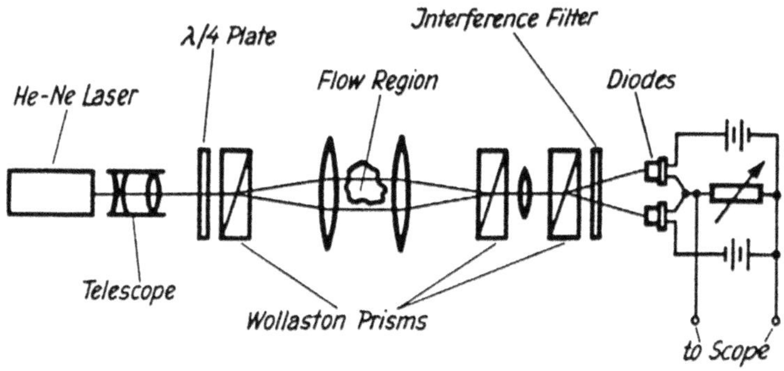

LDI using two separate diodes for complementary interference detection.

Figure 3. Laser Differential Interferometer

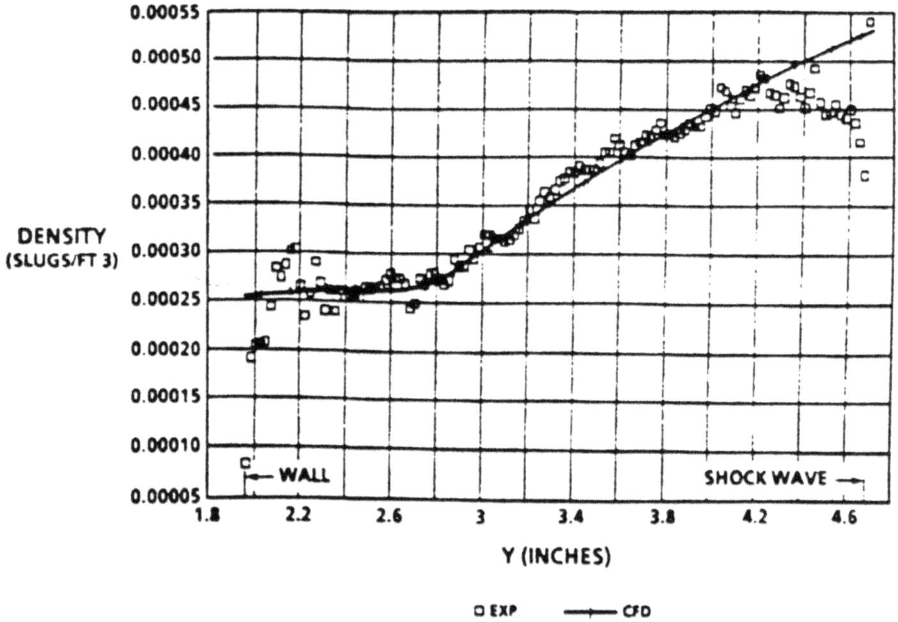

Figure 4. Measured Desnity Vs. CFD Solution for Blunt Cone.

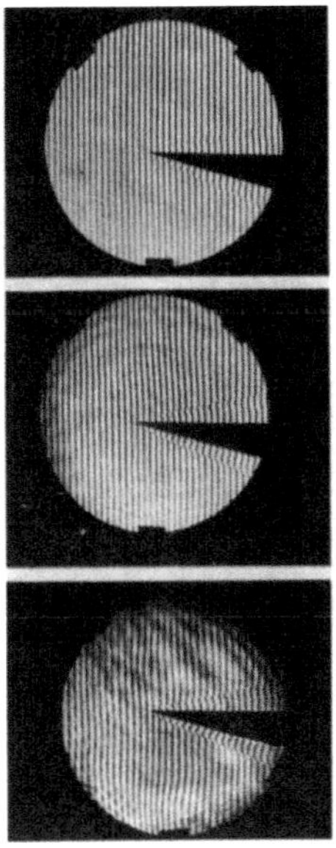

a) N = 1

b) N = 2

c) N = 3

Figure 5. Finite fringe interferograms

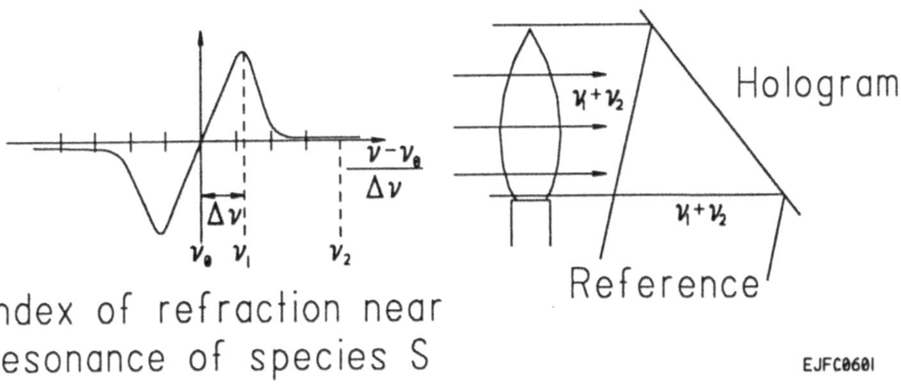

Figure 6. Holographic recording at frequencies ν_1 (near resonance) and ν_2 (off resonance)

LOCAL MEASUREMENT OF TEMPERATURES AND CONCENTRATIONS: A REVIEW FOR HYPERSONIC FLOWS

C. DANKERT, *DLR Göttingen / Germany*
R. CATTOLICA, *University of California at San Diego / USA*
W. SELLERS, *NASA Langley Research Center / USA*

ABSTRACT: The quality of reentry simulation for Shuttle, HERMES, Sänger, and NASP systematically suffers from the strong non-equilibrium of rotational and vibrational temperature due to the rapid acceleration of the test gas in the nozzle. Therefore the determination of temperatures is necessary and, if possible, preferable by a non-intrusive technique. The specific interests of this review are optical techniques such as electron beam fluorescence, laser-induced fluorescence, and coherent anti-Stokes Raman scattering. The capabilities available for local measurements with temporal resolution and quantitative accuracy are discussed for velocity, temperature, density, species concentrations, and fluctuations due to turbulence. The applicability of these methods of measurement is presented and discussed for the coming topic in aerothermodynamics: experimental techniques of hot gases in high enthalpy flows.

1. Introduction and Scope

The interest in hypersonic aerodynamic research is now increasing due to the new hypersonic space transport vehicles like the European HERMES, the German Sänger, and the U.S. NASP. The two classical major problems of such reentry vehicles are steering at high angle of attack with flaps or thrusters and the extreme heat loads in the nose area of the vehicle structure during aerodynamic breaking. Therefore the knowledge of gas- and wall-temperatures, wall-material characteristics, and species composition is essential. The latter point is also of interest for environmental studies of the upper atmosphere.

The design of hypersonic space vehicles depends on computational fluid dynamics codes (CFD) which simulate the hypersonic flow field. Another, somewhat historic, engineering tool is the wind tunnel with a flowing gas medium and a test model at rest.

Hypersonic research requires an integration between flight, CFD and wind tunnel methods:

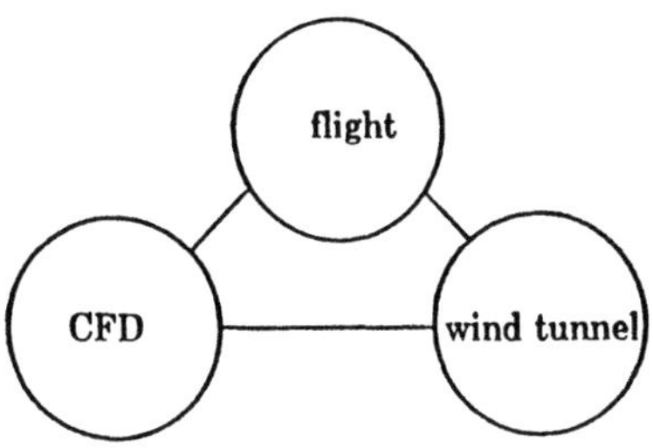

A. Boutier (ed.), *New Trends in Instrumentation for Hypersonic Research*, 563–581.

This paper focusses on wind tunnel aspects of this interaction, taking into account the different conditions in flight, and keeping in mind the validation of CFD codes.

Several optical methods are reviewed which are non-intrusive [1, 2] and have already been applied to hypersonic wind tunnel testing. These techniques are surveyed especially for possible use in two of the new European wind tunnels: the F4 at Le Fauga and the HEG at Göttingen. Both wind tunnels operate in intermittent modes at testing times of 100 ms in F4 and only 1 ms in HEG, which is very short compared to the low density hypersonic wind tunnels [5, 6] which run for some hours. The densities in the low density tunnels are $10^{12} - 10^{16}$ *molecules / cm*^{3} ($30\,K < T_{rot} < 2500\,K$, $800\,K < T_{vib} < 2500\,K$), and for the intermittent tunnels up to 10^{20} *molecules/cm*^{3} (T_{vib}, $T_{rot} < 10\,000\ K$).

The new tunnels use air as the test gas, and due to the high temperatures in the flow, the O_2 – and N_2 – molecules will dissociate and recombinate to NO. So we have to work with at least three molecular species (N_2, O_2, and NO) in a temperature range from 100 K to 10 000 K with non-equilibria in rotation, vibration and thermal dissociation. Furthermore, the state population distribution might not be of a Boltzmann type [3, 4] so the definition of temperatures becomes questionable.

2. Electron-Beam Fluorescence Diagnostics

2.1 DESCRIPTION

The electron-beam technique is well established for hypersonic wind tunnel testing since more than 25 years [7, 8]. Also, for flight-research diagnostic for hypersonic boundary layer measurements this method can be used to probe nitrogen [2, 9 – 16]; eight previous flight experiments with optical detection of electron-beam fluorescence are summarized by Cattolica [2].

Limitations of the electron-beam technique at high density concern quenching of the signal, reduction of spatial resolution by gas scattering, and possible interference from background emission. Muntz et al. [24] proposed the Electron-Photon-Fluorescence, a combined electron beam ionization/laser-induced fluorescence detection approach. In addition Muntz [46] has also proposed a pulse electron-beam technique to overcome some of these limitations.

Many types of electron guns were used for wind tunnel applications, with energies ranging from 5 kV up to 300 kV and beam currents ranging from 1 μA up to 1 mA [17]. Most of the scientists probed the N_2^+ -ion and measured rotational and vibrational temperatures [3, 5, 6, 7], NO [18, 19, 20], O_2 and O [21], CO [22], and CO_2 [23] were also investigated.

2.2 APPLICATION REGIMES

The density range of the e^--beam fluorescence is, according to Lewis [17], 10^{13} to 10^{16} *molecules/cm*^{3}. At higher densities quenching and beam scattering problems occur and make this technique difficult but not impossible; the pulsed e^--beam technique (10 ns) may extend the density range up to 10^{18} *molecules/cm*^{3} [46]. In **Fig. 1** the application densities are listed and compared to some other optical methods [17] like LIF (laser-induced fluorescence), LIPF (laser-induced predissociative fluorescence), Rayleigh scattering, CARS (coherent anti-Stokes Raman scattering), and 2D-LIF (planar laser-induced fluorescence).

2.3 ELECTRON BEAM AND WIND TUNNEL

The standard experimental set-up for temperature and density measurements is shown in **Fig. 2.** The electron gun is mounted in a separate chamber at low pressure ($< 10^{-4}$ mbar) differentially pumped by turbomolecular pumps [25]. Older systems used oil-diffusion pumps [5, 6, 7]; hence the chamber must be cleaned from time to time. The electron beam in Fig. 2 is pulsed rectangular (1 kHz ... 50 kHz) to increase the signal-to-noise ratio by means of the lock-in-technique. The emitted light from $N_2^+(0,0)$ band (at $\lambda \approx 400\ nm$) is focussed by a lens outside the wind tunnel on a monochromator and a photomultiplier. Therefore the wind tunnel must have either a window with a good transmission in the near ultraviolet or one can use a fiberoptic. The width of the entrance slit of the spectrograph is dependant on the optical resolution which is needed: for rotational spectra a resolution of $\Delta\lambda_{rot} = 0.05\ nm$, and for vibrational bands of $\Delta\lambda_{vib} = 0.5\ nm$. **Fig. 3** shows an unresolved N_2 fluorescence spectrum including bands from the first negative and second positive system. Usually, for temperature determination the $N_2^+(01)$ and $N_2^+(12)$ transitions are used.

2.4 RESULTS AND OUTLOOK

One example of applying the e^--beam technique to hypersonic flow is shown in **Fig. 4.** The density profile and the velocity gradient across a normal shock are measured [47] in a low density wind tunnel at a Mach number of 24. Both velocity and density show the expected change along the Rankine-Hugoniot relations by a factor of 6 known from continuum theory for high Mach numbers. The possibility of two-dimensional temperature-, density-, and velocity-fields is also given by this technique [18, 48, 49]. Some planar results taken in wind tunnels are published [47, 48].

Focussing on the two new wind tunnels F4 and HEG, a gun of about 20 - 50 kV and variable current output (low current for point, high current (1 mA) for planar-measurement) is favoured. Such a gun can be pulsed up to 50 kHz [18] which will provide several e^--shots even during the 1 ms running time of the HEG. In F4 some problems will occur due to charged particles generated by the arc-heater system. At higher densities the electron beam fluorescence suffers from quenching problems. In gas mixtures one has to take into account direct ionization, Penning ionization via collisions between target molecules and metastable atoms, and charge-transfer reactions by collisions of ionized atoms with target molecules. Using metastable atoms as a high energy source gives us many suggestions to improve the conventional electron beam technique [47].

3. Laser-Induced Fluorescence Diagnostics

3.1 DESCRIPTION

The principle of LIF is the use of laser-excited emission to provide local measures of the number density distribution functions [17]. The introduction of tunable lasers of high spectral power density has resulted in a manifold applicability of one-photon and multi-photon excitation processes of atoms and molecules. The LIF technique is on the way of becoming a standard tool for non-intrusive measurements in supersonic wind tunnel flow fields [26 – 30] in a wide density range. It can measure translational, vibrational and rotational temperatures by exciting selectively (in contrast to the electron beam) each molecular state even at very low densities. At higher densities quenching may occur [36, 37], but it is possible to modify this technique and detect predissociative molecular states [31] (LIPF). **Fig. 5** shows (in principle) the LIF- and LIPF-processes; of major interest is the repulsive curve crossing the potential curve on the upper right hand side: the laser excited molecule (O_2) can dissociate immediately after the stimulation process. This has the big advantage of very low quenching influence, but the disadvantage of low intensity compared to the conventional LIF-technique.

3.2 APPLICATION REGIMES

The density range of the LIF-technique is enormous, see Fig. 1 [17], but one should point out that the LIF-sensitivity is strongly dependant on the molecule which is stimulated: *NO* has best sensitivity [27, 29, 50, 51, 52, 53, 54], *CO* and N_2 are extremely difficult [32, 33] (two-photon process in the vacuum ultra violet) but possible, O_2 is about two orders of magnitude less sensitive than *NO* but still a good candidate at higher densities [31, 34, 50, 53, 54]. Two-dimensional temperature fields can also be measured by LIF [27, 29, 30, 50, 55] and LIPF [31, 35] if powerful UV-lasers are used.

3.3 LIF-APPARATUS AND WIND TUNNEL

The experimental set-up with a tunable ArF*-excimer laser in front of a supersonic vacuum wind tunnel is shown in **Fig. 6**. Three windows with a good transmission in UV (193 nm) are installed. For pointwise measurement the laser beam can be focused by a lens outside the tunnel [26]. The fluorescence light is collimated by two lenses on a photomultiplier. The maximum laser pulse energy of the ArF*-laser is 100 mJ at a pulse length of 13 ns and a maximum repetition rate of 250 Hz.

Fig. 7 shows a static *NO* excitation spectrum at room temperature [52] with equilibrium of rotational and vibrational energies. The measured spectrum is compared to the computer calculation. The $B^2\Pi_r(v' = 7) \leftarrow X^2\Pi_r(v'' = 0)$-band is the only one within the tuning range of the ArF*-excimer laser at $T = 300\,K$. Therefore the spectrum is very simple and easy to identify.

Planar images can be recorded by replacing the multiplier by a Newton telescope plus a CCD-camera and spreading the laser beam by a cylinder lens [27, 50].

3.4 RESULTS AND OUTLOOK

A good demonstration for the high spatial resolution is given in **Fig. 8.** Along the center line of a supersonic free jet (orifice diameter D = 3.1 mm) single point measurements of rotational temperature were conducted by laser-induced fluorescence of *NO*, conducted with a rapid-tuning ring dye layer [50]. The agreement with the isentropic temperature distribution calculated along the center line is quite good except at x/D greater than 2 where rotational freezing appears to set in. Also the position of the Mach disc at $x/D = 3.2$ is measured clearly and a temperature increase from 80 K to 300 K is detected. It is interesting that such quantitative results can be obtained with only 1 mW laser power and that the measurements can be executed at a repetition rate of of 4 kHz!

Another application of the LIF-technique to turbulent boundary layers can demonstrate the high temporal resolution when operating with laser pulses of 10 ns. **Fig. 9** shows a comparison of the root-mean-square (RMS) fluctuation amplitudes of temperature and density obtained by spectroscopic methods [54] and from hot wire data taken in a 12 mm thick, fully turbulent boundary layer that is adjointed by a relatively undisturbed flow of the same thickness, Mach number 2.

The laser results, *NO*-LIF and O_2-LIF/Raman, are in agreement. They both show a region in the boundary layer near $y/\delta = 0.6$ where the density fluctuations are exceptionally large. A histogram of the *NO*-LIF fluctuations displays a bimodal character which suggests that a weak wave was oscillating across the sample volume. These comparisons demonstrate that the O_2-LIF/Raman technique can provide accurate measurements in air flows at the low temperatures and densities which are prevalent in high-speed wind tunnels.

Focussing on the F4- and HEG-wind tunnels the *NO*-LIF- and O_2-LIPF-methods are applicable, see Fig. 1, for planar measurements. Due to the short running time (1 ms) of HEG and the low laser repetition rate two separate excimer lasers will be used which are tuned to two selected transitions. These two lasers are fired one after the other, and two camera systems are installed, which are synchronized with the lasers.

4. Cars Diagnostics

4.1 DESCRIPTION

CARS has been shown to be a versatile and powerful technique for spectroscopic studies in supersonic flows [1, 38, 39, 56] and in hot gases [4, 37, 40, 41]. It is ideally suited to determine density and temperature profiles with high spatial resolution and to study rotational and vibrational relaxation processes. Many species have been studied: H_2, D_2, OH, O_2, NH_3, N_2, CO_2..., so all species in hypersonic wind tunnels can be detected. The multi-parameter dependancy of CARS performance makes the applicability region complex to define. Single shot experiments (20 ns) are possible.

4.2 APPLICATION REGIMES

Fig. 1 shows the possible number density range for CARS-measurements having an optimum at 10^{17} to 10^{20} *molecules/cm*^{3} [17]. The spectroscopy experts from ONERA [42] checked the possible density range for 300 K and 3000 K:

multiline CARS $5 \cdot 10^{15}$ to $2 \cdot 10^{17}$ *molecules/cm*^{3}

multiplex CARS 10^{12} to $2 \cdot 10^{18}$ *molecules/cm*^{3}

4.3 CARS-APPARATUS AND WIND TUNNEL

A standard set-up incorporates a frequency-doubled Nd: YAG laser at 532 nm [38]. Its output is split into two parts, see **Fig. 10:** one beam is used to pump a tunable dye laser, the other part provides the pump beam for the CARS process. Both laser beams are carefully aligned in time and space and focussed by a lens outside the wind tunnel. The two optical windows of the tunnel are not critical (visible light). The anti-Stokes photons are detected via monochromator and photomultiplier; an explain OMA-system with photodiodes is also possible. **Fig. 11** shows a measured N_2 CARS spectrum from a flame [45]. It was taken in the narrow-band mode and is well resolved. For single shot experiments it is possible to detect only two representative lines of the spectrum.

4.4 RESULTS AND OUTLOOK

In a Mach 10 hypersonic air flow of a blow-down wind tunnel temperatures and densities of nitrogen were measured [56]. The data were recorded using a non-linear optical technique named dual-line CARS which allows single laser-shot spatially-resolved measurements. The flow was probed from the free stream to the vicinity of a two-dimensional ramp model (25°). Averaged temperature and density profiles are shown in **Fig. 12.** The measurements are compared with those predicted by a Navier-Stokes solver and show good agreement.

Dual-line CARS is a sensitive technique to probe molecules in hypersonic flow. Time-averaged temperature and density can be measured simultaneously in the vicinity of a two-dimensional model with a separation flow field region.

Focussing on the F4- and HEG-wind tunnels all scanning CARS methods are ruled out out due to a data acquisition time of several minutes. Single shot experiments remain [42]:

- dual-line CARS with high sensivity (10^{15} *molecules/cm*^{3})
- broad band CARS with minor sensivity (10^{17} *molecules/cm*^{3}).

The dual-line CARS-technique will be applied to the F4; it is also applicable to HEG.

5. Additional Non-Intrusive Methods

Raman- and Rayleigh scattering are two additional candidates for use in supersonic flows. These techniques and some derivations are listed by Bonnet and Gressilon [1] and Miles [43]. A new method should be pointed out based on the combination of Raman Excitation and Laser-Induced Electronic Fluorescence (RELIEF). Two or three lasers are used [43, 44], tagging O_2-molecules. The method provides two-dimensional temperature- and density-surveys of the flow field, but is limited to temperatures below 800 K.

Absorption measurements of O_2 and H_2O with tunable semi-conductor diode lasers offer high potential for diagnostic applications at low prize and compact size [50, 57]. The latter point is of major importance for future flight experiments.

6. Conclusion

6.1 DISCUSSION

Several non-intrusive techniques look promising for high-temperature hypersonic flows. These techniques include electron beam, LIF, laser absorption, and CARS. All four techniques suffer from the problem of spectrum overlapping at temperatures of 3000 K and above. This can be solved by spectra simulation and choice of transitions which are not influenced by other bands [26]. The expected results in F4 and HEG are planar temperature plots of NO and O_2 measured by LIF, temperature measurements at least pointwise and along a line at lower densities by electron beam fluorescence, and single laser-shot temperature measurements pointwise by CARS at higher densities. All three techniques have overlapping density ranges, see Fig. 1; the electron beam is (up to now) the only one providing continuous measurements during the tunnel running times.

A summary of the advanced optical diagnostic methods for temperature and density for hypersonic research is listed in **Table 1.**

The techniques listed above will provide temperature profiles near the nose of a reentry vehicle at high angle of attack and realistic flow velocities. This should help to validate CFD-data. The strong non-equilibrium of rotational and vibrational temperature is typical in hypersonic wind tunnels due to the rapid expansion in the nozzle. The chemical non-equilibrium (production of NO) is characteristic for wind tunnels with stagnation temperatures higher than 5000 K. Both effects will occur in F4 and HEG, and this is in strong contrast to the flight situation in a static atmosphere of a reentry vehicle. Therefore, flight experiments are necessary with an electron beam or a laser on board. Only flight data can simulate (because it is the flight reality) the true conditions to compare with CFD. Furthermore such flight data of temperatures and species concentrations are of great importance for environment studies of the upper atmosphere.

Finally, it should be recognized that many of the diagnostic techniques are not only species specific, but, in addition, measure the internal state distribution of the probed molecules [57]. Thus the newer diagnostics present a revolution of sorts in that they are leading to an understanding of aerodynamic phenomena from the molecular level rather than from the traditional bulk properties approach.

6.2 SUMMARY

1. The capabilities are available for local measurements with temporal resolution and quantitative accuracy of

    - velocity
    - temperature
    - density
    - species concentrations
    - fluctuations due to turbulence

2. Planar and point field measurements are demonstrated in laboratories for quantitative data.
3. Some techniques are now ready for wind tunnel application.
4. The potential for flight experiments is demonstrated for the electron beam for high altitudes; the laser systems require miniaturization.

6.3 RECOMMENDATIONS

Major funding is now required to move these techniques from the laboratory to the wind tunnels and to flight experiments. These techniques can supply data with resolution required for CFD development.

7. References

[1] J.P. Bonnet, D. Gressilon, *Advanced Non-Intrusive Experimental Techniques in Fluid and Plasma Flows.* Experiments in Fluids, Springer Verlag 1991, 12, pp. 131-135.

[2] R.J. Cattolica, *Modern Development in Electron-Beam Fluorescence.* Rarefied Gas Dynamics, 1990, Aachen. Ed. A.E. Beylich. Weinheim: VCH, 1991, pp. 1581-1592.

[3] A.K. Rebrov, R.G. Sharenfutdinov, *Relaxation Process in a Jet in the Presence of Phase Transition.* Proceedings of the Fifteenth International Symposium on Rarefied Gas Dynamics, Vol. II, 1986, p. 109. Stuttgart: B.G. Teubner, 1986.

[4] N. Herlin, M. Pealat, M. Lefebore, P. Alnot, J. Perrin, *Rotational Energy Transfer on a Hot Surface in a Low-Pressure Flow Studied by CARS.* Surface Science 258, 1991, pp.381-388.

[5] J.-C. Lengrand, J. Allegre, M. Raffin, *Electron Beam Fluorescence Technique at Rarefied Dynamics.* Proceedings of the 14th International Symposium on Rarefied Gas Dynamics, Vol. I (1984), p. 423. Ed. H. Oguchi. Tokyo: University of Tokyo Press, 1984.

[6] C. Dankert, K.-A. Bütefisch, *Influence of Nozzle Geometry on Vibrational Relaxation in an Expansion Flow.* Proceedings of the 9th International Symposium on Rarefied Gas Dynamics, Göttingen 1974, Vol. I, p. B.20-1.

[7] E.P. Muntz, *The Electron Beam Fluorescence Technique.* AGARDograph, 132, 1965.

[8] K.A. Bütefisch, D. Vennemann, *The Electron Beam Technique in Hypersonic Rarefied Gas Dynamics.* Progress in Aerospace Science, Vol. 15, p. 217, 1974.

[9] R.J. Cattolica, R.L. Schmitt, R.E. Palmer, *Feasibility of Non-Intrusive Optical Diagnostic Measurements in Hypersonic Boundary Layers for Flight Experiments.* AIAA 28th Aerospace Science Meeting, AIAA-90-0627, Reno, Nevada/USA, 1990.

[10] J.H. De Leeuw, W.E.R. Davies, *Measurement of Temperature and Density in the Upper Atmosphere Using an Electron Beam.* Canadian Journal of Physics, 50 (1972), pp. 1044-1052.

[11] R. Grandal, E.V. Thrane, J. Troim, *Polar 5 - An Electron Accelerator Experiment within an Aurora. 4.Measurements of the 391.4 nm Light Produced by an Artificial Electron Beam in the Upper Atmosphere.* Planet Space Sci., 28 (1980), pp. 309-319.

[12] R.R. O'Neil, W.R. Pendleton, Jr., A.M. Hart, A.T. Stair, Jr., *Vibrational Temperature and Molecular Density of Thermospheric Nitrogen Measured by Rocket-Borne Electron Beam Induced Luminescence.* Journal Geophys. Res., 79 (1974), pp. 1942-1958.

[13] A.A. Haasz, J.H. De Leeuw, W.E.R. Davies, *Thermophysics N_2 and O_2 Concentration with Rocket-Borne Electron Beam Fluorescence Probes.* Journal Geophys. Res., 81 (1976), pp. 2382-2391.

[14] G. Israelson, J.R. Winckler, *Measurements of 3914 Å Light Production and Electron Scattering from Beams Artificially Injected into the Ionosphere.* Journal Geophys. Res., 80 (1975), pp. 3709-3712.

[15] R.R. O'Neil, F. Bein, D. Burt, J.A. Sandock, A.T. Stair, Jr., *Summarized Results of the Artificial Aurora Experiment, Precede.* Journal Geophys. Res., 83 (1978), pp. 3275-3280.

[16] P.M. Banks, W.J. Raitt, A.B. White, R.I. Bush, P.R. Williamson, *Results from the Vehicle Charging and Potential Experiment on STS-3.* J. Spacecraft, 24 (1987), pp. 138-149.

[17] J.W.L. Lewis, *Optical Diagnostics of Low-Density Flow Fields.* Rarefied Gas Dynamics, 16th Int. Symp., Pasadena/USA, 1988. Edited by E.P. Muntz, D.P. Weaver, D.H. Campbell. Progress in Astronautics and Aeronautics, Vol. 117 (1989), pp. 107-132.

[18] A. Mohammed, *Electron Beam Probing of Blow-Down Hypersonic Flow.* Ph.D. Thesis of University of Paris Sud, 17th Dec 1991.

[19] S.L. Petrie, J.J. Komr, *Application of the Electron Beam Fluorescence Technique to the Measurement of the Properties of NO in Non-Equilibrium Flows.* AFFDL-TR-72-144, Dec. 1972.

[20] R.J. Cattolica, *Electron Beam Fluorescence Measurements of Nitric Oxide.* Rarefied Gas Dynamics, 16th Int. Symp., Pasadena/USA, 1988. Edited by E.P. Muntz, D.P. Weaver, D.H. Campbell. Progress in Astronautics and Aeronautics, Vol. 117 (1989), pp. 133-139.

[21] S.L. Petrie, A.A. Boiarski, S.S. Lazdinis, *Electron Beam Studies of the Properties of Molecular and Atomic Oxygen in Non-Equilibrium Flows.* 6th Aerodynamic Testing Conference, AIAA 71-271, 1971.

[22] P.J. Dagdigian, J.P. Doering, *Electron Impact Ionization Excitation of CO in a Supersonic Beam.* Journal of Chemical Physics, Vol. 78, No. 4, Feb. 1983.

[23] A.E. Beylich, *Experimental Investigation of Carbon Dioxide Jet Plumes.* The Physics of Fluids, Vol. 14, No. 5, May 1971, pp. 898-905.

[24] E.P. Muntz, J.A. Kunc, D.A. Erwin, *A Pulsed Electron-Photon Technique for Temperature and Specie Concentration Measurements at Points in Relatively Dense, Unseeded Air Flows.* AIAA-87-1526, 22nd AIAA Thermophysics Conference, July 8-10, 1987.

[25] C. Dankert, G. Koppenwallner, *An Experimental Study of Nitrogen Condensation in a Free Jet Expansion.* Rarefied Gas Dynamics, Eleventh Symposium. Edited by R. Campargue, Vol. II, Paris 1979: Commissariat a l'Energie Atomique (CEA), pp. 1107-1118.

[26] C. Dankert, G. Gundlach, *Laser Induced Fluorescence Apparatus for Application in Rarefied Hypersonic Flow Fields. Part I: Static Cell Experiments with NO.* DLR-IB 222-91 A 11, DLR Göttingen/Germany, 1991.

[27] W.H. Beck, C. Dankert, G. Eitelberg, G. Gundlach, *Preliminary Laser Induced Fluorescence Measurements in Several Facilities in Preparation for Application to Studies in the High Enthalpy Shock Tunnel Göttingen.* 30th Aerospace Sciences Meeting, AIAA 92-0143, Reno, Nevada/USA, 1992.

[28] T. Niimi, T. Fujimoto, N. Shimizu, *Planar Measurements of Temperature in Rarefied Gas Flow by LIF Images.* Rarefied Gas Dynamics, Proceedings of the 17th International Symposium on Rarefied Gas Dynamics, Aachen 1990. Edited by A.E. Beylich. Weinheim: VCH, 1991.

[29] R.K. Hanson, A.T. Chang, J.M. Seitman, M.P. Lee, P.H. Paul, B.E. Battles, *Laser-Induced Fluorescence Diagnostics for Supersonic Flows.* 28th Aerospace Sciences Meeting, AIAA-90-0625, Reno, Nevada/USA, 1990.

[30] B.K. Mc Millin, J.L. Palmer, R.K. Hanson, *Two-Dimensional Temperature Measurements of Shock Tube Flows Using Planar Laser-Induced Fluorescence Imaging of Nitric Oxide.* AIAA 22nd Fluid Dynamics, Plasma Dynamics & Laser Conference, AIAA-91-1670, Honolulu, Hawaii, 1991.

[31] P. Andresen, A. Bath, W. Graeger, H.W. Luelf, G. Meijer, J.J. Meulen, *Laser-Induced Fluorescence with Tunable Excimerlasers as a Possible Method for Instantaneous Temperature Field Measurements at High Pressure: Checks with an Atmospheric Pressure Flame.* Appl. Opt. 27 (1988), pp. 365-378.

[32] F. Aguillon, A. Lebehot, J. Roussean, R. Campargue, *Two-Rotor Resonant Third Harmonic Generation in the XUV Range Reflecting Internal State Populations in CW CO Free Jets.* Proceedings of the Fifteenth International Symposium on Rarefied Gas Dynamics, Grado/Italy, Vol. II, 1986, pp. 565-574. Stuttgart: B.G. Teubner, 1986.

[33] O. Faucher, F. Aguillon, A. Lebehot, R. Campargue, *Two-Photon Third Harmonic Generation in a CW Nitrogen Free Jet Zone of Silence Reflecting Quantitatively Internal State Populations.* Proceedings of the 17th International Symposium on Rarefied Gas Dynamics, Aachen/Germany 1990. Edited by A.E. Beylich. Weinheim: VCH, 1991, pp. 1088-1095.

[34] G. Lauter, D. Fletcher, R. Mc Kentic, *A Method for Measuring Tempertures and Densities in Hypersonic Wind Tunnel Air Flows Using Laser-Induced* O_2 *Fluorescence.* 28th Aerospace Sciences Meeting, 1990, Reno, Nevada/USA, AIAA-90-0626.

[35] M.P. Lee, P.H. Paul, R.K. Hanson, *Quantitative Imaging of Temperature Fields in Air Using Planar Laser-Induced Fluorescence of* O_2. Optical Letter 12 (1987), 75.

[36] K. Shibuya, F. Stuhl, *Fluorescence Lifetime and Collisional Quenching of the Predissociative* $NO\,B^2\Pi\,(v'=7)$ *State.* Chemical Physics 79, 1983, pp. 367-381.

[37] K. Kohse-Höinghaus, U. Meier, B. Attal-Tretout, *Laser-Induced Fluorescence Study of OH in Flat Flames of 1 - 10 bar Compared with Resonance CARS Experiments.* Appl. Opt. 29 (10), 1990, pp. 1560-1569.

[38] H.-D. Barth, C. Jackschath, T. Pertsch, F. Huisken, *CARS Spectroscopy of Molecules and Clusters in Supersonic Jets.* Appl. Phys. B45, 1988, pp. 205-214.

[39] H.-D. Barth, F. Huisken, A.A. Ilyokhin, *Coherent Raman Spectroscopy of Nitrogen Molecules and Clusters in Supersonic Jets.* Appl. Phys. B52, 1991, pp. 84-89.

[40] W. Stricker, M. Woyde, *CARS Temperature Measurements in High Pressure Flames.* Proceedings of the Joint Meeting of the German and Italian Section of the Combustion Institute, Ravelo/Italy, 1989, Paper 1.1.

[41] G. Marowsky, A. Slenczka, *Laser Spectroscopy VIII.* Edited by W. Persson and S. Scramberg, Springer Series Opt.Sci. 55, 1987, p. 334.

[42] J.P. Taran, M. Pealat, F. Grisch, *Private Communication.*

[43] R.B. Miles, D.M. Nosenchuck, *Three-Dimensional Quantitative Flow Diagnostics.* Lecture Notes in Engineering 45, M. Gad-el-Hak, Editor. Advances in Fluid Mechanics Measurements, Springer Verlag, 1969.

[44] R.B. Miles, J.J. Connors, E.C. Markovitz, P.J. Howard, G.J. Roth, *Instantaneous Profiles and Turbulence Statistics of Supersonic Free Shear Layers by Raman Excitation Plus Laser-Induced Electronic Fluorescence (Relief) Velocity Tagging of Oxygen.* Experiments in Fluids, 8, pp. 176-24, Springer Verlag 1989.

[45] W. Kreutner, W. Stricker, Th. Just, *Comparison of Spontaneous Raman and CARS Measurements in a Laminar Flame at Atmospheric Pressure.* Applied Spectroscopy, Vol. 41, No. 1, 1987, pp. 98-105.

[46] E.P. Muntz, D.A. Erwin, *Rapid Pulse Electron Beam Fluorescence for Flow Field Diagnostics.* Hypersonic ARW, Theoretical and Experimental Methods in Hypersonic Flows, 70th Fluid Dynamics Panel Meeting, ed. by Kluwer, 1992.

[47] E. Hirai, K.-A. Bütefisch, C. Dankert, *Velocity and Density Determination by Electron Beam Technique.* Hypersonic ARW, Theoretical and Experimental Methods in Hypersonic Flows, 70th Fluid Dynamics Panel Meeting, ed. by Kluwer, 1992.

[48] R.J. Cattolica, *Electron Beam Fluorescence Imaging for Hypersonic Research.* Hypersonic ARW, Theoretical and Experimental Methods in Hypersonic Flows, 70th Fluid Dynamics Panel Meeting, ed. by Kluwer, 1992.

[49] A.K. Mohammed, *Electron Beam Velocimetry.* Hypersonic ARW, Theoretical and Experimental Methods in Hypersonic Flows, 70th Fluid Dynamics Panel Meeting, ed. by Kluwer, 1992.

[50] R.K. Hanson, *Laser-Based Diagnostics for Hypersonic Flows.* Hypersonic ARW, Theoretical and Experimental Methods in Hypersonic Flows, 70th Fluid Dynamics Panel Meeting, ed. by Kluwer, 1992.

[51] G. Gundlach, C. Dankert, *Nonintrusive Temperature Determination by LIF in Hypersonic Wind Tunnels.* Hypersonic ARW, Theoretical and Experimental Methods in Hypersonic Flows, 70th Fluid Dynamics Panel Meeting, ed. by Kluwer, 1992.

[52] W.H. Beck, U. Koch, M. Müller, *Spectroscopic Diagnostic Techniques for the High Enthalpy Shock Tunnel in Göttingen (HEG): Preparatory LIF Studies on other Facilities.* Hypersonic ARW, Theoretical and Experimental Methods in Hypersonic Flows, 70th Fluid Dynamics Panel Meeting, ed. by Kluwer, 1992.

[53] F. Dionnet, J. Monard, D. Puechberty, *Broad Band Excimer Laser Induced Fluorescence Applied to Hypersonic and Supersonic Air Flows.* Hypersonic ARW, Theoretical and Experimental Methods in Hypersonic Flows, 70th Fluid Dynamics Panel Meeting, ed. by Kluwer, 1992.

[54] R.L. Mc Kenzie, D.G. Fletcher, *Laser-Spectroscopic Measurement Techniques for Hypersonic, Turbulent Wind Tunnel Flows.* Hypersonic ARW, Theoretical and Experimental Methods in Hypersonic Flows, 70th Fluid Dynamics Panel Meeting, ed. by Kluwer, 1992.

[55] U. Meier, I. Plath, K. Kohse-Höinghaus, *Determination of Minority Species Concentrations and Temperature in Combustion Systems by Laser-Induced Fluorescence.* Hypersonic ARW, Theoretical and Experimental Methods in Hypersonic Flows, 70th Fluid Dynamics Panel Meeting, ed. by Kluwer, 1992.

[56] F. Grisch, P. Bouchardy, M. Pealat, B. Chanetz, T. Pot, M.C. Coet, *Rotational Temperature and Density Measurements in a Hypersonic Flow by Dual-Line CARS*. Hypersonic ARW, Theoretical and Experimental Methods in Hypersonic Flows, 70th Fluid Dynamics Panel Meeting, ed. by Kluwer, 1992.

[57] R.J. Exton, *Absorption, Scattering, and Fluorescence Techniques for Hypersonic Flow Measurements*. Hypersonic ARW, Theoretical and Experimental Methods in Hypersonic Flows, 70th Fluid Dynamics Panel Meeting, ed. by Kluwer, 1992.

Tab. 1 **Summary of Advanced Optical Diagnostic Methods for Temperature and Density for Hypersonic Research**

Technique	Flow Properties	Spatial Type	Temporal Resolution	Lab Experience	Wind Tunnel Experience	Flight Experience	Comments on Status
Electronic Beam Fluorescence	ρ_i, v, T_{trans}, T_{rot}, T_{vib}	point, 1D, 2D	high, continuous	extensive	moderate	some (8 flights)	Needs modernization in WT and can provide flight experiments within 3 years
Laser Fluorescence	ρ_i, v, T_{trans}, T_{rot}, T_{vib}	point, 1D, 2D	high, pulsed	moderate	some	none	Needs more utilization in WT for CFD validation
Laser Absorption	ρ_i, v, T_{trans}, T_{rot}, T_{vib}	line-of-sight	moderate, continous	moderate	some	none	Primarily for internal flows, line-of-sight method
Coherent Laser Raman (Cars)	ρ_i, v, T_{trans}, T_{rot}, T_{vib}	point	moderate, pulsed	moderate	some	none	More development, WT applications at high density
Laser Rayleigh Scattering	ρ, v, T_{trans}	point, 1D, 2D	low, pulsed	some	minimal	none	WT applications at high density, not species specific
Laser Raman Scattering	ρ_i	point	very low, pulsed	minimal	none	none	WT applications only at high density

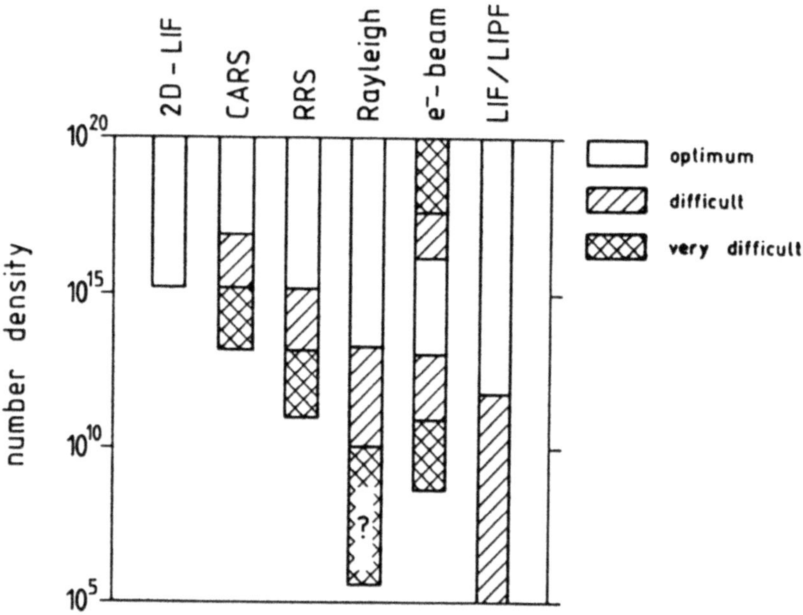

Fig. 1 Diagnostics application regimes [17]

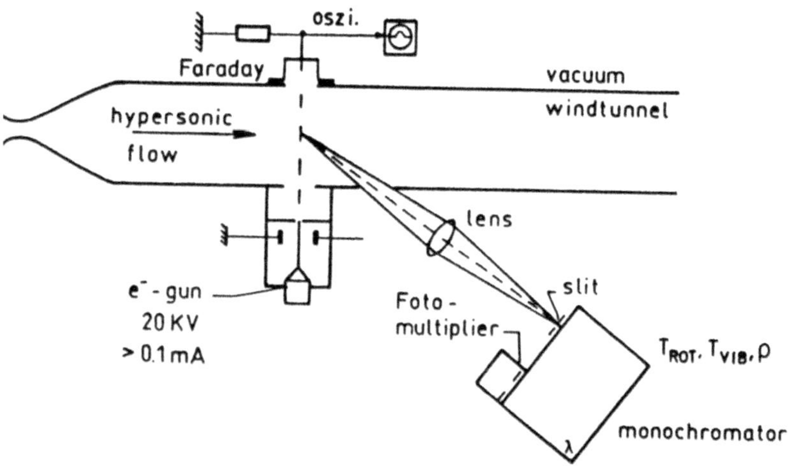

Fig. 2 Electron beam apparatus in a wind tunnel

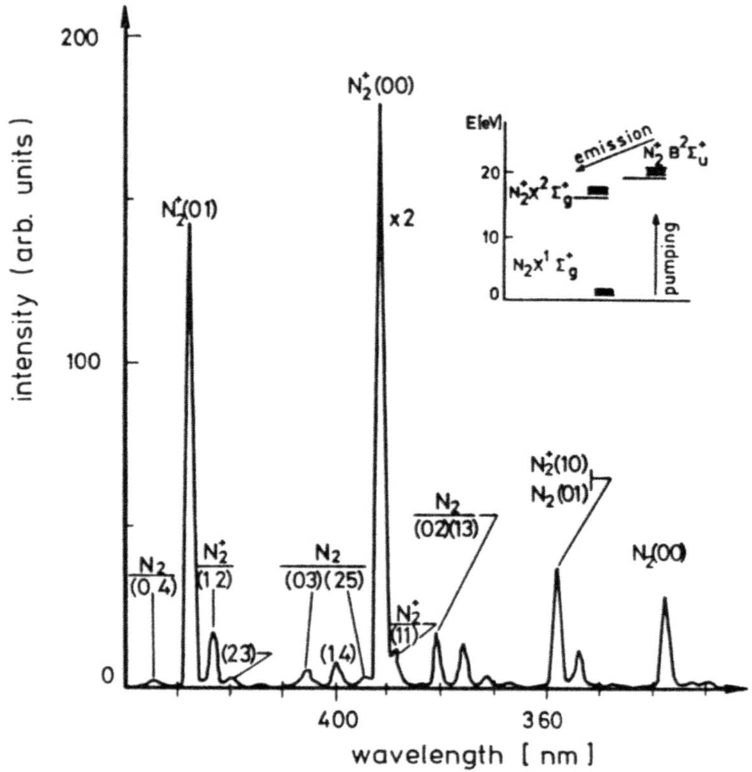

Fig. 3 Unresolved N_2 fluorescence spectrum including bands from the first negative and the second positive system. Most temperature measurements use the $N_2^+(01)$ and $N_2^+(12)$ transitions.

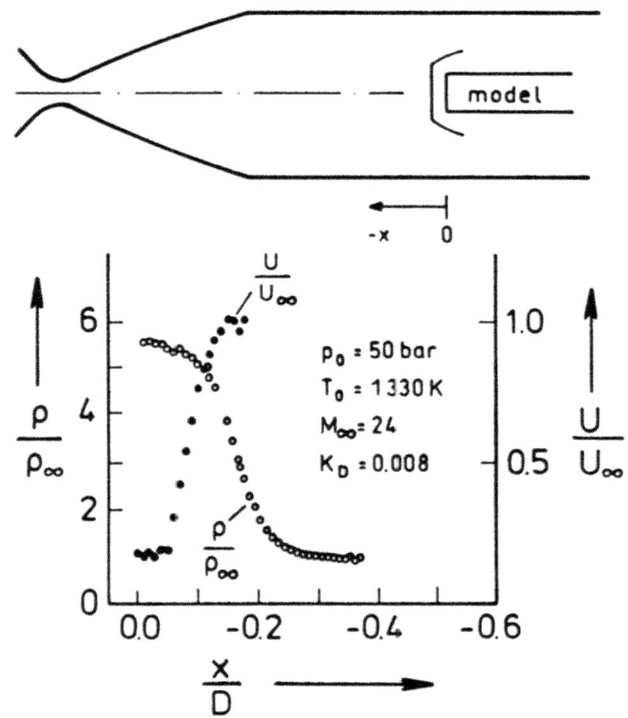

Fig. 4 Velocity measurements across a normal shock together with the corresponding density profile at $M_\infty = 24$ [47].

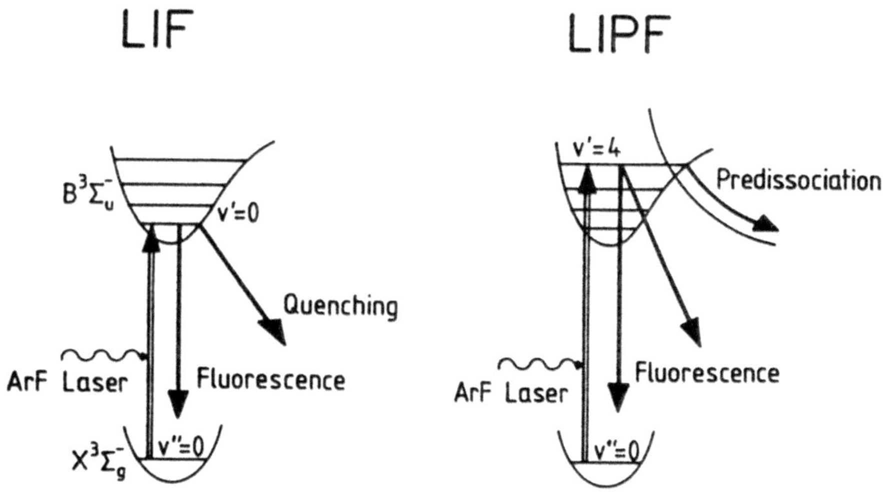

Fig. 5 Principle of LIF with laser excitation, fluorescence, quenching, and LIPF with the additional predissociation potential.

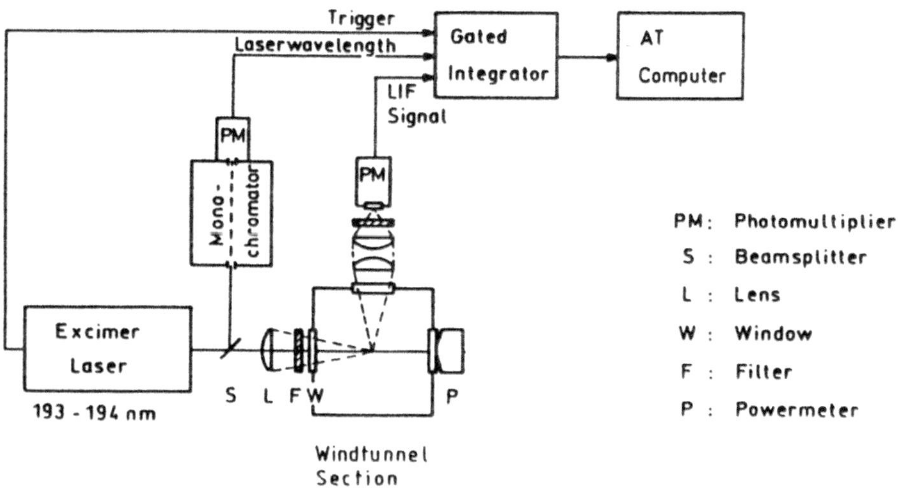

Fig. 6 LIF-apparatus installation in a wind tunnel.

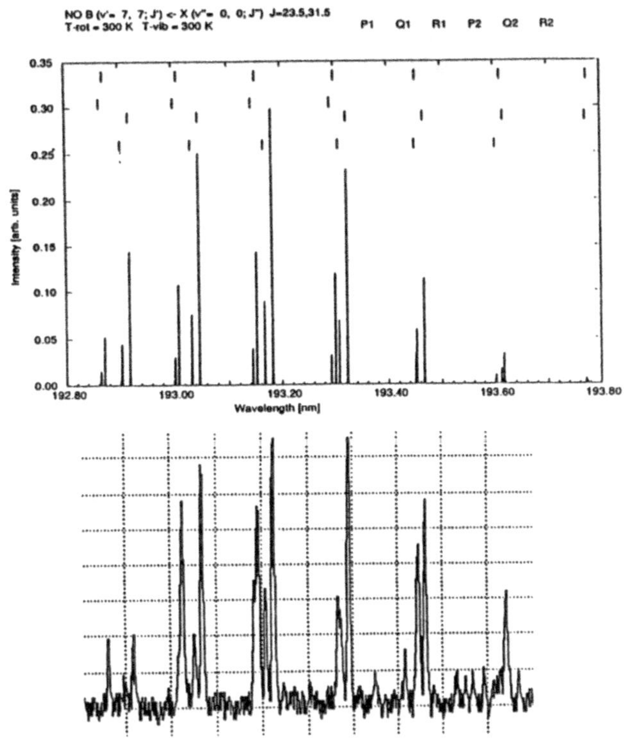

Fig. 7 Calibration measurement in comparison to calculated spectra at room temperature.

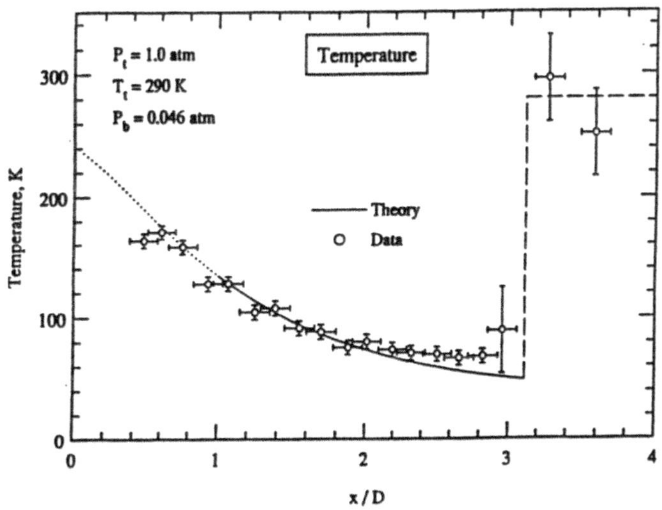

Fig. 8 Rotational temperature along free jet center line by LIF [50].

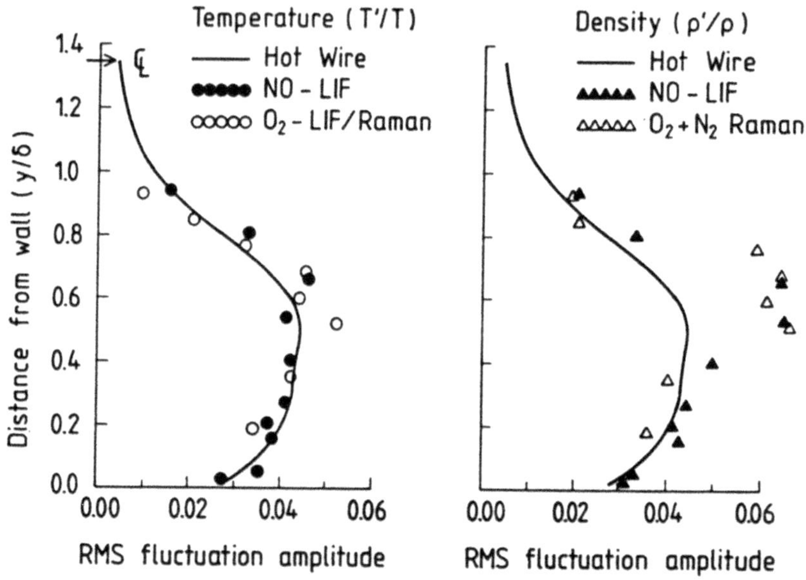

Fig. 9 Measurements of temperature and density fluctuation in turbulent Mach 2 channel flow [54].

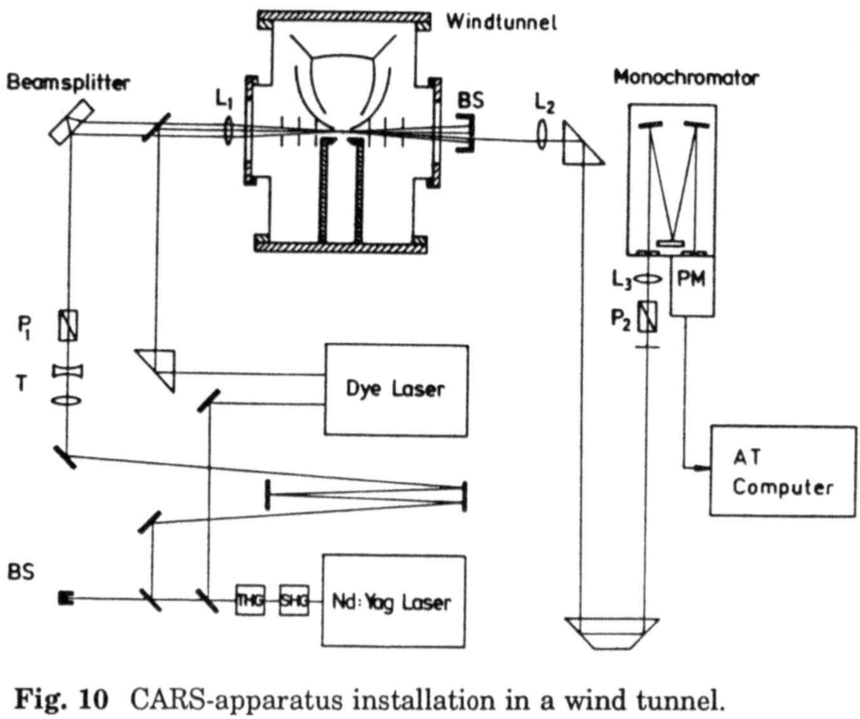

Fig. 10 CARS-apparatus installation in a wind tunnel.

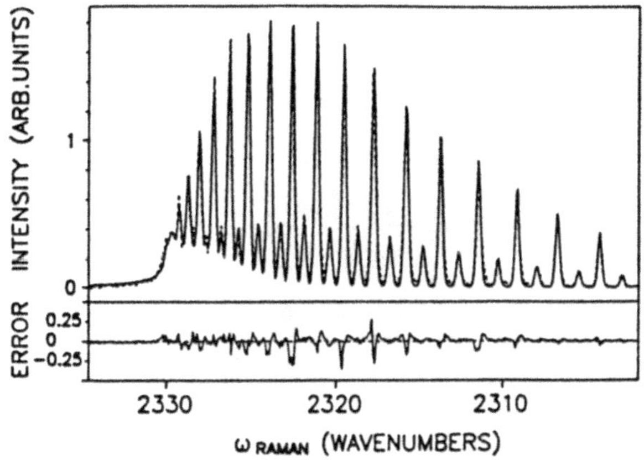

Fig. 11 Narrow-band N_2 CARS spectrum (Q-branch) from a propane/air flame above burner surface, flame center. Measurement (dashed), theoretical (full line), and error surve (lower track). Approximately 30 shots were averaged at each point; $T = 1983\ K$ [45].

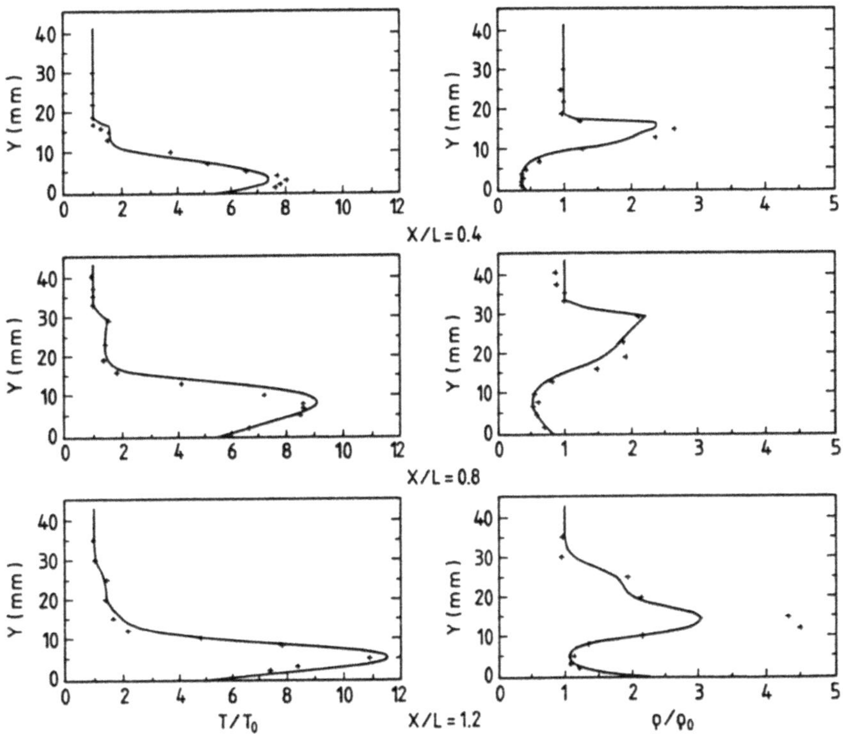

Fig. 12 Rotational temperature (left) and total density (right) profiles normalized by the free stream conditions at: a) X = 0.072 mm (X/L = 0.4), c) X = 0.144 m (X/L = 0.8), e) X = 0.214 m (X/L = 1.2): (-) HOMARD2 code; (+) experiment [56].

Pressure and Force Measurements on Models Set in Hypersonic Flows: A Review

Charles G. Miller
NASA Langley Research Center
16 Victory Street, MS 408
Hampton, VA 23665-5225 USA

K. W. Naumann
French-German Research Institute Saint-Louis ISL
Rue de l' industrie 5
F-68301 Saint-Louis
France

ABSTRACT. A review of measurement techniques used to obtain aerodynamic forces and moments and surface/flow field pressures for models tested in hypersonic-hypervelocity impulse facilities and in conventional-type hypersonic wind tunnels is presented. Although force and moment measurement techniques presently used in hypersonic wind tunnels are relatively unchanged from the 1960's and 1970's, significant advances have recently been made for impulse facilities. For both hypersonic wind tunnels and impulse facilities, the state-of-the-art has advanced via refinements, improved test techniques, and advances in semiconductor technology, data acquisition systems and computers. The introduction of electronically scanned pressure systems over a decade ago "revolutionized" pressure measurements in hypersonic wind tunnels and a second "revolution" is impending with the development and application of optical, two-dimensional, global pressure measurement techniques. The development and continued refinement of miniature piezoresistive transducers has provided the capability to perform detailed surface pressure measurements on relatively small, complex models in impulse facilities; these transducers also provide the capability for intrusive flow field pressure measurements with miniature survey rakes.

1. Preface

The original intent of this paper, as indicated by its title, was to provide a review of force and moment and of pressure measurement techniques used in hypersonic facilities, both conventional-type hypersonic wind tunnels and impulse-type hypersonic-hypervelocity facilities. This review was to be based primarily on reports presented at the NATO Advanced Research Workshop entitled "New Trends in Instrumentation for Hypersonic Research" and held at ONERA LeFauga-Mauzac, France, April 27 to May 1, 1992. Four papers were presented in session 4, which was entitled "Pressure-Forces", of the workshop and all addressed force and moment measurements in impulse facilities. The titles and authors of these papers are as follows:

(1) "Force Measurement in Hypervelocity Impulse Facilities" by J. M. Simmons, W. J. Daniel, D. J. Mee, and S. L. Tuttle, (paper 4A).

(2) "A Six Component Balance for Short Duration Hypersonic Facilities" by C. Jessen and H. Gronig (paper 4B).

(3) "Millisecond Aerodynamic Force Measurement Technique for High Enthalpy Test Facilities" by K. W. Naumann, H. Ende, and G. Mathieu (paper 4C).

(4)"Aerodynamic Force Measurements in the VKI LongShot Hypersonic Facility" by M. Carbonaro (paper 4D).

In the spirit of the original intent for this paper and for the sake of completeness, the synopses of the four papers presented at the workshop will be augmented by discussions on measurement

A. Boutier (ed.), New Trends in Instrumentation for Hypersonic Research, 583–592.

techniques routinely used to: (1) obtain aerodynamic forces and moments and pressure distributions in conventional-type (i.e., relatively low enthalpy, blowdown-to-vacuum) hypersonic wind tunnels and (2) to obtain pressure measurements in impulse facilities.

2. Introduction

The contribution of hypersonic/hypervelocity ground-based facilities to the aerodynamic and aerothermodynamic assessment of proposed aerospace vehicle concepts over the past four decades is well recognized. Aerodynamic assessments are typically performed via the measurement of forces and moments and of pressure distributions on stationary models subjected to hypersonic/ hypervelocity flow; similarly, aerothermodynamic assessments are performed via the measurement of heat transfer rate distributions. Models may be tested over a wide range of hypersonic simulation parameters (principally Mach number, Reynolds number, and normal shock density ratio or ratio of specific heats for continuum flow) and attitudes (angles of attack and sideslip) using conventional-type hypersonic wind tunnels and/or impulse facilities, some of which are capable of generating flow velocities representative of flight at high hypersonic Mach numbers. (A conventional-type hypersonic wind tunnel, referred to herein as hypersonic wind tunnel, is defined as: (1) operating at a relatively low enthalpy such that perfect or near-perfect gas behavior is achieved throughout the nozzle expansion process; (2) providing a given Mach number via a fixed nozzle area ratio; and (3) providing a range of Reynolds number via variation of reservoir pressure. The facility may be intermittent (i.e., blow-down to vacuum) or continuous, running for several hours. Impulse facilities may be of the shock tube family, such as shock tunnels (steady expansion process) and expansion tubes (unsteady expansion process), or hot-shot type whereby electrical energy is provided directly to the test medium. (A collection of presentations made on impulse facilities, which include descriptions of facilities and instrumentation, was recently edited by Rodgers, 1990.) Test times for such impulse facilities typically range from 0.1 to 100 millisecond. (Ballistic ranges, whereby the model is launched and allowed to free flight or is tracked, are not considered herein. Although arc-heated wind tunnels are important contributors to the study of materials at high temperatures, they are not generally considered amenable for aerodynamic/aerothermodynamic testing and thus are also not considered herein.) Both types (i.e., conventional-type and impulse) of facilities provide challenges to experimental aerodynamicists in pursuit of credible aerodynamic data. From a simplistic viewpoint, three interdependent factors require consideration in the measurement of forces and moments and of pressure distributions irrespective of the type of hypersonic facility; they are (1) magnitude, (2) time, and (3) the influence of temperature change.

The magnitude of the aerodynamic forces and moments and the surface/flow field pressures during all phases of the facility flow sequence (i.e., flow establishment, testing period, and flow breakdown) will naturally influence the measurement approach selected. Care must be exercised so to not exceed the acceptable range of the instrumentation, nor damage it, and to provide acceptable signal-to-noise ratios. Most hypersonic wind tunnels and impulse facilities are capable of at least an order of magnitude variation in dynamic pressure. This variation coupled to variations in model shape (very blunt to very slender) and attitude generally precludes a single measurement device from covering the entire range of magnitude possible in most tests.

The time required to establish flow within the facility and the period of steady or quasi- steady flow is quite short for impulse facilities. Because of this short test time, impulse facilities are attractive from an aerothermodynamic testing viewpoint (e.g., Schultz and Jones, 1973). Generally, the heat transfer rate to models may be measured to a greater accuracy than surface pressures in such facilities. Force and moment measurements in impulse facilities, particularly for those having run times less than 3-4 millisecond, are quite challenging due primarily to the finite mass of the model (model inertia; infinite mass corresponds to no model movement whereby zero mass corresponds to instantaneous response to the flow) and the associated dynamics. On the other hand, with the possible exception of extremely low levels of surface pressure such as those encountered in the base region of blunt bodies and on very slender configurations at high hypersonic Mach numbers, model surface pressures may be more accurately measured than heat transfer rates in hypersonic wind tunnels. Primarily because of differences in run time, both types of hypersonic facilities challenge experimentalists in different ways. One common denominator, however, is the detrimental effect of temperature change on the instrumentation commonly used for force and moment and for pressure measurements.

The relatively long run times, hence model exposure times to the flow, of hypersonic wind tunnels often allow the model and the support system to become quite hot (e.g., Mach 10 air tunnels require stagnation temperatures around 1800°R to avoid liquefaction during expansion of the nozzle flow). Commonly used techniques to measure model forces and moments and pressure distributions in these tunnels are sensitive to temperature change. Thus steps must be taken to isolate the instrumentation from temperature effects (e.g., water cooled and/or ceramic coated model support systems, glass-ceramic models, etc.) and/or accurately compensate for such effects. One example of recent changes in methodology to minimize, if not eliminate, the effect of temperature on pressure measurements in hypersonic wind tunnels having an injection/retraction system is the incorporation of onboard, small volume transducers even for relatively small models. The short tubing lengths possible with onboard transducers significantly reduces the lag time, allowing steady-state values of pressure to be measured after only a few seconds of exposure to the flow. The model may then be retracted from the flow prior to the occurrence of appreciable temperature changes. Although the short model exposure times for impulse facilities generally precludes large temperature variations of the model, pressure instrumentation must often be flush mounted to achieve acceptable response. Any volume between the sensing element of a transducer, which is often protected by a screen, and the model surface will induce a finite lag as the volume is filled and may be viewed as a pneumatic filter (e.g., Moore, 1979). Commonly used piezoelectric and piezoresistive pressure transducers are quite sensitive to temperature; and when mounted in close proximity to the high enthalpy test medium, they must be thermally protected or compensated.

Because of the demise of interest in hypersonics in the 1970's and well into the 1980's, as resources and personnel were diverted into other disciplines and a major portion of the hypersonic wind tunnels were deactivated or terminated, force and moment measurements are currently performed much as they were in the 1960's and 1970's. Although refinements have been made in the materials and adhesives used in strain gage balances, improvements made in test techniques, and advances made in data acquisition and reduction, the basic technique is relatively unchanged over the past two decades. The lack of interest in hypersonics coupled with the relatively good performance of existing methods simply did not provide an incentive in the 1970's and 1980's to develop other techniques. This was not the case for pressure measurements, however. About a decade ago, electronically scanned pressure (ESP) systems became commercially available which essentially "revolutionized" pressure measurements in hypersonic wind tunnels. These systems significantly reduced the cost per channel compared to variable capacitance or reluctance transducers; were quite compact, with 16, 32, or 48 sensors being housed in a small module that could be mounted close to the model surface and often within the model; and could be calibrated immediately prior to a test. ESP systems have continued to be refined and presently are standard equipment for many hypersonic wind tunnels world wide.

Similar trends were experienced by the impulse facility community. Because impulse facilities are primarily used for aerothermodynamic and fluid dynamic studies, there was little incentive in the 1970's and early-to-mid 1980's to improve methods for force and moment measurements in the few facilities that remained operational. Advances were made, however, in the measurement of model surface and flow field pressures with the introduction of commercially available miniature piezoresistance pressure transducers around the mid 1970's. Employing silicon diaphragms, these transducers could be made quite small (typically to a diameter of 0.050 inch) thereby allowing detailed pressure distributions to be made on relatively small, complex models. As for hypersonic wind tunnels, impulse facilities benefitted greatly from advances in data acquisition systems (DAS) and computers. Gone were the days of large banks of oscilloscopes with cameras and the laborious digitation of traces from film. Today, data acquisition systems for impulse facilities provide essentially the same rapid data reduction as those for hypersonic wind tunnels.

The purpose of this paper is to provide a review of techniques presently used in hypersonic wind tunnels and in impulse facilities to measure model forces and moments and surface/flow field pressures. It is not intended to provide a comprehensive, indepth discussion of any one technique, but rather a synopsis.

3. Measurement Techniques

3.1 MODEL SURFACE PRESSURE

3.1.1. *Impulse Facilities.* Representative test times for shock tunnels and expansion tubes range from 0.1 to 10 millisecond. For such short times, model surface pressures are generally measured with transducers based on the piezoelectric or the piezoresistive principal. Such transducers may also be used for impulse facilities having longer run times such as hotshot tunnels (typically 20 to 100 milliseconds), as may other concepts such as small volume variable reluctance diaphragm transducers. Piezoelectric and piezoresistive pressure transducers are commercially available and cover a wide range of size, pressure range, and response time. Because of their wide spread utilization in impulse facilities world wide, these transducers are reviewed in this section.

Peizo is derived from the early Greek word meaning "to squeeze." For piezoelectric transducers, the surface of certain crystals is charged electrically when the crystal is loaded mechanically or "squeezed." The three basic components of such transducers are: (1) the sensor housing which hermetically encloses the quartz elements; (2) quartz elements, and (3) the diaphragm welded to the sensor housing and transmitting the pressure exerted onto the quartz elements. Piezoelectric transducers produce an output only when subjected to a change in pressure and thus are ideally suited for measuring dynamic events; since the electrical charge cannot be stored for an indefinite period of time, piezoelectric transducers cannot perform truly static measurements. High impedance piezoelectric systems (charge mode) require charge amplifiers for signal conditioning; low impedance piezoelectric systems have built-in charge-to-voltage converters and are referred to as the voltage mode. Acceleration compensation consists of a tuned accelerator system connected in series, with opposite polarity, to the pressure measuring system. Commonly used piezo-electric transducers (e.g., those manufactured by Kistler Instrument Corporation and PCB Piezotronics) have 0.1 or 0.22 inch diameter diaphragms, quoted rise times of one microsecond and resonant frequencies from about 150 kHz to up to 500 kHz. These transducers are quite sensitive to change in temperature; that is, the transducer output varies (generally, nonlinearly) with temperature. Elimination of temperature effects can be quite challenging, particularly for flush mounted units which are necessary for run times less than a millisecond, or so. A thin layer of RTV or layer of electrical tape fitted to the diaphragm help to isolate piezoelectric transducers from temperature effects.

The piezoresistance effect is defined as the change in electrical resistivity with applied stress; this change is relatively large in semi-conductors. Piezoresistive transducers utilize a silicon pressure sensing diaphragm. Silicon is used primarily because the high sensitivity of this material allows very small diaphragms with high resonant frequencics; however, silicon is brittle and can crack or shatter on impact or under excessive loading. Newer transducers incorporate four piezoresistive silicon strain gages atomically diffused into the diaphragm. (Older designs used wire gages bonded to the thin diaphragm. Newer gages using silicon resistors are free of mechanical hysteresis and have greater sensitivity.) These strain gages serve as the arms of a Wheatstone Bridge. This fully active bridge converts the strain-induced resistance change of the gage to voltage changes which may be measured. The bridge is normally energized by applying a regulated voltage across two opposite corners; the output voltage is proportional to the product of the excitation voltage and the resistance changes of the strain gages. Piezoresistive transducers having outside diameters as small as 0.030 inch (e.g., Kulite Semiconductor Products, Inc. model CQ-030) are available, but diameters of 0.050 to 0.060 inch are more representative of commonly used transducers (e.g., those manufactured by Endevco Corporation, Entran International, and Kulite). As with piezoelectric gages, piezoresistive gages are temperature dependent. Temperature compensation is generally accomplished with simple circuit techniques using passive shunt or series resistor elements whose resistance is temperature independent. A protective coating of silicon rubber or RTV may be applied to the diaphragm for thermal protection and a screen is generally used to protect the diaphragm from particles incident upon the transducer. The rise time of a piezoresistive transducer is inversely proportional to the resonant frequency; this frequency for commonly used transducers having a diameter of 0.050 to 0.060 inch ranges from 100 to 1500 kHz depending on the pressure range (diaphragm resonant frequency increases with increasing pressure range). A resonant frequency of approximately 1200 kHz provides a rise time of one microsecond.

Piezoresistive transducers offer the advantage of dc (O Hz) operation, hence provide the capability to perform static measurements and yet have excellent dynamic behavior. This capability allows transducers installed in a model, which has been installed in the facility, to be calibrated in-place over the appropriate range of pressure. This calibration of the entire system (sensor, signal conditioning and recording) is usually performed by first evacuating the facility. The pressure level is increased in discrete increments and recorded via a precision pressure standard. The transducer output at each pressure level provides a calibration of the transducer. Because piezoresistive transducers are capable of static measurements, such transducers are used in both impulse facilities and hypersonic wind tunnels. Both absolute (i.e., with respect to vacuum) and differential (i.e., measured between two different levels) transducers are available.

To achieve accurate measurements, painstaking care must be exercised in the static and/or dynamic calibration and the installation. Caution must be exercised to avoid damage to the transducer; to avoid loading it such to influence its output; not providing adequate thermal or mechanical (vibration) protection; not providing a proper seal to avoid leaks around the gage; and maintaining a controlled and known reference pressure for differential transducers. With care and for pressure levels in excess of one psid, measurements accurate to within 5 percent are possible.

The miniaturization of piezoresistive pressure transducers provides the capability to measure detailed total pressure surveys in shock layers and in boundary/shear layers with survey rakes. The transducer is mounted, naturally, at or very close to the sensing end of the probe.

3.1.2. *Hypersonic Wind Tunnels*. Model surface pressures are measured in many hypersonic wind tunnels with either electronically scanned pressure (ESP) piezoresistive (silicon) sensors, or with high-volume, multi-range, variable-capacitance diaphragm-type transducers (e.g., Miller and Smith, 1986; Miller, 1990). ESP modules are commercially available with different pressure ranges and typically contain 16, 32, 48 or 64 sensors. The small size of the sensor modules allows for installation in the model or near the base of the strut to minimize tubing length (pressure lag) between the model surface and sensor. When installed externally to the model, modules are generally housed within an insulated box through which air may be passed to provide a thermally stable environment, if required. These modules combine pressure transducers, internal multiplexing, and amplification to provide a scanner for a high data rate. An integral, pneumatically controlled mechanism allows the transducers to be calibrated on-line. This on-line calibration consists of applying accurately known pressure (or vacuum) levels to the sensors (original ESP systems used a 3-point calibration whereas newer systems use 5 points); these levels are generally selected to cover the expected pressure levels for a run. Since calibration is performed using an absolute pressure standard, maintaining the same reference pressure between the calibration and the wind tunnel test will provide absolute pressure measurements. Such systems contain their own signal-processing and thus are interfaced directly to the controller (computer). Representative sampling rates for the ESP systems are 10/sec for each of 128 channels, or 5/sec for 256 channels.

The primary advantages of the ESP system over the variable-capacitance system are significantly lower (by about 1/5) cost per channel, miniaturization, and faster response. The primary disadvantages are limited range of application for a given module and a higher uncertainty in the measurement of low pressure levels, for example less than 0.1 psid.

As one example, most hypersonic wind tunnels at NASA Langley (Miller, 1990) have Pressure System Incorporated (PSI) model 8400 systems having 256 or 512 channels. Such systems operate well in pressure ranges associated with low speed (subsonic and transonic) wind tunnels, but modifications are required to meet the low pressure (vacuum) requirements of hypersonic wind tunnels (e.g., Hutchinson, 1991). These modifications, which include improvements to the vacuum and pneumatic system used in the calibration procedure, and the calibration standard, are discussed by Hutchinson, 1991. Also, sixty variable-capacitance transducers have been assembled into a "portable" system that may be moved from facility to facility. Most tunnels have 10 to 20 variable-capacitance transducers onsite. When using the ESP system, several tubes are often teed to variable-capacitance transducers to provide a check on the pressure level measured. Similarly, when using multiple ESP modules, transducers from different modules are teed together to determine the level of agreement (or disagreement) between modules.

As mentioned previously, piezoresistive pressure transducers provide the capability to perform static measurements and have good dynamic characteristics. Thus, the same pressure transducers used in impulse facilities are also used in hypersonic wind tunnels. (As a rule of thumb, manufacturers recommend the selection of a transducer having a resonant frequency that is five times the frequency of interest, i.e., the useful frequency range is linear (within ±5 percent) for 20 percent of the resonant frequency.) Their fast response characteristics are amenable to the study of hypersonic internal (e.g., inlets) and external flow separation phenomena, and these transducers are often used in traversing flow field survey probes.

Complex models tested over a range of Reynolds number and attitude provide a wide range of surface pressure. One disadvantage of ESP systems is they are not multi-ranged. Thus, to accommodate a large surface pressure range, modules must be changed or possibly teed together provided they can withstand the overloads. Primarily for this reason, a pressure measurement system has been developed that employs a number of miniature piezoresistive pressure transducers with different ranges to appear as a single transducer. The different range pressure transducers are connected to a comparator circuit which compares the output from the transducers, determines which gage is reading closest to full scale, biases the signals from the other transducers, and provides a continuous single output so to appear as a single transducer. This system, manufactured by Dynamic Engineering Incorporated and referred to as the ATS-1 Pressure Measurement System, is presently being evaluated in the NASA Langley Unitary Plan Wind Tunnel.

Heavily instrumented pressure models for conventional-type hypersonic wind tunnels require a relatively long time to design and fabricate (often in excess of 1 year), are thus quite expensive, challenging to fabricate to high fidelity primarily due to the tubing process, and often require considerable setup time in the wind tunnel (corresponding to considerable tunnel down time). Thus, it is not surprising that there is considerable interest in the hypersonic community concerning recent advancements in optical pressure measurement systems (OPMS) (e.g., DeMeis, 1992) based on photoluminescence which includes both phosphorescence (e.g., Kavandi, et al., 1990) and fluorescence (e.g., Morris, et al., 1992). Such systems avoid many of the previously mentioned disadvantages associated with conventional pressure model fabrication and tunnel setup while providing the significant advantage of global two-dimensional pressure mappings.

As discussed by Morris, et al., 1992, models are coated with a material that photoluminesce when excited by the appropriate light source. A probe molecule is excited via the absorption of a photon having the appropriate energy. Upon returning to the ground electronic state by emitting a lower energy photon, the molecule photoluminesce. (Fluorescence is the emission of light generally within a time on the order of 10^{-8} sec; in contrast, phosphorescence is a delayed emission generally within 10^{-3} to 100 sec.) An alternate transition to the ground state is provided by a collision with an oxygen molecule. Rather than emitting a photon, the excess energy of the molecule is absorbed by the oxygen during a collisional deactivation. As the number of oxygen molecules increases in a given volume, the frequency of collisional deactivation increases and lowers or quenches the luminescence. Thus, for a coated model, the greater the static air pressure (i.e., the greater the oxygen concentration), the less the phosphorescence or fluorescence. Advances in image processing hardware and computers now allow the concept of oxygen quenching to be extended to the quantitative measurement of pressure. Digital cameras with a scientific-grade CCD array having 16 bits of resolution and 4 million pixels can be used to measure luminescence. Computers can quickly reduce and store the large quantities of data generated by such cameras and image processors can be used to analyze and display the data.

The development and/or application of OPMS (e.g., University of Washington - NASA Ames Research Center (Kavandi, et al., 1990); McDonnell Douglas Research Laboratory (MDRL) (Morris, et al., 1992); and Central Aerohydrodynamics Institute (Ts AGI) of the Commonwealth of Independent States (marketed by Inteco, an Italian Soft- ware firm) (Vollon and Loreto 1991; and Engler, et al., 1992)) has provided new diagnostic capability to the lower speed (subsonic to supersonic) aerodynamic community via successful demonstrations in subsonic and supersonic wind tunnels and in flight. The extension of this technique to hypersonic wind tunnels using air as the test medium will require compensation for temperature effects. The oxygen quenching process is temperature dependent and the properties of the coating may also vary with temperature, hence time. The technique developed by MDRL has recently been tested in the NASA Langley 12-Inch Mach 6 High Pressure Tunnel and these data are presently being analyzed.

3.2. FORCES AND MOMENTS

3.2.1. *Impulse Facilities*. A few general comments will be made prior to discussing the four papers presented at the subject workshop. Most force transducers have an elastic sensing element whose deformation provides a measure of the acting force. Aerodynamic forces in impulse facilities are generally measured with: (1) piezoelectric force transducers, where the sensing element is the same as the transduction element that produces the electrical output signal from the acting force, (2) piezoresistive or semiconductor strain gages, and/ or (3) conventional foil strain gages. Generally, three component and six component balances are used in impulse facilities having useful run times of approximately 3 to 10 millisecond (i.e., excluding facility flow establishment time). These balances may employ one or a combination of the measurement techniques (i.e., piezoelectric, piezoresistive, or foil strain gage). Aerodynamic force measurements in impulse facilities are not straightforward and may require additional (typically 6) accelerometers on the model for inertial corrections (i.e., the effect of model vibration) to the balance output. The natural frequency of the model material is also an important consideration, and the model must be dynamically calibrated. Jessen and Gronig, 1992 provide an excellent discussion of the rationale for the design and construction of a balance for use in shock tunnels having run times of several milliseconds. A classical review of force and moment measurements in impulse facilities is provided by Bernstein, 1975.

A new technique that provides the capability to measure aerodynamic forces for sub-millisecond test times has been developed and demonstrated by Simmons, et al., 1992 for nonlifting bodies at zero incidence. This technique is based on the interpretation of transient stress waves propagating within the model and support. Sharp cones, having half angles of 5° and 15° (lengths of 7.9 and 16.7 inches respectively) and fabricated from aluminum, were attached to a relatively long (80 inches) brass sting having an outside diameter of 1.5 inches. The sting, hence model, was suspended in the test section of the University of Queensland piston-driven Shock Tunnel T4 via two strings; hence the model was free to move axially although the actual movement is negligible during a run. To protect the sting from external forces due to shear layer impingement and the wake flow in general, it was enclosed in a shroud. Two strain gages were attached to the sting 180° apart. Upon establishment of the flow about the model and during the useful test period, a large number of stress wave reflections occur within the model. The period of these reflections is an order of magnitude less than the test time. The model reacts against the mass of the sting and stress waves propagate along the sting and reflect from the end. The resulting outputs of the strain gages mounted on the sting, which are sampled every 5 microseconds, are averaged to cancel bending effects and used to provide a time history of the drag. This time history of the drag is determined numerically via the deconvolution process (inverse dynamic problem) and follows the pitot pressure time history (i.e., decays with time). The rise time for this technique, which is shown to be insensitive to the assumed spatial and temporal loading distributions on the model, is approximately 100 microseconds. A technique that extends this capability to include measurements of lift and moment is under study and discussed by Simmons, et al., 1992. Future plans include the utilization of this suspended model technique to measure the thrust of a combustor. The exhaust of a combustor would be captured by a nozzle having parallel side walls and 11° diverging top and bottom walls. This nozzle would be attached to stings and suspended via strings as discussed previously, thus be free to move relative to the fixed-position combustor. The impulsive pressure loading on the inner walls of the nozzle will send stress waves along the stings where they are monitored by strain gages. A positive thrust will move the nozzle closer to the combustor.

An excellent discussion of the design rationale for a new geometry, six-component balance for use in a shock tunnel having a 4 millisecond flow duration is presented by Jessen and Gronig, 1992. The primary driving factors in the design of this balance, which utilized finite element analysis, were test time, balance natural frequency, and model moment of inertia. Based on a geometry involving connecting cross arms having quadratic cross sections and being conical as the arms extend outward from a center inner plug to a 1.65-inch-diameter outer shell, the resulting balance is quite short (length is 0.87 inch without the model mounting adaptor). Semiconductor strain gages are attached to each side of each arm, resulting in a total of 16 gages. Although challenging to fabricate, the balance is easy to handle, robust, and quite accurate. The minimum natural frequency of the balance without the model is 2.4 kHz and all six force and moment components are measured without acceleration compensation. A third-order static calibration involving 500 load cases was performed with the balance. (The minimum number of individual load cases required is 400. One thousand cases are required for good results and 1500 for excellent results.) This new balance has been successfully demonstrated via tests at Mach 8 in the RWTH Aachen Shock Tunnel for a 15° half angle sharp cone and a 30° half angle sharp cone with a cylindrical afterbody. Future plans include compensation for sting oscillation, testing in a contoured (as opposed to conical) nozzle that will provide uniform flow, and the design of an advanced proto-type involving calibration with 1500 load cases.

A novel device for free-flight testing a model in a conventional-type shock tunnel is the subject of the paper by Naumann, et al., 1992. This technique is insensitive to disturbances resulting from the flow establishment process of the facility and compensates for variations in flow parameters during the useful test period. The model, instrumented with piezoelectric accelerometers, is positioned in the test section by a model support system consisting of a clamping device. This clamp is closed about the sting attached to the model thereby fixing the position of the model in the test section. Just prior to flow arrival at the test section, pyrotechnic charges are ignited to open the clamp, thereby allowing the model and sting to free flight. Following the useful test period, the clamp closes on the sting thereby securing the model. A minimum of 6 accelerometers (more for redundant information) are used to infer the forces and moments, and wires from these accelerometers are routed through the sting for protection. Aerodynamic coeffi-cients are calculated directly from the measured accelerations and the pitot pressure time history. This technique has been demonstrated for a 30° half-angle sharp cone with cylindrical afterbody model having 8 symmetrically mounted accelerometers to provide 5 components. This "release-capture" force measurement technique has also been demonstrated with a body-plate model having side jets. The body houses two pyrotechnic charges, a settling chamber, and 1 to 3 parallel nozzles which may have different shapes. The pyrotechnic charges provide a maximum settling chamber pressure of about 7000 psi. All accelerometers are mounted on this body. The plate-wing, together with the upper surface of the body, provides a flat surface that is 5.5 inches in length and aligned with the oncoming flow; this plate-wing is exchangeable. The forces imparted to the body upon discharge of the pyrotechnics are measured via accelerometers for the case of no flow and for the test flow. These tests revealed that the 3 millisecond test time of the ISL Shock Tunnel was just sufficient to establish quasi-steady flow; i.e., the limiting factor is the establishment of the complex flow structure, not the dynamic response of the model. Future plans include testing with reservoir-supplied side jets and measurement of lateral force due to external combustion. Also planned are scramjet model tests with a built-in supply.

Although having a longer run time (10 to 20 milliseconds) than piston-driven or conventional-type shock tunnels, the VKI Longshot Hypersonic Facility nevertheless presents experimental aerodynamicists with a challenge in obtaining accurate measurements of forces and moments. The techniques developed for and used in the longshot facility at hypersonic Mach number of about 15 are discussed by Carbonaro 1992. In this report by Carbonaro, the techniques and problems associated with the measurement of model accelerations using piezoelectric or piezoresistive accelerometers, including calibration and transverse sensitivity aspects, are discussed. Inertial corrections are optimized taking into account accelerometer transverse sensitivity effects, time lags, signal filtering or smoothing, and tuning for the inertia of the vibrating masses. Low frequency oscillations due to sting vibration are well corrected, but high frequency oscillations must be damped by suitable numerical filtering or smoothing. The viability of the subject measurement technique is demonstrated via tests of the shuttle orbiter and Hermes configurations.

3.2.2. *Hypersonic Wind Tunnels.* Internal strain gage balances used most frequently in hypersonic wind tunnels are sting or blade strut supported, range in diameter from 0.56 to 0.85 inch, and cover a wide range of maximum design loads applicable to blunt, high drag models and to slender, high lift models (e.g., Miller, 1990). Most balances are six-component, water cooled, and the excitation voltage is typically 5 volts. The general procedure used in most tunnels is to calibrate all six-components, or as a minimum the axial-and-normal-force and pitching moment components, of the balance by loading the balance with weights and recording the output. These onsite data are compared to the formal calibration data to assure the balance is operating within acceptable tolerances. Tare runs with the model mounted to the balance are performed over the desired angle of attack range prior to and during the test series. Following the establishment of steady flow in the wind tunnel, the model is injected into the flow from a sheltered position, the angle of attack varied via the pitch-pause model of operation, and the model retracted prior to flow break-down. The model injection, attitude variation, and retraction is generally controlled by a computer.

As mentioned previously, a major concern with force and moment tests in heated flows of relatively long duration is the potential conduction heating to the balances through the model wall and along the sting or strut; therefore, only water cooled balances are used in such facilities. Conduction of heat along the sting or blade strut is generally reduced via water cooling and/or ceramic coating; for severe heating problems, a water-cooled balance-to-sting adapter may also be used. Protecting the balance from conduction effects through the model wall is usually achieved with passive techniques such as insulated balance sleeves and air gaps. Whenever model weight is not a consideration, stainless steel is selected over aluminum as the model material so to reduce conduction effects. Force and moment models cast of ceramic or machined from Macor, a machinable glass ceramic manufactured by Corning Glass Incorporated, have been made to reduce conduction effects.

3.3. DATA ACQUISITION SYSTEMS

3.3.1. *Impulse Facilities.* Significant advances have been made in the last decade in high frequency waveform recording systems. Commercially available analog-to-digital (A/D) systems typically are 12 bit, provide sample rates in excess of 2 MHz and have in excess of 16K of memory per channel. Such systems provide data acquisition and reduction capabilities for impulse facilities that are not unlike those for conventional-type hypersonic wind tunnels.

The cost of such systems is relatively high, typically \$5K to \$10K per channel, depending on the capability (bit, sample rate, memory). Thus, the number of channels at various facilities may run from only a few channels to the 128 enjoyed by the Calspan Shock Tunnel.

3.3.2. *Hypersonic Wind Tunnels.* Again, as an example, all hypersonic wind tunnels at NASA Langley have stand-alone data acquisition systems. The heart of the system for most tunnels is a 256-channel, 15 bit, 50 kHz or 100 KHz through-put rate, amplifier per channel, analog-to-digital (A/D) unit manufactured by the NEFF Instrument Corporation and having programmable gains and filters per channel and internal clock (NEFF system 620/series 600). These systems are controlled via a Hewlett Packard (HP) 9000 series model 375 computer networked to a central HP model 750 computer. Peripheral equipment for each DAS includes dual-disc and optical disc mass storage units, 9-track tape systems, and printers. The recent addition of a portable 20-channel NEFF system 490 having 12-bit A/D, up to 1 MHz sampling rate, programmable amplifier filter and 2 megaword RAM provides the capability to routinely utilize fast response pressure transducers to examine unsteady surface flow phenomena.

4. REFERENCES

Bernstein, L., (1975) "Force Measurements in Short Duration Hypersonic Facilities," AGARDogr aph 214.

Carbonaro, M., (1992) "Aerodynamic Force Measurements in the VKI LongShot Hypersonic Facility," NATO Advanced Research Workshop, LeFauga, France.

DeMeis, R., (1992) "Paint Under Pressure" Aerospace America, pp. 40-41, March 1992.

Engler, R. H., et al., (199), "Aerodynamic Assessment of an Optical Pressure Measurement System (OPMS) by Comparison with Conventional Pressure Measurements in a High Speed Wind Tunnel."

Hutchinson, M., (1991) "Langley Research Center's Hypersonic Wind Tunnel Pressure Instrumentation," proceedings of the 37th International Instrumentation Symposium. Instrument Society of America paper 91-074.

Jessen, C., and Gronig, H., (1992) "A Six Component Balance for Short Duration Hypersonic Facilities," NATO Advanced Research Workshop, LeFauga, France.

Kavandi, J., et al, (1990) "Luminescent Barometry in Wind Tunnels," Rev. Sci. Instr., Vol. 61, No. 11, pp. 3340-3347, November 1990.

Miller, C. G., and Smith, F. M., (1986) "Langley Hypersonic Facilities Complex - Description and Application," AIAA 86-0741-CP.

Miller, C. G., (1990) "Langley Hypersonic Aerodynamic/Aerothermodynamic Testing Capability - Present and Future," AIAA 90-1376.

Moore, J. A., (1979) "A Study of Response Time of Pitot Pressure Probes Designed for Rapid Response and Protection of Transducer, NASA TM 80091.

Morris, M. J., et al, (1992) "Aerodynamic Applications of Pressure-Sensitive Paint," AIAA 92-0264.

Naumann, K. W., Ende, H., and Mathieu, G., (1992) "Millisecond Aerodynamic Force Measurement Technique for High Enthalpy Test Facilities," NATO Advanced Research Workshop, LeFauga, France.

Rogers, R. C. (edited by) (1990) "Workshop on the Application of Pulse Facilities to Hyperveloci ty Combustion Simulation," NASP Workshop Publication 1008. Restricted distribution.

Schultz, D. L,; Jones, T. V., (1973) "Heat-Transfer Measurements in Short-Duration Hypersonic Facilities," AGARD AG-163. February 1973.

Simmons, J. M., Daniel, W. J., Mee, D. J., and Tuttle, S. L., (1992) "Force Measurement in Hypervelocity Impulse Facilities," NATO Advanced Research Workshop, LeFauga, France.

Vollon, A., and Loreto, A., (1991) "A New Optical Pressure Measurement System (OPMS)," 14th ICIASF Congress, Rockville, MD, USA, October 27-31, 1991.

VELOCITY MEASUREMENTS IN HYPERSONIC FLOWS: A REVIEW

Alain BOUTIER
ONERA - BP 72
92 322 - Châtillon CEDEX - FRANCE

William J. YANTA
NSWC - 10901 New Hampshire Avenue - White Oak
Silver Spring - Maryland 20903 - 5000 - USA

Gunter SMEETS
ISL - 5 rue de l'Industrie - BP 34
68 301 - Saint Louis - FRANCE

ABSTRACT. An analysis of the different means to measure the velocity is done; the relative merits of the particle based and the molecular based methods are evaluated, leading to a set of recommendations for future research.

1. Introduction

With the renewed interest in hypersonics, as seen by the large number of flight programs in the international community, there is a real need for accurate, detailed measurements of fluid flow and thermodynamic properties of high speed, high temperature flows. Since the last major hypersonic activities took place in the late 1960's, major strides have taken place both in computational fluid dynamics (CFD) and experimental measurements. For the CFD community these include large high speed computers with efficient algorithms. These new techniques now make it possible to make accurate predictions of very complex flows. This requires accurate global data for the calibration/verification of these complex codes. For the experimentalist, new tools include high powered lasers, miniature electron beams, fast digital cameras, high speed image processing hardware and software, fast desktop computers, paints that are temperature, pressure and shear sensitive. All of these techniques have provided the experimentalist with a vast assortment of means of obtaining nonintrusive measurements of a variety of fluid properties. These new diagnostics can either replace or complement the traditional physical probes which have numerous problems such as probe interference, lag corrections and the requirement for subsidiary cooling systems. This paper is concerned with the review and summary of the papers presented in Session 5 (Velocity Measurements) at the NATO Advanced Research Workshop [1]. Eight papers were presented ([2] to [9]) in that session. Although this paper will focus on the results from those papers, other results will be described for completeness.

2. Background

The choice of velocity measurement systems is governed by a variety of parameters. These include: whether the experiment is a flight test or takes place in a ground test facility, is the test duration a few milliseconds or a few seconds, are real gas properties being measured, is the experiment carried out in a small research facility or is the test being made on a component of a full scale vehicle.

Velocity measurement techniques are generally broken down into two categories: particle based and molecular based. Both of these categories can be used to make either point or planar measurements; they can also be time-of-flight or Doppler shifted methods.

A. Boutier (ed.), New Trends in Instrumentation for Hypersonic Research, 593–602.

Particle based velocimetry has a 30 year history of development which has made it a very important diagnostics. Particle based systems generally depend upon micron size particles that have been placed in the flow and the Doppler shifted light scattered by these particles is measured to determine the velocity of the particle and, hence, the velocity of the fluid in which the particle is immersed. The Doppler shift has been directly got with optical spectrometers (Michelson and Fabry-Pérot) or through set-ups creating interference of two Doppler shifts yielding beat frequencies that can be detected by photomultiplier tubes and which are further processed with a "counter" type of electronics or the more recently developed FFT type of devices. These systems are generally referred to as Laser Doppler Velocimeters (LDV). Variants of this system measure the time for the particle to travel a known distance, two spot or two dash velocimeters sometimes called laser transit velocimeters; or, measure the distance the particles have gone in a known time, Particle Image Velocimetry (PIV). The problem of all the particle based systems is that the effect of the particle's inertia must be seriously considered. In low density flows or wherever there may be large velocity gradients, the particle's velocity may not accurately represent the flow velocity. The kind of particle chosen for the experiment is of vital importance. This will be described in more detail in the following section.

A spectrometric technique using a Michelson spectrometer is described in [10]. The velocity can be recorded in real time and with a μs time resolution provided the scattering particles are numerous enough for creating a continuous stream of scattered light from the measuring volume. As the Doppler shifted light is delivered by an ensemble of may be many particles, a good signal to noise ratio can be obtained also with rather small particles resulting in less problems from particle inertia. The technique is of special interest for impulse facilities.

Molecular based velocity measurements systems have not had the broad development efforts that the particle based systems have had. A small number of researchers have used the electron beam and laser systems to measure velocities. It is only in the last five years that molecular based systems have been developed to make accurate measurements in "lab-top" experiments. Again, these systems fall into two categories: time-of-flight and Doppler shifted. These will be discussed in more detail later.

3. Scattering particles

As was mentioned above, the choice of scattering particle for hypersonic measurements is very important. The 1985 conference held at NASA Langley [11] was devoted to various aspects of the use of aerosols for velocity measurements. In [4], which describes the application of the classical "fringe" type LDV system to several of ONERA's wind tunnels, a variety of particles have been used, including ZrO_2, SiO_2, alumina powder, and natural particles occurring in the flow system. The powders were generally fed through a cyclone generator to remove large particles or agglomerates. The nominal size for the ceramic powders was about 0.5 μm and their specific gravity varied from 2.2 to 5.5 g.cm^{-3}. Their melting temperatures vary from 1700°C to 2700°C, thus making them quite suitable for high temperature flows. Since these powders were maintained at the same pressure as the supply pressure of the wind tunnel, this generally required a complex plumbing system to introduce the aerosols into the flow. Tests were also carried out in an arc facility where naturally occurring particles were present. It is ascertained that the particles were probably copper or copper oxide particles with diameters between 15 and 30 μm. In [2], particles are used which when compared to predictions, appear to have a diameter of 0.5 μm and a specific gravity of 0.1 g.cm^{-3}. Results with these particles will be described later in this paper. No description is given as to the type of particles or the nature of the particle material. In [3] a fluidized bed was used to disperse Alumina (Al_2O_3) with a mean diameter of 0.3 μm into a Mach 6 wind tunnel supply system. All of the above systems were used in continuous running wind tunnels. For the case of impulse facilities, Avidor [12] used exploding aluminum wires in the supply section of a shock tunnel to generate aerosols. The wires were exploded shortly before the tunnel was run, providing a continuous source of scattering particles during the multi milli-second run. It was estimated that the particle size was less than 0.02 μm so this technique could not be used with a conventional LDV system for single particle scattering. If the gas temperatures are not significant (less than 250°C) then some of the organic aerosols such as polystyrene, latex, dioctylphthalate and other aerosols that have been in use over the years could be used.

4. Particle based results

Some of the significant results from [2] to [9] will now be discussed. In [3], both laser transit anemometer and LDV measurements were made in a Mach 6 wind tunnel. Velocity measurements of the normal velocity component using agglomerate free 0.3 micron alumina normalized to free stream reference velocity, with theoretical predictions for different particle sizes as a function of position behind a 40.8° oblique shock are shown in Figure 1. The effects of particle lag are seen in the figure. The measured velocity values converge to within 8% of the expected 133 m/s 8 millimeters behind the shock. The values were within 5% of the 133 m/s 12 millimeters behind the shock. Shown in Figure 2 are the results from [2]. An oblique shock wave is generated by a flat plate model at -10° incidence in a Mach 12.5 air flow. The measurements, which were made above the plate, about 125 mm from the leading edge, are shown in Figure 3. The probability density measurements shown, clearly point out the shift from the dominant freestream peak ahead of the shock as the probe volume is traversed towards the plate. It was concluded that the measured particle response indicates a particle of 0.5 μm with a specific gravity of 0.1: theoretically these particles have a 95% response to velocity fluctuations up to 50kHz. In [4] velocity results from three of ONERA's wind tunnels are presented. Specifically, the results from S4 Modane are shown. The Mach number was 12 and three different electronic processing systems were used: a digital oscilloscope Lecroy 7200, a Dantec counter and the IFA 750 from TSI. The results are shown in Figure 4. The measured results lie between the calculated values for a perfect gas and a real gas. It is concluded in [4] that for these conditions, the results from the different processors correlated well with each other and that velocities up to 1 700 m/s could be measured with a good confidence level.

5. Molecular based systems

As was said above, molecular based velocity measurements can be categorized into two groups: time-of-flight and Doppler shift techniques. We will first discuss the time-of-flight methods.

A pulsing electron beam is utilized in [5] (shown schematically in Figure 5) to determine the velocity in a hypersonic flow. The electron beam generates ions with long lifetimes. One can then measure their time of flight during their travel downstream among the surrounding neutral molecules along a known distance. The beam is pulsed for 1 μsec at a repetition rate of 2-8 kHz. Downstream in the flow an ion collector acting like a Langmuir probe is positioned. From the time delay of the ion signal at the probe with respect to the corresponding electron beam pulse the time-of-flight of the ions is measured. When the probe is moved a well defined distance downstream or upstream the time-of-flight is measured again. The velocity may then be computed readily. Shown in Figure 6 are the results of measurements in a variety of gases at many different speeds. The results are compared with theoretical isentropic expansion velocities. One sees that the results are in good agreement. A slight variant of this velocity measurement technique was described in [6]. An electron beam was again used to excite the gas molecules in a low density, Mach 10 airflow. A lens (see Figure 7) forms the image of the flowfield on a high speed gated image intensifier. The final image is recorded on a 400 ASA sensitive photographic plate. The image intensifier has two purposes, the first one is to act as high speed shutter in order to freeze the movement of the plasma columns; the second is to use the high gain to amplify the low intensity light coming from the plasma columns. A photograph of the plasma is shown in Figure 8. One effectively knows the time-of-flight and one determines the distance from the scaled photograph; thus, one can deduce the velocity. It was concluded in [6] that the optical spatial imaging of luminous plasma columns induced by a pulsed electron beam can be a useful tool to measure the velocity of low density hypersonic flows. The last time-of-flight paper [8] describes results obtained with the RELIEF method. Air is tagged by vibrationally exciting oxygen molecules and interrogating at a later time by laser-induced electronic fluorescence. Lines can be instantaneously (10ns) written into the flowfield, and their displacement after an accurately measured time interval gives the flow velocity. If a cross is written into a flow, then the displacement of the cross gives an accurate measure of the mean flow velocity; rotation of the cross and distortion of the lines can be used to have access to vorticity and turbulence. Experiments have been carried out in flows up to Mach 4 and the results look very promising, but the technique requires a shorter wavelength laser source in order to generate high fluorescence intensities to low density hypersonic flows.

For steady, nonreacting flows, planar velocity measurements can be made using planar laser-induced iodine fluorescence (PLIIF)[7]. Seeding the flow with iodine vapor and then exciting these molecules with a light sheet of the 514.5nm line of an argon laser with an air-spaced etalon, the resulting broadband planar fluorescence is imaged into a CCD camera (Fig. 9). Time averaged data (over 30 s) are then taken. By scanning the laser frequency of discrete frequencies, a series of images is obtained from which the Doppler shift is deduced. For application to high-speed reacting flows, velocity measurements can be made using planar laser-induced OH fluorescence. The same approaches that were used in PLIIF can be used for OH PLIF if a sufficiently narrow linewidth, tunable, UV laser is used. Pointwise measurements of velocity have been demonstrated in a small, combustor, jet facility. Planar velocity measurements have also been demonstrated in a reacting jet.

Velocities and temperature measurements in high speed flows which were based on spectrally resolved Rayleigh scattering were described in [9]. Time averaged measurements were carried out in a Mach 1.3 air jet. The experimental set-up is shown in Figure 10. The central item in the image processing system is a planar mirror Fabry-Pérot interferometer. The results are shown in Figure 11. The maximum velocity is about 422m/s with a maximum error of about 10-15m/s.

A new approach in time-domain CARS is proposed [13] to perform single shot measurements of the velocity (as well as of the static temperature) in low pressure supersonic flows. Instead of repetitive probing the macroscopic polarizibility coherently driven by short pump and Stokes pulses with variable delay times, the coherence is created with a long duration pump and a short duration Stokes pulse, and is continuously probed with the same pump pulse. The decay of the anti-Stokes signal reveals the static temperature; the beating between two signals which are created in opposite crossing directions with respect to the flow direction gives the flow velocity.

6. Conclusions and recommendations

The results from the papers presented in the Velocity session at the NATO workshop indicated that major strides have been made in being able to make velocity measurements in high speed, high temperature flows that may have chemical reactions present. It is also obvious that much work still remains to be carried out. Most of the experiments described in the above papers were carried out in carefully controlled, small scaled laboratory environments. Only a small number were attempted in large scaled test facilities; none were attempted in flight vehicles.

The following recommendations are offered for future research:

* Continued improvement in particle techniques for use with LDV systems. The development for particles with large aerodynamic diameters with low specific gravity, such as micro spheres, is important. The continued development of electronic processors capable of measuring frequencies with a range of 100's of megahertz is also important.
* A continued emphasis on molecular based techniques including both time-of-flight and Doppler techniques.
* Attempts to apply the Michelson spectrometer to molecular based Doppler measurements: recording the wavelengths of spectral lines stimulated by electron beam or laser radiation.
* The development of high powered, high repetition rate UV lasers.
* In order to supplement turbulence modeling, particularly in the analysis of high speed mixing, the density-velocity correlation should be emphasized. This will require multi-parameter measurements from such techniques as CARS velocimetry, LIF and the electron beam.
* Planar techniques in blow down facilities.

7. References

[1] NATO Advanced Research Workshop - New trends in instrumentation for hypersonic research - ONERA Le Fauga-Mauzac - 27 April-1 May, 1992.

[2] F.K. Owen - Measurement of hypersonic flows [1].

[3] W.W. Hunter Jr., L.R. Gartrell, W.M. Humphreys Jr., D.W. Witte, C.E. Nichols Jr.- Development and assessment of an LDV system for Mach 6 flow field measurements [1].

[4] A. Boutier, J. Lefèvre, F. Micheli - Laser velocimetry applied in hypersonics [1].

[5] F. Hiraï, K.A. Bütefisch, C. Dankert - Velocity and density determination by electron beam technique [1].

[6] A.K. Mohamed - Electron beam velocimetry [1].

[7] J.C. McDaniel, S.D. Hollo, K.G. Klavuhn - Planar velocimetry in high-speed aerodynamic and propulsion flowfields [1].

[8] R. Miles, W. Lempert, J. Forkey, B. Zhang, D. Zhou - Filtered Rayleigh and RELIEF imaging of velocity, temperature and density in hypersonic flows for the study of boundary layers, shock structures, mixing phenomena and the acquisition of in-flight air data [1].

[9] R.G. Seasholtz - 2D velocity and temperature measurements in high speed flows based on spectrally resolved Rayleigh scattering [1].

[10] G. Smeets, A. George - Michelson spectrometer for instantaneous Doppler velocity measurements. J. Phys. E: Sci. Instrum., Vol.14, 1981.

[11] W.W. Hunter, C.E. Nichols - Wind tunnel seeding systems for laser velocimeters - NASA Conference Publication 2393, Proceedings of a workshop held at NASA Langley research center, March 19-20, 1985.

[12] J.M. Avidor, A.M. Schneiderman - Experimental investigation of high Reynolds number compressible axisymmetric turbulent wakes - AIAA paper 74-576, presented at AIAA 7th Fluid and Plasma Dynamics Conference, Palo Alto, CA, June 17-19, 1974.

[13] M. Lefebvre, M. Péalat, J. Strempel - Single-shot Time-Domain CARS: application to temperature and velocity measurements in supersonic flows - Submitted to Optics Letters.

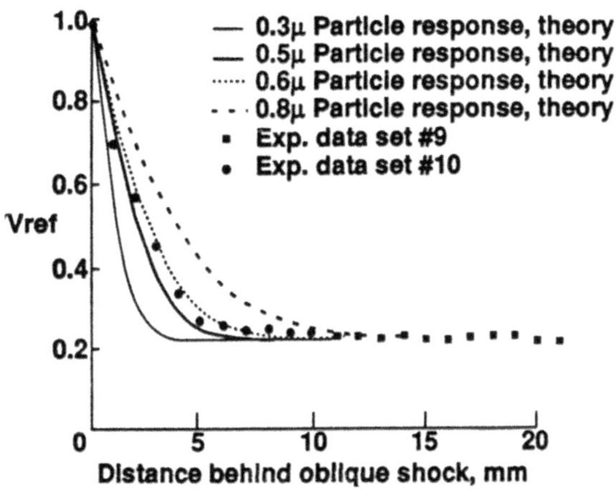

Fig.1: Comparison of LDV normal velocity component measurements using agglomerate free 0.3 μm Al_2O_3, normalized to freestream reference velocity, with theoretical predictions for different particle sizes as a function of position behind az 40.8° oblique shock [3].

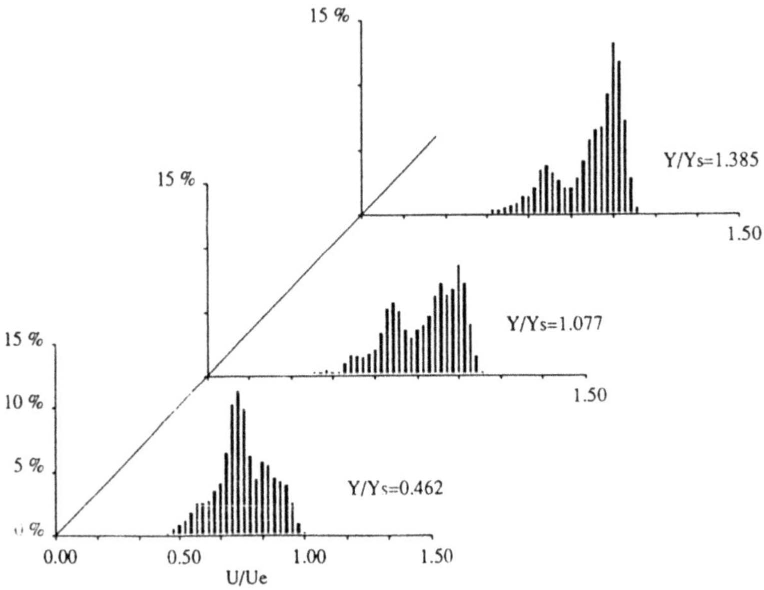

Fig.2: Particle response in hypersonic flow [2].

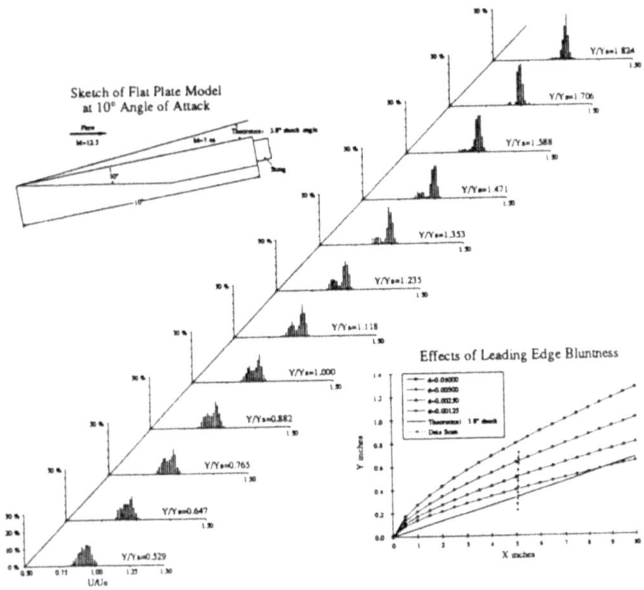

Fig.3: Particle response in the Wright Laboratory 20-inch hypersonic wind tunnel [2].

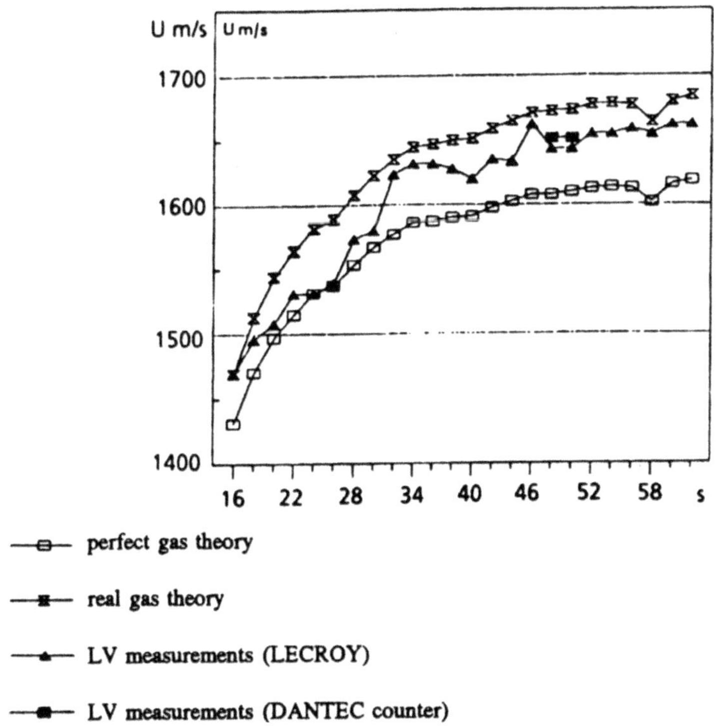

Fig.4: Velocity evolutions during run 1015 at S4 Modane [4].

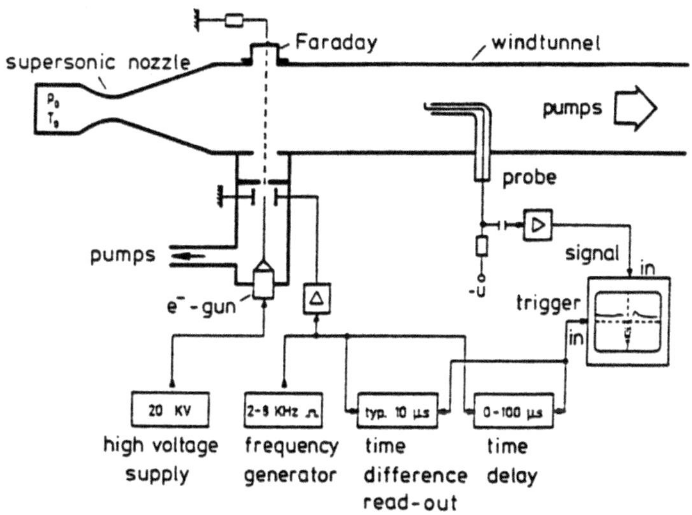

Fig.5: Experimental set-up for velocity measurements by means of a pulsed electron beam [5].

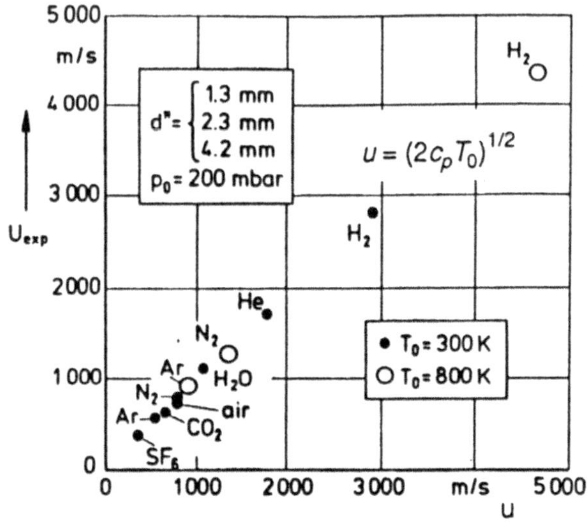

Fig.6: Measured flow velocities for different atoms and molecules; d* = orifice diameter [5].

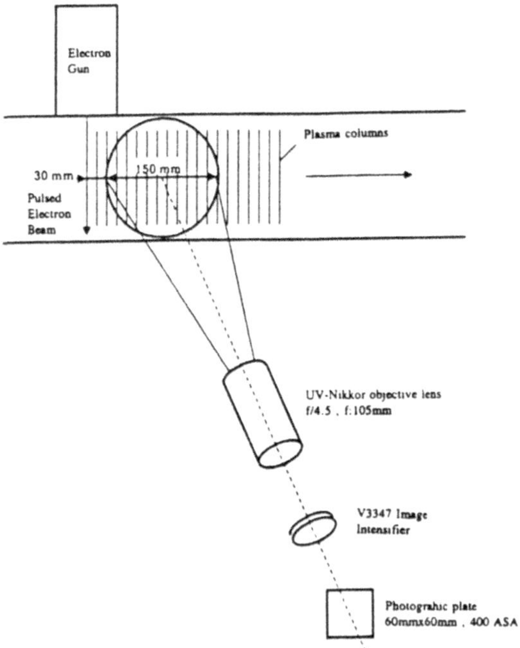

Fig.7: Experimental set-up [6].

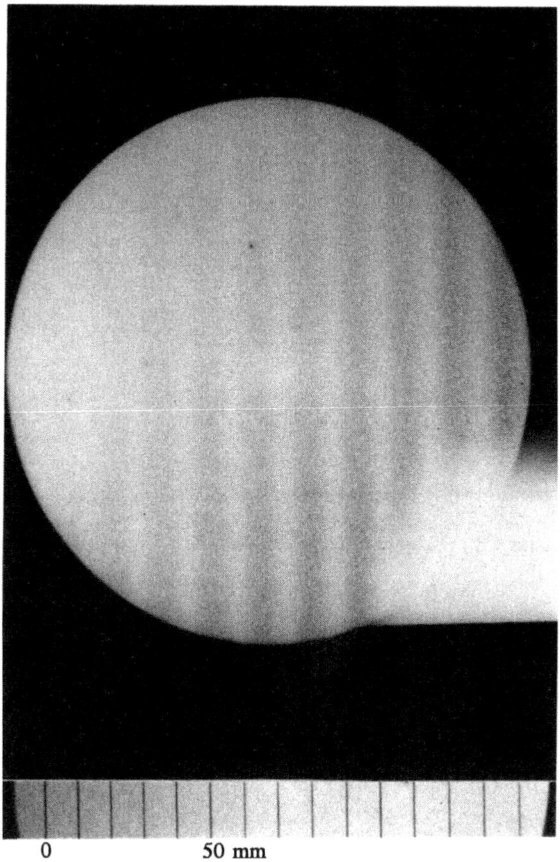

Fig. 8: Plasma columns in the free flow [6].

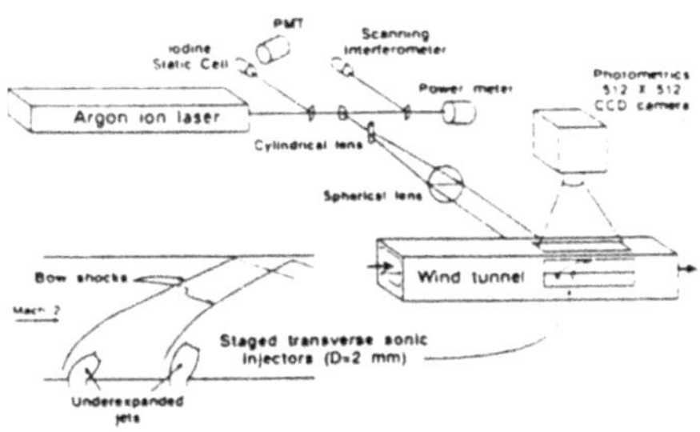

Fig. 9: Experimental setup for PLIIF velocity measurements [7].

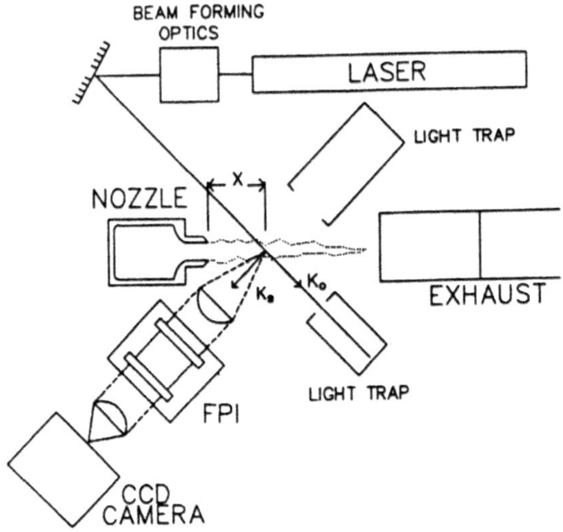

Fig.10: Experimental setup for measurement of velocity in Mach 1.3 free jet using imaging Fabry-Pérot interferometer [9].

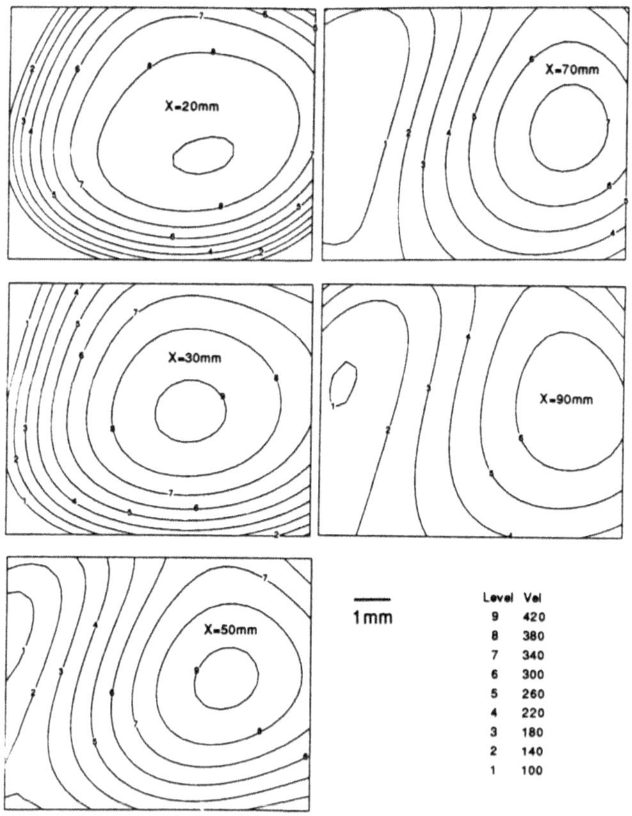

Fig.11: Measured constant axial velocity contours.
Distances from nozzle exit are X = 20,30,50,70,90 mm. Isentropic velocity = 422 m/s [9].

THERMAL FLUX MEASUREMENTS IN HYPERSONIC FLOWS - A REVIEW -

J.F. Wendt
von Karman Institute for Fluid Dynamics
Chaussée de Waterloo, 72
B-1640 Rhode Saint Genèse-Belgium

D. Balageas
O.N.E.R.A.
BP 72
F-92322 Châtillon, Cedex-France

R.D. Neumann
The Univerisity of Dayton Research Institute
300 College Park
Dayton, Ohio 45469, USA

ABSTRACT. A summary of the collection of papers from Session 6A "Heat Flux " and Session 6B "Infrared Thermography" is presented. The present status and problem areas associated with distinct methods are discussed and recommendations for future research and development are presented.

1. Introduction

One of the authors (R.D.N.) opened this session with a paper (6A) containing a statement which summarizes the reality faced by the designers and users of instruments to determine heat flux :
"The fundamental problem is that you and I would like to hear broad, sweeping, universal truths about thermal sensors, but the subject is dominated by the details of the particular application".

The tone of the session having been set by the above remark, the remaining presentations from a number of large national aerospace laboratories as well as a few smaller establishments touched on virtually all techniques to measure heat flux on models in hypersonic flows; the sole exception was the liquid crystal technique.

Clearly, the choice of technique by the user will depend, among other requirements,

A. Boutier (ed.), New Trends in Instrumentation for Hypersonic Research, 603–613.

on the objective of the experiment. Measurements are carried out in wind tunnels to provide detailed aerothermal design data where the need exists for a large quantity of data to be gathered on numerous complex shapes. Wind tunnel measurements are also carried out to validate or calibrate codes and in this case the spatial (and even temporal) resolution of the measurement technique is much more important than the quantity of data gathered. Flight measurements, likewise, generate or provide a check on design data; an equally important task on an operational vehicle may be "vehicle tending", i.e., monitoring certain flight parameters.

Once the objective as described above is clear, then the key parameters which will fix the choice of the instrument will be :

- the level of heat flux and surface temperature;
- the duration of the test;
- the material and geometry of the structure, i.e., the model or vehicle sub-element.

In earlier years, most if not all quantitative measurements of heat flux were performed with discrete gauges (thermocouples, thin films, etc.). Today, the user often has a choice between the discrete gauges which have developed, generally speaking, a certain maturity and survey techniques which provide an overall surface image (the equivalent of an infinite number of discrete gauges).

Our review is arranged according to the discrete or field nature of the technique. In each case, the present status, problem areas, and recommendations for future research and development are presented.

2. Discrete Techniques

2.1. THERMOCOUPLES AND THIN FILM GAUGES

2.1.1. *Present Status.* Thermocouples have been the active element in numerous heat flux techniques spanning many years of applications. The thin-skin method, which relies on an inner surface temperature measurement in conjunction with a mathematical model relating temperature, model geometry, and material properties to heat flux, seems to be less employed at present in wind tunnel applications in spite of the ease of fabrication. Surface thermocouples, which use the same mathematical model as thin-film gauges, find application in harsh environments and where heat fluxes and corresponding changes in surface temperature are large. Consigny et al [6B] described the use of a constantan jacket in which a series of chromel wires were embedded flush to the surface as a means of simplifying the thermal model. His paper also stressed the need to numerically simulate the data reduction process, particularly when materials of different thermal properties are employed.

Many specialized forms of gauges based on thermocouples and employing a suitable sub-surface model have found application in flight instruments [6A, 6D].

2.1.2. *Problem Areas.* The major problem area appears to be an understanding of non-uniform heating effects and non-uniform substrates; this is a common issue with all of the following techniques as well and will be discussed in 5.0.

2.2. THIN-FILM GAUGES

These gauges, using the thick-skin technique as the model relating heat flux to the surface temperature, determined through the change in resistance of a thin metallic film, are in an advanced state of development. Miller [6C] concludes that with suitable care they present the most accurate means of measuring heat flux on wind tunnel models (± 5 to 8 %).

2.2.1. *Problem Areas.* Erosion of the gauges in harsh environments which characterize most high enthalpy wind tunnels has always been a problem. Olivier et al. [6E] describe their method of manufacture including protective coatings which seems to have produced a very durable gauge.

The trend is to manufacture the models, or large sections of the models, with a single insulating material, such as MACOR, a machinable ceramic, so as to eliminate thermal discontinuities. However, this is a costly process and the models are easily damaged.

2.2.2. *Recommendations.* The development of more resistant coatings to protect thin film gauges from erosion and thermal shock should be accelerated.

Better substrate materials which are easily machinable to high tolerances, characterized by nearly temperature-independent thermal properties, and are shock-resistant should be the object of research by materials specialists.

2.3. GRADIENT HEAT FLUX GAUGES

2.3.1. *Present Status.* Two examples of micro-machined gradient-type heat flux gauges were mentioned; one was developed at ONERA [6F] and the other is a commercial product of the Vatell Corporation in the U.S. Both gauges are fabricated by alternately stacking layers of insulator and thermocouples on a test article. While the two examples differ in detail, they both represent applications of integrated circuit fabrication technology to heat flux instruments and they both suggest a fundamental shift in philosophy from instruments that are inserted into existing structures to instruments that are fabricated on or with the structure. These examples of aerospace heat flux instruments join parallel developments in the measurement of heat transfer in rotating machinery.

2.3.2. *Problem Areas.* The development problems for this class of heat gauges are somewhat different than traditional gauges. There still exist fabrication problems associated with the etching and deposition of films on a substrate. Gauge thickness is a problem that will be solved with more experience. The application of this class of thermal gauges to arbitrary model materials and calibration of the gauge at room and elevated surface temperatures are problem areas.

2.3.3. *Recommendations.* Encourage development of this promising technology.

2.4. GAUGE CALIBRATION

2.4.1. *Present Status.* There is a need for heat flux gauge calibration both at room temperature for aerodynamic ground test applications and at elevated temperatures for high temperature applications both on the ground and in flight. AEDC has developed very high heat flux calibration techniques which are essentially conducted at room temperature.

2.4.2. *Problem Areas.* Corresponding high temperature calibration devices do not exist. In-situ, high temperature, high heat flux calibration devices are increasingly needed as the heat flux sensors become smaller and more complex and are increasingly integrated into the high temperature structure as previously discussed. There are several problem areas; e.g., controlling the errors due to surface radiation, producing convective and radiative calibration sources and developing portability within the calibration device.

2.4.3. *Recommendation.* Develop a standardized, portable, thermal calibration device capable of calibration up to 1000°C at heating rates up to 1000 watts/cm^2.

3. Survey Techniques

3.1. INFRARED

Infrared thermography is the survey technique which appears most often in the papers presented.

3.1.1. *Present status.* Miller [6C] explains that due to recent advances providing high data acquisition rate radiometric measurements and computer based digitized image processors, infrared has received renewed interest at Langley. His group uses three single detector imagers with 6,25 and 30 frames per second in the 8-12 μm range. The models are viewed through ZnSe windows with antireflective coatings. The calibrations are made directly in the test section. Work is in progress for using emissivity calibration laws varying with angle of view (for three-dimensional models) and temperature. Curvature effects are taken into account by extrapolating to zero time the coefficient-time history.

Paper [6G], not orally presented, describes the application of infrared thermography to hypersonic blowdown wind tunnel tests using long-wave Agema cameras (782 and 880) with digitization by a microcomputer-based system. Original work is made which concerns image restoration for enhanced image resolution. This is obtained using an FFT inverse filter algorithm with a generalized system transfer function in which are cascaded an average sampled system modulation transfer function (imaging and sampling effect correction) and a temperature amplitude transfer function (lateral conduction effect correction). The corrections thus obtained in the case of Görtler vortex patterns are presented.

Paper [6H] presents the application of IR thermography in a low density hypersonic wind tunnel. The camera is an Agema 780 short wave (3-5 μm range) version. The calibration is also made in the test section using a model instrumented with thermocouples. The acquisition rate of the digitized images is rather low : 1 image/second.

Paper [6I] deals with a hot shot wind tunnel application with run durations as small as 20 to 100 ms (F4 wind tunnel). To follow the fast transients, two solutions are proposed :

i) line scanning using an Agema 880 long wave camera (one of the scanning movements is blocked) with $7^o \times 7^o$ objective. Through a ZnSe window, a line on the model is scanned 2500 times per second. A microcomputer system digitizes in real time, with 12 bits, 250 points per line.

ii) high frame rates (400 Hz) obtained by using a staring array camera built at ONERA. The camera is based on a 64×64 HgCdTe array working at 3-5 μm. A real-time system stores the digitized 12-bit data.

Paper [6K] presents results obtained with a long wave Agema 782 camera and a portable microcomputer-based system allowing analog-to-digital conversion (12 bits) and storage in real time of all the frames. As in paper [6G], the work presented deals with Görtler vortices. Here, the enhancement of the spatial resolution is obtained by a special arrangement using extra IR lenses. The Görtler phenomenon is taken as an illustration of the possibilities of IR survey techniques for identification of the governing parameters of the heat transfer : heat transfer coefficient and recovery temperature. This last parameter is rarely measured in spite of its importance, especially in an interaction zone where it can experience important variations. Use of spatial Fourier analysis is also presented for identification of the spanwise modulation of the heat transfer.

Most survey techniques, such as IR, will continue to use discrete gauges for continuous recalibration.

3.1.2. *Problem areas.* Neumann [6A] emphasizes the challenge of the technique : the ability to handle the data efficiently (density of information is 500 times higher than that of gauges).

For ground tests, there is a need :

i) to broaden the range of temperature response, for example, in shock-interaction regions (problem of peak heating where the peaks are chopped off).

ii) for transition location measurements by means of a survey technique such as IR thermography while the forces and moments are measured simultaneously.

In flight, the interaction regions must also to be studied but many difficulties are to be solved to successfully obtain infrared measurement surveys (see, for example, problems encountered with the Shuttle in-flight measurement).

In high enthalpy wind tunnels, energy emitted by hot parts of the facility (the throat, for instance) may perturb the measurements if reflection off the model is not avoided. High-emissivity diffuse surfaces are needed.

For many models, the three dimensional geometry is difficult to handle and must be taken into account when using a survey technique.

3.1.3. *Recommendations.* For Miller [6C], in the 90's optical video-based infrared emission and relative intensity phosphor thermography systems will provide the vast majority of aerothermodynamic information from conventional hypersonic wind tunnels. They will provide quantitative, global three-dimensional heating distributions (multiple cameras and/or mirror systems). There will be an extension of the technique to impulse facilities (a few milliseconds) thanks to high speed video cameras with fast response phosphors and staring array cameras as in [6I].

Active infrared thermography with two main techniques which use uniform extended radiant heat flux stimulations combined with classical IR scanners or laser flying-spot cameras are to be evaluated since they can be useful for continuous flow experiments and for eliminating spurious radiation artefacts which can be encountered in high enthalpy facilities.

Two and three-dimensional conduction solvers must be used routinely allowing better identification of the heat transfer parameter in the presence of model curvature and high aerodynamic gradients. The three-dimensionality of models is also to be taken into account by procedures which establish correlations between the pixels of the infrared images and the surface elements of the model. A possibility is the inlay of these IR images with a calculated projection of the model 3-D mesh. This operation allows to take into account the variation of emissivity with the angle of view, to eliminate the pixels which are not on the model for computer time and memory saving, and to obtain a more meaningful presentation of results (iso-coordinate distribution in relation to aerodynamic phenomena in place of line or column distributions rarely interesting for the aerodynamicist).

3.2. PHOTOLUMINESCENCE

3.2.1. *Present Status.* The temperature sensitivity of various phosphor compounds has been known for many years; recent advances in imaging techniques and associated digital image processing algorithms have made this method a practical reality.

Reports were presented by three different laboratories, ONERA, NASA Langley and AEDC/Calspan, on their experiences with the phosphor technique [6B, 6C, 6D]. It seems that the so-called "two-color" or relative intensity method exhibits better accuracy; it has become "the most widely used heat transfer measurement technique at (NASA) Langley" [6C]. Developments were noted in the choice of phosphor compound and binder material, the coating thickness, and model construction (which seems to be moving toward slip casting techniques). Processing software is being further refined.

Miller [6C] described a new system that is undergoing acceptance tests at the time of this workshop (May 1992). It possesses the hardware and software for target recognition capability. A frame rate of 30 frames per second in possible and a calibration procedure which includes the response of each pixel is included. Provision for a second camera, synchronized with the first, is provided for three-dimensional analysis.

3.2.2. *Problem Areas.* The temperature range for the application of phosphors is unduly narrow for some applications. Their use in impulse facilities will pose problems because of the need for a finite phosphor thickness which may result in a false indication of the model surface temperature; in addition, erosion of the phosphor coating will be a problem in some facilities. The phosphor surface tends to be too rough for certain applications. Finally, none of the authors commented on the spatial resolution of the technique.

3.2.3. *Recommendations.* Efforts should be made to obtain mixtures of phosphors which will extend the useful temperature range and improve signal strength. The influence of phosphor thickness and binders on signal-to-noise ratio should be quantified. Model materials, discussed in 5.0, should be analyzed for ease of machining, although as mentioned above the ceramic slip-casting technique appears promising.

The limitations of phosphors in short duration facilities should be identified. Since the IR technique may not be usable in flows with hot particles emanating from the reservoir, the photoluminescence method may prove a good candidate in this more challenging flow environment.

3.3. PAINTS

Paints for heat transfer measurement were mentioned by two experimenters : thermosensitive paint presented by Consigny from ONERA [6B] and phase change paint presented by Matthews AEDC/Calspan [6D].

3.3.1. *Present status.* The thermosensitive paint used at ONERA, presenting three color changes, allows quantitative measurement of the heat transfer coefficient. The calibration procedure is detailed in the paper. The main advantages of the technique are the possible achievement of very detailed maps of heat transfer coefficient and the detection of possible hot spots.

The phase change paint is used at Calspan and also provides results in reasonable agreement with thin skin data.

3.3.2. *Problem areas.* One of the major problems with the thermosensitive paint is the tedious data reduction needed. Consigny [6B] presented a digital image processing method based on segmentation of two-dimensional histograms.

3.3.3. *Recommendations.* No recommendations were formulated on this subject.

4. Model Materials

4.1. PRESENT STATUS

Three types of materials are used for models with survey techniques :

i) thin skin (0.5 mm) metallic models in stainless steel as in [6G,6H] or in nickel [6H] with black matte paint for high emissivity.

ii) very poor conductors such as Rhodorsil RTV 147 used for both thermosensitive paints and IR thermography in [6G] and Norcoat 4000 [6G,6K], which consists of hollow silica microspheres in a silicone elastomer, are used for small effusivity $(k\rho C)^{1/2}$ and high emissivity.

iii) ceramic materials like Macor used in [6B,6C] which are interesting because of their refractority and high emissivity.

4.2. PROBLEM AREAS
There is a need to obtain low effusivity, high emissivity, low-cost models; there is also a need to calibrate the emissivity dependence with temperature and angle of view (for angles higher than 60°). Finally, there is a need for the temperature dependence of the thermal properties (effusivity for semi-infinite models, conductivity and diffusivity for thin skin models). A step heating procedure is used for effusivity measurements at ONERA [6B].

4.3. RECOMMENDATIONS
Ceramic models will be the norm instead of the exception in the 90's [6C]. Sintered slip-cast ceramic techniques are needed.

5. Non-Isothermal Surface Phenomena

5.1. PRESENT STATUS
Several authors discussed the effects of surface temperature discontinuities which may distort the level of inferred heat flux through temperature measurements. Temperature mismatches between the measurement device and the measured structure distort both the flow of heat within the structure as well as the convective heat transfer to the surface through boundary layer distortion. Non-isothermal effects are found both in wind tunnel studies and more dramatically in flight applications where heating rates can be changed by a factor of two. Correction methodology is available for more simple flows, but not for the more practical cases; for example, the presence of shock interactions.

5.2. PROBLEM AREAS
Correction methodology based on analytical methods is limited to classical plate flows. Corresponding CFD-based correction methodology is not validated for complex situations and probably suffers the same resolution errors of other thermal CFD applications. There are two problem areas : a lack of enduring sensitivity to the problem and perceived deficiencies in existing CFD correction methodology.

5.3. RECOMMENDATIONS
Develop a series of validating experiments based on thermal surface mismatch and shock interaction processes to challenge and encourage the CFD community.

Upgrade AGARDograph 165 to provide a mechanism of continuing education.

6. Flight Heat Flux Gauges

6.1. PRESENT STATUS

Flight gauges differ from ground test gauges in several important ways; they are high temperature gauges, their fundamental measurement technique as well as construction materials are tied to ever-changing flight material systems and they must function in severe environments of long test duration or extremely complex material systems. The very character of these gauges, whether discrete sensors or sensors integrated into the structure at the time of fabrication creating a 'smart skin', is open.

6.2. PROBLEM AREAS

Since little was presented, there appears to be a lack of experience and/or perception within the instrumentation community.

6.3. RECOMMENDATIONS

Create increased awareness of the problems and possibilities through AGARD meetings or documentation.

7. Flight Measurements

7.1. PRESENT STATUS

This workshop skirted the fringes of flight measurement technology. 'Making instruments that fly' was a common theme. The difficulty of flight measurements is both technical (high temperatures and the integration of sensors into complex structures) and cultural in that many diverse groups with very different perspectives come together in a flight test project. The 'no holes' criterion is a reflection of the difficulty that the measurement engineer has in discussing instrumentation with a structures engineer. This interaction is not the only one occurring in a flight test discussion. It is important that the instrumentation community develop the hardware to make instruments that fly and, as well, it is important that this community carefully develop and articulate the philosophy of why the proposed measurements are required in flight. For the thermal flux example, the measurement of temperature is easy. The inference of heat flux is far more complex. Whether heat flux is required in flight is an open issue that must be carefully articulated to the broader flight test community.

7.2. PROBLEM AREAS

Flight instrumentation is discussed in too narrow a technical setting. There is a need for broader dialog between instrumentation engineers and those who would design the flight vehicle; i.e., flight test systems engineers and structures engineers.

7.3. RECOMMENDATIONS

Focus a discussion on high temperature, high-heating-rate instrumentation appropriate for both flight and elevated temperature ground test facilities (such as arc tunnels).

At some point (perhaps after the instrumentation people have developed further among themselves), create a discussion format between the instrumentation engineers ('everything is measurable') and flight test systems engineers ('here are the constraints') to determine what is most useful to measure in flight, given practical constraints and how that data is used.

8. Principal Recommendations

Heat flux instrumentation is an enabling technology, both in wind tunnels and in flight. A strong need exists today for greater accuracy due to code validation studies and for survey methods which can, e.g., illustrate transition front location with time. Future requirements indicate that surface temperatures and heat fluxes in flight may reach 2000 K and 10^4 to even 10^5 watts/cm^2, respectively. Modest investments in instrumentation development and in test planning at an early stage can reduce substantially future total vehicle costs.

Near-term actions which are recommended :

- exploration for phosphors with a larger temperature range;
- more emphasis on determination of recovery temperature for code validation;
- a concerted look at "active" IR;
- an up-date of the classic AGARDograph 165 by Schultz & Jones, so as to set down in one source the recent developments in heat flux measurements;
- efforts must be devoted to techniques suitable for the accurate determination of the reservoir conditions of high enthalpy facilities;
- the special instrumentation problems of the newly-emerging arc-heated facilities need more attention. (They are more closely allied with flight instrumentation.)

Longer-term actions which are recommended :

- development of model materials with smaller temperature dependence on ρck and easier machinability;
- IR array cameras in the kilohertz range for unsteady flows;
- instrumentation techniques for surface temperatures to 2000 K and heat fluxes to 10^4 or even 10^5 watts/cm^2.

9. Titles of Contributions

6A Neumann, R.D.: Temperature and heat flux measurement challenges for high temperature aerospace applications.

6B Consigny, H.; Le Sant, Y.; Bouvier, F.; Baumann, P.; Merienne, M.C.: Heat transfer measurement techniques used or in development at ONERA Chalais-Meudon.

6C Miller, C.G.: Aerothermodynamic measurement techniques employed in NASA Langley hypersonic facilities.

6D Matthews, R.K.: Aerothermal test methodology and techniques for the development of hypersonic vehicles.

6E Olivier, H.; Vetter, M.; Jessen, Grönig, H.: Measurements on models for hypersonic real gas conditions.

6F Kayser, P.; Rodrigues, M.; Godefroy, J.C.: Thin gradient heat fluxmeters developed at ONERA.

6G Carlomagno, G.M.; De Luca, L.; Cardone, G.: Hypersonic aerodynamics research with an infrared imaging system.

6H Westby, M.F.: Heat transfer measurements using infrared thermography in rarefied flows.

6I Barbe, S.; Boscher, D.; Deschamps, J.; Gauffre, G.; Girard, A.: Infrared thermography for hot-shot wind-tunnel.

6J Gülhan, A.; Jöres, V.; Kindler, K.: Experiments on the catalytic efficiency of heat protection materials in the arc heated facility wind-tunnel (LBK) in Cologne.

6K Boscher, D.; Baudouy, B.; Deom, A.; Coet, M.C.; Delery, J.; Balageas, D.: Infrared thermography characterization of Görtler vortex type patterns in hypersonic flows.

GPSR Compliance

The European Union's (EU) General Product Safety Regulation (GPSR) is a set of rules that requires consumer products to be safe and our obligations to ensure this.

If you have any concerns about our products, you can contact us on ProductSafety@springernature.com

In case Publisher is established outside the EU, the EU authorized representative is:

Springer Nature Customer Service Center GmbH
Europaplatz 3
69115 Heidelberg, Germany

Batch number: 10607320

Printed by Printforce, the Netherlands